Proceedings in Life Sciences

Biochemistry of Membrane Transport

FEBS – Symposium No. 42

Edited by
G. Semenza and E. Carafoli

With 392 Figures

Springer-Verlag
Berlin · Heidelberg · New York 1977

Professor Dr. Giorgio Semenza, Laboratorium für Biochemie, Eidgenössische Technische Hochschule Zürich, 8006 Zürich/ Switzerland

Professor Dr. Ernesto Carafoli, Laboratorium für Biochemie, Eidgenössische Technische Hochschule Zürich, 8006 Zürich/ Switzerland

Umschlagmotiv: Dr. Klaus Müller, ETH Zürich

ISBN-13: 978-3-642-66566-0 e-ISBN-13: 978-3-642-66564-6
DOI:10.1007/ 978-3-642-66564-6

Library of Congress Cataloging in Publication Data. Febs Symposium on the Biochemistry of Membrane Transport, Swiss Institute of Technology, 1976. Biochemistry of membrane transport. (FEBS symposium; no. 42) (Proceedings in life sciences). Includes bibliographical references and index. 1. Biological transport--Congresses. 2. Membranes (Biology)--Congresses. I. Semenza, G., 1928- II. Carafoli, Ernesto. III. Title. IV. Series: Federation of European Biochemical Societies. FEBS symposium; no. 42. [DNLM: 1. Biological transport--Congresses. 2. Cell membrane--Physiology--Congresses. QH601 F32b 1976] QH509.F3 1976 574.8'75 76-58327

Softcover reprint of the hardcover 1st edition 1977

Offsetprinting: Beltz Offsetdruck, D-6944 Hemsbach über Weinheim

2131/3130-543210

These proceedings are dedicated to Walther Wilbrandt on the occasion of his 70th birthday. He was among the first to shed light into the black box of membrane transport.

Preface

This volume contains the proceedings of the FEBS Symposium on the Biochemistry of Membrane Transport, which was held at the Swiss Institute of Technology, Zürich, July 18-23, 1976. Of the speakers invited originally, only five could not attend the meeting, and of the lectures given, all but one of the texts are published here. Thus, this volume gives a faithful account of the way the meeting was originally conceived and actually took place.

This Symposium on Biochemistry of Membrane Transport was the first Symposium sponsored by the FEBS outside the yearly FEBS-Meetings, after the Special Meeting on Industrial Biochemistry, which took place in Dublin in 1973, and it reflects the interest and the trend for gatherings of smaller size than the official FEBS Meetings. The topic of the Symposium was an easy choice, not only because membrane transport is becoming more and more important to biochemistry every year, but also because of the long-standing interest of Swiss Science in the field. In the choice of the topics and of the speakers, efforts were made to achieve as balanced a coverage of the area as possible. However, since some aspects of membrane biochemistry were dealt with extensively at the parallel 10th International Congress of Biochemistry in Hamburg, GFR, certain topics were given less emphasis than others. The Symposium was attended by about 400 participants (we expected 200-250); among them 48 were invited speakers, and some 200 contributed posters.

Were essential advances in the field presented at this Symposium? This is of course the important question, and it is not up to us to answer it. However, it is the organisers' impression that the original goal - that of presenting a rather balanced view of the state of the art in transport biochemistry at a time when membrane transport proteins are beginning to be isolated and characterised - was reached by and large.

It was perhaps more important that the Symposium brought together an impressive number of specialists from different backgrounds and using different, albeit parallel, systems. The possibilities for interchanges and cross-fertilisation were thus optimal, and it is the firm hope of the organisers that the Symposium will pay substantial dividends to people working in the area in the future.

In assembling the program on behalf of the Swiss Society of Biochemistry (and also on behalf of the Swiss Groups of Biomembranes and of Bioenergetics), we were assisted by the Advisory Board (H. Aebi, C. Bron, M.M. Burger, F. Gey, B. Jeanrenaud, H. Keberle, H. Moor, K. Mühlethaler, H. Reuter, G. Schatz, H.J. Schatzmann, R. Schwyzer, J. Seelig, W. Simon, R.W. Straub, P.G. Waser, J.P. v. Wartburg, E.R. Weibel, W. Wilbrandt (Honorary President), P. Zahler), who provided invaluable suggestions and to whom we would like to express our deepest appreciation. We would also like to acknowledge the essential financial support of FEBS, of the Swiss National Science Foundation, of the Swiss Federal Institute of Technology, the Swiss Society of Natural Sciences, as well as the support of the following firms: Interfarma, Bender & Hobein, Digitana AG, and Swissair. Particularly warm thanks are due to the European Committee of the Weizmann Institute, whose financial support rendered possible the Aharon-Katzir-Katchalsky Memorial Lecture (page 582).

We are deeply indebted to our colleagues in the staff of the Department of Biochemistry of the Swiss Federal Institute of Technology for their help in the organisation. An especial word of thanks we would like to express to Miss Ulrike Zilian for her impeccable secretarial work.

The volume appears some 7 months after the date of the Symposium. This is due to the cooperative effort of the invited speakers and of the staff of Springer-Verlag. To them also sincere thanks are also due.

G. Semenza
E. Carafoli

Contents

Kinetic Aspects of Membrane Transport

Transport in Epithelia and in Some Other Mammalian Cells

Regulation of Transport Processes

Ion Transport in Mitochondria

Bacterial Transport Systems

List of Contributors

AFFOLTER, H., Laboratorium für Biochemie der Eidgenössischen Technischen Hochschule, Universitätstraße 16, CH-8092 Zürich

ALPES, H., Fachbereich Biologie, Universität Konstanz, D-775 Konstanz

APELL, A.J., Fachbereich Biologie, Universität Konstanz, D-775 Konstanz

AQUILA, H., Institut für Physiologische Chemie und Physikalische Biochemie der Universität München, Goethestraße 33, D-8000 München 2

BABEL, W., Institut für Physiologische Chemie und Physikalische Biochemie der Universität München, Goethestraße 33, D-8000 München 2

BAKER, P.F., University of London, King's College, Strand, London WC 2R 2LS, England

BAMBERG, E., Fachbereich Biologie, Universität Konstanz, D-775 Konstanz

BARENHOLZ, Y., University of Virginia, Dept. of Biochemistry, Charlottesville, Va. 22901, USA

BAUER, P.J., Abt. Biophysikalische Chemie, Biozentrum der Universität Basel, Klingelbergstraße 70, CH-4056 Basel

BEIL, F.U., Max-Planck-Institut, Jahnstraße 29, D-6900 Heidelberg

BENZ, R., Fachbereich Biologie, Universität Konstanz, D-775 Konstanz

BLOK, M.C., Biochemisches Laboratorium der Rijksuniversiteit, Padualaan 8, De Uithof, Utrecht, The Netherlands

BOFFOLI, D., Istituto di Chimica Biologica, Facoltà di Medicina, Università di Bari, Via Amendola 165/A, I-70126 Bari

BOOS, W., Universität Konstanz, Fachbereich Biologie, Postfach 7733, D-775 Konstanz

BRAN, M.D., Dept. of Physiological Chemistry, Johns Hopkins University, School of Medicine, 725 N. Wolfe Street, Baltimore, Maryland 21295, USA

BRINLEY Jr., F.J., Department of Physiology, University of Maryland, Baltimore 21201, USA

BROCKLEHURST, R., Dept. of Biochemistry, University of Manchester, Institute of Science and Technology, Manchester M60 1QD, England

BRUNNER, J., Laboratorium für Biochemie, Eidgenössische Technische Hochschule, Universitätstraße 16, CH-8092 Zürich

BÜRKLI, A., Laboratorium für Biochemie, Eidgenössische Technische Hochschule, Universitätstraße 16, CH-8092 Zürich

CABANTCHIK, Z.I., Laboratory of Kidney and Electrolyte Metabolism, Building 10, Room 6N320, National Institutes of Health, Bethesda, Maryland 20014, USA

CAPUANO, F., Istituto di Chimica Biologica, Facoltà di Medicina, Università di Bari, Via Amendola 165/A, I-70126 Bari

CARAFOLI, E., Laboratorium für Biochemie der Eidgenössischen Technischen Hochschule, Universitätstraße 16, CH-8092 Zürich
CECCHINI, G., Dept. of Biological Chemistry, University of Michigan, Ann Arbor, Michigan 48109, USA
CHAPMAN, D., Chemistry Department, Chelsea College, University of London, Manresa Road, London SW3, England
CHERRY, R.J., Laboratorium für Biochemie, Eidgenössische Technische Hochschule, Universitätstraße 16, CH-8092 Zürich
CHIESI, M., Laboratorium für Biochemie der Eidgenössischen Technischen Hochschule, Universitätstraße 16, CH-8092 Zürich
CLAUSEN, T., Institute of Physiology, Aarhus University, DK-Aarhus
CORNELL, B.A., Chemistry Department, Chelsea College, University of London, Manresa Road, London SW3, England
CHRISTENSEN, H.N., Dept. of Biological Chemistry, University of Michigan, Ann Arbor, Michigan 48104, USA
CRANE, R.K., Dept. of Physiology, CMDNJ-Rutgers Medical School, Piscataway, NJ 08854, USA
CROMPTON, M., Laboratorium für Biochemie der Eidgenössischen Technischen Hochschule, Universitätstraße 16, CH-8092 Zürich
van DEENEN, L.L.M., Biochemisches Laboratorium der Rijksuniversiteit, Padualaan 8, De Uithof, Utrecht, The Netherlands
DENCHER, N.A., Abt. Biophysikalische Chemie, Biozentrum der Universität Basel, Klingelbergstraße 70, CH-4056 Basel
EARNSHAW, P., Dept. of Biochemistry, University of Manchester, Institute of Science and Technology, Manchester M60 1QD, England
EDDY, A.A., Dept. of Biochemistry, University of Manchester, Institute of Science and Technology, Manchester, M60 1QD, England
FECKL, J., Institut für Physiologische Chemie und Physikalische Biochemie der Universität München, Goethestraße 33, D-8000 München 2
FRIZZELL, R.A., School of Medicine, University of Pittsburgh, Dept. of Physiology, Pittsburgh, PA 15213, USA
GECK, P., Klinikum der Johann-Wolfgang-Goethe-Universität, Abt. f. Physikalische Biochemie, Theodor-Stern-Kai 70, D-6000 Frankfurt/M. 70
de GIER, J., Biochemisches Laboratorium der Rijksuniversiteit, Padualaan 8, De Uithof, Utrecht, The Netherlands
GINSBURG, H., The Hebrew University of Jerusalem, Biophysics Group, The Institute of Life Sciences, Dept. of Botany, Jerusalem, Israel
GOODMAN, D.P.B., University of Pennsylvania, The School of Medicine G3, Dept. of Biochemistry, Philadelphia, PA 19174, USA
GRELL, E., Max-Planck-Institut für Biophysikalische Chemie, D-3400 Göttingen-Nikolausberg
GROSS, E., Department of Health, Education and Welfare, Public Health Service, National Institute of Health, Bethesda, Maryland, USA

GUERRERI, F., Istituto di Chimica Biologica, Facoltà di Medicina, Università di Bari, Via Amendola 165/A, I-70126 Bari
HACKENBROCK, C.R., Dept. of Cell Biology, The University of Texas, 5323 Harry Hines Blvd., Dallas, Texas 75235, USA
HASSELBACH, W., Max-Planck-Institut, Jahnstraße 29, D-6900 Heidelberg
HAUSER, H., Laboratorium für Biochemie, Eidgenössische Technische Hochschule, Universitätstraße 16 , CH-8092 Zürich
HEINZ, E., Klinikum der Johann-Wolfgang-Goethe-Universität, Abt. Physikalische Chemie, D-6000 Frankfurt/M. 70
HEYN, M.P., Abt. Biophysikalische Chemie, Biozentrum der Universität Basel, Klingelbergstraße 70, CH-4056 Basel
HINKLE, P.C., Section of Biochemistry, Molecular and Cell Biology, Cornell University, Ithaca, N.Y. 14853, USA
HIRATA, H., Dept. of Biochemistry, Jichi Medical School, Minamikawachi, Jichi, Tochigi-ken, Japan 329-04
HÖCHLI, H., Dept. of Cell Biology, The University of Texas, 5323 Harry Hines Boulevard, Dallas, Texas 75235, USA
HOKIN, L.E., 383 Med. Sci. Bldg., 426 N. Charter Street, University of Wisconsin, Madison, Wis. 53706, USA
HORVAI, G., Laboratorium für Org. Chemie, Eidgenössische Technische Hochschule, Universitätstraße 16, CH-8092 Zürich
IVANOV, V.T., Shemyakin Institute of Bioorganic Chemistry, USSR Academy of Sciences, Ul. Vavilova 32, Moscow V-312, USSR
IZZO, G., Istituto di Chimica Biologica, Facoltà di Medicina, Università di Bari, Via Amendola 165/A, I-70126 Bari
JANKO, K., Fachbereich Biologie, Universität Konstanz, D-775 Konstanz
JILKA, R.J., School of Medicine, Dept. of Biochemistry, 1402 South Grand Boulevard, Saint Louis, Missouri 63104, USA
JIMENO, J., Institut de Biologie Molêculaire, Université de Paris VII, 2, Place Jussieu, F-75221 Paris-Cedex 05
KABACK, H.R., Roche Institute of Molecular Biology, Kingsland Street, Nutley, New Jersey 07110, USA
KAGAWA, Y., Dept. of Biochemistry, Jichi Medical School, Minamikawachi, Jichi, Tochigi-ken, Japan 320-04
KASAHARA, M., Section of Biochemistry, Molecular and Cell Biology, Cornell University, Ithaca, N.Y. 14853, USA
KEPES, A., Institut de Biologie Molêculaire, Université de Paris VII, 2, Place Jussieu, F-75221 Paris-Cedex 05
KESSLER, M., Laboratorium für Biochemie der Eidgenössischen Technischen Hochschule, Universitätstraße 16, CH-8092 Zürich

KEYNES, R.D., Physiological Laboratory, University of Cambridge, Downing Street, Cambridge CB2 3EG, England
KINNE, R., Max-Planck-Institut für Biophysik, Kennedy-Allee 70, D-6000 Frankfurt/M. 70
KLINGENBERG, M., Institut für Physiologische Chemie und Physikalische Biochemie der Universität München, Goethestraße 33, D-8000 München 2
KNAUF, P.A., The Hospital for Sick Children, 555 University Avenue, Toronto, Ontario, Canada
KOLB, H.A., Fachbereich Biologie, Universität Konstanz, D-775 Konstanz
KOTYK, A., Czechoslovak Academy of Sciences, Laboratory for Cell Membrane Transport, Budejovicka 270, Praha 4, Czechoslovakia
KRÄMER, R., Institut für Physiologische Chemie und Physikalische Biochemie der Universität München, Goethestraße 33, D-8000 München 2
LÄUGER, P., Fachbereich Biologie, Universität Konstanz, D-775 Konstanz
LEE, M., Dept. of Biological Chemistry, University of Michigan, Ann Arbor, Michigan 48198, USA
LEGRUM-SCHUHMANN, B., Max-Planck-Institut für Biophysik, Kennedy-Allee 70, D-6000 Frankfurt/M. 70
LEHNINGER, A., Dept. of Physiological Chemistry, Johns Hopkins University, School of Medicine, 725 N. Wolfe Street, Baltimore, Maryland 21205, USA
LENTZ, B.R., University of Virginia, School of Medicine, Dept. of Biochemistry, Charlottesville, Va. 22901, USA
LORUSSO, M., Istituto di Chimica Biologica, Facoltà di Medicina, Università di Bari, Via Amendola 165/A, I-70126 Bari
MALATHI, P., Dept. of Physiology, CMDNJ-Rutgers Medical School, Piscataway, NJ 08854, USA
MALMSTRÖM, K., Laboratorium für Biochemie der Eidgenössischen Technischen Hochschule, Universitätstraße 16, CH-8092 Zürich
MARTONOSI, A., School of Medicine, Dept. of Biochemistry, 1402 South Grand Boulevard, Saint Louis, Missouri 63104, USA
MAX, E., University of Pennsylvania, The School of Medicine G3, Dept. of Biochemistry, Philadephia, PA 19174, USA
MEURY, J., Institut de Biologie Moléculaire, Université de Paris VII, 2, Place Jussieu, F-75221 Paris-Cedex 05
MITCHELL, R., Dept. of Molecular Biology, Vanderbuilt University, Nashville, Tennessee 37235, USA
MOORE, P., Dept. of Biological Chemistry, University of Michigan, Ann Arbor, Michigan 48109, USA
MURER, H., Max-Planck-Institut für Biophysik, Kennedy-Allee 70, D-6000 Frankfurt/M. 70
MÜHLETHALER, K., Lab. f. Zellbiologie der ETH, CH-8093 Zürich/Hönggerberg
NAKAMURA, H., School of Medicine, Dept. of Biochemistry, 1402 South Grand Boulevard, Saint Louis, Missouri 63104, USA
OBERBÄUMER, I., Max-Planck-Institut für Biophysikalische Chemie, D-3400 Göttingen-Nikolausberg

OP DEN KAMP, J.A.F., Biochemisches Laboratorium der Rijksuniversiteit, Padualaan 8, De Uithof, Utrecht, The Netherlands

OVCHINNIKOV, Y., Shemyakin Institute of Bioorganic Chemistry, USSR Academy of Sciences, Ul. Vavilova 32, Moscow V-312, USSR

OXENDER, D.L., Dept. of Biological Chemistry, University of Michigan, Ann Arbor, Michigan 48109, USA

PAPA, S., Istituto di Chimica Biologica, Facoltà di Medicina, Università di Bari, Via Amendola 165/A, I-70126 Bari

PASSOW, H., Max-Planck-Institut für Biophysik, Kennedy-Allee 70, D-6000 Frankfurt/M. 70

PFEIFFER, E., Klinikum der Johann-Wolfgang-Goethe-Universität, Abt. für Physikalische Biochemie, Theodor-Stern-Kai 70, D-6000 Frankfurt/M. 70

PHILO, R., Dept. of Biochemistry, University of Manchester, Institute of Science and Technology, Manchester, M60 1QD, England

PIETRZYK, C., Klinikum der Johann-Wolfgang-Goethe-Universität, Abt. f. Physikalische Biochemie, Theodor-Stern-Kai 70, D-6000 Frankfurt/M. 70

PILARSKA, M., Dept. of Molecular Biology, Nashville, Tennessee 37235, USA

POST, R.L., Dept. of Physiology, Vanderbuilt University, Nashville, Tennessee 37232, USA

PREISER, H., Dept. of Physiology, CMDNJ-Rutgers Medical School, Piscataway, NJ 08854, USA

PRETSCH, E., Laboratorium f. Org. Chemie, Eidgenössische Technische Hochschule, Universitätstraße 16, CH-8092 Zürich

PRING, M., Max-Planck-Institut für Biophysik, Kennedy-Allee 70, D-6000 Frankfurt/M. 70

QUINN, P.J., Chemistry Department, Chelsea College, University of London, Manresa Road, London SW3, England

RASMUSSEN, H., University of Pennsylvania, The School of Medicine G3, Dept. of Biochemistry, Philadephia, PA 19174, USA

REPKE, K.R.H., Akademie der Wissenschaften der DDR, Forschungszentrum für Molekularbiologie und Medizin, Lindenberger Weg 70, DDR-1115 Berlin-Buch

REUTER, H., Universität Bern, Pharmakologisches Institut, Friedbühlstraße 49, CH-3000 Bern

REYNAFARJE, B., Dept. of Physiological Chemistry, Johns Hopkins University, School of Medicine, 725 N. Wolfe Street, Baltimore, Maryland 21295, USA

ROBIN, A., Institut de Biologie Moléculaire, Université de Paris VII, 2, Place Jussieu, F-75221 Paris-Cedex 05

ROLEOFSEN, B., Veterinär-pharmakologisches Institut, Universität Bern, Länggass-Straße 124, CH-3000 Bern

ROSEMAN, S., Dept. of Biology, Johns Hopkins University, McCollum Pratt Institute, Baltimore 21218, USA

ROTHSTEIN, A., The Hospital for Sick Children, 555 University Avenue, Toronto, Ontario, Canada

RUF, H., Max-Planck-Institut für Biophysikalische Chemie, D-3400 Göttingen-Nikolausberg

SAITO, A., Dept. of Molecular Biology, Vanderbuilt University, Nashville, Tennessee 37235, USA

SALZMANN, M., Laboratorium für Biochemie der Eidgenössischen Technischen Hochschule, Universitätstraße 16, CH-8092 Zürich

SCARPA, A., Johnson Research Foundation, University of Pennsylvania, Philadelphia, PA 19174, USA

SCHATZMANN, H.J., Veterinär-pharmakologisches Institut, Universität Bern, Länggass-Straße 124, CH-3000 Bern

SCHNEIDER, G., Laboratorium für Biochemie, Eidgenössische Technische Hochschule, Universitätstraße 16, CH-8092 Zürich

SCHULTZ, S.G., School of Medicine, University of Pittsburgh, Dept. of Physiology, Pittsburgh, PA 15213, USA

SEMENZA, G., Laboratorium für Biochemie, Eidgenössische Technische Hochschule, Universitätstraße 16, CH-8092 Zürich

SIGEL, E., Laboratorium für Biochemie der Eidgenössischen Technischen Hochschule, Universitätstraße 16, CH-8092 Zürich

SIMON, W., Laboratorium für Org. Chemie, Eidgenössische Technische Hochschule, Universitätstraße 16, CH-8092 Zürich

SONE, N., Dept. of Biochemistry, Jichi Medical School, Minamikawachi, Jichi, Tochigi-ken, Japan 329-04

TANNENBAUM, C., Laboratorium für Biochemie der Eidgenössischen Technischen Hochschule, Universitätstraße 16, CH-8092 Zürich

THOMA, A.P., Laboratorium für Org. Chemie, Eidgenössische Technische Hochschule, Universitätstraße 16, CH-8092 Zürich

TIFFERT, T., Dept. of Physiology, University of Maryland, Baltimore 21201, USA

TOGGENBURGER, G., Laboratorium für Biochemie der Eidgenössischen Technischen Hochschule, Universitätstraße 16, CH-8092 Zürich

THOMPSON, T.E., Dept. of Biochemistry, University of Virginia, School of Medicine, Charlottesville, Virginia 22901, USA

VANDERKOOI, J.M., Dept. of Biochemistry and Biophysics, University of Pennsylvania, School of Medicine, Philadelphia, Pennsylvania 19174, USA

VERKLEIJ, A.J., Biochemisches Laboratorium der Rijksuniversiteit, Padualaan 8, De Uithof, Utrecht, The Netherlands

WACKER, H., Laboratorium für Biochemie, Eidgenössische Technische Hochschule, Universitätstraße 16, CH-8092 Zürich

WANG, Ch. T., Dept. of Molecular Biology, Vanderbuilt University, Nashville, Tennessee 37234, USA

WILBRANDT, W., Pharmakologisches Institut der Universität, Friedbühlstraße 49, CH-3010 Bern

WOLOSIN, J.M., The Hebrew University of Jerusalem, Biophysics Group, The Institute of Life Sciences, Dept. of Botany, Jerusalem, Israel

YOSHIDA, M., Dept. of Biochemistry, Jichi Medical School, Minamikawachi, Jichi, Tochigi-ken, Japan 320-04

ZAKI, L., Max-Planck-Institut für Biophysik, Kennedy-Allee 70, D-6000 Frankfurt/M. 70

ZINGSBEHEIM, H.P., Max-Planck-Institut für Biophysikalische Chemie, D-3400 Göttingen-Nikolausberg

Electron Microscopy of Membranes

Electron Microscopic Studies on Membranes

K. MÜHLETHALER

In the discussions on membrane structure the electron microscopic contributions are of some importance because only with this technique can a direct image of these cell structures be obtained. The first observations were based on fixed and sectioned material and led to the so-called "unit membrane concept", postulated by Robertson in 1960. In the same year, work was going on in our laboratory to develop an alternative method for specimen preparation in the frozen state (Moor et al., 1961).

It was soon recognized that this method, which was called freeze-etching, was well suited for membrane studies, because fracturing occurred preferentially along cellular membrane systems (Moor and Mühlethaler, 1963; Moor, 1964; Branton and Moor, 1964). In contrast to the standard sectioning technique, in which only cross sections could be studied, extended surface views of the membranes were disclosed. It became apparent that these cell structures are composed of an amorphous lamellar sheet, decorated with particles having diameters between 60 and 140 Å. Some of these particles could be washed off in dilute buffer solutions, whereas the remaining ones seemed to be firmly embedded in the membrane matrix. Based on these findings, we postulated, in 1965, a model of the thylakoid membrane which consisted of a lipid bilayer in which globular proteins were embedded (Fig. 1; Mühlethaler et al., 1965). Although such a structural arrangement - called the "fluid mosaic model" - by Singer and Nicolson (1972) is now generally accepted, our proposal was not taken very seriously in 1965. At that time the Robertson-Danielli concept was at the height of its fame and controversial ideas had little chance of acceptance. The particles visible in our micrographs were said to represent lipid micelles formed during freezing (Glauert, 1968). The expected instability of the lipids during the quenching process became the main argument against the freeze-fracture technique and its application to membrane studies.

In order to deal with objections we searched for a method which would allow us to have a control for each of the successive preparatory steps used for freezing and replication of the object. An important requirement was that the molecular events at each step could be examined independently of the electron microscope.

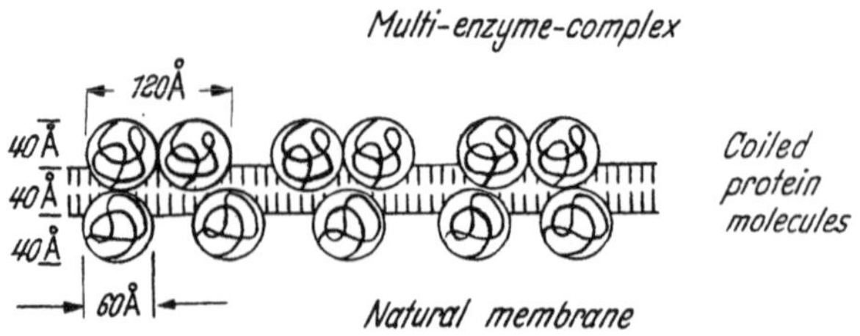

Fig. 1. First membrane model of a chloroplast thylakoid as derived from early freeze-etch observations (Mühlethaler et al., 1965)

The approach we chose was to use the Langmuir-Blodgett technique in combination with attenuated total reflection infrared spectroscopy, called the ATR-IR spectroscopy (Harrick, 1967) and electron microscopy. By spreading molecules on a water surface as indicated by Blodgett and Langmuir (1937) single- and multilayers can be produced for experimental use. These layers can be mounted on a trapezoidal germanium plate for spectroscopy or electron microscopy. Due to multiple internal reflections the beam entering the germanium plate is attenuated and by using polarized infrared light the intensity and directions of the group vibrations can be determined. In parallel investigations such lipid layers served as model systems to study the changes introduced by freeze-fracturing. Measurements done by Kopp et al. (1975a,b) indicated that the molecular structures of the layers before and after cooling to -100°C and before and after evacuation yielded no detectable effects.

Immediately after deposition, mono- or multilayers have properties similar to the liquid crystal state in the smectic phase. These layers appear to be intrinsically unstable and rearrange spontaneously into irregularly shaped crystallites (Fig. 2a,b). Kopp et al. (1975a,b) studied the successive stages from the freshly deposited Langmuir-Blodgett layer to ordered islands and finally to sharply defined crystals by ATR-IR spectroscopy and electron microscopy. As shown in Figure 3a and b, the spectrum of a freshly deposited set of four tripalmitin layers is different if the layers are measured several hours after deposition on the germanium plate. During aging the absorption peaks become sharper and more structured. It indicates that the various groups in the molecules lose the freedom of motion which they apparently have in the Langmuir-Blodgett layer.

The perfect agreement of the ATR-IR spectra with the freeze-etch micrographs proves that the structural framework as seen in the electron microscope corresponds to the original state. Similar results have been obtained in studies using X-ray diffraction and freeze-etching in combination (Deamer et al., 1970; Gulik-Krzywicki, 1975; Verkleij, 1975). These authors found that several lipid water associations do not transform during the freezing process. The structure of the paraffin chains was preserved and it was found that the X-ray repeat distances of several lipid phases agreed well with the freeze-etch measurements.

As demonstrated by Moor (1964) cells remain viable if the freezing procedure is the same as in the freeze-etch technique, which indicates that also the protein components are well preserved after shock freezing.

It is now generally agreed that fracturing occurs along the hydrophobic zone within the membrane. A special specimen holder has been constructed to replicate both sides of the fracture plane (Mühlethaler et al., 1970). With the two outer surfaces and fracture faces, four different views of the membrane can be studied. In the now standard procedure, fracturing is done in a vacuum of about 1.10^{-6} torr at a temperature of -100°C. Under these conditions, residual gases may still contaminate the freshly cleaved faces or structural distortions may be introduced during fracturing. In order to eliminate these artifacts a new ultra-high vacuum freeze-fracture apparatus has been built; it enables one to fracture at -196°C without contamination (Gross et al., 1976).

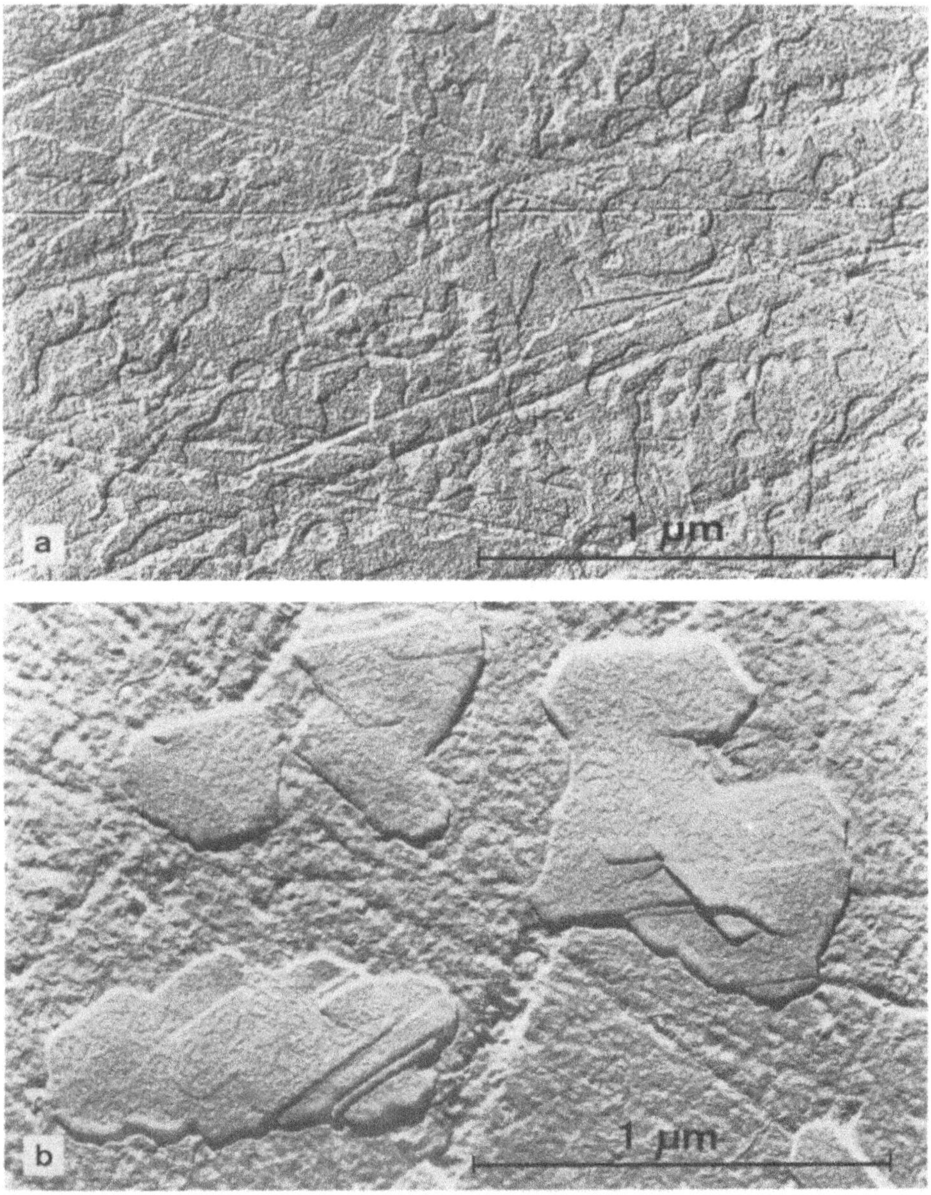

Fig. 2a,b. (a) Electron micrograph of four Langmuir-Blodgett (LB) layers of tripalmitin deposited on the optically plane polished surface of a germanium ATR reflection plate (Kopp et al., 1975b). (b) Tripalmitin microcrystals developed spontaneously from LB layers as shown above at room temperature (Kopp et al., 1975b)

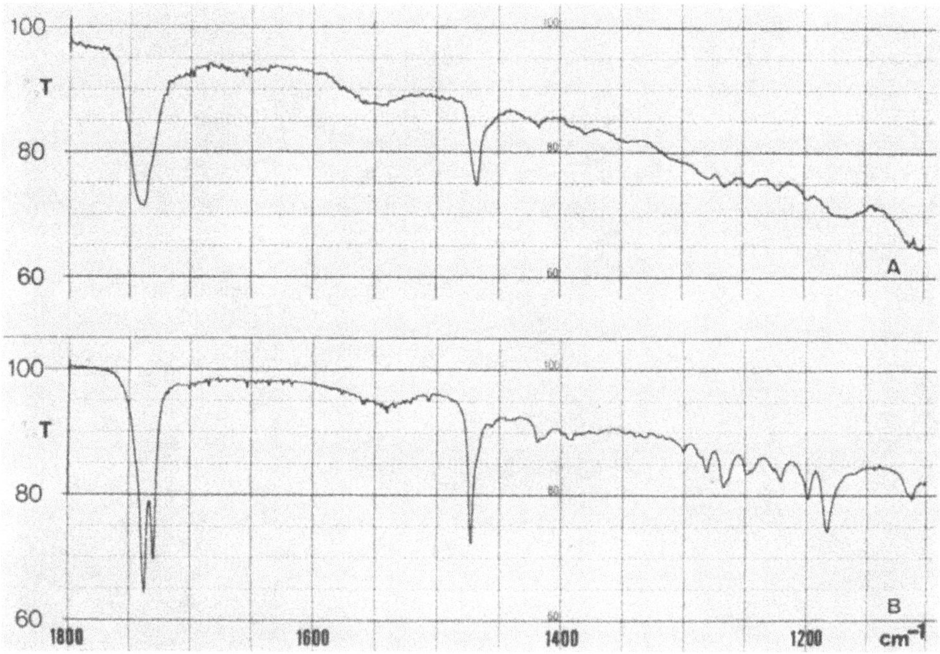

Fig. 3a,b. (a) IR-ATR spectrum of four layers of tripalmitin on germanium reflection plate in the LB state (Kopp et al., 1975b). (b) IR-ATR spectrum of the same four layers of tripalmitin as above but scanned at 164 h after deposition (Kopp et al., 1975b)

As shown in Figure 4, the replica of the fracture face of yeast plasmalemma reflects the structural detail with a clarity so far unattainable. An optical diffractogram was made of the area with the regular particle arrangement, and used for a computer reconstruction (Gross et al., 1976). The reconstructed image demonstrates clearly that each particle has a central depression. The significance of these depressions is not yet clear, however. It can be speculated that the pore is related to the enzyme activity or transport. On the complementary fracture face the imprints of these particles appear as shallow holes. This indicates that large complexes penetrate the outer half of the plasmalemma and probably reach the cell surface.

After this technical part I am going to discuss some examples that illustrate how freeze-etching can be combined with other procedures to gain new information about membrane structure and changes in that structure. Most examples are drawn from studies on plant membranes that we performed in our laboratory, although of course similar approaches can be applied to animal or bacterial membranes.

Each membrane system has its characteristic arrangement of particles (Fig. 4) which can be influenced by factors such as growth and development, seasonal alterations or osmotic effects. This can be demonstrated with chloroplast thylakoids, yeast cell membranes and tonoplasts. The processes causing these variations are not yet fully understood. Within limited areas the same membrane system may display different structures, as seen in stroma- and grana thylakoids. The large complexes expected to contain photosystem II (Arnitzen et al., 1969) located in the hydrophobic zone,

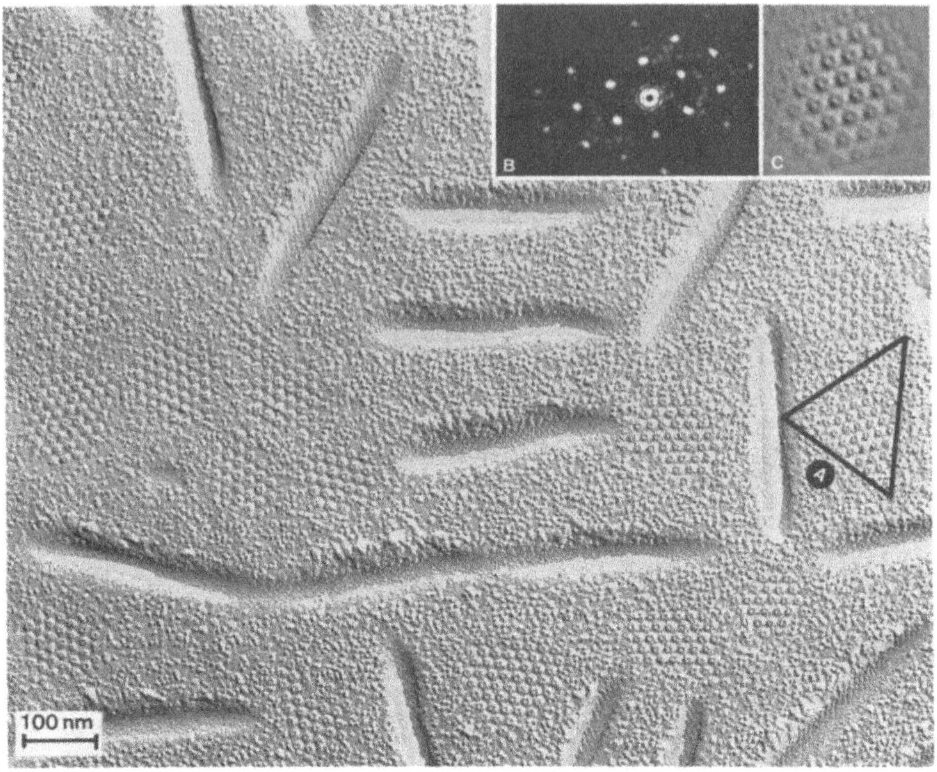

Fig. 4a-c. The plasmatic fracture face (PF) of yeast plasmalemma fractured at $-196^{o}C$ and Pt/C shadowed (20 Å) at 1.10^{-9} torr. Upper right corner of the micrograph: the optical diffractogram (B) of a crystal-like region (A) and the generated image (C) from these reflections reconstructed by filtered computer technique is shown (Gross et al., 1976)

are only present in fractures through grana thylakoids (Wehrli, 1975). Therefore, it must be concluded that the water splitting system is restricted to the grana areas only. On the other hand, the complexes seen at the outer surface of the thylakoids containing the coupling factor 1 (Berzborn et al., 1974; Miller and Staehelin, 1976) are only found on stroma thylakoids. Serological reactions indicate that these CF_1-particles are mobile within the membrane. If the thylakoids are incubated with antisera containing only antibodies against CF_1, the particles become aggregated in clusters. Since the mean distance between the particles in the untreated control was greater than the distance which can be bridged by an antibody, it can be proposed that the antibody molecule first reacts through only one of its binding sites with a particle. Subsequently this complex moves laterally within the surface of the thylakoid until it meets another particle where the antibody molecule reacts through its second binding site (Berzborn et al., 1974). This succession of events generates the aggregates seen in the electron micrographs by lateral movements.

In addition to the lateral mobility a transverse movement of the particles has also been observed. According to Murakami and Packer

(1970) the thickness of a pair of fused grana membranes decreases by 25% after illumination. This process is reversible. Using the freeze-fracture technique they found that the particles seen in the hydrophobic core of the grana thylakoids change their orientation upon illumination or after treatment by divalent ions.

As shown by Parish (1974) not only the location and distribution of the particles may change over a period of time but also their quantity and size. He investigated the number and size of the particles seen on the internal faces of the tonoplast in cells of the shoot cambial-zone of *Salix fragilis*. It was found that during the period from winter to spring the number of particles attached to the tonoplast lamella next to the cytoplasm (the PF face according to the new nomenclature proposed by Branton et al., 1975) is significantly reduced. On the complementary side the number of particles remained approximately constant. In winter both faces contained the same distribution of particle sizes but in spring the exoplasmic face (EF) exhibited significantly larger complexes than those on the opposite side (Parish, 1974). Parallel with the increase in particle size, a reduction in number was recorded. This would indicate that an aggregation of particles occurs in the period between winter and spring. These findings indicate that the tonoplast membrane contains predominantly complexes of uniform size during the dormant period, whilst in the active tonoplast these are replaced by two distinct populations of particles.

Another distinctive feature of the tonoplast membrane is its ability to expand during osmotic swelling of the vacuole and to contract during shrinking. Were the number of particles to remain approximately constant, their frequency would be expected to decrease if the vacuoles swell. In contrast their number should increase per unit area during shrinking. Investigations on yeast vacuoles by Niedermeyer (1976) have shown that these events are not as simple as that. If isolated vacuoles are transferred from 1.0 M to 0.1 M sorbitol the surface area swells 94% and the number of particles increases 203% at the extracellular fracture face and 26% at the plasmatic fracture face. The particle diameters on both fracture faces can be classified in three distinct classes with diameters of 7, 10 and 14 nm respectively. Under isotonic conditions the 7 nm particles at the extracellular fracture face are absent. Under these conditions 80% of the particles are 10 nm in diameter and the rest 14 nm. After either shrinking or swelling conditions the 7 nm particles appear and in addition the percentage of larger-sized particles increase on both fracture faces (Niedermeyer, 1976). A plausible explanation for these findings is that stretching of the tonoplast membrane may cause a transverse movement into the fracture plane of some proteins not previously visible. This shift of proteins to another zone within the lipid matrix would generate the "new" particles.

During swelling and shrinking of the vacuole the thickness of the tonoplast membrane remains unchanged. Therefore we have to suppose that stretching increases the mobility of the unsaturated fatty acid chains but still preserves the general state of the liquid crystalline (fluid) phase.

If the vacuoles are incubated in 0.1 M sorbitol containing protease, the surface increases 650% compared to normal. According to Niedermeyer (1976) the results of these investigations may indicate that membrane lipids in the fluid phase allow elasticity of the membrane but that the protein components determine the degree of elasticity.

The results discussed demonstrate the complexity of the various membrane systems as seen with the freeze-fracture and freeze-etch technique. It can be expected that with the technical improvements, such as high-pressure freezing (Riehle, 1968; Höchli, 1974) and the use of ultra-high vacuum as well as additional methods such as image reconstruction, antibody labeling, differential decoration and ATR-IR spectroscopy, new significant results on membrane structure will become available.

Summary

In the first part, the quality of the structural preservation by freeze-etching and freeze-fracturing is discussed. Control measurements by ATR-infrared spectroscopy and X-ray diffraction indicate that the original state of the lipid layers is well preserved.

In the second part, the general structural appearance of chloroplast thylakoids, yeast plasmalemmas and tonoplasts is described. It is shown that each membrane system has its specific arrangement of particles. This can be changed due to various influences: in thylakoid membranes local differences in the structure are related to special physiological activities (grana-stroma lamellae). In the vacuolar membrane seasonal changes have been observed and in the very elastic tonoplasts the membrane particles rearranged according to the state of swelling or shrinking.

These observations confirm the findings that membrane particles are very mobile and can be moved in lateral and vertical directions within the lipid layer, obviously without any loss of their physiological activities.

References

Arnitzen, C.J., Dilley, R.A., Crane, F.L.: A comparison of chloroplast membrane surfaces visualized by freeze-etch and negative staining techniques; an ultrastructural characterization of membrane fractions obtained from digitonin-treated spinach chloroplasts. J. Cell Biol. 43, 16-31 (1969)

Berzborn, R.J., Kopp, F., Mühlethaler, K.: Mobility of chloroplast coupling factor 1 (CF_1) at the thylakoid surface as revealed by freeze-etching after antibody labeling. Z. Naturforsch. 29c, 694-699 (1974)

Blodgett, K.B., Langmuir, I.: Built-up films of barium stearate and their optical properties. Phys. Rev. 51, 964-982 (1937)

Branton, D., Bullivant, S., Gilula, N.B., Karnovsky, M.J., Moor, H., Mühlethaler, K., Northcote, D.H., Packer, L., Satir, B., Satir, P., Speth, V., Staehelin, L.A., Steere, R.C., Weinstein, R.S.: Freeze-etching nomenclature. Science 190, 54-56 (1975)

Branton, D., Moor, H.: Fine structure in freeze-etched Allium cepa L. root tips. J. Ultrastr. Res. 11, 401-411 (1964)

Deamer, D.W., Leonard, R., Tardieu, A., Branton, D.: Lamellar and hexagonal lipid phases visualized by freeze-etching. Biochem. Biophys. Acta 219, 47-60 (1970)

Glauert, A.M.: Electron microscopy of lipids and membranes. J. Roy. Micros. Soc. 88, 49-70 (1968)

Gross, H., Bas, E., Kübler, O., Moor, H.: An experimental system for freeze-fracturing in ultra high vacuum at temperatures to -196°C. 6th Europ. Congr. Electron Microscopy Jerusalem, 1976. Vol. 1, p. 402, Brandon, D.G. (ed.), T.A.L.L. Intern. Publ. Co.

Gulik-Krzywicki, T.: Structural studies of the associations between biological membrane components. Biochim. Biophys. Acta (MR) 415, 1-28 (1975)

Harrick, N.J.: Internal Reflection Spectroscopy. New York: Interscience, 1967

Höchli, M.: Entwicklung einer neuen Druck-Gefriertechnik und ihre Anwendung in der Elektronenmikroskopie. Diss. ETH Nr. 5304 (1974)

Kopp, F., Fringeli, U.P., Mühlethaler, K., Günthard, Hs.H.: Instability of Langmuir-Blodgett layers of barium stearate, cadmium arachidate and tripalmitin, studied by means of electron microscopy and infrared spectroscopy. Biophys. Struct. Mechanism 1, 75-96 (1975a)

Kopp, F., Fringeli, U.P., Mühlethaler, K., Günthard, Hs.H.: Spontaneous rearrangement in Langmuir-Blodgett layers of tripalmitin studied by means of ATR infrared spectroscopy and electron microscopy. Z. Naturforsch. 30c, 711-717 (1975b)

Miller, K.R., Staehelin, A.: Analysis of the thylakoid outer surface. J. Cell Biol. 68, 30-47 (1976)

Moor, H.: Die Gefrier-Fixation lebender Zellen und ihre Anwendung in der Elektronenmikroskopie. Z. Zellforsch. 62, 546-580 (1964)

Moor, H., Mühlethaler, K.: Fine structure in frozen-etched yeast cells. J. Cell Biol. 17, 609-628 (1963)

Moor, H., Mühlethaler, K., Waldner, H., Frey-Wyssling, A.: A new freezing ultramicrotome. J. Biophys. Biochem. Cytol. 10, 1-13 (1961)

Mühlethaler, K., Moor, H., Szarkowski, J.W.: The ultrastructure of the chloroplast lamellae. Planta 67, 305-323 (1965)

Mühlethaler, K., Wehrli, E., Moor, H.: Double fracturing methods for freeze-etching. 7th Congr. Electron Microscopy, Grenoble. Favard, P. (ed.). Vol. I, pp. 449-450 (1970)

Murakami, S., Packer, L.: Protonation and chloroplast membrane structure. J. Cell Biol. 47, 332-351 (1970)

Niedermeyer, W.: The elasticity of the tonoplast membrane related to its ultrastructure and chemical composition. Diss. ETH Nr. 5649, 1976

Parish, G.R.: Seasonal variation in the membrane structure of differentiating shoot cambial-zone cells demonstrated by freeze-etching. Cytobiology 9, 131-143 (1974)

Riehle, U.: Über die Vitrifizierung verdünnter wässriger Lösungen. Diss. ETH Nr. 4271 (1968)

Robertson, J.D.: A molecular theory of cell membrane structure. Proc. 4th Intern. Congr. Electron Microscopy, Bd. 2, 159-171, 1960, Vol. II, pp. 159-171

Singer, S., Nicolson, G.: The fluid mosaic model of the structure of cell membranes. Science 175, 720-731 (1972)

Verkleij, A.J.: Architecture of biological and artificial membranes as visualized by freeze-etching. Diss. Utrecht 1975

Wehrli, E.: Struktur der Thylakoidmembran. Eine Gefrierätz- und Doppel-Abdruckstudie. Diss. ETH Nr. 5571 (1975)

Motional Freedom of Integral Proteins in the Mitochondrial Energy-Transducing Membrane

C. R. HACKENBROCK and M. HÖCHLI

A. Introduction

The energy-transducing membrane of the mitochondrion is distinct from all other membranes of eukaryote cells in its major role in the conservation of metabolic energy. Such a vital function requires a complex and comparatively high degree of membrane transport activity. Indeed the biochemistry of membrane transport may find its ultimate complexity in the energy-transducing membrane. Nevertheless, the progress which has occurred over the past twenty years in the elucidation of substrate, ion, and electron transport processes in this membrane has been overwhelming.

Of special interest to us is the critical consideration of the generally intuitive belief pervading the discipline of mitochondrial biology that this highly metabolically active membrane is a rather rigid, "solid state" membrane, crowded with integral proteins for the purpose of their precise transport and metabolic interactions. The sequential and rapid events inherent in electron transport and oxidative phosphorylation generally tend to support the inference that the proteins of the energy-transducing membrane are stabilized by an organized protein-protein lattice (Fleisher et al., 1967; Sjöstrand and Barajas, 1970; Capaldi and Green, 1972). Related to this inference is a further assumption that the specific metabolically active proteins are ordered laterally in the plane of the membrane with recurring intermolecular spacing (Klingenberg, 1968).

We believe, however, that conformational flexibility is inherent in the ion, substrate, electron transport and energy-transducing activities of the large number of metabolically active integral proteins contained in the energy-transducing membrane. Such conformational events require a high degree of fluidity in the bilayer lipid and lateral motional freedom in the integral proteins. We wish to present here new data and observations which clearly reveal that the energy-transducing membrane is much less crowded with integral proteins and is considerably less rigid and indeed more fluid in its molecular organization than has been previously recognized.

B. Thermotropic Lipid Phase Transitions and Lateral Motional Freedom of Integral Proteins

I. Mitochondrial Membrane Preparations and Experimental Approach

Lipid fluidity and protein diffusion were investigated in rat liver mitochondrial membranes by combining differential scanning calorimetry (DSC) with freeze fracture electron microscopy. Several types of preparations were used: freshly isolated whole mitochondria which routinely show a condensed configuration (Fig. 1); an inner membrane-matrix fraction which has a condensed configuration in 300 milliosmolar medium (Fig. 2); and an inner membrane-matrix fraction which has a spherical membrane configuration in 40 milliosmolar medium (Fig. 3; Hackenbrock, 1972). The spherical configuration of the energy-transducing membrane is useful for freeze-fracture studies in which large area fracture faces are required for adequate structural analysis of the hydrophobic interior of the membrane. All three preparations identified above are active in electron transport and oxidative phosphorylation. A purified preparation of the outer mitochondrial membrane (Parsons et al., 1966) was also studied and is included here for the purpose of completeness and comparison. The purities of the membrane fractions were determined by assessing the activities of typical membrane marker enzymes such as monoamine oxidase and cytochrome c oxidase after solubilization of the fractions with Lubrol (Table 1). Since the two mitochondrial membranes contain a distinct lipid and protein composition, highly purified membrane preparations are required if the calorimetric determinations of lipid phase transitions are to be accurate.

A Perkin-Elmer DSC-2 differential scanning calorimeter was used to determine the onset temperatures of the exotherm and endotherm lipid phase transitions. The difference between these two onset temperatures was used to define the extent of the phase transition region (Ladbrook and Chapman, 1969). Alternate cooling and heating runs were repeated twice on each membrane sample and at least two different samples of each membrane fraction were monitored.

All membrane samples were analyzed in the presence of 50% ethylene glycol, since it was determined early in our studies that mitochondrial membranes showed transition regions which extended below $0^{\circ}C$. As in some bacterial membranes (Steim et al., 1969), we determined that ethylene glycol depressed the transition temperatures in mitochondrial membranes. Such depressions have been corrected for, both in the text and in the DSC curves presented here, in order to compare directly the DSC data with the freeze fracture analysis where 30% glycerol was used as the membrane cryoprotector. The corrected as well as uncorrected onset temperatures and the extents of the transition regions for all membrane preparations are summarized in Table 2. In practice, lipid transition exotherms, rather than endotherms, proved to be of greater value for comparison with freeze-fracture observations since freeze-fracturing was carried out on membranes pre-cooled to various temperatures.

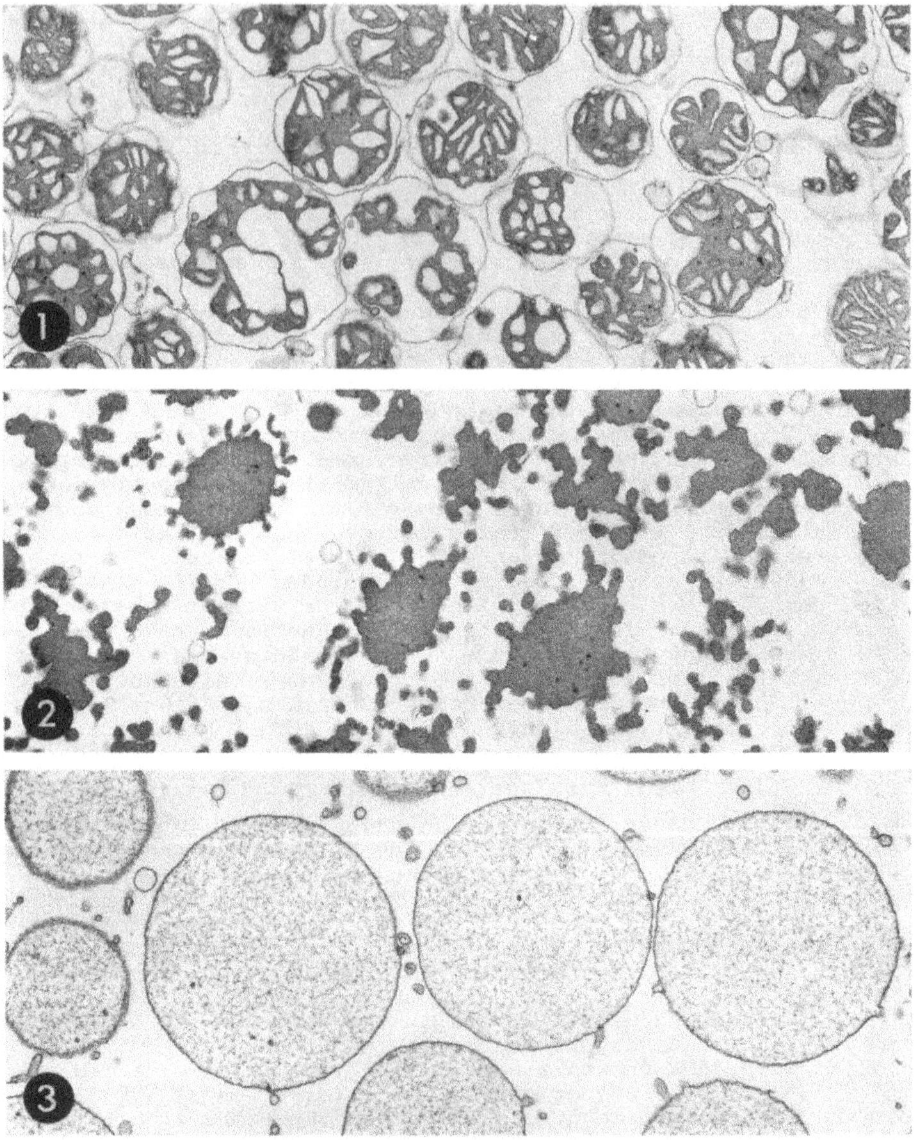

Figs. 1-3. Thin section electron micrographs of mitochondrial membrane preparations

Fig. 1. Condensed configuration of freshly isolated rat liver mitochondria

Fig. 2. Purified condensed inner membrane-matrix fraction

Fig. 3. Purified spherical inner membrane-matrix fraction. x 20,000

Table 1. Relative purity of fractionated mitochondrial membranes

	Specific activity[a]	
	Monoamine oxidase	Cytochrome c oxidase
Whole mitochondria	12.8	720
Condensed inner membrane-matrix fraction	0.257	1060
Spherical inner membrane-matrix fraction	0.014	1200
Outer membrane fraction	352.0	150

[a]Specific activity: for monoamine oxidase in nmoles benzylamine oxidized $\cdot$ min^{-1} $\cdot$ mg^{-1} protein; for cytochrome c oxidase in natoms oxygen reduced $\cdot$ min^{-1} $\cdot$ mg^{-1} protein

Table 2. Relative lipid phase transition temperatures of whole mitochondria and fractionated membranes[a]

	Onset temperature in 50% ethylene glycol (%)		Onset temperature corrected (°C)[b]		Extent of transition region (°C)
	Exothermal	Endothermal	Exothermal	Endothermal	
Whole mitochondria	+ 3, - 5	- 21, - 14	+ 9, + 1	- 15, - 8	24
Condensed inner membrane-matrix fraction	- 10	- 21, - 14	- 4	- 15, - 8	11
Spherical inner membrane-matrix fraction	- 10	- 21, - 14	- 4	- 15, - 8	11
Outer membrane fraction	+ 3, - 5	- 21	+ 9, + 1	- 15	24

[a]All temperatures given are averages of 2 or 3 heating or cooling runs and are ± 1°C

[b]Corrected for 6°C depressions induced by 50% ethylene glycol

II. Studies on Whole Mitochondria

DSC revealed that the transition exotherm of both whole mitochondria and the purified outer membrane fraction was biphasic with a slight onset temperature occurring at 9°C and a major onset temperature occurring at 1°C (Fig. 4a,b). At high recording sensitivity (0.2 mCal $\cdot$ sec^{-1}) it was determined that the onset temperatures of the transition exotherm of whole mitochondria resulted from phase transitions occurring in the purified outer membrane (Fig. 4c).

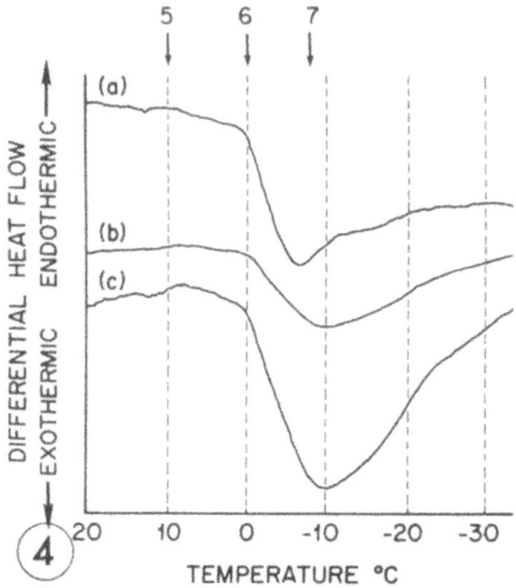

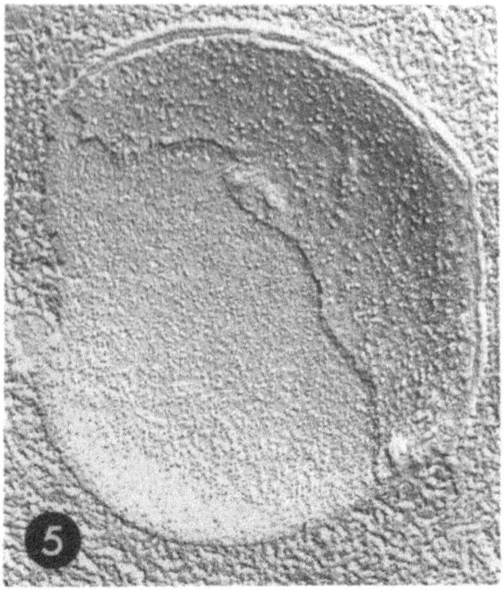

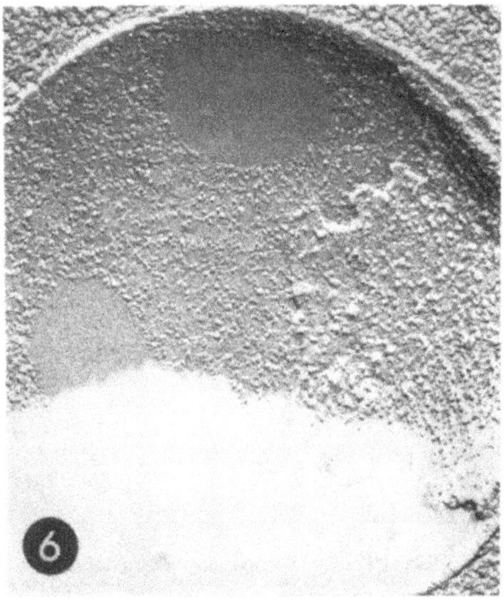

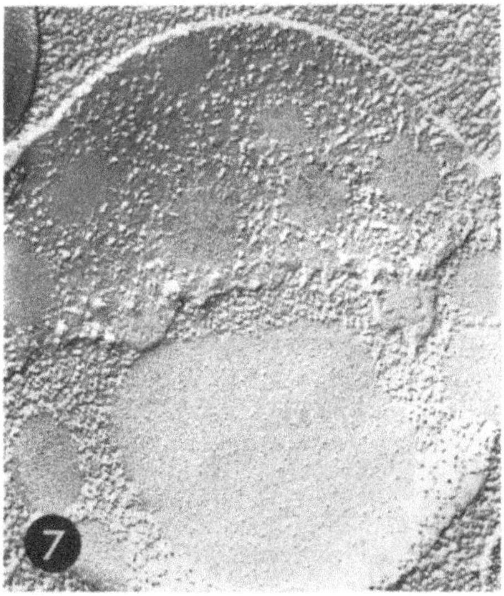

Figs. 4-7. Lipid phase transition and lateral mobility of integral proteins

Fig. 4a-c. DSC cooling runs of whole mitochondria (a) and purified outer membrane (b,c). Recording sensitivity, 0.5 mCal · sec^{-1} (a,b), 0.2 mCal · sec^{-1} (c). Arrows identify temperatures on the curve for whole mitochondria which correspond to Figs. 5-7

Fig. 5. Concave fracture faces of both membranes of a mitochondrion frozen from $10^{\circ}C$

Fig. 6. Cooled to $0^{\circ}C$, then frozen

Fig. 7. Cooled to $-8^{\circ}C$, then frozen. Note lateral displacement of integral proteins and smooth regions of gel state lipid. x 67,500

Freeze-fracturing was carried out on whole mitochondria which were cooled slowly (as in the DSC studies) to 10, 0, and -8°C prior to rapid freezing. When frozen from either room temperature or 10°C, i.e., from above the transition exotherm, whole mitochondria revealed a random distribution of intramembrane particles in the fracture faces of the hydrophobic interior of both the outer membrane and energy-transducing membrane (Fig. 5). Intramembrane particles which are embedded in a smooth continuum of the hydrophobic interior of both membranes show a frequency size distribution peak of 8 nm in diameter for the outer membrane and 10 nm for the energy-transducing membrane (Hackenbrock, 1973). The smallest resolved intramembrane particles in the energy-transducing membrane measure approximately 4.5 nm, while close to 70% of the particles measure between 10 and 18 nm in diameter. Thus the energy-transducing membrane contains a wide range of intramembrane particle sizes and an exceptionally high number of large intramembrane particles compared to other natural membranes. Such particles in the fracture face of the energy-transducing membrane can account for virtually all the integral membrane proteins. Essentially, this includes the heme proteins of the cytochrome chain other than cytochrome c, the membrane-bound dehydrogenases, the hydrophobic, non-catalytic proteins of the ATPase complex, and the various ion and substrate-permease and translocating integral proteins.

That intramembrane particles represent integral membrane proteins is well documented in the literature (Chen and Hubbell, 1973; Grant and McConnell, 1974; Kleeman et al., 1974; Papahadjopoulos et al., 1975). In the sections to follow, we will equate intramembrane particles with integral proteins. It has been estimated that only 50% of the protein of the energy-transducing membrane is integral (Capaldi and Tan, 1974; Harmon et al., 1974) and that this integral protein occupies only about one third of the bilayer lipid surface area of the membrane (Vanderkooi, 1974). Such estimates predict that the energy-transducing membrane may represent a less "crowded pool" than was heretofore assumed and that the integral proteins are afforded considerable lateral area in the bilayer lipid for translational mobility. We thus hypothesized that provided the bilayer lipid is fluid and the integral proteins are not stabilized or anchored by a continuous, rigid protein-protein lattice, a high potential exists for free lateral translational diffusion by the integral proteins in the energy-transducing membrane.

The potential for lateral translational diffusion by the integral proteins was studied by lowering the temperature of whole mitochondria to below the major onset temperature of the transition exotherm just prior to rapid freezing for freeze-fracturing. At 0°C obvious lateral separations between smooth, particle-free regions and particle-rich regions were observed in the hydrophobic fracture faces of the outer membrane (Fig. 6). Whole mitochondria cooled slowly to -8°C revealed dramatic lateral separations between smooth, particle-free regions and particle-rich regions in the hydrophobic fracture faces of both membranes (Fig. 7). We concluded that the striking lateral movement and aggregation of intramembrane particles is the result of a thermotropic growth of protein-excluding regions of liquid crystalline to gel state phase transitions in the bilayer lipid. Such phase transitions were clearly indicated in our DSC studies. The onset temperatures of the transition exotherm and the first indications of lateral separation between integral proteins and gel state bilayer lipid were in close agreement (Figs. 4-7). A reasonable assumption here is

Table 3. ADP:O and acceptor control ratios in mitochondria before and after low temperature-induced lateral separations in the inner membrane

Exp.	Pretreatment with glycerol medium[a]		ADP:O	A.C.
	Temp. in °C	Time in min		
1	No pretreatment		1.80	5.50
2	25	2	1.48	3.38
3	25	2	1.45	3.33
4	0	8	1.50	3.19
5	0	8	1.59	3.05
6	-8	8	1.42	3.43
7	-8	8	1.48	3.00
8	-8	15	1.56	3.05

[a]Mitochondria were pretreated in glycerol medium (30% glycerol; 250 mM sucrose; 10 mM Tris, pH 7.4) at temperatures and times indicated prior to polarographic analysis

that the smooth protein-free regions in the hydrophobic fracture faces of the membranes represent gel state bilayer lipid which develops below the onset temperature of the transition exotherm. It is the growth of gel state bilayer lipid that displaces the integral membrane proteins laterally.

The low temperature-induced lateral translational motion and resultant aggregation of the integral proteins in the energy-transducing membrane, as well as in the outer membrane, is rapidly reversible. When whole mitochondria were equilibrated at temperatures in the region of their transition exotherm for 15 min and then rapidly returned to 30°C for only a few seconds prior to rapid freezing, the freeze-fractured membranes exhibited complete randomization of integral proteins and no sign of smooth, protein-free, gel state regions. Thus, rapid disaggregation, free lateral translational diffusion, and complete randomization of integral proteins occurred in both membranes as the temperature was raised and a gel state to liquid crystalline phase transition occurred in the bilayer lipid.

Of major interest in these studies was the finding that there exists a high potential for free lateral translational diffusion by integral proteins in the fluid bilayer lipid of the energy-transducing membrane. Of considerable significance in this regard, was the determination that the thermotropic lateral separation between integral proteins and gel state bilayer lipid, and the reversal of this separation in the energy-transducing membrane, had no disrupting effect on either respiratory chain electron transport or on oxidative phosphorylation. Mitochondria equilibrated at temperatures well below the onset temperature of the transition exotherm for 8 to 15 min and then returned to room temperature were as efficient in oxidative phosphorylation as mitochondria which were not subjected to thermotropic lateral separations of membrane lipids and integral proteins (Table 3). Clearly the thermotropic transitions and separations of bilayer lipids and integral proteins demonstrated in these studies did not disrupt any organized molecular protein-protein lattice in the energy-transducing membrane necessary for the structural and functional integrity of the membrane required for electron transport and ATP synthesis. Such observations suggest strongly that no such general protein-protein lattice exists in the energy-transducing membrane.

III. Studies on the Purified Energy-Transducing Membrane

Unlike whole mitochondria and the purified outer membrane, both the purified condensed and spherical inner membrane-matrix preparations showed a monophasic onset temperature of the transition exotherm. The onset temperature occurred at -4°C irrespective of the overall configuration of the membrane (Fig. 8b,c). Clearly the onset temperature of the transition exotherm is lower in the energy-transducing membrane than in the composite two-membrane system of the whole mitochondrion (Fig. 8a) owing to the higher onset temperature characteristic of the outer membrane (Fig. 4b,c). Since the two energy-transducing membrane preparations were highly purified (i.e., free of outer membrane), slight but routinely superimposable exothermal deflections occurred above noise level at 21, 12, and 1°C (Fig. 8b,c). Such reproducible small deflections which were detected at an instrument sensitivity of 0.5 mCal·sec^{-1} may be related to discontinuities in Arrhenius activation energies which have been reported for several inner membrane enzymes over the temperature range of 8 to 27°C (Raison et al., 1972; Lee and Gear, 1974).

In freeze-fracture studies the purified spherical energy-transducing membrane exhibited integral membrane proteins in a dispersed distribution above the onset temperature of the transition exotherm (Fig. 9). However, at just below the onset temperature, large, smooth regions of gel state lipid were clearly separated laterally from integral protein-rich regions in the hydrophobic fracture face of the membrane (Fig. 10). At -13°C, which is well below the peak of the transition exotherm (Fig. 8c), such lateral separations were impressive, with the integral proteins often crowded together to occupy only one half or one third of the surface area of the lipid bilayer (Fig. 11). At -13°C the gel state bilayer lipid was extensive and clearly free of integral proteins in the hydrophobic interior of the membrane (Fig. 11). These observations are consistent with calculations that predict that the integral proteins of the energy-transducing membrane occupy only one third of the surface area of the membrane (Vanderkooi, 1974). The use of the purified, spherical energy-transducing membrane clearly reveals that a large surface area exists in the bilayer lipid of this membrane which represents considerable space for the lateral translational diffusion of integral proteins.

Such impressive thermotropic lateral separations between gel state lipid and integral proteins are completely reversible in the purified energy-transducing membrane, as was the case for membranes of the whole mitochondrion. When returned to above the transition temperature, complete randomization of integral proteins occurred in a few seconds by lateral translational diffusion through the fluid bilayer lipid (Fig. 12). Consistent with this finding were results of a calorimetric analysis during heating from -30°C. DSC revealed a transition endotherm for both the spherical and condensed

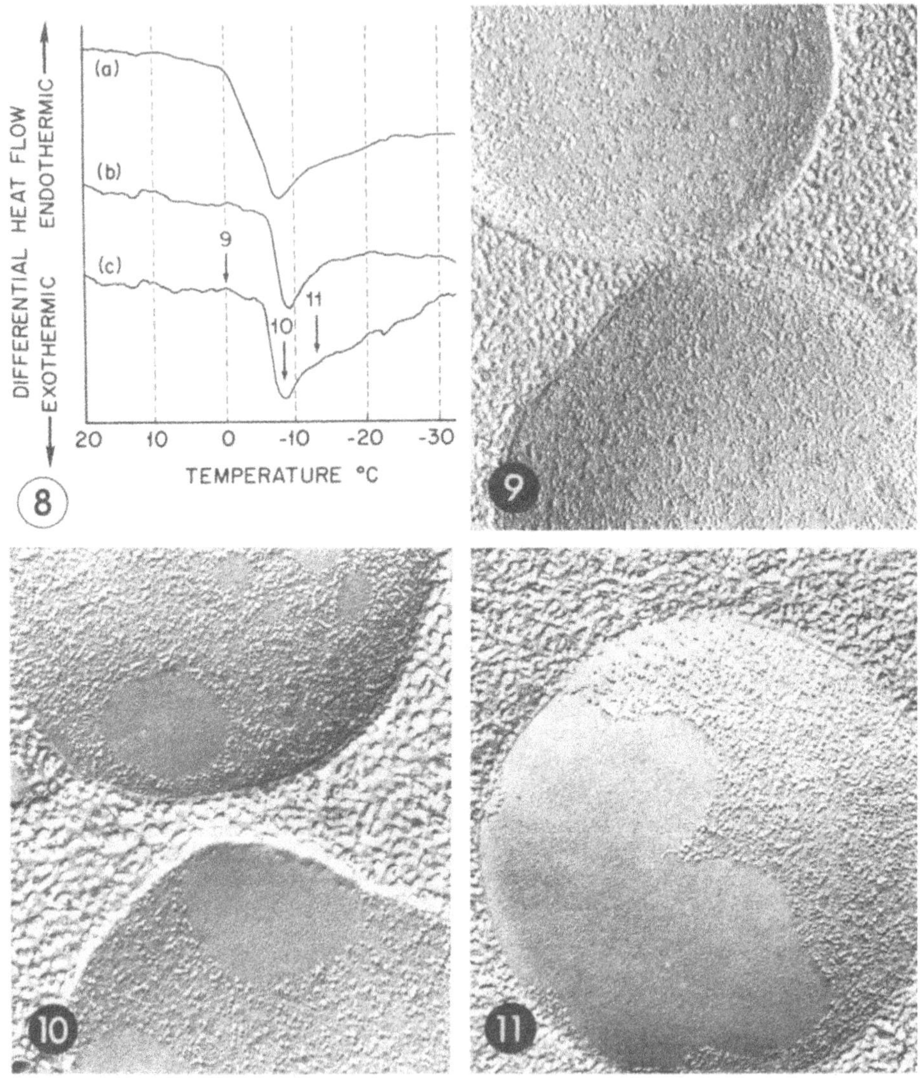

Figs. 8-11. Lipid phase transition and lateral mobility of integral proteins

Fig. 8a-c. DSC cooling runs of whole mitochondria (a), condensed inner membrane-matrix fraction (b), and spherical inner membrane-matrix fraction (c). Arrows identify temperatures on the curve for the spherical inner membrane-matrix fraction which correspond to Figs. 9-11

Fig. 9. Fracture faces of both halves of the spherical energy-transducing membrane cooled to 0 °C, then frozen

Fig. 10. Cooled to -8 °C, then frozen

Fig. 11. Cooled to -13 °C, then frozen. Note lateral displacement of integral proteins and large, smooth regions of gel state lipid. x 67,500

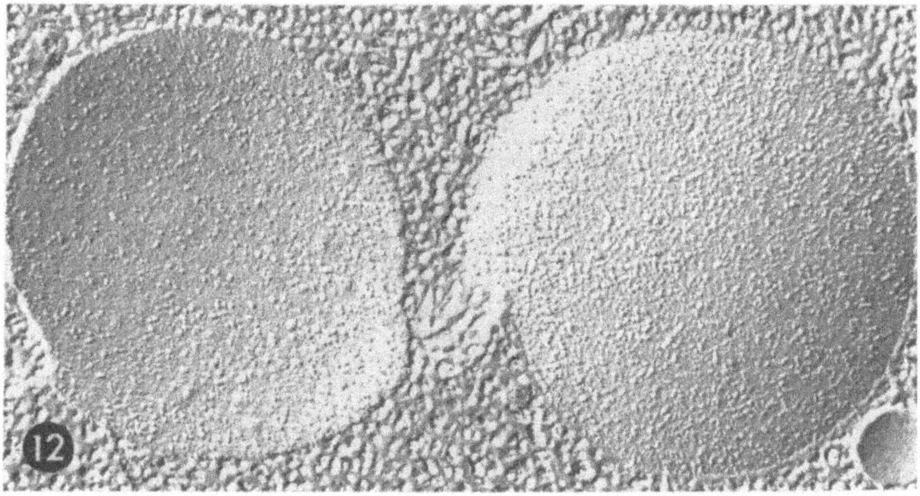

Fig. 12. Free lateral diffusion of integral proteins in the energy-transducing membrane. Membranes cooled to -10°C, returned to 25°C and then frozen show completely random integral proteins in both fracture faces (cf. membranes at -8 to -13°C, Figs. 10, 11). x 67,500

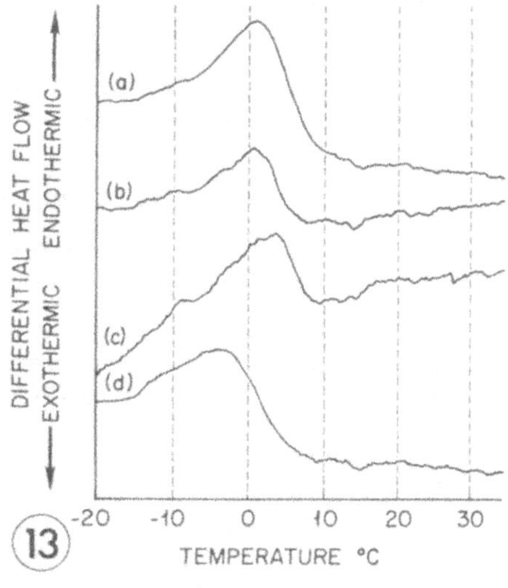

Fig. 13a-d. DSC of heating runs of whole mitochondria (a), condensed inner membrane-matrix fraction (b), spherical inner membrane-matrix fraction (c), and outer mitochondrial membrane fraction (d)

energy-transducing membrane (Fig. 13b,c), as well as for whole mitochondria (Fig. 13a), which is characteristic of the melting properties of bilayer lipid. The endotherm was slightly biphasic, with the first onset temperature appearing at -15°C and the second at -8°C (Fig. 13). Table 2 summarizes the onset temperatures of the exotherm and endotherm transitions as well as the extents of the transition regions for all mitochondrial membrane preparations used in our studies.

IV. Lipid and Protein Composition Related to the Fluid Nature of the Energy-Transducing Membrane

Our studies show that thermotropic lipid phase transitions, lipid-protein lateral separations, and free lateral translational diffusion of integral proteins can occur in the energy-transducing membrane. We have demonstrated further that such reorganizational events in the molecular structure of the membrane are not destructive to electron transport or oxidative phosphorylation. These findings reveal the fluid nature of the energy-transducing membrane. That the full extent of the transition exotherm occurs at subzero temperatures predicts a high degree of molecular motion in the bilayer lipid at physiological temperature. Such a high degree of fluidity is consistent with the composition of the inner membrane phospholipids. It is well known that model bilayer systems composed of unsaturated phospholipids have lower characteristic transition temperatures than model bilayer systems composed of saturated phospholipids. Cardiolipin represents approximately 20% of the phospholipid of the energy transducing membrane and is 90% unsaturated (Colbeau et al., 1971). Phosphatidylcholine and phosphatidylethanolamine make up the remaining 80% of the lipid and are collectively 53% unsaturated. The saturated to unsaturated membrane phospholipid ratio is 0.65. This compares to a ratio of 1.75 for the outer mitochondrial membrane. Thus our studies reveal that the relatively high content of unsaturated phospholipids, combined with the virtual absence of cholesterol in the energy-transducing membrane is consistent with the subzero phase transition, the narrow phase transition region, and the high degree of lipid fluidity in this membrane. We would point out further that these new findings are in agreement with earlier observations from this laboratory that the energy-transducing membrane displays a high degree of metabolically linked configurational plasticity (Hackenbrock, 1966, 1968a).

The effect of membrane proteins on the fluidity of the bilayer lipid is more difficult to appraise. The energy-transducing membrane is composed of approximately 75% protein, however, only 50% of the total protein is integral to the membrane (Capaldi and Tan, 1974; Harmon et al., 1974). The majority of these integral proteins are metabolically active, presumably globular proteins, possessing extensive apolar surfaces in hydrophobic association with the acyl chains of the membrane phospholipids. From studies on less complex membrane systems (McConnell et al., 1972; Trauble and Overath, 1973; Jost et al., 1973), it is to be expected that between 15 to 25% of the phospholipid component of the energy-transducing membrane may be immobilized by the integral protein. The relatively large integral protein content of the energy-transducing membrane may result in a low ordering efficiency of the phospholipid acyl chains which would cause a low transition temperature in the bilayer lipid. The effect of the large peripheral protein content may also add to the low transition temperature of the energy-transducing membrane through electrostatic interaction with the polar head groups of the bilayer lipids. Such an effect has been proposed from studies of protein-lipid electrostatic interactions in model membrane systems (Chapman et al., 1974).

C. Inhibition of Lateral Translational Motion of Integral Proteins by Artificial Protein-Protein Latticing

I. Covalent Latticing of Integral Membrane Proteins

It is our contention that no rigid protein-protein lattice occurs normally throughout the protein component of the energy-transducing membrane. We have developed several methods of artificial latticing to determine if such latticing would influence the thermotropic lipid-protein lateral separations and lateral translational diffusion of integral proteins as described above. Thus we wished to compare the native membrane with membranes containing artificially latticed integral proteins in terms of the motional freedom of the proteins.

For the strongest, most effective artificial protein-protein lattice, we utilized glutaraldehyde as a bifunctional reagent to crosslink the membrane proteins. Thermotropic separations between gel-state lipid and integral proteins were induced in the spherical energy-transducing membrane at -8 or -13°C (Figs. 10, 11). As noted earlier, when the temperature was returned to above the transition temperature, complete randomization of integral proteins occurred in a few seconds through lateral translational diffusion during the gel-state to liquid crystalline lipid phase transition (Fig. 12). However, crosslinking of the aggregated integral proteins at -13°C with 1% glutaraldehyde prevented free lateral translational diffusion of integral proteins when the membrane was returned to above the transition temperature (Fig. 14). Thus the low temperature-induced separations between integral proteins and gel-state lipid were maintained after the gel-state lipid was permitted to melt (Fig. 14). Under these conditions two types of smooth lipid regions were noted above the transition temperature. In addition to the usual smooth, "flat" lipid regions, smooth, "bulging lipid regions were observed (Fig. 14). We would speculate that the smooth, "flat" regions may be glutaraldehyde crosslinked phosphatidylethanolamine and the smooth "bulging" regions may be non-fixed phosphatidylcholine and cardiolipin. In any case, it is clear that the integral proteins were completely immobilized through a generalized covalent crosslinking. Thus it is highly unlikely that the native energy-transducing membrane contains any generalized protein-protein lattice comparable to the artificial, covalently crosslinked protein lattice that we have utilized in this study.

II. Electrostatic Latticing of Integral Membrane Proteins

We have used polycationic ferritin as a visually detectable macromolecular ligand to probe the distribution of anionic groups on the surface of the energy-transducing membrane (Hackenbrock, 1975; Hackenbrock and Miller, 1975). Electrostatic binding of the ligand to the membrane surface inhibits ascorbate-TMPD cytochrome c oxidase activity and succinate permease activity (Fig. 15). In the case of ascorbate oxidation, kinetic data revealed that polycationic ferritin binds to cytochrome c oxidase as a non-competitive inhibitor with respect to cytochrome c (Fig. 16). Thus polycationic ferritin which contains approximately 65 cationic charges at neutral pH binds to the anionic groups of cytochrome c oxidase, an

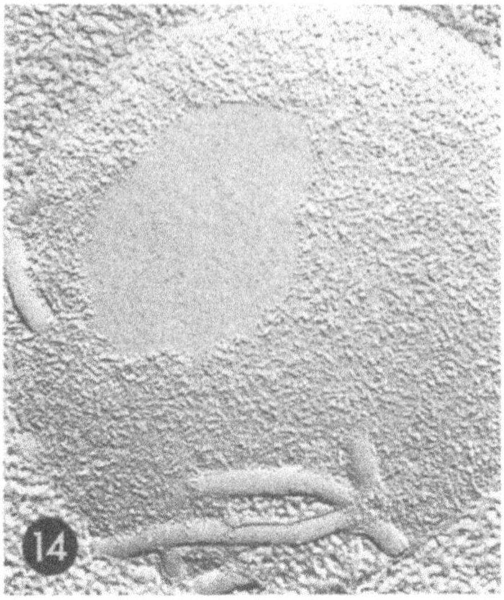

Fig. 14. Latticing and immobilization of integral proteins in the energy-transducing membrane through covalent crosslinking. Convex fracture face after cooling to -13°C, then treated with 1% glutaraldehyde, then warmed to 25°C, and finally frozen shows integral proteins unable to diffuse laterally into the bilayer lipid. x 67,500

integral protein exposed on both surfaces of the membrane (Hackenbrock and Miller-Hammon, 1975). In addition, it is clear that polycationic ferritin, a large molecule of approximately 11 nm in diameter, replaces cytochrome c, a naturally occurring peripheral membrane protein, in its binding to cytochrome c oxidase. Polycationic ferritin may therefore be regarded as an artificial peripheral protein, electrostatically bound to any integral membrane protein which contains anionic groups exposed at the membrane surface.

Polycationic ferritin bound at 25°C at a concentration of 30 µg/ml to the surface of the spherical energy-transducing membrane was observed to be impressed deeply into the membrane (Fig. 17). Such binding impressions permitted the simultaneous observation of the distribution of the bound ligand and the distribution of integral proteins in the fracture face of the membrane (Fig. 17). When the temperature was lowered to the region of the transition exotherm, it was observed that the bound polycationic ferritin co-migrated laterally with the integral proteins (Fig. 18). Thus we observed that at low concentrations, polycationic ferritin was bound tightly to the anionic groups of integral proteins exposed at the membrane surface and freely translated laterally with such proteins during thermotropic liquid crystalline to gel-state lipid phase transitions.

These observations led us to test the possible effects of electrostatic latticing of the integral membrane proteins on the motional freedom of such proteins. Polycationic ferritin was bound at 25°C to the spherical energy-transducing membrane, but at 90 µg/ml, and subsequently followed by the addition of native (anionic) ferritin. Examination of the deep-etched surface of the membrane after binding revealed a complete but not packed coverage by ferritin (Fig. 19). In this way native ferritin was used to electrostatically crosslink the polycationic ferritin, which was electrostatically bound to the integral membrane proteins exposed at the membrane surface. When cooled to the region of the transition exo-

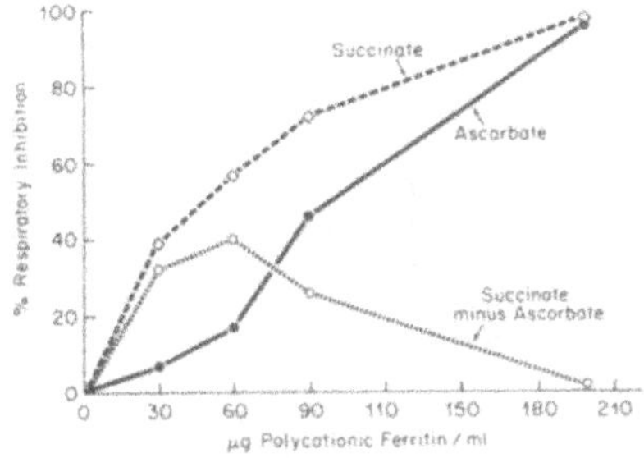

15

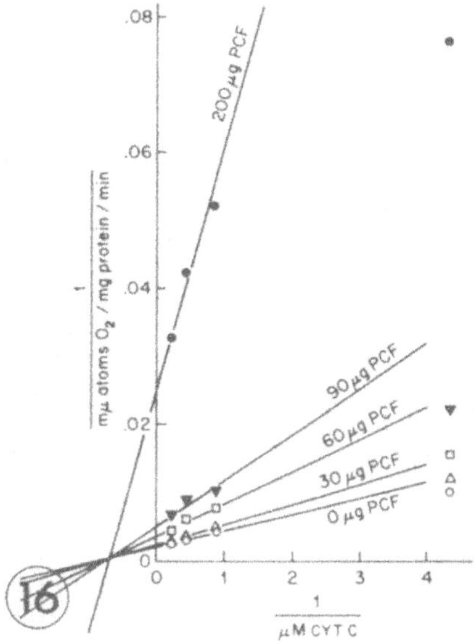

16

Figs. 15, 16. Metabolic inhibition in the energy-transducing membrane of the inner membrane-matrix fraction by electrostatic binding of polycationic ferritin

Fig. 15. Percent inhibition of succinate oxidase and ascorbate-TMPD cytochrome c oxidase activity

Fig. 16. Lineweaver-Burk plot of 1/ascorbate-TMPD cytochrome c oxidase activity vs. 1/cytochrome c concentration at progressively increasing concentrations of polycationic ferritin

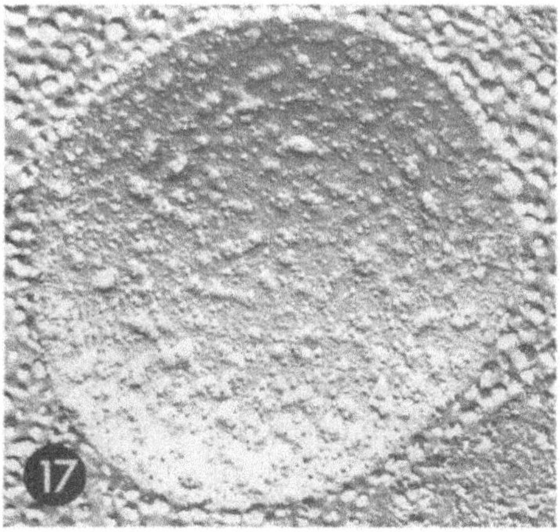

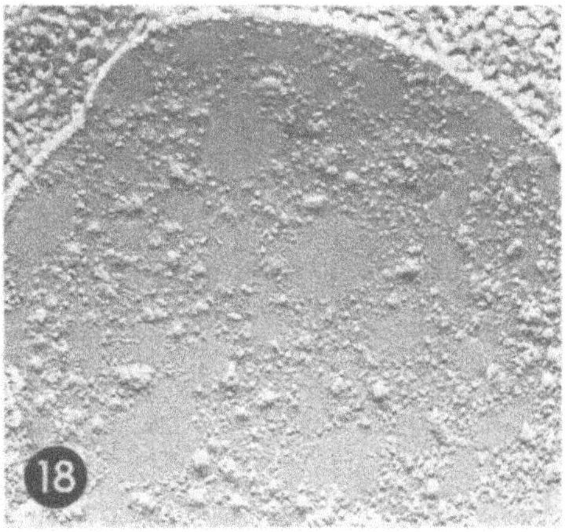

Figs. 17, 18. Concave fracture face of the energy-transducing membrane showing ferritin impressions after surface binding of polycationic ferritin at 30 μg/ml

Fig. 17. Membrane frozen from $30^{o}C$

Fig. 18. Membrane cooled to $-10^{o}C$, then frozen, shows ferritin impressions only in regions of integral proteins. x 67,500 therm after this general electrostatic crosslinking by artificial peripheral proteins, the thermotropic lateral translational mobility of the integral proteins was greatly diminished (Fig. 20). Only small, irregular regions of gel-state lipid could be detected (Fig. 20), compared to the dramatic gel-state lipid regions observed in the transition exotherm of non-latticed membranes (Figs. 10, 11). Clearly the electrostatic peripheral protein lattice prevented the lateral displacement of integral proteins during growth of the thermotropic gel-state bilayer lipid.

The free lateral translational diffusion of integral proteins which occurs in the transition endotherm during warming from subzero temperatures was virtually inhibited after electrostatic latticing. The spherical energy-transducing membrane was cooled to the transition exotherm to induce thermotropic lateral separations between integral proteins and gel-state bilayer lipid (Figs. 10, 11). This was followed by addition of polycationic ferritin at 90 μg/ml plus the subsequent addition of native ferritin at low temperature to electrostatically lattice the integral proteins in their low-temperature-induced aggregation. The deep-etched surface of the membrane revealed a highly packed coverage of ferritin with obvious ferritin-free, smooth regions which likely represented thermotropic gel-state lipid regions as viewed from the membrane surface (Fig. 21). The distribution of the highly packed bound ferritin on the membrane surface agreed with the distribution of the highly packed integral proteins observed below the onset temperature of the transition exotherm. The free lateral translational diffusion of integral proteins and the complete mixing of these proteins with the liquid crystalline bilayer lipid which occurs in unlatticed membranes when the temperature is raised to 25°C (Fig. 12), was inhibited after the electrostatic latticing by artificial peripheral proteins (Fig. 22). It may be noted that a small population of integral proteins of less than average diameter did manage to diffuse back into lipid-rich regions after raising the temperature. These small proteins represent integral proteins which presumably do not contain anionic groups exposed on the membrane surface (Fig. 22).

These data reveal that metabolically active integral proteins, such as cytochrome c oxidase and succinate permease, contain anionic groups which are exposed on the surface of the energy-transducing membrane. These proteins and most other integral proteins in this membrane can translate laterally by free diffusion in the fluid bilayer lipid of the membrane and can, as we have demonstrated, be inhibited from doing so by immobilization through electrostatic latticing with artificial peripheral proteins. Thus it is highly unlikely that the native energy-transducing membrane contains any generalized electrostatically stabilized protein-protein lattice comparable to the artificial electrostatic peripheral protein lattice that we utilized in this study.

III. Immunoglobulin Latticing of a Specific Integral Membrane Protein

1. Orientation and Site-by-Site Distribution of Cytochrome c Oxidase

We have used an affinity purified immunoglobulin monospecific for cytochrome c oxidase (cyt ox Ab) as a membrane-impermeable probe

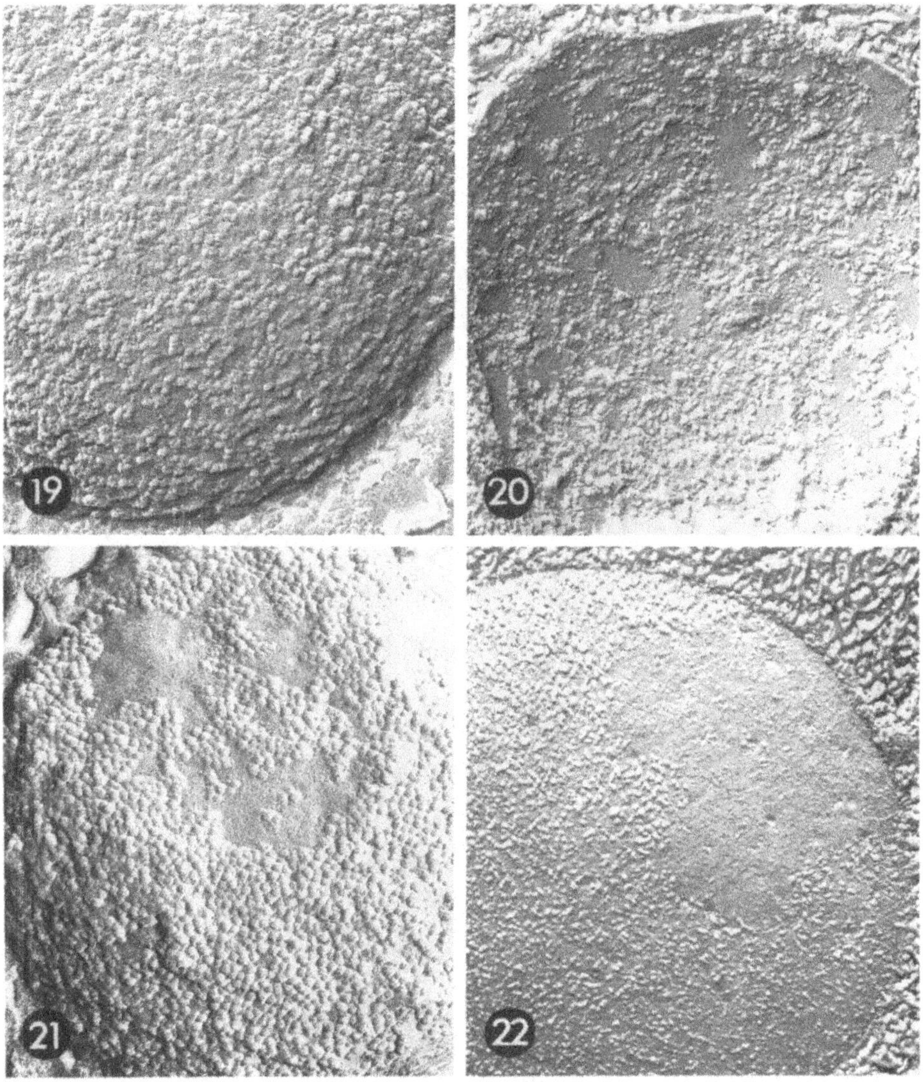

Figs. 19-22. Latticing and immobilization of integral proteins in the energy-transducing membrane through polycationic ferritin

Fig. 19. Deep-etched membrane surface after latticing with polycationic followed by native ferritin, then frozen from 25 °C

Fig. 20. Concave fracture face after latticing, then cooled to -10 °C, then frozen

Fig. 21. Deep-etched membrane surface after cooling to -10 °C, then latticed, then frozen

Fig. 22. Convex fracture face after cooling to -10 °C, then latticed, then warmed to 25 °C, and finally frozen shows a large protein-free region.
Figs. 19 and 21, x 115,000; Figs. 20 and 22, x 67,500

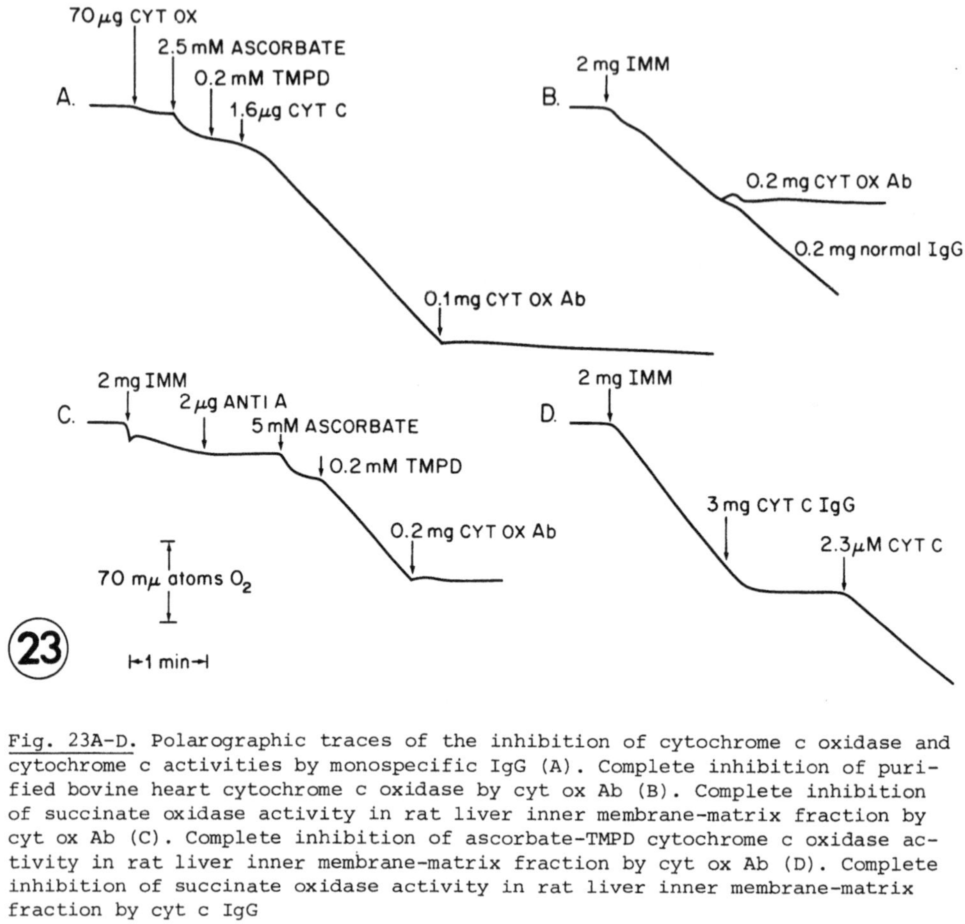

Fig. 23A-D. Polarographic traces of the inhibition of cytochrome c oxidase and cytochrome c activities by monospecific IgG (A). Complete inhibition of purified bovine heart cytochrome c oxidase by cyt ox Ab (B). Complete inhibition of succinate oxidase activity in rat liver inner membrane-matrix fraction by cyt ox Ab (C). Complete inhibition of ascorbate-TMPD cytochrome c oxidase activity in rat liver inner membrane-matrix fraction by cyt ox Ab (D). Complete inhibition of succinate oxidase activity in rat liver inner membrane-matrix fraction by cyt c IgG

to determine the distribution and orientation of the enzyme on the two surfaces of the energy-transducing membrane (Hackenbrock and Miller-Hammon, 1975). In addition, we have prepared a ferritin conjugate of the cyt ox Ab, an immunoglobulin monospecific for cytochrome c (cyt c IgG), and an affinity-purified goat immunoglobulin monospecific for rabbit immunoglobulin (goat anti-rabbit IgG). The cyt ox Ab was developed against purified beef heart cytochrome c oxidase while the cyt c IgG was developed against purified horse heart cytochrome c. All IgG used in our studies have been carefully characterized for their monospecific cross-reactivity with the cytochromes of completely solubilized rat liver mitochondria (Hackenbrock and Miller-Hammon, 1975).

Cyt ox Ab inhibited the activity of the purified bovine heart cytochrome c oxidase at a ratio of 100 µg cyt ox Ab to 70 µg oxidase (Fig. 23a). Cyt ox Ab also inhibited succinate oxidase activity (Fig. 23b) and ascorbate-TMPD cytochrome c oxidase activity (Fig. 23c) immediately and completely when it was restricted in its binding to the outer surface of the condensed energy-transducing membrane. Likewise, the binding of cyt c IgG inhibited electron transport when reacted with the outer surface of the membrane (Fig. 23d). Cytochrome c is generally believed to be a peri-

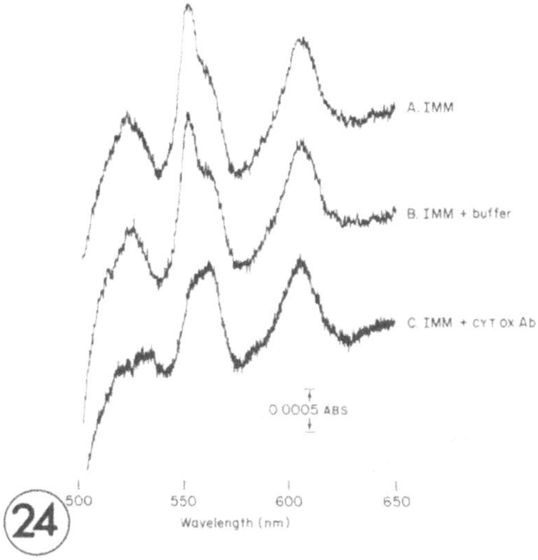

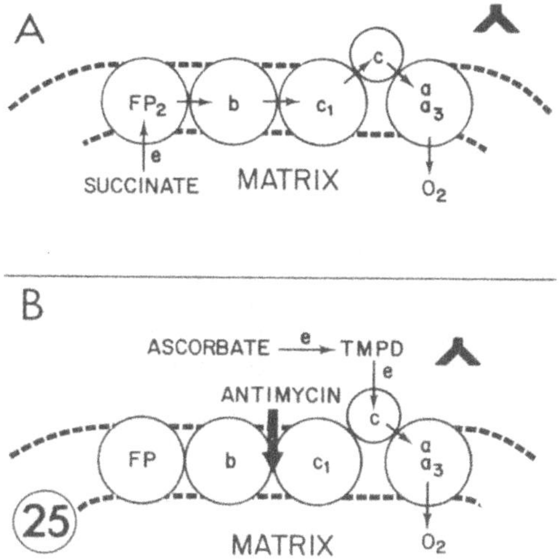

Fig. 24A-C. Reduced minus oxidized spectra of cytochromes in the energy-transducing membrane of the inner membrane-matrix (IMM) fraction (A). Spectra without additions (B). Spectra after addition of 0.1 M sodium phosphate (C). Spectra after addition of 0.1 M sodium phosphate plus cyt ox Ab

Fig. 25A,B. Pathways of electron flow and accessibility of membrane bound cytochrome c oxidase and cytochrome c to monospecific IgG (⅄) in the intact energy-transducing membrane. (A) succinate oxidation; (B) ascorbate-TMPD oxidation

pheral protein confined to the outer surface of the energy-transducing membrane. When cyt ox Ab became bound to the cytochrome c oxidase exposed on the outer surface of the membrane, a loss of cytochrome c occurred as determined by the reduced minus oxidized difference spectrum of the membrane cytochromes (Fig. 24). No loss of the integral heme proteins occurred i.e., cytochrome c oxidase (604 nm band), cytochrome b (562 nm band) and cytochrome c_1 (554 nm band).

These results clearly reveal that all functional cytochrome c oxidase is exposed on the outer surface of the energy-transducing membrane. The pathways of electron transport utilized in these experiments are shown graphically in Figure 25.

We have shown previously that sonication of the spherical energy-transducing membrane yields completely inverted membrane vesicles (Hackenbrock and Miller-Hammon, 1975). We determined that neither cytochrome c nor high concentrations of cyt c IgG affected succinate or ascorbate-TMPD supported electron transport in the completely inverted energy-transducing membrane (Fig. 26a). Low concentrations of cyt ox Ab, however, immediately and completely inhibited succinate oxidase and ascorbate-TMPD cytochrome c oxidase activity in the inverted membrane (Fig. 26b,c).

These results reveal that all functional cytochrome c oxidase is exposed on the inner surface of the energy-transducing membrane. The pathways of electron transport utilized in our studies on the inverted energy-transducing membrane are shown graphically in Figure 27.

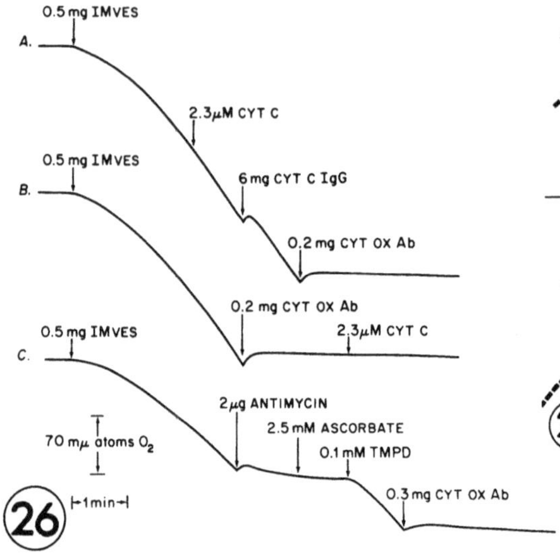

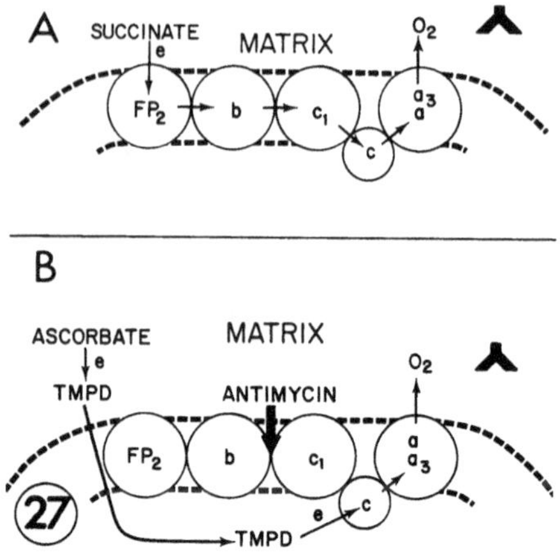

Fig. 26A-C. Polarographic traces of the inhibition of cytochrome c oxidase in inverted vesicles of the energy-transducing membrane (A,B). Insensitivity of succinate oxidase activity in inverted vesicles to cytochrome c or cytochrome c IgG and complete inhibition by cyt ox Ab (C). Complete inhibition of ascorbate-TMPD cytochrome c oxidase activity by cyt ox Ab

Fig. 27A,B. Pathways of electron flow and accessibility of membrane bound cytochrome c oxidase and cytochrome c to monospecific IgG (⅄) in the completely inverted vesicular energy-transducing membrane. (A) succinate oxidation; (B) ascorbate-TMPD oxidation

In summary, when taken collectively our data reveal that all functional cytochrome c oxidase occupies a transmembrane position as an integral protein since it is totally accessible on both surfaces of the energy-transducing membrane to an inhibiting affinity-purified immunoglobulin monospecific for the oxidase. Our data also show that all functional cytochrome c occurs only on the outer surface of the energy-transducing membrane.

The site-by-site distribution of cytochrome c oxidase on the surface of the purified energy-transducing membrane in the condensed configuration was determined with use of the ferritin conjugate of the cyt ox Ab. After complete inhibition by the ferritin-cyt ox Ab conjugate, microsamples were prepared for thin section electron microscopy. A typical site-by-site distribution of cytochrome c oxidase on the outer surface of the intact energy-transducing membrane is demonstrated in Figure 28. The oxidase clearly exhibits a nonequidistant intermolecular spacing in the plane of the membrane and thus a disordered lateral distribution. Such a distribution is consistent with our observations, which reveal the bilayer lipid of this membrane to be highly fluid. Important for our purposes in the following sections is the determination that the shortest intermolecular distance often appeared to be 100 nm.

2. Motional Freedom of Cytochrome c Oxidase

We have combined the use of thermotropic lateral translational motion of integral proteins with the use of cyt ox Ab to visually

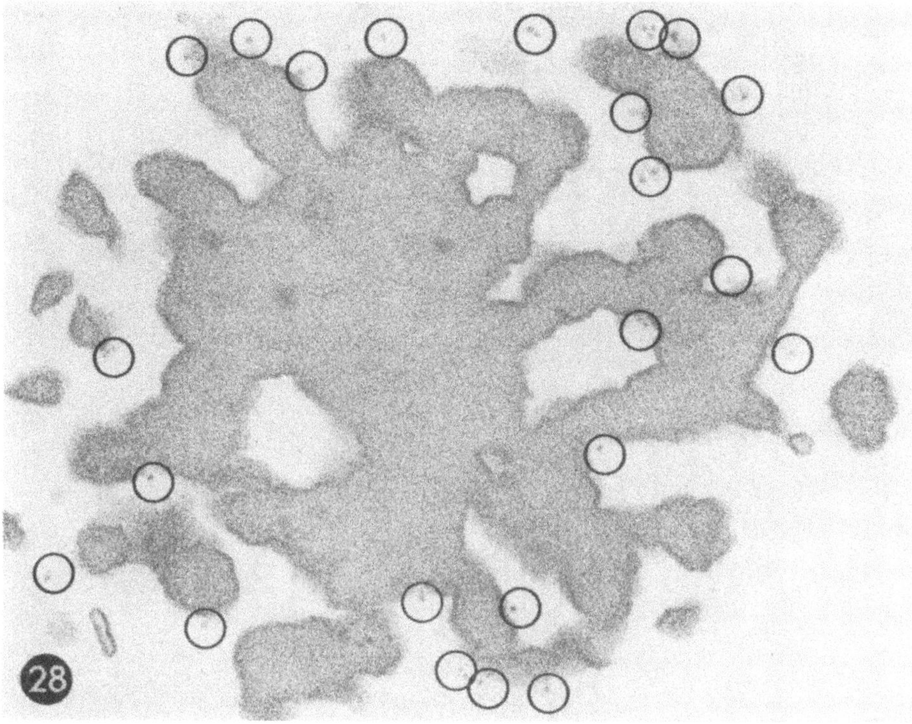

Fig. 28. Site-by-site distribution of cytochrome c oxidase on the outer surface of the energy-transducing membrane in the condensed configuration. Fixed after complete oxidase inhibition by the ferritin conjugate of cyt ox Ab. x 113,000

follow the lateral movements specifically of cytochrome c oxidase in the hydrophobic interior of the energy-transducing membrane. Rabbit cyt ox Ab was first bound to the spherical energy-transducing membrane at 25°C at concentrations which completely inhibited cytochrome c oxidase activity. Subsequently, goat anti-rabbit IgG was added in order to crosslink, at least partially, the cyt ox Ab which was bound to the cytochrome c oxidase exposed at the membrane surface. It will be recalled that we determined the most common shortest intermolecular distance between oxidases to be approximately 100 nm (Fig. 28). Thus we anticipated that the lattice, goat anti-rabbit IgG bound to cyt ox Ab bound to cytochrome c oxidase on the membrane surface, would only partially immobilize the oxidase, since fully extended IgG can span only 15 to 20 nm. Irrespective of the degree of latticing, however, integral membrane proteins would be partially latticed by immunoglobulins, representing peripheral proteins, specific for cytochrome c oxidase.

Without such latticing, it will be recalled that the integral membrane proteins aggregate through exclusion from rather extensive gel-state lipid regions in the hydrophobic interior of the energy-transducing membrane at temperatures in the transition exotherm (Figs. 10, 11). After partial immunoglobulin latticing, however, a significant inhibition of the thermotropic lateral translational movement of integral proteins was observed (Fig. 29). Under these

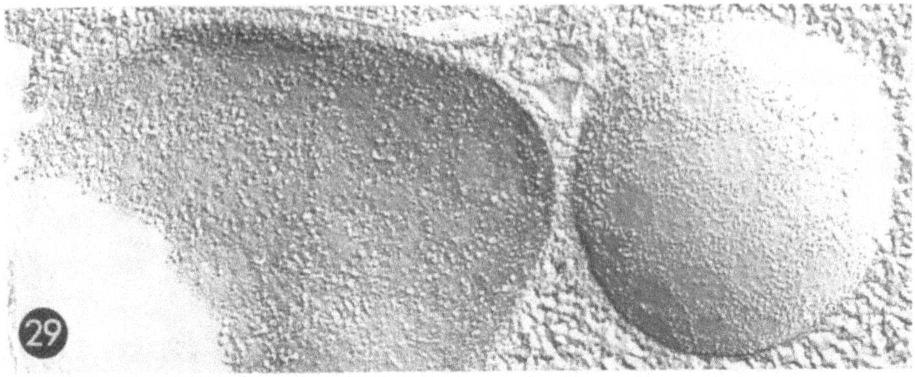

Figs. 29, 30. Lateral mobility of cytochrome c oxidase in the energy-transducing membrane and its hindrance by immunoglobulin latticing. Rabbit cyt ox Ab was permitted to bind to the membrane surface followed by addition of goat anti-rabbit IgG. After latticing, membranes were cooled to -13°C and then frozen

Fig. 29. Both fracture faces reveal only slight lateral displacement of intramembrane particles, x 67,500 (cf. non-latticed membranes, Fig. 11)

Fig. 30. High magnification reveals immunoglobulin lattice in the etched ice surface in register with the integral proteins and no immunoglobulin in regions of smooth gel state bilayer lipid. x 132,000

conditions a large number of rather small gel-state bilayer lipid regions was observed in the membrane fracture face (Fig. 29) but the dramatic separations between large patches of aggregated integral proteins and gel-state lipid found in unlatticed membranes (Fig. 10, 11) were uncommon.

After such limited thermotropic separation of membrane components the immunoglobulin lattice which was bound specifically to cytochrome c oxidase could be resolved on the membrane surface directly over areas rich in aggregated integral proteins, while membrane regions rich in gel-state lipid contained no surface-bound immunoglobulin (Fig. 30). Thus the immunoglobulin lattice, monospecific for cytochrome c oxidase, was determined to partially inhibit lateral motional freedom of the oxidase. Further, since the IgG did co-migrate laterally with the oxidase in such limited translational

Fig. 31. Graphical representation of experimental procedure and rationale utilizing immunoglobulin latticing of cytochrome c oxidase (cytox) after low temperature-induced lipid phase transition

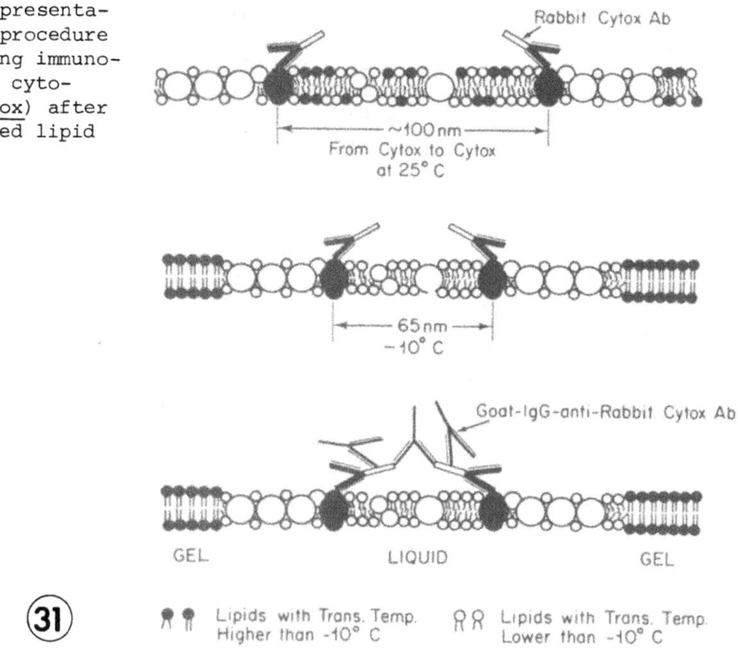

motion, it was thus clearly determined that cytochrome c oxidase in fact can translate laterally in the hydrophobic interior of the energy-transducing membrane.

3. Lateral Translational Motion of Cytochrome c Oxidase Independent of other Integral Proteins

We have seen that in the transition exotherm dramatic separations occur between regions of gel-state bilayer lipid and regions of aggregated integral proteins (Figs. 10, 11). We have also presented evidence that this thermotropic separation of membrane components is completely and immediately reversible upon raising the temperature above the transition endotherm (Fig. 12). Thus the low-temperature-induced, aggregated integral membrane proteins randomize rapidly by free lateral translational diffusion as the temperature rises and the gel-state to liquid crystalline phase transition occurs in the bilayer lipid. Cytochrome c oxidase occupies a completely transmembraneous position and has been suggested to occur as a large, hexameric integral protein in the native energy-transducing membrane (Hackenbrock and Miller-Hammon, 1975). In the present study we determined that the largest integral membrane proteins can be completely immobilized through a continuous membrane surface lattice composed of rabbit cyt ox Ab and goat anti-rabbit IgG, while many small integral proteins continue to display free lateral translational diffusion after such latticing. Figure 31 explains graphically the rationale of the experimental procedure. The spherical energy-transducing membrane was first cooled to the transition exotherm to induce aggregation of integral proteins through exclusion from the gel-state bilayer lipid (Figs. 10, 11). This was followed by addition of rabbit cyt ox Ab and subsequently by goat anti-rabbit IgG. Since in the region of the transition exotherm the intermolecular distance between the integral membrane proteins is considerably decreased, such complex latticing by immunoglobulin was expected to crosslink and immobilize the crowded cytochrome c oxidase molecules completely (Fig. 31). Any integral proteins not naturally bound to cytochrome c oxidase would be free to continue to diffuse laterally in the bilayer lipid.

After such immunospecific latticing, it was observed that the largest integral proteins (up to 18 nm in diameter) in the energy-transducing membrane did not disaggregate or randomize through free lateral translational diffusion when the temperature was raised and the gel-state to liquid crystalline phase transition occurred in the bilayer lipid (Figs. 32, 33). Nonspecific IgG did not inhibit diffusion of integral proteins of any size. Of particular interest was the observation that at 25°C the largest integral proteins remained aggregated in both the outer half (Fig. 32) and inner half (Fig. 33) of the membrane. At 25°C the outer half of the membrane revealed virtually no integral proteins in the large patches of liquid crystalline bilayer lipid (Fig. 32). The inner half of the membrane showed similar large patches of liquid crystalline bilayer lipid free of large integral proteins; however, these lipid patches contained a large number of small (4.4 to 7.4 nm in diameter) integral proteins (Fig. 33). Thus we determined that numerous small integral proteins can diffuse laterally in the energy-transducing membrane independent of cytochrome c oxidase.

On the basis of these data we conclude that cytochrome c oxidase can translate laterally in the bilayer lipid of the native energy-transducing membrane. A large number of other integral proteins

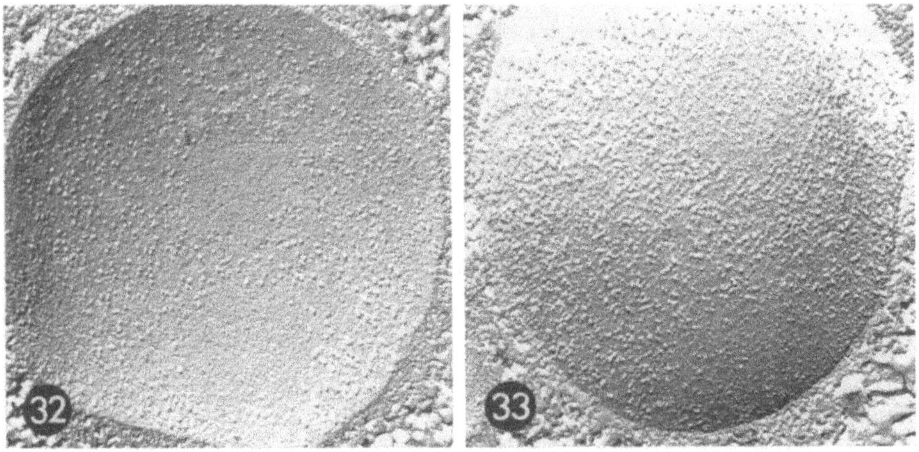

Figs. 32, 33. Free lateral diffusion of integral proteins in the energy-transducing membrane independent of cytochrome c oxidase mobility. Membranes were cooled to -10°C, then latticed with rabbit cyt ox Ab and goat anti-rabbit IgG, then returned to 25°C and finally frozen

Fig. 32. Concave fracture face shows smooth, protein-free regions at 25°C

Fig. 33. Convex fracture face shows large areas containing small but not large integral proteins at 25°C. x 67,500 can diffuse laterally in the membrane independent of cytochrome c oxidase. Diffusion of the oxidase can clearly be inhibited by crosslinking with an immunoglobulin lattice monospecific for the oxidase. Thus it is highly unlikely that the native energy-transducing membrane contains any stabilizing inter-cytochrome c oxidase or indeed any stabilizing inter-respiratory chain lattice comparable to the artificial immunospecific peripheral protein lattice that we have utilized in this study. We would note, however, that the number of large integral proteins inhibited in their free lateral translational diffusion by the specific immunoglobulin lattice is greater than can be accounted for by the number of cytochrome c oxidase molecules or molecule hexamers located in the membrane (Hackenbrock and Miller-Hammon, 1975). Thus we would conclude further that other as yet unidentified integral proteins may diffuse laterally with the oxidase in the native membrane. These may consist of cytochrome c_1, cytochrome b, and/or the hydrophobic portion of the ATPase complex. Those integral proteins which can diffuse independently of cytochrome c oxidase appear to be confined to the inner half of the energy-transducing membrane. These may consist of the various membrane-bound dehydrogenases, ion- and substrate-transporting proteins, cytochrome b, and/or the hydrophobic portion of the ATPase complex. Of considerable importance is to determine whether or not the ATPase complex can diffuse laterally in the energy-transducing membrane independently of cytochrome c oxidase and other respiratory integral proteins, since the spatial relationship between ATPase and the redox components, i.e., whether in molecular complex or spatially separated, is intimately related to the mechanism of energy conservation in the mitochondrion.

D. Fluidity and Function in the Energy-Transducing Membrane

The combined data of our DSC and freeze-fracture studies reveal that the bilayer lipid of the structurally and functionally intact energy-transducing membrane of the mitochondrion is highly fluid. Our finding is consistent with the known lipid composition of the membrane, namely a high percentage of unsaturated phospholipids and the lack of cholesterol. Such a lipid composition can explain the unusually low viscosity of approximately 0.1 P at 30°C determined for the lipid component of mitochondria (Keith et al., 1970) and the high mobility of phospholipid acyl chains detected by Carbon-13 NMR spectra (Keough et al., 1973). Both the low viscosity and high degree of motional activity in the bilayer lipid component of the membrane provide for the long-range lateral translational diffusion of integral proteins observed in our studies. There are, however, other requirements for lateral translational diffusion of membrane proteins. The integral proteins must be free of self-latticing forces and especially free of anchoring and latticing through peripheral proteins. In the energy-transducing membrane this appears to be the case.

What functional advantage is offered by the fluidity of the bilayer lipid? To answer this question we propose two levels of lateral motional freedom for integral proteins in the energy-transducing membrane, a slow, long-range lateral diffusion related to membrane biogenesis and a rapid, short-range lateral diffusion related to membrane metabolic and transport activities.

In the biogenesis of the energy-transducing membrane apparently all the integral proteins, other than several hydrophobic polypeptide subunits of cytochrome c oxidase, the ATPase complex, and a cytochrome b component, are nuclear gene products. The nuclear encoded polypeptides destined for residence in the energy-transducing membrane are most likely incorporated into the mitochondrion at the 100 or more inner membrane-outer membrane contact sites (Hackenbrock, 1968b; Hackenbrock and Miller, 1975). Supporting this idea of protein incorporation is the observation that cytoplasmic ribosomes anchor to the outer mitochondrial membrane surface preferentially at contact sites (Kellems et al., 1975). An earlier study revealed that polypeptides transcribed by the mitochondrial genome and destined for the energy-transducing membrane are identified first in the inner boundary membrane portion (thus in the vicinity of the contact sites) rather than in the cristal membrane portion of the energy-transducing membrane (Werner and Neupert, 1972). We would speculate that integral proteins are incorporated into the outer membrane as nuclear gene products or into the inner membrane as mitochondrial gene products at the membrane-membrane contact sites. Once incorporated, the proteins may assume their ultimate functional location in the energy-transducing membrane by a slow, long range, free lateral translational diffusion in the highly fluid bilayer lipid.

We believe that the fluid nature of the bilayer lipid of the energy-transducing membrane provides for the motional freedom of conformational events in integral proteins inherent in metabolic and transport activities. Lateral translational motion of integral proteins during metabolic activity is most likely rapid and short range. The "paristaltic" mechanism for the ion transport event requires protein conformational motion and lipid fluidity for such motion (Singer, 1974).

With regard to the time structure of the metabolic proteins in the energy-transducing membrane, it is difficult to reconcile the half time of oxidation of some redox proteins with the half time of ATP synthesis in terms of a physical coupling of these proteins with ATPase. For example, at the third energy-transducing segment in the respiratory chain the half time oxidation of cytochrome c oxidase is 0.5 msec (Chance et al., 1967) whereas ATP synthesis coupled to this oxidation is approximately 100 msec. This represents a considerable delay in rate and half time of ATP synthesis (Lemasters, 1975; Thayer and Hinkle, 1975; Lemasters and Hackenbrock, in press, 1976). Clearly such delays may occur as a function of the time required for conformational changes to occur in redox and ATPase components during conformational coupling in the membrane, or as a function of the time required to generate a protonmotive gradient across the membrane adequate to synthesize ATP. These delays, however, may also be accounted for by some component of protein diffusion, in addition to conformational motion, in the overall mechanism during which oxidative energy is transduced.

Fluidity in the bilayer lipid of course would permit such diffusion. Lateral translational diffusion of redox components and ATPase may well occur in the energy-transducing membrane if the three energy-transducing segments of the respiratory chain transduce oxidative energy to the same ATPase complex. Lateral translational diffusion by cytochrome c appears to occur during interchain electron transport (Wohlrab, 1970). Extended half times of the oxidation of cytochrome b and of flavoproteins in oxygen pulse experiments are also consistent with lateral diffusion of membrane components (Chance et al., 1967). Delay times in the transfer of reducing equivalents from membrane-bound dehydrogenases to the cytochrome chain appear to occur through the lateral diffusion of ubiquinone between the dehydrogenases and b cytochromes (Kröger et al., 1973a, 1973b).

The fluid nature of the bilayer lipid of the energy-transducing membrane can provide for the rapid, short-range lateral motion required in all of these metabolic events. The lateral translational diffusion of integral proteins occurs through Brownian motion with diffusion coefficients centering at approximately 5×10^{-9} $cm^2 \cdot sec^{-1}$ at 20°C (Edidin and Fambrough, 1973; Poo and Cone, 1974; Liebman and Entine, 1974). This is equal to an approximate molecular displacement of 100 nm in one second. Therefore, as a first approximation, the potential for free lateral translational diffusion of integral proteins in the energy transducing membrane can be greater than 10 nm $\cdot$ 100 $msec^{-1}$ at physiological temperature. Such lateral diffusion rates in redox proteins and ATPase would be consistent with the time structure of the various respiratory chain segments as well as with the time structure of the redox-ATPase couple.

E. Summary and Conclusions

The mitochondrial energy-transducing membrane is distinct from all other membranes in the eukaryote cell owing to its complex activities related to the conservation of metabolic energy. This membrane is also distinctive in its molecular composition. It possesses a high content of both peripheral and integral proteins, a high degree of unsaturated phospholipids, and is free of cholesterol. The result of this unique molecular composition is a highly effective concentration of integral proteins partitioned in a polar bilayer phospholipid environment of relatively low viscosity and high fluidity. The peripheral polar environment of the bilayer phospholipid provides for the precise vertical orientation of the metabolically active integral proteins. The fluid environment of the bilayer phospholipid provides for the conformational and lateral translational motion of the integral proteins inherent in metabolic activity. The integral proteins of the energy-transducing membrane are not immobilized, anchored, or organized by any general, rigid protein-protein lattice throughout the membrane. Some integral proteins, such as cytochrome c oxidase show a high potential for lateral translational diffusion independent of some and in association with other integral proteins. Slow, long-range lateral translational diffusion may permit integral proteins, which have been incorporated into the energy-transducing membrane during biogenesis, to reach their sites of functional activity in the membrane. In transport and energy-transducing activities, rapid, short-range lateral translational diffusion of the energy-transducing proteins of the membrane may play a rate-limiting role in the mechanism of electron transport in various segments of the respiratory chain as well as in the mechanism of oxidative phosphorylation. Estimates of the rates of lateral translational diffusion of integral membrane proteins are consistent with the time structure of various segments of the respiratory chain as well as with the time structure of the redox-ATPase couple.

Acknowledgments. We wish to thank our associates Katy Hammon and Mary Tobleman who have contributed significantly to the research presented here. Dr. Höchli is a Fellow of the Swiss National Foundation and the Muscular Dystrophy Association of America. These studies were supported by research grants from the U.S. National Science Foundation and the National Institutes of Health.

References

Capaldi, R.A., Green, D.E.: FEBS Lett. 25, 205 (1972)
Capaldi, R.A., Tan, P.-F.: Federation Proc. 33, 1515 (1974)
Chance, B., DeVault, D., Legallais, V., Mela, L., Yonetani, T.: In: Nobel Symp. 5, Fast Reactions and Primary Processes in Chemical Kinetics. Claesson, S. (ed.). New York: Interscience 1967, p. 437
Chapman, D., Urbina, J., Keough, K.M.: J. Biol. Chem. 249, 2512 (1974)
Chen, Y.S., Hubbell, W.L.: Exp. Eye Res. 17, 517 (1973)
Colbeau, A., Nachbaur, J., Vignais, P.M.: Biochim. Biophys. Acta 249, 462 (1971)
Edidin, M., Fambrough, D.: J. Cell Biol. 57, 27 (1973)
Fleisher, S., Fleisher, B., Stoechenius, W.: J. Cell Biol. 32, 193 (1967)
Grant, C.W.M., McConnell, H.M.: Proc. Natl. Acad. Sci. U.S. 71, 4653 (1974)
Hackenbrock, C.R.: J. Cell Biol. 30, 269 (1966)
Hackenbrock, C.R.: J. Cell Biol. 37, 345 (1968a)
Hackenbrock, C.R.: Proc. Natl. Acad. Sci. U.S. 61, 598 (1968b)
Hackenbrock, C.R.: J. Cell Biol. 53, 450 (1972)
Hackenbrock, C.R.: In: Mechanisms in Bioenergetics. Azzone, G.F., Ernster, L., Papa, S., Quagliariello, E., Siliprandi, N. (eds.). New York: Academic Press 1973, p. 77
Hackenbrock, C.R.: Arch. Biochem. Biophys. 170, 139 (1975)
Hackenbrock, C.R., Miller, K.J.: J. Cell Biol. 65, 615 (1975)
Hackenbrock, C.R., Miller-Hammon, K.: J. Biol. Chem. 250, 9185 (1975)
Harmon, H.J., Hall, J.D., Crane, F.L.: Biochim. Biophys. Acta 334, 119 (1974)
Jost, P.C., Griffith, O.H., Capaldi, R.A., Vanderkool, G.: Proc. Natl. Acad. Sci. U.S. 70, 480 (1973)
Kellems, R.E., Allison, V., Butow, R.A.: J. Cell Biol. 65, 1 (1975)
Keith, A., Bulfield, G., Snipes, W.: Biophysic. J. 10, 618 (1970)
Keough, K.M., Oldfield, E., Chapman, D., Beynon, P.: Chem. Phys. Lipids 10, 37 (1973)
Kleeman, W., Grant, C.W.M., McConnell, H.M.: J. Supramol. Struct. 2, 609 (1974)
Klingenberg, M.: In: Biological Oxidations. Singer, T.P. (ed.). New York: John Wiley & Sons 1968, p. 3
Kröger, A., Klingenberg, M., Schweidler, S.: Europ. J. Biochem. 34, 358 (1973a)
Kröger, A., Klingenberg, M., Schweidler, S.: Europ. J. Biochem. 39, 313 (1973b)
Ladbrooke, B.D., Chapman, D.: Chem. Phys. Lipids 3, 304 (1969)
Lee, M.P., Gear, A.R.L.: J. Biol. Chem. 249, 7541 (1974)
Lemasters, J.J.: Federation Proc. 34, 596 (1975)
Lemasters, J.J., Hackenbrock, C.R.: Europ. J. Biochem. (in press)
Liebman, P.A., Entine, G.: Science 185, 457 (1974)
McConnell, H.M., Wright, K.L., McFarland, B.G.: Biochem. Biophys. Res. Commun. 47, 273 (1972)
Papahadjopoulos, D., Vail, W.J., Moscarello, M.: J. Membrane Biol. 22, 143 (1975)
Parsons, D., Williams, G.R., Chance, B.: Ann N.Y. Acad. Sci. 137, 643 (1966)
Poo, M.-M., Cone, R.A.: Nature (London) 247, 438 (1974)
Raison, J.K., Lyons, J.M., Thomson, W.W.: Arch. Biochem. Biophys. 142, 83 (1971)
Singer, S.J.: Ann. Rev. Biochem. 43, 805 (1974)
Sjöstrand, F.S., Barajas, L.: J. Ultrastruct. Res. 32, 293 (1970)
Steim, J.M., Tourtellotte, M.E., Reinert, J.C., McElhaney, R.N., Rader, R.L.: Biochemistry 63, 104 (1969)
Thayer, W.S., Hinkle, P.C.: J. Biol. Chem. 250, 5336 (1975)
Trauble, H., Overath, P.: Biochim. Biophys. Acta 307, 491 (1973)
Vanderkooi, G.: Biochim. Biophys. Acta 344, 307 (1974)
Werner, S., Neupert, W.: Europ. J. Biochem. 25, 369 (1972)
Wohlrab, H.: Biochemistry 9, 474 (1970)

Lipid-Lipid and Lipid-Protein Interactions in Membranes

Some Aspects of Lipid-Phase Transition on Membrane Permeability and Lipid-Protein Association

M. C. Blok, L. L. M. van Deenen, J. de Gier, J. A. F. Op den Kamp, and A. J. Verkleij

A. Introduction

Because the barrier properties of biological membranes are provided by a lipid bilayer, it has been useful to establish correlations between membrane characteristics and the chemical and physical properties of their constituent lipids by comparative studies on permeability properties of liposomes. Studies on liposomes demonstrated that the bilayer is a selective barrier, the permeability being strongly dependent on the polarity and molecular size of the permeant. Water molecules easily penetrate the bilayer, but the permeability for ions is many orders of magnitude less. Small polar non-electrolytes, such as glycol and glycerol, have an intermediate permeability whereas a number of hydrophobic ionophores have been demonstrated to penetrate rather easily. In natural membranes chemical variations in lipids occur both in the polar headgroups and in the hydrophobic regions. Studies on liposomes prepared from synthetic phospholipids gave information on the influence of the chemical structure of the lipid molecule on the barrier function of the bilayer. These studies showed that parameters such as size and charge of the polar headgroup, chain length and degree of unsaturation of the fatty acid chains and interactions of phospholipid with sterol are of great importance for the permeability behavior of the artificial membranes. Similar results were obtained when variations were induced in the chemical composition of the lipids of natural membranes. The comparative permeability experiments on liposomes and biological membranes showed an excellent correlation, not only for penetration of non-electrolytes, but also for the effect of lipid composition on valinomycin-induced transport (de Gier, 1973; van Deenen et al., 1975).

B. Effect of Lipid-Phase Transition on Water Permeation

The present discussion is limited to some recent studies on the effect of the lipid-phase transition (gel to liquid-crystalline) on permeability properties of membranes. Until now, most studies concerning the effect of membrane lipid composition on permeability properties of membranes have been carried out on systems that contain the lipids in the liquid-crystalline state. However, arguments can be forwarded that at least in a number of membranes, local regions may exist which contain lipids in the gel state. Therefore, a comparative study on the permeation of water through liposomal membranes above and below the gel liquid-crystalline phase transition temperature was carried out (Blok et al., 1976). It was found that liposomes prepared of dimyristoyl lecithin and dipalmitoyl lecithin behaved in glucose solutions as perfect osmometers, both above and below the gel to liquid-crystalline phase

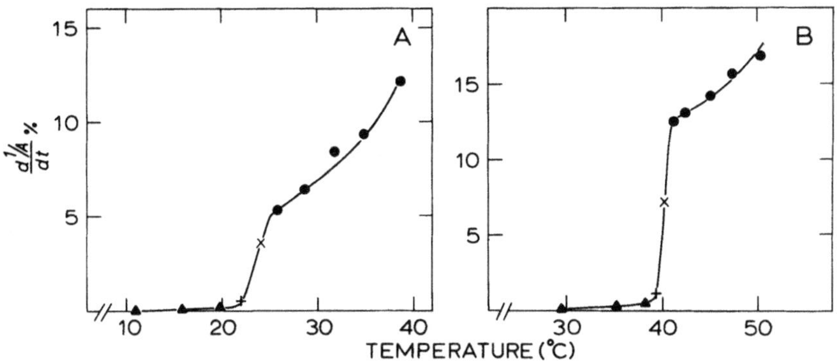

Fig. 1A,B. Temperature dependence of the osmotic shrinkage of liposomes; dimyristoyl lecithin (A) and dipalmitoyl lecithin (B)

transition temperature. In as much as glucose can be considered incapable of permeating through the lipid bilayer under the conditions of these experiments, it is concluded that liposomal membranes with their lipids in the gel-state are still permeable to water. This conclusion was supported by directly measuring the diffusion of $^{3}H_2O$ through bilayers made of dipalmitoyl lecithin (Blok et al., 1976).

As demonstrated in Figure 1, the rate of water permeation through lipid bilayers changes drastically on passing the transition temperature. Calculation of the activation energy demonstrates increase in the energy barrier of water permeation across lipid bilayers in the gel state when compared with those in the liquid-crystalline state; the water permeation has activation energies of 9.5 ± 1.8 and 26.4 ± 0.9 kcal $\cdot$ mol^{-1} above and below the transition temperature, respectively.

Cholesterol has been shown to decrease the rate of permeation of non-electrolytes across lipid bilayers in the liquid-crystalline state. It is well-known that cholesterol reduces the energy content of the lipid phase transition and it was found of interest to investigate the temperature dependency of water permeation across liposomes of mixtures of synthetic phospholipid with cholesterol. In a previous study from this laboratory indications were already obtained that the presence of cholesterol in a phospholipid bilayer may increase the rate of permeation of non-electrolytes below the transition temperature (de Gier et al., 1969).

The effect of cholesterol on water permeation in liposomes of dipalmitoyl lecithin is demonstrated in Figure 2. The temperature dependency of the osmotic shrinkage of these liposomes shows that above the transition temperature the rate of permeation of water is decreased by the presence of cholesterol. Increasing amounts of cholesterol cause a gradual disappearance of the abrupt change in the rate of water permeation near the gel to liquid-crystalline phase transition of liposomes of synthetic saturated lecithins. At cholesterol concentrations above 30 mol% no further discontinuity in the rate of water permeation exists.

Cholesterol was found to produce an increase in the activation energy of the water permeation above the transition temperature of the saturated lecithins at about 15 mol% cholesterol. Below

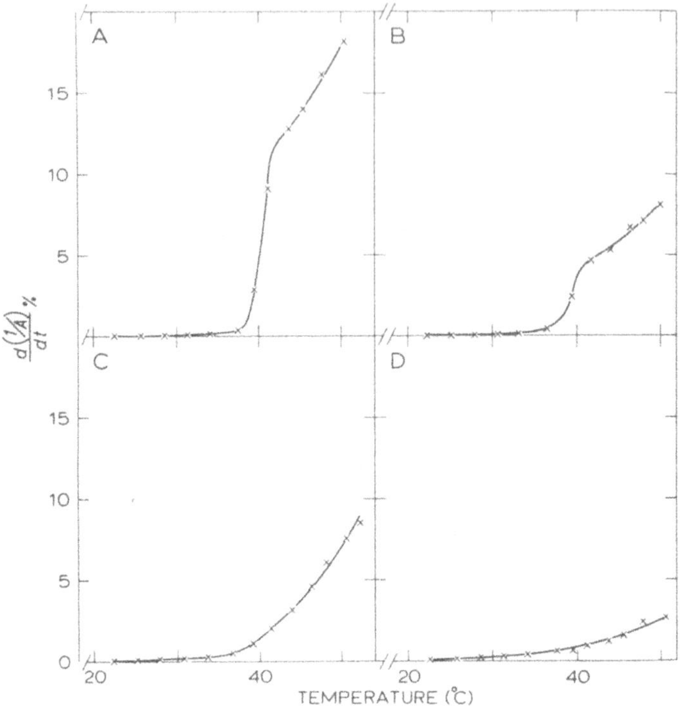

Fig. 2A-D. Temperature dependence of the osmotic shrinkage of liposomes of dipalmitoyl lecithin containing 0 (A), 10 (B), 20 (C) and 33 (D) mol% of cholesterol

the transition temperature a gradual decrease in the activation energy of water permeation was observed in the region of 0 to 33 mol% of cholesterol. In systems containing unsaturated phosphatidylcholines, cholesterol enhanced the activation energy of the water permeation to a lesser extent. Furthermore, it was found that the position of the cis double-bond in the fatty acid chain of the phospholipid was important.

C. Effects of Lipid-Phase Transition on Ionic Permeability

Lipid bilayers made of phospholipids occurring in natural membranes are rather impermeable to ions. Using synthetic phospholipids an exception to this rule can be observed when the chain length of the fatty acid constituents of the phospholipid is reduced below a certain critical value. The barrier properties of artificial membranes of dilauroyl lecithin differ significantly from those of higher lecithin analogs since the former bilayers are not capable of retaining ions. It is of interest to note that synthetic phospholipids of intermediate chain length have been demonstrated to be lytic against erythrocytes, their activity being comparable to lysolecithins. A significant decrease in bilayer thickness, depending on the chain length of diacyl-lecithins or on the presence of lysolecithins leads to a decreased interaction between the lipid molecules and is believed to be responsible for the inability of these bilayers to act as an effective permeability barrier (Mandersloot et al., 1975). By contrast liposomes

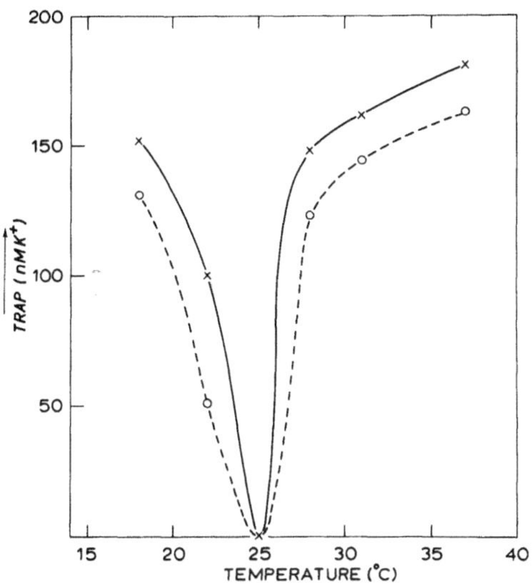

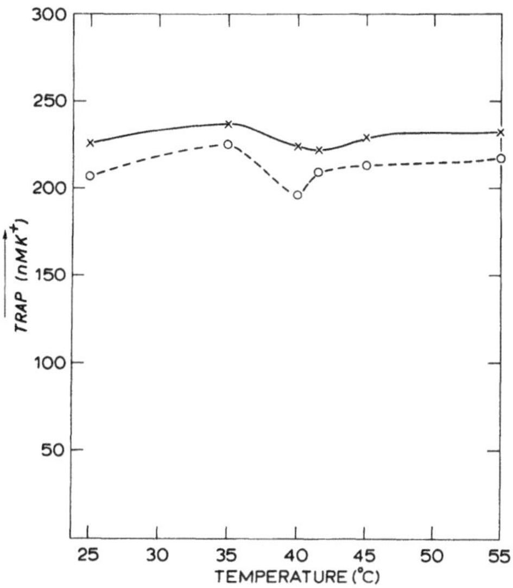

Fig. 3. Temperature dependence of the quantity of potassium trapped by liposomes after an incubation of 30 sec (X-X) and 5 min (O---O). Left: dimyristoyl lecithin. Right: dipalmitoyl lecithin

prepared from saturated lecithins with 14 or more carbon atoms per acyl chain exist as stable bilayers which are practically impermeable to ions above the gel to liquid-crystalline phase transition temperature.

Previous studies of Haest et al. (1972) showed a fragility of the permeability barrier of Escherichia coli when these cells were cooled at, or below the transition temperature of the membrane lipids, resulting in a spontaneous loss of e.g. intracellular K^+. It was found that liposomes prepared from dimyristoyl lecithin incubated in the vicinity of the transition temperature also revealed a release of trapped potassium. To see whether the length of the paraffin chain in the phospholipid is important for the increased release of ions from liposomes near the transition temperature, the potassium leak was measured from liposomes of dimyristoyl lecithin and dipalmitoyl lecithin as a function of temperature (Fig. 3). Indeed, it was found that the release of potassium at the transition temperature where liquid and solid domains coexist in the plane of the membrane is much more pronounced for liposomes of dimyristoyl lecithin than for those of dipalmitoyl lecithin.

The increased ionic permeability of bilayers in which there is an equilibrium between the liquid-crystalline and gel-state appeared to decrease with increasing diameter of the permeating ion (Blok et al., 1975). Comparative osmotic shrinkage experiments in hypertonic media indicate that liposomes of dimyristoyl lecithin have a very high permeability for NaCl, KCl and RbCl, but less for LiCl. Liposomes of dipalmitoyl lecithin showed shrinkage in all four alkaline chlorides in the vicinity of the transition temperature, although the shrinkage in LiCl is still most pronounced; liposomes of distearoyl lecithin did not show significant differences in one of the four media under identical conditions.

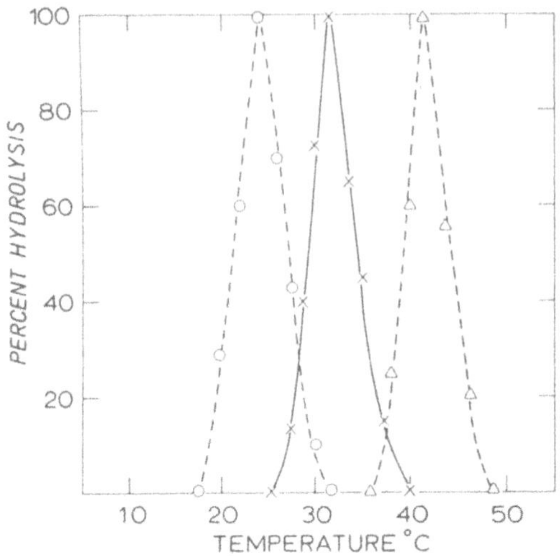

Fig. 4. Temperature dependence of the hydrolysis of liposomes of synthetic phospholipids by pancreatic phospholipase A_2; dimyristoyl lecithin (O---O), dipalmitoyl lecithin (Δ---Δ) and an equimolar mixture of both lecithins (X-X)

These experiments support the conclusion that the increased permeability in the bilayers at the temperature of the phase transition depends on the diameter of the permeating ion and the length of the hydrocarbon chain of the lecithins. Similar results were obtained in experiments using a series of potassium salts where the anions differed in size. The findings on the selectivity of the enhanced permeability for ions support the idea that a macroscopic rupture and resealing of the liposomal membrane is not responsible for the observed alterations in the barrier properties of the bilayers under conditions of phase separation. It has been suggested that a dynamic equilibrium at the boundaries of liquid and solid domains is important for the formation of statistical pores which allow the permeation of solutes (Blok et al., 1975).

D. Lipid-Phase Transition and Lipid-Protein Association

Various investigations indicate that gel to liquid-crystalline phase transition of phospholipids may affect protein-lipid interaction. A facilitated interaction of a protein with lipid bilayers, where liquid and solid domains coexist, was observed in studies on the action of phospholipase A_2 from pancreas (Op den Kamp et al., 1974, 1975). This enzyme can act at a slow rate at substrate molecules of unsaturated lecithins in liposomes above the transition temperature, but the rate of enzymatic hydrolysis is greatly enhanced near the transition temperature of the substrate concerned. However, fully saturated lecithins are hydrolyzed only at the transition temperature as demonstrated by experiments on dilauroyl-, dimyristoyl- and dipalmitoyl-lecithin; both above and below the transition temperature of the phospholipid the rate of hydrolysis of the substrate in the bilayers is negligible (Fig. 4). This figure also demonstrates that an equimolar mixture of dimyristoyl lecithin and dipalmitoyl lecithin, which shows co-crystallization of the paraffin chains, is hydrolyzed between 25 and 40°C with a maximum at 32°C, this in agreement with the calorimetric scan. Somewhat different results have been obtained with a mixture of two saturated lecithins which show a monotectic behav-

Fig. 5. Hydrolysis of lecithin-cholesterol liposomes by phospholipase A_2 at different temperatures. Liposomes of dimyristoyl lecithin containing 0 (X-X), 10 (O-O), 20 (∇-∇) and 35 (□-□) mol% of cholesterol

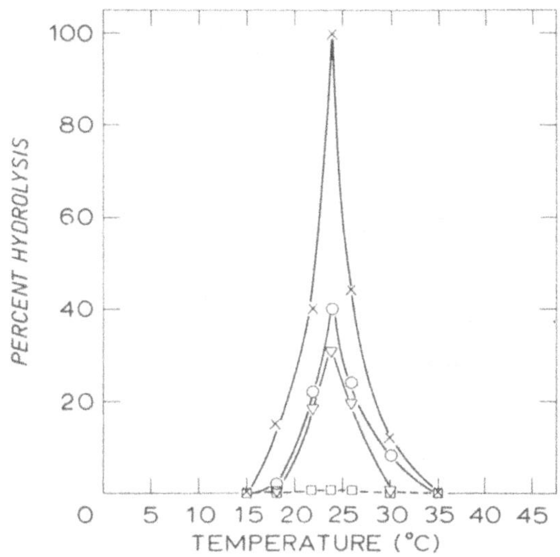

ior. An equimolar mixture of dilauroyl lecithin and distearoyl lecithin is hydrolyzed at all temperatures, but two optima in the rate of hydrolysis were found at 0 and 40°C, which temperatures coincide with the phase transition of dilauroyl- and distearoyl-lecithin respectively in this mixture. This result suggests that not only the cooperative phase transition stimulates hydrolysis, but that already the coexistence of two phases in a bilayer promotes the penetration of the pancreatic phospholipase A_2 into the interface. It was found that in this monotectic system the more fluid lecithin species is hydrolyzed at a faster rate over the whole temperature range. Also in the cocrystallizing system - at the temperature region of the phase transition - the species with the shorter fatty acid chains are attacked preferentially, although after continued incubation all components are completely hydrolyzed, though at a variable rate.

The results obtained on the action of pancreatic phospholipase A_2 on the long-chain saturated lecithin molecules, suggest that the enzymatic hydrolysis is enhanced by perturbations in the packing of the lipid molecules in the bilayer, which are created at the temperature of the phase transition. It is proposed that a penetration of the enzyme into the lipid layer is facilitated at the border of the domains with lipid in the gel and liquid-crystalline state. Because of the effect of cholesterol on the cooperative phase transition of lecithin molecules, it was of interest to test whether the introduction of this sterol into the bilayer will result in an alteration of the action of phospholipase A_2 from pig pancreas. As was demonstrated in Figure 5, in the presence of 10 and 20 mol% of cholesterol the hydrolysis of dimyristoyl lecithin proceeds at a slower rate than in the absence of cholesterol; maximal hydrolysis still occurs at 23°C and a similar temperature profile was obtained in the absence and presence of cholesterol. However, in the presence of about 35 mol% of cholesterol, hydrolysis is completely inhibited at all temperatures. Identical results have been obtained with mixtures of cholesterol and other saturated lecithins. Furthermore, the restoration of the phase transition by the addition of polyene antibiotics, such as filipin and amphotericin B to the liposomes of saturated lecithin-cholesterol re-

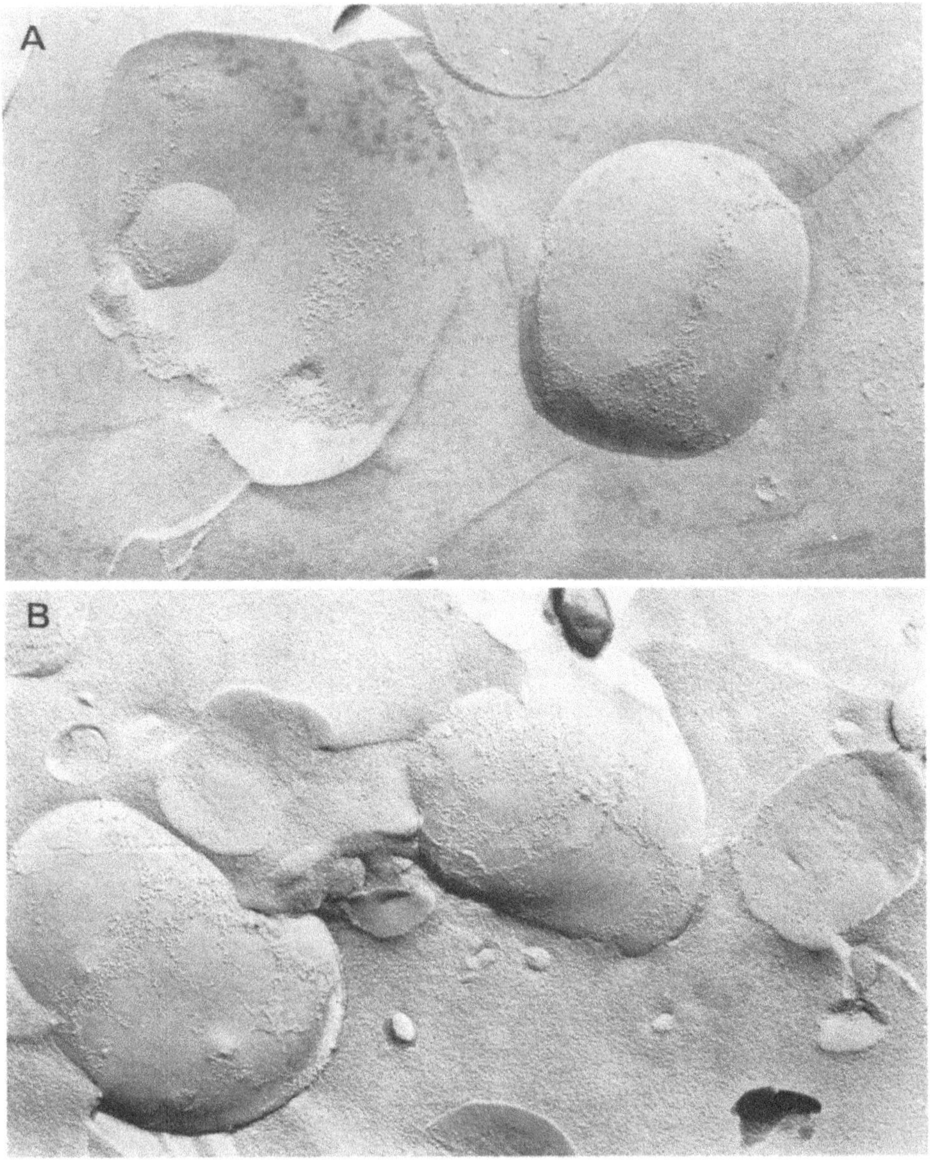

sults in a restored susceptibility towards the enzyme (Op den Kamp et al., 1975). Under these conditions, the substrate in the liposomes, which contains the cholesterol-antibiotic complexes are accessible for the enzyme not only in, but also below and above the transition temperature. It can be argued that in this case the packing of the lipid molecules is disturbed, because of the presence of these complexes, thereby facilitating the penetration of the enzyme into the perturbed bilayer.

Freeze-etching has well documented that lipid phase transition effects the topographical distribution of intrinsic membrane pro-

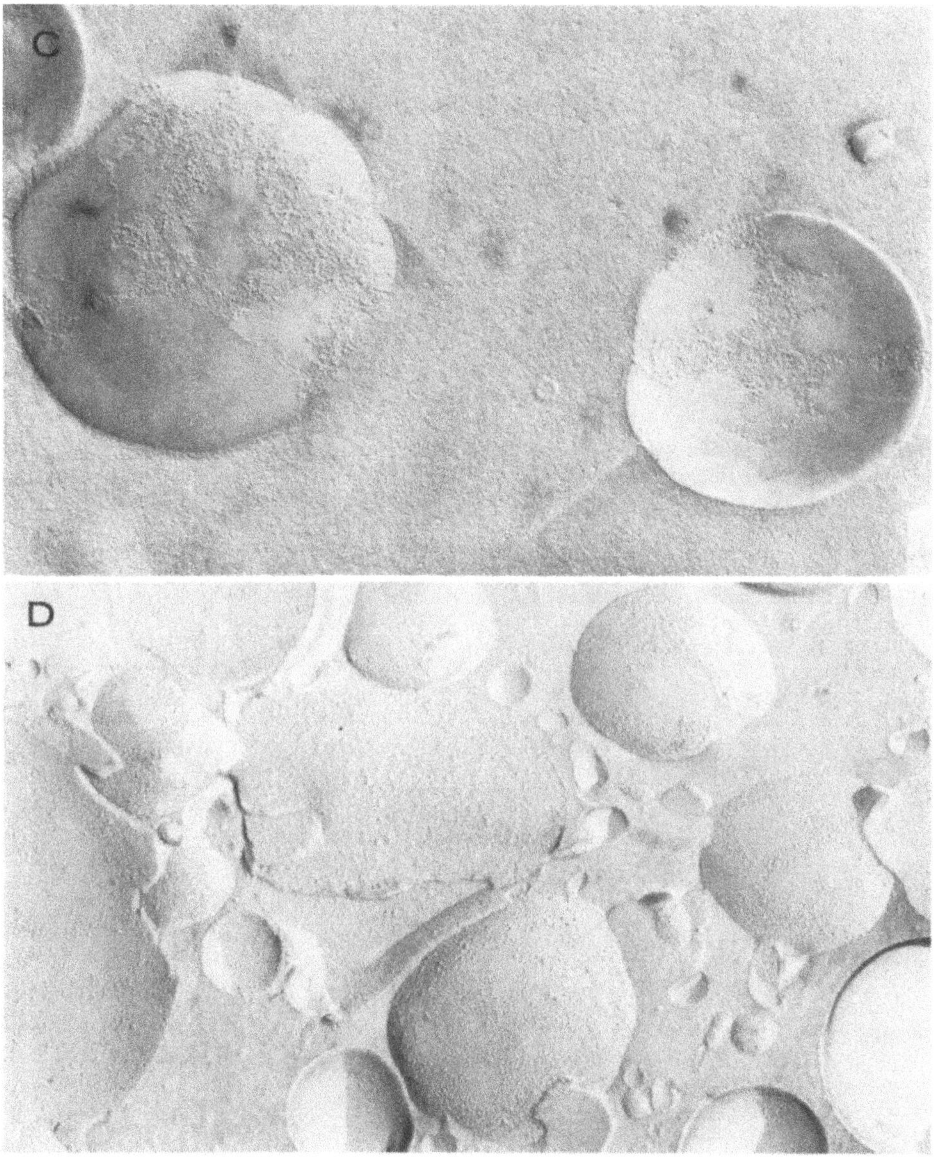

Fig. 6A-D. Inner and outer fracture faces of Acholeplasma laidlawii grown on different fatty acids. (A) grown on elaidic acid; (B) grown on an equimolar mixture of palmitic and oleic acid; (C) grown on myristic acid; (C) grown on oleic acid. Quenched from -15 $^{\circ}$C

teins. (For a recent review see Verkleij and Ververgaert, 1975). Of particular interest in this respect are studies on Acholeplasma laidlawii, because the region of the lipid phase transition can be varied by supplementing the growth medium with particular fatty acids. A fairly homogeneous distribution of protein particles on the fracture faces of A. laidlawii is observed above the transition temperature. Aggregation of the intercalated particles occurs after cooling the membranes to the point where phase transition of the lipids commences (Fig. 6). The temperature at which this aggregation of particles starts, was found to be dependent on the nature of the fatty acid constituents of the membrane lipids and to be related to their transition temperatures. The particle-free areas of the fracture faces of the cell membranes are considered to represent lipid regions in the gel-state. Below the lipid phase transition the protein particles are aggregated in different domains. It was recently found that these domains are different in size for cells grown on various fatty acids and their mixtures (Fig. 6).

References

Blok, M.C., Neut-Kok, E.C.M., van der, Deenen, L.L.M., van, Gier, J., de: The effect of chain length and lipid phase transition on the selective permeability properties of liposomes. Biochim. Biophys. Acta 406, 187-196 (1975)

Blok, M.C., Deenen, L.L.M., van, Gier, J., de: Effect of the gel to liquid-crystalline phase transition on the osmotic behavior of phosphatidylcholine liposomes. Biochim. Biophys. Acta 433, 1-2 (1976)

Deenen, L.L.M., van, Gier, J., de, Demel, R.A., Kruyff, B., de, Blok, M.C., Neut-Kok, E.C.M., van der, Haest, C.W.M., Ververgaert, P.H.J.Th., and Verkleij, A.J.: Lipid-lipid and lipid-protein interaction in model systems and membranes. Ann. N.Y. Acad. Sci. 264, 124-141 (1975)

Gier, J., de, Mandersloot, J.G., Deenen, L.L.M. van: The role of cholesterol in lipid membranes. Biochim. Biophys. Acta 173, 143-145 (1969)

Gier, J., de: Comparative permeability studies on liposomal and biological membranes with various lipid compositions. Biochem. Soc. Trans. 1, 331-333 (1973)

Haest, C.W.M., Gier, J., de, Es, G.A., van, Verkleij, A.J., Deenen, L.L.M., van: Fragility of the permeability barrier of Escherichia coli. Biochim. Biophys. Acta 288, 43-53 (1972)

Mandersloot, J.G., Reman, F.C., Deenen, L.L.M., van: Barrier properties of lecithin/lysolecithin mixtures. Biochim. Biophys. Acta 382, 22-26 (1975)

Op den Kamp, J.A.F., Gier, J., de, Deenen, L.L.M., van: Hydrolysis of phosphatidylcholine liposomes by pancreatic phospholipase A_2 at the transition temperature. Biochim. Biophys. Acta 345, 253-256 (1974)

Op den Kamp, J.A.F., Kauerz, M.Th., Deenen, L.L.M., van: Action of pancreatic phospholipase A_2 on phosphatidylcholine bilayers in different physical states. Biochim. Biophys. Acta 406, 169-177 (1975)

Verkleij, A.J., Ververgaert, P.H.J.Th.: The architecture of biological and artificial membranes as visualized by freeze-etching. Ann. Rev. Phys. Chem. 26, 101-120 (1975)

A Calorimetric and Fluorescent Probe Study of Phase Transitions in Phosphatidylcholine Liposomes

T. E. THOMPSON, B. R. LENTZ, and Y. BARENHOLZ

Summary

Fluorescence depolarization of the hydrophobic probe 1,6 diphenyl-1,3,5 hexatriene and differential scanning calorimetry are used to monitor gel-liquid crystalline phase transitions in aqueous dispersions of synthetic phosphatidylcholines. Parameters of the phase transitions are determined for both small single-lamellar vesicles and large multilamellar liposomes. Dimyristoyl, dipalmitoyl, distearyl, and dioleyl phosphatidylcholine are studied in single component systems and in binary mixtures. Phase diagrams constructed from these data indicate that all binary systems exhibit unmixing characteristics. The miscibilities of the lipid pairs are markedly different in small, single-lamellar vesicles and large multilamellar liposomes. These results are discussed in terms of the relations between function, composition, and curvature of the bilayer component of biological membranes.

A. Introduction

Much of the current research on biological membranes is based on the fluid mosaic hypothesis formulated by Singer and Nicolson (1972). This hypothesis, which is a refinement of the Danielli concept dating from the 1930s (Danielli and Davson, 1935), is composed of two essential elements. The first of these derives directly from Danielli's suggestion that the lipid component has the structural form of a bilayer, which provides the basic barrier properties displayed by the membrane. In its simplest form, this set of properties is expressed as a generalized membrane impermeability to small water-soluble molecules. The lipids comprising the bilayer, while confined to this structure, are free to exhibit a variety of modes of motion including rotation and translation. The second major element comprising the fluid mosaic hypothesis concerns the protein components of the membrane. These molecules are immersed to varying degrees in the bilayer. Some are deeply imbedded in the hydrophobic core, others span the bilayer while still others may be associated only with the polar faces of the bilayer. The various polypeptide components are free to move within the membrane to form a variety of complex structures that may endure for very long or very short periods of time. The protein components confer on the membrane its basic physiological functions. In their simplest form they are expressed as specific transport processes.

Although the essential physiological properties of biological membranes must derive primarily from protein components, there is much evidence to suggest that these functions are strongly dependent on lipid composition. In general the relations between function and lipid composition have been formulated in terms of membrane fluidity and lipid-protein as well as protein-protein interactions, which are in turn governed by the distribution of different types of lipids both between the two opposing monolayers that comprise the bilayer, and within the plane of the bilayer.

There is considerable evidence to suggest that in the membrane of the human erythrocyte there is a marked asymmetric distribution of lipids between the inner and outer faces of the membrane (Bretscher, 1972; Verkleij et al., 1973; Gordeskey and Marinetti, 1973). Essentially all of the sphingomyelin and most of the phosphatidylcholine are located in the outer surface while most of the phosphatidylethanolamine and all phosphatidylserine are located mainly in the inner membrane face. Associated with this lipid transmembrane compositional asymmetry is an asymmetry of protein components (Bretscher, 1973). Recent studies on bovine retinal rod outer segment disc membranes indicate a marked transmembrane asymmetry in the distribution of phosphatidylserine and phosphatidylethanolamine (Smith and Litman, unpublished data). It seems certain that these asymmetries have functional significance, although at the present time the relations between functional properties and compositional asymmetries are unknown.

The fluidity of the lipid bilayer component in biological membranes has been shown to influence a variety of membrane functions (Singer and Nicolson, 1972). This property is dependent on both lipid composition and temperature. In addition, bilayer fluidity can be markedly altered by a phase transition from the liquid crystalline to the crystalline state. Differential calorimetry, X-ray measurements, fluorescence studies, nuclear magnetic resonance (NMR) and electron spin resonance (ESR) studies carried out on biological membranes (Chapman and Dodd, 1971; Lee, 1975) indicate that the lipid in many membranes undergoes a thermotropic phase transition. This phase transition from crystalline to liquid crystalline is accompanied by a loss of order in the two dimensional structure and by an abrupt increase in membrane fluidity. It is also accompanied by a change in bilayer thickness reflecting a change in the partial specific volume of the lipid in the bilayer. These changes in structure have been measured by several different techniques and found to be very similar in biological membranes and in liposomes (Wilkins et al., 1971; Ashe and Steim, 1971; Blazyk and Steim, 1972; Abramson and Pisetsky, 1972; Rand and Pangborn, 1973; Lee, 1975). In microorganisms the composition of the lipid is adjusted biosynthetically by the organism to cause the transition temperature to be just slightly below the temperature at which the organisms are grown (Rottem et al., 1973a).

The distribution of lipid components in the plane of the membrane may be affected by a phase transition in two ways. If the component lipids are similar in type, for example phosphatidylcholines differing only in the degree of unsaturation of their acyl chains, then solidification may not grossly alter the distribution of components. On the other hand, if the thermotropic behavior of the lipids is very different, phase separation may occur. In this case a phase consisting predominantly of the lipid with the higher transition temperature forms a compositionally different domain within the plane of the membrane (McConnell et al., 1972; Phillips et al., 1972; Lee, 1975).

Phase separation has been studied in liposomes prepared from synthetic lipids (Shimshick and McConnell, 1973) and from lipids extracted from the inner membranes of *E. coli* (Linden et al., 1973) using mobile spin labels. The triphasic Arrhenius plots obtained for the fluidity were directly related to the phospholipid composition of the membrane and resembled the Arrhenius plots for glucoside transport. Phase separation was also found in the sarcoplasmic reticulum of rabbit muscle (McConnell et al., 1972). There are many indications that the membrane activities reflect phase transition phenomena. The work of Overath and co-workers and Fox and co-workers (Tsukagoshi et al., 1971; Overath et al., 1971; Overath and Träuble, 1973; Schairer and Overath, 1969) shows that there is a direct correlation between carrier-mediated lactose uptake and the phase transitions of the lipid components of E. coli fatty acid auxotrophs. Similar results have been obtained with Mycoplasma. Verkleij and co-workers (1972) and Rottem and co-workers (1973b), using this organism, found that the fluidity of the membrane as well as the transition temperature affected both the transport of α-methyl glucoside and the activity of the membrane-bound ATPase. Elter and co-workers (1973) found correlations between the activities of two membrane-bound microsomal enzymes, UDP-glucoronyl transferase and glucose 6-phosphatase and the fluidity of the membrane. The activity of Na^+, K^+-dependent ATPases was found to be strongly dependent on the fluidity of the membrane by Grisham and co-workers (Grisham and Barnett, 1973; Tanaka and Ternya, 1973). It seems reasonable to believe that these effects on physiologically important membrane functions may be the result of changes in membrane protein conformation and aggregation caused by alteration in lipid bilayer fluidity and phase separation (Shimshick and McConnell, 1973; Linden et al., 1973).

The purpose of this paper is to discuss studies of fluidity and phase transition characteristics in small, single-lamellar vesicles and large, multilamellar liposomes prepared from pure synthetic phospholipids and binary mixtures of these compounds. Particular attention is focussed on the thermal characteristics of small, single-lamellar phospholipid vesicles characterized by Huang (1969), which have radii of curvature between 100 and 120 Å. The small radius of curvature of bilayers in this type of system is responsible for a set of physical properties that differ markedly from those of the planar bilayers of multilamellar liposomes. Since the plasma membranes of cells and the membranes of subcellular organelles frequently are characterized by regions exhibiting radii of curvature in this range, the studies on small, single-lamellar vesicles suggest that very small radii of curvature regions of biological membranes may be expected to have special properties that may be of physiological importance (Thompson et al., 1974).

The work to be discussed is based on fluorescence depolarization studies using the fluorescence probe 1,6 diphenyl-1,3,5 hexatriene (DPH) (Shinitzky and Barenholz, 1974) and on thermal work employing a differential scanning calorimeter of the heat conduction type (Ross and Goldberg, 1974; Suurkuusk et al., 1976).

B. Liposome Systems

The use of multilamellar liposome dispersions for the study of phospholipid bilayers was introduced by Bangham and co-workers (1965). The phospholipid in these structures is organized into concentric bilayers, each separated from its neighbor by an in-

Table 1. Physical properties of homogeneous fraction II vesicles[a,b]

Sedimentation coefficient	$s^{o}_{20,w}$	$(2.63 \pm 0.10) \times 10^{-13}$ sec
Diffusion coefficient	$D^{o}_{20,w}$	$(2.03 \pm 0.4) \times 10^{-7}$ cm^{2} sec^{-1}
Partial specific volume of vesicle wall	$\bar{v}$	0.9848 ± 0.0007 ml g^{-1}
Intrinsic viscosity	$[\eta]$	0.041 dl g^{-1}
Vesicle shell weight		2.0×10^{6} daltons
Vesicle diameter		210 ± 8 Å

[a]Aqueous phase: 0.1 M KCl + 0.01 M Tris, pH 8.0, 20°C

[b]Taken from Huang and Thompson, 1974

terspersed water lamella 25 to 50 Å in thickness. Each bilayer is a closed surface in most systems so that, in effect, each interspersed water lamella is a separate aqueous compartment. These multilamellar liposomes are generally large enough to be visible in the light microscope. The lipid bilayers comprising each liposome, although curved, are characterized by large radii of curvature in the range of 0.1 to 10 μ, that is, the bilayers are essentially planar (Bangham et al., 1967; Bangham, 1968). In contrast to these large, multilayer structures, liposomes can be prepared that are small spherical vesicles each composed of a single continuous lipid bilayer enclosing a volume of aqueous solution. The preparation of aqueous dispersions of these vesicles, which are homogeneous in size and have radii of about 100 Å, was first described by Huang (1969). The two essential steps involved in vesicle preparation are dispersion followed by fractionation according to size. Although dispersion is usually accomplished by sonication, it may be achieved by other means (e.g., Batzri and Korn, 1973). Fractionation was originally carried out by Huang by molecular sieve chromatography on large-pore agarose beads (Huang, 1969; Huang and Thompson, 1974), but may more conveniently be achieved by differential centrifugation (Barenholz et al., 1976). Some of the important physical parameters that characterize such vesicles prepared from hen egg phosphatidylcholine are summarized in Table 1 (Huang and Thompson, 1974). The most interesting characteristic of these vesicles is their very small size. The diagrammatic cross section through a vesicle shown in Figure 1 illustrates two consequences of this extreme bilayer curvature. First, if the bilayer is taken to be 40 Å in thickness (Wilkins et al., 1971), then the radius of the inner bilayer face is about half that of the outer surface. Second, relative to the bilayer wall, the outer surface is defined by a positive radius while the radius of the inner face is negative. This situation implies molecular packing constraints on the inner monolayer that are quite different from those on the outer monolayer (Thompson et al., 1974).

As might be expected, several of the physical properties of these vesicles are distinctly different from the corresponding properties of liposomes of larger radius. For example, the proton magnetic resonance spectra obtained from aqueous dispersions of large liposomes and large multilamellar structures of the type studied by Bangham exhibit resonance lines that are so broad that they can

Fig. 1. Diagrammatic cross section through minimum radius vesicle. Insert illustrates molecular arrangement in bilayer

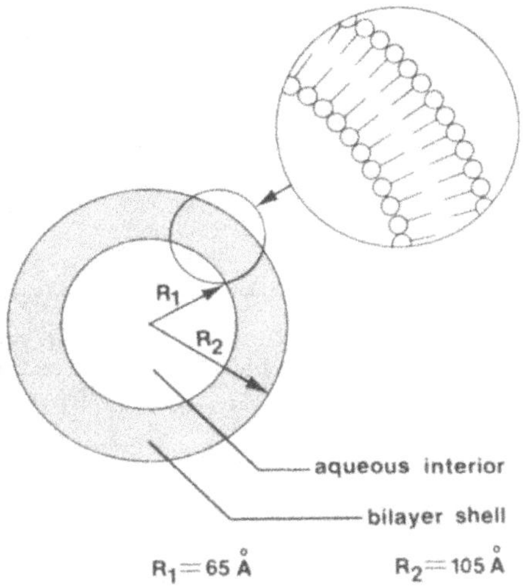

be resolved only by special techniques (Seiter and Chan, 1973). However, upon sonication, the resonance lines sharpen dramatically and are easily resolved by low-field-strength equipment. In fact, the vesicles having a minimum radius of curvature are characterized by lines of maximum sharpness (Chapman et al., 1968; Plenkett et al., 1968; Sheard, 1969). Although the origin of this dependence of line width on liposome size has been the subject of considerable controversy, recent work strongly supports the view that the marked decrease of proton NMR line widths in small-radius vesicles is due to increased molecular motion in these structures (Sheetz and Chan, 1973; Seiter and Chan, 1973).

This interpretation is directly supported by the fact that the partial specific volume of the bilayer in the small vesicles can be shown experimentally to be larger than the partial specific volume of the lipid in bilayers of larger radius of curvature (Sheetz and Chan, 1973).

Perhaps the most interesting special feature of these minimum radius of curvature vesicles is the transbilayer compositional asymmetry in vesicles formed from certain binary lipid mixtures. Table 2 is a summary of the results obtained for three such systems, each with a total vesicle mol fraction ratio equal to 1 for the two components. It is apparent that the relative concentrations of the two components in the apposing halves of the vesicle bilayer are markedly different in these three systems. Whatever the detailed molecular basis for this transbilayer compositional asymmetry may be, it is clear that the explanation must rest in the dissimilarities in the interactions between the two lipids of each pair and in the smallness of size and disparity in size of the radii of curvature of the two bilayer faces. Thus, in these small-vesicle systems a relation exists between bilayer composition and morphology. The implications of this situation for biological membranes rest on the widely accepted idea that the bilayer is the structural form of most of the lipid in biological membranes. These implications take on greater significance in view of the fact that the functional properties at any locus in a biological membrane must be directly related to the organization and properties of the molecules composing the locus. Thus, in regions

Table 2. Molar composition of small vesicles[a]

	Total vesicle	Outer surface	Inner surface
1.	PG/PC = 1[b]	PG/PC = 2.0	PG/PC = 0.33
2.	PE/PC = 1[c]	PE/PC = 0.65	PE/PC = 3.0
3.	Chol/PC = 1[d]	Chol/PC = 0.89	Chol/PC = 1.3

[a]Abbreviations: PG: phosphatidylglycerol; PC: phosphatidylcholine; PE: phosphatidylethanolamine; Chol: cholesterol

[b]Michaelson et al., 1973

[c]Litman, 1973

[d]Huang et al., 1974

of biological membranes exhibiting very small radii of curvature, a relation may exist between membrane barrier properties, as they are a reflection of the properties of a lipid bilayer, and the curvature of the membrane. If the lipid compositional asymmetry is generalized to include protein components, a relation may exist between many other membrane properties and curvature.

It seems reasonable to suppose that the cause-effect nature of this relation could operate in either of two ways. On the one hand, changes in curvature driven by physiological mechanisms such as microtubule or microfilament formation could cause changes in composition and hence in function. On the other hand, changes in the relative compositions of components in opposite bilayer faces, accomplished in the cell by biosynthetic mechanisms, could result in alterations of membrane curvature, which in turn could have functional significance (Thompson et al., 1974).

Whether such relations actually exist in biological membranes is not known at the present time. It is certain, however, that minimum radii of curvature structures of the order of 100 Å are exhibited by many biological membranes of diverse origin (Fawcett, 1966).

C. Phase Transition Characteristics in Phosphatidylcholine Bilayers

Pure phospholipids in multilamellar liposomes undergo gel-liquid crystalline phase transitions of relatively high cooperativity (for a summary, see Lee, 1975). Single component phosphatidylcholines in this type of system usually show transition-temperature ranges of 1-3°C. Two-component bilayer phase transition characteristics are considerably more complex, but generally fall within a range defined by the following extremes. If the component lipids are similar in generic type, for example, phosphatidylcholines differing only in acyl chain length by two methylene groups, solidification does not alter the distribution of components. In both the liquid crystalline and gel phases the two components are miscible in all proportions. This type of system behaves as an ideal mixture with values of characteristic parameters given by a mol fraction-weighted average of the corresponding parameters for the pure components. A good approximation of this type of behavior is exhibited by liposomes comprised of dimyristoyl phosphatidylcholine and dipalmitoyl phosphatidylcholine (Shimshick and McConnell, 1973; Lentz et al., 1976b). At the other extreme are systems comprised of lipids of dissimilar type with phase transition temperatures differing by tens of degrees. In this case phase separation may occur with gel and liquid crystalline phases consisting predominantly of one component. In a system of this type, the coexistence of gel and liquid crystalline phases of very different composition may occur over relatively large compositional and temperature ranges. Examples of this type of system are: 1-palmitoyl, 2-oleyl phosphatidylcholine mixed with dipalmitoyl phosphatidylcholine (Lentz et al., 1976b; McConnell, 1975, Lee, 1975).

The important point to be developed in the discussion that follows is that the phase transition characteristics of all types of bilayer systems, of unitary as well as binary composition, are strongly dependent on bilayer curvature. Thus, in small single-lamellar vesicles, in contrast to large liposomes, the transitions occur at lower temperatures over a wider temperature range. These characteristic differences between small and large radii of curvature bilayers are well illustrated by phosphatidylcholine systems.

I. Dipalmitoyl Phosphatidylcholine Liposomes

1. Calorimetry

The calorimetric studies discussed in this section have been carried out in a newly developed differential scanning calorimeter of the heat conduction type (Ross and Goldberg, 1974; Suurkuusk et al., 1976). The instrument is designed for measuring heat capacities and heat effects accompanying thermally induced transitions of solutes in dilute solution over the temperature range 0-70°C at scanning rates from 3 to 50 deg h^{-1}. Measurements are carried out in the heating mode after a 3-6 h equilibration at 4°C.

Figure 2 shows a typical heat capacity vs. temperature scan of a suspension of large, multilamellar liposomes prepared from dipalmitoyl phosphatidylcholine (Suurkuusk et al., 1976). The scan is characterized by two transitions with T_m values of 35.4 and 41.2°C and corresponding enthalpy changes of 1.6 and 8.2 kcal · mol^{-1}, respectively. T_m is defined as the position of a maximum in the heat capacity function, and the corresponding ΔH is the integral of the C_p function over the appropriate temperature range calculated as described below. These thermodynamic quantities are in reasonable agreement with T_m values of 34.0 ± 0.2 and 41.75 ± 0.06°C and the corresponding values of ΔH = 2.3 ± 0.2 and 9.7 ± 0.2 kcal · mol^{-1} obtained by Hinz and Sturtevant (1972) on a very similar preparation of dipalmitoyl phosphatidylcholine liposomes. The value of 8.2 kcal · mol^{-1} for ΔH of the high-temperature transition is in better agreement with values reported by other workers (Ladbroke and Chapman, 1969; de Kruijff et al., 1973, 1975; Klopfenstein et al., 1974; Vaughn and Keough, 1974).

Although the temperature-dependent behavior of small vesicle preparations of varying degrees of homogeneity has been described using NMR spectroscopy (Sheetz and Chan, 1972; Lee et al., 1972; Uhing, 1975), ESR (Hubbell and McConnell, 1971, dilatometry

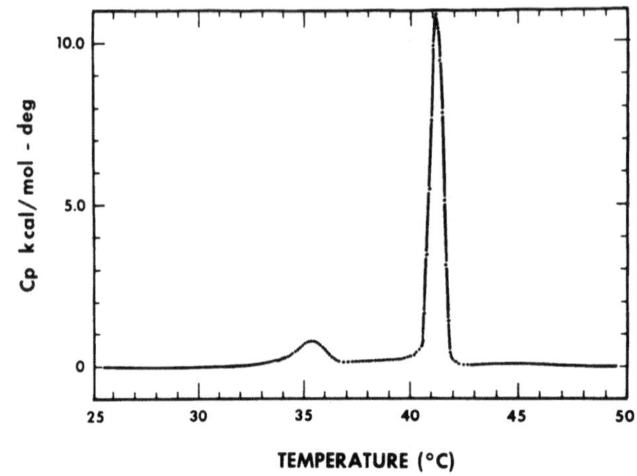

Fig. 2. Differential calorimetric temperature scan of large, multilamellar liposomes prepared from dipalmitoyl phosphatidylcholine dispersed in 50 mM aqueous KCl solution

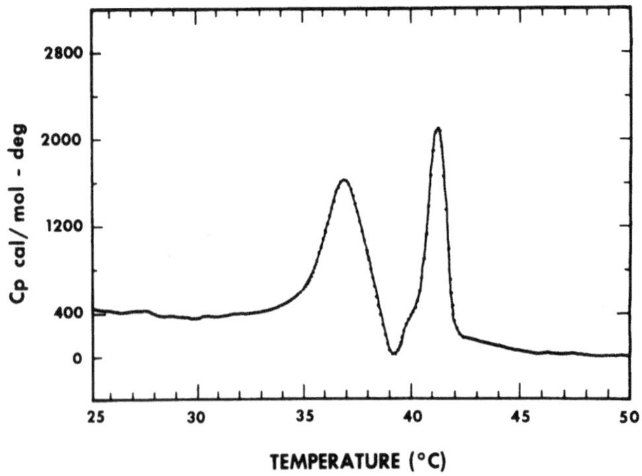

Fig. 3. Differential calorimetric temperature scan of small, single-lamellar vesicles prepared from dipalmitoyl phosphatidylcholine dispersed in 50 mM aqueous KCl. Vesicle dispersion was introduced into calorimeter immediately after preparation at 60°C and allowed to equilibrate at about 0°C for 3 h before beginning scan

(Sheetz and Chan, 1972; Melchior and Morowitz, 1972), light scattering (Yi and MacDonald, 1973), and fluorescence (Lussan and Faucon, 1971; Vanderkooi and Chance, 1972; Cogan et al., 1973; Papahadjopoulos et al., 1973), the transition characteristics of a homogeneous population of small, single-lamellar phospholipid vesicles, prepared as described in the preceding section, have not previously been determined by direct calorimetric measurements. A calorimetric scan of such vesicles freshly prepared from dipalmitoyl phosphatidylcholine is shown in Figure 3 (Suurkuusk et al., 1976). Two distinct heat capacity maxima at 36.9 and 41.2°C are observed. The enthalpy changes associated with each are 3.9 and 2.9 kcal · mol^{-1}, respectively. It is clear that the thermal characteristics of the small vesicles are quite different from those displayed by multilamellar liposomes.

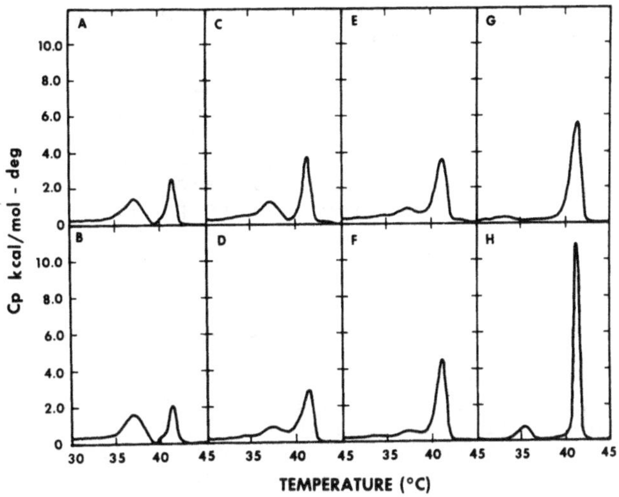

Fig. 4A-H. Differential calorimetric scans of small vesicle dispersions prepared from dipalmitoyl phosphatidylcholine as function of time under different conditions: aqueous phase composition 50 mM KCl: (A) sample held at 48-50°C for 1 week before scanning; (B) identical with Figure 2, sample was introduced into calorimeter immediately after preparation and allowed to equilibrate at about 0°C for 3 h before scanning; (C, D, E, and F) samples after 2, 7, 13, and 19 cycles from 4 to 50°C; (G) a small vesicle preparation stored in frozen state at -10°C for 90 h; (H) identical with Figure 1, scan of fresh dispersion of multilamellar liposomes

The thermal characteristics of multilamellar liposomes are stable with respect to time; the basic features of the transitions, T_m and ΔH, are reproducible after several temperature cycles. This, however, is not the case for the small vesicles whose thermotropic behavior is highly dependent upon the history of the sample. With successive calorimetric scans of the same vesicle preparation, it is observed that the heat associated with the transition at 41.2°C increases while that associated with the 36.9°C transition decreases.

The changing thermotropic behavior of the small vesicles as a function of time under different conditions is illustrated in Figure 4 (Suurkuusk et al., 1976). Scan A was performed with a sample stored above the highest transition temperature for one week. Scan B, which is identical with Figure 3, was initiated within 3 h after preparation. The thermal characteristics of the two samples are very similar, indicating that the small vesicle preparation is essentially stable by this criterion if maintained above the highest transition temperature. However, a distinct increase in turbidity is observed after prolonged storage above the transition temperature.

Scans C, D, E, and F were performed after increasing numbers of temperature cycles from 0 to 70°C with the same sample used in scan B. It is evident that the enthalpy change of the transition at 41°C grows at the expense of that associated with the transition at 37°C. Scan G was performed on a sample of the small vesicle preparation stored frozen at -10°C for approximately 90 h. The thermotropic behavior of this sample is similar to that shown in scan F, and to that of a fresh multilamellar preparation shown in scan H and Figure 2. These results demonstrate that small vesicle preparations stored at temperatures below T_m are unstable. While it is clear that the thermotropic behavior of small vesicles is substantially different from that of multilamellar liposomes, it appears that upon exposure to temperatures below the T_m, species slowly form spontaneously, which resemble in some respects multilamellar liposomes in their calorimetric characteristics. The absence of a well-defined low-temperature transition in scan G clearly indicates that this species is not, however, identical with multilamellar liposomes.

2. Fluorescence Depolarization of DPH

Fluorescence depolarization studies outlined in this and the following sections, utilized a modified Perkin-Elmer MPF-3-Spectrofluorimeter operating in the ratio mode. The modifications include addition of two polarizers, one in the excitation beam and the other in the emission beam, and installation of a multiwedge plate depolarizer in the emission beam. This latter modification eliminates the correction for preferential phototube sensitivity. The instrument is best suited for use in the cooling mode with scanning rates of 10 to 30 deg h^{-1}.

The general theory of the fluorescence probe depolarization technique as applied to membrane structure has been outlined in detail by Cogan et al. (1973). The use of DPH as a fluorescent probe* specific for the hydrophobic region of lipid bilayer has been described by Shinitzky and Barenholz (1974). The technique is based on determination of the rotational motion of the probe molecule. This motion is characterized by the anisotropy parameter, r, as defined by:

$$r = \frac{(I_{\parallel} / I_{\perp}) - 1}{(I_{\parallel} / I_{\perp}) + 2} = \frac{I_{\parallel} - I_{\perp}}{I_{\parallel} + 2I_{\perp}}$$

where $I_{\parallel}$ is the fluorescence intensity parallel to and $I_{\perp}$ is the intensity perpendicular to the plane of polarization of the excitation beam. Total fluorescence is given by $I_{\parallel} + 2I_{\perp}$. The theoretical upper limit for r when the probe is in a medium of infinitely high viscosity is $r_0 = 0.40$. The lower limit in a medium of very low viscosity is zero. With DPH the experimentally determined upper limit for r is 0.362.

Fluorescence anisotropy is not related in a simple way to the absolute temperature. However, the anisotropy can be related to the apparent microviscosity, η, of the probe environment (Shinitzky et al., 1971). For a homogeneous phase, the natural logarithm of the viscosity should be a linear function of 1/T. The anisotropy may be related to the apparent microviscosity experienced by the probe by means of a modified version of the Perrin equation,

$$r_0/r = 1 + [k\ T\ \tau] \ / \ [\eta\ v(r)]$$

which relates the ratio r_0/r to the lifetime of the excited state,

---
*DPH, 1,6-diphenyl-1,3,5-hexatriene

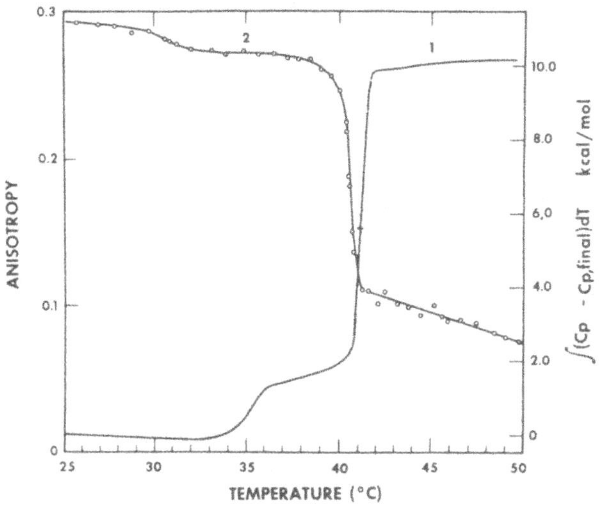

Fig. 5. Comparison of calorimetric and fluorescence results for liposomes: (curve 2) fluorescence anisotropy of DPH as function of temperature; (curve 1) calorimetrically determined excess enthalpy as function of temperature. These data are the integral under curve in Figure 3

τ, the apparent microviscosity, η, and the effective rotational molar volume of the probe, v(r), which is itself a function of r. k is the Boltzmann constant and T is the absolute temperature. The functional dependence of v on r for DPH has been determined by Shinitzky and Barenholz (1974). Values of the fluorescence lifetimes for DPH in liposome dispersions at various temperatures can be estimated from a limiting low temperature value of 11.4 nsec following the procedure of Shinitzky and Barenholz (1974). This procedure gives a value of $\tau = 10.1 \pm 0.5$ nsec at 21°C for dimyristoyl phosphatidylcholine small vesicles and a value of $\tau = 10.6 \pm 0.5$ nsec at 20°C for dipalmitoyl phosphatidylcholine multilamellar liposomes. These values are in good agreement with a value of 9.8 nsec obtained for small, single-lamellar dimyristoyl phosphatidylcholine vesicles at 21°C by the phase modulation technique (Spencer, 1970) in the laboratory of G. Weber. They are also in reasonable agreement with $\tau = 9.6$ nsec obtained at 20°C for large, multilamellar dipalmitoyl phosphatidylcholine liposomes determined by single photon counting (Yguerabide, 1972) in the laboratory of L. Brand. Both independent measurements gave a single lifetime for the excited state of the DPH in these systems.

Fluorescence depolarization measurements have been carried out on preparations similar to those examined calorimetrically. The resulting DPH fluorescence anisotropies and the calorimetrically determined excess enthalpies (the integral of the C_p curve in Figure 3 corrected for baseline heat) are plotted as a function of temperature in Figures 5 and 6 (Suurkuusk et al., 1976).

The correspondence between calorimetric and fluorescence data for multilamellar liposomes is excellent in the region of the main phase transition as shown in Figure 5. The transition temperature determined from fluorescence depolarization measurements, 40.1 ± 0.5°C, is in good agreement with the value of 41.2°C determined calorimetrically. However, a significant difference between the calorimetrically and fluorometrically measured range of the lower temperature transition in multilamellar liposomes is apparent in Figure 5. While the temperature range for the lower tem-

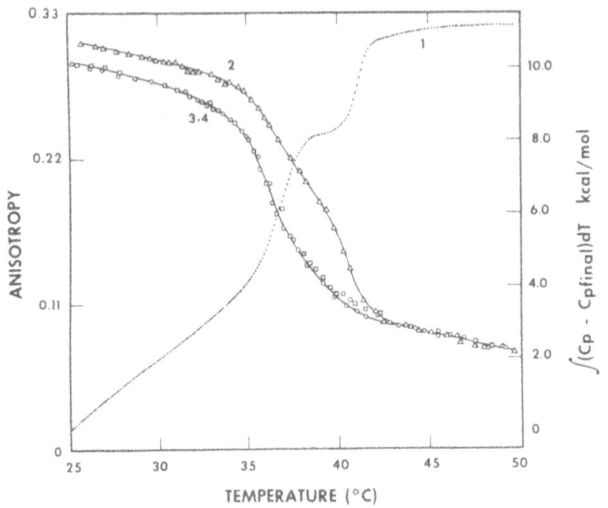

Fig. 6. Comparison of calorimetric and fluorescence results for small-vesicle preparations: (curve 1) calorimetrically determined excess enthalpy as function of temperature; (curve 2) fluorescence anisotropy of DPH vs. temperature after small-vesicle preparation was held below transition temperature for 6 h; (curves 3,4) fluorescence anisotropy of DPH vs. temperature for freshly prepared small vesicles (o) and for small vesicles held at 48-51°C for 5 days (□)

perature transition measured by the fluorescence polarization technique was 25.2-33.9°C, a range of 33.2-36.7°C was obtained by the calorimetric method. Instrumental uncertainties cannot account for this difference, nor can the fact that in the fluorometric studies sucrose was present in the aqueous phase and the vesicles contained DPH but both were absent in the calorimetric experiments. Identical calorimetric scans were obtained in systems with and without 15% sucrose or DPH. Since the calorimetric data were taken during a heating scan, and the fluorometry data during a cooling scan, this difference could be due to a real hysteresis associated with the low-temperature transition. It is interesting that the values of this transition temperature range obtained by studies of tempo partitioning into dipalmitoyl phosphatidylcholine dispersions (Shimshick and McConnell, 1973) (25-31°C) are also lower than those obtained calorimetrically.

A comparison of the calorimetric and fluorometric results for small vesicles is shown in Figure 6. Curve 1 is the integrated calorimetric scan for a small vesicle preparation held below the transition temperature for 6 h. Curve 2 is a plot of the fluorescence anisotropy as a function of temperature for a similar preparation maintained at 4°C for 6 h. The similarity between curves 1 and 2 is apparent. Both show clear transitions at 37 and 41-42°C. Curves 3 and 4, which are completely coincident, are fluorescence anisotropy data obtained for freshly prepared small vesicles and for small vesicles held at 48-51°C for 5 days. Curve 3,4 shows a single transition centered at 37°C with no indication of the high-temperature transition at 41-42°C seen in curves 1 and 2.

The following conclusions can be drawn from the data in Figure 6. (1) Comparison of curve 1 with curve 2 indicates that the fluorescence probe accurately monitors the phase transition characteristics of small vesicles as defined by the calorimetric scan. (2) Comparison of either curve 1 or 2 with curve 3,4 clearly shows that storage at 4°C alters the transition characteristics of small vesicles, while storage at temperatures above the transition region (48-51°C) produces no alteration in these characteristics. (3) The fact that curve 3 shows only a single transition at 37°C strongly

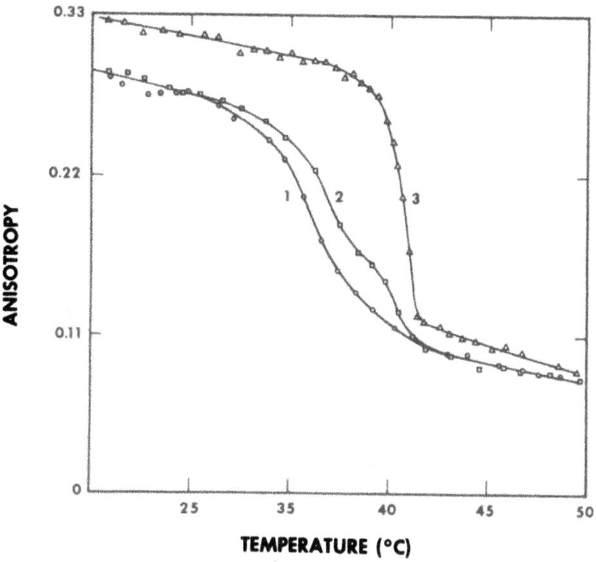

Fig. 7. Plot of fluorescence anisotropy of DPH vs. temperature: (curve 1) freshly prepared small-vesicle dispersion; (curve 2) same small-vesicle preparation shown in curve 1 but after 4.3 h in temperature range 4-50°C (cycled 13 times); (curve 3) same sample shown in curve 2 after freezing at -10°C followed by 12 h at 48-50°C

suggests that pure small vesicles exhibit a single distinctive transition. A calorimetric scan of small vesicles immediately after preparation and thus completely equivalent to the fluorescence data of curve 3 was impossible to obtain. For technical reasons the scan in the calorimeter employed in the study must be made from low to high temperature. Since the equilibration time in the calorimeter below T_m is about 3 to 6 h, it is clear that at the initiation of the scan, the phase transition characteristics of the small-vesicle preparations have already been altered. In contrast to this situation, the fluorescence spectrometer is best suited for a high- to low-temperature scan. Thus, in this instrument the sample is above the transition range during the initial temperature equilibration, a condition under which alteration in the thermal characteristics does not occur, as is shown by the identity of curves 3 and 4, Figure 6.

The effect of repeated cycling of a sample of small vesicles from 4 to 50°C is illustrated in Figure 7 (Suurkuusk et al., 1976). Curve 1 in this figure is a plot of the fluorescence anisotropy of the DPH probe vs temperature obtained for freshly prepared vesicles. Curve 2 is a similar plot of data obtained after the same preparation has been cycled 13 times between 4 and 50°C. The appearance of a transition at about 41°C is in agreement with the calorimetric data for the same sample shown in Figure 4E. Curve 3, Figure 7, is a plot of data obtained after the cycled sample was frozen at -10°C and then gently stirred at 48-50°C for 12 h before fluorescence depolarization was measured (Suurkuusk et al., 1976).

It is apparent that this treatment of small vesicles produces a species whose temperature-dependent behavior is similar to that displayed by multilamellar liposomes as shown in Figure 5, curve 2. Curve 3, Figure 7, however, shows no evidence of the low-temperature transitions displayed by liposomes. These results are in agreement with the calorimetric data discussed above.

These observations strongly suggest that the time- and temperature-dependent behavior of small-vesicle dispersions reflects a fusion process occurring at temperatures below the gel-liquid crystalline phase transition. The relation between vesicle fusion and the time-dependent thermal characteristics of the system can be formulated as a model with the following characteristics. (1) A pure dispersion of pure small vesicles, S, exhibits a single-phase transition at 37°C with an enthalpy change ΔH_S. (2) A dispersion of multilamellar structures, L, containing no small vesicles, has a phase transition at 41°C with an enthalpy change ΔH_L. (3) At temperatures below the phase transition, the L forms are thermodynamically more stable than small vesicles. In this temperature range the small vesicles slowly associate until an equilibrium mixture of S and L of minimum free energy is formed. (4) At temperatures above the transition, where both L and S are in the liquid crystalline state, a kinetic barrier prevents interconversion between forms. (5) The calorimetric scans obtained for small-vesicle preparations maintained below the transition are a composite of contributions from both S and L forms.

The following scheme summarizes this model (Suurkuusk et al., 1976):

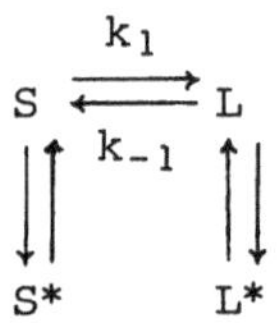

Here S and L represent the gel forms of the small vesicles and aggregates and S* and L*, the liquid-crystal states. k_1 and k_{-1} are the forward and reverse apparent first-order rate constants for the interconversion of S and L. It should be stressed that, although the thermal characteristics of the L form are similar to the main transition of multilamellar liposomes, the two types of liposomes are not identical. Both the fluorescence anisotropy measurements and the calorimetric studies clearly show that the low-temperature transition at $T < 36^\circ C$ observed in multilamellar liposomes is absent in the L form. In addition the width at half-height of the main transition at 41.2°C is 1.3 ± 0.3°C for the L type vesicles whereas freshly prepared multilamellar liposomes exhibit a width at half-height of 0.7°C.

On the basis of this model, the calorimetric scans of the various vesicle preparations can be resolved into three separate transitions to yield estimates of ΔH_1, ΔH_2, and ΔH_3, the enthalpies associated with the transitions at 34.6, 36,9, and 41.2°C. The inverse correlation between ΔH_2 and ΔH_3 is shown in Figure 8 (Suurkuusk et al., 1976). The solid curve is the linear least-squares represion line. The value of ΔH_3 extrapolated to $\Delta H_2 = 0$ is 8.4 kcal/mol. This is very similar to the value of 8.3 kcal/mol experimentally obtained for a fresh preparation of multilamellar liposomes. These results suggest that ΔH_2 and ΔH_3 are in fact independent measures of the relative amounts of small vesicles, S, and multilamellar structures, L, in each preparation. The results further suggest that the pure small vesicles are characterized by a single phase transition at 36.9°C, whereas the L form is characterized by a phase transition at 41.2°C. These results are consistent with the basic assumptions of the model.

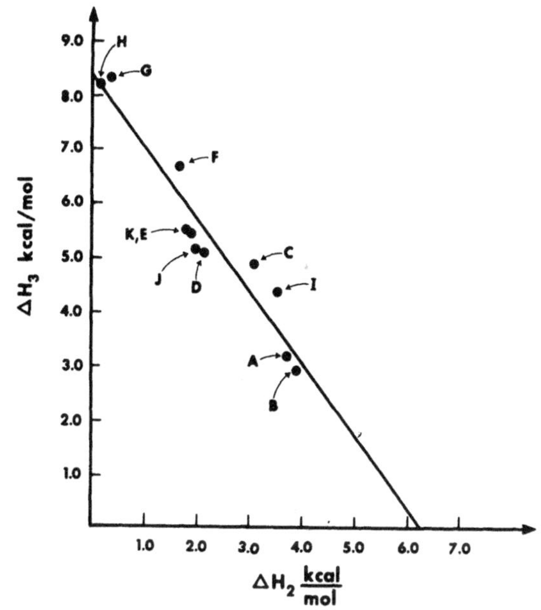

Fig. 8. Plot of ΔH_2 vs. ΔH_3 calculated from calorimetric scans obtained for vesicles under different conditions. Letters identifying individual points are not germaine to the discussion

Within the terms of this model the results in Figure 8 can be represented by

$$\Delta H_3 = \Delta H_L - (\Delta H_L/\Delta H_S)\Delta H_2$$

Values of $\Delta H_L = 8.4$ kcal/mol and $\Delta H_S = 6.3$ kcal $\cdot$ mol^{-1} are estimated from the parameters of the least-squares regression line.

These results clearly show that the thermotropic behavior of the small, single-lamellar vesicle comprised of pure dipalmitoyl phosphatidylcholine is distinctly different from that of the large multilamellar liposomes. This is quite apart from the problem of the intrinsic instability of the small vesicle prepared from this phospholipid. The data presented above also demonstrate that fluorescence depolarization measurements utilizing DPH faithfully monitor the major phase transition characteristics in this system, as defined by the primary method of differential scanning calorimetry. Using this technique, the marked differences in thermotropic behavior between large, multilamellar liposomes and small, single-lamellar vesicles can be demonstrated for other single-component phospholipid systems. These differences are clearly apparent in a comparison of the data in Figures 9 and 10 for multilamellar and single-lamellar vesicles respectively prepared from dioleyl (DOPC), dimyristoyl (DMPC), dipalmitoyl (DPPC), and distearoyl (DSPC) phosphatidylcholines (Lentz et al., 1976a).

II. Liposomes Composed of Binary Mixtures of Phosphatidylcholines

In general biological membranes are comprised of mixtures of a large number of different lipids. Thus, in contrast to the studies on single-component bilayers discussed previously, an examination

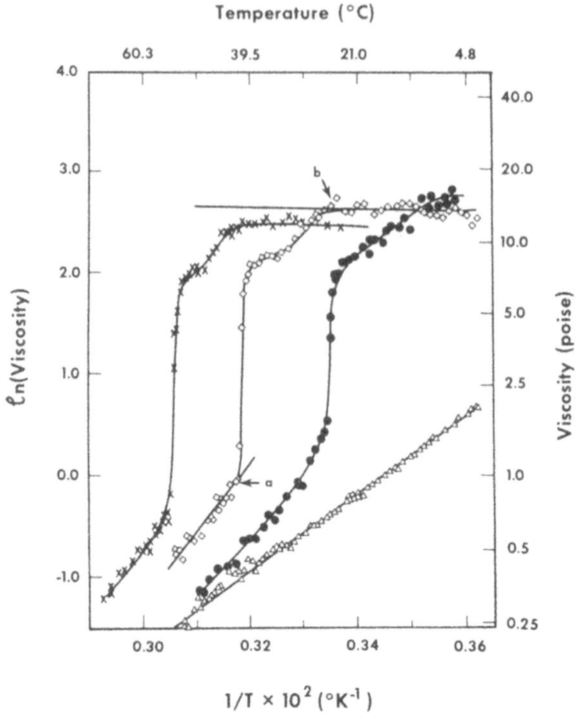

Fig. 9. Plot of natural logarithm of apparent microviscosity vs. 1/T for dispersions of large, multilamellar vesicles prepared from: Δ: DOPC; ●: DMPC; ◊: DPPC; ×: DSPC. Arrows a and b designate limits of transition range

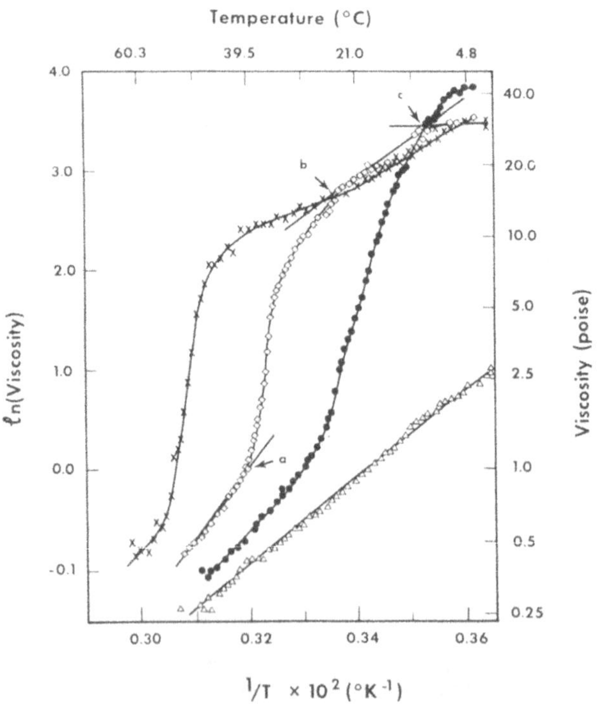

Fig. 10. Plot of natural logarithm of apparent microviscosity vs. 1/T for dispersions of small, single-lamellar vesicles prepared from: Δ: DOPC; ●: DMPC; ◊: DPPC; ×: DSPC. Arrows a and b designate limits of transition range, while arrows b and c delimit a possible low temperature structural change

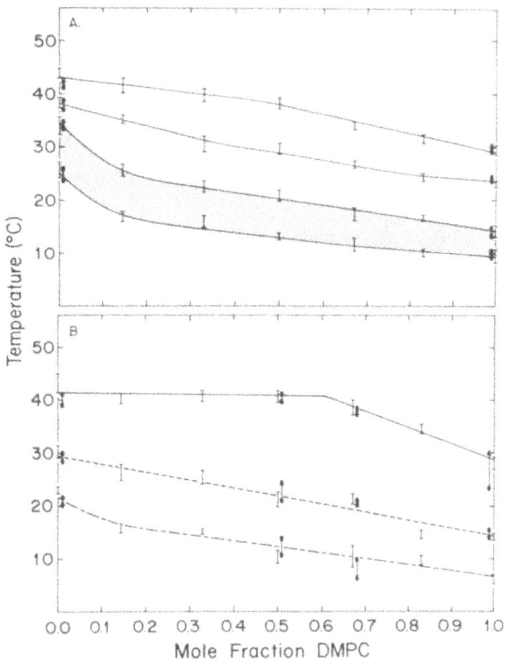

Fig. 11A,B. Phase diagrams for dispersions of DMPC/DPPC mixtures: (A) large, multilamellar liposomes; two upper lines delimit high-temperature phase transition range, lower lines, low-temperature transition range. (B) small, single-lamellar vesicles. ⁎I⁎: repeated data points

of thermotropic properties in multicomponent bilayer systems may be expected to be of direct relevance to structure-function relations in biological membranes. Several recent papers have reported studies of phase transitions and lateral phase separation in multicomponent bilayers and have discussed the possible functional significance of this phenomenon in biological membranes (Phillips et al., 1970; Linden et al., 1973; Chapman et al., 1974; Shimshick and McConnell, 1973; Wu and McConnell, 1973; McConnell, 1975). These studies have utilized multilamellar liposomes or undefined mixtures of these structures and small single-lamellar vesicles. Since it is apparent that the phase transition characteristics in single-component bilayers are markedly different in small vesicles and multilamellar liposomes, it is important to examine the phase characteristics in well-defined dispersions of both types. The following discussion is a comparison of the thermal characteristics of bilayers of both small and large radii of curvature formed from binary mixtures of phosphatidylcholines. The primary data are generated by monitoring the fluorescence depolarization of diphenylhexatriene in the manner described above. The characteristics of each system are best presented by generating phase diagrams from the primary data. This is done by plotting the delimiting temperatures defining a phase transition versus composition (for example, points a and b, Fig. 10). The following phase diagrams are based on the work of Lentz et al. (1976b).

The phase diagram for DMPC and DPPC in multilamellar liposomes is shown in Figure 11A. The compositional behavior of both the high- and low-temperature (with cross-hatching) transitions can be obtained, but the range of the low-temperature transition is less well defined than the range of the high-temperature transition. The shape of the phase diagram for the high-temperature transition suggests that these two very similar lipids mix nearly ideally both

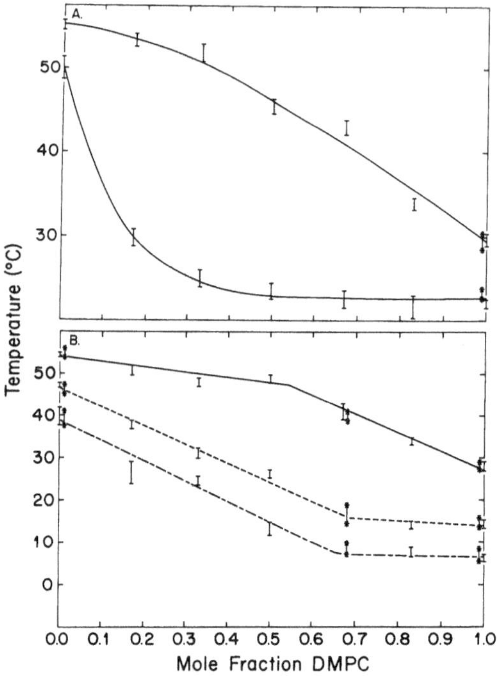

Fig. 12A,B. Phase diagrams for dispersions of DMPC/DSPC mixtures: (A) large, multilamellar liposomes; (B) small, single-lamellar vesicles. Ɪ: repeated data points

in the gel and liquid crystalline phases in the bilayers of multilamellar liposomes. A calorimetric study of this system, although subject to uncertainties in interpretation due to instrumental difficulties, yielded a phase diagram in qualitative agreement with that in Figure 11A (Chapman et al., 1974).

The composition dependence of the low temperature transition for DMPC/DPPC mixtures is also recorded in Figure 11A. This binary mixture is the only one for which this transition can be consistently observed at all compositions. As is the case for the high-temperature transition, this transition shows little, if any, broadening in the mixed-lipid vesicles. The low-temperature transition is separated from the main transition by about the same temperature span at all compositions above 10 mol% DMPC. The general characteristics of the phase diagram in DMPC/DPPC mixtures suggest that the mixing of these two phospholipids in multilamellar bilayers is essentially ideal.

Figure 12A is the phase diagram for the high-temperature phase transition in multilamellar liposomes composed of mixtures of DMPC and DSPC, two phosphatidylcholines with saturated acyl chains differing in length by four carbon atoms. This phase diagram is markedly different from that shown in Figure 11A obtained for two phosphatidylcholines with saturated chains differing in length by only two carbons. Figure 12A indicates that DMPC and DSPC mix nonideally. The horizontal section of the solidus curve (lower line) in the mol fraction range 0.3 to 1.0 suggests that in this range two solid phases of different composition coexist. This same conclusion was reached by Shimshick and McConnell (1973) based on tempo partitioning, although the detailed shape of their phase diagram was somewhat different from that in Figure 12A. Also, since Shimshick and McConnell presented no data in the range 0 to 0.38 mol fraction

Fig. 13A,B. Phase diagrams for dispersions of DOPC/DPPC mixtures. (A) large, multilamellar liposomes; (B) small, single-lamellar vesicles. ⁑: repeated data points

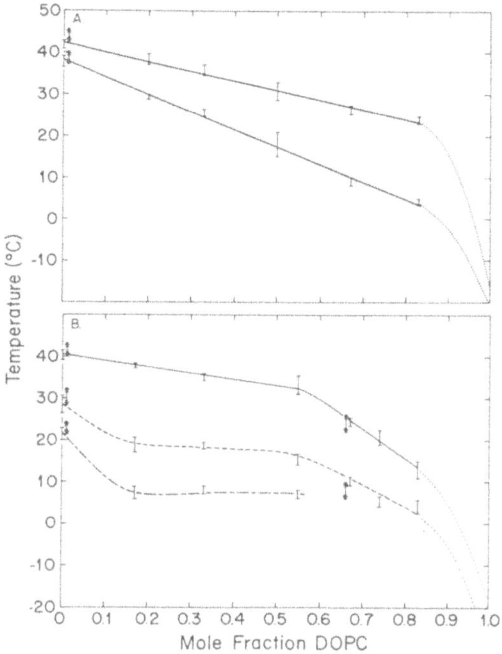

DMPC, it is not possible to specify the concentration at which the horizontal solidus line terminates. However, the fact that it must be within this range is consistent with Figure 12A. An earlier scanning calorimetric study (Phillips et al., 1970), however, indicated that the horizontal segment terminates below 0.22 mol fraction DMPC. This is clearly inconsistent with the results in Figure 12A.

The phase diagram for mixtures of DOPC and DPPC in large multilamellar liposomes is presented in Figure 13A. This figure clearly indicates that a <u>cis</u> double bond in the acyl chains of one component markedly alters the mixing properties of these lipids. Oleic acid (18:1Δ9) and palmitic acid (16:0) differ in length by only two carbon atoms, as do myristic (14:0) and palmitic acids, the mixture of Figure 11A. However, DOPC and DPPC obviously do not mix in the apparently ideal manner of DMPC and DPPC (Fig. 11A). In contrast to the very slight effect of composition on the breadth of the transition in the latter system, Figure 13A shows increased broadening of the transition with increasing content of DOPC in multilamellar liposomes, up to 0.83 mol fraction DOPC. Above this concentration, the low-temperature limit of the phase transition is very close to or below $0^{\circ}C$, and the transition becomes so poorly defined that it is difficult to determine the limits. The dotted line in Figure 13A is drawn on the basis of the results of Phillips et al. (1970), who demonstrated that hydrated DOPC (50% water, wt/wt) has a gel-liquid crystalline phase transition at $-22^{\circ}C$. A recently published differential calorimetry scan of DOPC multilamellar vesicles (de Kruijff et al., 1975) suggests that a phase transition occurs in the neighborhood of $-20^{\circ}C$. Within these limits, the shape of the phase diagram of Figure 13A (especially the rapid drop in transition temperature as pure DOPC is approached) clearly implies that DOPC and DPPC do not mix ideally. Such a result was not observed for any mixture of two saturated lecithins of similar chain-length incorporated into large, multilamellar liposomes.

A calorimetric study of multilamellar liposomes composed of a 1/1 mixture of DOPC and DPPC (i.e., one point on the phase diagram of Fig. 13A) showed two phase transitions, one in the region of the DPPC transition and one in the region of the DOPC transition (Phillips et al., 1972). A transition in the range of -20°C must be measured either in a buffer containing a high concentration of some organic antifreeze or in ice. Within the uncertainty imposed by these conditions, however, it appears that these phosphatidylcholines segregate into regions rich in DPPC and DOPC, respectively. Thus, the phase diagram of Figure 13A corresponds only to the transition of the DPPC-rich regions, that of the DOPC-rich regions being inaccessible without inclusion of antifreeze in the aqueous medium. These observations suggest that the unusual shape of the apparent solidus line of Figure 13A should be interpreted in terms of coexisting gel and liquid crystalline phases at temperatures below this line. As noted previously, the fact that the liquidus line of Figure 13A does not have a more negative slope at high DPPC mol fraction does suggest that domains rich in each of the lipids may exist in the lipid crystalline state. The observation of two phase transitions in 1/1 liposomes (Phillips et al., 1972) would seem to support this conclusion.

Phase diagrams derived for small, single-walled vesicle systems presented in Figures 11B, 12B, and 13B are quite different from those for large multilamellar liposome dispersions shown in Figures 11A, 12A, and 13A. As is evident in 11B, 12B, and 13B, the phase diagrams are more complex. The phase transitions are broader in single-lamellar vesicles, and at least two straight lines are needed to fit the data in microviscosity Arrhenius plots at low temperatures. For this reason, the phase diagrams in Figures 11B, 12B, and 13B have two solidus lines. The higher temperature line delimits the main phase transition in small vesicles. The significance of the lower temperature line is unclear.

Figure 11B is the phase diagram derived from the fluorescence data for small single-lamellar vesicles composed of DMPC/DPPC mixtures. Its most striking feature is the nearly horizontal liquidus line, suggesting that these two very similar lipids mix nonideally in the liquid-crystalline state. This is particularly surprising since these two lipids mix nearly ideally in the multilamellar liposomes described above. There are several possible explanations for this behavior: (1) the vesicles formed by the sonication process are a compositionally heterogeneous population, with some vesicles very rich in DPPC and others very rich in DMPC; (2) the lipids are distributed unequally between the nonequivalent inner and outer monolayers of the small vesicles; (3) because of the special packing requirements introduced by the small radius of curvature of sonicated vesicles (Sheetz and Chan, 1972; Thompson et al., 1974; Lentz et al., 1976a), these two similar lipids really do mix imperfectly nearly to the point of immiscibility, in the inner or outer monolayer of the small vesicle bilayer. The first explanation seems unlikely, for three reasons. First, complete or nearly complete segregation of the two lipids into separate vesicle populations would have resulted in an Arrhenius plot of a much different shape than was observed (Lentz et al., 1976b). Second, determination of the fatty acid composition of the sonicated vesicle preparations indicates no preference for either lipid of a mixture to appear in large as opposed to small liposomes. If there were no preferential partitioning of lipids between vesicles of different radii of curvature, it would be difficult to imagine a reason for preferential partitioning between vesicles of very similar radii of curvature. Third, the existence of a population of vesicles composed almost entirely of DPPC would necessarily imply a coexisting population composed almost exclusively of DMPC. This should result in an apparent nonmixing in the gel phase as well as in the liquid crystalline phase. However, no such behavior is observed. For a similar reason, the second explanation is unlikely, that is, enrichment of the other monolayer with DMPC necessarily would imply a concomitant enrichment of the inner monolayer with DPPC. No evidence for this is observed. Thus, it appears most likely that the shape of Figure 11B indicates that packing constraints in these small vesicles cause DMPC and DPPC to mix nonideally in the liquid crystalline state. That such imperfect mixing occurs in the liquid crystalline phase and not the gel phase is surprising. On the other hand, it is, in fact, reasonable to expect that the extended, all-trans configuration of the phospholipid acyl chains in the gel state (Seelig and Seelig, 1974) would be less affected by the unusual packing requirements of a small radius of curvature vesicle than would the disordered, gauche-containing acyl chains of the liquid crystalline state.

The phase diagram for mixtures of DMPC and DSPC incorporated into small single-lamellar vesicles is shown in Figure 12B. Again, there is a clear indication of nonideal mixing in the liquid crystalline state in these vesicles that is not observed in multilamellar liposomes. As is the case with multilamellar liposomes, the phase diagram of this lipid mixture also suggest immiscibility in the gel phase. However, immiscibility occurs in small vesicles only at concentrations of DMPC above 0.6 mol fraction, whereas in multilamellar liposomes this occurs above 0.3 mol fraction (Fig. 12A). It is interesting to note that for single-lamellar vesicles (as for multilamellar liposomes) the Arrhenius plots of the apparent microviscosity for 0.83 and 0.67 mol fraction DMPC show two distinct straight lines in the phase transition region (Lentz et al., 1976b). This suggests two phases of different composition in the gel state. In summary, small, single-lamellar vesicles composed of DMPC/DSPC show the same gel-phase immiscibility as do multilamellar liposomes, but over a narrower range of composition. In addition, they display a nonideality of mixing in the liquid crystalline phase not observed in the multilamellar liposomes. The interpretation of this behavior is similar to that suggested for the results obtained for small vesicles composed of DMPC/DPPC; the disordered acyl chains of the liquid crystalline phase mix considerably less ideally, while the ordered acyl chains of the gel phase mix more nearly ideally under the packing constraints imposed by the small radius of curvature of a single-lamellar vesicle.

The DOPC/DPPC phase diagram shown in Figure 13B illustrates the effect of a cis-double bond in the acyl chains of one phospholipid component in the small, single-lamellar vesicle system. The 9,10 double bond causes a fixed gauche-like kink at this point in the acyl chain. By analogy to the behavior of large-multilamellar liposomes, this phase diagram probably represents the phase behavior of only the DPPC-rich regions of the lipid bilayer. Although there is no direct evidence in small, single-bilayer vesicles, another phase change probably occurs in the range of -20°C, which is presumed to be the phase-transition range for pure DOPC small vesicles. The phase diagram in Figure 13B is interpreted on the basis of these assumptions. As in the case of the saturated systems already discussed, the shape of the liquidus line in Figure 13B suggests nonideal mixing of DPPC and DOPC in the liquid crystalline phase. The existence of a nearly horizontal section of this line below 0.55 mol fraction DOPC suggests liquid crystalline phase immiscibility even more convincingly than in compositionally similar multilamellar liposomes (Fig. 13A). It is interesting that the solidus line in Figure 13B also has a distinctly horizontal section (0.17 to 0.55 mol fraction DOPC). This probably reflects a limited miscibility of DPPC and DOPC in the DPPC-rich gel state. This phenomenon was not observed for DPPC/DOPC mixtures in large, multilamellar liposomes (Fig. 13A). The Arrhenius plots of microviscosity for mol fractions of DOPC from 0.17 to 0.55 are fit well by two straight lines in the temperature range of the phase transitions (correlation coefficient of the slope > 0.986, standard deviation from a straight line < 6.5%) (Lentz et al., 1976b). This is analogous to the behavior of the other systems studied for which gel phase immiscibilities were noted (DMPC/DSPC multilamellar liposomes and small vesicles). Thus, the phase diagram in Figure 13B is consistent with the hypothesis that, in small radius of curvature bilayers, lipids with fully extended acyl chains (those of saturated lipids below the phase transition) mix more nearly ideally than do lipids with kinked acyl chains (unsaturated chains, or any chain above the phase transition).

It is worthwhile noting that the small vesicle, mixed lipid systems so far discussed are examples of lipid bilayers for which the term "phase separation" is not an entirely appropriate description of the gel-liquid crystalline phase transition. This is so because lipid immiscibility seems often to be associated with the liquid crystalline phase of the small radius of curvature bilayers, which should result in some degree of phase separation (or lipid-domain formation) even above the phase transition temperature. It is this liquid crystalline phase bilayer heterogeneity (rather than gel phase heterogeneity) that results in the phase transition broadening in mixed-lipid, small vesicles. It is not clear whether this heterogeneity should increase on cooling through the phase transition, as implied by the common use of the term "phase separation" (McConnell et al., 1972).

D. Conclusions

The results discussed in this paper indicate that the mixing of phosphatidylcholine molecules within a bilayer is affected not only by the nature of the acyl chains, but also by the geometry of the bilayer. Two different phosphatidylcholine species mix less ideally in small single-walled vesicles that have a radius of curvature of 100-120 Å than in multilamellar liposomes of very large radius of curvature. The phase diagrams obtained for small vesicles are most simply interpreted in terms of nonideal mixing of components in the plane of the bilayer as opposed to nonideal mixing across the plane of the bilayer. In most cases, too, the range of temperatures for which a gel phase exists in equilibrium with a liquid crystalline phase is increased in small vesicles relative to large liposomes. Thus, the effect of bending a lipid bilayer to give a small radius of curvature is to increase the range of temperature and composition over which different phases may coexist in the plane of the bilayer. There has been considerable speculation recently about the importance of lateral phase separation of lipids in biomembranes (Shimshick and McConnell, 1973; Linden et al., 1973). If these speculations are correct, then the results on phosphatidylcholine vesicles suggest that regions of high curvature in biomembranes may be associated with plane formation of lipid domains of differing composition. Of even greater significance is the fact that these results suggest the possibility of lateral segregation of lipids into different domains in highly curved bilayers even above the gel-liquid crystalline phase transition. This is not "lateral phase separation" in the sense originally used by McConnell et al. (1972), since it is induced by changes in bilayer curvature rather than by passage below the liquid crystalline phase transition temperature. Perhaps this phenomenon would more appropriately be referred to as lateral lipid domain formation. Since most biomembranes exist at temperatures above the gel-liquid crystalline phase transition of the component lipids, this phenomenon provides a means of introducing lateral inhomogeneities into the bilayers of biomembranes. Such inhomogeneities should result in the same special compressibility properties that have been speculated to occur in bilayers undergoing lateral phase separation (Shimshick and McConnell, 1973; Linden et al., 1973).

Biological membranes, of course, are comprised of many different lipid species. The saturated phosphatidylcholines discussed above are rarely found as components of biological systems. There is evidence, however, that sphingomyelins, common major components of most mammalian membranes, do exhibit phase transitions in the biological temperature range (Shipley et al., 1974; Barenholz et al., 1976b). It is thus possible that lipids of this class may be of prime importance in lateral phase separation phenomena in biological membranes.

References

Abramson, M.B., Pisetsky, D.: Biochim. Biophys. Acta 282, 80-84 (1972)

Ashe, G.B., Steim, J.M.: Biochim. Biophys. Acta 233, 810-814 (1971)

Bangham, A.D.: Prog. Biophys. Mol. Biol. 18, 29-95 (1968)

Bangham, A.D., De Gier, J., Greville, G.D.: Chem. Phys. Lipids 1, 225-246 (1967)

Bangham, A.D., Standish, M.M., Watkins, J.C.: J. Mol. Biol. 13, 238-252 (1965)

Barenholz, Y., Litman, B.J., Goll, J., Thompson, T.E.: manuscript in prepara- (1976a)

Barenholz, Y., Suurkuusk, J., Mountcastle, D., Thompson, T.E., Biltonen, R.L.: Biochemistry 15, 2241-2247 (1976b)

Batzri, S., Korn, E.D.: Biochim. Biophys. Acta 298, 1015-1019 (1973)

Blazyk, J.F., Steim, J.M.: Biochim. Biophys. Acta 266, 737-741 (1972)

Bretscher, M.S.: Nature (New Biol.) 236, 11-12 (1972)

Bretscher, M.S.: Science 181, 622-629 (1973)

Chapman, D., Dodd, G.H.: In: Structure and Function of Biological Membranes. Rothfield, L.I. (ed.). New York: Academic Press 1971, pp.13-81

Chapman, D., Fluck, D.J., Plenkett, S.A., Shipley, G.G.: Biochim. Biophys. Acta 163, 255-261 (1968)

Chapman, D., Urbina, J., Keough, K.M.: J. Biol. Chem. 249, 2512-2521 (1974)

Cogan, U., Shinitzky, M., Weber, G., Nishida, T.: Biochemistry 12, 521-528 (1973)

Danielli, J.F., Davson, H.: J. Cellular Comp. Physiol. 5, 495-508 (1935)

Elter, S., Zakin, D., Vessey, S.: J. Mol. Biol. 78, 351-362 (1973)

Fawcett, D.: An Atlas of Fine Structure: The Cell. Philadelphia: W.B. Saunders Co. 1966
Gordeskey, S.E., Marinetti, G.V.: Biochem. Biophys. Res. Commun. 50, 1027-1031 (1973)
Grisham, C.M., Barnett, R.E.: Biochemistry 12, 2635-2637
Hinz, H., Sturtevant, J.M.: J. Biol. Chem. 247, 6071-6075 (1972)
Huang, C.: Biochemistry 8, 344-352 (1969)
Huang, C., Sipe, J., Chow, S.T., Martin, B.: Proc. Natl. Acad. Sci. U.S. 71, 359-362 (1974)
Huang, C., Thompson, T.E.: Methods Enzymol. 32, 485-489 (1974)
Hubbell, W.L., McConnell, H.M.: J. Am. Chem. Soc. 93, 314-326 (1971)
Klopfenstein, W.E., de Kruyff, B., Verkleij, A.J., Demel, R.A., van Deenen, L.L.M.: Chem. Phys. Lipids 13, 215-222 (1974)
Kruijff, B., de, Cullis, P.R., Radda, G.K.: Biochim. Biophys. Acta 406, 6-20 (1975)
Kruyff, B., de, De Greef, W.J., Van Eyk, R.V.W., Demel, R.A., Van Deenen, L.L.M.: Biochim. Biophys. Acta 298, 479-499 (1973)
Ladbrooke, B.D., Chapman, D.: Chem. Phys. Lipids 3, 304-367 (1969)
Lee, A.G.: Prog. Biophys. Mol. Biol. 29, 3-56 (1975)
Lee, A.G., Birdsall, N.J.M., Levine, Y.K., Metcalf, J.C.: Biochim. Biophys. Acta 255, 43-56 (1972)
Lentz, B.R., Barenholz, Y., Thompson, T.E.: Biochemistry (submitted) (1976a,b)
Linden, C.D., Wright, K.L., McConnell, H.M., Fox, C.F.: Proc. Natl. Acad. Sci. U.S. 70, 2271-2275 (1973)
Litman, B.J.: Biochemistry 12, 2545-2554 (1973)
Lussan, C., Faucon, J.F.: FEBS Lett. 19, 186-188 (1971)
McConnell, H.M.: In: Functional Linkage in Biomolecular Systems. Schmitt, F.O., Schneider, D.M., Crothers, D.M. (eds.). New York: Raven Press 1975, pp. 123-131
McConnell, H.M., Devaux, P., Scandella, S.C.: In: Membrane Research. Fox, C.F. (ed.). 1st ICN-UCLA Symp. Mol. Biol. New York: Academic Press 1972, pp. 27-37
Melchior, D.L., Morowitz, H.J.: Biochemistry 11, 4558-4562 (1972)
Michaelson, D.M., Horwitz, A.F., Klein, M.P.: Biochemistry 12, 2637-2645 (1973)
Overath, P., Hill, F.F., Lamnek-Hirsch, I.: Nature (New Biol.) 234, 264-267 (1971)
Overath, P., Träuble, H.: Biochemistry 12, 2625-2634 (1973)
Papahadjopoulos, D., Jacobson, K., Nir, S., Isac, T.: Biochim. Biophys. Acta 311, 330-348 (1973)
Phillips, M.C., Hauser, H., Paltauf, F.: Chem. Phys. Lipids 8, 127-133 (1972)
Phillips, M.C., Ladbrooke, B.D., Chapman, D.: Biochim. Biophys. Acta 196, 35-44 (1970)
Plenkett, S.A., Flook, A.G., Chapman, D.: Chem. Phys. Lipids 2, 273-290 (1968)
Rand, R.P., Pangborn, W.A.: Biochim. Biophys. Acta 318, 299-305 (1973)
Ross, P.D., Goldberg, R.N.: Thermochim. Acta 10, 143-150 (1974)
Rottem, S., Yashouv, J., Neeman, Z., Razin, S.: Biochim. Biophys. Acta 323, 495-508 (1973a)
Rottem, S., Cirillo, V.P., De Kruyff, B., Shinitzky, M., Razin, S.: Biochim. Biophys. Acta 323, 509-519 (1973b)
Schairer, H.U., Overath, P.: J. Mol. Biol. 44, 209-214 (1969)
Seelig, A., Seelig, J.: Biochemistry 13, 4839-4845 (1974)
Seiter, C.H.H., Chan, S.: J. Am. Chem. Soc. 95, 7541-7558 (1973)
Sheard, D.: Nature (London) 223, 1057-1059 (1969)
Sheetz, M.P., Chan, S.I.: Biochemistry 11, 4573-4581 (1972)
Shimshick, E.J., McConnell, H.M.: Biochemistry 12, 2351-2360 (1973)
Shinitzky, M., Barenholz, Y.: J. Biol. Chem. 249, 2652-2657 (1974)
Shinitzky, M., Dianoux, A.C., Gitler, C., Weber, G.: Biochemistry 10, 2106-2113 (1971)
Shipley, G.G., Avecilla, L.S., Small, D.M.: J. Lipid Res. 15, 124-131 (1974)
Singer, S.J., Nicolson, G.L.: Science 175, 720-731 (1972)
Spencer, R.: Ph.D. dissertation, University of Illinois, Urbana (1970)

Suurkuusk, J., Lentz, B.R., Barenholz, Y., Biltonen, R.L., Thompson, T.E.: Biochemistry 15, 1393-1401 (1976)
Tanaka, P., Ternya, A.: Biochim. Biophys. Acta 323, 584-591 (1973)
Thompson, T.E., Huang, C., Litman, B.J.: In: The Cell Surface in Development. Moscona, A.A. (ed.). New York: Wiley 1974, pp. 1-16
Tsukagoshi, N., Fielding, P., Fox, C.F.: Biochem. Biophys. Res. Commun. 44, 497-502 (1971)
Uhing, M.C.: Chem. Phys. Lipids 14, 303-308 (1975)
Vanderkooi, J.M., Chance, B.: FEBS Lett. 22, 23-26 (1972)
Vaughn, D.J., Keough, K.M.: FEBS Lett. 47, 158-161 (1974)
Verkleij, A.J., Verveagort, P.H.J., Van Deenen, L.L.M., Elbers, P.F.: Biochem. Biophys. Acta 288, 326-332 (1972)
Verkleij, A.J., Zwaal, R.F.A., Roelofsen, B., Comfurius, P., Kasteijn, D., Van Deenen, L.L.M.: Biochim. Biophys. Acta 323, 178-193 (1973)
Wilkins, M.H.F., Blaurock, E.E., Engleman, D.M.: Nature (New Biol.) 230, 72-76 (1971)
Wu, S.H., McConnell, H.M.: Biochem. Biophys. Res. Commun. 55, 484-491 (1973)
Yi, P.N., MacDonald, R.C.: Chem. Phys. Lipids 11, 114-134 (1973)
Yguerabide, J.: Methods Enzymol. 26, 498-578 (1972)

Phase Transitions, Protein Aggregation and a New Method for Modulating Membrane Fluidity

D. CHAPMAN, B. A. CORNELL, and P. J. QUINN

A. Introduction

It is well known that hydrated phospholipids can undergo a thermotropic phase transition in which the hydrocarbon chains change from a relatively rigid and ordered arrangement to a phase in which they are fluid and disordered. In the rigid, so-called crystal phase, the lateral diffusion of lipids in the bilayer is restricted compared with diffusion rates in the fluid, liquid-crystal conformation. The temperature (T_c) at which lipids undergo a phase change and the dependence of this temperature on hydrocarbon chain length, unsaturation, polar group, and metal ion interactions have been documented for many lipid systems (Chapman, 1975). In mixed lipid dispersions the phase transition may involve a cocrystallisation of the constituent lipids or conversely their segregation within the lamellar plane giving rise to broad melting endotherms or a number of individual, relatively narrow endotherms, respectively. The phase diagrams of several mixed lipid systems have been established (Ladbrooke and Chapman, 1969; Phillips et al., 1970; Shimshick and McConnell, 1973), particularly those of binary phospholipid-cholesterol mixtures (Ladbrooke et al., 1968). These studies have shown that cholesterol broadens the phase transition of the phospholipid hydrocarbon chains and restricts the lateral diffusion of lipids in the bilayer. In mixed lipid systems containing a protein or polypeptide, crystallisation of the higher melting point lipid effects a redistribution of the constituent proteins and lipids into aggregated patches within the lamellar plane (Cornell et al., 1976).

Phase transition and phase separation behaviour characteristic of pure lipids and lipid-polypeptide systems has also been observed in biological membranes. This is illustrated in Figure 1, which shows electron micrographs of freeze-fracture replicas from *Mycoplasma mycoides* var. *capri* membranes (kindly supplied by Dr. Razin). The replica in Figure 1A is from an adaptive strain of the microorganism containing relatively low amounts of cholesterol that have been equilibrated at 4°C prior to quenching to low temperatures. The fracture face of the membrane can be seen to be divided into two separate regions, one possessing a high density of particles (believed to represent membrane proteins), the other a smooth area devoid of particles. Figure 1B is the corresponding freeze-fracture surface of the cholesterol-rich parent strain. The protein appears to remain dispersed more or less randomly throughout the plane of the membrane (Rottem et al., 1973). In this case the crystallisation of the lipid is prevented by the relatively high concentration of cholesterol in the membrane. Similar temperature-induced protein aggregation has been reported in mitochondrial membranes (Hackenbrock, 1976) and lipid-reconstituted Ca^{2+}-activated ATPase of sarcoplasmic reticulum (Kleemann and McConnell, 1976). Not all membranes, however, exhibit these effects. Haest et al. (1974), for example, noted that proteins of some bacterial mem-

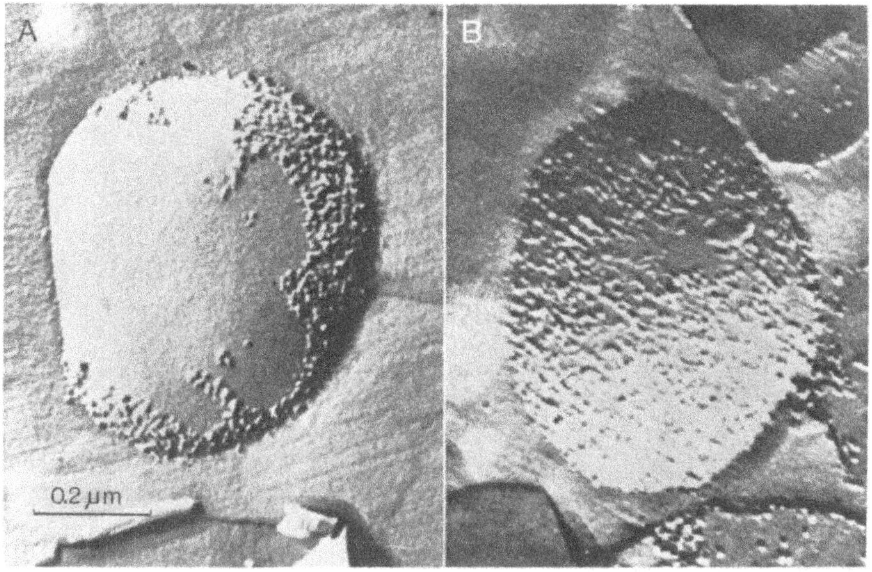

Fig. 1A,B. Platinum-carbon replica of convex fracture faces of freeze-cleaved *Mycoplasma mycoides* var. *capri* membranes suspended in 20% glycerol and incubated at 4°C prior to freezing. (A) membranes of the adapted strain; (B) membranes of the cholesterol-rich native strain (electron micrographs by Rottem et al., 1973)

branes remain dispersed even when the membranes are cooled to below the lipid phase transition temperature. Erythrocyte glycophorin also appears to be dispersed both above and below the phase transition temperature (Kleemann et al., 1974).

The retinal thylakoid protein, rhodopsin, provides another very interesting example. Studies of reconstituted rhodopsin membranes have shown that dark-adapted rhodopsin is aggregated below the lipid transition temperature but is dispersed randomly at higher temperatures. When the chromophore is bleached, however, the rhodopsin molecules remain dispersed at all temperatures (Hong and Hubbell, 1973).

Protein aggregation within the fluid lipid bilayer triggered by mechanisms other than lipid crystallisation has been related to various cell processes including pinocytosis (Orci and Perrelet, 1973), viral budding (Bose and Brundige, 1972), patching of surface antigens (de Petris and Raff, 1972), the cell growth cycle (Scott et al., 1971), and to membrane interactions (McDonough and Lilien, 1975; Furcht and Scott, 1975) and fusion (Lucy, 1976). The factors that determine the state of aggregation of protein within a lipid membrane above and below the lipid transition temperature remain however a subject of speculation. In this paper we examine the various factors that may effect the distribution of lipids and intrinsic membrane proteins in simple dispersions. We also introduce a new technique for modulating membrane fluidity that we have recently developed and that should prove useful in further studies of the dependence of protein aggregation processes on membrane fluidity. We envisage many additional applications of this new technique for the investigation of membrane structure and function.

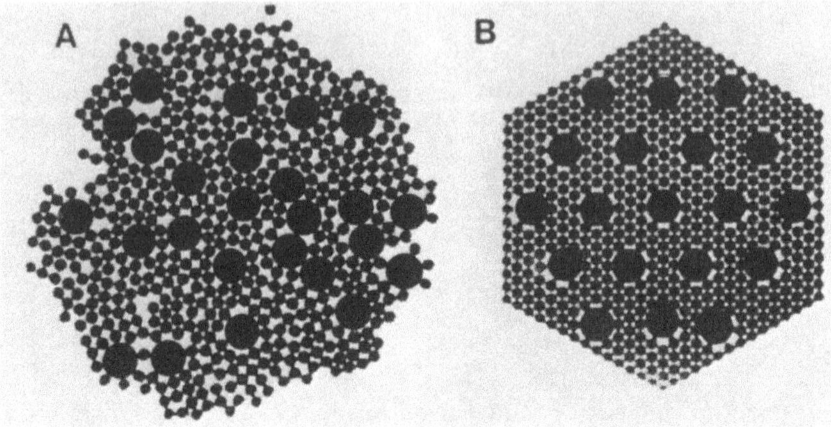

Fig. 2A,B. Two-dimensional array of lipids and proteins (penetrating polypeptide segments) based on assumption that (A) the lipids and proteins (one polypeptide unit) are randomly packed; (B) the proteins (one polypeptide unit) are ordered on a hexagonal super-lattice

B. Spontaneous Lipid-Protein Distribution

In order to illustrate what we understand by the spontaneous or natural distribution of lipid and protein molecules (and lipid-cholesterol molecules) in the plane of a cell membrane, we will begin by considering the arrangement of lipid and protein molecules in a fluid membrane at a temperature well above the main lipid phase transition temperature. We will simplify the situation by assuming that there are no specific electrostatic or hydrogen bonding attractions between the various constituents other than the self-association of the water molecules required to maintain the stability of the fluid bilayer structure. An important concept related to the protein distribution in cell membranes is the essential disorder of the fluid state. Thus broad non-crystalline wide-angle X-ray diffraction bands are observed in many biomembrane systems. This concept of disorder may be considered in a simple model involving the hydrocarbon chains of the membrane phospholipids and the penetrating polypeptide segments of intrinsic membrane proteins (Chapman and Cornell, 1976).

If the penetrating polypeptide segment of an intrinsic membrane protein is represented by a circle of diameter 120 Å and the lipid chains by a smaller circle of 48 Å diameter, we can arrange the components either in an ordered or, alternatively, in a disordered manner. This is schematically illustrated in Figure 2A, in which the components are arranged in a randomly disordered but closely packed array dictated by a simple Monte-Carlo technique and in Figure 2B, in which the polypeptide segments are displayed on a hexagonal super-lattice.

It is immediately apparent that more protein-protein contacts are made when the components are presented in a disordered array. These correspond to spontaneous protein-protein contacts within the fluid or disordered membrane structure. It is possible, using this simple model system, to predict the statistical probability of any particular size of protein-protein cluster as a function of the relative protein-to-lipid concentration. The model system can also be used to describe polypeptide units of larger size within the lipid bilayer, allowing the simulation of intrinsic proteins, which comprise a number of linked helical subunits. We recognize that using a small circle of diameter 48 Å to represent the fluid lipid chains is an oversimplification of the situation at low protein contents, but the situation will become increasingly more realistic at higher protein-to-lipid ratios, because the gauche isomers of the lipid chains are progressively inhibited by contact with the protein structures.

The spontaneous protein aggregates that occur in the fluid or disordered structure of cell membranes, illustrated in Figure 2B, may play an important role in certain membrane functions including transport processes. It may be expected that where membrane functions are influenced by protein dispositions, such spontaneous protein aggregates would be especially prevalent in membranes containing a relatively high protein-to-lipid content or where, as a result of a trigger mechanism, the local concentration of protein is increased.

C. Protein Distribution Affected by Lipid Crystallisation

As we noted previously, cooling certain membranes to the point where the lipid crystallises can cause the intrinsic proteins to aggregate. We have illustrated this process in the electron micrographs shown in Figure 1. Thus the formation of a crystalline array of lipids often excludes the protein from large areas of the membrane. It is important to appreciate that it is the melting of these extensive crystalline lipid regions from which the protein has been excluded that gives rise to the endotherms observed by scanning calorimetry with cell membrane systems. This observation and cautionary comment on the interpretation of calorimetric data was made some time ago by Chapman and Urbina (1971).

We now propose to examine the mechanisms underlying this type of lipid-protein distribution. For convenience we will consider three separate situations in which membranes have (1) a low protein content, (2) a high protein content, and (3) an intermediate protein content.

I. Low Protein Content

Consider a single protein molecule situated in a plane of pure lipid and consider what happens when the temperature is reduced below the phase transition temperature, T_c, so that the lipid crystallises and gives sharp wide-angle X-ray diffraction peaks. When an integral protein is interpolated into this layer a number of factors will influence the crystallisation process.

If the protein does not fit comfortably into a space normally occupied by an integral number of lipid chains, the crystallised array will be subject to considerable lateral stress resulting in a

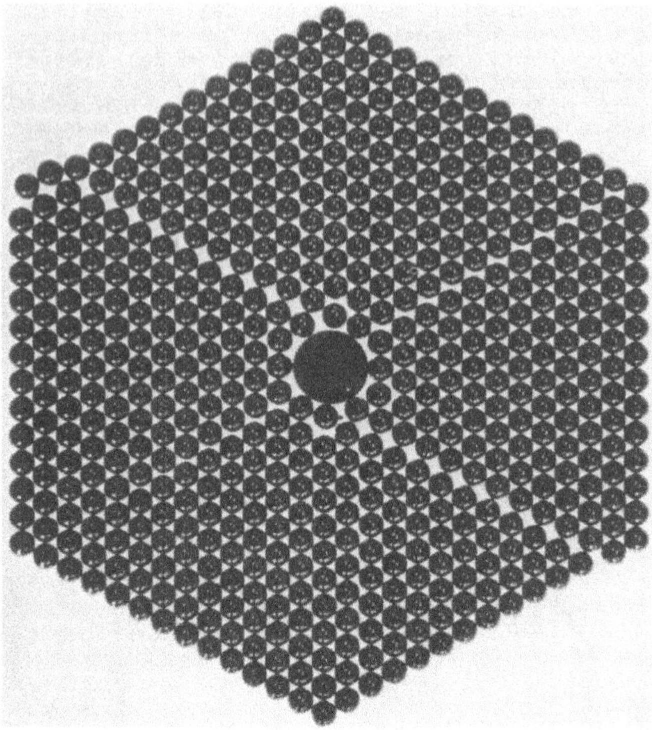

Fig. 3. Typical pattern of packing faults resulting from introduction of cylinder of noncritical size into ball bearing raft. Ball bearings represent hydrocarbon chains of 48 Å diameter. Cylinder is 120 Å in diameter. Packing faults lie along hexagonal axes of closely packed array of ball bearings

disordering of the lipid molecules adjacent to and extending away from the protein. In terms of our simplified model this disorder will take the form of packing faults radiating outward from the protein as illustrated in Figure 3. These packing faults will close after a number of molecular distances (dependent on the elasticity assumed for the lipid matrix) to form dislocations that ultimately vanish, permitting a return to the original hexagonal packing. A number of studies of other solid state systems have shown that the directions along which such dislocations occur are invariably along a path where the atoms are most closely packed. In a two-dimensional hexagonal matrix these are the directions of the three hexagonal axes, mutually separated by 120°C.

At low concentrations of protein relative to lipid, the disruptive effect of imperfect packing arrangements may be absorbed as dislocations. However, as the proportion of protein increases the width of the dislocations will begin to approach the average spacing of the proteins. If the sample is cooled evenly throughout the transition, the hydrocarbon chains will crystallise and the protein will become trapped within the dislocations that form spontaneously due to thermal fluctuations during crystallisation. Until the concentration of the protein is high enough to produce a dislocation density in excess of that which occurs spontaneously in the pure lipid, the "zone refining" of the protein into aggregates is prevented by its entrapment in packing faults. With increased protein concentrations, the effect of the lipid crystallisation will be to produce localised regions with a relatively high protein-to-lipid ratio distributed throughout the plane of the bilayer. It follows that if the size of the protein cannot permit regular hexagonal packing of the surrounding lipid chains,

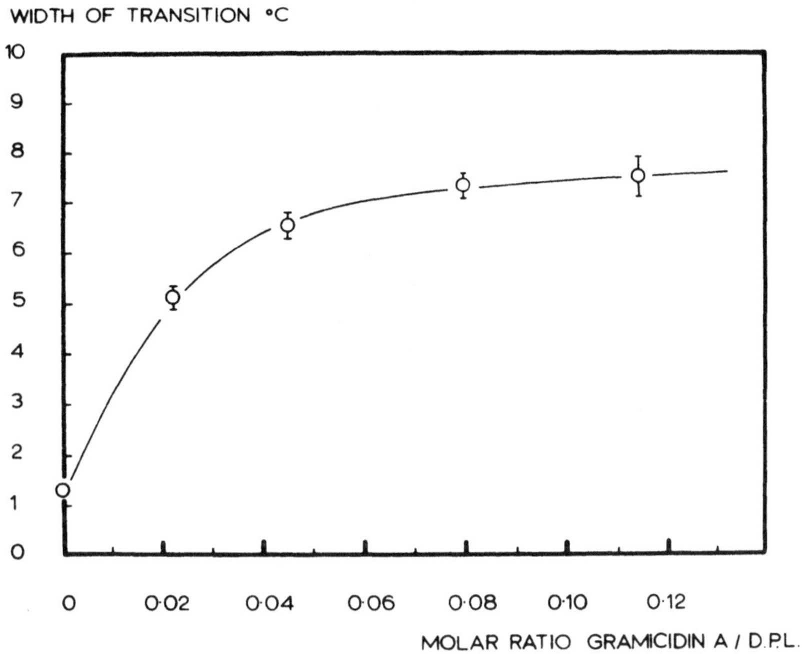

Fig. 4. Effect of adding increasing amounts of intrinsic hydrophobic polypeptide (Gramicidin A) to bilayer dispersions of dipalmitoyl lecithin on width of endothermic phase transition

regardless of the cross-sectional shape of the protein, the system will be unstable to lipid phase transitions. Moreover, if fluid hydrocarbon chains are crystallised in such a system by cooling below the phase transition temperature, both the protein and the lipid will be aggregated.

II. High Protein Content

As the proportion of protein to lipid increases, a limiting case will eventually be reached when the protein content is so high that the lipid phase transition will be smeared out and removed. The fact that cholesterol can abolish the endothermic phase transition of phospholipid hydrocarbon chains is well known (Ladbrooke et al., 1968). Intrinsic polypeptides and proteins can produce the same effect as cholesterol in this respect. This has been demonstrated recently in studies of the intrinsic trans-bilayer channel polypeptide, gramicidin A (Cornell et al., 1976). The manner in which the phase transition of dipalmitoyl lecithin is broadened when increasing amounts of polypeptide are incorporated into the lipid bilayer is shown in Figure 4. Once the phase transition of the lipid is removed, the driving force responsible for redistributing the components will no longer exist and only minor changes in the disposition of the lipids and proteins are to be expected.

III. Intermediate Protein Content

In the intermediate range of protein concentrations the regions of the membrane most affected by the crystallising lipid will be those that spontaneously possess very few protein-protein or protein-lipid-protein contacts. The crystallising lipid will exclude the protein and at least part of the non-crystalline lipid surrounding the protein into regions in which the protein concentration is relatively high. The extent to which these protein-rich areas now run together will depend upon the rate at which the system is cooled. The various processes involved in the lipid crystallisation and protein aggregation or protein-lipid clustering are illustrated schematically in Figure 5. This shows an initially random distribution of lipids and proteins (Fig. 5A) in which areas of lipid nucleation appear (Fig. 5B), producing domains of crystalline lipid intersected by packing faults. As crystallisation of the lipid proceeds, protein molecules are forced to aggregate (Fig. 5C), achieving an equilibrium distribution at low temperatures depicted in Figure 5D.

D. Crystallisation of Lipids in Biomembranes

In a natural membrane as distinct from a reconstituted or model membrane there will be a range of lipid classes and lipid acyl chains. When such membranes are cooled slowly, the higher melting point lipid will tend to crystallise out first. Depending upon the relative amount of intrinsic membrane protein, the crystallising lipid will exclude the proteins into regions of lower melting point lipids, producing a localised increase in protein-to-lipid ratio, thus giving rise to protein aggregates or protein-lipid clusters (Cornell et al., 1976).

If the protein is easily accommodated by virtue of size within a hexagonally packed array of lipid hydrocarbon chains or alternatively is present in sufficiently low amounts so that it becomes trapped within the crystallising lipid, the effect of lowering the temperature on protein redistribution will be negligible. This could explain some of the anomalies in which the distribution of certain proteins appears to be unaffected by equilibrium at low temperature.

E. Lipid Phase Transitions and Transport Processes

The packing arrangements of pure synthetic lecithins above and below the phase transition temperature have been shown in many experiments to influence the passive diffusion of solutes across bilayers, e.g., studies of water permeation in bilayer vesicles of saturated lecithins above and below the phase transition temperature have shown a decrease in the energy of activation from 110 $kJ \cdot mol^{-1}$ in the crystalline phase to 40 $kJ \cdot mol^{-1}$ above the phase transition temperature, suggesting that the diffusion process itself depends on the packing arrangement of the lipid (Blok et al., 1976). The permeability of phospholipid bilayers at temperatures corresponding to the phase transition temperature ap-

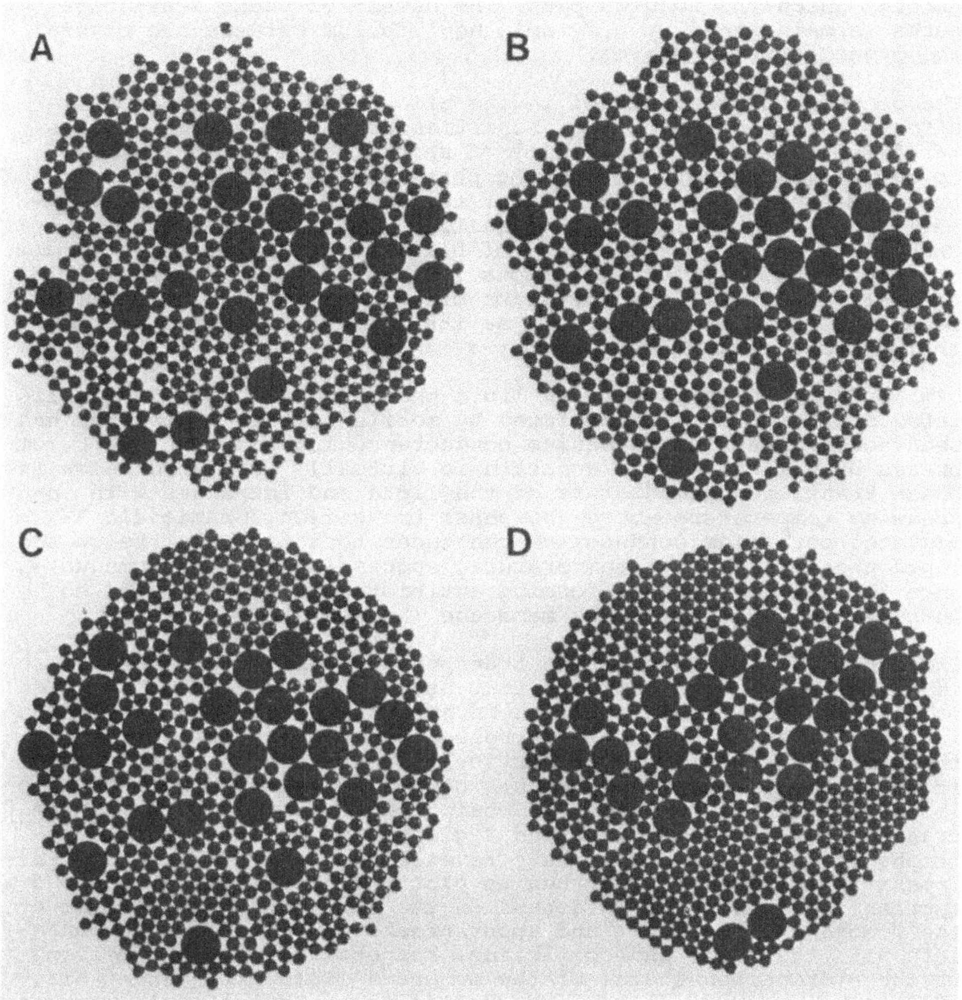

Fig. 5A-D. Schematic diagram showing how lipid crystallisation can lead to protein aggregation. (A) random distribution of lipid and protein; (B) more extensive areas of lipid nucleate-producing domains of crystalline lipid intersected by packing faults seen in (C) the crystallisation of certain areas of lipid forces protein molecules to aggregate; (D) equilibrium low temperature distribution of lipid and protein

pears to be increased markedly compared to membranes in purely crystalline or liquid-crystalline (fluid) configuration. The apparent activation energy of 8-anilino-1-naphthalene sulphonate (ANS) transport in dimyristoyl lecithin bilayers, for example, is large in both the crystalline and liquid-crystalline conformation but is reportedly zero at the phase transition temperature (Tsong, 1975). This enhanced permeability has been shown to depend on the size of the permeant solute as well as the length of the hydrocarbon chain of the saturated lecithin (Blok et al., 1975). The usual interpretation of this phenomenon is that discontinuities originating at the boundaries of the crystal (rigid) and fluid (disordered) domains in the bilayer give rise to pores through which the solutes pass. The number of these statistical pores is maintained by a dynamic equilibrium between the crystal and liquid-crystal phases.

The presence of cholesterol in the bilayer produces predictable effects on the permeability properties of the structure. Cholesterol enhances the permeability of saturated lecithin bilayers to glycerol and glycol below the phase transition temperature but depresses the permeability of the membrane towards these solutes above the lipid phase transition (De Gier et al., 1969). Comparison of the permeability of bilayers of saturated lecithins containing cholesterol to various ions and glucose indicates that cholesterol affects the partition of solutes between the aqueous phase and the membrane as well as the rate of diffusion through the membrane (Papahadjopoulos et al., 1971).

Similar relationships between lipid phase and the rate of facilitated diffusion of ions mediated by mobile carrier ionophores has been noted. Thus the potassium conductance induced in bilayer membranes by valinomycin or nonactin is virtually abolished below the phase transition temperature of the lipid and increases with increasing temperature above the phase transition. Gramicidin A-mediated potassium conductance continues to some extent below the lipid phase transition temperature, suggesting that the conductance through the channel-forming antibiotic is not affected so much by the fluidity of the membrane (Krasne et al., 1971).

The dependence of many active transport processes on membrane lipid fluidity has been demonstrated. Usually the evidence for this is derived from discontinuities in Arrhenius relationships, which in many instances appear to correspond to a phase transition of the membrane lipid. This is particularly evident in those cases in which the transport system has been reconstituted into phospholipid bilayers of defined acyl chain composition. Studies of the transport of β-galactosides and β-glucosides in fatty acid auxotrophs of *Escherichia coli* have revealed as many as three distinct breaks in the slopes of Arrhenius plots. The position of these discontinuities was shown to depend on the fatty acid composition of the membranes; the lower and upper breaks were observed to correlate with the onset and completion, respectively, of the melting of the hydrocarbon chains of the membrane lipids (Linden et al., 1973). The active transport of proline across cytoplasmic membrane vesicles of *E. coli* has been related to different energies of activation at temperatures above and below the lipid phase transition temperature (Esfahani et al., 1971; Shechter et al., 1974). Other examples where such behaviour has been observed include ATPase-dependent functions of *Mycoplasma* cells.

At this stage we may enquire as to what happens when, due to lipid crystallisation or phase separation, proteins are forced to cluster together and as a result trap lipid between them. Recent nuclear magnetic and electron spin resonance spectroscopic studies of hydrophobic polypeptide in fluid lipid bilayers are pertinent to this question. These studies show that when the fluid lipid becomes trapped between the proteins the lipid chains become relatively immobile (Cornell et al., 1976). This effect has also been observed with the lipids of the "purple membrane" where spectroscopic studies have shown that the "trapped" lipid in the membrane is rather rigid in character (Esser and Lanyi, 1973). Thus we can predict that when lipid is trapped between enzyme proteins this will cause the lipid to become less fluid and in some cases rather immobile. In turn this will (1) cause the lipid to become less compressible, (2) produce a loss in sensitivity of the lipid to temperature effects (due to smearing of the lipid phase transition), (3) cause a reduction in permeability in this region, and (4) produce a restriction on conformational changes in the protein. This restriction on protein conformational changes is expected to arise from the less compressible nature of the trapped lipid and probably affects the activation energy of processes mediated by membrane-bound proteins. In turn this may be related in some instances to the discontinuities observed in the Arrhenius plots, which are coincident with the lipid phase transition temperature. Thus we envisage that, when a membrane or reconstituted model-membrane is heated and where protein aggregates or lipid-protein clusters exist, the lipid trapped between the protein will at first be insensitive to temperature. However, as the bulk of the remaining lipid becomes fluid, the lateral diffusion of this lipid will begin to separate the clusters (the reverse of the process shown in Fig. 5). The lipid will then begin to attain a more normal fluidity, and compressibility and enzyme conformational possibilities will be restored to more active forms.

F. A New Method for Modulating Membrane Fluidity

We have seen that the fluidity characteristics of biological membranes have important consequences for lateral disposition and mobility of proteins, and this determines the way membranes perform their vital functions. The detailed chemical analyses of biological membranes show that they consist of a wide range of phospholipid classes (lecithins, sphingomyelins, etc.) and their associated fatty acids, the nature of which dominates the fluidity of the membrane. Within each phospholipid class present, the range of acyl chains is usually considerable, consisting of various chain lengths and a number of unsaturated bonds. Phospholipids extracted from brain and retinal tissues, for example, possess a high proportion of very long and highly unsaturated fatty acids ($C_{22:6}$) while those from lung tissues are relatively shorter and more saturated (van Deenen, 1965).

The relationship between membrane lipid fluidity and biochemical and biological processes has been actively studied (Raison, 1973; Chapman and Quinn, 1976) and transport processes, membrane-bound enzyme activities, malignant transformation, membrane adhesion, membrane fusion, protein rotation and diffusion and patch and cap formation have all been related to membrane lipid fluidity (Cherry, 1976). The idea (Chapman, 1966) that some biological systems, e.g., the thermophilic microorganisms, have a feedback mechanism designed to maintain membrane fluidity at a constant level by changing the length or degree of unsaturation of the associated fatty acids has been supported by a number of experiments (Lyons, 1972).

Because of this link between membrane lipid fluidity and various cellular processes, a number of experimental techniques have been devised to alter the lipid fluidity of cell membranes, including genetic, nutritional, and temperature manipulation (Fourcans and Jain, 1975; Silbert, 1975; Cronan and Gelmann, 1975). Recently, we have invented an entirely original approach to this idea of modulating membrane lipid and membrane fluidity. The method consists of hydrogenating membrane lipids in situ, i.e., within the membrane structure.

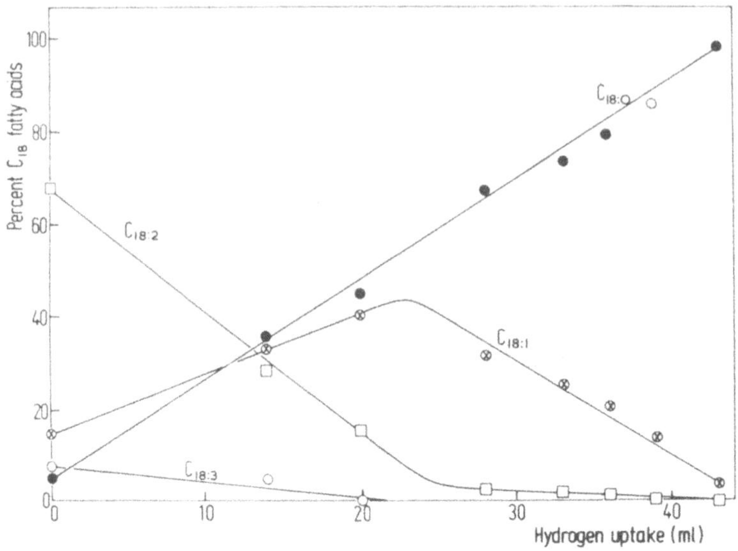

Fig. 6. Change in proportion of C_{18} fatty acids during hydrogenation of model biomembrane prepared by dispersing 0.5 g of soya lecithin and catalyst in water

The hydrogenation of triglycerides has considerable and well-documented exploitation in the production of hardened fats. All natural oils (glycerides) possess a spectrum of acyl chains containing different numbers of double bonds analogous to the acyl chains of membrane phospholipids. Studies of the heterogeneous catalytic hydrogenation of glycerides show that the fatty acyl residues with one double bond are hydrogenated more slowly or at a later stage than those with two or more double bonds, also that the hydrogenation of a polyunsaturated acyl residue once started tends to stop or slow down considerably at the mono-enoic stage (Coenen, 1970).

The hydrogenation of natural oils is usually performed in the presence of catalysts such as nickel, copper, platinum, or palladium, which are generally referred to as heterogeneous catalysts. The catalytic reaction in these cases is thought to take place entirely on the surface of the metal. Hydrogenation, however, can also be catalysed by transition metals in the atomic form, in which case the process is referred to as homogeneous catalysis. Homogeneous catalysts have a number of advantages over heterogeneous catalysts, including the fact that all the metal atoms of homogeneous catalysts are potentially available as catalytic centres.

We have succeeded in hydrogenating phospholipids in the presence of heterogeneous catalysts only when the substrate is dispersed in organic solvents in which the phospholipid is readily accessible to the catalytic surface of the metal. However, using homogeneous catalysts, we have been successful in hydrogenating phospholipids in completely aqueous systems, multibilayer lipid structures or sonicated dispersions consisting of single bilayer vesicles. The changes in proportion of C_{18} fatty acids produced as a result of hydrogenation of soya lecithin when presented as multibilayer lipid structures in water in the presence of a homogeneous catalyst are shown in Figure 6.

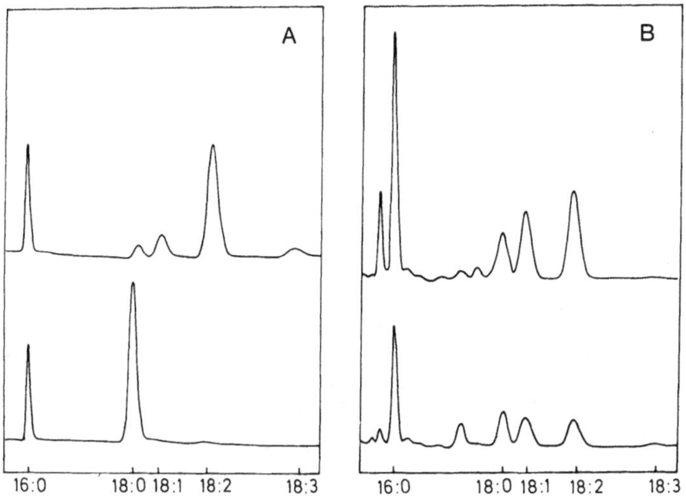

Fig. 7A,B. Fatty acid methyl esters prepared from (A) soya lecithin and (B) sarcoplasmic reticulum analysed by gas liquid chromatography. Upper traces show major fatty acids of unhydrogenated preparations and lower traces, after complete or partial hydrogenation of respective samples

We have demonstrated that the technique can also be applied directly to cell membranes. Figure 7 shows a gas liquid chromatographic trace of fatty acid methyl ethers prepared from soya lecithin (A) and rabbit muscle sarcoplasmic reticulum (B) before and after hydrogenation. This technique therefore provides a satisfactory method for hydrogenating the double bonds of the constituent lipids of cell membranes and hence modulating membrane fluidity. In the future we will report on our studies using this technique for the study of protein aggregation as a function of the amount of hydrogenation of the lipids of particular membranes. This hydrogenation technique may have many future biological and biochemical applications for the study of membranes and cell systems.

Acknowledgments. We wish to acknowledge a Senior Wellcome Trust Research Fellowship (to D.C.) and support from the Nuffield Foundation (to B.A.C.). We also wish to thank Professor S. Razin for supplying us with electron micrographs of *Mycoplasma* cells.

References

Blok, M.C., van Deenen, L.L.M., De Gier, J.: Effect of the gel to liquid crystalline phase transition on the osmotic behaviour of phosphatidylcholine liposomes. Biochim. Biophys. Acta 433, 1-12 (1976)

Blok, M.C., Van Der Neut-Kok, E.C.M., van Deenen, L.L.M., De Gier, J.: The effect of chain length and lipid phase transitions on the selective permeability properties of liposomes. Biochim. Biophys. Acta 406, 187-196 (1975)

Bose, H.R., Brundige, M.A.: Selective association of sindus virion proteins with different membrane fractions of infected cells. J. Virol. 9, 785-791 (1972)

Chapman, D.: Liquid crystals and cell membranes. Ann. N.Y. Acad. Sci. 137, 745-754 (1966)

Chapman, D.: Phase transitions and fluidity characteristics of lipids and cell membranes. Quart. Rev. Biophys. 8, 185-235 (1975)

Chapman, D., Cornell, B.A.: Phase transitions, protein aggregation and membrane fluidity. Nobel Symposium 34. New York - London: Plenum Press 1976, pp.

Chapman, D., Quinn, P.J.: The control of cell membrane fluidity by biochemical and catalytic hydrogenation processes. Chem. Phys. Lipids 16, (1976)

Chapman, D., Urbina, J.: Phase transitions and bilayer structure of Mycoplasma Laidlawii B. FEBS Lett. 12, 169-172 (1971)

Cherry, R.J.: Protein and lipid mobility in biological and model membranes. In: Biomembranes, Vol. 3. Chapman, D., Wallach, D.F.H. (eds.). London: Academic Press, 1976, pp. 48-97

Coenen, J.W.E.: Hydrogenation of oils and fats. In: Die Margarine heute. Leiden: E.J. Brill 1970, pp. 62-91

Cornell, B.A., Chapman, D., Eliasz, A.W.: The interactions of helical polypeptide segments which span lipid bilayers. J. Mol. Biol. (submitted for publication)

Cronan, J.E., Gelmann, E.P.: Physical properties of membrane lipids: biological relevance and regulation. Bacteriol. Rev. 39, 232-356 (1975)

Deenen, L.L.M., van: Phospholipids and biomembranes. In: Progress in the Chemistry of Fats and Other Lipids. Vol. VIII, Part 1. Holman, R.T. (ed.). Oxford: Pergamon Press 1965, pp. 1-127

Esfahani, M., Limbrick, A.R., Knutton, S., Oka, S., Wakil, S.J.: The molecular organisation of lipids in the membrane of Escherichia coli: phase transitions. Proc. Natl. Acad. Sci. U.S. 68, 3180-3184 (1971)

Esser, A.F., Lanyi, J.K.: Structure of the lipid phase in cell envelope vesicles from Halobacterium cutirubrum. Biochemistry 12, 1933-1939 (1973)

Fourcans, B., Jain, M.K.: Role of phospholipids in transport and enzymic reactions. Adv. Lipid Res. 12, 147-226 (1974)

Furcht, L.T., Scott, R.E.: Modulation of the distribution of plasma membrane intramembranous particles in contact inhibited and transformed cells. Biochim. Biophys. Acta 401, 213-220 (1975)

Gier, J., de, Mandersloot, J.G., van Deenen, L.L.M.: The role of cholesterol in lipid membranes. Biochim. Biophys. Acta 173, 143-145 (1969)

Hackenbrock, C.R.: Molecular organization and the fluid nature of the mitochondrial energy transducing membrane. Nobel Symposium 34. New York-London: Plenum Press 1976, pp.

Haest, C.W.M., Verkleij, A.J., De Gier, J., Scheek, R., Ververgaert, P.H.J., van Deenen, L.L.M.: The effect of lipid phase transitions on the architecture of bacterial membranes. Biochim. Biophys. Acta 356, 17-26 (1974)

Hong, K., Hubbell, W.L.: Lipid requirements for rhodopsin regenerability. Biochemistry 12, 4517-4523 (1973)

Kleemann, W., Grant, C.W.M., McConnell, H.M.: Lipid phase separations and protein distribution in membranes. J. Supramol. Struct. 2, 609-616 (1974)

Kleemann, W., McConnell, H.M.: Interactions of proteins and cholesterol with lipids in bilayer membranes. Biochim. Biophys. Acta 419, 206-222 (1976)

Krasne, S., Eisenman, G., Szabo, G.: Freezing and melting of lipid bilayers and the mode of action of nonactin, valinomycin and gramicidin. Science 174, 412-415 (1971)

Ladbrooke, B.D., Chapman, D.: Thermal analysis of lipids, proteins and biological membranes. Chem. Phys. Lipids 3, 304-367 (1969)

Ladbrooke, B.D., Williams, R.M., Chapman, D.: Studies on lecithin-cholesterol-water interactions by differential scanning calorimetry and X-ray diffraction. Biochim. Biophys. Acta 150, 333-340 (1968)

Linden, C.D., Wright, K.L., McConnell, H.M., Fox, C.F.: Lateral phase separations in membrane lipids and the mechanism of sugar transport in Escherichia coli. Proc. Natl. Acad. Sci. U.S. 70, 2270-2275 (1973)

Lucy, J.A.: The membrane of the hen erythrocyte as a model for studies on membrane fusion. Nobel Symposium 34. New York-London: Plenum Press 1976, pp.

Lyons, J.M.: Phase transitions and control of cellular metabolism at low temperatures. Cryobiology 9, 341-350 (1972)

McDonough, J., Lilien, J.: Inhibition of mobility of cell surface receptors by factors which mediate specific cell-cell interactions. Nature (London) 256, 416-417 (1975)

Orci, L., Perrelet, A.: Membrane-associated particles: Increase at sites of pinocytosis demonstrated by freeze-etching: Science 181, 868-869 (1973)

Papahadjopoulos, D., Nir, S., Ohki, S.: Permeability properties of phospholipid membranes: effect of cholesterol and temperature. Biochim. Biophys. Acta 266, 561-583 (1971)

Petris, S., de, Raff, M.C.: Distribution of immunoglobulin on the surface of mouse lymphoid cells as determined by immunoferritin electron microscopy. Antibody-induced, temperature dependent redistribution and its implications for membrane structure. Europ. J. Immunol. 2, 523-535 (1972)

Phillips, M.C., Ladbrooke, B.D., Chapman, D.: Molecular interactions in mixed lecithin systems. Biochim. Biophys. Acta 196, 35-44 (1970)

Raison, J.K.: The influence of temperature-induced phase changes on the kinetics of respiratory and other membrane-associated enzyme systems. Bioenergetics 4, 285-309 (1973)

Rottem, S., Yashouv, J., Ne'eman, Z., Razin, S.: Cholesterol in mycoplasma membranes. Composition, ultrastructure and biological properties of membranes from *Mycoplasma mycoides* var. *capri* cells adapted to grow with low cholesterol concentrations. Biochim. Biophys. Acta 323, 495-508 (1973)

Scott, R.E., Carter, R.L., Kidwell, W.R.: Structural changes in membranes of synchronized cells demonstrated by freeze-cleavage. Nature (New Biol.) 233, 219-220 (1971)

Shechter, E., Letellier, L., Gulik-Krzywicki, T.: Relations between structure and function in cytoplasmic membrane vesicles isolated from an *Escherichia coli* fatty acid auxotroph. High-angle X-ray diffraction, freeze-etch electron microscopy and transport studies. Europ. J. Biochem. 49, 61-76 (1974)

Shimshick, E.J., McConnell, H.M.: Lateral phase separation in phospholipid membranes. Biochemistry 12, 2351-2360 (1973)

Silbert, D.F.: Genetic modification of membrane lipid. Ann. Rev. Biochem. 44, 315-339 (1975)

Tsong, T.Y.: Effect of phase transition on the kinetics of dye transport in phospholipid bilayer structures. Biochemistry 14, 5409-5414 (1975)

Rotational Diffusion of Proteins in Membranes

R. J. CHERRY, A. BÜRKLI, M. BUSSLINGER, and G. SCHNEIDER

Using the technique of flash photolysis, we have developed a method for measuring rotational relaxation times in the microsecond-millisecond time range. The principal aim of developing such a method is to investigate protein rotation in membranes. From considerations of membrane viscosity, it is clear that a globular "intrinsic" protein (as defined by Singer and Nicolson, 1972) would have a rotational relaxation time at least as slow as microseconds (e.g., see Cherry, 1976). Rhodopsin, which is the only membrane protein whose rotational diffusion has been quantitatively determined, has a relaxation time of 20 μs at 20°C (Cone, 1972).

Measurement of the rotational diffusion of membrane proteins should test to what extent proteins undergo free diffusion, as envisaged in currently popular models of membrane structure (Singer and Nicolson, 1972). Where free diffusion is restricted, diffusion measurements may provide a useful method of investigating interactions that are responsible for the restriction. Finally, it may be possible to link the mobility (or immobility) of proteins to functional properties of the membrane.

A. Measurement of Slow Rotational Diffusion by Flash Photolysis

Triplet states of organic molecules typically have lifetimes of milliseconds at room temperature. This long lifetime may be exploited to measure rotational motion that is too slow to be measured by the conventional method of fluorescence polarization.

Triplet states are conveniently detected by observing transient absorbance changes of a given sample following excitation by a pulse of light. Absorbance changes arising from either triplet-triplet absorption or from ground state depletion may be observed. When the exciting light is plane polarized, only those molecules whose transition moment for absorption lies in or near the plane of polarization are excited. Hence a partially oriented population of excited molecules is photoselected from the initial random distribution. As a result, the transient absorbance changes are in general dichroic. Dichroism decays as the excited molecules again become disordered due to Brownian rotation: the rate of decay enables the rate of rotation to be calculated. Because of the long lifetime of the triplet state, relaxation times as slow as milliseconds may be determined in this way. Similar measurements can in principle also be made using other long-lived photoproducts, such as free radicals or stereoisomers.

The flash photolysis apparatus used in the experiments described here is shown schematically in Figure 1. The exciting source is an Electro Photonics model 43 dye laser. The dye solution is coumarin 6 (10^{-4} M in ethanol), which has an untuned emission at 540 nm; the light pulse has a half width of 1-2 μs and an energy

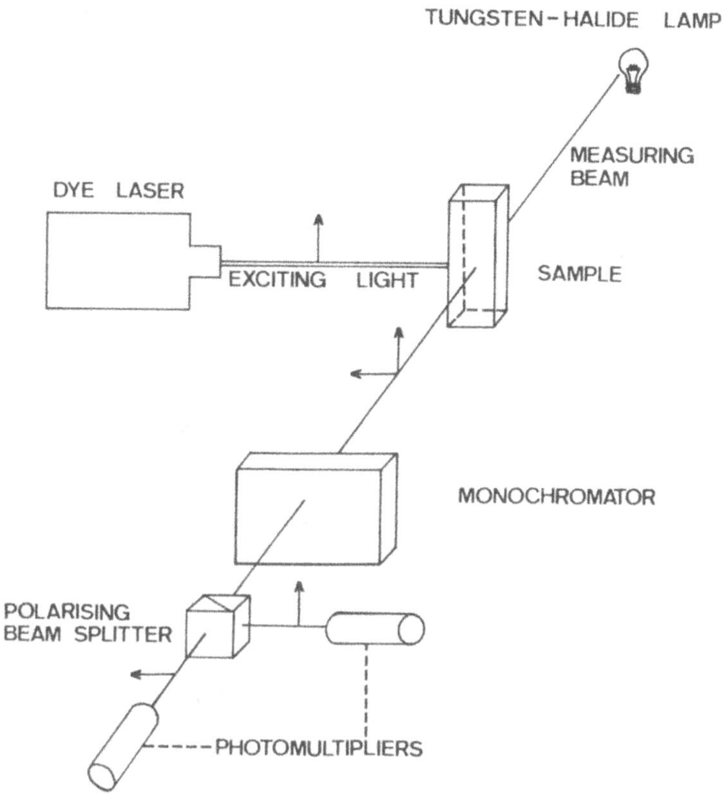

Fig. 1. Schematic diagram of flash photolysis apparatus

of 100-200 mJ. The exciting light is vertically polarized by a Glan Taylor prism. The sample is contained in a 1 cm fluorimeter cell and the measuring source is a 100 W tungsten-halide lamp. The measuring beam passes through the sample at right angles to the exciting beam. The wavelength of the transient absorption is selected by a Leitz in-line mirror monochromator. After emerging from the monochromator, the measuring beam is split into vertically and horizontally polarized components by a Barr and Stroud BC6 polarizing beam splitter. The intensities of the two components are then measured separately by two photomultipliers (EMI 9683 QB) and displayed on a Tektronix 5103 dual beam oscillioscope. For measurement purposes, it is often convenient to display the difference between the two signals on one trace, using the differential input of the oscilloscope amplifier.

The data are analyzed by calculating the anisotropy parameter r(t) defined by

$$r(t) = \frac{A_{\parallel}(t) - A_{\perp}(t)}{A_{\parallel}(t) + 2A_{\perp}(t)} \qquad (1)$$

where $A_{||}(t)$, $A_{\perp}(t)$ are the absorbance changes at time t for light polarized parallel and perpendicular to the exciting flash. For a spherical molecule, r(t) decays exponentially with time according to the expression (Jablonsky, 1961)

$$r(t) = r_o \exp(-t/\phi) \tag{2}$$

where ϕ is the rotational correlation time and r_o is the value of r at t = 0. For a sphere of radius a immersed in a medium of viscosity η

$$\phi = \frac{4\pi a^3 \eta}{3kT} = \frac{1}{6D} \tag{3}$$

where D is the rotational diffusion coefficient (Einstein, 1906; Jablonsky, 1961).

For an irregular body the situation is more complicated. In the general case the right-hand side of eq. (2) is replaced by the sum of five exponentials (Ehrenberg and Rigler, 1972; Chuang and Eisenthal, 1972). For a body (such as an ellipsoid of revolution) that possesses an axis of symmetry, there are three exponentials.

B. Rotational Diffusion of Protein-Eosin Conjugates in Viscous Solutions

In order to measure rotational diffusion by the methods outlined above, it is of course necessary to have a chromophore with appropriate spectroscopic properties. Eosin was previously demonstrated to be a suitable probe molecule (Razi-Naqvi et al., 1973); recently we have prepared the reactive isothiocyanate derivative, which enables eosin to be coupled covalently to proteins (Cherry et al., 1976b). We have investigated the validity of the method by measuring the rotational diffusion of protein-eosin conjugates in viscous glycerol-water mixtures (Cherry and Schneider, 1976).

Table 1 contains the results of experiments carried out with proteins of varying sizes. From the measured rotational diffusion coefficient and the solution viscosity, the radius of the protein was calculated using Equation (3). As can be seen, the experimental values agree reasonably well with the expected values calculated by assuming the proteins to be spherical. Differences between experimental and theoretical values are probably due to deviations from spherical shape and to hydration effects. We could also demonstrate that the measured rate of rotation varied in the predicted manner with the solution viscosity. These experiments clearly demonstrate that the method reliably measures rotational motion in the microsecond-millisecond time range.

Table 1. Determination of particle size from rotational diffusion measurements[a]

	Mol wt	Theoretical radius[b]	Radius from rotational diffusion
β lactoglobulin A (monomer)	18,400	17.6 Å	22.4 Å
Ovalbumin	43,500	23.5 Å	34 Å
Sucrase-isomaltase	221,000	40 Å	52 Å
Qβ virus	4.2×10^6	118 Å[c]	143 Å

[a]Protein measurements made in 94% glycerol. Value for Qβ virus is mean of measurements made in 40% and 50% glycerol, 40% and 50% sucrose

[b]Theoretical radii of proteins calculated from mol wt assuming proteins to be spherical and unhydrated

[c]Radius determined by negative stain electron microscopy

C. Protein Rotation in the Human Erythrocyte Membrane

The first application of the method to membrane proteins has been carried out with the human erythrocyte. This system was chosen because the composition and properties of the membrane proteins are relatively well characterized. Major constituents are spectrin, a fibrous protein composed of two polypeptides of molecular weights 220,000 and 240,000, and band 3 (otherwise known as component a). Band 3 consists of hydrophobic proteins that span the membrane, have a molecular weight of about 90,000, and comprise about 24% of the total membrane protein (Steck, 1974). They are implicated in anion and other transport functions (Cabantchik and Rothstein, 1974; Ho and Guidotti, 1975; Passow et al., 1975; Lin and Spudich, 1974; Brown et al., 1975). The major sialoglycoprotein, glycophorin, also spans the membrane and has been characterized extensively (Tomita and Marchesi, 1975). It is thought that glycophorin in association with band 3 constitutes the membrane particles revealed by freeze-fracture electron microscopy (Pinto da Silva and Nicolson, 1974; Guidotti, 1972).

I. Labeling of Human Erythrocytes with Eosin Isothiocyanate

Red blood cells obtained from fresh or recently outdated human blood were washed three times with 310 mosM sodium phosphate buffer, pH 7.4. Fifteen ml of packed cells were incubated with 3 mg eosin-isothiocyanate for 3 h at 22°C. The cells were then washed twice more with isotonic buffer to remove any unreacted label and subsequently hemolyzed in 40-50 vol of 20 mosM sodium phosphate buffer, pH 7.4. The ghosts were sedimented by centrifugation for 20 min at 20,000 g and washed three to four times with 20 mosM buffer. All operations except the labeling step were performed at 0-4°C.

The amount of bound eosin was determined spectrophotometrically as described previously (Cherry et al., 1976a), after first solubilizing ghosts with sodium dodecyl sulphate (SDS). Protein was determined by the method of Lowry et al. (1951). Typically the ghosts contained 0.5-1.5 μg eosin per mg protein.

II. Identification of Labeled Components

1. Gel electrophoresis

The erythrocyte membrane proteins were analyzed on SDS-polyacrylamide gels using methods based on those of Fairbanks et al. (1971). However, instead of the 5.6% acrylamide gels used by these authors, gels were prepared from a solution containing 7.5% acrylamide (wt/vol), 2.73% bis acrylamide (with respect to acrylamide). Gel dimensions were 0.85 x 8 cm. Ghosts were solubilized by mixing the suspension with an equal volume of a solution containing 2% SDS, 2% mercaptoethanol, and 8 M urea and incubating for 5 min at 100°C. A sample of the solubilized ghosts containing 0.5 mg protein, 1 drop glycerin, and 5 μl of 0.5% bromophenol blue was loaded onto the gels which were run for 8 h at a current of 20 mA per gel.

When electrophoresis was complete, the gels were cut into slices approximately 1 mm thick. Each slice was mounted in a specially designed holder in the sample space of a Perkin-Elmer 204 fluorimeter. The excitation wavelength was 470 nm and emission was scanned from 490-600 nm. Eosin was detected by its fluorescent emission at about 545 nm. Because of the turbid nature of the sample, there was also a strong scattering peak from the exciting light. The eosin emission was normally superimposed on the tail of this scattering band.

After examination for fluorescence, the gel slices were fixed and stained for carbohydrate using the periodic acid-Schiff (PAS) procedure of Fairbanks et al. (1971). The stain was detected spectrophotometrically at 560 nm. Finally the slices were stained with coomassie blue, again as described by Fairbanks et al. (1971), and the absorption of each slice was measured at 530 nm.

The results of this analysis are shown in Figure 2. The resolution of the protein bands is not optimal, both because of the large diameter of the gels and because of the detection methods. This is unimportant for the purpose of locating eosin since the main components of the membrane can be clearly distinguished. Further, smaller diameter 5.6% acrylamide gels gave exactly the same protein distribution as that described by Fairbanks et al. (1971).

Figure 2 shows that most of the eosin fluorescence is associated with band 3. Some fluorescence is also detected in the region of bands 1 and 2 (spectrin). Only trace amounts are associated with other proteins. A particular advantage of the 7.5% gel system is that glycophorin is clearly separated from band 3. It can be seen that no eosin fluorescence is detectable in the position of the PAS-1 stain.

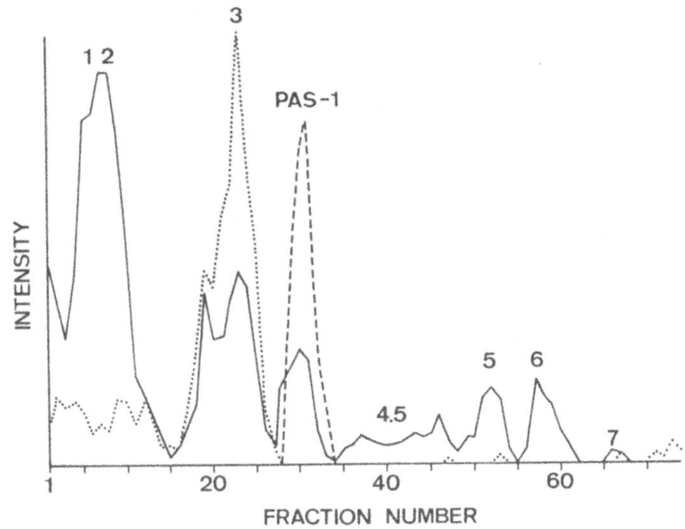

Fig. 2. SDS-polyacrylamide gel electrophoresis of eosin-labeled erythrocyte membranes: ——: coomassic blue stain; ---: PAS stain; ...: eosin fluorescence. Bands are numbered according to nomenclature of Steck (1974)

2. Selective Extraction

Selective extraction of proteins from the ghost membrane was used to confirm the preceding findings and to provide some quantitative estimation of the extent to which proteins other than band 3 are labeled. Selective release of bands 1, 2, and 5 was achieved by incubating 1 vol ghosts with 9 vol 0.1 mM EDTA for 25 min at 37°C (Fairbanks et al., 1971). Under these conditions we find that only small amounts of eosin are released from the membrane. For example, in one typical experiment, 25 ± 7% of the total membrane protein appeared in the supernatant following centrifugation of ghosts subjected to the foregoing procedure. However, only 7 ± 3% of the total eosin was solubilized. SDS-polyacrylamide gels confirmed that nearly all of bands 1, 2, and 5 were selectively released from the membrane. The conclusion from a number of such experiments is that less than 10% of bound eosin is associated with these components.

Band 3 proteins were solubilized by incubating 1 vol ghosts with 7 vol of a solution containing 0.5% Triton X-100 (wt/vol), 61 mM sodium tetraborate, pH 8, for 20 min at 0-4°C (Yu et al., 1973). Since the method is only partially selective and not all of band 3 is released, these experiments are at best semiquantitative. We find that 60-80% of bound eosin together with about 40% of the membrane protein is solubilized by the foregoing procedure.

Finally, we extracted labeled ghosts with 2:1 chloroform-methanol as described by Hamaguchi and Cleve (1972). No eosin was detectable in the organic phase, which contains lipids. A small amount of eosin (~ 4% of the total) appeared in the aqueous phase. SDS-polyacrylamide gels confirmed that the aqueous phase contained only PAS-staining glycoproteins; none of these components could be detected at the interface, which contains the remainder of the membrane proteins.

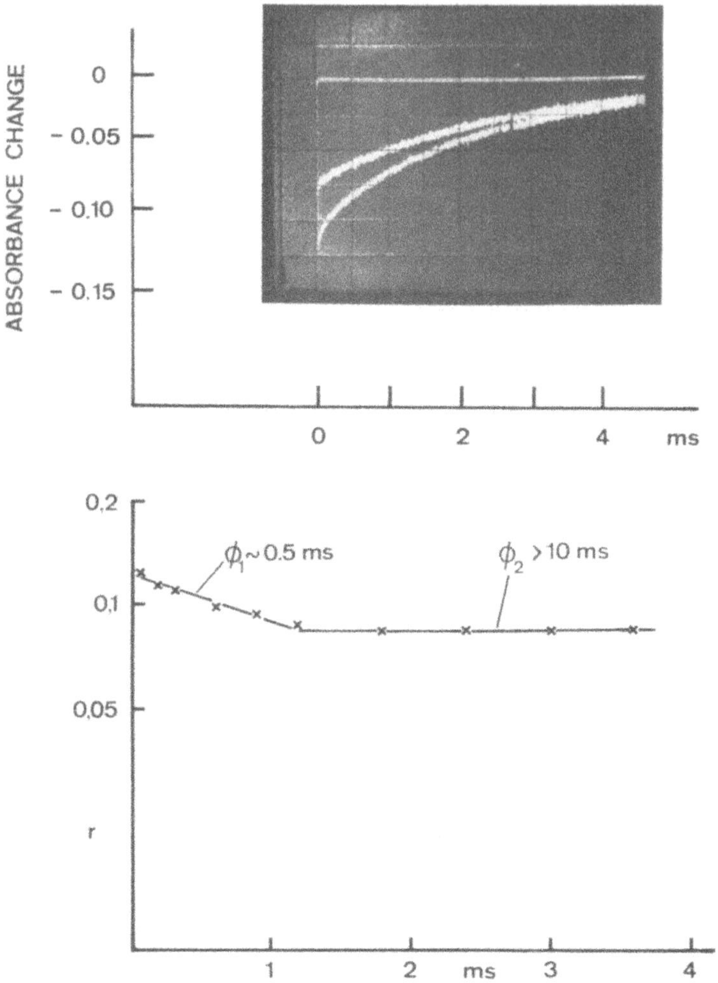

Fig. 3A,B. Results of flash photolysis experiments with eosin-labeled erythrocyte membranes. (A) Ground state depletion signals measured at 520 nm. Stronger (lower) signal is obtained with measuring beam polarized parallel to exciting flash, weaker signal with perpendicular polarization. Ghosts suspended in 20 mosM phosphate buffer, pH 7.4, 22°C. Prior to measurement samples were bubbled with argon for 10 min to displace oxygen. (B) Time dependence of anisotropy parameter

Taken together the extraction experiments provide good confirmation of the result obtained fluorimetrically with SDS-polyacrylamide gels, namely that the labeling procedure leads to a selective attachment of eosin to band 3. Such a selective labeling is in accord with other authors, who find that only band 3 and sialoglycoproteins are available at the outer surface of the red cell (Steck, 1974; Bretscher, 1973; Morrison et al., 1974; Shin and Carraway, 1974). Since some of our label is associated with spectrin, we conclude that there is some permeation of eosin isothiocyanate through the membrane. We find little labeling of glycophorin; a similar observation was made by Cabantchik and Rothstein (1974) with DIDS (4,4'diisothiocyano-2,2' stilbene disulphonic acid). DIDS (added to intact cells) is found to bind preferentially to band 3 with less than 5% of the label associated with glycophorin.

III. Rotational Diffusion Measurements

Flash photolysis signals obtained from eosin-labeled ghosts are shown in Figure 3A. It can be seen immediately that the transient absorbance change is dichroic and the dichroism is long-lived. Quantitative analysis of the data is performed by calculating the anisotropy parameter (eq. (1)). A semilogarithmic plot of the time dependence of r is shown in Figure 3B.

Figure 3B demonstrates that the decay of dichroism contains at least two components. At long time, r(t) is, as far as can be ascertained, independent of time so that the curve has the form

$$r(t) = r_1 \exp(-t/\phi_1) + r_2$$

Replotting $\log[r(t) - r_2]$ against t gives $\phi_1 \approx 0.5$ ms.

A full discussion of the interpretation of this result is given elsewhere (Cherry et al., 1976a). In summary, the observed time dependence of r is probably due to the anisotropic motion of band 3. The experimental result is explicable if rotation is confined to an axis normal to the plane of the membrane. The time constant ϕ_1 is related to the diffusion coefficient for rotation about this axis. The time independent component r_2 is due to the negligible rate of rotation about axes in the plane of the membrane. Possibly there is some aggregation of band 3 that may also contribute to r_2.

The only previous quantitative determination of the rotation of a membrane protein is Cone's measurement with rhodopsin (Cone, 1972), which yielded a relaxation time of 20 μs at 20°C. Rotation of band 3 proteins in the human erythrocyte membrane is clearly much slower than this.

Band 3 is probably involved in anion transport across the erythrocyte membrane (Cabantchik and Rothstein, 1974; Passow et al., 1975; Ho and Guidotti, 1975), although as pointed out by Lepke et al. (1976), the evidence is not completely conclusive. Lepke et al. find that sulphate exchange is completely inhibited when 1.7×10^6 molecules of the irreversible inhibitor H_2DIDS are bound to band 3. This figure implies a turnover number for chloride exchange of about $2 \times 10^4 \cdot sec^{-1}$ per site (at 25°C) if all these sites are involved in transport. If only some of these sites are responsible for transport, or if the transport system resides in a component of the membrane other than band 3 (Lepke et al. find that only about 70% of the total bound H_2DIDS is located on band 3), then the turnover number must be higher. Hence a rotation time in the order of tens of microseconds or faster would be required if transport occurred by a rotational mechanism. Irrespective of any particular interpretation, the present experiments do not detect any component of rotational motion of band 3 in this order of magnitude. Thus if band 3 is responsible for anion transport, a mechanism involving rotation of the whole protein is ruled out. This is hardly a surprising conclusion since arguments based on the structure of band 3 also indicate that rotation <u>across</u> the membrane is improbable (Bretscher, 1973; Ho and Guidotti, 1975). Nevertheless, the present studies demonstrate how rotation measurements may be useful in elucidating transport mechanisms in other systems in which the outcome is not so easily anticipated.

D. Summary

Slow rotational diffusion may be measured by flash photolysis using eosin-isothiocyanate as a probe. Measurements with protein-eosin conjugates in viscous solutions demonstrate that the method provides quantitatively reliable information.

When eosin-isothiocyanate is reacted with intact human erythrocytes, the probe binds selectively to band 3 proteins. Transient absorption signals from eosin-labeled membranes exhibit dichroism; the decay of dichroism contains one component of time constant about 0.5 ms and another time-independent component. This finding is consistent with rotational motion of band 3, which occurs only about an axis normal to the plane of the membrane.

These experiments demonstrate that the flash photolysis method can be successfully used to investigate protein rotation in the erythrocyte membrane. Further application of the method should provide useful information about protein mobility in other membranes.

Acknowledgments. We are grateful to Prof. G. Semenza for his continual support of the work described here, to Prof. H. Passow for a helpful discussion of anion transport, and to the Swiss National Science Foundation for financial assistance.

References

Bretscher, M.S.: Membrane structure: some general principles. Science 181, 622-699 (1973)

Brown, P.A., Feinstein, M.B., Sha'afi, R.L.: Membrane proteins related to water transport in human erythrocytes. Nature (London) 254, 523-525 (1975)

Cabantchik, Z.I., Rothstein, A.: Membrane proteins related to anion permeability of human red cells. I. Localisation of disulfonic binding sites in proteins involved in permeation. J. Membrane Biol. 15, 207-226 (1974)

Cherry, R.J.: Protein and lipid mobility in biological and model membranes. In: Biological Membranes, Vol. III. Chapman, D., Wallach, D.F.H. (eds.). London-New York: Academic Press 1976, pp. 47-102

Cherry, R.J., Schneider, G.: A spectroscopic technique for measuring slow rotational diffusion of macromolecules. Part II: Determination of rotational correlation times of proteins in solution. Biochemistry 15, 3657-3661 (1976)

Cherry, R.J., Bürkli, A., Busslinger, M., Schneider, G., Parish, G.R.: Rotational diffusion of band 3 in the human erythrocyte membrane. Nature (Lond.) 263, 389-393 (1976)

Cherry, R.J., Cogoli, A., Oppliger, M., Schneider, G., Semenza, G.: A spectroscopic technique for measuring slow rotational diffusion of macromolecules. Part I: Preparation and properties of a triplet probe. Biochemistry 15, 3653-3656 (1976b)

Chuang, T.J., Eisenthal, K.B.: Theory of fluorescence depolarisation by anisotropic rotational diffusion. J. Chem. Phys. 57, 5094-5097 (1972)

Cone, R.A.: Rotational diffusion of rhodopsin in the visual receptor membrane. Nature (New Biol.) 236, 39-43 (1972)

Ehrenberg, M., Rigler, R.: Polarised fluorescence and rotational Brownian motion. Chem. Phys. Lett. 14, 539-544 (1972)

Einstein, A.: On the theory of Brownian motion. Ann. Phys. 17, 549-562 (1905)

Fairbanks, G., Steck, T.L., Wallach, D.F.H.: Electrophoretic analysis of the major polypeptides of the human erythrocyte membrane. Biochemistry 10, 2606-2616 (1971)

Guidotti, G.: Membrane proteins. Ann. Rev. Biochem. 41, 731-752 (1972)
Hamaguchi, H., Cleve, H.: Solubilisation and comparative analysis of mammalian erythrocyte membrane glycoproteins. Biochem. Biophys. Res. Comm. 47, 459-464 (1972)
Ho, M.K., Guidotti, G.: A membrane protein involved in anion exchange. J. Biol. Chem. 250, 675-683 (1975)
Jablonski, A.: Über die Abklingungsvorgänge polarisierter Photolumineszenz. Z. Physik 16a, 1-4 (1961)
Lepke, L., Fasold, H., Pring, M., Passow, H.: A study of the relationship between anion exchange and binding to the red blood cell membrane of 4,4'-diiisothiocyano stilbene disulfonic acid (DIDS) and its dihydro derivative (H_2DIDS). J. Membrane Biol. (in press)
Lin, S., Spudich, J.A.: Biochemical studies on the mode of action of cytocholasin B. J. Biol. Chem. 249, 5778-5783 (1974)
Lowry, O.H., Rosenbrough, N.J., Farr, A.L., Randall, R.J.: Protein measurement with the folin phenol reagent. J. Biol. Chem. 193, 265-275 (1951)
Morrison, M., Mueller, T.J., Huber, C.T.: Transmembrane orientation of the glycoproteins in normal human erythrocytes. J. Biol. Chem. 249, 2658-2660 (1974)
Passow, H., Fasold, H., Zaki, L., Schuhmann, B., Lepke, S.: Membrane proteins and anion exchange in human erythrocytes. In: Biomembranes: structure and function. Gardos, G., Szász, I. (eds.). (FEBS Symp. Ser. Vol. XXXV, Proc. 9th meeting, Budapest, 1974). Budapest: Publ. House Hung. Acad. Sci. 1975, pp. 197-214
Pinto da Silva, P., Nicolson, G.L.: Freeze-etch localisation of concanavalin A receptors to the membrane intercalculated particles of human erythrocyte ghost membranes. Biochim. Biophys. Acta 363, 311-319 (1974)
Razi Naqvi, K., Gonzalez-Rodriguez, J., Cherry, R.J., Chapman, D.: A spectroscopic technique for studying protein rotation in membranes. Nature (New Biol.) 245, 249-251 (1973)
Shin, B.C., Carraway, K.L.: Lactoperoxidase labelling of erythrocyte membranes from the inside and outside. Biochim. Biophys. Acta 345, 141-153 (1974)
Singer, S.J., Nicolson, G.L.: The fluid mosaic model of membrane structure. Science 175, 720-731 (1972)
Steck, T.L.: The organisation of proteins in the human red blood cell membrane. J. Cell Biol. 62, 1-19 (1974)
Tomita, M., Marchesi, V.T.: Amino-acid sequence and oligosaccharide attachment sites of human glycophorin. Proc. Natl. Acad. Sci. U.S. 72, 2964-2968 (1975)
Yu, J., Fischman, D.A., Steck, T.L.: Selective solubilisation of proteins and phospholipids from red blood cell membranes by non-ionic detergents. J. Supramol. Struct. 1, 233-248 (1973)

Protein-Protein Interactions in the Purple Membrane

M. P. HEYN, P.-J. BAUER, and N. A. DENCHER

A. Introduction

The purple membrane from Halobacterium halobium is a specialized area of the bacterial membrane. The purified purple membranes are flat oval sheets with an average diameter of 0.5 μm. They contain only a single type of protein, bacteriorhodopsin, of mol wt 26,000. The proteins make up 75% of the total mass of the membrane, the lipids, 25%. The intense purple color is due to the chromophore all-trans or 13-cis retinal, which is covalently bound to a lysine residue of the opsin. In the light-adapted state of the membrane, the longest wavelength absorption band occurs at 570 nm. When light is absorbed in this band, a photochemical cycle is initiated. Via a number of intermediates, the system returns to the 570 nm complex. The intermediate with the longest lifetime absorbs at 412 nm and has a lifetime of several msec. In contrast to the 570 nm complex, the 412 nm complex reacts with hydroxylamine to form retinaloxime. By illuminating the purple membrane with 570 nm light in the presence of hydroxylamine, it is thus possible to produce a colorless chromophore-free membrane, the apomembrane (Oesterhelt et al., 1974). A great deal is known about the structure of the membrane from X-ray diffraction and electron microscopy studies (Blaurock, 1975; Blaurock and Stoeckenius, 1971; Henderson, 1975; Henderson and Unwin, 1975; Unwin and Henderson, 1975). The membrane is 45 Å thick, with the approximately globular protein molecules extending across it. Within the membrane the proteins are arranged in a two-dimensional hexagonal array, an almost perfect crystal lattice of space group P3. The protein contains seven helical rods of lengths varying from 35 to 40 Å. Three are perpendicular to the plane of the membrane, four are slightly inclined. Seventy to eighty percent of the protein is in the helical form, the remaining mass consists mostly of sequences connecting the helices. A projection of the electronic density in the plane of the membrane is shown in Figure 1. At the upper left vertex of the unit cell, three protein molecules are arranged in a cluster. The three helices that are perpendicular to the plane of the membrane are viewed end-on and can be clearly identified. Due to the tilting the other four helices, which are on the outside of the trimer, cannot be clearly discerned in this projection. In between these trimers the lipids are supposed to be arranged in bilayer form. The three proteins within the trimer are in direct contact. There are about 10 lipid molecules per bacteriorhodopsin. The almost close packing of the proteins leads to rigid flat sheets and an unusually high thermal stability of the proteins. In accordance with the function of the purple membrane as a light-driven proton pump, all the bacteriorhodopsin molecules are oriented in the same way.

Since each bacteriorhodopsin molecule contains a strongly absorbing chromophore, one may attempt to use this naturally occurring label to learn more about the protein-protein interactions via the

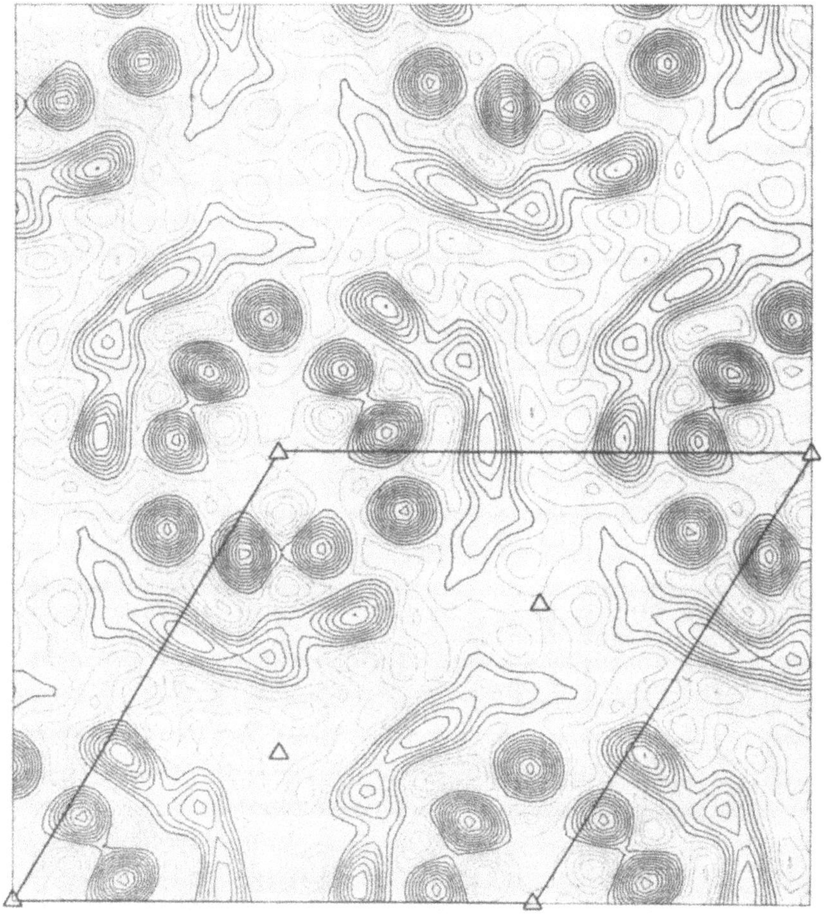

Fig. 1. Contour map of projected structure of purple membrane at 7 Å resolution. Dimensions of unit cell shown are 62 Å × 62 Å. Low density areas contain the lipids. Reproduced with permission from Unwin and Henderson (1975)

interaction between their chromophores. We will show in the next section that such interactions can indeed be observed, and that they are strongly dependent on the geometrical arrangement of the chromophores. Since the chromophores are tightly bound to the proteins, these signals are thus also dependent on the spatial arrangement of the proteins with respect to each other. Chromophore-chromophore interactions are well known in the spectroscopy of nucleic acids and proteins. There one usually deals with intramolecular interactions, such as between the bases of DNA or between the peptide bond transitions in proteins. These interactions lead to hypochromism and to the characteristic exciton bands in the optical activity. Such effects occur when identical or almost identical chromophores interact via their transition charges. The necessary conditions for their occurrence are: a fixed or preferred chromophore-chromophore geometry, a high extinction coefficient, reasonably close proximity, and a favorable orientation. In the case of the 570 nm band of the purple membrane, the first two requirements are met. The location and orientation of the chromophore within the protein is not known at present. From Figure 1, a lower bound for the chromophore-chromophore distance of about 12 Å may be estimated when the chromophores are located on the inner side of the trimer. In this case the interactions occuring between the chromophores within one trimer will dominate over inter-trimer interactions.

B. Interprotein Exciton Coupling Effects

When two mechanical oscillators, e.g., two pendulums, are mechanically coupled, we may observe the familiar effect of resonance: if we initially excite only one of the oscillators, the oscillation is transferred to the other and vice versa. In these nonstationary states the energy is localized on one or the other of the oscillators. Two stationary states exist, the in- and out-of-phase oscillations, in which both oscillators are in motion all the time, in which the excitation energy is delocalized. The frequency or energy of these two coupled modes is slightly different from those of the uncoupled oscillators, and depends on the interaction between them. In many respects we may consider the array of 570 nm transition dipoles in the purple membrane as an assembly of coupled oscillators. The coupling is due to the interaction between the transition charges when absorption occurs. To a good approximation we may neglect coupling with the other chromophore transitions, since they are far removed in energy (little resonance). Due to the coupling we expect the 570 nm absorption band to split up in energy, the so-called exciton splitting. For the interaction between two retinal chromophores, we obtain two stationary states labeled with + and -. To a first approximation, the splitting depends in the following way on the geometry:

$$\lambda_{\pm} = \lambda_o \mp \frac{\lambda_o^2}{h\,c\,R_{12}^3}\left[\vec{\mu}_1 \cdot \vec{\mu}_2 - 3\,\frac{(\vec{\mu}_1 \cdot \vec{R}_{12})\,(\vec{\mu}_2 \cdot \vec{R}_{12})}{R_{12}^2}\right] \quad (1)$$

$\vec{\mu}_1$ is the transition dipole moment vector of the 570 nm transition of the first chromophore; $\vec{\mu}_2$ the corresponding quantity for the second chromophore. $\vec{R}_{12}$ is the distance vector from 1 to 2 and λ_o is the wavelength in the absence of coupling. From Eq. (1) it is clear that the splitting depends in a sensitive way on the distance between the chromophores and on their orientation. The splitting calculated on the basis of Eq. (1), assuming the most favorable geometry, is, however, very much smaller than the bandwidth of the 570 nm absorption band, so that only very small effects are to be expected in the absorption spectrum. Due to the coupling the split bands also become optically active. Optical activity is most conveniently measured using circular dichroism (CD). For the dimer the + and - stationary states acquire optical activity with rotational strengths R_+ and R_-:

$$R_{\pm} = \mp \frac{\pi}{2\lambda_o}\,\vec{R}_{12} \cdot \vec{\mu}_1 \times \vec{\mu}_2 \quad (2)$$

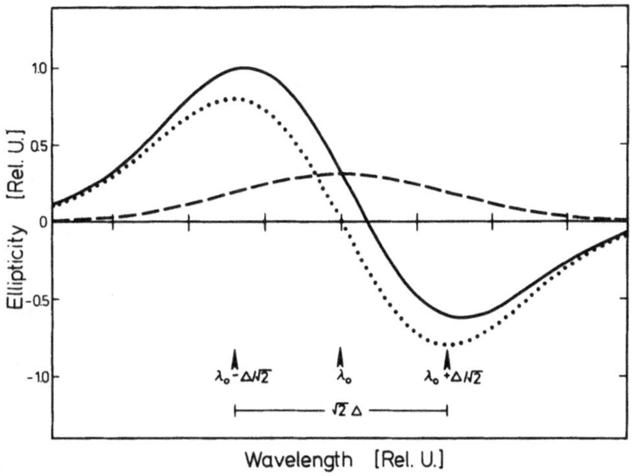

Fig. 2. ——: Simulated purple membrane CD spectrum obtained by superposition of symmetric exciton CD spectrum (...) and positive CD band proportional to absorption spectrum (---). λ_0: wavelength of absorption maximum. Δ: absorption band width

Since the amplitude of the optical activity in the two bands is of the same magnitude but of opposite sign, considerable effects may be observed even if the splitting is small. The expected CD curve thus consists of a positive and a negative lobe of equal amplitude, symmetrically located with respect to λ_0. Note the sensitive dependence of $R_\pm$ on the transition dipole geometry. The amplitude of the CD coupling bands decreases with R_{12}^{-2}, since it is proportional to the product of $R_\pm$ and $(\lambda_+ - \lambda_-)$. The coupling effect has therefore a short range. These considerations can be easily generalized to the case of the trimer and the hexagonal lattice. The appropriate quantity to use in studying the chromophore-chromophore interactions is thus circular dichroism rather than absorbance.

Of course we also have to consider the optical activity of the chromophore in the absence of chromophore-chromophore coupling. The interaction between the chromophore and the opsin is expected on theoretical grounds to lead to a CD band proportional to the 570 nm absorption band. Both for detergent solubilized rhodopsin and for rhodopsin within disc membranes, such a broad positive CD band is observed. Due to the high rotational and translational mobility of rhodopsin within these membranes, no exciton coupling effects are expected because such motions will lead to a complete averaging out of the coupling amplitudes. Both in rhodopsin and bacteriorhodopsin, the retinal chromophore is bound to a lysine residue of the opsin via a Schiff base. For bacteriorhodopsin we may therefore expect a superposition of an exciton band due to the chromophore-chromophore interactions and a positive band centered at the absorption maximum due to the interaction of the chromophore with the protein (Fig. 2). As a result of this superposition, the two CD extrema will be asymmetric in amplitude and the crossover will be shifted from λ_0 to the red (Fig. 2). These and other predictions are in accordance with the experimental spectra (Heyn et al., 1975; Bauer et al., 1976). A number of experiments were performed to provide stronger evidence for this interpretation of the spectra:

1. The purple membrane can be solubilized in Triton X-100 without denaturing the protein. Since micelles with single bacteriorhodopsin molecules are formed, one expects to observe the CD spectrum of uncoupled chromophores consisting of the positive band at λ_0. This is indeed the case.

2. By working at low temperatures and high light intensities, the photoequilibrium between the 570 nm and 412 nm complexes can be shifted to the side of the 412 nm complex. The exciton coupling bands at 570 nm should now disappear, since the isolated remaining 570 nm complexes have no unbleached neighbors left with which they could couple. In accordance with this argument a broad positive band centered at 570 nm remains.

3. It is possible to reconstitute the chromophore-free apomembrane by the successive addition of aliquots of all-trans retinal (Oesterhelt and Schumann, 1974). In the initial stages of such a reconstitution experiment, only single isolated retinal binding sites will be occupied and one expects to observe only the positive band. The coupling bands, which are due to the interaction between pairs of neighboring chromophores, will make their appearance at a later stage in the titration and are expected to increase in amplitude quadratically with the percentage reconstitution. The observed spectra are in good quantitative agreement with these considerations.

Once the coupling nature of the CD spectra is established, one may use this effect to learn more about protein-protein interactions in the membrane, in conjunction with other techniques such as X-ray diffraction, electron microscopy, and flash photolysis. In the next section an example is discussed in which the state of aggregation of the proteins within the membrane is changed. The hexagonal lattice is disassembled by organic solvents. Removing them leads to reassociation and reassembly of the proteins in the lattice. Several other possible applications of the CD coupling effects to the study of protein assembly within the purple membrane will be briefly mentioned.

C. Rotational Mobility of Bacteriorhodopsin Within the Purple Membrane Induced by Organic Solvents

The organic solvents dimethylsulfoxide, diethylether, and dimethylformamide cause a reversible blue shift of the visible absorption maximum of the purple membrane from 570 nm to about 470 nm (Oesterhelt et al., 1973). At a given percentage of organic solvent these changes can also be induced by raising the temperature. We showed (Heyn et al., 1975) that the change in the absorption spectrum for dimethylsulfoxide (DMSO) was accompanied by a large change in the CD. The exciton bands at 570 nm disappeared and were replaced by a broad positive band at 470 nm. Since the latter band showed no exciton features, and since in the presence of DMSO the X-ray diffraction rings from the hexagonal lattice disappeared, we suggested that DMSO induces a disordered state of the membrane, permitting some rotational and translational motion of the bacteriorhodopsin molecules within the membrane. It was assumed that the membranes remained intact since they could still be easily centrifuged down and since the CD effects were rapidly reversible by diluting out the DMSO or by lowering the temperature. It thus appeared that we

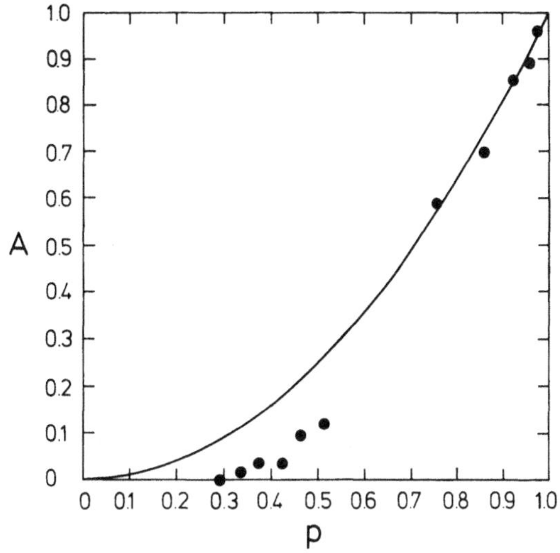

Fig. 3. Exciton coupling amplitude A plotted vs. fraction p of chromophores that remain in 570 nm state. p is varied from 1 to 0.29 by increasing ether percentage (vol/vol) from 0 to 6%. Continuous curve is expected dependency $A = p^2$ when no rotation occurs

had a method to disassemble and reassemble the protein lattice within the membrane. Whereas at room temperature at least 35% (vol/vol) DMSO is required to have observable color shifts, with diethylether only 3 or 4% is required. Diethylether can also be very simply removed by evaporation. The experiments with DMSO were followed up with more detailed work with ether.

In general, averaging over a number of different transition dipole moment geometries will lead to a reduction of the exciton CD amplitude. If there is rotational mobility of the bacteriorhodopsin molecules in the purple membrane, we expect to observe such a reduction. Rotation around axes perpendicular to the membrane is the most likely motion. It can be shown that this motion is already sufficient to cause a complete disappearance of the exciton amplitude. If a simple equilibrium exists between the 570 nm and 470 nm complexes, as suggested by Oesterhelt et al. (1973), then the CD amplitude will decrease when organic solvents are added, due not only to rotation but also to the shift in equilibrium toward the 470 nm state. In order to obtain evidence from the CD spectra in favor of bacteriorhodopsin rotation, it is thus necessary to show that in the presence of ether the CD amplitude decreases faster than expected on the basis of this equilibrium. If p is the fraction of chromophores that is in the 570 nm state, then the CD exciton coupling amplitude will be proportional to p^2 in the absence of rotation. In Figure 3 the experimental CD coupling amplitude A, normalized with respect to its value in the absence of ether, is plotted versus p for various percentages of ether. The determination of p was made by measuring the absorbance at 570 nm with respect to that of the apomembrane. In this way the considerable light scattering contribution to the effective absorbance at 570 nm could be subtracted. With decreasing p values the ether percentages varied from 0% to 6%. The continuous curve is the expected dependence $A = p^2$ when no rotation occurs. Up to 3% ether, the decrease in A follows this curve, and is presumably just due to a decrease in the number of 570 nm chromophores. Between 3 and 4% a clear deviation sets in, with A falling to 12% of its maximum value at a p value of 0.52. A reaches zero when the absorbance at 570 nm is still quite high, showing clearly that its decrease is faster than expected on the basis of the equilibrium alone. If we had determined p without correcting for the light scattering, the deviation from the p^2 curve would have been even larger. Removal of the ether by evaporation led to the original exciton bands. If the bacteriorhodopsin molecules possess some rotational and translational mobility, the rigidity of the membrane may be decreased and the membranes need no longer be flat sheets. If the membranes curl up, the radius of gyration becomes smaller and the light scattering increases. Whereas from 0 to 3% and from 4 to 6% ether the light scattering at 750 nm did not change, a substantial abrupt increase was observed between 3 and 4%.

In parallel with these CD experiments, flash experiments were performed to see if an ether-induced rotational diffusion of the bacteriorhodopsin molecules could be detected in the decay of the linear dichroism (R.J. Cherry, M.P. Heyn and U. Müller, to be published). Linear dichroism can be transiently induced when the chromophore is bleached by a short linearly polarized flash of 540 nm light. The anisotropy of the linear dichroism, r, at time t after the exciting pulse is defined by:

$$r(t) = \frac{A_{||}(t) - A_{\perp}(t)}{A_{||}(t) + 2A_{\perp}(t)} \qquad (3)$$

where $A_{||}$ and $A_{\perp}$ are the absorbance changes for 560 nm light polarized parallel and perpendicular, respectively to the flash polarization (see Cherry, 1976, for more details). If no rotational diffusion of the bacteriorhodopsin molecules within the membrane can occur, r should be independent of time. If rotational diffusion of the bacteriorhodopsin molecules occurs within the lifetime of the transient, the anisotropy r will decay with characteristic time constants determined by the rotational diffusion constants. In agreement with the crystalline arrangement of the protein molecules, it was concluded from the persistence of the dichroism during the lifetime of the 412 complex that the rotational relaxation time of the bacteriorhodopsin molecules is slower than 20 msec (Razi Naqvi et al., 1973). The results of the flash experiments with ether are summarized in Figure 4, in which the anisotropy is plotted semilogarithmically vs. time. In the absence of ether, a single exponential decay is observed with a rotational relaxation time of 63 msec, presumably due to the slow rotation of whole purple membrane sheets. For a thin disc of 5000 Å diameter and 45 Å thickness, the relaxation times for rotation around the long and short axes are equal, so that only one exponential is expected. The order of magnitude of the relaxation time is correct for a disc of these dimensions. At 5% ether, a rapid initial decrease occurs, with a rotational relaxation time of about 2.6 msec, followed by a slow decay with a relaxation time which is about equal to that in the absence of ether. Upon removal of the ether a single exponential slow decay is observed again with a relaxation time that is somewhat shorter than the initial one. The simplest interpretation of these experiments is that in the presence of ether, rotation of the bacteriorhodopsin molecules within the membrane is induced, with the whole membrane sheets rotating slowly as before. In agreement with the CD experiments this fast component first makes its appearance at about 4% ether. The alternative explanation for the observed increase in light scattering, based on aggregation of the membranes, can be ruled out, since this would lead to an increase in the longest relaxation time. The trivial explanation for the CD

Fig. 4. Time dependence of anisotropy r of 560 nm depletion signal. ×: before addition of ether; •: in presence of 5% ether

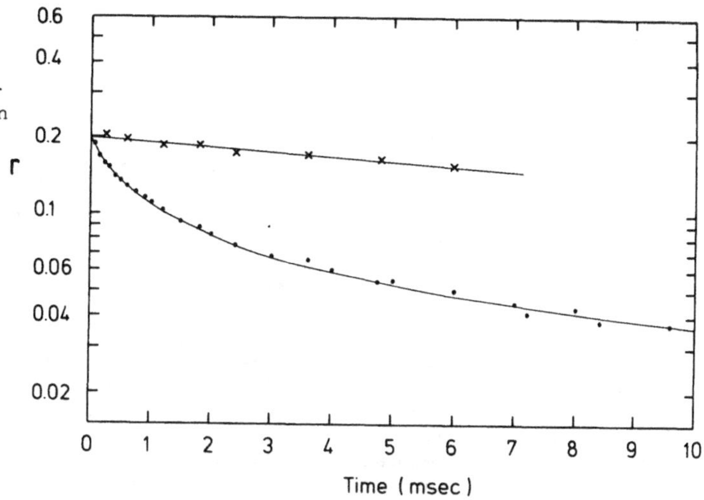

experiments, namely a complete fragmentation of the membrane, seems to be ruled out by the fact that in 5% ether the slow rotation due to the whole membrane is still present and by the observation of an increase in light scattering.

It thus appears that the protein-protein interactions in the purple membrane are clearly changed by ether: rotational mobility and disorder are induced. The original state of assembly of the membrane can be restored by removing the ether. The detailed molecular mechanism of this ether effect remains unclear. Presumably the ether enters the membrane and leads to a swelling of the purple membrane and a concomitant change in the protein-protein contacts. The ether also affects the chromophore, as is evident from the color shift and the fact that the 470 nm complex reacts with hydroxylamine (Oesterhelt et al., 1973). A possible denaturation or conformational change in the protein must therefore be considered. It could be shown by CD measurements in the 190-230 nm range that the secondary structure of bacteriorhodopsin is not significantly affected by the presence of 5% ether.

D. Conclusions

Exciton coupling effects provide a sensitive way to monitor protein-protein interactions in the purple membrane. Changes in the state of assembly of the bacteriorhodopsin molecules within the membrane can be detected and a clear differentiation is possible on the basis of the CD spectra between aggregated and nonaggregated bacteriorhodopsin molecules. Finally we wish to mention two potentially useful applications. The state of aggregation of bacteriorhodopsin molecules that are incorporated in artificial systems (liposomes, vesicles) may be determined in this way. Another interesting possibility, presently being studied, occurs in the biogenesis of the purple membrane where one wants to find out at which stage the proteins are assembled into the hexagonal lattice.

References

Bauer, P.-J., Dencher, N.A., Heyn, M.P.: Evidence for chromophore-chromophore interactions in the purple membrane from reconstitution experiments of the chromophore-free membrane. Biophys. Struct. Mech. 2, 79-92 (1976)

Blaurock, E.A.: Bacteriorhodopsin: a trans-membrane pump containing α-helix. J. Mol. Biol. 93, 139-158 (1975)

Blaurock, E.A., Stoeckenius, W.: Structure of the purple membrane. Nature (New Biol.) 233, 152-154 (1971)

Cherry, R.J.: Protein and lipid mobility in biological and model membranes. In: Biological Membranes, Vol. III. Chapman, D., Wallach, D.F.H. (eds.). London: Academic Press 1976, pp. 47-102

Henderson, R.: The structure of the purple membrane from Halobacterium halobium: analysis of the X-ray diffraction pattern. J. Mol. Biol. 93, 123-138 (1975)

Henderson, R., Unwin, P.N.T.: Three-dimensional model of purple membrane obtained by electron microscopy. Nature (London) 257, 28-32 (1975)

Heyn, M.P., Bauer, P.-J., Dencher, N.A.: A natural CD label to probe the structure of the purple membrane from Halobacterium halobium by means of exciton coupling effects. Biochem. Biophys. Res. Commun. 67, 897-903 (1975)

Oesterhelt, D., Meentzen, M., Schuhmann, L.: Reversible dissociation of the purple complex in bacteriorhodopsin and identification of 13-cis and all-trans retinal as its chromophores. Europ. J. Biochem. 40, 453-463 (1973)

Oesterhelt, D., Schuhmann, L.: Reconstitution of bacteriorhodopsin. FEBS Lett. 44, 262-265 (1974)

Oesterhelt, D., Schuhmann, L., Gruber, H.: Light-dependent reaction of bacteriorhodopsin with hydroxylamine in cell suspensions of Halobacterium halobium: demonstration of an apo-membrane. FEBS Lett. 44, 257-261 (1974)

Razi-Naqvi, K., Gonzalez-Rodriguez, J., Cherry, R.J., Chapman, D.: Spectroscopic technique for studying protein rotation in membranes. Nature (New Biol.) 245, 249-251 (1973)

Unwin, P.N.T., Henderson, R.: Molecular structure determination by electron microscopy of unstained crystalline specimens. J. Mol. Biol. 94, 425-550 (1975)

Incorporation of Pure Hydrolases Isolated from Brush Border Membranes in Single-Bilayer Lecithin Vesicles

J. Brunner, H. Hauser, G. Semenza, and H. Wacker

Summary

1. Both hydrolases, sucrase-isomaltase and aminopeptidase, when in their integral form interact with lecithin to form stable lipoprotein complexes.

2. In the reconstituted lipoprotein complex the hydrophobic part of the amphipathic hydrolase is incorporated in the lecithin bilayer. The total protein mass in contact with lecithin is less than 5%.

3. With sucrase-isomaltase the hydrophobic part incorporated in the lecithin bilayer belongs to the isomaltase subunit.

4. The sucrase-isomaltase is incorporated in the lecithin bilayer such that the bulk of the protein is not in contact with the lipid surface but sitting on a stalk of a minimum length of ~ 10 Å.

5. The reconstituted lipoprotein/membrane system is a lecithin single-bilayer vesicle (diameter ~ 300 Å) with all the enzyme incorporated being oriented towards the external aqueous phase.

6. Lecithin forms a single closed bilayer surrounding a water-filled cavity. The properties of the lipid bilayer - molecular packing, segmental and cooperative motion, and ion permeability - are unchanged compared to the reference state of a pure lecithin bilayer vesicle.

7. No hydrolysis-related sugar transfer is measured in the reconstituted membrane system.

8. The following properties of the reconstituted system resemble those observed in "native" brush border membranes: (1) the asymmetric incorporation of the enzyme; (2) the attachment of sucrase-isomaltase to the membrane is accomplished by the hydrophobic "anchor", which belongs to the isomaltase subunit; (3) the susceptibility of membrane-bound hydrolases to detachment by papain treatment: the proteins released from either system are indistinguishable in their sedimentation behavior and in their electrophoretic mobility in polyacrylamide gels.

Table 1. Brush border membrane hydrolases used in the present study

Hydrophobic glycoprotein	Sucrase-isomaltase (SI)	Aminopeptidase (A)
Origin	Rabbit small intestine	Pig kidney
Hydrophobic part	< 10000	< 10000
Aggregation in H_2O	$\sim 10^6$	$\sim 10^6$
Subunits	2	2
Stokes radius	70 ± 5 Å	
Detergent solubilisation:		
Triton X-100	Monomers	Monomers
Na-cholate		
Protease solubilisation:		
Papain	Monomers (mol wt 220000)	
Trypsin		Monomers (mol wt 280000)

A. Introduction

Artificial membranes have been reconstituted from egg lecithin and pure hydrolases isolated from brush border membranes. The two-component system is simple and allows the successful application of physical techniques, particularly of high resolution nuclear magnetic resonance (NMR). The reconstituted membranes are used to study the nature of the lipid-protein interaction and to shed light on the question of whether or not the hydrolases are involved in the transport of their substrates or their hydrolytic products.

B. Method of Reconstitution

Pure egg lecithin and highly purified hydrolases (see Table 1) were used for the reconstitution. Sucrase-isomaltase and aminopeptidase were extracted from brush border membrane of rabbit small intestine (Cogoli, et al., 1972; Sigrist et al., 1975; Takesue, 1969; Takesue et al., 1973) and hog kidney (Wacker et al., 1976a), respectively.

Some physico-chemical properties of the two enzymes, relevant to reconstitution, are summarized in Table 1. Both proteins are amphipathic glycoproteins (sugar content ~ 15% and 20%, respectively) with the hydrophobic part amounting to less than ~ 5% of the total protein mass. The hydrophilic portion amounting to ≥ 95% of the total protein mass has characteristics typical for water-soluble globular proteins. Due to their amphipathic character both proteins tend to polymerize in aqueous solvents to aggregates $\sim 10^6$ daltons (Wacker et al., 1976a; Sigrist et al., 1975). In order to solubilize the integral proteins in their monomeric form, detergents such as Triton X-100 and sodium cholate are required. The hydrophilic water-soluble part of the protein is separated from the hydrophobic tail by papain and trypsin treatment, respectively, of the detergent form of the proteins (Wacker et al., 1976a). The same water-soluble form ("papain form") of sucrase-isomaltase is released by papain digestion of the brush border membrane.

To reconstitute an artificial membrane system, 2.5% (0.06 M) sodium cholate recrystallized as the acid from aceton/H_2O 4:1 (vol/vol) was added to 1-2% unsonicated egg lecithin dispersions in buffer (0.1 M NaCl, 0.01 M Tris/HCl, pH 7.0, 0.02% NaN_3) and to protein solutions containing 2-5 mg protein in the same buffer. The amount of detergent was sufficient to disintegrate the multilamellar lecithin structures to small mixed micelles of lecithin and cholate and also to dissociate the aggregated (detergent free) form of the proteins to monomeric protein-cholate complexes. The lecithin/cholate micelles and protein-cholate complex were mixed at a weight ratio of (1.5-4):1 and the mixture was passed over a Sephadex G-50 column (26 × 1.5 cm, equilibrated with the same buffer, flow rate 7 ml · h^{-1}) to remove 99.9% of the cholate. During gel filtration a stable lipoprotein complex was formed. The advantages of the foregoing reconstitution procedure are: (1) the quick and efficient removal of the detergent; (2) the reproducible formation of a stable lipoprotein complex, the morphology of which is well suited for permeability measurements and for the application of physical techniques such as high resolution NMR.

It is important to stress that this reconstitution occurs only when the integral proteins containing the hydrophobic portion are incubated with lecithin. After removal of the hydrophobic tail by enzymatic treatment, the hydrophilic water-soluble part of the protein does not interact with lecithin vesicles (Pattus et al., 1976; Wacker et al., 1976b).

C. Physico-chemical Properties of the Reconstituted Membrane System

The best test of whether or not the protein has interacted with lecithin to form a stable lipoprotein complex is to subject the reaction product to density gradient centrifugation. Both proteins give similar results. Figure 1 shows the density gradient pattern of the reconstituted aminopeptidase-lecithin system. Besides some free lipid and protein at least three lipoprotein complexes characterized by apparent hydrated densities d = 1.048-1120 g · ml^{-1} are present. Hence the lipoprotein complexes formed are operationally defined as low density lipoproteins. Ultracentrifugal analysis of the reconstituted lipoprotein complex confirms the inhomogeneity (Fig. 2). Consistent with the density gradient centrifugation, free (uncomplexed) lipid and protein in the aggregated form are identified in addition to a series of Schlieren peaks representing lipoprotein complexes (Fig. 2a). A simple calculation shows that these lipoprotein complexes represent lecithin vesicles (see below) containing an integral number of protein molecules. This is demonstrated in Table 2. In column 2 of Table 2 the apparent sedimentation coefficients of the four lipoprotein fractions of Figure 2a (see legends) are given and compared with those calculated in column 3. In columns 4 and 5, respectively, the ratios of the experimental sedimentation coefficients are compared with those calculated from Eq. (1)

$$M_2(1 - \bar{v}_2\rho) / M_1(1 - \bar{v}_1\rho) = s_2/s_1 \quad (1)$$

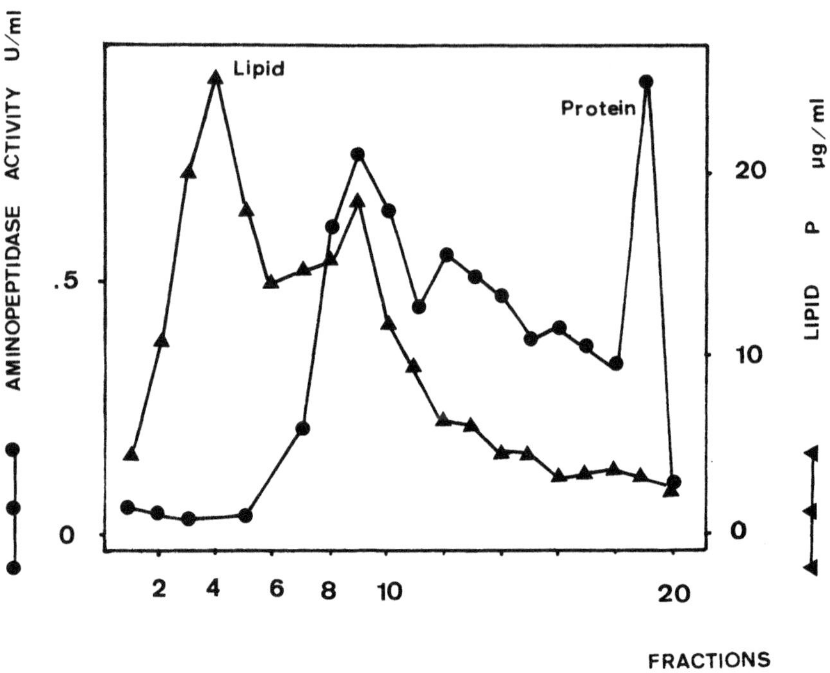

Fig. 1. Density gradient centrifugation of lipoprotein complexes reconstituted from lecithin and aminopeptidase (lecithin concentration 6.7 mM, weight ratio 10) using a linear (0-12%) sucrose gradient. After removal of larger lipoprotein particles on Sepharose 4B, 0.15-0.25 ml of the sample in buffer (0.1 M NaCl, 0.01 M Tris-HCl, pH 7.2, 0.1 mM EDTA, 0.02% NaN_3) was layered on top of the gradient. After centrifugation at ~60,000 g for 20 h the content of the centrifuge tube was displaced upward with 40% sucrose. 0.25 ml fractions were collected and the sucrose concentration was determined with an Abbé hand refractometer. The first peak of the effluent represents pure lecithin vesicles; the peak in the last few fractions showing aminopeptidase activity represents free uncomplexed protein. The fractions 6 to 18 contain lipoprotein complexes with a hydrated density d = 1.048-1.120 g/ml (Wacker et al., 1976b)

where M_i, $\bar{v}_i$, s_i are the particle weight, the partial specific volume, and the sedimentation coefficient, respectively, of the ith lipoprotein complex. There is good agreement between the experimental and calculated values. Figure 2b shows that the number of peaks and thus of lipoprotein complexes increases when more sucrase-isomaltase is incubated with lecithin indicating that the average number of protein molecules associated with each lecithin vesicle is increased. In Figure 2b at least 6 Schlieren peaks representing vesicles containing 1-6 protein molecules can be detected besides the Schlieren peak on the left of the meniscus with $s_{20,w} = 2.9 \times 10^{-13}$ s representing pure lecithin vesicles.

Figure 3 is a schematic diagram summarizing the main features of the reconstituted sucrase-isomaltase-lecithin membrane system:

1. An asymmetric membrane is formed with the bulk of the protein together with the fully active centers being oriented towards the aqueous phase. The arrangement of the protein molecule is similar to the in vivo situation (Louvard et al., 1975a,b) in the intes-

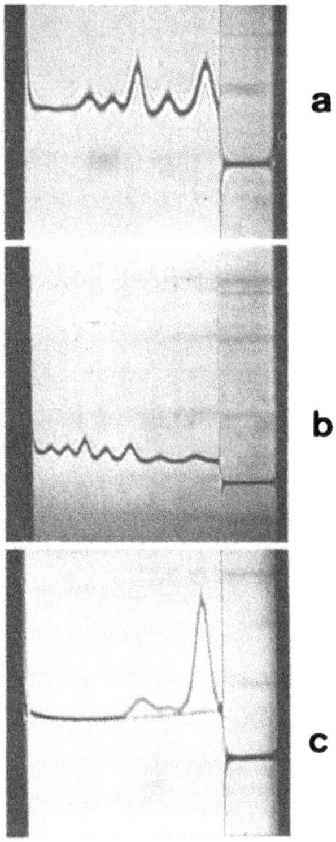

Fig. 2a-c. Ultracentrifugal Schlieren pattern of lipoprotein complex reconstituted from egg lecithin and sucrase-isomaltase (lecithin conc 0.037 M, lecithin/protein wt ratio 4, solvent used: 0.1 M NaCl); sedimentation (from right to left) at 50,740 rpm and 20°C; phase plate angle 70°, time 41.3 min; the peak at the extreme right represents pure lecithin vesicles with $s_{20,w} = 2.9$ S; the other 4 peaks represent lipoprotein complexes (I, II, III, and IV, cf. Table 2) in order of increasing apparent sedimentation coefficients; I: $s_{20,app} = 6.6$ S; II: $s_{20,app} = 10.3$ S; III: $s_{20,app} = 13.1$ S; IV. $s_{20,app} = 15.7$ S (a). Schlieren pattern of lipoprotein complex reconstituted from egg lecithin and sucrase-isomaltase (lecithin conc 0.016 M, lecithin/protein wt ratio 1.4, solvent: 0.1 M NaCl); sedimentation at 50,740 rpm and 20°C, phase plate angle 75°, time 42 min (b). Schlieren pattern of lipoprotein complex shown in Figure 2a after papain treatment; sedimentation at 50,740 rpm and 20°C; phase plate angle 60°, time 39.9 min. In addition to free lipid with a sedimentation coefficient $s_{20,w} = 2.9$ S (right hand peak), the "papain" form of sucrase-isomaltase is identified as fastest migrating peak with an apparent sedimentation coefficient $s_{20,app} = 10.0$ S

Table 2. Sedimentation behavior of the lipoprotein complex (Fig. 2a) reconstituted from lecithin and sucrase-isomaltase

Lipoprotein fraction[a]	$s_{20,app} \times 10^{13}$ (s)	$s \times 10^{13}$ (s)[b]	s_x/s_I[c]	s_x/s_I[d]	$M \times 10^{-6}$ (dalton)[e]	$\bar{v}$ (ml/g)[e]
I	6.6	6.7	1.0	1.0	4.35	0.972
II	10.3	9.8	1.56	1.46	4.57	0.961
III	13.1	12.8	1.99	1.91	4.79	0.950
IV	15.7	15.7	2.40	2.35	5.01	0.941

[a] Schlieren peaks 2-5 of Fig. 2a (see legends)

[b] Sedimentation coefficient $s_{20,w}$ calculated from the following equation

$$s_{20,w} = \frac{M(1 - \bar{v}\,\rho_{20})}{6\pi N r \eta}$$

where M and $\bar{v}$ are given in the table, ρ_{20} is the density of water at 20°C, N is Avogadro's number, r is the Stokes radius determined from gel filtration (r = 170 Å) and η is the viscosity of water at 20°C (η = 0.01 poise)

[c] Ratio of experimental sedimentation coefficients, x = I, II, III, IV

[d] Calculated ratios of sedimentation coefficients using values in 3rd column

[e] Particle weight of lipoprotein complex calculated on the basi that 1-4 sucrase-isomaltase molecules are present per lecithin vesicle (see below) of 5500 lipid molecules.
the partial specific volumes were calculated from the lipid-protein weight ratio in each lipoprotein complex, using the partial specific volume of lecithin volume of lecithin $\bar{v} = 0.984\ ml \cdot g^{-1}$ and of sucrase isomaltase $\bar{v} = 0.793\ ml \cdot g^{-1}$

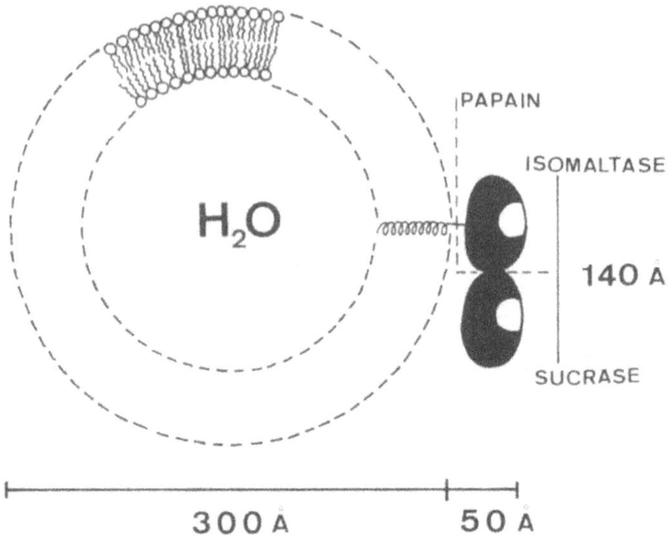

Fig. 3. Schematic diagram summarizing the main features of the lipoprotein/ membrane system reconstituted from lecithin and sucrase-isomaltase. The isomaltase subunit contains a hydrophobic peptide segment, which is inserted in the lecithin bilayer. It is assumed that this peptide chain is α-helical and spans the lipid bilayer. Horizontal dashed line: citraconylation of the protein in the reconstituted artificial membrane system liberates the sucrase subunit. Vertical dashed line: point of attack of papain, which cleaves off all hydrophilic portions of the sucrase-isomaltase complex

tinal brush border membrane where the enzyme is oriented towards the lumen of the intestine. The asymmetry seems to be a general feature of the reconstitution process since the same is true for the membrane system reconstituted from aminopeptidase and lecithin. The molecular details underlying the mechanism of the reconstitution procedure are not understood.

2. Papain digestion of the reconstituted membrane removes all but the hydrophobic part of the sucrase-isomaltase molecule inserted in the lecithin bilayer. The resulting water-soluble "papain-form" of the protein shows a sedimentation behavior in the analytical ultracentrifuge very similar to that observed by papain treatment of the detergent form of the protein (cf. Fig. 2c). Furthermore the two "papain" forms obtained from the reconstituted system and the detergent form of sucrase-isomaltase, respectively, give identical polyacrylamide gel electrophoresis patterns.

3. The lipoprotein particle is essentially a lecithin vesicle with an internal, water-filled cavity surrounded by a single continuous bilayer. The particle diameter determined by gel filtration on Sepharose 4B is about 340 Å, only 40 Å larger than that of pure lecithin vesicles (Brunner et al., 1976). Furthermore the particle diameter of the lipoprotein complex does not vary significantly with the number of proteins incorporated.

^1H and ^{31}P-NMR measurements at 360 and 36.4 MHz, respectively, together with electron spin resonance (ESR) spin label experiments using fatty acid spin labels

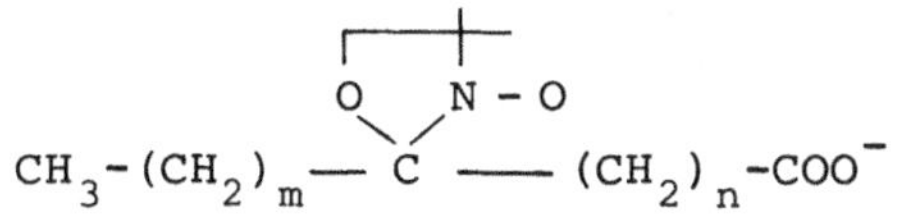

(m, n) = 12, 3; 5, 10; 1, 14 incorporated in the lecithin bilayer show no spectral changes when the protein is incorporated in the bilayer. Chemical shift and line width (relaxation) measurements and the order parameters S_3 (Hubbell and McConnell, 1971) derived from the ESR spin label experiments do not show any difference between the pure lecithin bilayer and the reconstituted membrane system. This indicates that the molecular packing and motion of the lecithin molecules are unaffected by the incorporation of the peptide segment. Apparently the average protein mass penetrating the lipid bilayer is insufficient to affect the cooperative motion of the lipid.

4. Since the size of the lecithin vesicle in the reconstituted system (Fig. 3) is the same as that of a pure lecithin bilayer vesicle, the distribution of the lecithin molecules between the outer and inner layers of the bilayer is also expected to remain the same. This can be demonstrated by measuring the spectral perturbations in the ^1H and ^{31}P NMR spectra of lecithin vesicles and the reconstituted system (Fig. 3) when paramagnetic shift reagents are added. With both systems, $K_3[Fe(CN)_6]$ shifts about two-thirds of the $-\overset{+}{N}(CH_3)_3$ proton signal upfield to the same extent, indicating that the ratio of lecithin molecules on the outer and inner layers of the bilayer is about 2. The same is true for the shift changes of the ^{31}P-signal induced by $Pr(NO_3)_3$ or $Eu(No_3)_3$.

5. The system depicted in Figure 3 lends itself to permeability studies because the internal water-filled cavity is surrounded by a single, closed bilayer of known dimensions. Consistent with the notion that the molecular packing and motion of the lipid in the bilayer are not affected by the incorporation of polypeptide chains is the finding that the bilayer permeability for Na^+ is hardly changed in the reconstituted system. No hydrolysis-related transport is observed, i.e., sucrose added to the reconstituted system is hydrolyzed, and the products of hydrolysis - fructose and glucose - diffuse across the bilayer with permeability coefficients identical to those observed when pure fructose or glucose is added to the reconstituted system. (The values for the permeability coefficients of Na^+ and various sugars will be the subject of a separate publication (Hauser, H., Brunner, J., Graham, D., Semenza, G., to be submitted).

Previous work showed that papain-solubilized sucrase-isomaltase incorporated into black lipid membranes produced an increase in the permeability for glucose and fructose stemming from sucrose (Storelli et al., 1972). The apparent discrepancy between the black lipid membranes and the single bilayer-lipoprotein system described above is not understood at the moment. More work is required to clarify the discrepancy between the two systems (see also Semenza, 1976).

D. Attachment of the Protein to the Lipid Bilayer

By labeling the amino groups of sucrase-isomaltase in the detergent form with methyl-maleic acid anhydride (citraconic acid anhydride) according to

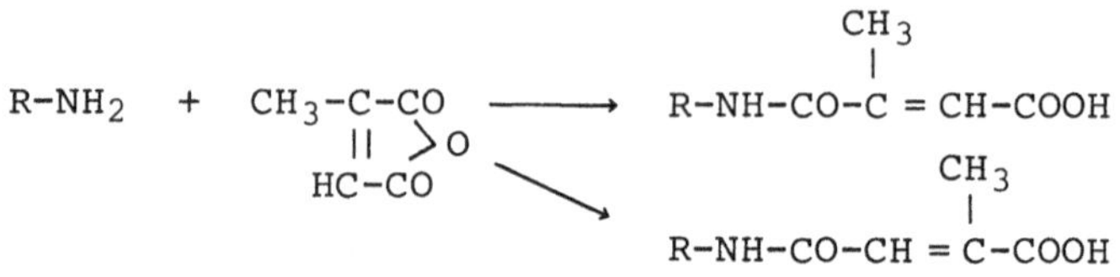

a positively charged amino group on the protein is replaced by a negative charge. As a consequence, the sucrase-isomaltase complex dissociates into its two subunits (Braun et al., 1975), which can be identified by polyacrylamide gel electrophoresis. When the reconstituted system (Fig. 3) is treated in the same way, only the sucrase subunit is released from the lipoprotein complex, indicating that the sucrase-isomaltase is attached to the lecithin bilayer by the isomaltase subunit (Fig. 3). Using the same labeling procedure with "native" brush border membranes isolated according to Storelli et al. (1977), it can be shown that the attachment of the enzyme in vivo is similar to that in the reconstituted system in the sense that it is the isomaltase subunit that is associated with the membrane. Hence the citraconylation technique provides evidence for the hydrophobic "anchor" (by which sucrase-isomaltase is linked to the lipid bilayer) belonging to the isomaltase subunit. The small hydrophobic tail inserted in the bilayer is likely to be α-helical, and work to shed light upon this question is presently in progress.

The bulk of the protein does not appear to be in contact with the lipid surface, but seems to be sitting on a stalk of a minimum length of ~ 10 Å. Evidence for this comes from ^{1}H-NMR work. The shift changes in the ^{1}H-NMR spectrum of the reconstituted system induced by the paramagnetic anion $[Fe(CN)_6]^{-3}$ are identical with those observed with pure lecithin vesicles. The upfield shifts observed with the signals from the choline group of lecithin depend on r^{-3} where r is the distance between the paramagnetic center and the choline proton under consideration. The finding that the magnitude of the induced shifts, both with respect to the shift change and the proportion of the signal intensity shifted, is identical in pure lecithin bilayers and in the reconstituted system (Fig. 3) can only be reconciled by postulating that the lipid surface is readily accessible to the large $[Fe(CN)_6]^{-3}$ anion in both systems. Since the approximate diameter of the $[Fe(CN)_6]^{-3}$ anion is 9 Å, the bulk of the protein must be at a distance > 9 Å. Such a gap between the lipid surface and the bulk of the sucrase-isomaltase could also possibly account for the accessibility of the protein to enzymatic attack by papain, which removes all but the hydrophobic "anchor". If, however, the lipid surface was covered by sucrase-isomaltase it would not be accessible to $[Fe(CN)_6]^{-3}$, and both the magnitude of the shift changes induced and the proportion of the signal intensity shifted would be expected to be significantly decreased since the surface of one protein molecule amounts to ~ 1/8 of the total lipid surface.

Acknowledgements. Our work was partially supported by the SNSF, Berne, and by Nestlé Alimentana, Vevey. We also acknowledge the expertise of Mrs. B. O'Neill and of Mr. K. Howell.

References

Braun, H., Cogoli, A., Semenza, G.: Dissociation of small-intestinal sucrase-isomaltase complex into enzymatically active subunits. Europ. J. Biochem. 52, 475-480 (1975)

Brunner, J., Skrabal, P., Hauser, H.: Single bilayer vesicles prepared without sonication: physico-chemical properties. Biochim. Biophys. Acta 455, 322-331 (1976)

Cogoli, A., Mosimann, H., Vock, C., v. Balthazar, A.K., Semenza, G.: A simplified procedure for the isolation of sucrase-isomaltase complex from rabbit intestine. Its aminoacid and sugar composition. Europ. J. Biochem. 30, 7-14 (1972)

Hubbell, W.L., McConnell, H.M.: Molecular motion in spin-labelled phospholipids and membranes. J. Am. Chem. Soc. 93, 314-326 (1971)

Louvard, D., Maroux, S., Desnuelle, P.: Topological studies on the hydrolases bound to the intestinal brush border membrane. II. Interactions of free and bound aminopeptidase with a specific antibody. Biochim. Biophys. Acta 389, 389-400 (1975a)

Louvard, D., Maroux, S., Vannier, C., Desnuelle, P.: Topological studies on the hydrolases bound to the intestinal brush border membrane. I. Solubilization by papain and Triton X-100. Biochim. Biophys. Acta 375, 236-248 (1975b)

Pattus, F., Verger, R., Desnuelle, P.: Comparative study of the interactions of the trypsin and detergent form of the intestinal aminopeptidase with liposomes. Biochem. Biophys. Res. Commun. 69, 718-723 (1976)

Semenza, G.: Small intestinal disaccharidases: their properties and role as sugar translocators across natural and artificial membranes. In: The Enzymes of Biological Membranes, Vol. 3. Martonosi, A. (ed.). New York-London: Plenum Press 1976, pp. 349-383

Sigrist, H., Ronner, P., Semenza, G.: A hydrophobic form of the small intestinal sucrase-isomaltase complex. Biochim. Biophys. Acta 406, 433-446 (1975)

Storelli, C., Kessler, M., Müller, M., Murer, H., Semenza, G.: (1977) to be submitted

Storelli, C., Vögeli, H., Semenza, G.: Reconstitution of a sucrase-mediated sugar transport system in lipid membranes. FEBS Lett. 24, 287-292 (1972)

Takesue, Y.: Purification and properties of rabbit intestinal sucrase. J. Biochem. (Tokyo) 65, 545-552 (1969)

Takesue, Y., Yoshida, T.O., Akaza, T., Nishi, Y.: Localisation of sucrase in the microvillus membrane of rabbit intestinal mucosal cells. J. Biochem. (Tokyo) 74, 415-423 (1973)

Wacker, H., Lehky, P., Vanderhaege, F., Stein, E.A.: On the subunit structure of particulate aminopeptidase from pig kidney. Biochim. Biophys. Acta 429, 546-554 (1976a)

Wacker, H., Müller, F., Semenza, G.: Incorporation of hydrophobic aminopeptidase from hog kidney into egg lecithin liposomes: number and orientation of aminopeptidase molecules in the lecithin vesicles. FEBS Lett. 68, 145-152 (1976b)

Ionophores

Chiral Recognition of Biogenic Amines by Synthetic Neutral Ionophores in Membranes

A. P. THOMA, E. PRETSCH, G. HORVAI, and W. SIMON

Extensive reviews by Ovchinnikov et al. (1974) as well as by Koryta (1975) provide clear evidence that certain lipophilic, electrically neutral compounds exhibit cation transport properties in liquid membranes. Such molecules are responsible for the outstanding electrochemical cation selectivity (Stefanac and Simon, 1966; Mueller and Rudin, 1967) (see also De Levie, 1976) of these systems. Among different synthetic ligands that behave as ionophores (Ovchinnikov et al., 1974; Koryta, 1975), derivatives of the 3,6-dioxaoctanedioic acid diamide (Fig. 1) show carrier properties for alkali and alkaline earth metal cations coupled with a pronounced selectivity when used in electrochemical cells (Simon et al., 1975; Morf et al., 1976). Measurements of ^{13}C nuclear magnetic resonance (NMR) chemical shifts as well as spin-lattice relaxation times T_1 (Table 1) clearly indicate an interaction of the ligand in Figure 1 with α-phenylethylammonium ions. Since the chemical shifts change only up to a molar ratio of 1:1 when adding the ammonium chloride to the ligand (in $CDCl_3$), evidence for the existence of a 1:1 carrier-cation complex is corroborated (see Büchi et al., 1976). The introduction of chirality into ligands of the type shown in Figure 1 should therefore lead to diastereomeric interactions with enantiomeric cations and thus to corresponding chiral recognition. Extensive studies of this phenomenon in complexation, using conformationally chiral ligands, have been reported by Cram et al. (Newcomb et al., 1974; Cram et al., 1975). Although remarkable enantiomer differentiation has been achieved in conventional liquid-liquid extraction devices (Newcomb et al., 1974) chiral recognition in electrochemical membrane systems has been realized only recently with configurationally chiral ionophores (Thoma et al., 1975). In the meantime ligands of the latter type of chirality have been reported to be capable of chiral recognition in extraction experiments (Curtis et al., 1975).

In view of models for stereospecific transport processes in biological systems, the behavior of membranes incorporating the compounds shown in Figure 2 has been studied. The ligands 1-3 have been prepared according to general principles described earlier (Ammann et al., 1975). Compounds 4 and 5 are representatives of a series designed and prepared in the laboratories of Prelog (Bedeković, 1976; Bedeković and Prelog, 1976). Through electromotive force (EMF) measurements on the cell,

$$\text{Hg; } Hg_2Cl_2\text{, KCl (sat.) / } 10^{-1}M\ NH_4NO_3 \text{ / sample solution // membrane // reference solution, AgCl; Ag} \quad (1)$$

the selectivity coefficients K^{Pot}_{XY} were obtained using the expression

$$\log K^{Pot}_{XY} = \frac{\Delta E \cdot F}{2.303 \cdot R \cdot T} - \log a^{1/z}_{Y^{z+}} + \log a_{X^+} \quad (2)$$

Fig. 1. Ionophore of type 3,6-dioxaoctanedioic acid diamide

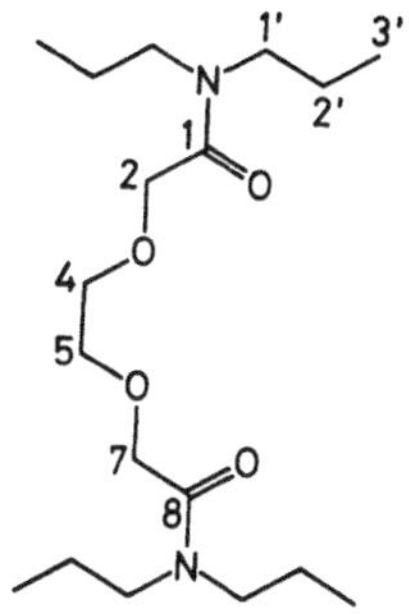

Table 1. ^{13}C-Chemical shifts (in ppm relative to tetramethylsilane, standard deviation < 0.05 ppm) and spin-lattice relaxation times (T_1 in sec, standard deviation ≤ 5%) of the ligand shown in Fig. 1 (L) and of its complex with α-phenylethylammonium chloride (L · PEA) (solvent: $CDCl_3$, concentration ~0.3 M)

Carbon atom[a]	Chemical shifts [ppm]		Spin-lattice relaxation times T_1 [sec]	
	L	L · PEA	L	L · PEA
1,8	168.7	169.2	15.3	6.1
2,7	70.2	68.8	1.0	0.3
4,5	70.7	70.2	0.9	0.3
1' cis[b]	47.4	47.6	} 0.9	0.3
1' trans[b]	48.7	48.5		
2' cis[b]	20.8	20.8	} 1.3	0.5
2' trans[b]	22.2	21.9		
3' cis[b]	11.4	11.4	} 2.5	1.7
3' trans[b]	11.3	11.2		

[a] See Fig. 1

[b] cis and trans relative to carbonyl oxygen atom

where ΔE is the potential difference between 0.1 M solutions of the chlorides, and z is the charge of the interfering ion Y; F is the Faraday constant, T the absolute temperature, R the gas constant, and a, the ion activities. Figure 3 contains selectivity sequences obtained with a liquid membrane free of carrier (DOA: dioctyladipate) as well as in the presence of ligands 2 and 6 (valinomycin), respectively. Through the incorporation of ligands the selectivity range increases substantially. As expected (Ovchinnikov et al., 1974), the valinomycin-containing membrane considerably prefers K^+ over Na^+. In the presence of 2, however, (RS)-α-phenylethylammonium is singled out. Such EMF studies indicate interactions of ligands 1-6 with the biogenic ammonium ions shown in Figure 3. Since all of the molecules 1-6 are chiral, they are theoretically capable of inducing enantiomer selectively in membranes. Table 2 contains results of EMF measurements on cell (1) in the presence of 0.1 M solutions of (R)- and (S)-α-phenyl-

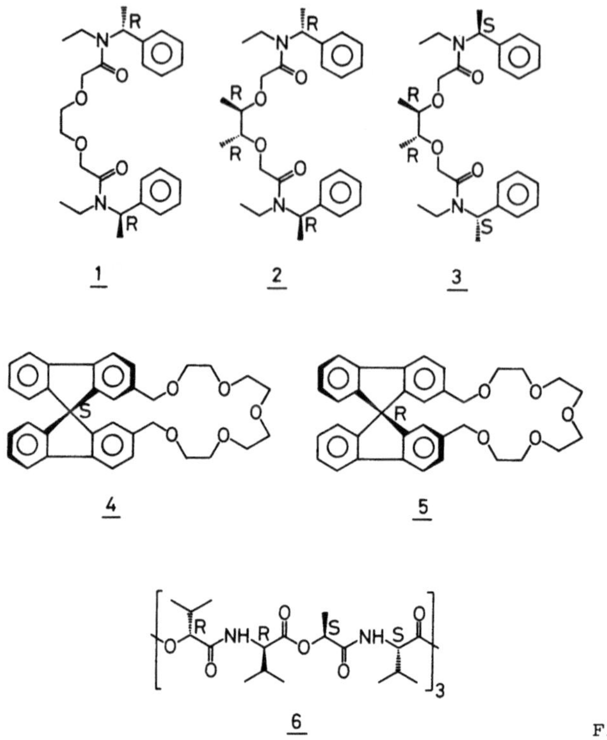

Fig. 2. Chiral ionophores

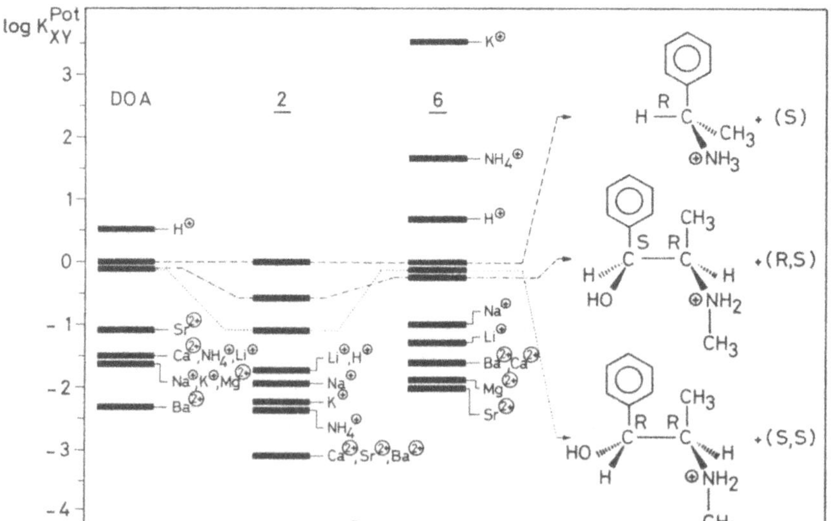

Fig. 3. Results of potentiometric determination of selectivity coefficients using 0.1 M chloride solutions. Membranes consist of 66% (wt/wt) membrane solvent (here dioctyladipate was used), 33% (wt/wt) poly-(vinylchloride) and 1% (wt/wt) of ligand; if no ligand is incorporated in membrane, weight ratio membrane-solvent/poly-(vinyl chloride) is 67/33 (see column on left in this figure)

Table 2. Enantiomer selectivity of ligands 1-6 determined by potentiometry ($\Delta EMF = EMF_R - EMF_S$ in mV; T = 25 ± 0.1 °C) and by electrodialysis (racemate X: ^{14}C-(R)-, ^{3}H-(S)-α-phenylethylammonium chloride; racemate Y: ^{3}H-(R)-, ^{14}C-(S)-α-phenylethylammonium chloride; T = 21 °C); confidence limits are 95%. Membrane solvent is o-nitrophenyloctylether

Ligand	Potentiometry		Transport $\frac{t_R}{t_S}$	
	ΔEMF	$\frac{K_{RL}}{K_{SL}}$	racemate X	racemate Y
1	-0.3 ± 0.4	0.99 ± 0.02		
2	2.3 ± 0.1	1.09 ± 0.003	1.067 ± 0.015	1.070 ± 0.026
3	1.2 ± 0.5	1.05 ± 0.02		
4	-3.5 ± 0.7	$(1.15 \pm 0.03)^{-1}$	$(1.093 \pm 0.027)^{-1}$	$(1.094 \pm 0.012)^{-1}$
5	2.9 ± 0.8	1.12 ± 0.04	1.124 ± 0.017	1.121 ± 0.012
6[a]	0.0 ± 0.4	1.00 ± 0.02	1.00 ± 0.02	1.00 ± 0.02

[a] Membrane solvent: dioctyladipate

ethylammonium chloride, respectively. The enantiomer selective EMF response may be described by

$$\Delta EMF = EMF_R - EMF_S = \frac{RT}{F} \ln \frac{K_{RL}}{K_{SL}} \tag{3}$$

with: EMF_R, EMF_S: potential differences measured over the cell (1) in equimolar solutions of the monovalent enantiomeric cations R and S, respectively; R: gas constant; T: absolute temperature; F: Faraday constant; K_{RL}, K_{SL}: complex formation constants of the enantiomeric cations R and S with the chiral ligand L in the liquid membrane (1:1 stoichiometry). The values given in column 3 of Table 2 were obtained from the data in column 2. A comparison of the results for 1 and 2 indicates that the introduction of methyl groups at sites 4 and 5 (see Fig. 1), leading to central chirality R at both carbon atoms, is the main cause of enantiomer selectivity. This is in agreement with the only slightly different enantiomer selectivity of 3 relative to 2. As expected, the effect of 4 is reversed by 5. Although valinomycin 6 is chiral, no enantiomer selectivity was detectable (see Table 2). The potentiometrically determined enantiomer selectivity ΔEMF is correlated to the transport selectivity (Morf et al., 1976)

$$\frac{t_R}{t_S} = \frac{K_{RL}}{K_{SL}} \tag{4}$$

with t_R, t_S: transport numbers of the enantiomeric cations R and S through the liquid membrane.

By applying a potential difference across neutral carrier membranes (see legend to Fig. 3) in contact with similar solutions of racemic α-phenylethylammonium chloride, a carrier-mediated

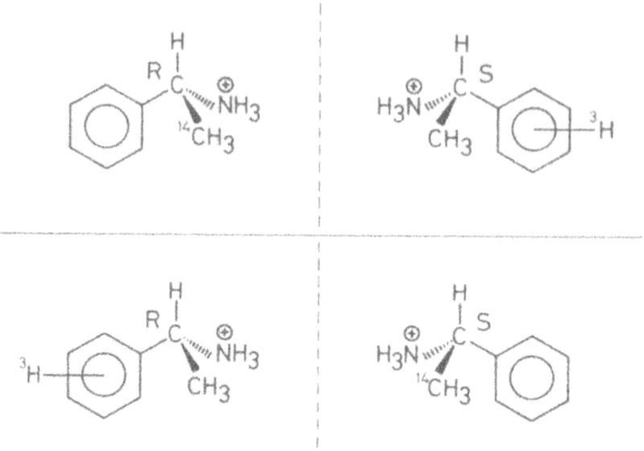

Fig. 4. Labeling of enantiomeric cations for transport studies

cation transport through the membrane is expected. According to Eq. (4) a ratio $t_R/t_S \neq 1$ has to be expected for ionophores with $\Delta EMF \neq 0$. Using the double labeling of racemates shown in Figure 4 and described in detail in Thoma et al. (1975), transport numbers may easily be determined for the enantiomers and may be confirmed by reversing the labeling. Results of these electrodialytic transport experiments are shown in the last two columns of Table 2 and are in good agreement with the data obtained potentiometrically (see column 3).

To obtain information on the transport mechanism, five membranes (thickness: 40 μm) were stacked, and the concentration of ligand and cation in each membrane was measured before and immediately after the transport experiment as well as 5 days after restacking the membranes. Since a concentration gradient of valinomycin develops (Fig. 5, t = 3 h), which decays almost completely after a relaxation period (Fig. 5, t = 5 days), the α-phenylethylammonium cation has obviously been transported by a carrier mechanism. During the transport process a cation profile builds up (Fig. 6, t = 3 h), which has the same trend as the ligand profile. This cation gradient disappears after some time (Fig. 6, t = 5 days).

In a similar electrodialysis experiment only the carrier in section A was labeled. After the transport experiment only about 20% of the labeled ligand drained off section A reached section E. This proves an exchange of ligands during the transport process so that a carrier-relay mechanism must be operative (Wipf et al., 1969). A detailed discussion of transport mechanisms will be given elsewhere.

Acknowledgments. This work was partly supported by the Schweizerischer Nationalfonds zur Förderung der wissenschaftlichen Forschung. We thank Prof. Dr. Y.A. Ovchinnikov for a supply of ^{14}C-labeled valinomycin. The ligands 2 and 3 are part of a larger program conducted by M. Güggi to correlate structure with ion selectivity. We thank him for his contribution. One of us (G.H.) thanks the Scientific Exchange Agreement for financial support.

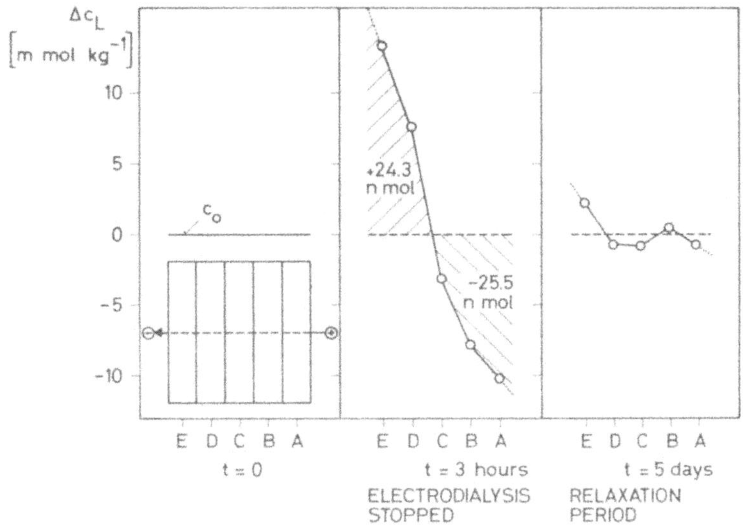

Fig. 5. Study of mobility of valinomycin within stack of membranes before (t = 0), immediately after (t = 3 h) electrodialysis as well as after 5-day period of restacking membranes (t = 5 days). c_o: initial ligand concentration; Δc_L: change of total ligand concentration. Radius of circles corresponds to 95% confidence limits

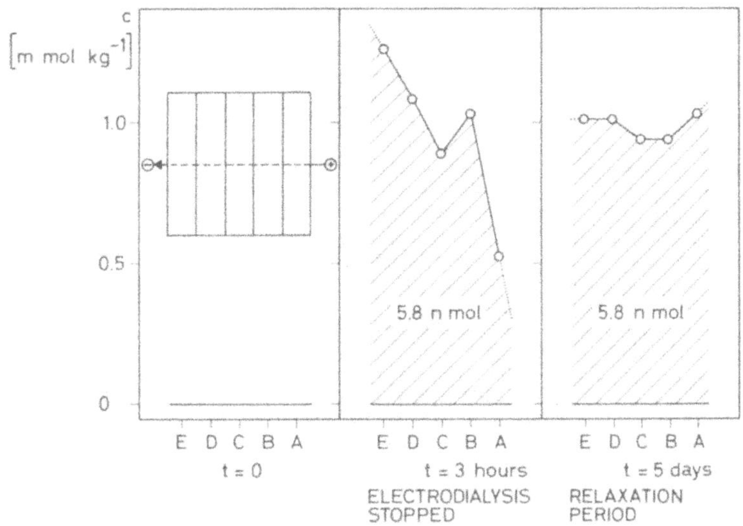

Fig. 6. Study of mobility of α-phenylethylammonium ions within stack of membranes before (t = 0), immediately after (t = 3 h) electrodialysis as well as after 5-day period of restacking membranes (t = 5 days). c: cation concentration in membrane sections

References

Ammann, D., Bissig, R., Güggi, M., Pretsch, E., Simon, W.: Preparation of neutral ionophores for alkali and alkaline earth metal cations and their application in ion selective membrane electrodes. Helv. Chim. Acta 58, 1535-1548 (1975)

Bedeković, D.: Dissertation ETH Nr. 5777 (1976)

Bedeković, D., Prelog, V.: in preparation, 1976

Büchi, R., Pretsch, E., Simon, W.: Information on the coordinating sites of ionophores obtained by carbon-13 NMR relaxation time measurements. Tetrahedron Lett. 20, 1709-1712 (1976)

Cram, D.J., Helgeson, R.C., Sousa, L.R., Timko, J.M., Newcomb, M., Moreau, P., De Jong, F., Gokel, G.W., Hoffman, D.H., Domeier, L.A., Peacock, S.C., Madan, K., Kaplan, L.: Chiral recognition in complexation of guests by designed host molecules. Pure Appl. Chem. 43, 327-349 (1975)

Curtis, W.D., Laidler, D.A., Stoddart, J.F., Jones, G.H.: Chiral Recognition by Configurationally Chiral Cryptands. J. Chem. Soc. Chem. Comm. 835-837 (1975)

De Levie, R.: Electrochemistry of artificial ultrathin lipid membranes; a review. J. Electroanal. Chem. 69, 265-297 (1976)

Koryta, J.: Ion-Selective Electrodes. Cambridge, London, New York: Cambridge Univ. Press 1975

Morf, W.E., Wuhrmann, P., Simon, W.: Transport properties of neutral carrier ion selective membranes. Anal. Chem. 48, 1031-1039 (1976)

Mueller, P., Rudin, D.O.: Development of $K^+ - Na^+$ discrimination in experimental bimolecular lipid membranes by macrocyclic antibiotics. Biochem. Biophys. Res. Commun. 26, 398-404 (1967)

Newcomb, M., Helgeson, R.C., Cram, D.J.: Enantiomer differentiation in transport through bulk liquid membranes. J. Am. Chem. Soc. 96, 7367-7369 (1974)

Ovchinnikov, Y.A., Ivanov, V.T., Shkrob, A.M.: BBA Library 12, Membrane-Active Complexones. Amsterdam: Elsevier 1974

Simon, W., Pretsch, E., Ammann, D., Morf, W.E., Güggi, M., Bissig, R., Kessler, M.: Recent developments in the field of ion selective electrodes. Pure Appl. Chem. 44, 613-626 (1975)

Štefanac, Z., Simon, W.: In-vitro-Verhalten von Makrotetroliden in Membranen als Grundlage für hochselektive kationenspezifische Elektrodensysteme. Chimia 20, 436 (1966)

Thoma, A.P., Cimerman, Z., Fiedler, U., Bedeković, D., Güggi, M., Jordan, P., May, K., Pretsch, E., Prelog, V., Simon, W.: Enantiomerenselektives Verhalten in Membranen eines chiralen, elektrisch neutralen Ionophors. Chimia 29, 344-346 (1975)

Wipf, H.K., Pache, W., Jordan, P., Zähner, H., Keller-Schierlein, W., Simon, W.: Mechanism of alkali cation transport in bulk membranes using macrotetrolide antibiotics. Biochem. Biophys. Res. Commun. 36, 387-393 (1969)

Wuhrmann, P., Thoma, A.P., Simon, W.: Calcium carrier properties of neutral synthetic ligands in bulk membranes. Chimia 27, 637-639 (1973)

Recent Developments in the Structure-Functional Studies of Peptide Ionophores

YU. A. OVCHINNIKOV and V. T. IVANOV

The last decade witnessed the birth and tempestuous development of a new field in bioorganic chemistry. Its objects of study are membrane-active complexones (see Ovchinnikov et al., 1974b and references therein), often called ionophores because of their ability to bind metal ions in solution and to transport them in one or the other stage of the transmembrane ion transporting process. It is this property that has secured a place for the ionophores among modern biochemical and biophysical techniques as highly effective tools for studying ion movements across membranes.

The ionophores constitute a highly varied class of compounds, both natural and synthetic, cyclic and linear, differing greatly in molecular weights and possessing diverse functional groups. In their mode of action, they can be divided into two main classes: (1) carriers (sometimes called "cage carriers") and (2) substances forming ion-conducting pores or "channels" (Fig. 1). Among the carriers, a preeminent place is occupied by macrocyclic depsipeptides of the groups of valinomycin and the enniatins. Of these, valinomycin

$\lfloor$(D-Val-L-Lac-L-Val-D-HyIv)$_3\rfloor$ valinomycin (1)

$\lfloor$(L-MeVal-D-HyIv)$_3\rfloor$ enniatin B (2)

(Vm) itself is unsurpassed in its potassium ion transporting capacity and the K/Na selectivity of its membrane action. The enniatins are interesting mainly because of their broad spectrum of action - members of this group being capable of binding and transporting ions of diverse sizes and valences. The most representative of the "channel" type of ionophore are the 15-membered linear peptide, gramicidin A (Gr A) and the structurally closely allied gramicidins B and C.

```
L     L   D   L   D   L   D   L   D   L   D   L   D   L
HCO-Val-Gly-Ala-Leu-Ala-Val-Val-Val-Trp-Leu-Trp-Leu-Trp-Leu-Trp-NH(CH2)2OH
```

gramicidin A (3)

Systematic synthesis of new ionophores and subsequent structural, mechanistic, and physicochemical studies have been carried out for a number of years in the Shemyakin Institute of Bioorganic Chemistry of the USSR Academy of Sciences.

The present paper presents new findings that have resulted from the latest study of the valinomycin, enniatin B (En), and gramicidin A groups of peptides.

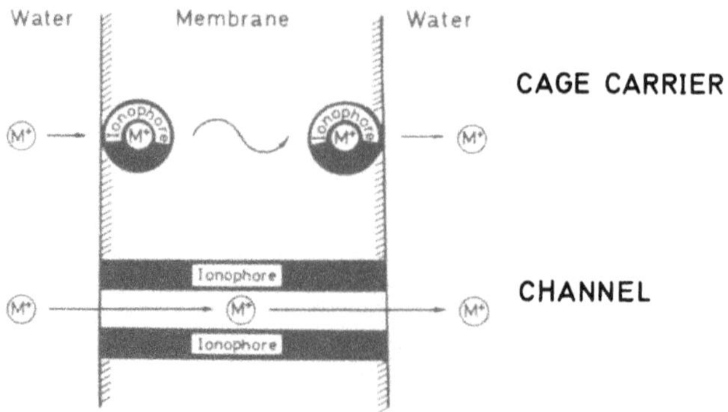

Fig. 1. Schematic illustration of mechanism of cation transport through membranes

A. Valinomycin, Enniatin B and Their Analogs

Our previous studies on the structure-membrane-affecting relations in the valinomycin and enniatin ionophores were centered on the effect of ring size, backbone polar groups, side chain bulkiness, and asymmetric center configuration upon the stability of the alkali metal complexes, complexing kinetics, and lipophilicity and surface activity of both the free and complexed depsipeptides and the mode of their transmembrane ion transport (Ovchinnikov et al., 1974b). This approach left unclear, however, the question of the effect on the ionophore properties of the side chain polarity and the electrical charge on the side chains. To shed light on this aspect of the problem, a series of valinomycin and enniatin B analogs (4, 5) and (6-9), respectively, which contained either a carboxyl or amino function in the side chain were synthesized (Ivanov et al., 1974) and the ability of these compounds to form complexes depending on the ionic state of the side chain functions and their behavior in artificial membrane systems was investigated.

The valinomycin analogs Vm(Lys) and Vm(Glu) differed from the parent compound in the replacement of a L-Val residue by a L-Lys or L-Glu residue and the enniatin analogs, En(MeLys) and En(MeGlu), by the replacement of a L-MeVal residue by a L-MeGlu or N-MeLys residue.

In contrast to the uni-charged complexes Vm · M^+ and En · M^+ (M^+ = alkali metal), the Vm(Lys) · M^+ and En(MeLys) · M^+ complexes with their ionized ε-NH_2 groups carry a charge of 2^+ while the complexes Vm(Glu) · M^+ and En(MeGlu) · M^+ with an ionized γ-carboxyl of the glutamic acid residue are electrically neutral. This circumstance can be expected to affect the stability of the metal ion complexes, the magnitude of the effect serving as a measure of the screening of the metal ion in the molecular cavity from the environment, so that the larger the effect, the less the amount of screening. On the other hand the Vm(Glu) and En(MeGlu) should have properties in common with the nigericin antibiotics, which are characterized by the presence of carboxylic groups and the ability to form electrically neutral complexes with metal ions (see Ovchinnikov et al., 1974b, pp. 198-210). Finally Vm(Lys) and En(MeLys) with their ionized amino groups should be interesting tools for analysis of the modes of action of ionophores in membranes, capable of revealing novel, hitherto inaccessible information.

No.	Abbrev.	Formula
4	Vm(Lys)	(D-Val-L-Lac-L-Val-D-HyIv)$_2$-D-Val-L-Lac-L-Lys-D-HyIv (cyclic)
5	Vm(Glu)	(D-Val-L-Lac-L-Val-D-HyIv)$_2$-D-Val-L-Lac-L-Glu-D-HyIv (cyclic)
6	En(MeLys)	(L-MeVal-D-HyIv)$_2$-L-MeLys-D-HyIv (cyclic)
7	En(Lys)	(L-MeVal-D-HyIv)$_2$-L-Lys-D-HyIv (cyclic)
8	En(MeGlu)	(L-MeVal-D-HyIv)$_2$-L-MeGlu-D-HyIv (cyclic)
9	En(Glu)	(L-MeVal-D-HyIv)$_2$-L-Glu-D-HyIv (cyclic)

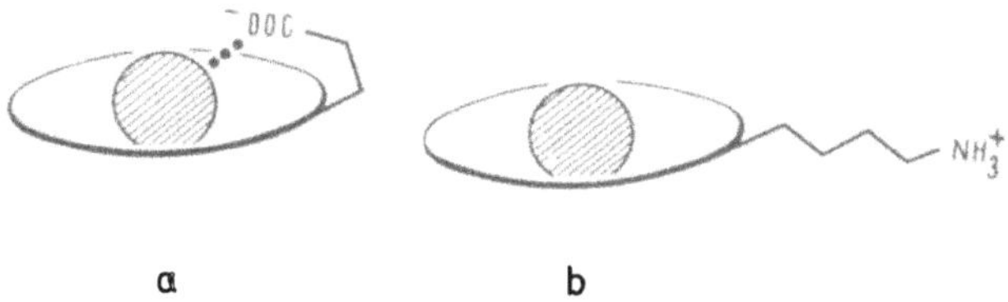

Fig. 2a,b. Schematic representation of conformation of equimolar complexes of (a) En(MeGlu) and En(Glu) and (b) En(MeLys) and En(Lys) with alkaline metal ions

The metal-binding properties of compounds (5-9) were investigated with the side chains in both the neutral and charged states. For this purpose the circular dichroism (CD)-monitored salt titration of the cyclodepsipeptide was carried out in the presence of 1.2 equivalents of hydrogen chloride or triethylamine. The results obtained are summarized in Table 1. They show that the majority of the analogs both with charged or with electrically neutral side chains form complexes with potassium ions that in general are of somewhat lower stability than complexes of valinomycin or enniatin B. Contrary to the enniatin analogs, the valinomycin derivatives reveal no signs of complexing with Na^+, thereby retaining the high K/Na selectivity of the parent antibiotic. Noteworthy is the fact that the potassium complexes of the charged form of the enniatin analogs En(MeGlu) and En(Glu) (in the presence of 1.2 eq. Et_3N) are markedly more stable than the corresponding complexes in which the carboxyl group is nonionized (i.e., in the presence of 1.2 eq. HCl). Most likely the ionized carboxyl complexes are additionally stabilized by electrostatic interaction of the bound cation with the carboxylate anion (Fig. 2a). This effect confirms an earlier conclusion that the bound metal ion in the enniatin complexes is not completely screened from the anion or the solvent (Ovchinnikov et al., 1974a). The En(MeLys) and En(Lys) lysine side chains are apparently stretched (Fig. 2b), so that their ionization only weakly affects the complex stability. The side chain also has almost no effect on the K^+-complex stability of the valinomycin derivatives, in harmony with the effective screening of the centrally located cation from the counterion due to their well-known rigid "bracelet" structure (Fig. 3).

Table 1. Stability constants (M^{-1}) of Na^+ and K^+ complexes of valinomycin and enniatin B derivatives as compared with the parent compounds

Compound	Solvent	Na^+		K^+		Na^+	K^+	Reference
		HCl 1.2 eq.	Et_3N 1.2 eq.	HCl 1.2 eq.	Et_3N 1.2 eq.			
1 Vm							$1.6 \cdot 10^4$	Andreev et al., 1971
4 Vm(Lys)	80% aqueous ethanol	10	10	1900	3600			Sumskaya et al., 1976a
5 Vm(Glu)		10	10	3300	4300			Sumskaya et al., 1976a
2 En						340	2100	Ovchinnikov et al., 1974a
6 En(MeLys)		50	50	60	110			Sumskaya et al., 1976a
7 En(Lys)	96% aqueous ethanol	150	150	80	450			Sumskaya et al., 1976a
8 En(MeGlu)		370	360	300	3400			Sumskaya et al., 1976a
9 En(Glu)		270	600	130	1200			Sumskaya et al., 1976a

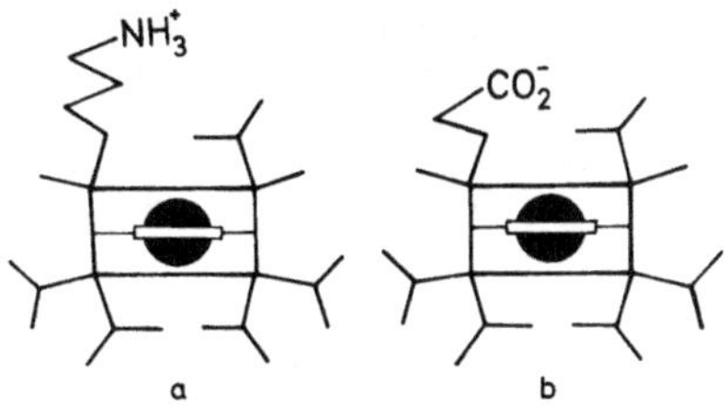

Fig. 3a,b. Schematic representation of equimolar complexes of (a) Vm(Lys) and (b) Vm(Glu)

Table 2. Effect of valinomycin, enniatin B, and their analogs on efflux of potassium ions from liposomes

No.	Compound	K^+ outflow 5 min after ionophore addition (in % of total K^+ content in liposomes)
1	Vm	34
4	Vm(Lys)	29
5	Vm(Glu)	31
2	En	15
6	En(MeLys)	18
8	En(MeGlu)	9

Judging from the stability of the complexes, these valinomycin and enniatin B derivatives have all the prerequisites for ionophoric function. In fact, in the experiments with the lecithin liposomes, Vm(Lys), Vm(Glu), En(MeLys), and En(MeGlu) mediate potassium ion transport along the salt concentration gradient, the rate of potassium chloride efflux from the liposomes in the presence of these compounds (Table 2) being commensurate with the corresponding parameters for valinomycin or enniatin B. The liposome experiments were carried out at pH 6.5, i.e., under conditions where one could expect ionization (at least in aqueous solution) of both the lysine ε-NH_2 group (pK $\approx$ 9) and the glutamic acid γ-COOH group (pK $\approx$ 3).

One of the characteristic properties of the ionophores is their ability to transfer the cations they bind from the aqueous to the organic phase (see Ovchinnikov et al., 1974b, pp. 102-109). In the example of Vm(Lys) and Vm(Glu), the effect of the ionic state of the side chains on this process was investigated by measuring by an extraction technique the ability of these compounds to mediate the transition of potassium ions from water to methylene chloride at pH 5.7 and 9. Picrate (Pi^-) was selected as the lipophilic anion because of its high extinction coefficient and the possiblity of determining spectrophotometrically its concentration in the organic phase. The Pi^- in the organic phase is due to the complex cation $LM^+ \cdot Pi^-$ where L is the ionophore and is, therefore, equal to the amount of potassium transferred from the aqueous phase. Two series of experiments (A and B) were carried out. In series A, conditions for the salt "saturation" of valinomycin, i.e., for the maximum entrance of the potassium ion into the organic phase, were established by making high potassium chloride

Table 3. Valinomycin- and analogs-mediated potassium picrate extraction from water into methylene chloride

No.	Compound ($c = 10^{-4}$M)	pH	Concentration of picrate anion in methylene chloride M($\times 10^4$)		Extractable form of complex (schematic)	Degree of complexation	
			A	B		A	B
		5	1.01	0.75		Quantitative formation of equimolar complex	0.75
1	Vm	7	1.02	0.74			
		9	0.99	0.76			
		5	1.98	0.82			0.41
4	Vm(Lys)	7	1.01	0.46			
		9	1.00	0.45			0.45
		5	0.98	0.51	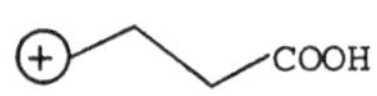		0.50
5	Vm(Glu)	7	0.99	0.49			
		9	0.01	0.01	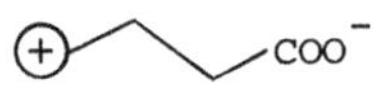		0.56

concentrations in the aqueous phase; in series B, the concentrations and solution volumes were so selected that while there was, as before, excess K^+ and Pi^- with respect to ionophore, their concentrations were still insufficient for attaining complete complexation. The series A experiments yielded the stoichiometry of the process. Series B gave the relative capacity of the ionophore to transfer potassium ions into the organic phase. From the results obtained (summarized in Table 3), it follows that for the very high excesses of salt (series A) the ionophore K^+-complexes in the methylene chloride turned out in every case to be equimolar; the pH of the medium had no effect on the valinomycin-mediated extraction of potassium picrate, which was always one picrate anion per ionophore molecule. The series B data for Vm(Lys) at pH 7 and 9 and Vm(Glu) at pH 5 and 7 showed that the efficiency of the potassium salt extraction by valinomycin and its analogs diminishes when a polar group is incorporated into the side chain. The sharp increase in Pi^- concentration of the extracted Vm(Lys) on passing from pH 7 to 5 is apparently due to protonation of its amino group and formation of a di-charged complex cation of the type  coupled with two equivalents of Pi^-.

With Vm(Glu) at pH 9, on the other hand, no signs of picrate ion transition into the organic phase were observed, but that a K^+-complex did form was established from its breakdown in the chloroform solution on shaking it with aqueous hydrochloric acid (pH 5) and determining the K^+ in the aqueous phase by flame photometry. From the data obtained it followed that at pH 9, i.e., with the carboxyl group in an ionized state, Vm(Glu) extracts K^+ from water into methylene chloride in the form of a 1:1 macrocycle:cation complex and that the extractable form of the complex is electrically neutral, since the positive charge of the potassium ion is compensated by the negative charge of the carboxylate anion.

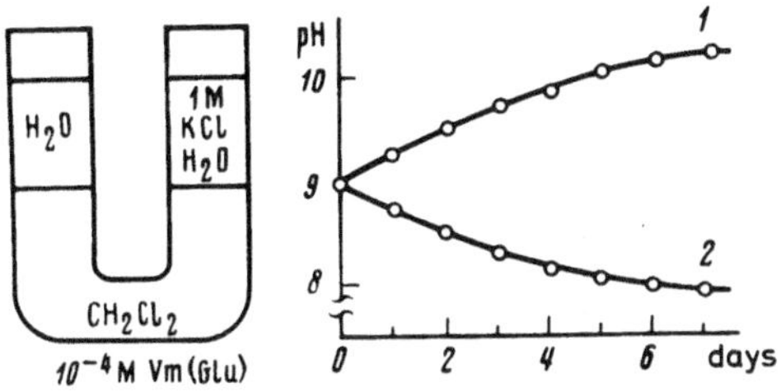

Fig. 4. Rate of $M^+ \longrightarrow H^+$ exchange across Vm(Glu)-containing liquid membrane. Curve 1: change in pH in left arm and Curve 2, in right arm of U tube

Thus, under certain conditions Vm(Glu) acquires properties characteristic of nigericin, monensin, and allied antibiotics, a large group of carboxyl-containing ionophores that carry out nonelectronogenic cation transport in membrane systems according to the following scheme:

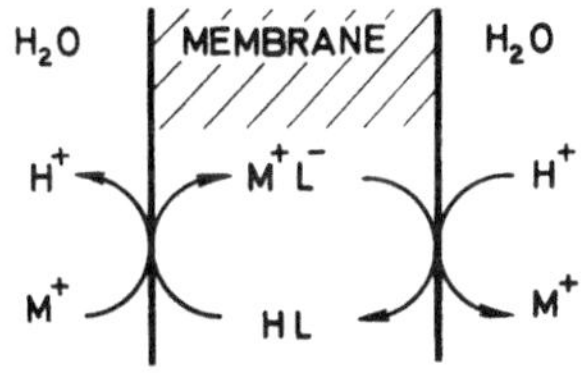

A typical test for the ability of nigericin ionophores to mediate $M^+ \rightleftharpoons H^+$ exchange in artificial membrane systems is to carry out conductance experiments in a U tube (see Ovchinnikov et al., 1974b, p. 276). In such a system neutral ionophores transport metal ions only in the presence of lipophilic anions while the nigericin group of antibiotics is capable of exchanging metal cations for hydrogen. In view of previous statements on Vm(Glu), we tested it by the U-tube method, covering the bottom with a solution of Vm(Glu) in methylene chloride; water was added to this in the left arm, and 1 M aqueous potassium chloride solution in the right arm. Runs were made at differing initial pH values of the aqueous phase (pH 5.7 and 9). No changes in pH occurred for a week at pH 5 and 7. At pH 9 the right-hand side of the U tube became acidified and the left-hand side, alkaline (Fig. 4), i.e., K^+ ions were replaced by H^+. Consequently the U-tube experiment provided additional proof of the observation that in neutral and acid solutions Vm(Glu) is a conventional neutral ionophore (of the valinomycin, enniatin or nactin type) performing only electrogenic cation transport, whereas, as the pH is increased, nonelectrogenic transport enters the scene and the analog starts behaving like an antibiotic of the nigericin group.

Thus, a unique type of membrane-active complexones has been synthesized and characterized. The compounds are in many respects - spatial structure, metal-binding properties, lipophilicity of the K^+-complexes, etc. - similar to the naturally occurring depsipeptide ionophores; but at given pH values they differ from the latter in being electrically non-neutral. As a result they have widened the area of application of the metal-binding complexones for studying ion transport processes across biological and artificial membranes.

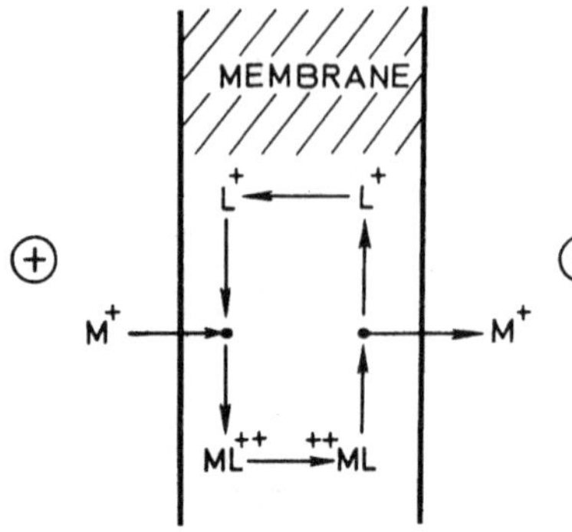

Fig. 5. Model of transmembrane cation transport by dicharged complexes

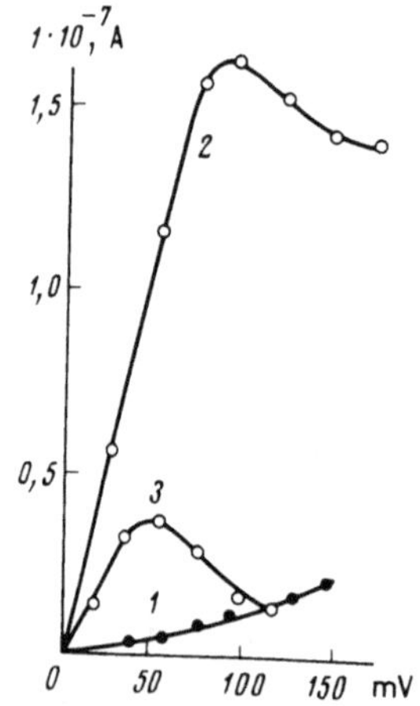

Fig. 6. Effect of 1-methyltetrachlorobenzimidazole (MTB) on the current-voltage characteristics of membranes in solutions containing $1.3 \cdot 10^{-6}$ M Vm(Lys), $1 \cdot 10^{-5}$ M $CuSO_4$, $5 \cdot 10^{-5}$ M hydroquinone and differing amounts of KCl. 1: 1M KCl (reference); 2: 1M KCl + 1M MTB; 3: 3M KCl + 10^{-5} M MTB; pH 3.5 ± 0.2

The behavior of Vm(Lys) in a bilayer under ε-amino protonating conditions serves to illustrate the possibilities uncovered here. In conformity with the theoretical model of Markin et al. (1969), a negative slope in some region of the steady state current voltage curve could have been expected because with the increase in electric field intensity, the NH_3^+ cations are "pressed in" toward the cathodically polarizable boundary of the membrane and are thereby excluded from the potassium ion transport cycle (Fig. 5). Hence the intramembrane flow of K^+-free ionophore molecules should diminish as the applied voltage is increased, and this should decrease the rate of ion transport across the anodically polarizable boundary of the bilayer. Actually in conventional membranes from bovine brain phospholipids, one does not observe such "falling" characteristics, perhaps because the protonated molecules contribute little to the process owing to the high positive potential jump across the membrane boundary, but they do appear when a "dipolar modifier" (1-methyltetrachlorobenzimidazole + Cu^{2+}) is added to the membrane in order to lower the positive potential difference between the membrane and the aqueous bathing solution (Fig. 6, Sumskaya et al., 1976a). Thus this derivative served as the means for experimental substantiation of the model.

Table 4. Effect of valinomycin, enniatin B, and their spin- and fluorescence-labeled analogs on the efflux of potassium ions from egg lecithin liposomes

No.	Compound ($c = 10^{-4}$ M)	Amount of released K^+ in % of initial KCl content after 5 min at 25°C
1	Vm	34
10	Vm(LysSL)	35
11	Vm(LysDns)	32
2	En	15
12	En(MeLysSL)	16
13	En(LysSL)	17
14	En(MeLysDns)	14
15	En(LysDns)	15

The behavior of the enniatins and especially of valinomycin in various membrane systems has been extensively investigated. However, these investigations were performed primarily with electrochemical methods, which do not give direct answers to such mechanistically important questions as location of the ionophore in the membrane, its manner of interaction with the lipid, and the effect on this interaction of cations, of the nature and state of the lipid, etc. Such questions could be resolved by spin or fluorescence labeling of the ionophores, thereby permitting a straightforward determination of their mobility and their immediate environment by electron spin resonance (ESR) or fluorescence spectroscopy. It is of considerable consequence that these methods permit the use of high dilutions, and the experiments can be made with heterogeneous, in particular, biological systems over a wide range of experimental conditions. Veatch et al. (1975) convincingly demonstrated the power of such an approach by preparing the dansyl derivative of gramicidin C and studying its behavior in bilayers.

With this in mind we synthesized spin- and fluorescence-labeled valinomycin derivatives (10 and 11) and enniatin B derivatives (12-15) issuing from analogs (4, 6 and 7) containing a free amino group (Ivanov et al., 1974).

Experiments with liposomes showed that compounds (10-15) were on a par with valinomycin or enniatin B in their transmembrane potassium ion transport capacities (Table 4), i.e., the insertion of spin or fluorescence labels into the molecules does not interfere with their ionophoric activity.

Taking advantage of the spin labels in compounds (10, 12, 13) we used ESR spectroscopy to study their metal-binding properties. In all cases when potassium thiocyanate was added to a solution of the depsipeptide in 96% aqueous alcohol there was a fall in the signal intensity (I, see Fig. 7) and an increase in the correlation time (τ_c), indicating that complexing lowers the mobility of the radical-containing fragment (Table 5). The stability constants

No.	Abbrev.	Formula
10	Vm(LysSL)	[(D-Val-L-Lac-L-Val-D-HyIv)$_2$-D-Val-L-Lac-L-Lys(SL)-D-HyIv]
11	Vm(LysDns)	[(D-Val-L-Lac-L-Val-D-HyIv)$_2$-D-Val-L-Lac-L-Lys(Dns)-D-HyIv]
12	En(MeLysSL)	[(L-MeVal-D-HyIv)$_2$-L-MeLys(SL)-D-HyIv]
13	En(LysSL)	[(L-MeVal-D-HyIv)$_2$-L-Lys(SL)-D-HyIv]
14	En(MeLysDns)	[(L-MeVal-D-HyIv)$_2$-L-MeLys(Dns)-D-HyIv]
15	En(LysDns)	[(L-MeVal-D-HyIv)$_2$-L-Lys(Dns)-D-HyIv]

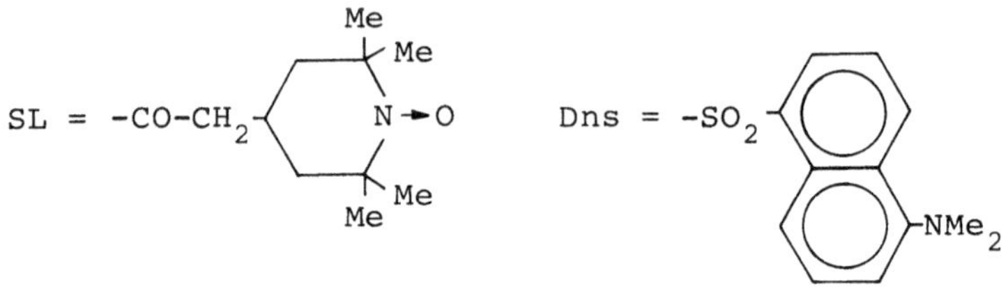

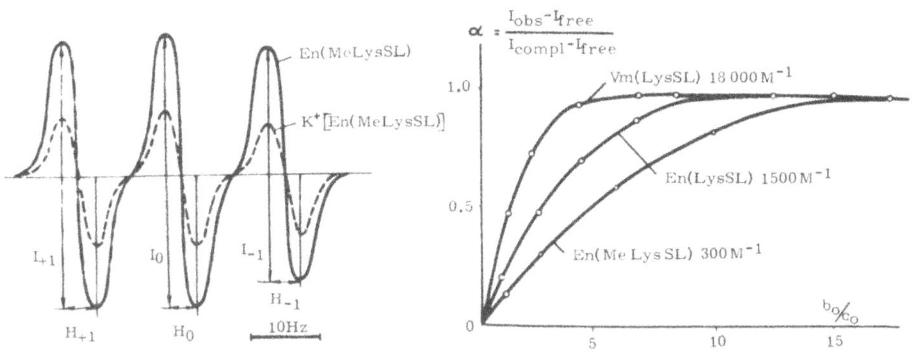

a b

Fig. 7a,b. (a) ESR spectra of compound (12) (——) and of its K^+-complex (---) in 96% aqueous ethanol and (b) potassium thiocyanate titration of spin-labeled analogs in 96% aqueous ethanol

determined from the titration curves for the complexes (Fig. 7b) are 18,000 M^{-1} for Vm(LysSL), 300 M^{-1} for En(MeLysSL), and 1500 M^{-1} for En(LysSL). Judging from the τ_c value, the nitroxyl radical is more immobilized in the En(MeLysSL) and particularly in the En(LysSL) complexes than in the Vm(LysSL) complex. In the case of En(MeLysSL), this result could be taken as indicating possible interaction of the weakly screened cation in the enniatin complexes with the nitroxyl O atom [by analogy with the En(MeGlu)·K^+-complex, see Fig. 2a)].

Table 5. Rotational correlation time (τ_c) for spin-labeled cyclodepsipeptides and their K^+-complexes

No.	Compound	$\tau_c \cdot 10^{11}$ sec Free cyclodepsipeptide	K^+-complex
10	Vm(LysSL)	2.5	3.0
12	En(MeLysSL)	2.0	4.0
13	En(LysSL)	2.3	4.5

Table 6. Fluorescence spectral data for Vm(LysDns) and DnsLys (λ_{exc} = 340 nm)

Solvent[a] (solute concentration 1.3 · 10^{-6} M)	Vm(LysDns)[a]			DnsLeu		
	λ_{max}	Q	P	λ_{max}	Q	P
Heptane-ethanol (99:1)	460	0.23	< 1%	460	0.22	< 1%
Ethanol	500, 520	0.14	1.5%	500, 520	0.18	1.5%
Water-ethanol (99:1)	460	0.016	31-37%	550	0.20	1.5%
Egg lecithin liposomes	500, 520	0.13	10-15%	-	-	-

[a]On adding 1 M KCl solution to aqueous Vm(LysDns) solution P = 50%; in all other cases, salt addition did not affect λ_{max}, Q, and P values

In the Vm(LysSL) · K^+-complex the cation is much more effectively screened from the environment and so does not interact with the nitroxyl moiety. In En(LysSL), accessibility of the nitroxyl radical to the cation is apparently additionally favored by the replacement of the N-methyl group by an NH group. This can also explain the greater stability of the En(LysSl) complex as compared to the En(MeLysSL) complex.

The high metal-binding and ionophoric activity of the spin-labeled derivatives combined with the high sensitivity of their ESR spectra to subtle changes of molecular structure demonstrate their considerable potentiality in the study of membrane systems.

The dansyl derivative of valinomycin (11) was investigated by fluorescence spectroscopy. The position of λ_{max}, the quantum yield (Q), and the fluorescence polarization (P) were measured in solvents of varying polarities for Vm(LysDns), both free and its K^+-complex. The results were compared with those for the simple model, dansyl-L-leucine (DnsLeu) in the same solutions. They are summarized in Table 6.

As might have been expected, on passing from heptane to ethanol, the fluorescence peak shifts toward longer wavelengths, and the quantum yield of fluorescence diminishes to some extent. The behavior of Vm(LysDns) was always very similar to that of the model leucine derivative.

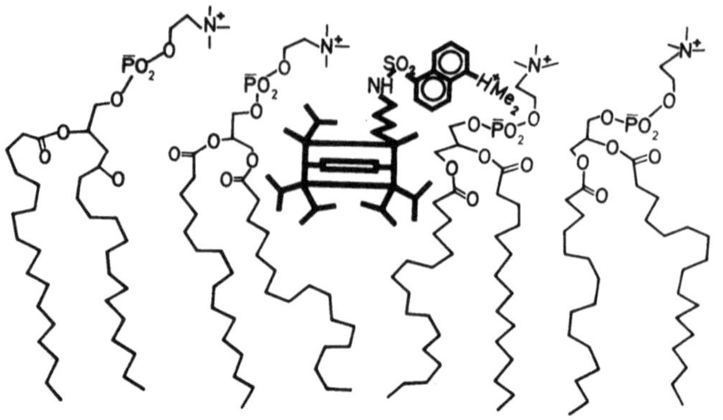

Fig. 8. Preferred disposition of Vm(LysDns) in liposome membranes (schematically)

Further augmentation of solvent polarity (aqueous solutions) leads in the case of DnsLys to still larger shifts of λ_{max}, to longer wavelengths, and to considerable decrease in the Q values. With Vm(LysDns), although the quantum yield was normal for water, the observed λ_{max} was characteristic for heptane, indicating, we believe, association of the depsipeptide under such conditions with the formation of a micellar solution. Support for such an assumption can be seen in the anomalously high polarization of fluorescence.

The Vm(LysDns) was further studied in an artificial membrane system composed of egg lecithin liposomes. Its λ_{max} and Q values indicated that the dansyl label under such conditions must be in a highly polar environment, i.e., it must be localized in the polar head region of the lipid rather than in the interior of the membrane. Hence, it may be considered likely that the ionophore backbone must also be situated near the membrane surface. One such possible disposition is shown in Figure 8. Taking into account the similar ion-carrying properties of Vm(LysDns) and of the parent valinomycin, it would be only natural to conclude that the latter should be similarly located in the membrane. This would be in harmony with the high surfactant properties of valinomycin observed earlier in this laboratory (Shemyakin et al., 1969) and with the CD measurements on this compound in dimyristyllecithin liposomes (Grell et al., 1975). The fluorescence spectra of Vm(LysDns) proved to exhibit little sensitivity to the K^+ complexing in ethanol (Table 6), indicating the invariability of the label's environment under such conditions. In water, the degree of polarization was found to increase, apparently manifesting further association of the depsipeptide and providing additional evidence of the high lipophilicity of the complex cation. As in alcohol, no significant spectral changes were observed in the liposome suspensions on adding potassium chloride, i.e., under conditions of intensive transmembrane ion transport (see Table 4). Consequently the diffusion time of the complex ion of Vm(LysDns) within the membrane must be small compared with the sojourn of the ion near the surface, a conclusion that must be borne in mind when analyzing the shapes and magnitudes of the energy barriers in the transmembrane transport process.

It was recently discovered in this laboratory that valinomycin can form 2:1 macrocycle: cation complexes in the form of "sandwiches", as well as the conventional 1:1 complex, the former much less stable than the latter. A hypothetical structure was suggested for

Fig. 9. Possible structure of the $(Vm)_2 \cdot K^+$-complex

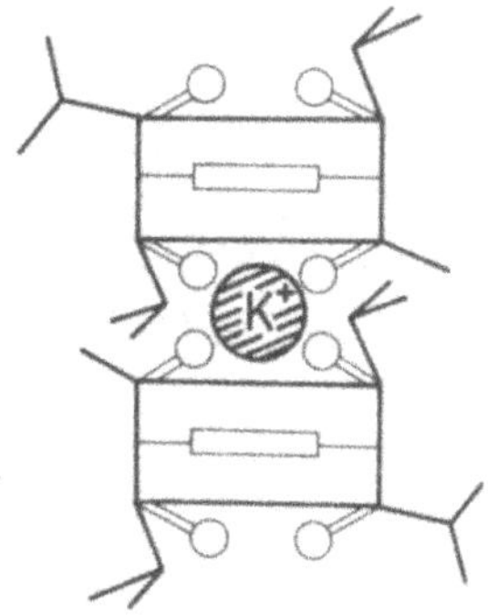

Table 7. Extraction of potassium picrate from water into methylene chloride containing 10^{-4} M ionophore

No.	Compound	Picrate anion concentration in methylene chloride, $M \cdot 10^4$	
		1 M KCl	$5 \cdot 10^{-4}$ M KCl
16	Vm(Glu)-Vm(Lys)	2.01	2.08
17	Vm(LysPht)	1.04	1.05
22	En(MeLys)-En(MeGlu)	0.99	-

these sandwiches (Fig. 9), and it was conjectured that they may participate in the transmembrane ion transport process (Ivanov, 1975; Ivanov et al., 1975). It occurred to us that "sandwiching" might be facilitated if two depsipeptide rings were joined together by a sufficiently large chain so that folding would bring the rings into a position favorable for both to interact with a single ion, so to say, intramolecular "sandwiching".

Accordingly, from the functional derivatives Vm(Lys) and Vm(Glu) we prepared the bis-valinomycin analog Vm(Lys)-Vm(Glu) (16) by acyl chloride condensation (Sumskaya et al., 1976b). However, analysis of its potassium chloride titration curve in ethanol (controlled by CD spectroscopy) showed that the rings behave independently, each binding its own ion to form a di-ionic complex of low stability ($370 \pm 50\ M^{-1}$).

The same holds true for extraction experiments. The analog (16), like the monomeric analog Vm(LysPht) (17), taken for comparison, mediated the stoichiometric (per ion binding center) extraction of potassium picrate from water into methylene chloride over a large range of concentrations. In other words both ion binding centers of bis-valinomycin are capable of solubilizing potassium ions in a nonpolar medium with an efficiency indistinguishable under the experimental conditions from the efficiency of the nearest analog with a single center (Table 7).

An explanation for this is provided by an examination of molecular models, from which it can be seen that the two rings of the bis-analog are arranged in a conformation like that shown in Figure 9.

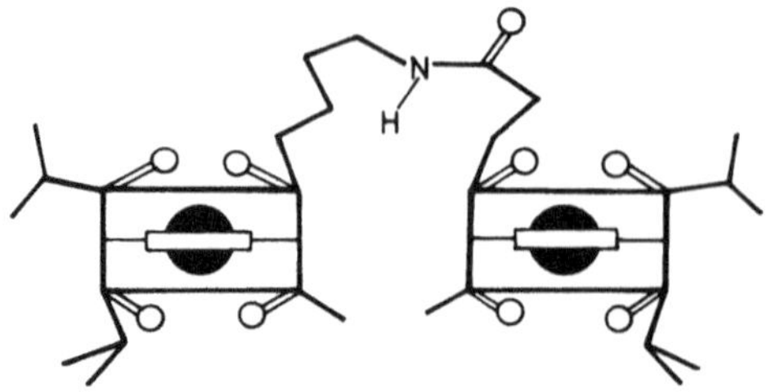

Fig. 10. Proposed structure of [Vm(Lys)-Vm(Glu)] · $(K^+)_2$-complex

Thus it seems impossible for the rings to assume such a position that the cation undergoing complexation could interact with the carbonyl groups on the sterically least hindered "lactyl" ends of the "cylinders" where the lactic acid residues are located (lower ends in Fig. 10).

Just recently we synthesized a new series of valinomycin analogs (18-21), (Fonina et al., 1976)

$$\overline{\underline{|(\text{D-Val-L-Pro-L-Val-D-HyIv})_n|}}$$

n = 3 (18)
n = 4 (19)

$$\overline{\underline{|(\text{D-Val-L-Lac-L-Val-D-Pro})_n|}}$$

n = 3 (20)
n = 4 (21)

The CD-salt dependence of the compounds and the resultant titration curves are indicative of the formation of exceptionally stable 2:1 complexes with potassium and cesium ions in 96% ethanol. Figure 11 illustrates the course of change in the CD curves of compounds (18) and (20) on titration with CsCl and the ensuing titration curves. With compound (20) the curve runs much higher than the theoretical one calculated for an equimolar complex with an infinitely high stability constant. Similar results were obtained on titrating with KCl. Compound (18) displays a different CD-salt titration pattern: initially $\Delta\varepsilon$ increases, reaches a maximum, and then begins to diminish, leveling off at large excess of salt. A similar picture holds for the changes in the CD curves for compound (19) on titration with KCl and CsCl and for compound (21) on titration with KCl. Tentatively, we consider the most probable sandwich structure to be such that each ligand retains the valinomycin-like bracelet system of six intramolecular $H \longrightarrow 1$ hydrogen bonds, while the role of ligands is played by the carbonyls of the X-Pro tert-amide groups (Fig. 12). Thus the formation of double complexes should be considered a common property of the members of the valinomycin group of depsipeptides. We are now in the progress of a quantitative study of this phenomenon for different solutions. Its implications for the transmembrane transport of the ions as yet remain obscure.

The opposite holds for the enniatin antibiotics, where there is strong evidence for the "nonstoichiometric" complex playing a

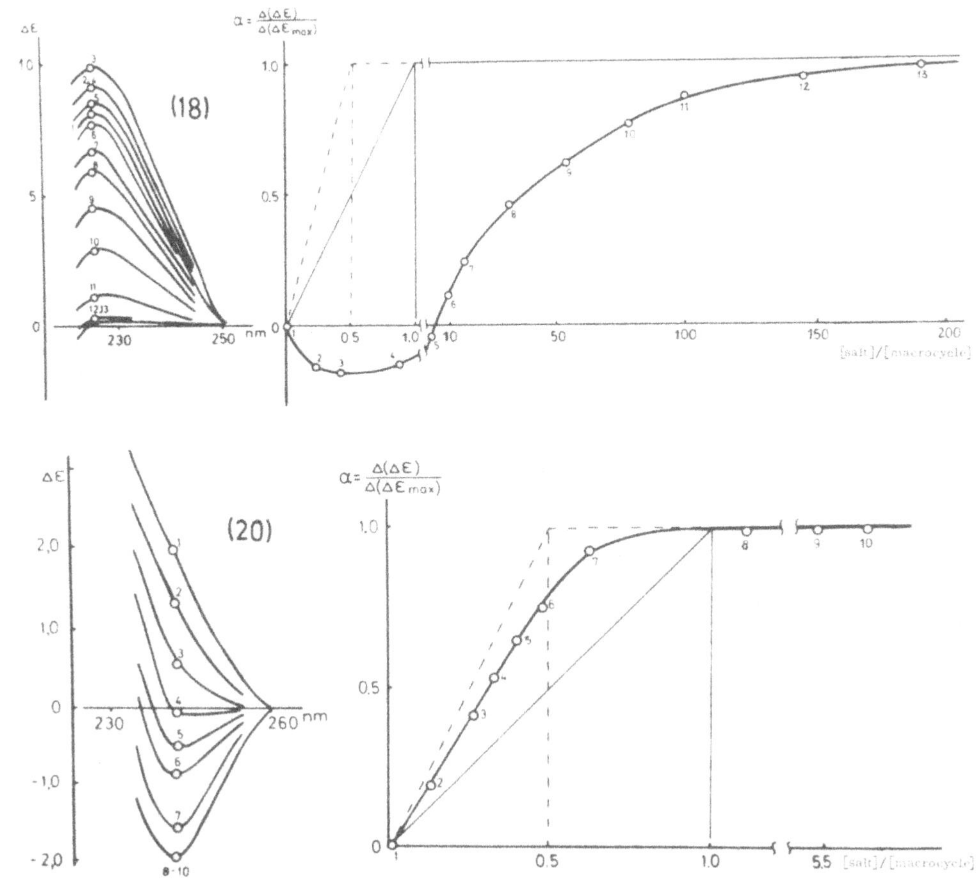

Fig. 11. Change in CD Curve and titration curve for compounds (18) and (20) on titrating with cesium chloride in 96% aqueous alcohol. Initial concentration 10^{-4} M, b_o/c_o cyclodepsipeptide/salt molar ratios

major role in their mediation of transmembrane alkali ion transport (Ovchinnikov, 1974; Ovchinnikov et al., 1974a; Ivanov et al., 1973). The advent of the dimeric complexes provides new support for such a mode of transport, as well as more information concerning the conceptual aspects of the problem. The dimers we used in the enniatin series were En(MeLys)-En(MeGlu) (22) and the desmethyl derivative En(Lys)-En(Glu) (23) synthesized in this Institute (Sumskaya et al., 1976b). Because of the low lipophilicity of (23), extraction experiments were carried out with (22), whereas (23) was subjected to titration with potassium chloride; the small quantities of the more difficultly synthesised (22) precluded its use for titration. Both compounds, however, exhibited very similar properties, giving grounds for treating them together.

The stability constant of the 1:1 "intramolecular sandwich" complex of (23) (Fig. 13) in 96% aqueous alcohol proved to be $(1.4 \pm 0.3) \cdot 10^4$ M^{-1} from the titration experiments, showing the complex to be of exceptionally high stability. The same holds for

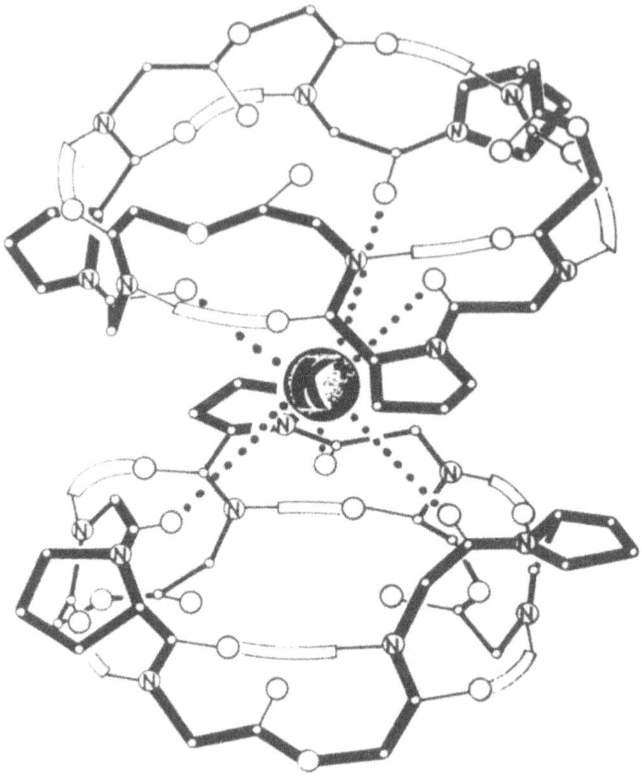

Fig. 12. Hypothetic structure of "sandwich" complex $(18)_2 \cdot K^+$

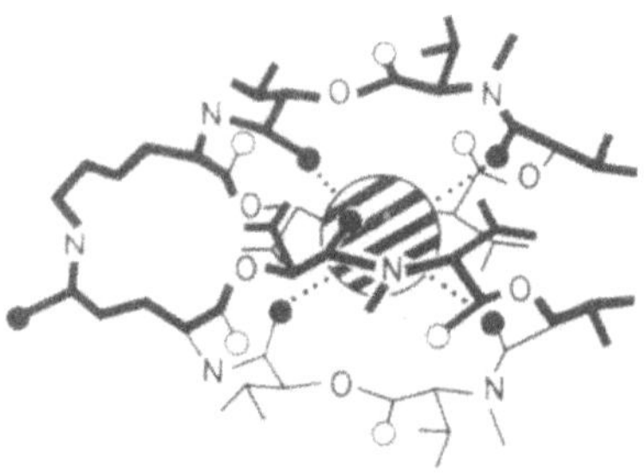

Fig. 13. Conformation of $[En(Glu)\text{-}En(Lys)] \cdot K^+$-complex in solution

(22), which even at very high potassium chloride concentrations carried over the potassium ions into the organic phase only in equimolar quantities, i.e., in the form of an intramolecular sandwich (Table 7).

Of considerable import is that under the same conditions no signs of sandwich formation were observed with sodium ions. It is well known that in contrast to the enniatin complexing reaction with its practically imperceptible K/Na selectivity, enniatin-mediated transmembrane ion transport is markedly K/Na selective (Ivanov et al., 1973). Since sodium ions are not sandwiched even by dimers, one might well conjecture that this ion is transported across membranes by enniatin B in the equimolar form.

The opposite is the case with potassium ions. The high stability of their 2:1 complexes with the enniatin dimers is evidence of a much greater tendency for the potassium ions to form such complexes, a conclusion that should be valid also for the monomers. This, together with the marked K/Na selectivity of the enniatin-mediated transmembrane transport process, in contrast with the low selectivity of the monomer complexing reaction, indicates the high probability of potassium transport of the enniatins across membranes in the form of the double (2:1) complex.

B. Gramicidin A and Its Analogs

It was assumed initially that gramicidin A (Gr A) functions as an ion carrier, but Haydon and Hladky (1972) showed that in membranes Gr A forms ion-conducting channels. A fluorescence study by Veatch et al. (1975) of the membrane-bound Gr A analog, dansyl Gr C, and of its mediated ion conductivity in the membrane led to the conclusion that two Gr A molecules are required for the formation of each "active" ion-conducting channel.

It seemed logical to find some relationship between this mechanism and the fact that Gr A exists in solution as a dimer chromatographically separable into four conformationally different species (1, 2, 3 and 4) (Veatch and Blout, 1974; Veatch et al., 1974). It was, of course, clear that the membrane affecting properties of the antibiotic must be governed by its conformation. However, the membrane conformation of Gr A has not been unequivocally established, at least two models, the $\pi_{(L,D)}$-helix (Urry et al., 1971) and the β-double helix (Veatch et al., 1974; Fossel et al., 1974) having been proposed for it (Fig. 14). Dimerization of the $\pi_{(L,D)}$-helix was postulated as occurring by head-to-head association (although other possibilities were not excluded) (Urry et al., 1971); in the β-double helix, all hydrogen bonds were assumed to be intermolecular and the chains oriented either parallel or antiparallel.

Species (1) and (2) have very similar CD spectra (see Fig. 15) (Veatch et al., 1974), that of (4) is approximately their mirror image, while species (3) displays a spectrum differing considerably from the others. All four exhibit a strong amide I band in the IR spectrum at 1633 cm^{-1} whereas the IR spectrum of (3) has an additionally resolved component at 1680 cm^{-1}. Because the elementary cells of the $\pi_{(L,D)}$-helix and the β-double helix differ little in symmetry, they should display similar IR bands, a fact which made the authors retain both possible structures without choosing between them.

Other points that are obscure in the Gr A behavior are the relation between the primary structure of the antibiotic and its associative ability, conformational states, and membrane-affecting properties.

Thus, although much progress has been achieved in elucidation of the Gr A properties, much yet awaits an answer. Considerable responsibility for the uncertainties lies in the dearth of experimental data that could lend unequivocal support to or provide sound reason for rejecting the models proposed.

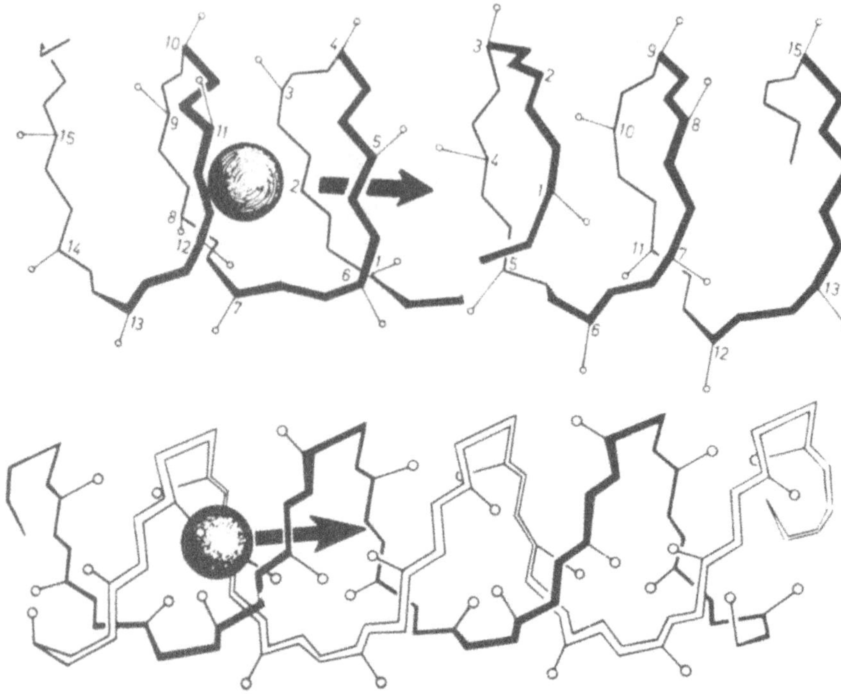

Fig. 14. Schematic representation of $\pi_{(L,D)}$-helix (upper structure) and β-double helix of Gr A

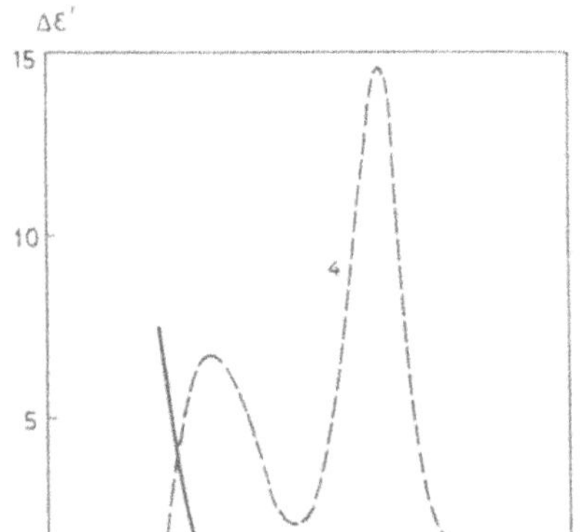

Fig. 15. CD curves of individual Gr A dimeric forms in dioxane (Veatch et al., 1974). Numbers on curves: corresponding species

Table 8. Stability constants (K_{dimer}, M^{-1}) of Gr A anlog dimers

	No.	Solvent	Dimerization constant CD	FP
Species 3	25	ethanol	$(3.0 \pm 0.4) \cdot 10^2$	$(3.5 \pm 1) \cdot 10^2$
		dioxane	-	$(2.0 \pm 0.5) \cdot 10^3$
	28	ethanol	$(1.4 \pm 0.4) \cdot 10^2$	$(2 \pm 4) \cdot 10^2$
		dioxane	$(9.6 \pm 0.5) \cdot 10^4$	$\sim 10^5$
Species 4	26	ethanol	$(3 \pm 1) \cdot 10^2$	$\sim 10^3$
		dioxane	$\sim 10^4$	$\sim 10^4$
	29	ethanol	$(1 \pm 0.5) \cdot 10^2$	$\sim 10^2$
		dioxane	$(3.5 \pm 2) \cdot 10^3$	$(2.6 \pm 2) \cdot 10^3$
	31	ethanol	$(3 \pm 2) \cdot 10^4$	$(9 \pm 4) \cdot 10^3$
		dioxane	$\sim 10^5$	$(4.0 \pm 1.5) \cdot 10^4$
See text	24	ethanol	$(2.5 \pm 0.5) \cdot 10^2$	$(3 \pm 1) \cdot 10^2$
		dioxane	-	$(5.5 \pm 0.5) \cdot 10^2$
	34	ethanol	$(3.2 \pm 0.7) \cdot 10^2$	$(3.8 \pm 0.5) \cdot 10^2$
		dioxane	-	$(2 \pm 1) \cdot 10^2$
See text	27	ethanol	$< 10^2$	$\lesssim 10^2$
		dioxane	$\sim 10^3$	$\sim 3 \cdot 10^3$
	30	ethanol	$\lesssim 10^2$	$\lesssim 10^2$
		dioxane	-	-
	33	ethanol	-	$< 10^3$
		dioxane	-	-

As a contribution to filling in this gap, we have synthesized a number of Gr A analogs (Shepel et al., 1976), differing from the parent antibiotic in chain length [compounds (24) - (30), (33) and (34)] in the L and D sequences of the amino acid residues [compounds (32) and (33)] and in some of the residues themselves [compound (31)]. The analogs were then investigated with respect to their associative capacity by determining the concentrational dependence of the CD curves and the fluorescence polarization (FP) over the maximum possible concentration range (10^{-2} - 10^{-5} M) in ethanol and dioxane (see Table 8).

The CD spectral dependence on concentration for some of the analogs is shown in Figures 16 and 17, the spectra always being obtained under equilibrium conditions. The figures also show the differential dichroic absorption at λ = 228 nm vs. lg c, where c is the overall concentration of the antibiotic.

No.	Formula
3	HCO-Val-Gly-Ala-Leu-Ala-Val-Val-Val-Trp-Leu-Trp-Leu-Trp-Leu-Trp-NH$(CH_2)_2$OH
24	HCO-Val-Gly-Ala-Leu-Ala-Val-(- -)-Trp-Leu-Trp-Leu-Trp-Leu-Trp-NH$(CH_2)_2$OH
25	HCO-Val-Gly-Ala-Leu-Ala-Val-Val-Val-(- -)-Trp-Leu-Trp-Leu-Trp-NH$(CH_2)_2$OH
26	HCO-Val-Gly-(- - - -)-Val-Val-Trp-Leu-Trp-Leu-Trp-Leu-Trp-NH$(CH_2)_2$OH
27	HCO-Val-Gly-Ala-Leu-Ala-(- - - -)-Leu-Trp-Leu-Trp-Leu-Trp-NH$(CH_2)_2$OH
28	HCO-Val-Gly-Ala-Leu-Ala-Val-Val-Val-(- - - -)-Trp-Leu-Trp-NH$(CH_2)_2$OH
29	HCO-Val-Gly-(- - - - - -)-Trp-Leu-Trp-Leu-Trp-Leu-Trp-NH$(CH_2)_2$OH
30	HCO-Val-Gly-Ala-Leu-Ala-(- - - - - -)-Leu-Trp-Leu-Trp-NH$(CH_2)_2$OH
31	HCO-Val-Gly-Val-Leu-Val-Leu-Val-Leu-Trp-Leu-Trp-Leu-Trp-Leu-Trp-NH$(CH_2)_2$OH
32	HCO-Val-Gly-Ala-Leu-Ala-Val-Val-Val-Gly-Trp-Trp-Leu-Trp-Leu-Trp-NH$(CH_2)_2$OH
33	HCO-Val-Gly-Ala-Leu-Ala-Val-Val-Val-Gly-Trp-Leu-Trp-Leu-Trp-Leu-Trp-NH$(CH_2)_2$OH
34	HCO-Leu-Val-Gly-Ala-Leu-Ala-Val-Val-Val-Trp-Leu-Trp-Leu-Trp-Leu-Trp-NH$(CH_2)_2$OH

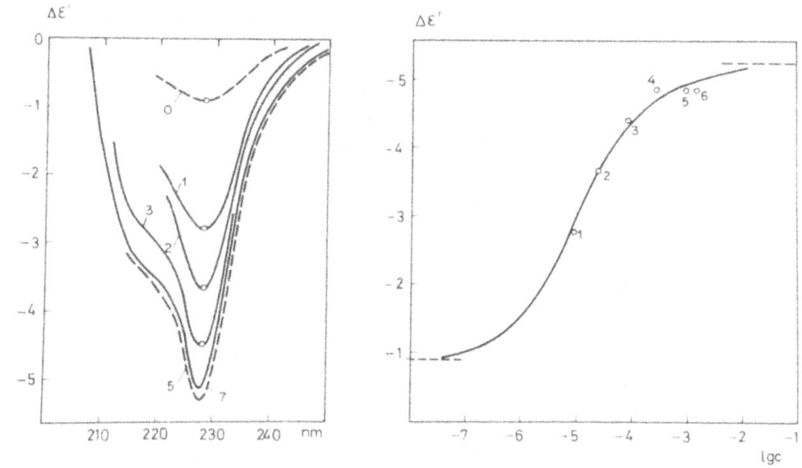

Fig. 16a,b. Concentrational dependence of (a) the CD curves and (b) Δε', for λ = 228 nm of analog (28) in dioxane. (---): monomer (O) and dimer (7)

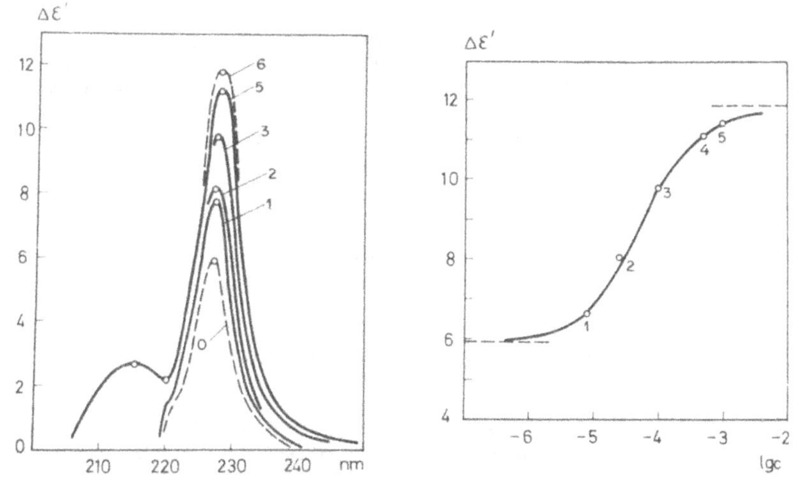

Fig. 17a,b. Concentrational dependence of (a) CD curves and (b) Δε' at λ = 228 nm for analog (26) in dioxane. (---): monomer (O) and dimer (6)

In the majority of cases, an increase in the dimer concentration is accompanied by changes in the CD curves, thus providing the means for following the dimerization process.

The CD parameters also provide a clue to both the monomer and the dimer conformations. Generally speaking, a clear-cut relationship between the CD curves and the chain length or the nature of the displaced residue can be observed only for some, not all the analogs. A comparison of the CD spectra of the analog dimers (the most favorable dimerizing conditions being concentrated solutions in dioxane) with those of the separate Gr A dimeric species showed that in certain analogs the conformational equilibrium is shifted toward one or the other of these species. From this standpoint, the analogs can be divided into four groups (Table 8). In one, consisting of analogs (25) and (28), the CD spectra are close to that of the Gr A species (3) (Fig. 16). Of particular interest here is analog (28) since its CD spectrum at concentrations for complete dimerization in dioxane can be described in terms of a combination of all four individual Gr A species, with species (3) amounting to 65% of the total.

A second group whose CD spectra closely resemble that of Gr A species (4) comprises analogs (26, 29, and 31), the most interesting being analog (31), which possesses the species (4) conformation to the extent of approximately 80%.

The two analogs (24) and (34) comprise a third group whose CD spectra resemble those of the Gr A species (1) and (2) but are about half their amplitude. These spectra cannot be described as a linear combination of the spectra of the four Gr A species.

The last group contains analogs (27, 30, and 33), characterized by the fact that the CD amplitudes are less than 0.2 of those characteristic of the Gr A dimers.

A comparison of the shift in the analog dimer equilibrium with the structure of the monomers uncovers a number of interesting facts. Thus, decrease in the relative Trp content is paralleled by augmentation of the species (3) content. In group 2 (Table 8) with predominant species (4) conformations, on the contrary, the Trp content is higher than for the other analogs. Moreover, in analog (31) Ala has been replaced by Val and Val by Leu, i.e., the less bulky hydrophobic residues have been replaced by more bulky ones. This all evidences that an essential part in stabilization of species (4) is played not only by H-bonds, but also by hydrophobic interaction.

Let us now, in the light of the new data presented here, consider the effect of Gr A and its synthesized analogs on the ion conductance of lipid bilayers.

It is well known that in the presence of very small amounts of Gr A the membrane current at constant voltage undergoes jumps, indicating the "opening" and "closing" of the conductance channels. At the same time the channels have definite conductance values, of which some are realized more often than others. We found (Fig. 18) that in egg lecithin membranes, Gr A forms four basic types of channels differing in both conductance and dwell times. It would, therefore, seem logical to relate these types to the four dimer species discussed previously. Grounds for this may be found in the fact that with the 13-membered analog (25), which in solution exists mainly in the form of a single type of dimer, we have observed mainly a single type of channel with a conductance of 7 pmho. At the same time, with another 13-membered analog (24) whose dimeric forms in solution differ from the Gr A species, the histogram characterizing the conductance distribution of the channel displays a wide range of values. The 15-membered analog (31) was also found to be active in bilayers, but its individual channels are characterized by a much larger range of conductance than might have been expected since its solution contains predominantly a single dimer species. True, whereas there are several different conducting types of this analog in the bilayer, the differences between them are much less than in the case of

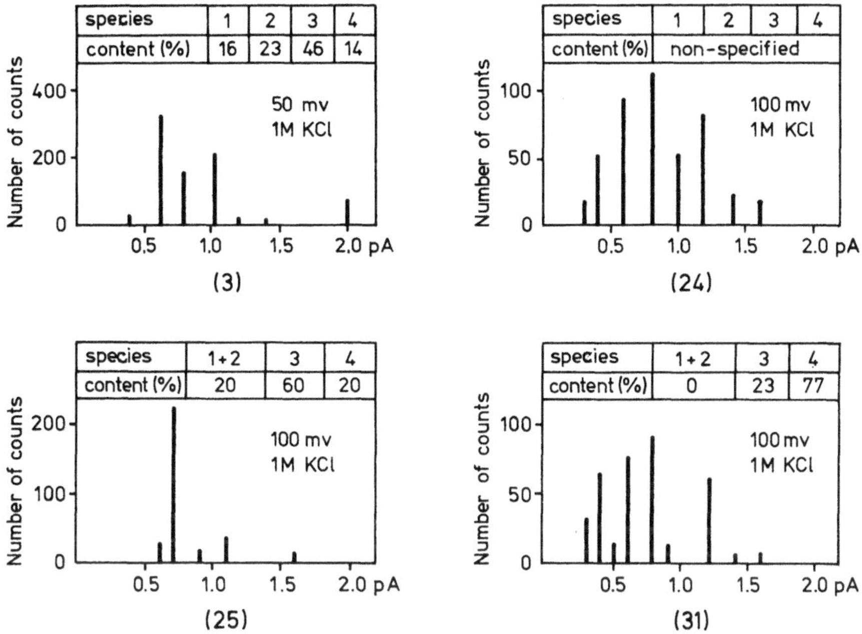

Fig. 18. Histograms of single channel conductances induced by Gr A (3) and its analogs (24), (25), and (31)

the naturally occurring Gr A, and we are inclined to ascribe them to small conformational differences that could have escaped detection in solution.

Other indirect support of the proposal advanced here may be seen in the fact that the CD curves of analog (33), which displays no conductance channels in bilayer membranes, are of low intensity, evidence of its lack of preferable conformations in solution. The fact that in this analog the L,D alteration of the asymmetric centers is violated shows that such ordering must be required for the formation of the Gr A-type channels.

Thus, the correlation that apparently exists between the number of different solution dimers and the number of different types of conductance channels serves as argument in favor of the concept that the conductance channels of the antibiotic are formed by means of its dimers and allows one to penetrate deeper into the conductance mechanism. Apparently, the dimer species does not seem to be crucial, but this requires further study.

References

Andreev, I.M., Malenkov, G.G., Shkrob, A.M., Shemyakin, M.M.: Conductometric study of complex formation of the cyclic depsipeptides and peptides with alcali metal ions. Mol. Biol. (USSR) 5, 614-622 (1971)

Fonina, L.A., Savelov, I.S., Avotina, G.Ya., Ivanov, V.T., Ovchinnikov, Yu.A.: Proc. XIV Eur. Pept. Symp. (Wepion, April 11-16, 1976), in press (1976)

Fossel, E.T., Veatch, W.R., Ovchinnikov, Yu.A., Blout, E.R.: A ^{13}C nuclear magnetic resonance study of gramicidin A in monomer and dimer forms. Biochemistry 13, 5264-5275 (1974)

Grell, E., Funck, T., Eggers, F.: Structure and dynamic properties of ion specific antibiotics. In: Membranes, Vol. III. Eisenmann, G. (ed.). New York-Basel: Marcel Dekker Inc. 1975, pp. 2-216

Haydon, D.A., Hladky, S.B.: Ion transport across thin lipid membranes: a critical discussion of mechanisms in selected systems. Quart. Rev. Biophys. 5, 187-282 (1972)

Ivanov, V.T.: "Sandwich" complexation in cyclopeptides and its implications in membrane processes. Ann. N.Y. Acad. Sci. 264, 221-243 (1975)

Ivanov, V.T., Evstratov, A.V., Sumskaya, L.V., Melnik, E.I., Chumburidze, T.S., Portnova, S.L., Balashova, T.A., Ovchinnikov, Yu.A.: Sandwich complexes as a functional form of the enniatin ionophores. FEBS Lett. 36, 65-71 (1973)

Ivanov, V.T., Sumskaya, L.V., Mikhaleva, I.I., Laine, I.A., Ryabova, I.D., Ovchinnikov, Yu.A.: Synthesis of valinomycin and enniatin B analogs containing charged, spin-labeled and fluorescent groups. Chim. Prirodn. Soed. (USSR), 346-358 (1974)

Ivanov, V.T., Fonina, L.A., Uvarova, N.N., Koz'min, S.A., Kropotnitskaya, T.B., Chekhlayeva, N.M., Balashova, T.A., Ovchinnikov, Yu.A.: Sandwich complexing of metal binding cyclopeptides and its biological implications. In: Peptides: Chemistry, Structure and Biology. Walter, R., Meienhofer, J. (eds.). Ann Arbor: Ann Arbor Sci. Publ. 1975, pp. 195-201

Markin, V.S., Pastushenko, V.F., Krishtalik, L.I., Liberman, E.A., Topaly, V.P.: Membrane potential and short circuit current of artificial phospholipid membranes in the presence of uncouplers of oxidative phosphorylation. Biophysika (USSR) 14, 462-473 (1969)

Ovchinnikov, Yu.A.: Membrane Active Complexones. Chemistry and Biological Function. FEBS Lett. 44, 1-21 (1974)

Ovchinnikov, Yu.A., Ivanov, V.T., Evstratov, A.V., Mikhaleva, I.I., Bystrov, V.F., Portnova, S.L., Balshova, T.A., Meshcheryakova, E.A., Tulchinsky, V.M.: The enniatin ionophores. Conformation and ion binding properties. Intern. J. Pept. Prot. Res. 6, 465-498 (1974a)

Ovchinnikov, Yu.A., Ivanov, V.T., Shkrob, A.M.: Membrane Active Complexones. Amsterdam: Elsevier 1974b

Shemyakin, M.M., Ovchinnikov, Yu.A., Ivanov, V.T., Antonov, V.K., Vinogradova, E.I., Shkrob, A.M., Malenkov, G.G., Evstratov, A.V., Ryabova, I.D., Laine, I.A., Melnik, E.I.: Cyclodepsipeptides as chemical tools for studying transport through membranes. J. Membrane Biol. 1, 402-430 (1969)

Shepel, E.N., Iordanov, St., Ryabova, I.D., Miroshnikov, A.I., Ivanov, V.T., Ovchinnikov, Yu.A.: Synthesis and antimicrobial activity of new gramicidin A analogs. Bioorgan. Chim. (USSR) 2, 581-593 (1976)

Sumskaya, L.V., Chekhlayeva, N.M., Barsukov, L.I., Terekhov, O.P., Djomin, V.V., Shkrob, A.M., Ivanov, V.T., Ovchinnikov, Yu.A.: Valinomycin and enniatin B functional derivatives as complexones and ionophores. Bioorgan. Chim. (USSR) 2, 351-364 (1976a)

Sumskaya, L.V., Balashova, T.A., Michaleva, I.I., Chumburidze, T.S., Melnik, E.I., Ivanov, V.T., Ovchinnikov, Yu.A.: Bis-derivatives of valinomycin and enniatin B as complexones and ionophores. Bioorgan. Chim. (USSR) 3, 5-21 (1977)

Veatch, W.R., Blout, E.R.: The aggregation of gramicidin A in solution. Biochemistry 13, 5257-5264 (1974)

Veatch, W.R., Fossel, E.T., Blout, E.R.: The conformation of gramicidin A. Biochemistry 13, 5249-5257 (1974)

Veatch, W.R., Mathies, R., Eisenberg, M., Styer, L.: Simultaneous fluorescence and conductance studies of planar bilayer membranes containing a highly active and fluorescent analog of gramicidin A. J. Mol. Biol. 99, 75-92 (1975)

Urry, P.W., Goodall, M.C., Glickson, J.D., Mayers, D.F.: The gramicidin a transmembrane channel: Characteristics of head-to-head dimerized $\pi_{(L,D)}$-helices. Proc. Natl. Acad. Sci. U.S. 68, 1907-1911 (1971)

Elementary Steps and Dynamic Aspects of Carrier-Mediated Cation Transport Through Membranes: the Streptogramin Antibiotics (Group B)

E. GRELL, I. OBERBÄUMER, H. RUF, and H. P. ZINGSHEIM

Summary

Certain streptogramin antibiotics of the B group strongly interact with membranes, a fact that was hitherto unknown. According to electrical measurements on phospholipid bilayer membranes the peptide-lactone virginiamycin S facilitates the transport of protons, alkali ions, and alkaline earth ions across these membranes, thus acting as a general cation mediator. The cation specificity is $H^+ >> Rb^+ > K^+ > Cs^+ > Na^+ \approx Ca^{2+}$. The pH-dependent interaction between phospholipid membranes (dispersions) and the antibiotic has been investigated employing spectroscopic methods. Virginiamycin S is bound predominantly to the surface region of the membrane; the dissociation constant of the membrane-bound compound is 7×10^{-6} M for the monodeprotonated neutral state and 3.7×10^{-5} M for the corresponding zwitterionic state whereas it is only 1×10^{-4} M for the negatively charged, twice deprotonated state of the antibiotic. About 250 phospholipid molecules form a binding site for one antibiotic molecule. The affinity to the membrane is caused mainly by interaction of the cyclic peptide-lactone ring of the antibiotic with the membrane.

The elementary steps as well as the dynamic aspects that are relevant to the mechanism of action of the proton and metal ion transport mediated by virginiamycin S have been investigated employing spectroscopic techniques and chemical relaxation methods. The kinetics of the proton transfer reactions of virginiamycin S have been studied in 33.8% water in methanol (vol/vol) and required a quantitative differentiation between the various protolytic states of the 3-hydroxypicolinyl residue of the antibiotic, which acts as proton acceptor and donor. In this medium its protolytic properties are similar to those shown when it is membrane-bound. At about pH 7, the overall proton transfer occurring at the 3-hydroxypicolinyl residue of the antibiotic is due mainly to an intramolecular process; the rate constant for proton transfer from the ring nitrogen to the phenolic oxygen is 850 sec^{-1} and that for the reverse process is 670 sec^{-1}. Evidence was found which suggested that slow conformational rearrangements of the peptide residue of virginiamycin S take place with rate constants of about 10^2 sec^{-1}. In nonpolar solvents and those of intermediate polarity, virginiamycin S forms stable complexes with cations. Complex formation between alkaline earth ions, and to a certain extent also between alkali ions and the antibiotic, is attributed to the twice and three times deprotonated protolytic states of the 3-hydroxypicolinyl residue; this is consistent with similar studies on simple model compounds.

A. Introduction

Many biological regulations of the function of biological membrane systems are coupled to their permeabilities to cations. Since the biological permeation systems exhibit a remarkable functional and structural complexity, much interest has been devoted to model membrane systems which show a cation specificity similar to that of biological membranes. In the model systems, the effect of antibiotics and synthetic compounds such as valinomycin, macrotetrolides, macrocyclic polyethers, X-537 A, 2,4-dinitrophenol, carbonylcyanide-3-chlorophenyl hydrazone (CCCP), and o-methyl red on the alkali ion and proton permeability of lipid bilayer membranes has been studied extensively in recent years by electrical measurements (Mueller and Rudin, 1967; Tosteson et al., 1967; Eisenman et al., 1968; Pedersen, 1967; Szabo et al., 1973; Schadt and Haeusler, 1974; Célis et al., 1974; Lea and Croghan, 1969; Hopfer et al., 1970, Le Blanc, 1971; Ruf, 1974, 1977a,b). The ability of valinomycin and other antibiotics to form complexes with alkali ions with selectivities similar to those of the corresponding membrane-bound compounds could also be detected in homogeneous solution and has been related to the special structural and conformational properties of these compounds (Shemyakin et al., 1969; Ovchinnikov et al., 1974; Grell and Funck, 1973; Grell et al., 1975).

The well-known action of the streptogramin antibiotics (Vazquez, 1975; Tanaka, 1975) is related to the interaction of these compounds with bacterial ribosomes, which leads to an inhibition of protein biosynthesis. This inhibition is in some cases sensitive to the presence of certain cations (Ennis, 1971, 1974). Much interest has been devoted to this family of antibiotics due to the synergistic effect of certain mixtures of streptogramins of group A and group B. Inhibition of respiration has also been reported in rat liver mitochondria by mikamycin (Dixon et al., 1971). This observation as well as the following structural considerations of the streptogramin antibiotics of group B suggested to us that these compounds could act as carriers for protons and metal ions through membranes. The general structure of the streptogramin antibiotics of the B group is shown in Figure 1a. The cyclic peptide-lactone ring consists of six to seven L- and D-amino acid residues with side chains exhibiting predominantly hydrophobic properties. Common to all the known compounds is a L-threonine residue that connects the peptide-lactone ring with the 3-hydroxypicolinyl residue. According to stereochemical considerations on models, the peptide-lactone ring can in principle adopt a conformation whereby a cavity is formed that is suitable for the coordination of cations and that is similar to that of cyclic peptides or depsipeptides exhibiting the well-known alkali-ion specificities. On the other hand, the 3-hydroxypicolinyl residue, which must exhibit the properties of a dibasic acid existing in four protolytic states, resembles several known proton carriers. Among these compounds are the weak acids such as nitrophenols and related compounds in which the negative charge of the deprotonated protolytic state is delocalized due to the π-electron system of the aromatic residue. In addition it is expected that the 3-hydroxypicolinyl residue acts as a suitable chromophore which permits the direct investigation of the possible elementary steps involved in the binding to the membrane and the cation transport processes in the membrane by sensitive optical methods. In order to test this hypothesis, electrical measurements of phospholipid bilayer membranes have been

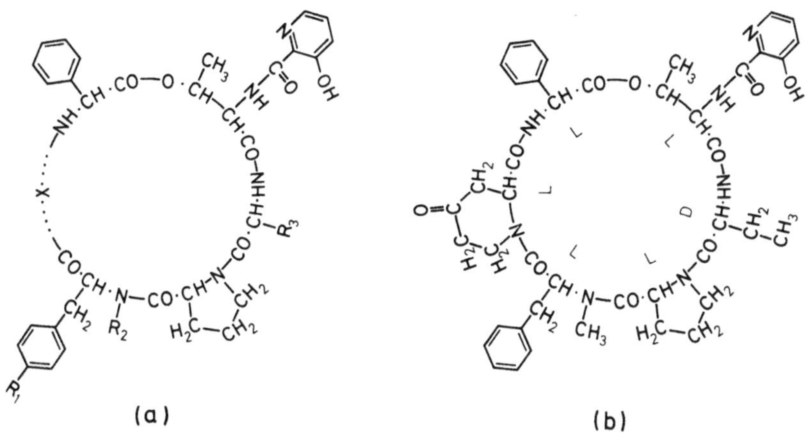

Fig. 1a,b. (a) General structure of streptogramin antibiotics of group B; (b) chemical formula of virginiamycin S

carried out in the presence of virginiamycin S (Vanderhaeghe and Parmentier, 1960), a typical representative of the streptogramin antibiotics of group B.

B. Conductance in Phospholipid Bilayer Membranes Mediated by Virginiamycin S

In order to test whether virginiamycin S facilitates the transport of cations, electrical conductivity was measured on lecithin bilayer membranes in the presence of tetramethylammonium chloride which does not affect membrane conductance. As shown in Figure 2, the membrane conductance under the conditions of zero current (G_o) increases remarkably upon the addition of a concentrated solution of the antibiotic dissolved in methanol. It is therefore concluded that virginiamycin S facilitates the transport of protons through membranes. For low virginiamycin S concentrations, linear dependence of G_o as a function of the concentration of the antibiotic is shown (cf. Fig. 2), which is consistent with a 1:1 stoichiometry; i.e., one proton is transported across the membrane by one antibiotic molecule. However, the corresponding slopes depend on pH, which indicates that the conductance must be sensitive to the population of the protolytic states of the 3-hydroxypicolinyl residue of the antibiotic. The dependence of G_o as a function of c_{Vir} in Figure 2a exhibits saturation behavior for the larger antibiotic concentrations. This implies that the number of free binding sites of the membrane cannot be considered large compared with those occupied by virginiamycin S. From the curvature it can be estimated that at pH 7.48 the binding constant for the antibiotic must be smaller than 10^{-4} M and larger than 10^{-5} M.

The pH dependence of the membrane conductance at constant antibiotic concentration may be seen from Figure 3. These measurements were carried out under conditions where a partial saturation of the free membrane binding sites occurs (cf. Fig. 2) Since the current-voltage dependence of the membrane in the presence of the

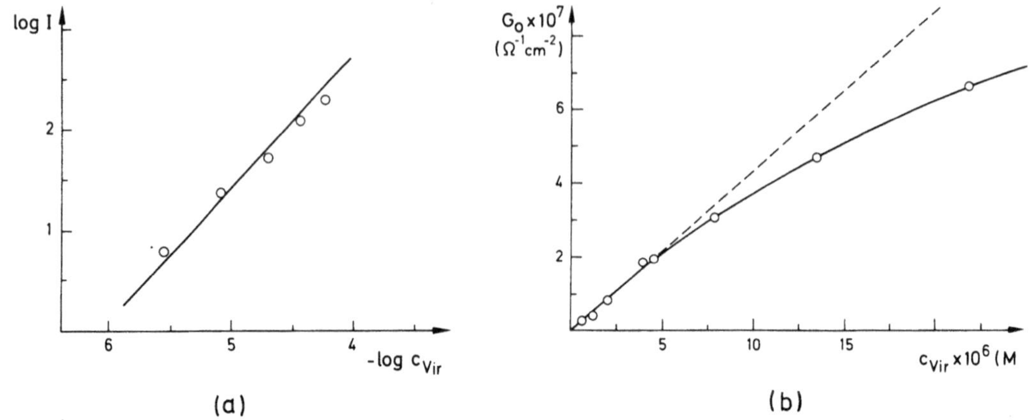

Fig. 2a,b. Dependence of conductance of virginiamycin S on antibiotic concentration: (a) log/log plot of current [I(pA)] of membrane made from egg-lecithin/cholesterol in n-decane (area ≈ 1.2 mm^2) at potential of 50 mV vs. concentration of virginiamycin S in presence of 3×10^{-3} M bis-tris/HCl pH 3.9 containing 0.1 M tetramethylammonium chloride at 25°C; (b) plot of conductance at zero-current (G_o) of membrane made from α,β-dioleoyl-L-lecithin in n-decane vs. concentration of virginiamycin S in presence of 5×10^{-3} M bis-tris/HCl pH 7.48 containing 0.1 M tetramethylammonium chloride at 25°C

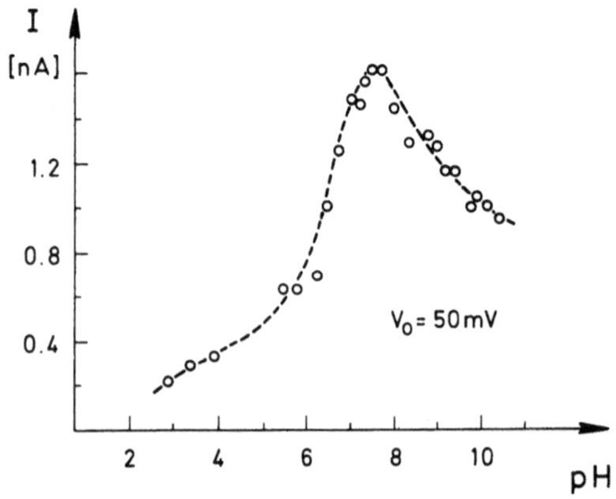

Fig. 3. Dependence of current (I) of membrane made from egg-lecithin/cholesterol in n-decane (area ≈ 1.2 mm^2) at a potential of 50 mV as function of pH in presence of 1.2×10^{-4} M virginiamycin S in 3×10^{-3} M bis-tris/HCl containing 0.1 M tetramethylammonium chloride at 25°C

antibiotic is linear in the potential range ± 60 mV, the current plotted in Figure 3 is directly proportional to the zero-current conductance. The conductance at low and high pH values is relatively small and reaches a maximum value [$(G_o)_{max}$] at pH 7.7, which is close to the pK_{II} value of the membrane-bound antibiotic (cf. Table 3). Taking into account the effect of the saturation of the binding sites and the change in proton concentration, it is possible to estimate that virginiamycin S is about 10^5 times more effective with respect to proton conductance at pH 7 than at pH 3.

The pH dependence of the membrane conductance in the presence of virginiamycin S can be explained according to the model considerations for deprotonated and negatively charged proton carrier molecules (Le Blanc, 1971; Haydon and Hladky, 1972; Ruf, 1974, 1977b). Two types of mechanism must be considered for the charge transport through the membrane. At high pH values (pH >> 7.7) the antibiotic in its negatively charged protoylytic state (cf. Fig. 8) is simply transferred from one aqueous compartment to the other according to the directed transport of a lipophilic anion. This transport mechanism predominates at high pH values (pH >> 7) and is consistent with the observation that the conductance at pH 9 is considerably greater than that at pH 6 (cf. Fig. 3). Below pH 9 a cyclic carrier mechanism must be introduced in order to explain the observed pH dependence of the conductance, which is characterized by a maximum at pH 7.7. This implies that the negatively charged, twice deprotonated state of the carrier is responsible for the flux of current through the membrane in the presence of a potential. At the interface of the membrane the negatively charged carrier, which is still bound to the membrane, can accept a proton from the aqueous phase. The loaded carrier can then diffuse to the opposite side of the membrane where the proton can be released. In the case of virginiamycin S, which represents the only known peptide compound that acts as a proton carrier in lipid bilayer membranes, the twice deprotonated 3-hydroxypicolinyl residue (protolytic state P_4^- in Fig. 8) acts as the proton acceptor whereas the monodeprotonated residue (protolytic states $P_2^{\pm}$ and P_o^3 in Fig. 8) acts as the proton donor. It must be expected that the permeability of the neutral protolytic state (P_o^3) of the antibiotic is considerably greater than that of the charged, zwitterionic state ($P_2^{\pm}$). Thus the loaded carrier in the protolytic state P_3^o preferentially penetrates the bilayer (cf. Figs. 8 and 29). A comparison with other proton carriers indicates that virginiamycin S is one of the most powerful to have been found so far. The maximum conductance at the pH optimum relative to the total concentration of the antibiotic [$(G_o)_{max}/c_{Vir}$] is about $7.5 \times 10^1\ \Omega^{-1}$ cm mole^{-1} under circumstances were no saturation of the binding sites at the membrane is observed. The corresponding values for CCCP, 2,4-dinitrophenol, and o-methyl red are $6.8 \times 10^3\ \Omega^{-1}$ cm mole^{-1}, $1.4 \times 10^1\ \Omega^{-1}$ cm mole^{-1}, and $5.6 \times 10^1\ \Omega^{-1}$ cm mole^{-1} (cf. Ruf, 1974, 1977a,b).

The conductance of bilayer membranes in the presence of virginiamycin S is also increased upon the addition of salts from alkali or alkaline earth ions (cf. Fig. 4). The slopes of the logarithmic plots of conductance vs. cation concentration can exceed the value of unity, and the pH dependence of the conductance differs also from that found in the absence of the cations mentioned previously. This will, however, not be discussed here. The cation specificity of virginiamycin S in lecithin membranes has been studied by measuring the membrane potential at zero current (V_o) as a function of the concentration of an alkali or an alkaline earth salt added to one of the compartments only (cf. Fig. 5). The permeability ratio (β) of a cation (I) can be determined relative to the permeability (P) of the proton according to Eq. (1) (cf. Szabo et al., 1973):

$$V_o = \frac{RT}{F} \ln \frac{c_H'' + \beta c_I''}{c_H'} \quad \text{where } \beta = \frac{P_I}{P_H} \tag{1}$$

c' denotes the concentration in one of the aqueous compartments and c" that in the other. The values of the relative permeabilities

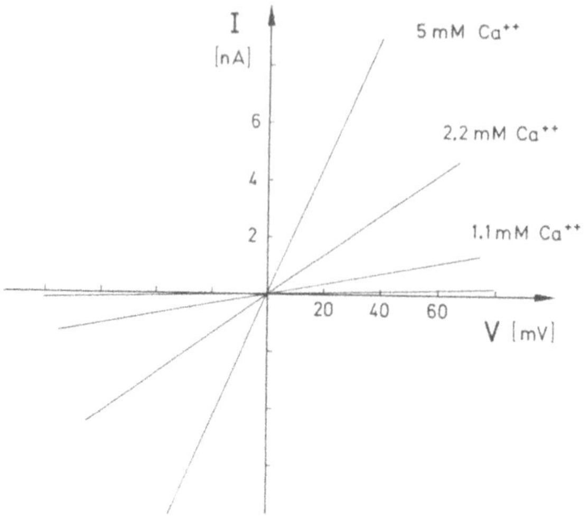

Fig. 4. Current-voltage dependence of membrane made from α,β-dioleoyl-L-lecithin in n-decane (area ≈ 1.2 mm^2) in presence of 3.4 × 10^{-5} M virginiamycin S, 5 × 10^{-3} M bis-tris/HCl pH 7.5, 0.1 M tetramethylammonium chloride, with and without several additions of $CaCl_2$ at 25°C

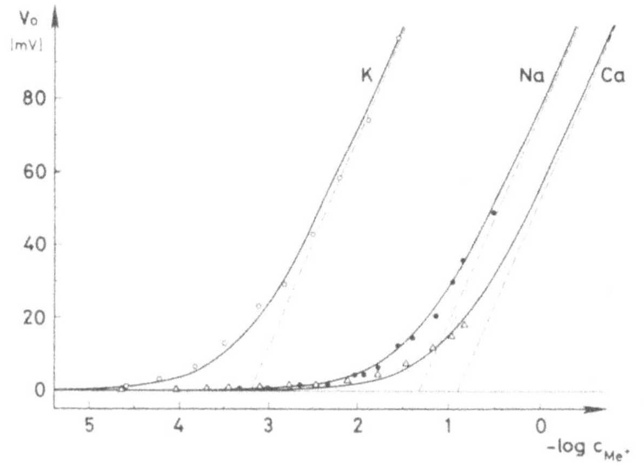

Fig. 5. Cation specificity of virginiamycin S: Plot of potential (zero-current) of membranes made from α,β-dioleoyl-L-lecithin in n-decane in presence of 3.4 × 10^{-5} M virginiamycin S in 5 × 10^{-3} M bis-tris/HCl pH 7.5, 0.1 M tetramethylammonium chloride as function of negative logarithm of NaCl, KCl, and $CaCl_2$ concentrations at 25°C

Table 1. Cation specificity of virginiamycin S at pH 7.5: Permeability ratio (P_I/P_H) relative to permeability of H^+ for α,β-dioleoyl-L-lecithin membranes at 25°C

Cation	Na^+	K^+	Rb^+	Cs^+	Ca^{2+}
$\frac{P_I}{P_H}$	$6.9 \pm 3 \times 10^{-7}$	$5.2 \pm 3 \times 10^{-5}$	$2.0 \pm 1.5 \times 10^{-4}$	$7.1 \pm 1.5 \times 10^{-6}$	$2.6 \pm 3 \times 10^{-7}$

that have been determined for several cations in the presence of virginiamycin S are given in Table 1. The cation specificity of virginiamycin S at pH 7.5 is thus $H^+ >> Rb^+ > K^+ > Cs^+ > Na^+ \approx Ca^{2+}$. A corresponding plot of the permeability ratio vs. the ionic radius of the cation is given in Figure 6a together with a similar plot for valinomycin (Fig. 6b).

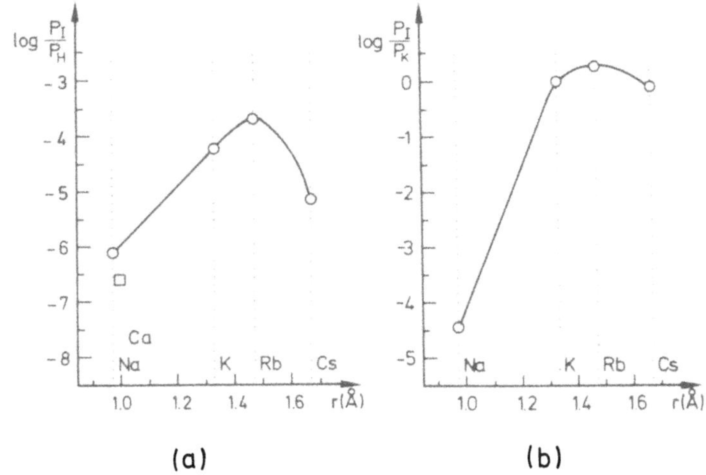

Fig. 6a,b. Cation specificity of (a) virginiamycin S at pH 7.5 and (b) of valinomycin (after Szabo et al., 1973): Plot of logarithm of permeability ratios relative to H^+ and K^+ permeability, respectively, vs. ionic radius of cations

Although the K^+/Na^+ selectivity of valinomycin is considerably larger than that of virginiamycin S, the latter exhibits a remarkable $Rb^+/K^+/Cs^+$ selectivity. This can be easily rationalized only if we assume that alkali ions are coordinated in a cavity that is formed by the peptide-lactone ring. Even though Ca^{2+} is not transported more efficiently than Na^+, we have to realize that virginiamycin S represents the only known peptide compound that facilitates the transport of alkaline earth ions through bilayer membranes. It is therefore concluded that virginiamycin S acts as a general cation carrier, being most efficient in the transport of protons, which is consistent with results obtained for other streptogramin antibiotics of the B group; this is shown in a schematic illustration in Figure 7. Since virginiamycin S facilitates the transport of more than one type of cation, this opens the possibility of carrying out interesting model studies under conditions where the coupling between the various fluxes is of importance, such as the transport of a cation against its concentration gradient induced by the existence of another concentration gradient. The present results also raise the question of whether the action of these antibiotics in the biological systems is restricted to the inhibition of protein biosynthesis or to the inhibition of mitochondrial respiration. In the following, the elementary steps as well as the dynamic aspects that are relevant to the mechanism of action of the cation transport mediated by these streptogramin antibiotics have been investigated employing spectroscopic techniques and chemical relaxation methods.

C. States and Elementary Steps Involved in Virginiamycin S-Mediated Cation Transport

I. Protolysis and Proton Transfer in Homogeneous Solution

Since the antibiotic virginiamycin S acts as a proton carrier in lipid bilayer membranes, proton transfer must be considered as an important elementary step in the overall transport process. The four protolytic states of the 3-hydroxypicolinyl residue of virginiamycin S are shown in Figure 8a. The 3-hydroxypicolinyl residue exists in two different monodeprotonated states, $P_2^{\pm}$ and P_3^{o}.

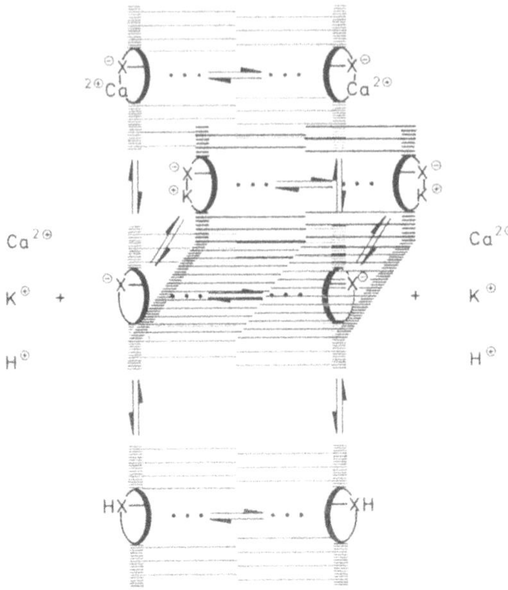

Fig. 7. Schematic illustration of virginiamycin S-induced H^+ and metal ion transport through phospholipid membrane. Residue XH represents monoprotonated protolytic states of 3-hydroxypicolinyl residue. Stoichiometry shown in this figure is not expected to correspond to the real one

The two states differ in the location of the proton, which can be bound either to the phenolic oxygen or to the ring nitrogen. Prior to any kinetic studies the thermodynamic parameters of the protolytic reactions of virginiamycin S must be determined.

The two apparent macroscopic dissociation constants, K_I and K_{II}, are determined by spectrophotometric titrations in the wavelength region 250 to 450 mμ where the absorption spectrum is related only to the 3-hydroxypicolinyl chromophore in a variety of solvents (Oberbäumer, 1975). An example of such a spectrophotometric pH titration of virginiamycin S in water is shown in Figure 9. The evaluation of this titration for the determination of the two macroscopic dissociation constants is illustrated in Figure 10. In order to calculate the four microscopic dissociation constants K_{12}, K_{13}, K_{24}, and K_{34} (cf. Fig. 8a), the equilibrium constant K_{23} must be known:

$$K_I = K_{12} + K_{13} \qquad K_{II}^{-1} = K_{24}^{-1} + K_{34}^{-1} \tag{2}$$

$$K_{23} = \frac{k_{32}}{k_{23}} = \frac{K_{12}}{K_{13}} = \frac{K_{34}}{K_{24}} \tag{3}$$

The equilibrium constant K_{23} is evaluated from the areas of the absorption bands of the states $P_2^{\pm}$ and P_3^{o}, which predominate between pH 3 and 5. This is done by computer simulation of the absorption spectra in the form of gaussian curves. The large spectral changes observed in this pH range upon the transfer of virginiamycin S from methanol to water (Fig. 11) reflect the sensitivity of the equilibrium constant K_{23} to the polarity of the medium. This implies that the spectroscopic properties of the antibiotic can also be utilized for studying the polarity of the medium surrounding the membrane-bound compound, which may provide

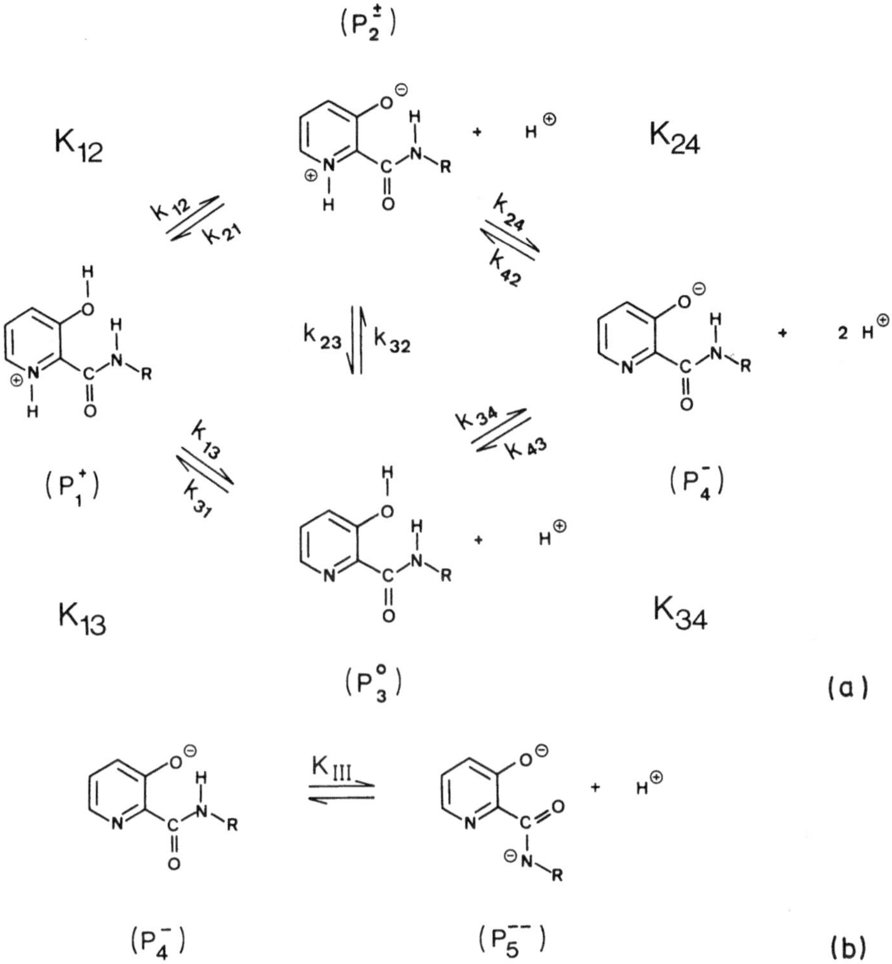

Fig. 8a,b. Protolytic states of virginiamycin S (R: peptide-lactone residue) (a) below pH 9.5 [$K_{12} = k_{12}/k_{21}$, $K_{13} = k_{13}/k_{31}$, $K_{24} = k_{24}/k_{42}$, $K_{34} = k_{34}/k_{43}$] and (b) above pH 9.5

information about the localization of the antibiotic in the membrane. As shown in Figure 12, a linear dependence results from the plot of -log K_{23} vs. the mole fraction of water (γ_{H_2O}) in the solvent system methanol/water. Thus the microscopic dissociation constants can be calculated. Some representative data are given in Table 2, together with data for the model compound 3-hydroxypicolinamide. In addition the dependencies of the various pK values of virginiamycin S upon γ_{H_2O} in the solvent system methanol/water are shown in Figure 13. These data will be of interest in the interpretation of the corresponding parameters for the membrane-bound antibiotic.

As shown in Figure 9c an increase in the hydroxyl ion concentration above pH 9 leads to further changes in the absorption spectrum that are characterized by a small blue shift (≈ 8 mμ) of the absorption band of the protolytic state P_4^- at about 339 mμ. Ac-

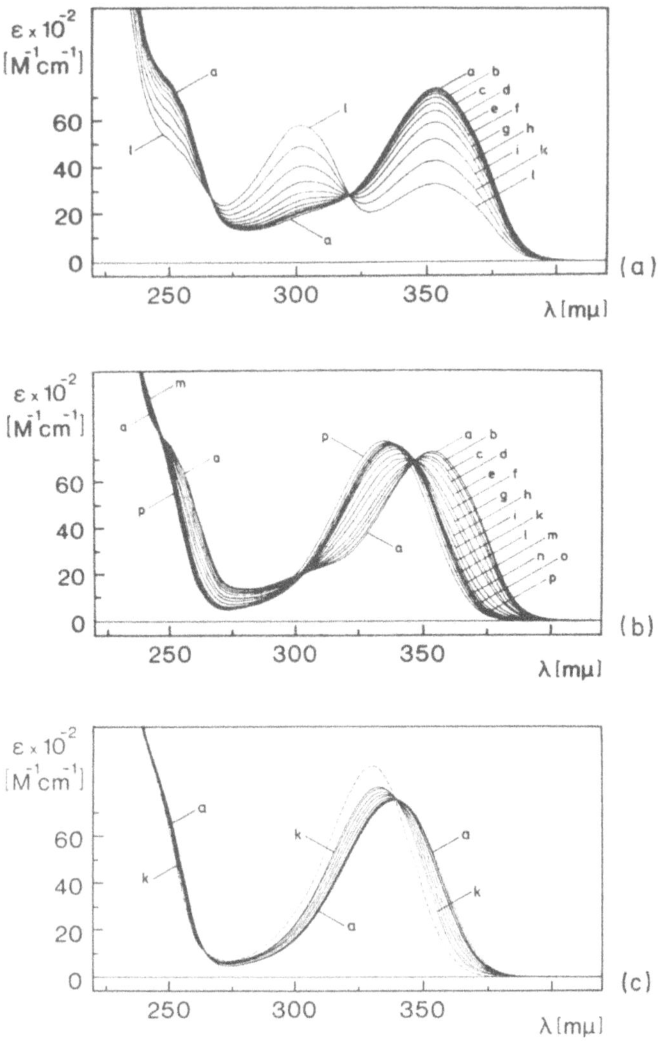

Fig. 9a-c. Spectrophotometric pH-titration of virginiamycin S in water containing 5 × 10^{-2} M tetramethylammonium chloride and 6 × 10^{-5} M HCl (a) with 2.29 M perchloric acid from pH 4.91 (spectrum a) to pH 1.22 (spectrum l), (b), with 0.35 M tetramethylammonium hydroxide from pH 4.48 (spectrum a) to pH 9.50 (spectrum m) and to pH 11.42 (spectrum p), (c), with 0.5 M sodium hydroxide from pH 8.88 (spectrum a) to pH 11.97 (spectrum k). Broken line in (c): spectrum of virginiamycin S extrapolated to end of titration

cording to similar titrations with simple model compounds, this spectral change represents the dissociation of the amide proton of the L-threonine residue, which is attached to the 3-hydroxy-picolinyl moiety, thus leading to the protolytic state $P_5^{=}$ (cf. Fig. 8b). The corresponding pK_{III} value is about 11.7 (cf. Fig. 10, Table 2).

The stationary conductance measurements on α,β-dioleoyl-L-lecithin bilayer membranes in the presence of the antibiotic indicated that the protolytic state P_4^- acts as a proton acceptor at the membrane interphase. Since the permeability of the membrane-bound antibiotic is expected to be larger in state P_3^o than in state $P_2^{\pm}$, the proton is preferentially carried across the membrane by the state P_3^o. After the unloading process the carrier action is completed by back diffusion of the antibiotic in the protolytic state P_4^-. Since the state P_3^o predominantly carries the proton across the membrane, the transport process must be coupled to a pre-equilibrium of the membrane-bound antibiotic that brings about proton

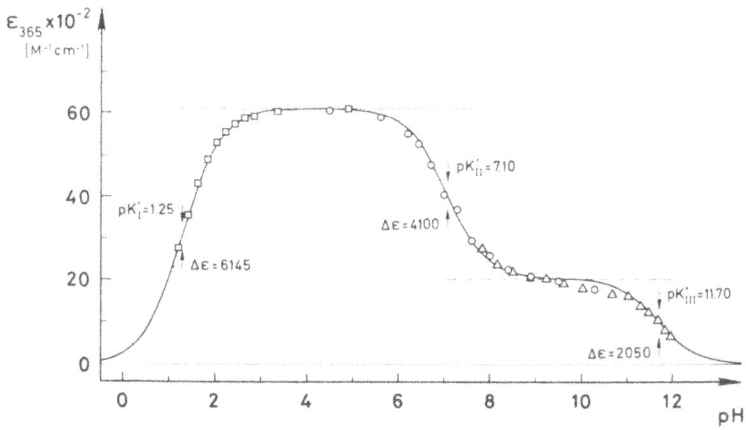

Fig. 10. Spectrophotometric pH-titration of virginiamycin S in water: Plot of extinction coefficient at 365 mμ vs. pH. Theoretical titration curves have been calculated for indicated pK-values (experimental conditions correspond to those specified in Fig. 9)

Fig. 11. Absorption spectra of virginiamycin S (2.31×10^{-5} M) in water (———), 70% (------), 40% (— — —), 20% (·········), 10% (—·—·—·—·), water in methanol (vol/vol) and in methanol (━━━━) containing 5×10^{-2} M tetramethylammonium chloride and 6×10^{-5} M HCL at 25°C

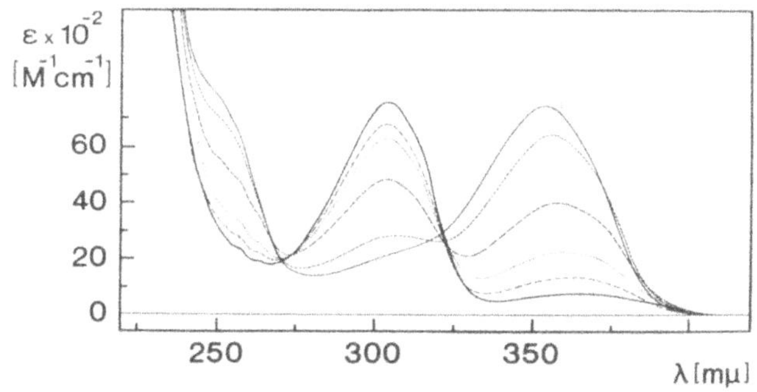

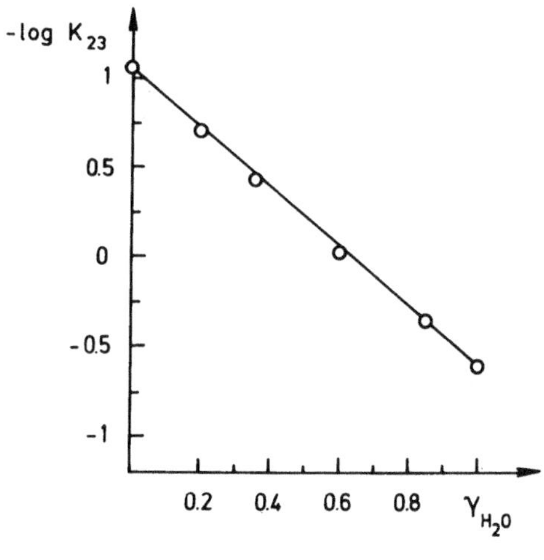

Fig. 12. Plot of negative logarithm of K_{23} vs. mole fraction of water in solvent system methanol/water

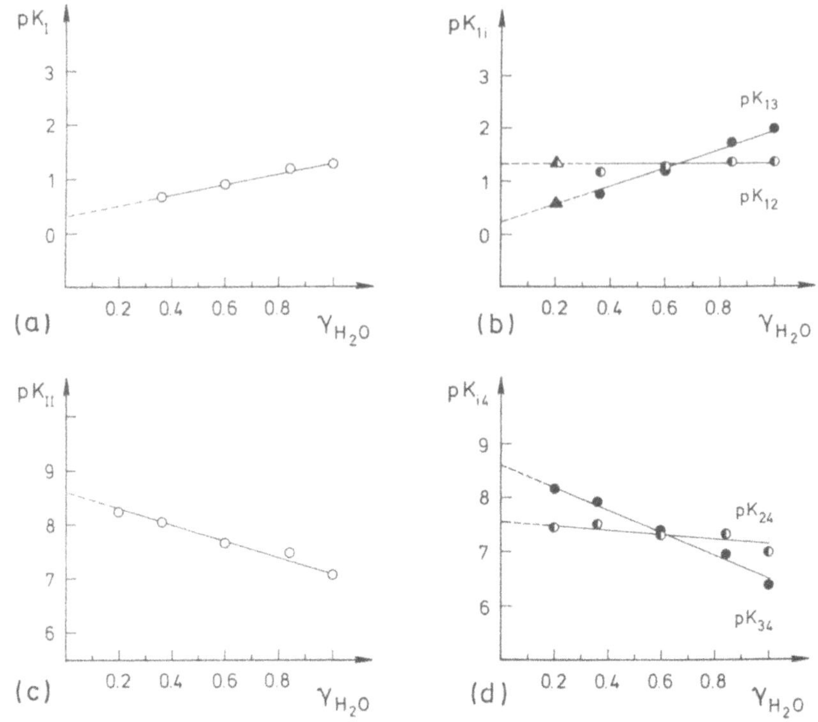

Fig. 13a-c. Dependencies of dissociation constants of virginiamycin S on solvent composition: Plots (a,c) of the macroscopic dissociation constants pK_I and pK_{II} and (b,d) of the microscopic dissociation constants pK_{1i} and pK_{i4} ($i = 2,3$) of virginiamycin S vs. mole fraction of water in solvent system methanol/water at 25°C

transfer between states $P_2^{\mp}$ and P_3^{o}. This proton transfer could occur via the states P_1^{+} or P_4^{-} by H^+ or OH^- catalysis, respectively, via intermolecular proton transfer with participation of the buffer molecules, or by intramolecular proton transfer. It is therefore important to determine the rate constants of the various reaction pathways. With respect to the reaction at the membrane surface it could be that the proton transfer occurs in a region of the membrane interface where the local concentrations of H^+, OH^-, and buffer molecules are small compared to their concentrations in the bulk solution. This implies that intramolecular proton transfer dominates.

Since the protolytic reactions of the 3-hydroxypicolinyl residue of virginiamycin S can be followed by absorption spectroscopy (cf. Fig. 9), absorption detection can also be used for studying the kinetics of proton transfer of virginiamycin S by temperature jump relaxation spectrometry (this method has been described by Eigen and De Maeyer, 1973). Three relaxation processes can be detected. The relaxation time of the fastest process ($\tau_1 \leq 3$ μsec) cannot be resolved and is attributed to the diffusion controlled recombinations of the populated protolytic states with H^+ and OH^-. The longer of the two resolvable relaxation times ($\tau_3 \approx 10$ msec, cf. Fig. 14c) is due to a pH-independent, conformational rearrangement of the peptide part of the antibiotic. This relaxation time cannot be attributed to a slow proton transfer reaction; it is

Table 2. Thermodynamic parameters related to protolytic reactions of virginiamycin S in different solvents and in membrane-bound state and of 3-hydroxypicolinamide in water; 25-27°C

Compound	Solvent	pK_I	pK_{12}	pK_{13}	K_{23}	pK_{II}	pK_{24}	pK_{34}	pK_{III}
Virginiamycin S	Methanol	0.33	1.43	0.36	0.09	8.60	7.50	8.56	-
	33.8% water in methanol (vol/vol)	0.84	1.20	1.09	0.79	7.80	7.44	7.54	-
	Water	1.27	1.37	1.98	4.07	7.08	6.98	6.37	11.70
	Membrane-bound state	-	-	-	0.77	7.80	7.44	7.55	-
3-Hydroxypicolinamide	Water	2.22	2.72	2.29	0.46	6.48	5.98	6.32	9.68

independent of the concentration of virginiamycin S and pH, does not depend on the presence of buffer, and cannot be observed with the model compound 3-hydroxypicolinamide. The sum of the forward and backward rate constants of the conformational rearrangement is $1.5 \pm 1 \times 10^2$ sec^{-1}. It has already been suggested that compounds related to virginiamycin S can undergo slow conformational rearrangements (Urry and Ohnishi, 1970; Anteunis et al., 1975).

The second relaxation time ($\tau_2 \approx 500$ µsec at a pH of approximately 7; cf. Fig. 14b) will be considered in more detail since it can be attributed to the overall proton transfer reaction. Within the pH range 3.5 to 9 the relaxation time τ_2 is independent of the virginiamycin S concentration but is sensitive to pH as shown by

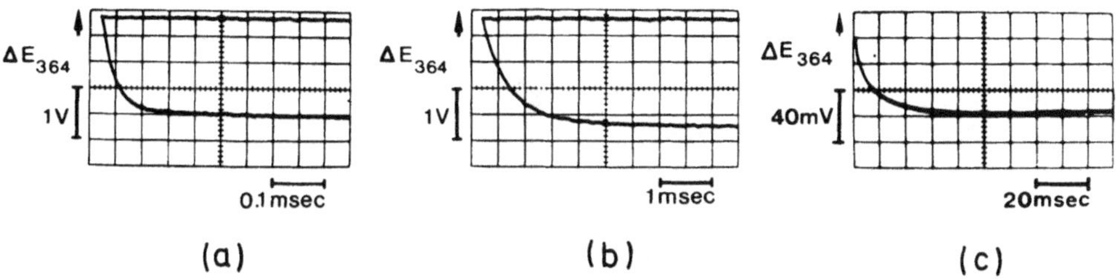

Fig. 14a-c. Kinetics of proton transfer and conformational rearrangement of virginiamycin S: Oscillograms of temperature jump relaxation experiments ($\delta T \approx 5$°C; total signal ≈ 25 V, 23°C) with (a) 6.25×10^{-5} M virginiamycin S in 33.8% water in methanol (vol/vol) in presence of 0.2 M tetramethylammonium chloride, 5×10^{-2} M acetic acid/tetramethylammonium hydroxide pH 6.42; (b) same experimental conditions but in absence of acetate buffer; (c) 8×10^{-5} M virginiamycin S in 40% water in methanol (vol/vol) in presence of 0.2 M tetraethylammonium chloride, 5×10^{-2} M acetic acid/tetraethylammonium hydroxide pH 5.45

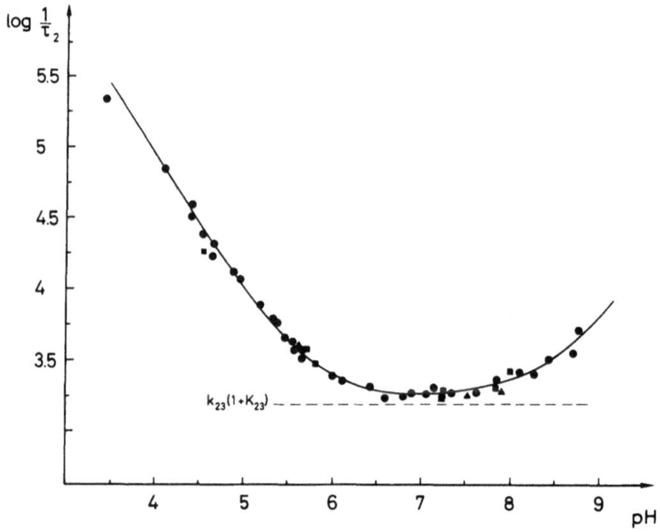

Fig. 15. Kinetics of proton transfer of virginiamycin S: Plot of reciprocal relaxation time τ_2 vs. pH

Figure 15. However, τ_2 is smaller in the presence of buffer (cf. Fig. 14a) indicating that the buffer molecules participate in the proton transfer reactions.

In the absence of buffer, several parallel reactions contribute to the overall proton transfer process. In addition to the intramolecular proton transfer, which is characterized by the rate constants k_{23} and k_{32}, there are intermolecular proton transfer reactions in which H^+ or OH^- participate:

$$H^+ + P_2^{\pm} \underset{k_{12}}{\overset{k_{21}}{\rightleftharpoons}} P_1^+ \underset{k_{31}}{\overset{k_{13}}{\rightleftharpoons}} P_3^{\circ} + H^+ \tag{4}$$

$$HO^- + P_1^+ \underset{k_{21}^*}{\overset{k_{12}^*}{\rightleftharpoons}} P_2^{\pm} \underset{k_{42}}{\overset{k_{24}}{\rightleftharpoons}} P_4^- + H^+ \tag{5}$$

$$OH^- + P_1^+ \underset{k_{31}^*}{\overset{k_{13}^*}{\rightleftharpoons}} P_3^{\circ} \underset{k_{43}}{\overset{k_{34}}{\rightleftharpoons}} P_4^- + H^+ \tag{6}$$

$$HO^- + P_3^{\circ} \underset{k_{43}^*}{\overset{k_{34}^*}{\rightleftharpoons}} P_4^- \underset{k_{24}^*}{\overset{k_{42}^*}{\rightleftharpoons}} P_2^{\pm} + OH^- \tag{7}$$

Furthermore, intermolecular proton transfer reactions can be expected to take place as a consequence of collisions between antibiotic molecules present in different protolytic states:

$$P_2^{\pm} + P_1^{+} \underset{k_{-A}}{\overset{k_A}{\rightleftharpoons}} P_3^{o} + P_1^{+} \tag{8}$$

$$P_2^{\pm} + P_4^{-} \underset{k_{-D}}{\overset{k_D}{\rightleftharpoons}} P_3^{o} + P_4^{-} \tag{9}$$

The contributions of the latter processes to the mathematical expression for the relaxation time τ_2 should lead to a dependence of τ_2 on the antibiotic concentration. However, within the experimental error of the present investigations, τ_2 is observed to be independent of the virginiamycin S concentration between pH 3.5 and 9. Thus the contributions of reactions (8) and (9) can be neglected and only upper limits for the rate constants k_A, k_{-A}, k_D, and k_{-D} can be given (Table 3). Under steady state conditions with respect to the protolytic states P_1^+ and P_4^-, a general expression for the second relaxation time τ_2, which can be attributed to the remaining proton transfer reactions, is given by Schuster et al. (1976). A similar, but less general treatment has also been derived by Ahrens (1973). Except for k_{23} and k_{32}, the magnitudes of the individual terms of the general expression can be estimated from the known values of general protolytic recombination rate constants and by using the microscopic dissociation constants (Table 2) to calculate the rate constants of the reverse processes (cf. Grell and Oberbäumer, 1976). For pH ranging from 3 to 6 (pH < 7) $k_{13}^* c_{OH^-}$, $k_{12}^* c_{OH^-}$, $k_{24}^* c_{OH^-}$, $k_{34}^* c_{OH^-}$, and the rate constants k_{21}^*, k_{31}^*, k_{43}^*, and k_{42}^* can be neglected compared to other terms in the general expression for τ_2^{-1}. This leads to the following approximation:

pH < 7

$$\tau_2^{-1} = k_{23} + k_{32} + \frac{k_{13}k_{21} + k_{12}k_{31}}{k_{12} + k_{13}} c_{H^+} + \frac{k_{24}k_{43} + k_{34}k_{42}}{k_{42} + k_{43}} \tag{10}$$

$$C_1 = \frac{k_{24}k_{43} + k_{34}k_{42}}{k_{42} + k_{43}}$$

The constant term, C_1, in Eq. (10) is estimated to be about 1.0×10^3 sec^{-1} and cannot be neglected.

In the alkaline region ranging from pH 8 to 10 (pH > 7) $k_{13}^* c_{OH^-}$, $k_{12}^* c_{OH^-}$, $k_{43} c_{H^+}$, $k_{42} c_{H^+}$, and the rate constants k_{21}^*, k_{31}^*, k_{24}, k_{34} are small compared to the other terms of the general expression for τ_2^{-1} and can also be neglected. Thus an approximation, Eq. (11), is obtained:

pH > 7

$$\tau_2^{-1} = k_{23} + k_{32} + \frac{k_{13}k_{21} + k_{12}k_{31}}{k_{12} + k_{13}} c_{H^+} + \frac{k_{24}^*k_{43}^* + k_{34}^*k_{42}^*}{k_{42}^* + k_{43}^*} c_{OH^-} \qquad (11)$$

The third term in Eq. (11) decreases with increasing pH and is about 10^2 sec^{-1} at pH 8. It is therefore small compared to the other terms and can be neglected so that the following approximation, in which the reciprocal relaxation time τ_2 is linearly dependent on c_{OH^-}, results:

pH > 7

$$\tau_2^{-1} = (1 + K_{23})\ (k_{23} + k_{OH}c_{OH^-}) \qquad k_{OH} = \frac{k_{24}^*\ k_{43}^*}{k_{42}^*+k_{43}^*} \qquad k_{-OH} = \frac{k_{34}^*\ k_{42}^*}{k_{42}^*+k_{43}^*} \qquad (12)$$

According to Eq. (10), for the acidic pH range the following is obtained:

pH < 7

$$\tau_2^{-1} = (1 + K_{23})\ (k_{23} + k_H c_{H^+}) + C_1 \qquad k_H = \frac{k_{13}\ k_{21}}{k_{12}+k_{13}} \qquad k_{-H} = \frac{k_{12}\ k_{31}}{k_{12}+k_{13}} \qquad (13)$$

Eq. (12) and (13) allow an approximate evaluation of the present experimental data under the limiting conditions mentioned previously. The corresponding dependence of the reciprocal relaxation time τ_2 upon c_{H^+} or c_{OH^-} in the acidic or alkaline pH ranges, respectively, is shown in Figure 16. Since K_{23} is known, the intercept of the plot in Figure 16a yields k_{23}. Thus C_1 can be cal-culated from the intercept of the plot in Figure 16b and is found to be 1.1×10^3 sec^{-1}, which is consistent with the estimated value. The overall rate constants, k_H and k_{OH}, are calculated from the slopes in the relevant figure. The kinetic parameters obtained in the absence of buffer are summarized in Table 3.

According to this kinetic data, it is evident that in the acidic and alkaline pH regions intermolecular proton transfer occurs with dominant participation of H^+ or OH^-, respectively. On the other hand, at about pH 7 the overall proton transfer is mainly due to the intramolecular process. The rate constants of the intramolecular proton transfer k_{23} and k_{32} are comparable to those of the estimated protolytic dissociation rate constants k_{24} and k_{34}. With respect to the intramolecular process this indicates that, e.g., the proton bound to the phenolic oxygen must first dissociate before it can be transfered to the ring nitrogen, implying that no

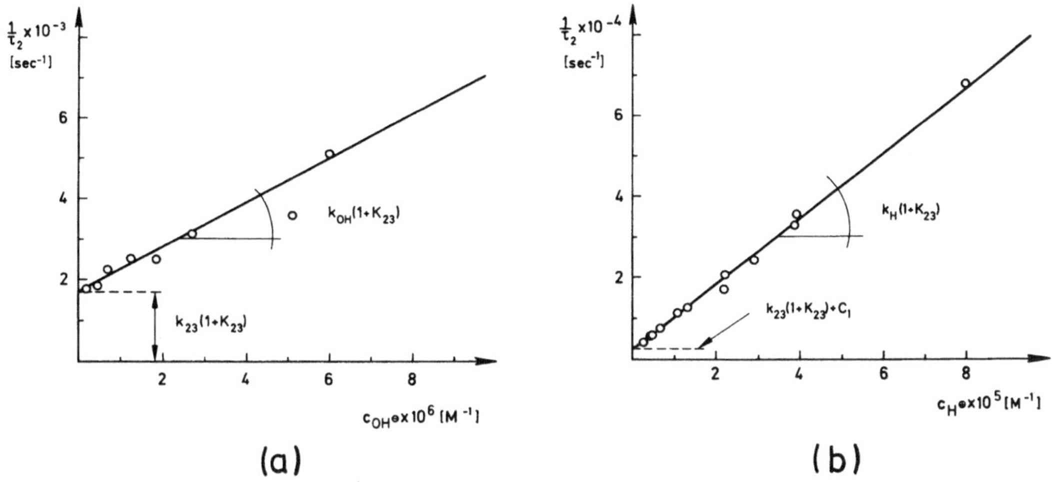

Fig. 16a,b. Evaluation of kinetic parameters for proton transfer of virginiamycin S: Plot of reciprocal relaxation times vs. concentration of (a) hydroxyl ions and (b) protons

special intramolecular proton transfer mechanism, such as participation of the neighboring carbonyl group, exists. It has already been mentioned that τ_2 does not depend on the virginiamycin S concentration as is observed for the model compound 3-hydroxypicolinamide. It is therefore concluded that the corresponding rate constants k_A, k_{-A} and k_D, k_{-D} [cf. Eqs. (8,9)] of the antibiotic must be considerably smaller than those of the model compound. This is probably due to the fact that steric hindrance leads to a decrease in the rate constant for the direct proton transfer between the 3-hydroxypicolinyl residues of two antibiotic molecules. The unexpectedly low value of k_H can also be explained on the basis of steric hindrance.

II. Complex Formation with Metal Ions in Homogeneous Solution

Since virginiamycin S facilitates the transport of cations through phospholipid bilayer membranes, it is expected that the antibiotic can also coordinate metal ions in homogeneous solution. Due to experimental limitations it is more suitable to clarify the principal aspects of cation coordination in homogeneous solution than under conditions where the ligand is bound to the membrane. Two different possibilities for complex formation with cations must be distinguished. The cations could be coordinated to the peptide-lactone ring to give structures similar to those of cation complexes of other ion-specific antibiotics. This behavior would be consistent with the measured alkali ion specificity (cf. Fig. 6a). In addition, cations could also be coordinated to one of the protolytic states of the 3-hydroxypicolinyl residue. In this case the affinity would depend mainly on electrostatic interactions. However with respect to cation specificity one would expect that the resulting free energy would be characterized by a monotonic dependence on the cation radius. This is not consistent with the

Table 3. Kinetic parameters of proton transfer reactions of virginiamycin S; 25 °C

Solvent	k_{23} (sec^{-1})	k_{32} (sec^{-1})	k_H ($M^{-1}sec^{-1}$)	k_{-H} ($M^{-1}sec^{-1}$)	k_{OH} ($M^{-1}sec^{-1}$)	k_{-OH} ($M^{-1}sec^{-1}$)	k_A; k_{-A} ($M^{-1}sec^{-1}$)	k_D; k_{-D} ($M^{-1}sec^{-1}$)
Water	9.8×10^2	4×10^3						
33.8% water in methanol (vol/vol)	8.5×10^2	6.7×10^2	4.4×10^8	3.5×10^8	2.8×10^8	2.2×10^8	$<5 \times 10^8$	$<3 \times 10^6$

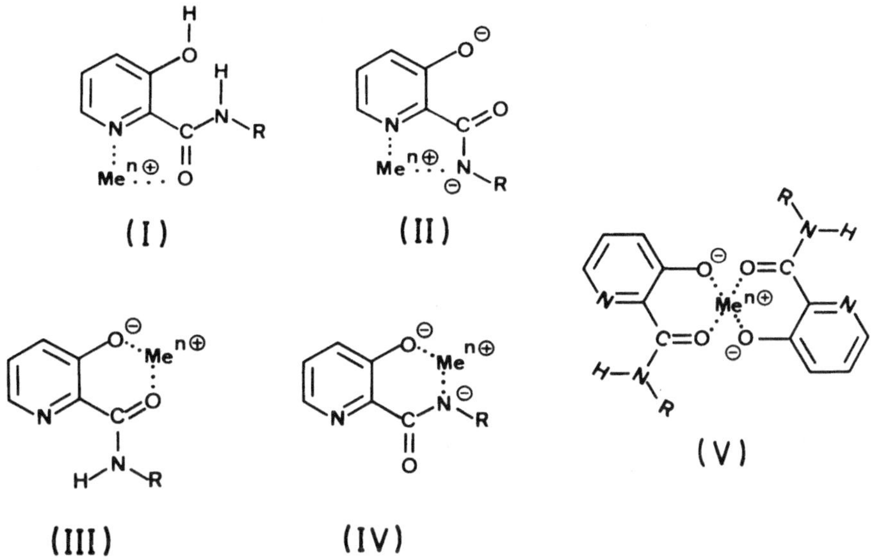

Fig. 17. Hypothetical structures of metal ion complexes of virginiamycin S (R = peptide-lactone residue) and related compounds

observed cation specificity for the membrane-bound virginiamycin S. A further complex structure could be characterized by a coordination of the cation to the peptide-lactone ring as well as to the 3-hydroxypicolinyl residue. This indicates that a variety of structures of cation complexes have to be considered. As far as coordination to the 3-hydroxypicolinyl residue is concerned, apart from the possible participation of the various protolytic states we also have to consider the formation of 2:1, 1:1, and possibly 1:2 ligand-cation complexes. Some selected examples for these types of hypothetical complex structures are shown in Figure 17. It is evident that among the various protolytic states the negatively charged states P_4^- and P_5^{--} should exhibit the largest affinities for the cation. The possibility can also not be excluded that the state P_5^{--} is capable of coordinating two cations which would lead to the formation of a 1:2 complex. In order to differentiate between coordination of the cations to the peptide-lactone ring or to the 3-hydroxypicolinyl group, circular dichroism and absorption spectroscopy are being applied. It is hoped that the circular dichroism will provide evidence mainly for conformational changes occurring upon complex formation between a cation and the peptide-lactone ring; the absorption spectrum, on the other hand, is most sensitive to processes in which the different protolytic states are involved, e.g., coordination to one particular protolytic state of the antibiotic.

In Figure 18 the circular dichroism of virginiamycin S in the protolytic state P_4^- is shown for ethanol as solvent together with the circular dichroism observed after the addition of several metal ions. Under similar experimental conditions, in the medium 20% water/methanol (vol/vol) absorption spectra of virginiamycin S in the presence of increasing $CaCl_2$ (Fig. 19a) and RbCl concentrations, respectively, (Fig. 19b) have been measured. These results

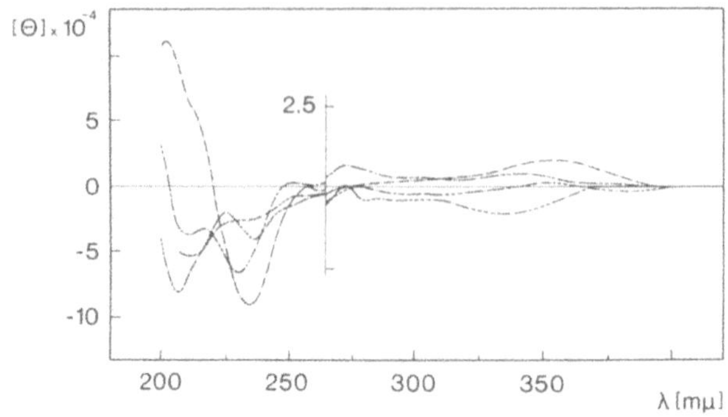

Fig. 18. Circular dichroism of virginiamycin S (8×10^{-5} M) in ethanol with 7.3×10^{-4} M tetramethylammonium hydroxide (·–·–·–), containing in addition 1 M $NaClO_4$ (— — — ·—), 0.1 M RbBr (— ---- — · —), 0.5 M $CaCl_2$ (— ---- — · —), respectively, at 25°C

tentatively suggest that virginiamycin S can form a 1:1 complex between the protolytic state P_4^- of the 3-hydroxypicolinyl residue of the antibiotic and the Ca^{2+} ion (cf. state III in Fig. 17). This results from a comparison of the absorption spectra of the different protolytic states (Oberbäumer, 1975) with that of the Ca^{2+} complex (e.g., spectrum g in Fig. 19a) and is consistent with similar titrations carried out at lower pH values. It is evident that a similar spectrophotometric titration with RbCl (Fig. 19b) does not lead to similar changes in the absorption spectra although complex formation is expected to occur according to the corresponding titration curve (Fig. 20). In contrast to this observation the change of the circular dichroism is comparatively larger in the presence of Rb^+ and Na^+. This implies a possible participation of

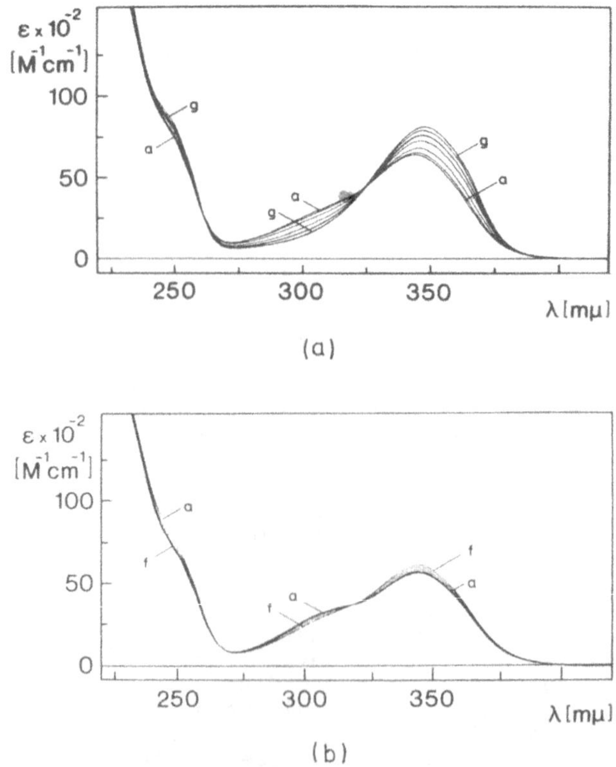

Fig. 19a,b. Spectrophotometric titration at 25°C of virginiamycin S (3.23×10^{-5} M) in 20% water/methanol (vol/vol) at pH 9.2 containing 2×10^{-2} M N-ethylmorpholine, 2×10^{-2} M tetramethylammonium chloride (a) with $CaCl_2$ from 1.38×10^{-4} M $CaCl_2$ (spectrum a) to 5.89×10^{-3} M $CaCl_2$ (spectrum g) and (b) with RbCl from 3.8×10^{-3} M RbCl (spectrum a) to 6.76×10^{-2} M RbCl (spectrum f)

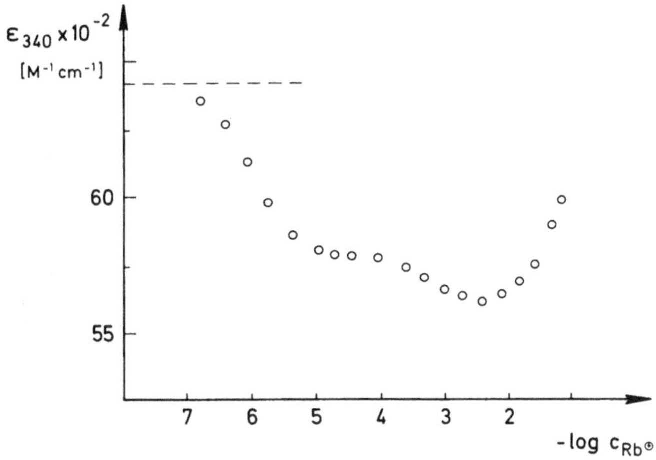

Fig. 20. Spectrophotometric titration of virginiamycin S in 20% water/methanol (vol/vol) with RbCl: Plot of extinction coefficient at 340 mμ vs. negative logarithm of Rb^+ concentration (experimental conditions correspond to those specified in Fig. 19b)

the peptide-lactone ring. The results of these titration studies exhibit a considerable complexity and cannot be directly evaluated (cf. Fig. 20). They indicate, however, that several types of one particular cation complex must exist, which occur in different states. For the complex that is formed upon titration with $CaCl_2$ under the conditions shown in Figure 19a, a binding constant of about 5×10^{-4} M is evaluated. A corresponding Rb^+ complex would be expected according to the spectral changes in Figure 19b to be about 500 times weaker.

Further changes in the absorption spectrum of the Ca^{2+} complex of virginiamycin S with respect to spectrum g in Figure 19a are observed upon a large additional increase of the calcium concentration, which is characterized by a small blue shift of the band at 348 mμ similar to that found for the alkaline titration shown in Figure 9c. A similar behavior is observed upon titration with $Cu(ClO_4)_2$ even at considerably lower pH values than in the case of the Ca^{2+} titrations. All these results suggest that Cu^{2+} and Ca^{2+} can also form weak 1:2 complexes where two cations are coordinated to the 3-hydroxypicolinyl residue in the protolytic state $P_5^{=}$ (cf. Fig. 8b). If this interpretation is correct, it would be expected that simple model compounds such as 3-hydroxypicolinamide should exhibit similar behavior. Two titrations of 3-hydroxypicolinamide with $Cu(ClO_4)_2$ at pH 4.8 are shown in Figure 21. The first main binding step leads to the formation of the 1:1 complex (Fig. 21a) (at the beginning of the titration small amounts of the 2:1 3-hydroxypicolinamide-Cu^{2+} complex are detected). The spectral change shown in Figure 21a is closely related to that observed upon the formation of the protolytic state P_4^- during a pH titration. Thus the resulting 1:1 complex must have the general structure of state (I) shown in Figure 23, where the metal ion is coordinated to the protolytic state P_4^-. The second spectral transition observed in the course of the Cu^{2+}-titration (Fig. 21b) corresponds to that found for the conversion of the protolytic state P_4^- to $P_5^{=}$, which is also similar to the one observed for virginiamycin S (Fig. 9c). It is therefore concluded that a second Cu^{2+} ion is coordinated to the ligand in the state (II) in Figure 23. The evaluation of the titrations illustrated in Figure 21 is given in Figure 22. The binding constant of the 1:1 complex at pH 4.8 is smaller than 10^{-5} M and that of the 1:2 complex is 2×10^{-3} . This implies that the Cu^{2+} ion must have an extremely high affinity for the nitrogens

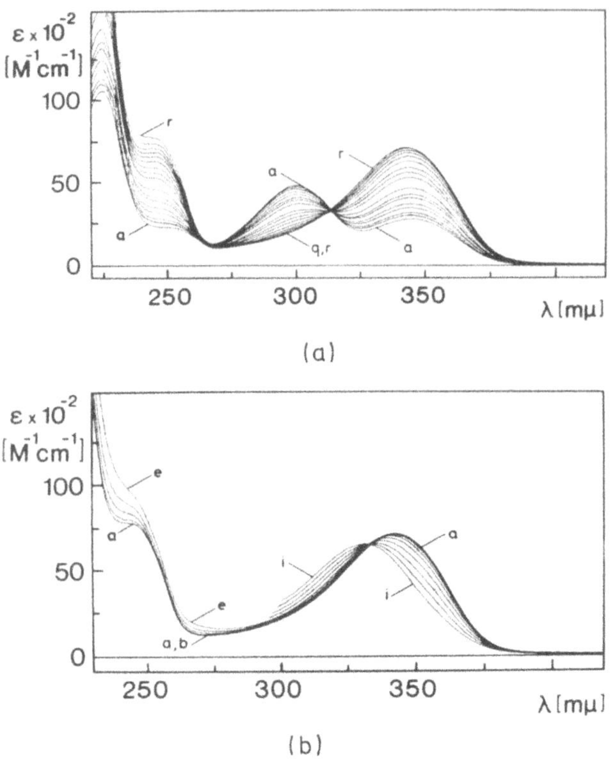

Fig. 21a,b. Spectrophotometric titration with $Cu(ClO_4)_2$ of 3-hydroxypicolinamide (1.63×10^{-5} M) at 25°C in water containing 10^{-3} M Na_2HPO_4, 0.05 M tetramethylammonium perchlorate, pH 4.80. Titration shown in (a) begins in absence of $Cu(ClO_4)_2$ (spectrum a) and ends with 1.12×10^{-4} M $Cu(ClO_4)_2$. Second titration (b) begins with 1.37×10^{-4} M $Cu(ClO_4)_2$ (spectrum a) and ends with 5.18×10^{-3} M $Cu(ClO_4)_2$

of the aromatic and the deprotonated amide residue so that the corresponding gain in free energy due to complex formation is considerably larger than the free energy required for the deprotonation of the weakly acidic amide group at pH 4.8. These results support the earlier suggestions based on the results obtained for virginiamycin S. The general complex formation behavior of these types of ligands is shown in Figure 23, which, however, must be considered as an oversimplification. Further efforts will be required in order to elucidate the properties of complex formation between cations and ligands such as virginiamycin S and related compounds in homogeneous solution as well as in the membrane-bound

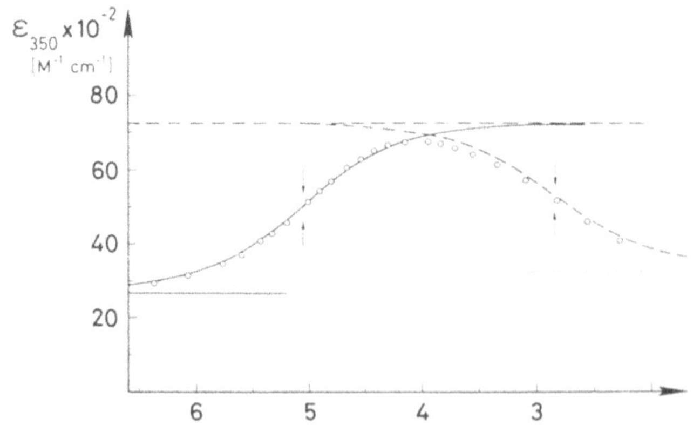

Fig. 22. Spectrophotometric titration of virginiamycin S with $Cu(ClO_4)_2$: Plot of extinction coefficient at 350 mμ vs. negative logarithm of Cu^{2+} concentration. Solid and broken lines represent theoretical titration curves (experimental conditions correspond to those specified in Fig. 21)

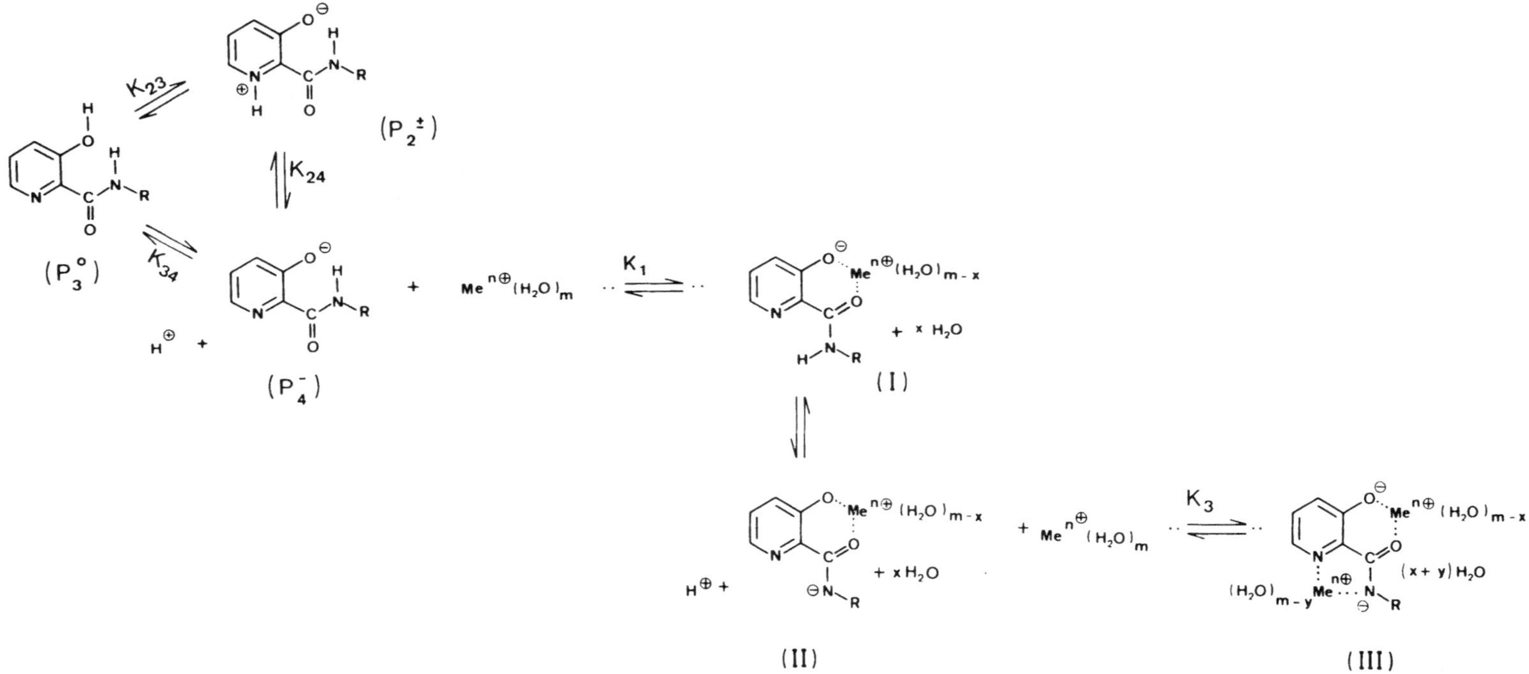

Fig. 23. Schematic illustration of 1:1 and 1:2 complex formation of 3-hydroxypicolinamide (R = H) and related molecules with metal ions (Me^{n+})

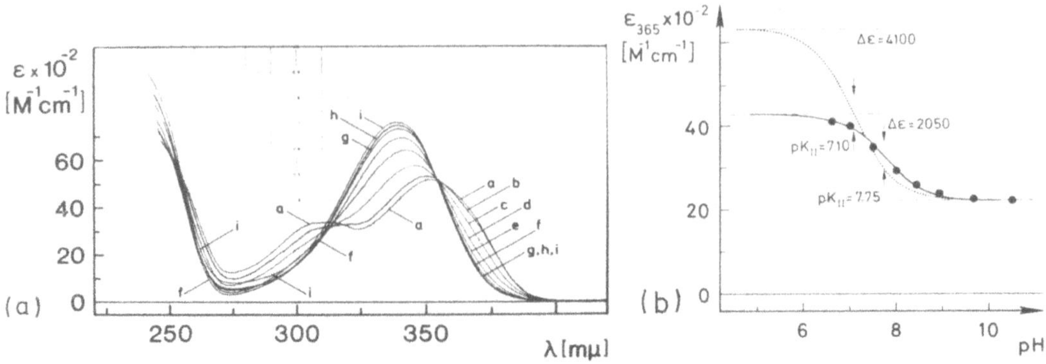

Fig. 24a,b. (a) Spectrophotometric pH-titration of virginiamycin S (2.31×10^{-5} M) with 0.35 M tetramethylammonium hydroxide ranging from pH 6.62 (spectrum a) to pH 10.50 (spectrum i) in a lecithin dispersion (2.24×10^{-2} M α,β-dimyristoyl-L-lecithin) containing 2×10^{-2} M tetramethylammonium chloride and 5×10^{-3} M N-ethylmorpholine; (b) plot of extinction coefficient at 365 mμ vs. pH in absence of lipid [....] and in presence of 2.24×10^{-2} M α,β-dimyristoyl-L-lecithin [●——●]. Theoretical titration curves have been calculated for indicated pK-values; (experimental conditions correspond to those specified in Fig. 24a)

state. Nevertheless, these results clearly indicate that a variety of cation complexes must be considered as possible candidates for carrying a cation across the membrane independently of whether transport according to a carrier mechanism or a different mechanism is responsible for the membrane conductance in the presence of the cations.

III. Interaction of Virginiamycin S with Membranes and Protolysis of the Membrane-Bound Antibiotic

In order to interpret the conductance data together with dynamic studies, in terms of elementary steps, it is necessary to investigate the interaction of virginiamycin S and related compounds with phospholipid bilayer membranes and to determine the relevant properties of the membrane-bound compounds. Since some of the thermodynamic parameters, such as K_{II} and K_{23}, depend on solvent composition, according to the studies in homogeneous solution, it is hoped that information is obtained about the polarity of the membrane region which contains the binding sites for the antibiotic.

For the determination of the protolytic dissociation constant of the membrane-bound virginiamycin S, spectrophotometric pH-titrations have been carried out in the presence of different concentrations of sonicated α,β-dimyristoyl-L-lecithin. An example of such a pH-titration is shown in Figure 24a. The evaluation of the data follows from Figure 24b. Although a certain fraction of free antibiotic contributes to the absorption spectra of the pH-titration in Figure 24a, it is evident that the spectra at low pH differ considerably from those observed for virginiamycin S in water (cf. Fig. 9b). This must reflect a change in the equilibrium constant K_{23} (cf. Fig. 8a) and indicates that the antibiotic is bound

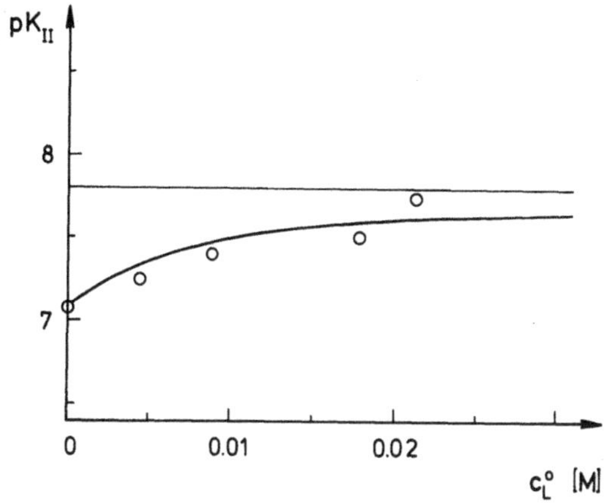

Fig. 25. Dependence of the pK_{II} value of virginiamycin S on lipid concentration: Plot of pK_{II} values vs. total concentration of α,β-dimyristoyl-L-lecithin (C_L^o). Thin solid line reflects extrapolated pK_{II}^L value according to theoretical curve (thick solid line), which has been calculated employing parameters given in Table 4

in an environment less polar than water. In Figure 25 the pK_{II} value of virginiamycin S determined in the presence of various lecithin concentrations is plotted vs. the total lipid concentration c_L^o. For the fit between the experimental data and the calculated dependence, which enables the determination of the protolytic dissociation constant of the membrane-bound antibiotic (pK_{II}^L) to be carried out, the binding constants given in Table 4 have been used. The pK_{II}^L value thus obtained is 7.80 (cf. Table 2). For the calculation of the microscopic dissociation constants it is necessary to determine the equilibrium constant between the two monodeprotonated protolytic states $P_2^{\pm}$ and P_3^o (cf. Fig. 8a) of the membrane-bound virginiamycin S (K_{23}^L). This is achieved by a spectrophotometric titration of virginiamycin S with sonicated α,β-dimyristoyl-L-lecithin at pH 4 (cf. Fig. 26a). This titration provides information about the affinity of the antibiotic for the lipid membrane, which is discussed later. Knowing the corresponding thermodynamic parameters (Table 4), it is possible to calculate the spectrum of the membrane-bound antibiotic according to the data shown in Figure 26a. The calculated spectrum is illustrated in Figure 26b together with the simulation of the absorption bands that is required for the calculation of K_{23}^L using the areas of the bands determined in homogeneous solution. The corresponding value at 27°C is 0.77 for membrane-bound virginiamycin S, implying that the populations of the two states are nearly equal. This value is about five times smaller than the one determined in water (Table 2). The microscopic dissociation constants that can now be calculated are also given in Table 2. Thus it is possible to compare the thermodynamic parameters of the protolytic reactions of the membrane-bound virginiamycin S with those obtained in homogeneous solution (Table 2, Figs. 12 and 13). The parameters of the protolytic reactions of virginiamycin S in 33.8% water in methanol (vol/vol) are remarkably different from those obtained in methanol or water alone; they are, however, nearly identical to those of the membrane-bound antibiotic.

If the antibiotic were preferentially bound to the inner, nonpolar part of the bilayer membrane, one would expect according to the spectroscopic studies in nonpolar solvents that the protolytic state P_3^o predominates. This is, however, not consistent with the experimental results. It is therefore concluded that the antibiotic is preferentially located in the surface region of the lipid bilayer membrane (cf. Fig. 29). The polarity of the surface region

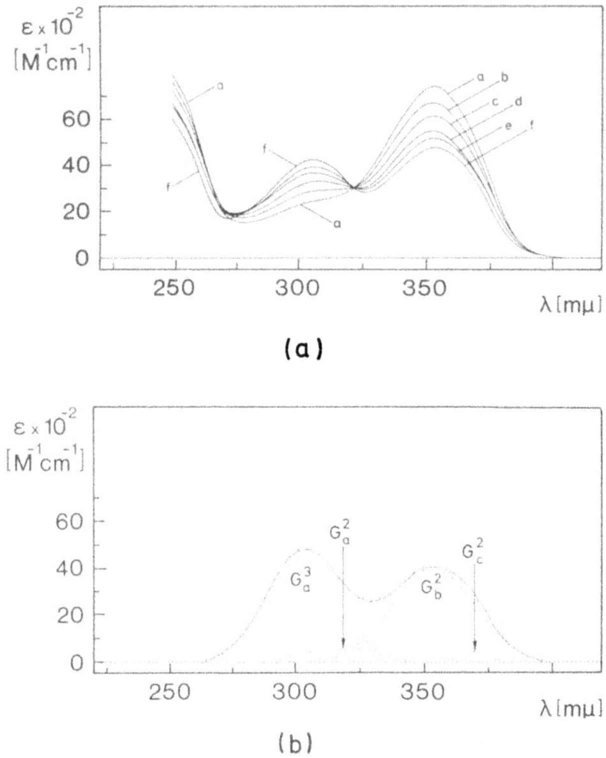

Fig. 26a,b. (a) Spectrophotometric titration of virginiamycin S (2.31×10^{-5} M) at 27°C with sonicated α,β-dimyristoyl-L-lecithin in 2×10^{-2} M tetramethylammonium chloride, 5×10^{-3} M N-ethylmorpholine/HCl pH ≈ 4.0 beginning with antibiotic alone (spectrum a) and in presence of 4.47×10^{-3} M to 2.24×10^{-2} M α,β-dimyristoyl-L-lecithin (spectrum b to spectrum f); (b) absorption spectrum of membrane-bound virginiamycin S at pH 4 and 27°C calculated by employing thermodynamic parameters given in Table 4 according to computer simulation of absorption bands in form of gaussian curves

of the membrane must be lower than the polarity of water and correlates well with the nearly identical data found in 33.8% water/methanol (vol/vol). Similar dependences have been reported for membrane-active molecules such as valinomycin and enniatin B under conditions where these molecules are bound to the membrane surface region (Grell et al., 1975).

Based on the titration shown in Figure 26a as well as on the plot of the corresponding data in Figure 27, the thermodynamic parameters related to the binding of the antibiotic to the membrane can be evaluated as follows. If n lipid molecules (L) form a binding site (B) for one antibiotic molecule (B = nL), the chemical equilibrium for the binding is given by Eq. (14):

$$\mathrm{Vir} + n\mathrm{L} \rightleftharpoons \mathrm{Vir} - n\mathrm{L} \tag{14}$$

The corresponding binding constant is

$$\frac{c_{\mathrm{VirL}}}{c_{\mathrm{Vir}} \; c_{\mathrm{B}}} = \frac{1}{K^{\mathrm{L}}} \tag{15}$$

where c_B denotes the concentration of the free and c_{VirL} that of the loaded binding sites. If we consider that

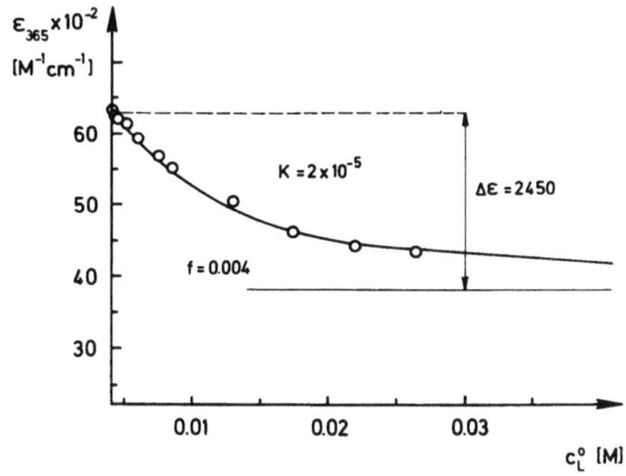

Fig. 27. Titration of virginiamycin S with sonicated α,β-dimyristoyl-L-lecithin: Plot of extinction coefficient at 365 mμ vs. α,β-dimyristoyl-L-lecithin concentration. Theoretical titration curve (thick solide line) has been calculated according to parameters given in Table 4, where $n = f^{-1}$

$$c^{\circ}_{Vir} = c_{Vir} + c_{VirL} \tag{16}$$

$$fc^{\circ}_{L} = c_{B} + c_{VirL} \tag{17}$$

where f denotes the reciprocal of the number of lipid molecules forming a binding site for one virginiamycin S molecule ($f = n^{-1}$) and c° the total concentrations, we obtain from Eqs. (15, 16, and 17):

$$\frac{c_{VirL}}{c_{Vir}} = (fc^{\circ}_{L} - c_{VirL})\,\frac{1}{K^{L}} = \frac{c_{VirL}}{c^{\circ}_{Vir} - c_{VirL}} \tag{18}$$

Eq. (18) leads to the quadratic equation:

$$c^{2}_{VirL} - c_{VirL}\,(K^{L} + c^{\circ}_{Vir} + fc^{\circ}_{L}) + c^{\circ}_{Vir}\,fc^{\circ}_{L} = 0 \tag{19}$$

In the case of a weak binding process the approximation can be introduced that the quadratic term in Eq. (19) is small compared to the other terms. Thus one obtains for the concentration of the membrane-bound compound:

$$c_{VirL} = \frac{c^{\circ}_{Vir}\,fc^{\circ}_{L}}{K^{L} + c^{\circ}_{Vir} + fc^{\circ}_{L}} \tag{20}$$

The spectrophotometric titration (Fig. 26a) allows differentiation between the free and the bound antibiotic. If the absorption due to the lipid is compensated by means of a difference measurement, we obtain for the extinction E (path length = 1 cm):

$$E = \varepsilon_{Vir}\,c_{Vir} + \varepsilon_{VirL}\,c_{VirL} \tag{21}$$

where the initial extinction in the absence of the lipid (E°) is:

$$E^{\circ} = \varepsilon_{Vir}\, c^{\circ}_{Vir} \tag{22}$$

According to Eqs. (21) and (22) we obtain for the difference in extinction,

$$\Delta E = E - E^{\circ} = c_{VirL}\,(\varepsilon_{VirL} - \varepsilon_{Vir}) = c_{VirL}\,\Delta\varepsilon \tag{23}$$

and thus together with Eqs. (20) and (21)

$$\frac{c^{\circ}_{Vir}}{\Delta E} = \frac{1}{c^{\circ}_{L}\,\Delta\varepsilon}\left(\frac{K^{L}}{f} + \frac{c^{\circ}_{Vir}}{f}\right) + \frac{1}{\Delta\varepsilon} \tag{24}$$

A plot of $c^{\circ}_{Vir}/\Delta E$ vs. the reciprocal total lipid concentration yields the parameters $1/\Delta\varepsilon$ and $f^{-1}(K^{L} + c^{\circ}_{Vir})$ (cf. Ruf, 1974, 1977b). In the case of a weak binding process ($K^{L} >> c^{\circ}_{Vir}$) an approximate value for the parameter K^{L}/f is obtained from this plot. If the experimental data are plotted according to Eq. (24), the value for $1/\Delta\varepsilon$ is obtained (cf. Fig. 28a), which allows the calculation of c_{VirL}. In order to determine K^{L} and $f = n^{-1}$, a modification of the procedure of Klotz (1953) is used. Eq. (18) leads to

$$\frac{c_{Vir}\, f\, c^{\circ}_{L}}{K^{L}} = c_{VirL}\,\left(1 + \frac{c_{Vir}}{K^{L}}\right) \tag{25}$$

or

$$\frac{c^{\circ}_{L}}{c_{VirL}} = \frac{K^{L}}{f c_{Vir}} + \frac{1}{f} \tag{26}$$

A plot of c°_{L}/c_{VirL} vs. the reciprocal of the free virginiamycin S concentration at pH 4 yields K^{L} and f (Fig. 28b, Table 4).

Since the titration has been carried out at pH 4, the binding constant K^{L} only reflects the affinity of the protolytic states $P^{\pm}_{2}$ and P°_{3}. If the macroscopic protolytic dissociation constant for the membrane-bound antibiotic K^{L}_{II} and the equilibrium constant K^{L}_{23} are known, the binding constants of the three relevant protolytic states (K^{L}_{2}, K^{L}_{3}, K^{L}_{4}) shown in Figure 29 can be calculated. If we consider the following equilibrium relations for the membrane-bound antibiotic (the c^{L} denote the corresponding concentrations)

$$\frac{c^{L}_{2}}{c_{2}\;c_{B}} = \frac{1}{K^{L}_{2}}\;;\quad \frac{c^{L}_{3}}{c_{3}\;c_{B}} = \frac{1}{K^{L}_{3}}\;;\quad \frac{c^{L}_{4}}{c_{4}\;c_{B}} = \frac{1}{K^{L}_{4}} \tag{27}$$

and

$$K^{L}_{23} = \frac{K^{L}_{34}}{K^{L}_{24}} \tag{28}$$

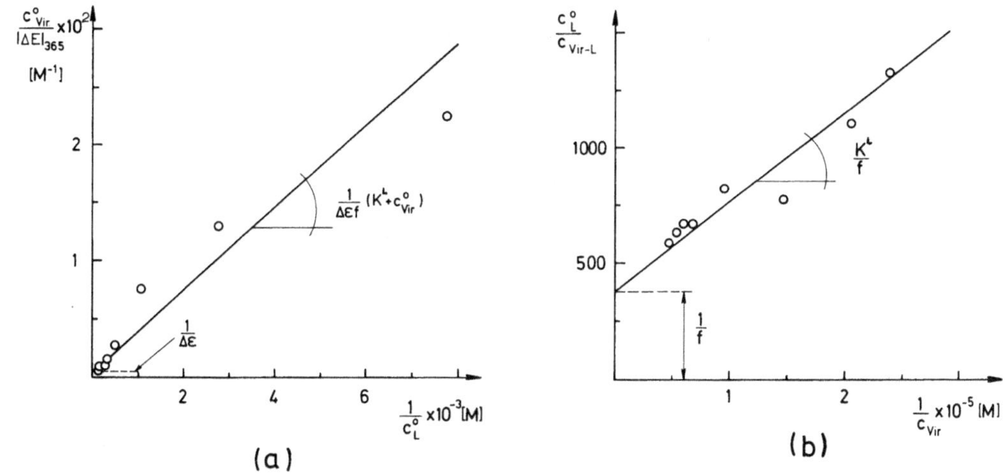

Fig. 28a,b. Determination of thermodynamic parameters related to binding of virginiamycin S to sonicated α,β-dimyristoyl-L-lecithin (experimental conditions correspond to those in Fig. 26a): (a) Plot of $c^{o}_{Vir}/|\Delta E_{365}|$ vs. reciprocal lipid concentration; (b) plot of c^{o}_{L}/c_{Vir-L} vs. reciprocal virginiamycin S concentration

we obtain, making use of the microscopic dissociation constants for the membrane-bound antibiotic K^{L}_{24} and K^{L}_{34},

$$K^{L}_{24} = \frac{c^{L}_{4} \; c_{H^+}}{c^{L}_{2}} \; ; \qquad K^{L}_{34} = \frac{c^{L}_{4} \; c_{H^+}}{c^{L}_{3}} \tag{29}$$

the following three relations:

$$\frac{K^{L}_{3}}{K^{L}_{2}} = \frac{K^{L}_{23}}{K_{23}} \; ; \qquad \frac{K^{L}_{4}}{K^{L}_{2}} = \frac{K_{24}}{K^{L}_{24}} \; ; \qquad \frac{K^{L}_{4}}{K^{L}_{3}} = \frac{K_{34}}{K^{L}_{34}} \tag{30}$$

Since the previously determined binding constant, which is valid at pH 4, is a function of the binding of the states $P^{\pm}_{2}$ and P^{o}_{3}, we must find a relation between K^{L} and K^{L}_{2} and K^{L}_{3}. If we consider the relations

$$c_{Vir} = c_{2} + c_{3} \; ; \qquad c_{VirL} = c^{L}_{2} + c^{L}_{3} \tag{31}$$

which are valid under the experimental conditions discussed here, we obtain for the equilibrium constant K^{L} [cf. Eq. (15)]

$$\frac{(1 + K^{L}_{23}) \; c^{L}_{3}}{c_{2}c_{B} + c_{3}c_{B}} = \frac{1}{K^{L}} \tag{32}$$

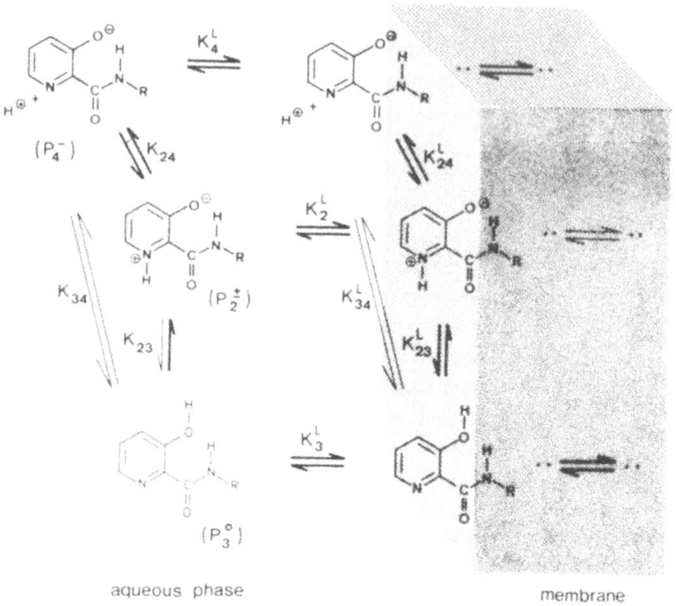

Fig. 29. Schematic illustration of binding of various protolytic state of virginiamycin S to lipid bilayer membrane

Table 4. Thermodynamic parameters related to binding of virginiamycin S to α,β-dimyristoyl-L-lecithin membranes at 27 °C

n	K^L[pH 4] (M)	K_2^L (M)	K_3^L (M)	K_4^L (M)
250	2×10^{-5}	3.7×10^{-5}	7×10^{-6}	1×10^{-4}

Thus, according to the previous considerations, we derive

$$K_2^L = \frac{K^L\ (1 + K_{23}^L)}{K_{23}^L\ (1 + K_{23}^{-1})} \ . \tag{33}$$

This implies that Eqs. (30) and (33) allow the determination of K_2^L, K_3^L and K_4^L (cf. Fig. 29 and Table 4) since the additional thermodynamic parameters required have already been determined.

The present results imply that virginiamycin S is bound more tightly to lecithin membranes than the H^+-carrier o-methyl red (Ruf, 1974, 1977a), but less tightly than the K^+-carrier valinomycin (Stark et al., 1971). The differences between the individual binding constants (K_2^L, K_3^L, K_4^L) are remarkably smaller than the differences observed for the corresponding states of o-methyl red (cf. Ruf, 1974, 1977a). This is attributed to the interaction of the large peptide-lactone ring with the phospholipid membrane. The determination of the thermodynamic parameters described previously is necessary for a quantitative interpretation of the conductance data (cf. Fig. 3) in terms of the permeabilities of the individual protolytic states, a suitable model for which is discussed by Ruf (1974, 1977b).

Although further experimental evidence is needed for the clarification of the various aspects discussed in this contribution, the present results indicate that the large peptide-lactone ring is required for a suitable interaction of this carrier with the membrane, whereas many of the observed specificities are attributed to the 3-hydroxypicolinyl residue. In this respect, the functional and structural properties of virginiamycin S may be close to the basic features of natural proteins acting as cation carriers in membranes.

Acknowledgments. We are very grateful to Professor H. Vanderhaeghe for making available to us the virginiamycin S used in his work. We wish to thank Professor M. Eigen, Professor G. Gottschalk, Professor P. Schuster and Privat-Dozent Dr. H. Diebler for helpful and stimulating discussions. We are very grateful to Dr. A. McConnell for reading the manuscript and to Miss E. Wittschieber for expert technical assistance.

References

Ahrens, M.-L.: Proton transfer kinetics of 5'-deoxypyridoxal. Biochim. Biophys. Acta 320, 86-96 (1973)

Anteunis, M.J.O., Callens, R.F.A., Tavernier, D.K.: Solution conformation of virginiamycins (staphylomycins). Eur. J. Biochem. 58, 259-268 (1975)

Cêlis, H., Estrada-O., S., Montal, M.: Model translocators for divalent and monovalent ion transport in phospholipid membranes. I. The ion permeability induced in lipid bilayers by the antibiotic X-537A. J. Membrane Biol. 18, 187-199 (1974)

Dixon, H., Kellerman, G.M., Mitchell, C.H., Towers, N.H., Linnane, A.W.: Mikamycin, an inhibitor of both mitochondrial protein synthesis and respiration. Biochem. Biophys. Res. Commun. 43, 780-786 (1971)

Eigen, M., DeMaeyer, L.: Theoretical basis of relaxation spectrometry. In: Techniques of Chemistry, Vol. VI, part 2. Weisberger, A., Hammes, G. (eds.). New York: Wiley 1973, pp. 63-146

Eisenman, G., Ciani, S.M., Szabo, G.: Some theoretically expected and experimentally observed properties of lipid bilayer membranes containing neutral molecular carriers of ions. Fed. Proc., Fed. Am. Soc. Exp. Biol. 27, 1289-1304 (1968)

Ennis, H.L.: Interaction of vernamycin A with Escherichia coli ribosomes. Biochemistry 10, 1265-1270 (1971)

Ennis, H.L.: Binding of the antibiotic vernamycin B_{α} to Escherichia coli ribosomes. Arch. Biochem. Biophys. 160, 394-401 (1974)

Grell, E., Funck, T.: Dynamic properties and membrane activity of ion specific antibiotics. J. Supramol. Struct. 1, 307-335 (1973)

Grell, E., Funck, T., Eggers, F.: Structure and dynamic properties of ion-specific antibiotics. In: Membranes, Vol. 3. Eisenman, G. (ed.). New York: Dekker 1975, pp. 1-128

Grell, E., Oberbäumer, I.: Dynamic aspects of carrier-mediated cation transport through membranes. In: Chemical Relaxation in Molecular Biology. Rigler, R., Pecht, I. (eds.). Berlin-Heidelberg-New York: Springer 1976

Haydon, D.A., Hladky, S.B: Ion transport across thin lipid membranes: a critical discussion of mechanisms in selected systems. Quart. Rev. of Biophysics 5, 187-282 (1972)

Hopfer, U., Lehninger, A.L., Lennarz, W.J.: The effect of the polar moiety of lipids on bilayer conductance induced by uncouplers of oxidative phosphorylation. J. Membrane Biol. 3, 142-155 (1970)

Klotz, I.M.: Protein Interactions. In: The Proteins, Vol. 1, part B. Neurath, H., Bailey, K. (eds.). New York: Academic Press 1953, pp. 727-806

Lea, E.J.A., Croghan, P.C.: The effect of 2,4-dinitrophenol on the properties of thin phospholipid films. J. Membrane Biol. 1, 225-237 (1969)

Le Blanc, O.H.: The effect of uncouplers of oxidative phosphorylation on lipid bilayer membranes: Carbonyl-cyanide m-chlorophenylhydrazone. J. Membrane Biol. 4, 227-251 (1971)

Mueller, P., Rudin, D.O.: Development of K^+ - Na^+ discrimination in experimental bimolecular lipid membranes by macrocyclic antibiotics. Biochem. Biophys. Res. Commun. 26, 398-404 (1967)

Oberbäumer, I.: Versuche zur Charakterisierung der Wechselwirkungen des Antibiotikums Virginiamycin S mit Lipid-Membranen. Diplomarbeit Göttingen 1975

Ovchinnikov, Yu.A., Ivanov, V.T., Shkrob, A.M.: Membrane Active Complexones. Amsterdam: Elsevier 1974

Pedersen, C.J.: Cyclic polyethers and their complexes with metal salts. J. Am. Chem. Soc. 89, 7017-7036 (1967)

Ruf, H.: Mechanismus des Carrier-induzierten Protonentransports durch Lipid-Membranen. Dissertation Göttingen (1974)

Ruf, H.: The mechanism of carrier-mediated proton transport across lipid bilayer membranes: I. The thermodynamics and kinetics of the binding of the model compound o-methyl red to liposomes. Submitted for publication (1977a)

Ruf, H.: The mechanisms of carrier-mediated proton transport across lipid bilayer membranes: II. A theory for the effects of o-methyl red on the electric properties of bilayer membranes. Submitted for publication (1977b)

Schadt, M., Haeusler, G.: Permeability of lipid bilayer membranes to biogenic amines and cations: Changes induced by ionophores and correlations with biological activities. J. Membrane Biol. 18, 277-294 (1974)

Schuster, P., Tortschanoff, K., Winkler, H.: Protonenübertragungsreaktionen zweibasischer Säuren in wässriger Lösung: 3-Hydroxypyridin. Z. Naturforsch. 31c, 219-224 (1976)

Shemyakin, M.M., Ovchinnikov, Yu.A., Ivanov, V.T., Antonov, V.K., Vinogradova, E.I., Shkrob, A.M., Malenkov, G.G., Evstratov, A.V., Laine, I.A., Melnik, E.I., Ryabova, I.D.: Cyclodepepsipeptides as chemical tools for studying ionic transport through membranes. J. Membrane Biol. 1, 402-430 (1969)

Stark, G., Ketterer, B., Benz, R., Läuger, P.: The rate constants of valinomycin-mediated ion transport through thin lipid membranes. Biophys. J. 11, 981-994 (1971)

Szabo, G., Eisenman, G., Laprade, R., Ciani, S.M., Krasne, S.: Experimentally observed effects of carriers on the electrical properties of bilayer membranes-equilibrium domain. In: Membranes, Vol. 1. Eisenman, G. (ed.). New York: Dekker 1973, pp. 179-328

Tanaka, N.: Mikamycin. In: Antibiotics III. Mechanism of Action of Antimicrobial and Antitumor agents. Corcoran, J.W., Hahn, F.E. (eds.). Berlin-Heidelberg-New York: Springer 1975, pp. 487-497

Tosteson, D.C., Cook, P., Andreoli, T., Tiefenberg, M.: The effect of valinomycin on potassium and sodium permeability of HK and LK sheep red cells. J. Gen. Physiol. 50, 2513-2525 (1967)

Urry, D.W., Ohnishi, M.: Nuclear magnetic resonance and the conformation of cyclic polypeptide antibiotics. In: Spectroscopic Approaches to Biomolecular Conformation. Urry, D.W. (ed.). Chicago: Am. Med. Assoc. 1970, pp. 263-301

Vanderhaeghe, H., Parmentier, G.: The structure of factor S of staphylomycin. J. Am. Chem. Soc. 82, 4414-4422 (1960)

Vazquez, D.: The streptogramin family of antibiotics. In: Antibiotics III. Mechanism of Action of Antimicrobial and Antitumor Agents. Corcoran, J.W., Hahn, F.E. (eds.). Berlin-Heidelberg-New York: Springer 1975, pp. 521-534

Studies on the Gramicidin Channel

E. BAMBERG, H. ALPES, H.-J. APELL, R. BENZ, K. JANKO, H.-A. KOLB, P. LÄUGER, and E. GROSS

A. Introduction

There is now considerable evidence that the ion permeability of biological membranes is not a diffuse property of the whole membrane but depends on the presence of specialized components in the membrane. Such a functional component may be a mobile carrier molecule that mediates the transport of ions in a sequence of complexation, translocation, and decomplexation steps. The second limiting case of an ion transport mechanism consists of a pore-like channel that offers to the ion a hydrophilic pathway through the apolar core of the membrane. The formation of channels in lipid membranes and the mechanism of ion transport through channels may be studied in experiments with certain hydrophobic peptides such as gramicidin.

Gramicidin is a linear pentadecapeptide (Fig. 1) that has been isolated from Bacillus brevis (Sarges and Witkop, 1965; Ovchinnikov et al., 1974). This molecule has several remarkable structural features. With the exception of glycine in position 2, it consists of strongly hydrophobic amino acids. As both end groups are blocked (by a formyl residue at the amino terminal and by an ethanolamine residue at the carboxyl terminal), the molecule is electrically neutral. Furthermore, the optical configuration of the amino acids is alternately D and L (glycine in position 2 may stand for a D amino acid).

If gramicidin A is introduced into a biological membrane (Pressman, 1965; Henderson et al., 1969; Podleski and Changeux, 1969; Wenner and Hackney, 1969; Johnstone, 1975) or into an artificial lipid bilayer (Liberman and Topaly, 1965; Mueller and Rudin, 1967; Tosteson et al., 1968; Hladky and Haydon, 1970, 1972; Goodall, 1970, 1971, 1973; Krasne et al., 1971; Urry et al., 1971; Meyers and Haydon, 1972; Bamberg and Läuger, 1973, 1974; Zingsheim and Neher, 1974; Kolb et al., 1975; Neher, 1975; Bamberg et al., 1976; Bamberg and Benz, 1976; Veatch et al, 1976; Eisenman et al., 1977), the membrane becomes cation permeable. In this respect gramicidin A is similar to the macrocyclic ion carriers such as valinomycin or monactin. It has become clear, however, that the action mechanism of gramicidin is quite different from that of the macrocyclic carriers and that gramicidin probably forms channels in a lipid membrane. One piece of evidence came from the experiments of Krasne et al., 1971. They were able to freeze lipid bilayers by reducing the temperature and observed that below the freezing point of the membrane the valinomycin conductance dropped to a very low value but that the gramicidin conductance still persisted in the frozen state. This finding favors the view that valinomycin acts as a mobile carrier whereas gramicidin creates channels in the lipid membrane. Further support for a channel mechanism comes from the observed transport rates (see below), as well as from bulk phase experiments (Byrn, 1974).

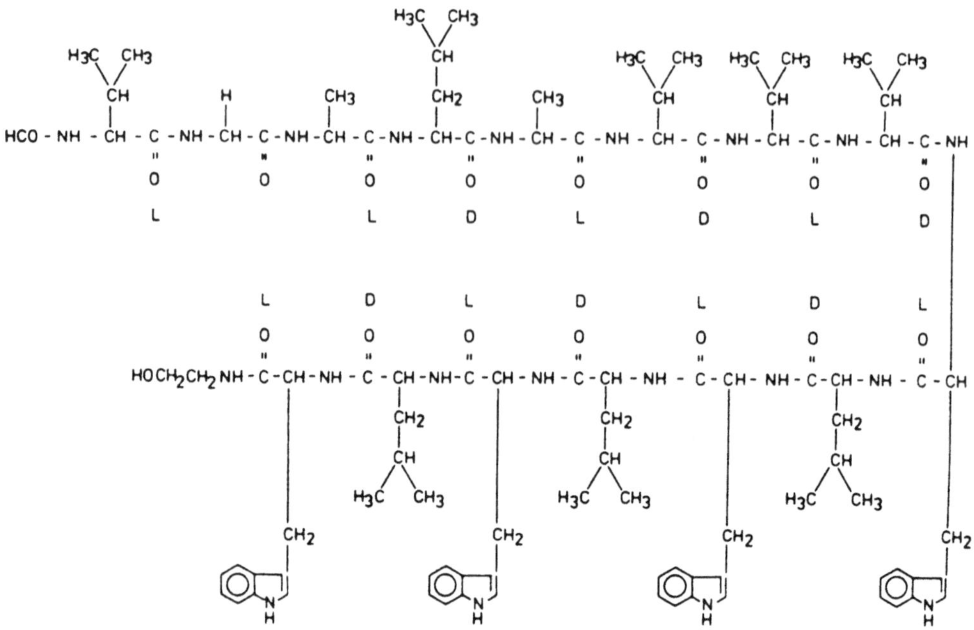

Fig. 1. Structure of valine-gramicidin A

A detailed model of the gramicidin A channel has been proposed by Urry (1971, 1972a,b) and independently by Ramachandran and Chandrasekaran (1972), on the basis of conformational energy considerations. The model of Urry consists of a helical dimer that is formed by head-to-head (formyl end to formyl end) association of two gramicidin monomers and that is stabilized by intra- and intermolecular hydrogen bonds (Fig. 2). The central hole along the axis of the π^6(L,D)-helix has a diameter of about 4 Å and is lined with the peptide C-O moieties, whereas the hydrophobic residues lie on the exterior surface of the helix. The total length of the dimer is about 20-30 Å, which is the lower limit of the hydrophobic thickness of a lipid bilayer. The length of the dimer would thus be sufficient to bridge the membrane if a local thinning of the lipid structure is assumed. Dimerization of gramicidin A in organic solvents has recently been studied by fluorescence and circular dichroism methods (Byrn, 1974). Further support for the dimer model comes from membrane experiments, described below.

Recently, Veatch and Blout (Veatch et al., 1974; Veatch and Blout, 1974; Fossel et al., 1974) proposed an alternative model of the gramicidin channel. They postulated a new class of double-helical dimers in which the two chains are coiled about a common axis and in which all hydrogen bonds are intermolecular (Fig. 3). These double helices have a cylindrical hole along the axis with a diameter of greater than 3 Å. The rise per turn is about twice that of the π(L,D)-helices of comparable diameter. Undoubtedly, the model of Veatch and Blout is a possible candidate for the structure of the gramicidin channel. The observation that desformylation of gramicidin A leads to a virtually complete loss of activity (Urry, 1971) is easily explained on the basis of Urry's model, but is less obvious for a double-stranded helix. At the moment the

Fig. 2. The structure of the π^6(L,D)-helix of gramicidin A, as proposed by Urry et al. (1971). The hydrophilic hole along the helix axis is lined with oxygen atoms of amide carbonyls alternately pointing up and down. The hydrophobic residues located at the periphery of the helix are not shown

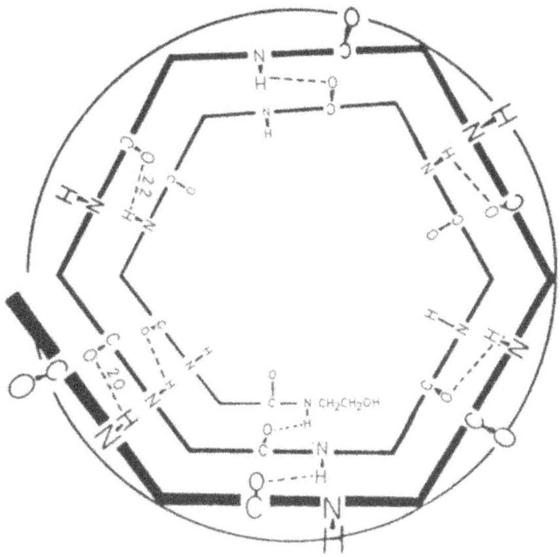

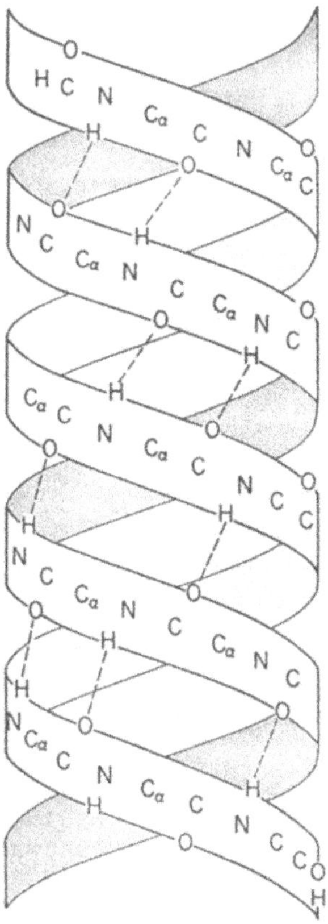

Fig. 3. Structure of the antiparallel-β-double-helical gramicidin dimer according to Veatch et al. (1974)

available experimental data do not seem sufficient, however, to exclude one or the other model.

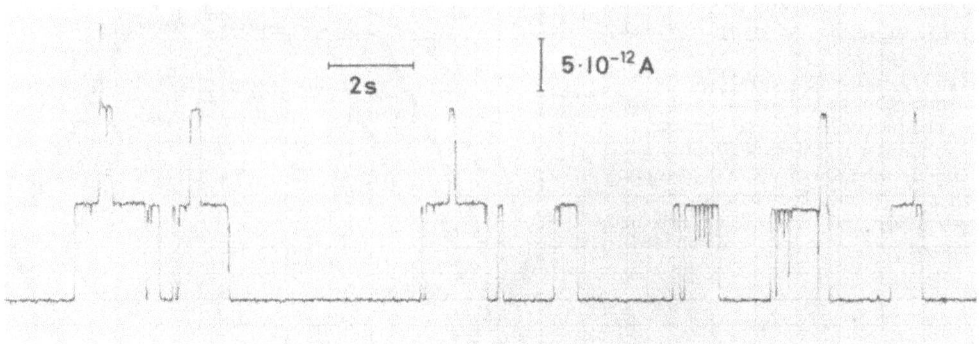

Fig. 4. Fluctuations of the membrane current in the presence of very small amounts of gramicidin A. 1 M CsCl, 25°C, V_m = 100 mV. The membrane has been formed from glycerylmonooleate. The baseline corresponds to the membrane conductance in absence of gramicidin

B. Kinetics of Channel Formation

Many details about the nature of the gramicidin-mediated cation permeability have been revealed by studying the electrical properties of gramicidin-doped artificial bimolecular lipid films. An instructive experiment, which was first performed by Hladky and Haydon, consists of adding extremely small amounts of gramicidin A to the membrane and recording the time course of the electric current at a given voltage (Hladky and Haydon, 1970, 1972; Bamberg and Läuger, 1974). Under these circumstances the current shows discrete fluctuations with more or less the same amplitude (Fig. 4). If increasing amounts of gramicidin are added to the membrane, more and more fluctuations build up on top of each other and eventually fuse into an average macroscopic current. This experiment can be explained by the assumption that the single current fluctuation comes about by the formation and the disappearance of a single channel. The conductance of the channel, which is calculated from the current fluctuations, is about 90 pS (1 pS = 10^{-12} Ω^{-1}) in 1 M CsCl at 25°C (Bamberg et al., 1976). This corresponds to a transport of about $6 \cdot 10^7$ cesium ions per sec through the single channel at a voltage of 100 mV. This value is larger by a factor of 10^3-10^4 than the turnover number of a translational carrier such as valinomycin (Läuger, 1972). The high transport rate of the single conductance unit of gramicidin A would therefore be difficult to reconcile with a carrier model but can easily be explained by a pore mechanism (Hladky and Haydon, 1972).

The probability distribution $p(\Lambda)$ for the occurrence of a given value of the single-channel conductance Λ is different in different lipids. As shown in Figure 5, $p(\Lambda)$ is relatively narrow in the case of a glycerylmonooleate membrane, but considerably broader for phosphatidylcholine membranes. Interestingly, fluctuations of Λ within the lifetime of the channel are not observed. In other words, a newly formed channel starts with a given conductance Λ, which stays constant throughout its lifetime τ^*. The next channel may show a significantly different Λ, which again remains constant during τ^*.

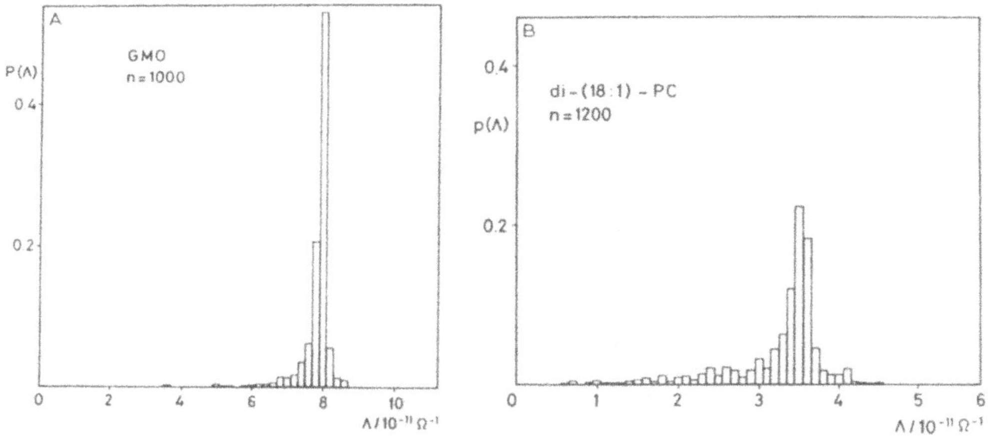

Fig. 5. Probability P(Λ) of the occurrence of conductance fluctuations of the magnitude Λ. P(Λ) is the number of events within an interval of the width $\Delta\Lambda = \pm 2 \times 10^{-13}\ \Omega^{-1}$ centered at Λ, divided by the total number n of events. Natural gramicidin B, 1 M CsCl, 25°C. Di-(18:1)-PC, dioleoylphosphatidylcholine; GMO, glycerylmonooleate (Bamberg et al., 1976)

The fact that Λ is distributed over a considerable range in certain lipids may have a number of different explanations. A first possibility would be that the sample of gramicidin B (see below) used in the experiments represented in Figure 5 is chemically heterogeneous. We consider this explanation unlikely because it would imply that the heterogeneity is expressed differently (in terms of Λ) in different lipids. Second, there may not be a single form of the channel, but a range of different hydrogen-bonded structures, and one of these structures may be "frozen in" when the channel forms. Still another possibility is that channels form in different lipid environments. If this explanation is correct, the patches of constant lipid environment must be rather large. This follows from the long lifetime of the channel (of the order of 0.1-1 sec), which, together with an estimated diffusion coefficient of the channel of 10^{-9}-10^{-8} $cm^2 \cdot sec^{-1}$, gives a mean diffusion-length of several hundred nanometers.

An obvious test for the hypothesis that the channel is a dimer would consist of measuring the dependence of membrane conductance λ on the aqueous gramicidin concentration c. Provided that a partition equilibrium of gramicidin between aqueous phase and membrane exists, the dimer model requires that λ should be proportional to c^2. Such conductance measurements turned out to be very inaccurate, however, apparently because a true partition equilibrium between membrane and water could never be reached (Tosteson et al., 1968; Hladky and Haydon, 1972; Veatch et al., 1976). This difficulty may be circumvented by studying the kinetics of the gramicidin system (Bamberg and Läuger, 1973; Bamberg and Benz, 1976).

If the dimer hypothesis is correct, then the channel-formation process may be described in the same way as a bimolecular chemical reaction between two gramicidin monomers G:

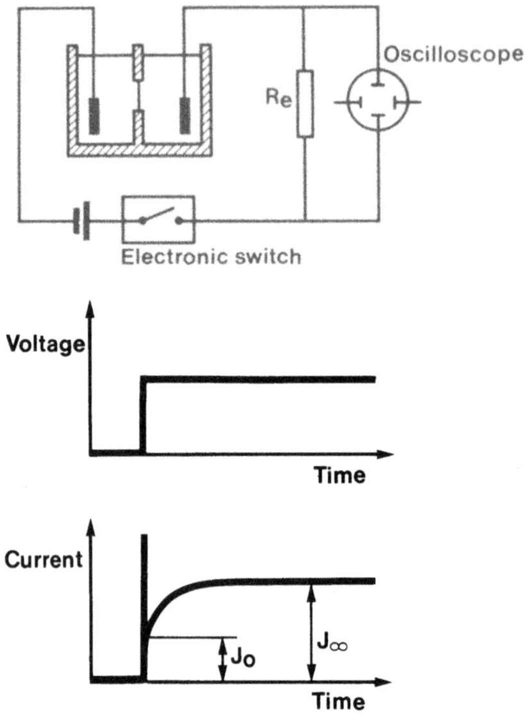

Fig. 6. Principle of the electrical relaxation method

$$G + G \underset{k_D}{\overset{k_R}{\rightleftarrows}} G_2 \qquad (1)$$

where k_R and k_D are the rate constants of association and dissociation. Information about the kinetics of such a reaction may be obtained by relaxation experiments (Bamberg and Läuger, 1974). The system is disturbed by a sudden displacement of an external parameter, and the time required for the system to reach a new stationary state is measured. In the case of a bilayer it is convenient to choose as the variable external parameter the electric field strength in the membrane (Fig. 6). Immediately after a sudden displacement of the voltage, a capacitive current is observed that decays with a time constant equal to the product of the circuit resistance and the membrane capacitance. This capacitive spike limits the time resolution of the method, which under favorable conditions is of the order of 1 μsec. After the decay of the capacitive transient, the membrane current approaches a stationary value J_∞ with a time constant τ_c, which contains information about the conductance mechanism under study.

Because the conductance of the bare membrane is negligibly small and because the monomers are likely to be nonconducting, the current is proportional to the instantaneous concentration of dimers G_2 in the membrane. For the foregoing case of an association-dissociation reaction, the relaxation time τ_c of the membrane current is then related to the rate constants k_R and k_D, the single-channel conductance Λ, and the stationary membrane conductance λ_∞ in the following way (Bamberg and Läuger, 1973):

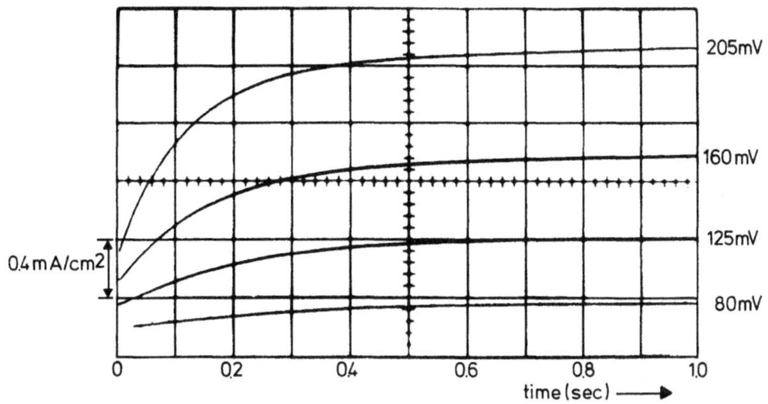

Fig. 7. Relaxation of membrane current after a sudden displacement of voltage (Bamberg and Läuger, 1973). 1 M NaCl, 2 × 10^{-9} M gramicidin, 25°C, dioleoyllecithin membrane. The amplitude of the voltage jump is indicated at the right side of the figure. The capacitative transient of the current is not resolved

$$\frac{1}{\tau_c} = k_D + 4\sqrt{\frac{k_R k_D \lambda_\infty}{L \Lambda}} \tag{2}$$

(L is Avogadro's constant). Thus, if τ_c is measured at different gramicidin concentrations, and $1/\tau_c$ is plotted as a function of the square root of λ_∞, a straight line should result.

The result of a typical relaxation experiment is shown in Figure 7. After the voltage jump at time zero, the membrane current starts at a relatively low level and increases asymptotically towards a new stationary value. According to the channel model, this increase of current reflects a rise in the equilibrium concentration of channels under the action of the electric field.

If the transients shown in Figure 7 are analyzed, it is found that the time course of the current is described by a single relaxation time τ_c. A plot of $1/\tau_c$ as a function of $(\lambda_\infty)^{1/2}$ gives a straight line in accordance with Eq. (2). The results of the relaxation experiments are thus consistent with the dimer model. This, of course, does not mean that more complicated models may be excluded at this stage of analysis. For instance, channel formation may be a two-step process, the first step being the formation of an inactive dimer, which in the second step is transformed into the conducting channel by a monomolecular reaction. Or there may be a preequilibrium between unreactive and reactive monomers (Urry, 1972b), only the latter being capable of forming dimers. While these possibilities require further investigation, the main result of the relaxation experiments is the conclusion that the overall reaction in channel formation has a molecularity that is higher than one.

Accepting for the moment the simple dimer model, we may analyze the experimental result in terms of Eq. (2) and determine k_D and k_R from the intercepts and the slope of the $1/\tau_c$ vs. $(\lambda_\infty)^{1/2}$ plot, using the Λ value that has been obtained from the single-channel experiments. For a dioleoyllecithin membrane at 25°C this gives

Table 1. Temperature dependence of gramicidin A channel parameters. Dioleoyllecithin/n-decane membrane, 1 M NaCl, 25°C (data from Bamberg and Läuger, 1973, 1974; Bamberg and Benz, 1976). k_R: association rate constant; k_D: dissociation rate constant; Λ: single-channel conductance; τ^*: mean lifetime

T (°C)	k_R ($cm^2 mol^{-1} sec^{-1}$)	k_D (sec^{-1})	Λ (pS)	τ^* (sec)	$1/k_D$ (sec)
10	$2.3 \cdot 10^{13}$	0.25	6.5	3.8	4.0
25	$20 \cdot 10^{13}$	1.6	12	0.72	0.63
40	$68 \cdot 10^{13}$	4.5	23	0.15	0.22

$k_D = 1.6\ sec^{-1}$ and $k_R = 2.0 \cdot 10^{14}\ cm^2 \cdot mol^{-1}\ sec^{-1}$ (Bamberg and Läuger, 1973, 1974). According to the dimer model, $1/k_D$ should be equal to the mean lifetime τ^* of the conducting channel, which can be independently determined from a single-channel experiment. τ^* and $1/k_D$ have been measured at 10, 25 and 40°C. As seen from Table 1, τ^* varies about 20-fold in this temperature range, but $1/k_D$ and τ^* agree within about 40% at all temperatures.

The value of the association rate constant k_R may be illustrated in the following way. If the monomers are present in the membrane in such a concentration that the mean distance is 10^3 Å, then a given monomer will associate with another monomer with a frequency of 2 per sec. As may be expected, this rate is considerably less than the limiting value of a diffusion-controlled reaction (Bamberg and Läuger, 1973). It may be instructive to calculate from the equilibrium constant of association, $K = k_R/k_D = 1.2 \cdot 10^{14}$ $cm^2 \cdot mol^{-1}$ (25°C), the ratio of monomers and dimers in the membrane under the conditions of the conductance experiments. If the actual numbers of monomers and dimers in a membrane of area A are n_m and n_d, respectively, then the ratio n_m/n_d is given by:

$$\frac{n_m}{n_d} = \sqrt{\frac{LA}{n_d K}} \tag{3}$$

For a membrane of area $A = 1\ mm^2$, the ratio n_m/n_d in a single-channel experiment ($n_d = 1$) is calculated to be about $7 \cdot 10^3$. The finding that a large number of monomers is present in the membrane under the conditions of a single-channel experiment is consistent with the observation (Hladky and Haydon, 1972) that at low conductance levels the probability for the simultaneous occurrence of exactly n channels follows the Poisson statistics. On the other hand, at high conductance levels, 10^8 or more channels may be present in a membrane of $A = 1\ mm^2$, and under these conditions, n_m/n_d becomes of the order of unity.

The mechanism by which the electric field influences the number of conducting channels is not known with certainty. One possibility discussed previously (Bamberg and Läuger, 1973) consists of the assumption that the decrease in membrane thickness brought about by the compressing force of the electric field tends to stabilize the channel. This explanation is suggested by the fact that the estimated length of the channel is less than the thickness of the

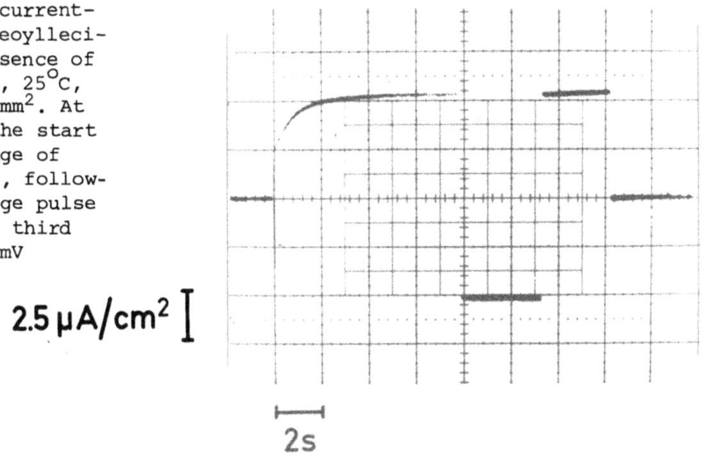

Fig. 8. Voltage-jum current-relaxation of a dioleoyllecithin membrane in presence of gramicidin. 1 M NaCl, 25°C, membrane area A = 8 mm^2. At time t = 2 s after the start of the trace a voltage of + 150 mV was applied, followed by a second voltage pulse to - 150 mV and by a third pulse back to 9 150 mV

membrane. Recent experiments (Bamberg and Benz, 1976) have shown, however, that the relaxation amplitude in the voltage-jump experiment is not correlated with the compressibility of the membrane. For instance, membranes made from lipid monolayers according to the method of Montal and Mueller, which are virtually incompressible, nevertheless show a pronounced voltage-dependence of channel formation. This finding indicates that there is a direct influence of electric field on the gramicidin molecule in the membrane.

A possible explanation for a direct field-effect on the channel formation kinetics has been proposed by Urry (1972b) who postulated an equilibrium between a nonconducting anti-β_2^6-helix and a conducting $\pi_{L,D}^6$-helix. Urry also presented arguments that the anti-β_2^6-helix has a smaller net dipole moment than the $\pi_{L,D}^6$-helix and that the anti-β_2^6-helix is not able to form stable head-to-head dimers. The action of an electric field in the membrane would then consist of shifting the equilibrium toward the form with the higher dipole moment, which is the conducting form.

Further information on the nature of the field effect may be obtained from the following experiment (Bamberg and Benz, 1976). A voltage V is applied to the membrane for a time sufficiently long to reach a steady-state current J, and then the voltage is quickly switched to -V (Fig. 8). At the switching of the voltage the current immediately jumps to -J and thereafter stays constant. This means that the channel does not possess an intrinsic structural polarity with respect to the direction of the electric field. Otherwise, the switching from V to -V should cause a disappearance of some of the channels followed by the formation of new channels of opposite polarity.

The finding that gramicidin channels do not possess an intrinsic polarity is consistent with the model of a head-to-head linked dimer. On the other hand, this model leads to difficulties with the foregoing explanation of the field effect. A dimeric channel formed by head-to-head association of two monomers exhibits mirror symmetry with respect to the central plane of the membrane, and therefore a field change that is energetically favorable to the active form of the monomer in one half of the bilayer is unfavor-

Valine-Gramicidin A

```
        1     2     3     4     5     6     7     8
HCO-L-Val-D-Gly-L-Ala-D-Leu-L-Ala-D-Val-L-Val-D-Val
                                                  |
HOCH2CH2NH-L-Trp-D-Leu-L-Trp-D-Leu-L-Trp-D-Leu-L-Trp
              15    14    13    12    11    10    9

                                -L-Phe  (B)
                                -L-Tyr  (C)
```

Fig. 9. Gramicidin A, B, and C

able in the other half. A more likely explanation is a quadratic field-effect on the free energy of the channel. For instance, if the active $\pi_{L,D}^{6}$-helix has a larger polarizibility than the inactive anti-β_{2}^{6}-helix, an increase of field strength would favor the former structure irrespective of the orientation of the helix with respect to the field. Another possibility (Gordon, 1973) would be to assume that inactive gramicidin monomers are adsorbed at the membrane surface. As the average dielectric constant of the monomer is larger than that of the hydrocarbon interior of the membrane, the monomer would be drawn into the membrane under the action of the electric field and thereby could be converted into the active form.

C. Studies with Analogs and Derivatives of Gramicidin

One of the main goals in studying model channels is a better understanding of the relationship between the chemical structure of the channel and its transport properties. A possible approach to this problem is to investigate chemical analogs of channel-forming compounds (Urry et al., 1971; Ovchinnikov, 1972; Goodall, 1973; Goodall and Urry, 1973; Bamberg et al., 1976; Apell et al., 1976; Bamberg and Janko, 1976).

I. Gramicidin A, B, and C

Natural gramicidin contains three main components named gramicidin A, B, and C (Gross and Witkop, 1965; Noda and Gross, 1972). Gramicidins B and C differ from A in that L-tryptophan in position 11 is replaced by L-phenylalanine and L-tyrosine, respectively (Fig. 9).

It may be seen from Figure 10 that the single-channel conductance Λ of gramicidin A is significantly higher than that of gramicidin B (Bamberg et al, 1976); gramicidin C shows similar Λ values as gramicidin A. A tentative explanation of the observed variation in Λ may consist of a change of electric field strength in the channel brought about by the different dipolar moments of tryptophane, phenylalanine, and tyrosine.

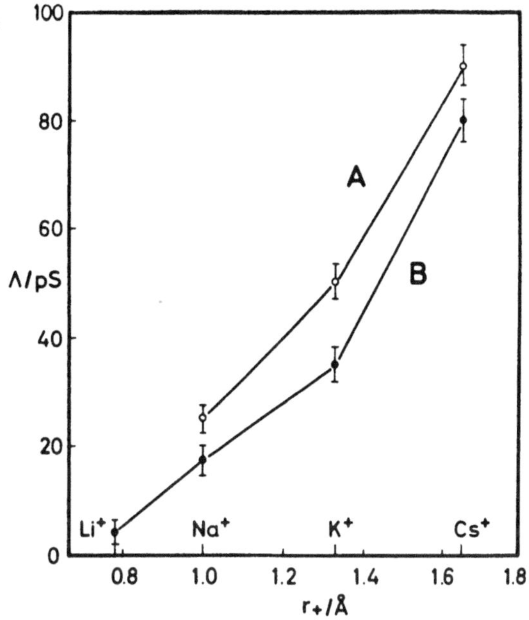

Fig. 10. Single channel conductance Λ of natural gramicidins A and B in glycerolmonooleate membranes; 25°C, ion concentration 1 M ($1\ pS = 10^{-12}\ S = 10^{-12}\ \Omega^{-1}$). Values of Λ for gramicidin C are similar to those for gramicidin A (Bamberg et al., 1976)

II. O-Pyromellitylgramicidin

Another interesting modification of the channel is the introduction of electric charges at the channel mouth. For instance, recent studies of the effect of pH on the conductance of sodium channels in nerve indicate that the access of ions to the channel is controlled by a negatively charged group (Woodhull, 1973; Drouin and Neumcke, 1974; Hille et al., 1975). A negatively charged analog of gramicidin may be obtained by introducing a pyromellityl residue at the hydroxyl end of the peptide (Fig. 11). O-pyrommellitylgramicidin forms ion channels in lipid membranes in a way similar to that of gramicidin (Apell et al., 1976). The records of conductance fluctuations at very low peptide concentrations have qualitatively the same appearance as those obtained with gramicidin. Also the current relaxation after a voltage jump is similar for the neutral and the negatively charged peptide. In other properties, however, both compounds differ markedly.

If O-pyromellitylgramicidin is added to one side of a lipid membrane, only a small membrane conductance develops (Fig. 12). If the compound is then added also to the other side, a drastic increase of conductance is observed; under the experimental conditions given in Figure 12 the conductance changes by a factor of about 5×10^3. This finding is in sharp contrast to the behavior observed with unmodified gramicidin, which yields nearly the same membrane conductance, irrespective of whether the peptide is added to only one or to both aqueous compartments. These observations can be explained on the basis of the dimer model if it is assumed that the two halves of the channel can combine only if they are present on opposite sides of the bilayer. Unmodified gramicidin, which is strongly hydrophobic, is likely to cross easily the membrane, whereas the negatively charged O-pyromellitylgramicidin presumably stays on the side of the bilayer where it has been added.

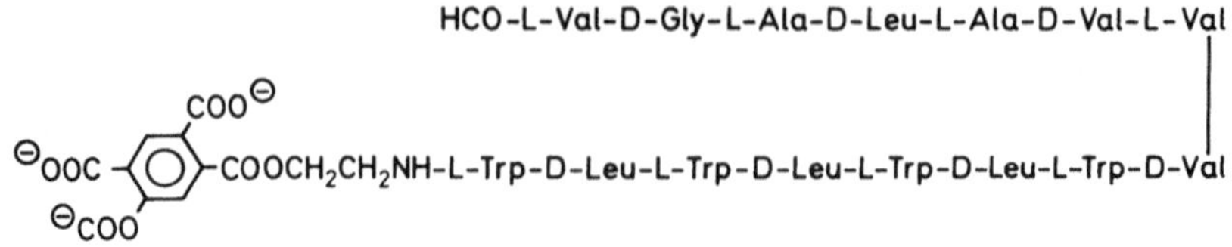

Fig. 11. O-pyromellitylgramicidin

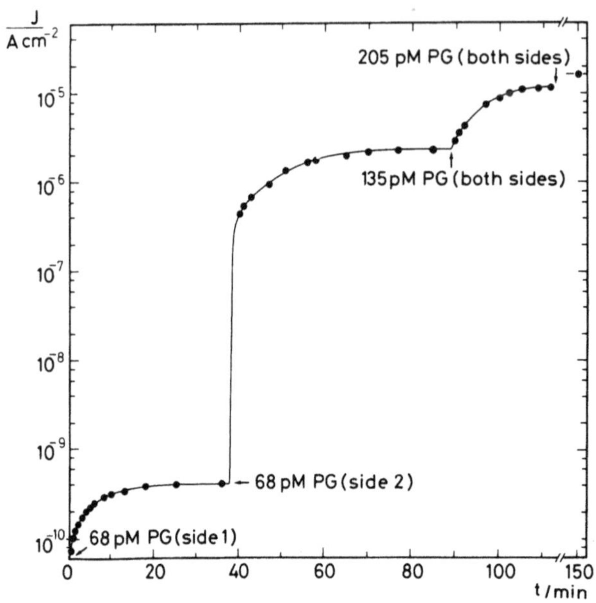

Fig. 12. Membrane current J after asymmetric and symmetric addition of O-pyromellityl-gramicidin (PG). The membrane was formed from dioleoyllecithin in n-decane. Both aqueous solutions contained 0.2 M CsCl, pH $\simeq$ 6.0 (unbuffered), $T = 25\,^{\circ}C$. The applied voltage was 25 mV. At time $t = 0$, 5 µl of a 2×10^{-7} M methanolic solution of PG were added to one aqueous compartment to give a concentration of 6.8×10^{-11} M. At $t = 38$ min, the same amount of PG was added to the other compartment; additional amounts were added later symmetrically to both sides up to a final concentration of 2.05×10^{-10} M (Apell et al., 1976)

As mentioned previously, it has proved rather difficult to determine the dependence of membrane conductance λ_o on the aqueous concentration of gramicidin. The difficulty of obtaining reproducible values of λ_o with given amounts of gramicidin added to the aqueous solutions apparently results from the strongly hydrophobic nature and the poor water-solubility of gramicidin. On the other hand, with O-pyromellitylgramicidin, which has a much higher water-solubility, reproducible conductance measurements can easily be performed. It can be seen from Figure 13 that the membrane conductance λ_o is proportional to the square of the aqueous concentration of O-pyromellitylgramicidin over a concentration range of at least 1:60. This result, of course, has to be expected if the channel consists of a dimer.

The analysis of the conductance fluctuations observed in the presence of small amounts of the peptide shows that the mean lifetime of the negatively charged channel and of the neutral channel are rather similar. There are striking differences, however, in the amplitudes of the fluctuations. It may be seen from Figure 14 that at low ionic strength (0.01 M) the conductance Λ_{PG} of the negatively charged channel is between five and six times higher than the conductance Λ_G of the neutral channel. On the other hand, at high ionic strength ($\geq$ 1 M), Λ_{PG} becomes equal to (or slightly less)

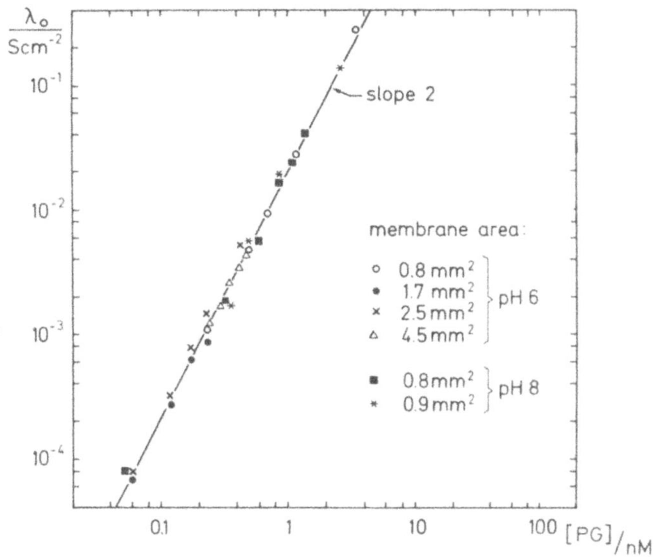

Fig. 13. Membrane conductance λ_o, expressed in S cm^{-2} (1 S = 1 Ω^{-1}, as a function of the concentration (PG) of O-pyromellitylgramicidin in water (Apell et al., 1976). Dioleoyllecithin/n-decane membrane, 0.2 M CsCl, T = 25°C. The applied voltage was 25 mV. Values of λ_o were obtained from different membranes with areas between 0.8 and 4.5 mm^2. In each experiment, the PG concentration was varied by successive symmetrical additions to both aqueous solutions. The pH of the solutions was either ~ 6 (unbuffered) or 8.0 (0.1 mM Tris)

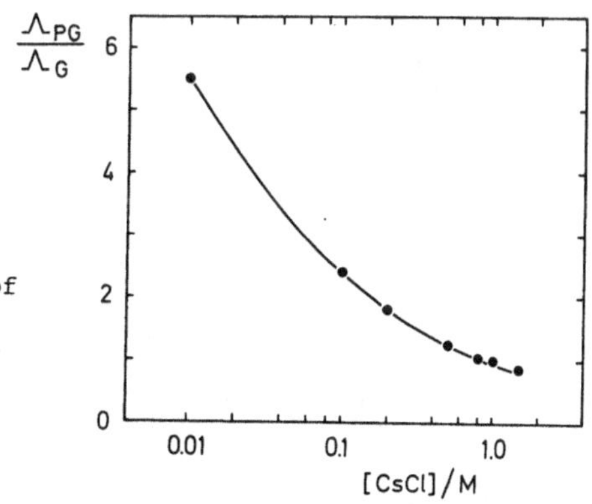

Fig. 14. Single-channel conductance Λ_{PG} of O-pyromellitylgramicidin divided by the single-channel conductance Λ_G of gramicidin. Monoolein/n-hexadecane, 25°C. The aqueous phase contained CsCl at various concentrations (Apell et al., 1976)

than Λ_G. These findings are indeed to be expected from the presence of fixed negative charges at the mouth of the O-pyromellityl-gramicidin channel. The electrical potential created by the negative charges leads to an accumulation of permeable positive ions near the opening of the channel; according to the Debye-Hückel theory of interionic attraction, this effect should become stronger at low ionic strength.

If a mixture of gramicidin and O-pyromellitylgramicidin is added to the aqueous phases, the analysis of the conductance fluctuations yields three peaks in the probability distribution (Fig. 15). The peaks at low and high Λ coincide with the peaks observed in pure gramicidin or pure O-pyromellitylgramicidin, respectively. A likely explanation of the intermediate peak is that it results from hybrid pores that are formed by association of a gramicidin monomer with a O-pyromellitylgramicidin. Such hybrid pores have also been observed by W. Veatch (personal communication) with a chemically modified gramicidin. The finding that a mixture of two different

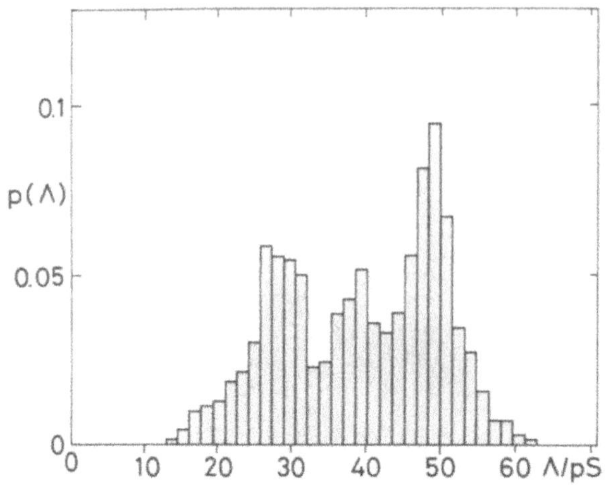

Fig. 15. Probability p(Λ) of conductance fluctuations in presence of mixture of gramicidin A and O-pyromellitylgramicidin A (Apell et al., 1976). Monoolein/n-hexadecane membrane in 0.2 M CsCL, pH ≃ 6 (unbuffered), T = 25°C. The total number of events was n = 700

gramicidin species yields, in addition to channels corresponding to the pure compounds, a single probability peak at intermediate conductances further supports the dimer model of the gramicidin channel.

If the channels of intermediate conductance are indeed hybrid dimers composed of a neutral and a negatively charged gramicidin monomer, then the current-voltage characteristic of the individual hybrid channel should be asymmetric. This expectation has been tested in experiments in which a small amount of gramicidin was added to side 1 and a much larger amount of O-pyromellitylgramicidin to side 2 (Fig. 16, curves 2 and 3). Under these circumstances the predominant form of the channel should be a hybride with the neutral monomer pointing to side 1 and the negatively charged monomer pointing to side 2. It was found that if cations were driven through the channel from the neutral side (side 1) to the negative side (side 2), the conductance was smaller by about 20% than for the opposite polarity (Fig. 16). Thus, even at the relatively low voltages applied in these experiments (± 50 mV), the single channel exhibits a distinct asymmetry in the current-voltage characteristic. The sign of the asymmetry is in accordance with expectations for a cation-permeable channel.

III. Malonyl-bis-Desformylgramicidin

Two gramicidin monomers may be covalently linked at the formyl ends by the bifunctional malonyl residue ($-COCH_2CO-$):

$$HO \text{———————} NH\text{-}COCH_2CO\text{-}NH \text{———————} OH$$

The malonyl dimer (malonyl-bis-desformylgramicidin) forms cation-selective channels of a conductance similar to that observed with normal gramicidin (Urry et al, 1971; Goodall, 1973; Bamberg and Janko, 1976). In contrast to normal gramicidin, the lifetime of the malonyl-bis-desformylgramicidin channel is very long (of the order of minutes) (Bamberg and Janko, 1976). If small amounts of the compound are added to the aqueous solution, the conductance

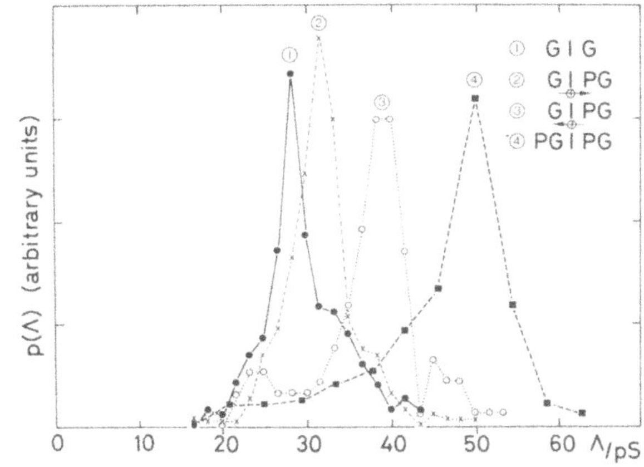

Fig. 16. Probability P(Λ) of conductance fluctuations (in arbitrary units) after symmetric and asymmetric additions of gramicidin A(G) and O-pyromellityl-gramicidin A(PG) to aqueous solutions (Apell et al., 1976). Monooleine/n-hexadecane membrane, 0.2 M CsCl, T = 25°C. Gramicidin A was added to give a nominal concentration of ~ 10^{-14} M and O-pyromellitylgramicidin A to give nominal concentration of ~ 10^{-11} M. The applied voltage was ± 50 mV. In the experiment represented by curve 2, the polarity of voltage was such that positive ions passed through the membrane from the gramicidin side to the O-pyromellityl-gramicidin side; curve 3 was obtained with recerse polarity. In symmetrical systems (curves 1 and 4), the conductance was independent of the polarity.For each system, values of p(Λ) have been multiplied by an arbitrary scale factor in order to obtain comparable peak heights

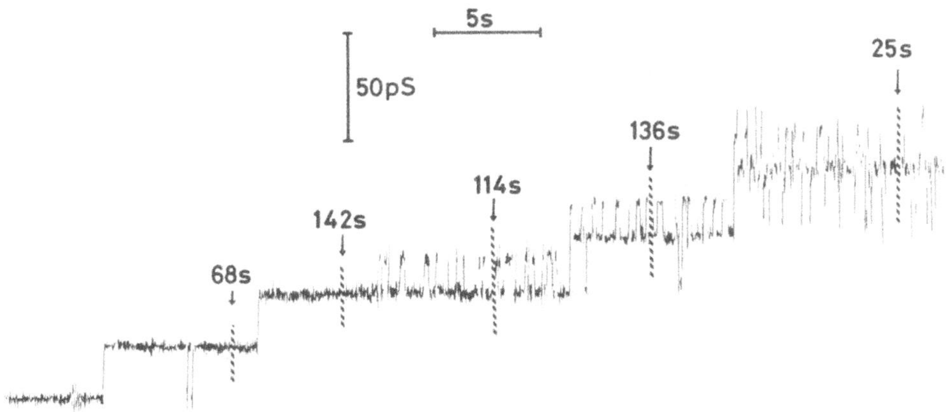

Fig. 17. Conductance fluctuations of a lipid membrane after addition of small amounts of malonyl-bis-desformylgramicidin to the aqueous phase (Bamberg and Janko, 1976). 1 M CsCl, 25°C. The membrane has been formed from a solution of 0.9% (wt/vol) of diphytanoyllecithin and 0.1% cholesterol in n-decane. Under same conditions, the channel lifetime of gramicidin A is of the order of 1 s

increases in a stepwise fashion (Fig. 17) whereas events corresponding to the closing of a channel are rarely seen. In addition to the long-lasting channels, the records show a second type of channel that frequently switches on and off (Fig. 17).

In the voltage-jump relaxation experiment, the covalently linked gramicidin dimer behaves completely different from normal gramicidin. In the whole concentration range, the relaxation amplitude is virtually zero (Bamberg and Janko, 1976). This finding shows that in the case of the covalently linked dimer, channel formation is a voltage-independent process.

D. Analysis of Channel Noise

In the past the kinetics of fast reactions (including transport processes in membranes) has been studied mainly by relaxation methods. An entirely different method is based on the principle that any reacting system exhibits random fluctuations around an equilibrium or stationary state (Feher and Weissman, 1973). By analyzing the time correlation of the fluctuations, information about the kinetic parameters of the system may be obtained. The fluctuation method has the advantage that the measurement is carried out while the system is in a stationary state.

The method of fluctuation analysis may be applied to the kinetic analysis of gramicidin channels in lipid bilayer membranes (Zingsheim and Neher, 1974; Kolb et al., 1975). A thin lipid membrane is, in the thermodynamic sense, a small system and is thus subject to relatively large fluctuations. In the case of a gramicidin-doped membrane, conductance fluctuations arise from the random formation and decay of channels. As described previously, from experiments with only one or a few channels present in the membrane, the single channel conductance and the mean lifetime of the channel may be obtained. It is not possible, however, to evaluate also the formation rate constant of the channel from single-channel experiments. A more complete analysis, yielding both k_D and k_R, becomes possible if the fluctuations of the macroscopic membrane conductance at different gramicidin concentrations are measured. Thus, the main difference between single-channel experiments and the study of macroscopic conductance fluctuations is that in the latter case additional information may be obtained from the variation of a further experimental parameter, the gramicidin concentration in the membrane.

If a constant voltage is applied to a bilayer membrane containing gramicidin A, the current J is found to fluctuate in time t around a mean value $\overline{J}$:

$$J(t) = \overline{J} + \delta J(t) \tag{4}$$

A straightforward way to analyze the fluctuating component $\delta J(t)$ of the membrane current is to compute the so-called autocorrelation function $C(\tau)$:

$$C(\tau) = \overline{\delta J(0) \cdot \delta J(\tau)} \tag{5}$$

$C(\tau)$ describes the average rate by which a current fluctuation decays that has occurred at time $\tau = 0$. The fluctuation-dissipation theorem states that the decay of a spontaneous fluctuation follows (on the average) the same time-law as the relaxation from a macroscopic perturbation. This means that for the dimerization reaction

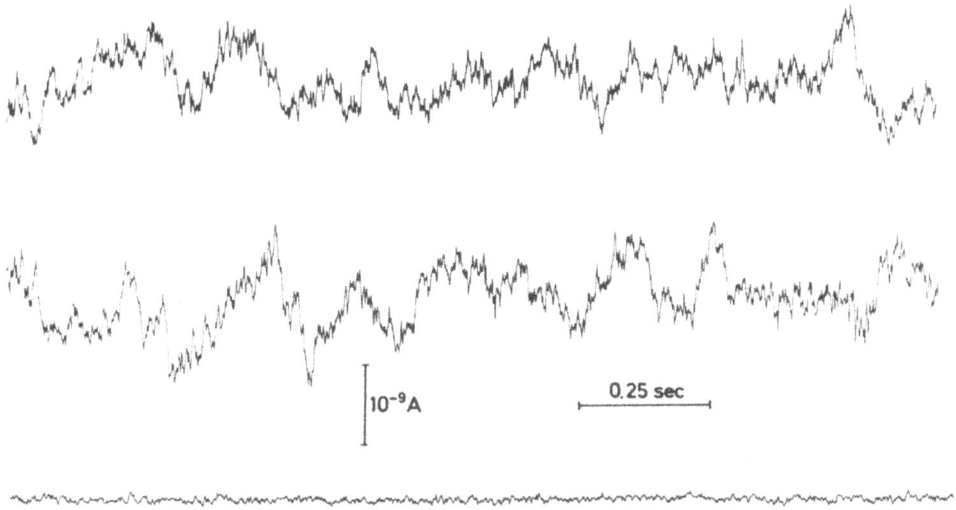

Fig. 18. Current noise from a dioleoyllecithin membrane in the presence of gramicidin A (upper traces) The mean current $\bar{J}$ was 390 nA, corresponding to a mean membrane conductance λ = 3.1 mS cm^{-2} (membrane area A = 0.68 mm^2, external voltage V_m = 18.5 mV). The aqueous phases contained 1 M NaCl. The lower trace represents a control experiment in which the noise was recorded in the same way as above but from a gramicidin-free membrane with an external resistor of 47 kΩ simulating the gramicidin-induced conductance (Kolb et al., 1975)

considered here, the autocorrelation function has the form:

$$C(\tau) = \overline{(\delta J)^2} \exp(-\tau/\tau_c) \quad (6)$$

where τ_c is the macroscopic relaxation time given by Eq. (2).

The autocorrelation function may be computed directly from a record of J(t) with the aid of suitable instruments (Zingsheim and Neher, 1974; Kolb et al., 1975). Figure 18 contains an example of the current fluctuations observed with a dioleoyllecithin membrane in the presence of gramicidin A. The autocorrelation function $C(\tau)$ of the current noise (Fig. 19) is found to be very nearly exponential with a single time constant τ_c. From here on the analysis proceeds in the same way as in the case of the relaxation experiments. The correlation time τ_c is plotted as a function of the square root of the mean membrane conductance; again, it is found that τ_c varies linearly with $\lambda_\infty^{1/2}$. The rate constants k_R and k_D that are obtained from the τ_c vs. $\lambda_\infty^{1/2}$ plot agree within a factor of two with the values calculated from the relaxation experiments (Table 2).

As the theoretical analysis shows, the mean square of the current fluctuations is related to the single-channel conductance Λ. Thus, from $C(0) = \overline{(\delta J)^2}$ information on Λ may be obtained. As shown in Table 2, the values of Λ that are calculated from noise analysis and from single-channel experiments are in satisfactory agreement.

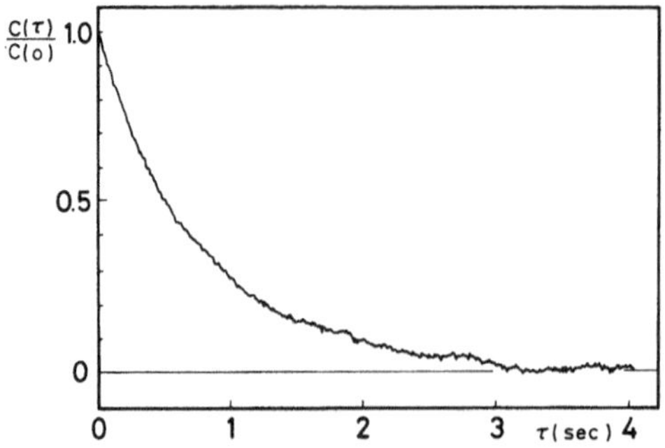

Fig. 19. Autocorrelation function $C(\tau)$ of current noise, divided by initial value $C(0) \simeq 7.8 \times 10^{-22}$ A^2. Membrane area $A = 0.70$ mm^2, external voltage $V_m = 18.5$ mV. The aqueous phase contained 1 M NaCl (Kolb et al., 1975)

Table 2. Kinetic analysis of the gramicidin A channel: comparison between noise analysis, relaxation method, and single-channel experiments (Kolb et al, 1975). Dioleoyllecithin membranes, 1 M NaCl, 25 °C

Method	k_R ($cm^2 mol^{-1} sec^{-1}$)	k_R ($cm^2 mol^{-1} sec^{-1}$)	k_D (sec^{-1})	k_D (sec^{-1})	Λ (pS)
	18.5 mV	98 mV	18.5 mV	98 mV	
Noise analysis	0.62×10^{14}	1.4×10^{14}	1.7	1.9	9
Relaxation method	1.3×10^{14}	1.5×10^{14}	-	1.6	-
Single-channel experiments	-	-	-	1.4	12

A second independent method for noise analysis consists of measuring the spectral density of the current noise (Kolb and Bamberg, 1976). The spectral density $S_J(f)$ is connected with the autocorrelation function $C(\tau)$ through the Wiener-Khintchine theorem; in the case considered here, $S_J(f)$ is given by

$$S_J(f) = \frac{4\,\tau_c\,\overline{(\delta J)^2}}{1 + (2\,\pi\,\tau_c\,f)^2} \tag{7}$$

where f is the frequency. Thus, $S_J(f)$ yields the same information as $C(\tau)$, namely the correlation time τ_c [Eq. (2)] and the variance $\overline{(\delta J)^2}$ of the current from which the single-channel conductance may be obtained. An experimental example of $S_J(f)$ is given in Figure 20; it is seen that the observed values of $S_J(f)$ may be represented rather accurately by the theoretical expression (Kolb and Bamberg, 1976). The values of τ_c and $\overline{(\delta J)^2}$ determined from $S_J(f)$ closely agree with those obtained from $C(\tau)$.

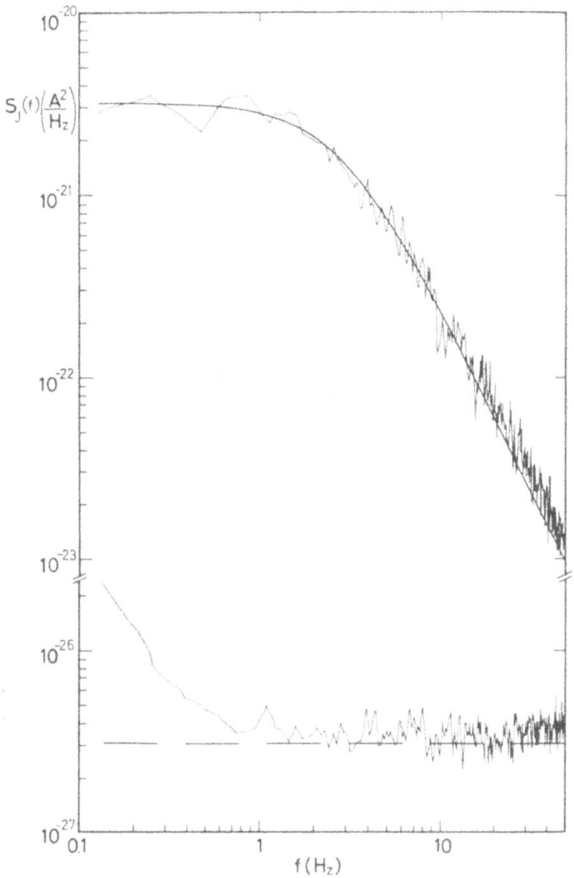

Fig. 20. Spectral density $S_J(f)$ of current noise vs. frequency f (Kolb and Bamberg, 1976). Glycerylmonoerucin/n-hexadecane, 1 M CsCl, 25° C, V_m = 30 mV. The upper trace represents the spectral density for a gramicidin-doped membrane with a mean macroscopic conductance of λ_o = 24 µS cm^{-2}. The theoretical curve was drawn according to Eq. (7) using the values $(\delta J)^2 = 1.4 \quad 10^{-20}$ A^2 and τ_c = 59 msec, which were determined from the simultaneously processed autocorrelation function. The lower trace is from a control experiment in which the current noise was recorded from a gramicidin-free membrane with an external resistor of 5.4 MΩ simulating the gramicidin-induced membrane conductance of the foregoing experiment. The conductance of the bare membrane was less than 25 nS cm^{-2}. The corresponding thermal noise level of the 5.4 MΩ resistor was computed from the Nyquist relation and is represented by the dashed line

E. Influence of Membrane Thickness and Ion Concentration on the Channel Properties

I. Kinetics of Channel Formation

It is known from the experiments of Hladky and Haydon (1972) that the lifetime of the gramicidin channel strongly decreases with increasing membrane thickness. A detailed study of the dependence of $k_D = 1/\tau^*$ and k_R on membrane composition has been carried out recently using autocorrelation, spectral density, relaxation, and single-channel measurements (Kolb and Bamberg, 1976). With membranes made from a series of monoglycerides with monounsaturated fatty acids, $\tau^* = 1/k_D$ was found to decrease by more than two orders of magnitude by increasing the chain length of the fatty acid from C_{16} to C_{24} (Fig. 21); at the same time k_R decreases about ten-fold between C_{18} and C_{24}. The variation of chain length from C_{16} to C_{24} corresponds to a variation in the hydrocarbon thickness d of the membrane from 2.8 to 6.9 nm (with n-hexadecane as solvent), or, with a channel length of 3.0 nm (Urry et al., 1971), to a variation in 1/d between 1.1 and 0.4. These studies further showed that the lifetime τ^* is not strictly correlated

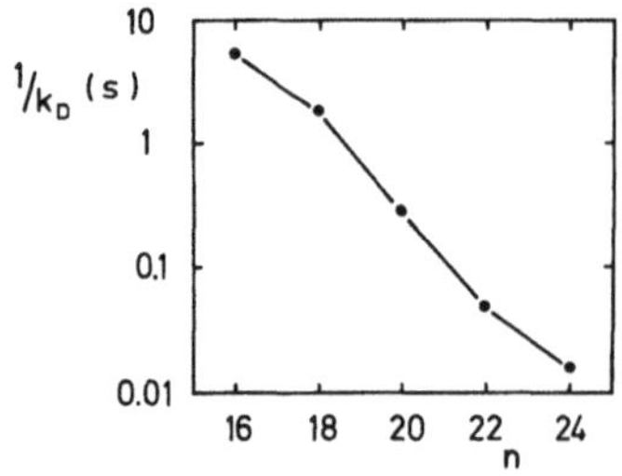

Fig. 21. Mean lifetime $\tau^* = 1/k_D$ of the gramicidin channel in membranes of different monoglycerides (Kolb and Bamberg, 1976). n is number of carbon atoms of the cis-monounsaturated fatty acid of the monoglyceride. The solvent was n-hexadecane. The k_D values were measured by the autocorrelation method in 1 M CsCl, except for membranes from monopalmitolein (n = 16) where 1 M NaCl was used; T = 25°C

with membrane thickness d. This is seen, for instance, by comparison of monoolein (C_{18})/n-hexadecane membranes ($d \approx 3.2$ nm, $\tau^* \simeq$ 2 sec) with solvent-free monoerucin (C_{22}) membranes formed by the Montal-Mueller technique ($d \approx 3.2$ nm, $\tau^* \simeq 0.7$). Despite the fact that the hydrocarbon thickness is approximately the same for both membranes, the values of τ^* differ by a factor of about 3.

The same studies revealed an interesting effect of aqueous ion concentration c_M on the kinetics of channel formation. Increasing c_M from 0.1 M to 1 M results in an decrease of k_D by a factor of about 3 and an increase of k_R by a factor of about 5 (monoolein/n-hexadecane membranes). This means that the equilibrium constant $K = k_R/k_D$ of dimerization is appreciably increased at higher ion concentrations. At the moment only a tentative explanation for the effect of c_M on K can be given. The concentration dependence of channel conductance indicates that at c_M around 1 M the channel becomes saturated, i.e., occupied by at least one cation (Hladky and Haydon, 1972). The influence of c_M on the dimerization could then simply be an electrostatic stabilization of the channel (if an ion is located in one half of the dimeric channel, then the replacement of the other half by an equivalent amount of hydrocarbon increases the free energy because the dielectric constant of the hydrocarbon is less than the average dielectric constant of the channel).

II. Single-Channel Conductance

In previous studies (Hladky and Haydon, 1972) it was found that the single-channel conductance Λ was largely independent of membrane thickness d. Measurements with monopalmitolein (C_{16}) and monoolein (C_{18}) membranes with either n-decane or n-hexadecane as solvent confirm these observations (Kolb and Bamberg, 1976). Despite a variation of d between 2.2 nm (monopalmitolein/n-hexadecane) and 4.8 nm (monoolein/n-decane), Λ remains virtually constant. However, with membranes made from monoglycerides with longer chain length n, the single-channel conductance strongly decreases with increasing n. The experimental results also indicate that Λ is determined mainly by the nature of the lipid and to a much lesser degree by the solvent used for membrane formation, despite a strong dependence of membrane thickness on the solvent. For instance, the Λ values (obtained by autocorrelation analysis) for monoerucin membranes formed from n-decane solution (d = 5.84 nm), n-hexadecane solution (d = 4.85 nm), and solvent-free monoerucin membranes (d = 3.15 nm) are not much different

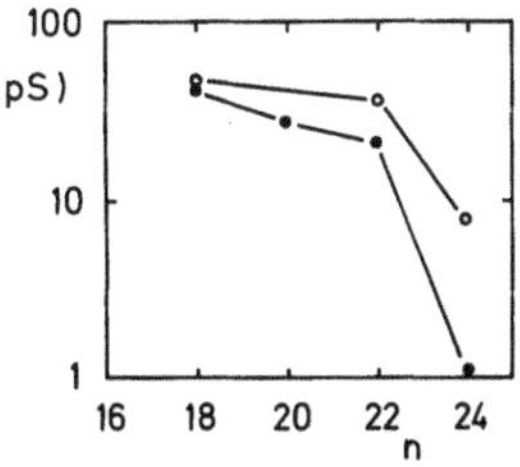

Fig. 22. Channel conductance Λ as obtained from single-channel studies (o) and autocorrelation analysis (•). n is the number of carbon atoms in the cis-monounsaturated fatty acid of the monoglyceride. The solvent was n-hexadecane, the aqueous phase contained 1 M CsCl ($25^{\circ}C$). The hydrocarbon thickness d varied between 3.18 nm (n=18) and 6.86 nm (n=24) (Kolb and Bamberg, 1976)

(5.8, 8,0, and 7.8 pS in 1 M NaCl). A similar conclusion holds for monopalmitolein and monoolein membranes in either n-decane or n-hexadecane.

These findings can be explained by the assumption that a membrane formed from a monoglyceride in a hydrocarbon solvent is inhomogeneous with respect to solvent content and thickness (Requena et al., 1975). If the membrane contains regions of different thicknesses, channels may form preferentially in the thinner regions. The properties of a region that contains less solvent than the average membrane is likely to be determined mainly by the nature of the lipid and much less by the solvent. This conclusion agrees with recent observations on the kinetics of carrier-mediated ion transport in membranes of different compositions (Benz, 1976).

The conductance Λ of the channel may be obtained either from single-channel analysis (Λ_{sc}) or from the autocorrelation function (Λ_a). In a number of systems Λ_a and Λ_{sc} were found to agree within the limits of experimental error (Zingsheim and Neher, 1974; Kolb et al., 1975; see also Table 2). On the other hand, in the case of the monoglycerides with longer fatty-acid chains (C_{22}, C_{24}), a marked discrepancy between Λ_{sc} and Λ_a was observed in 1 M CsCl, Λ_a being smaller than Λ_{sc} (Kolb and Bamberg, 1976; see Fig. 22). This discrepancy was found only at high ion concentration; it almost disappears in 0.1 M CsCl. The origin of the difference between Λ_a and Λ_{sc} is not yet clear. The main difference between the autocorrelation and the single-channel experiment is that the former is carried out with a membrane containing a very large number of channels. If the channels interact with each other, then Λ_a would be the conductance of a channel modified by the presence of the other channels, and Λ_{sc}, the conductance of an isolated channel. Because the discrepancy between Λ_a and Λ_{sc} persists even at a very low average conductance in the autocorrelation experiment, the foregoing explanation is feasible only if a local clustering of channels is assumed. Interactions between channels within a cluster may occur by mutual perturbation of the geometrical structure of the peptide helices, or, more likely, by the space charge built up by ions occupying the channels. The hypothesis that the reduction of the channel conductance in a multichannel experiment originates from electrostatic interactions between channels is consistant with the finding that the difference between Λ_a and Λ_{sc} vanishes at lower ion concentrations.

Acknowledgment. This work has been financially supported by the Deutsche Forschungsgemeinschaft (Sonderforschungsbereich 138).

References

Apell, H.J., Bamberg, E., Alpes, H., Läuger, P.: J. Membrane Biol. (in press)
Bamberg, E., Benz, R.: Biochim. Biophys. Acta 426, 570-580 (1976)
Bamberg, E., Janko, K.: Biochim. Biophys. Acta (in press)
Bamberg, E., Läuger, P.: J. Membrane Biol. 11, 177-194 (1973)
Bamberg, E., Läuger, P.: Biochim. Biophys. Acta 367, 127-133 (1974)
Bamberg, E., Noda, K., Gross, E., Läuger, P.: Biochim. Biophys. Acta 419, 223-228 (1976)
Benz, R.: in preparation (1976)
Byrn, S.R.: Biochemistry 13, 5186-5193 (1974)
Drouin, H., Neumcke, B.: Pflügers Arch. 351, 207-229 (1974)
Eisenman, G., Sandblom, J., Neher, E.: In: Metal-Ligand Interactions in Organic and Biochemistry, Pullman, B. (ed.). Dortrecht: D. Reidel, Publ. Co. (in press)
Feher, G., Weissman, M.: Proc. Natl. Acad. Sci. U.S. 70, 870-875 (1973)
Fossel, E.T., Veatch, W.R., Ovchinnikov, Yu.A., Blout, E.R.: Biochemistry 13, 5264-5275 (1974)
Goodall, M.C.: Biochem. Biophys. Acta 219, 471-478 (1970)
Goodall, M.C.: Arch. Biochem. Biophys. 147, 129-135 (1971)
Goodall, M.C.: Arch. Biochem. Biophys. 157, 514-519 (1973)
Goodall, M.C., Urry, D.W.: Biochim. Biophys. Acta 291, 317-320 (1973)
Gordon, L.G.M.: J. Membrane Biol. 12, 207-216 (1973)
Gross, E., Witkop, B.: Biochemistry 4, 2495-2501 (1965)
Henderson, P.J.F., McGivan, J.D., Chappel, J.B.: Biochem. J. 111, 521-535 (1969)
Hille, B., Woodhull, A.M., Shapiro, B.J.: Phil. Trans. Roy. Soc. Ser. B, 270, 301-318 (1975)
Hladky, S.B., Haydon, D.A.: Nature (London) 225, 451-453 (1970)
Hladky, S.B., Haydon, D.A.: Biochim. Biophys. Acta 274, 294-312 (1972)
Johnstone, R.M.: Biochim. Biophys. Acta 413, 252-264 (1975)
Kolb, H.-A., Bamberg, E.: submitted for publication (1976)
Kolb, H.-A., Läuger, P., Bamberg, E.: J. Membrane Biol. 20, 133-154 (1975)
Krasne, S., Eisenman, G., Szabo, G.: Science 174, 412-415 (1971)
Läuger, P.: Science 178, 24-30 (1972)
Liberman, E.A., Topaly, V.P.: Biochim. Biophys. Acta 163, 125-136 (1965)
Montal, M., Mueller, P.: Proc. Natl. Acad. Sci. U.S. 69, 3561-3566 (1972)
Mueller, P., Rudin, D.O.: Biochem. Biophys. Res. Commun. 26, 398-404 (1967)
Myers, V.B., Haydon, D.A.: Biochim. Biophys. Acta 274, 313-322 (1972)
Neher, E.: Biochim. Biophys. Acta 401, 540-544 (1975)
Noda, K., Gross, E.: In: Chemistry and Biology of Peptides. Proc. 3rd Am. Peptide Symp. Meienhofer, H. (ed.). Ann Arbor, Michigan: Ann Arbor Science Publishers 1972, p. 671
Ovchinnikov, Yu.A.: In: Federation of European Biochemical Societies. 8th Meeting, Vol. 28. Amsterdam: North Holland 1972, pp. 279-306
Ovchinnikov, Yu.A., Ivanov, V.T., Shkrob, A.M.: Membrane-Active Complexones. Amsterdam: Elsevier 1974
Podleski, T., Changeux, J.-P.: Nature (London) 221, 541-545 (1969)
Pressman, B.C.: Proc. Natl. Acad. Sci. U.S. 53, 1076-1083 (1965)
Ramachandran, G.N., Chandrasekaran, R.: Prog. Peptide Res. 2, 195-215 (1972)
Requena, J., Haydon, D.A., Hladky, S.B.: Biophys. J. 15, 77-81 (1975)
Sarges, R., Witkop, B.: J. Am. Chem. Soc. 87, 2011-2020 (1965)
Tosteson, D.C., Andreoli, T.E., Tiefenberg, M., Cook, P.: J. Gen. Physiol. 51, 373S-384S (1968)
Urry, D.W.: Proc. Natl. Acad. Sci. U.S. 68, 672-676 (1971)
Urry, D.W.: Biochim. Biophys. Acta 265, 115-168 (1972a)
Urry, D.W.: Proc. Natl. Acad. Sci. U.S. 69, 1610-1614 (1972b)
Urry, D.W., Goodall, M.C., Glickson, J.S., Mayers, D.F.: Proc. Natl. Acad. Sci. U.S. 68, 1907-1911 (1971)
Veatch, W.R., Blout, E.R.: Biochemistry 13, 5257-5264 (1974)

Veatch, W.R., Fossel, E.T., Blout, E.R.: Biochemistry 13, 5249-5256 (1974)
Veatch, W.R., Mathies, R., Eisenberg, M., Stryer, L.: J. Mol. Biol., in press (1976)
Wenner, C.E., Hackney, J.H.: Biochemistry 8, 930-938 (1969)
Woodhull, A.M.: J. Gen. Physiol. 61, 687-708 (1973)
Zingsheim, H.P., Neher, E.: Biophys. Chem. 2, 197-207 (1974)

Kinetic Aspects of Membrane Transport

The Asymmetry of Sugar Transport in the Red Cell Membrane

W. WILBRANDT

Among the transport systems interpreted in terms of carrier mediation, the glucose transport in the erythrocyte membrane is the most extensively studied system. It is not "active" in the sense of transport against chemical gradients, but there is ample evidence for the participation of binding to a membrane component in the translocation. The transport kinetics have been worked out for the steady state, and a considerable number of predictions have been derived from the kinetic equations (Widdas, 1952; Rosenberg and Wilbrandt, 1955; Wilbrandt and Rosenberg, 1961; Stein, 1967). It is fair to say that the great majority of these predictions has been confirmed by experiment. However, there are a few discrepancies that have led to rejection or modification of the basic concept.

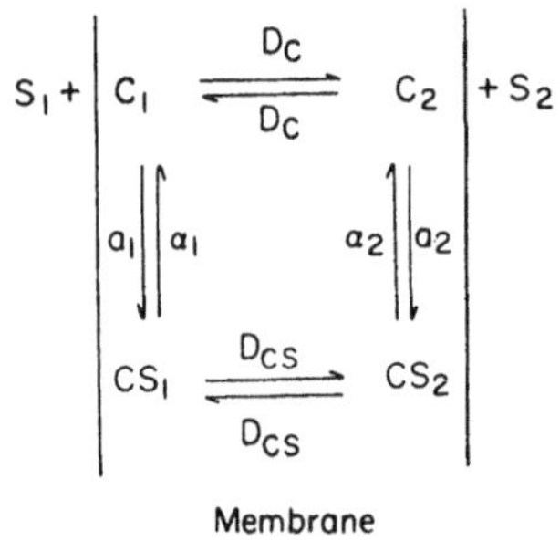

Fig. 1. Scheme of carrier mediation

The carrier system is shown schematically in Figure 1. Substrate S reacts reversibly with carrier C to form a transient transport complex CS. The chemical nature of the carrier C has been visualized by some authors (Rosenberg et al., 1956; Csaky and Ho, 1966) as lipid (in view of the lipid nature of the diffusion barrier) and by others (Katchalsky and Spangler, 1968; Lieb and Stein, 1970) as protein (in view of the chemical specificity of the transport resembling that of enzymatic reactions). According to the latter concept, by changes of conformation involving alternating access of the binding site to the two membrane surfaces rather than by actual transmembrane movement of the complex.

For specified conditions ($D_{CS} = D_C = D$; $\frac{1}{\alpha_1}\ \frac{1}{\alpha_2}\ \frac{1}{a_1}\ \frac{1}{a_2} \rightarrow 0$) the rate equation of the transport takes the simple form

$$v = v_{max}\ (\frac{S_1}{S_1 + K} - \frac{S_2}{S_2 + K}), \tag{1}$$

Partly supported by Schweiz. Nationalfonds, Projekt-Nr. 3.605-0.75.

with v = transport rate, v_{max}[= D.C_t] = maximum rate; K = Michaelis constant; C_t = total carrier concentration; other symbols as in Fig. 1).

According to this equation, which has been used extensively, the transport would be characterized by two parameters: the maximum rate v_{max} and the Michaelis constant K. Among the discrepancies mentioned, the most important seems to be a substantial difference in the numerical values of K as determined by different approaches (Wilbrandt and Rosenberg, 1961; Levine and Stein, 1966; Miller, 1968a,b). Thus, the evaluation of "zero trans" exit experiments according to Lineweaver and Burk (1934) in our laboratory yielded a value for K of 0.185, that according to Sen and Widdas (1962), a value of 0.019. This discrepancy will be referred to as "K paradox".

Table 1. D-glucose exit: parameters from exit experiments

Method	Temp. °C	m_o	m_i	b	K	n
Osmot. direct opacimetry	20	1.07	49.0	0.441	0.0050	2
Isotope	20	1.24	82.0	0.72	0.0070	3

It was shown previously (Wilbrandt, 1972; Fuhrmann, 1973) that the paradox can be resolved by the assumption of finite, but different reaction rate coefficients ($\frac{1}{\alpha_1} \neq \frac{1}{\alpha_2} > 0$) and unequal mobilities of free and loaded carrier ($D_{CS} \neq D_C$). Under these conditions the rate equation can be written in the form

$$v = \frac{S_1 - S_2}{S_1(m_2 + b + 2\frac{b}{K} S_2) + (m_1 + m_2)K + S_2(m_1 + b)} \tag{2}$$

(with $m_1 = (\frac{1}{\alpha_1} + \frac{1}{2D_{CS}}) \frac{1}{C_t}$, $m_2 = (\frac{1}{\alpha_2} + \frac{1}{2D_{CS}}) \frac{1}{C_t}$, $b = \frac{1}{2D_C C_t}$).

From experimentally determined exit rates of glucose from human red cells into different external concentrations, it was possible to evaluate numerically the four parameters m_i, m_o, b, and K. The striking feature of the results was that m_i (= m_1) appears to be larger than m_o (= m_2) by a factor of about 50, as shown in Table 1.

It was further shown that inserting these numerical values into the terms replacing the Michaelis constant in the formalisms of the Lineweaver-Burk and the Sen-Widdas procedures resolves the "K paradox": the numerical values of K as evaluated by the two procedures then become identical. It was concluded, therefore, that an asymmetry in the reaction rates at the two sides of the membrane ("reaction asymmetry") might underly the "K paradox".

There appeared, however, a possibility for a different kind of asymmetry. This emerged from observations on equilibrium exchange. If a substrate S in trace concentrations moves across the system, following its own gradient, in the presence of a second substrate R in equilibrium, the movement of S will be inhibited by R accord-

Table 2. Equilibrium exchange mixed

$R_1 = R_2 = R$

$S_1 \neq S_2$

$S << R$

R	S	"K_R" (isoton. = 1.0)
D-glucose	D-arabinose	0.041 ± 0.002
	D-ribose	0.053 ± 0.005
	D-xylose	0.157 ± 0.023
	D-galactose	0.183 ± 0.017

ing to affinity and concentration of R. A simple kinetic evaluation appears possible for the determination of the Michaelis constant K_R of the equilibrium substrate R. This constant, then, should of course be independent of the identity of the moving substrate S. Actually, however, such determinations yielded values that did depend on the identity of the moving substrate S in a systematic way: the higher the affinity and maximum rate of the moving substrate S, the higher the apparent Michaelis constant of the equilibrium substrate R (see Table 2).

It could be shown (see Fig. 2) that such behavior is to be expected in the case of a model first discussed by Miller (1968b) and termed by him diffusion limited system. In this model the carrier system proper is embedded within the membrane between two layers in which the movement of the substrate occurs without carrier mediation by thermal movement ("diffusion"). Quantitatively the system can be characterized by the resistance δ of the two layers (in terms of diffusion: δ = thickness of layer divided by diffusion constant). They will be referred to as "δ-resistances". The difference of the two concentrations in direct contact with the carrier is then changed from $(S_1 - S_2)$, as in Eqs. (1) and (2), to $[(S_1 - S_2) - (\delta_1 + \delta_2)\,v]$. The numerical value of $(\delta_1 + \delta_2)$ can be obtained from the apparent values of K_R obtained in experiments with two different substrates S_1 and S_2 by the plot shown in Figure 2 and a differential plot of $[(\frac{1}{K'_D})_1 - (\frac{1}{K'_D})_2]$ vs. R. For glucose at 20°C a value of 0.087 was obtained.

It appeared possible that an asymmetry in the δ resistances ($\delta_1 \neq 0$, $\delta_2 = 0$) could affect the kinetics in a way similar to that of a reaction asymmetry and, thus, could also be the basis of the K-paradox. The molecular structure underlying such an asymmetry could be visualized in terms of the well-known suggestion of Singer and Nicolson (1972) for the molecular membrane structure, if the carrier is assumed to be a nonpenetrating protein molecule that reaches the membrane surface on the outside but not on the inside.

The question of whether such a "structural asymmetry" could affect the kinetics in a way similar to that of the reaction asymmetry discussed previously can be approached in the following way. First it can be tested whether a structural asymmetry can likewise lead to a K paradox and, second, if this is so, which numerical value

$$S_1 \neq S_2$$
$$R_1 = R_2 = R$$
$$S \ll R$$

$$v_S = V_S \frac{(S_1 - S_2)K_R}{(R + K_R)\,K_S} = (S_1 - S_2)K_D$$

$$\downarrow \qquad \delta \neq 0$$

$$v_S = V_S \frac{(S_1 - S_2 - \delta v_S)}{(R + K_R)} \frac{K_R}{K_S}$$

$$\downarrow$$

$$v_S = (S_1 - S_2)K'_D$$

$$\frac{1}{K'_D} = \frac{K_S}{V_S} + \delta + \frac{K_S}{V_S} \cdot \frac{1}{K_R} \cdot R$$

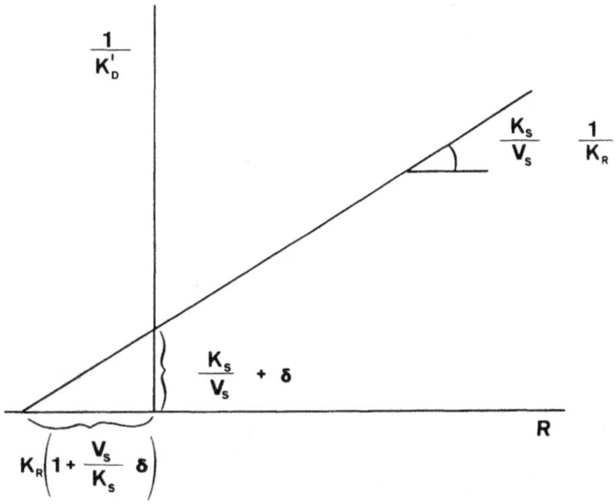

Fig. 2. Mixed equilibrium exchange with δ-layer

of δ would have to be assumed to account for the paradox actually observed. This value could then be compared with the value obtained from mixed equilibrium exchange.

In order to assess the effect of a δ resistance at the inside of the membrane on the result of the two evaluation procedures, the actual inside concentrations S_i in exit experiments were replaced by $(S_i - \delta v)$ and the evaluation repeated. The Sen-Widdas method was not affected by this change, since the inside concentrations do not appear numerically in the equations used. The Lineweaver-Burk plot remained linear but the slope decreased with increasing δ, yielding smaller values of K. By interpolation a value of δ could be found that equalized the K with that from the Sen-Widdas plot. It was $\delta = 0.070$, as compared to $\delta = 0.087$ as obtained from equilibrium exchange. Thus, two independent methods yielded reasonably similar results for the value of δ, and the introduction of this value into the evaluation procedures could abolish the K paradox equally as well as reaction asymmetry could. Thus, from this point of view, there is no reason to prefer one or the other interpretation.

There are, however, two observations that are not explained equally well by reaction asymmetry and structural asymmetry. The first is the equilibrium exchange paradox mentioned previously. In the formalism of mixed equilibrium exchange, the two reaction rate constants appear in reciprocal form in the reciprocal maximum rate. They do not, however, affect the value of K. Therefore, a difference between these rate constants (in other words, a reaction asymmetry) cannot underly the equilibrium exchange paradox, while, as shown above, the structural asymmetry can do so.

The second situation in which the two asymmetries give different results concerns the unidirectional fluxes (or, monofluxes). For the unidirectional flux from inside to outside, with the assumptions used in the derivation of Eq. (2), equations can be derived that show the following. A double reciprocal plot of flux rate M_1 against internal concentration S_i is expected to give a straight line under two conditions: (1) for an external concentration equal to the inside concentration ($S_2 = S_1$) and (2) for an external concentration equal to zero ($S_2 = 0$). The ordinate intercepts I of these two straight lines are

$$I = \left(\frac{1}{\alpha_1} + \frac{1}{D_{CS}} + \frac{1}{\alpha_2}\right) \quad \text{for } S_2 = S_1$$

and

$$\left(\frac{1}{\alpha_2} + \frac{1}{2D_{CS}} + \frac{1}{2D_C}\right) \quad \text{for } S_2 = 0.$$

Thus, here $\frac{1}{\alpha_1}$ is in the intercept for $S_2 = S_1$ but not in that for the line $S_2 = 0$. In terms of the resistances m_i, m_o, and b the intercepts are $(m_i + m_o)$ and $(m_o + b)$, respectively. Therefore, in the case of reaction asymmetry, if, according to the results given in Table 1, m_i is 50 times higher than m_o and b is relatively small, the intercept for $S_2 = S_1$ should be much higher than that for $S_2 = 0$, as shown in Figure 3. Actually the experiments show unequivocally that the intercept for $S_2 = 0$ is higher than that for $S_2 = S_1$, in accordance with previously (Levine et al., 1965) published results and with the assumption that $D_{CS} > D_C$. Thus, the results of the tests used in this analysis indicate that the assumption of reaction asymmetry not only can be replaced but has to be replaced by that of structural asymmetry.

Fig. 3. Double reciprocal plots of unidirectional flux M_1 vs. internal concentration S_i. Left: prediction based on reaction asymmetry; Right: observed

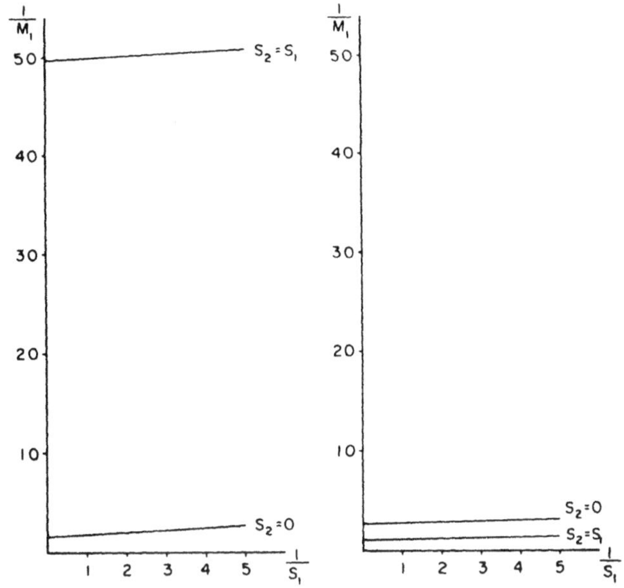

With respect to the question touched briefly in the introduction as to whether the assumed carrier should be visualized as a lipid or a protein molecule, the assumption of structural asymmetry would favor the protein concept.

We have approached this question experimentally from the following angle. The transmembrane movement of a lipid complex would be expected to be influenced by the fluidity of the membrane lipid. This is affected by a number of factors including the cholesterol content of the membrane. High cholesterol content is known to decrease fluidity. We have varied the cholesterol content of human erythrocytes by the method of Cooper et al. (1975) and compared the dependence of glucose transport on cholesterol content with that of valinomycin-mediated potassium transport, the latter being a well-established case of true mobile carrier operation with a lipid soluble transport complex.

For valinomycin-mediated potassium transport we found, as expected, a decrease of transport rate with increasing cholesterol content (Fig. 4). For glucose transport, however, there appeared a biphasic dependence (Fig. 5) with an inverse effect of cholesterol content (transport rate increasing with increasing cholesterol content) in a large part of the concentration range studied. These experiments, then, appear not to favor the assumption of a mobile lipid soluble complex in glucose transport (in accordance with Read and McElhaney, 1976) and, therefore, to be compatible with the assumption of structure asymmetry.

Acknowledgment. This work was partially supported by the Swiss National Science Foundation, Berne (Grant No. 3.605.075).

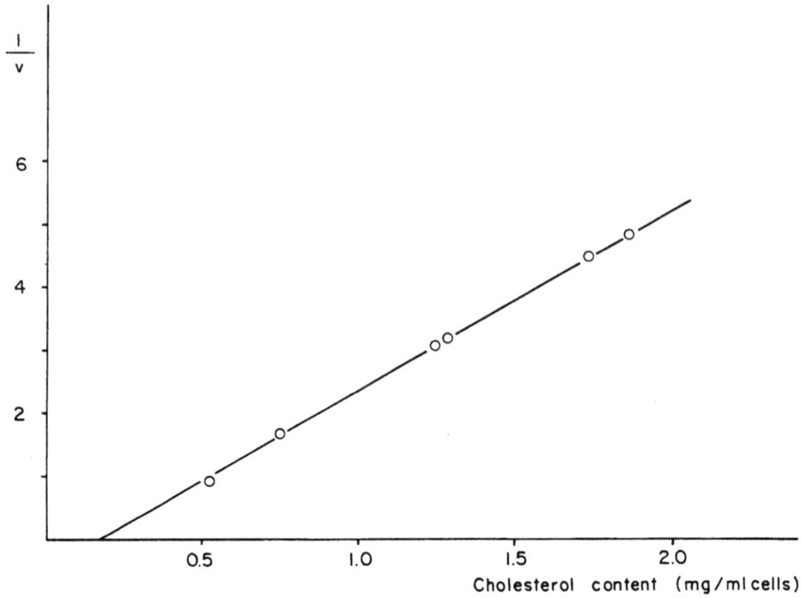

Fig. 4. Rate of K^+ loss from erythrocytes treated with 10^{-7} M valinomycin (v) as function of cholesterol content

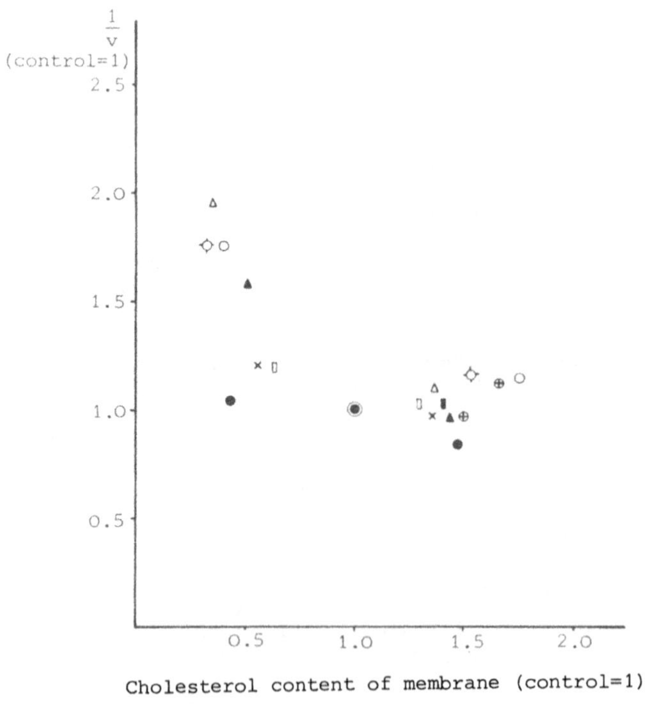

Fig. 5. Rate of glucose exit from erythrocytes (v), as function of cholesterol content

References

Cooper, R.A., Arner, C.E., Wiley, J.S., Shattil, S.J.: J. Clin. Invest. 55, 115-126 (1975)

Csaky, T.Z., Ho, P.M.: Proc. Soc. Exp. Biol. Med. 122, 860-865 (1966)

Fuhrmann, G.F.: Erythrocytes, Thrombocytes, Leucocytes; 2nd Intern. Symp., Vienna 1972. Stuttgart: Georg Thieme-Verlag 1973, pp. 102-105

Katchalsky, A., Spangler, R.: Quart. Rev. Biophys. 1, 127-175 (1968)

Levine, M., Oxender, D.L., Stein, W.D.: Biochim. Biophys. Acta 109, 151-163 (1965)

Levine, M., Stein, W.D.: Biochim. Biophys. Acta 127, 179-193 (1966)

Lieb, W.R., Stein, W.D.: Biophys. J. 10, 585-609 (1976)

Lineweaver, H., Burk, D.: J. Am. Chem. Soc. 56, 658-666 (1934)

Miller, D.M.: Biophys. J. 8, 1329-1338 (1968a)

Miller, D.M.: Biophys. J. 8, 1339-1352 (1968b)

Read, B.D., McElhaney, R.N.: Biochim. Biophys. Acta 419, 331-341 (1976)

Rosenberg, T., Vestergaard-Bogind, B., Wilbrandt, W.: Helv. Physiol. Acta 14, 334-341 (1956)

Rosenberg, T., Wilbrandt, W.: Exp. Cell Res. 9, 49-67 (1955)

Sen, A.K., Widdas, W.F.: J. Physiol. 160, 392-403 (1962)

Singer, S.J., Nicolson, G.L.: Science 175, 720-730 (1972)

Stein, W.D.: The Movement of Molecules Across Cell Membranes. New York-London: Academic Press 1976

Widdas, W.F.: J. Physiol. 118, 23-39 (1952)

Wilbrandt, W.: In: Biomembranes. Passive Permeability of Cell Membranes, Vol. 3. Kreuzer, F. (ed.). New York: Plenum Press 1972, pp. 79-99

Wilbrandt, W., Rosenberg, T.: Pharmacol. Rev. 13, 109-183 (1961)

Nonlinearity of Reciprocal Plots in Transport Kinetics

A. Kotyk

Determination of binding affinities of isolated transport proteins remains one of the methods generally used for identifying the protein with a part of the transport system in the intact cell. The dissociation constant K_D of the binding protein-ligand complex is then compared with the half-saturation constant K_T of transport and usually found to disagree by one or two orders of magnitude. Some of the reasons for this possible disagreement follow from kinetic considerations.

1. For example, the transmembrane rate of movement of the liganded carrier is substantially different from that of the free carrier. With the equilibrium approach (cf. Kotyk and Janáček, 1975), when the transmembrane movement is so slow as to permit virtual attainment of the association equilibrium at the membrane surfaces, we have

$$K_T = 2k_c K_{CS}/(k_c + k_{cs}) \tag{1}$$

where k_c and k_{cs} are the first-order rate constants of transmembrane movement of the free and the liganded carriers, respectively. Hence for $k_{cs} \gtreqless k_c$, we have $K_{CS} \gtreqless K_T$.

With the steady state approach (assuming that k_c and k_{cs} are each equal in both directions), we have

$$K_T = \frac{2k_c[K_{CS}(k_{cs} + k_{-2}) + k_{cs}k_{-2}k_1]}{2k_c k_{cs} + k_{-2}(k_c + k_{cs})} \tag{2}$$

where k_1 and k_{-2} are the association rate constant at side I and the dissociation rate constant at side II, respectively. If k_c and k_{cs} are of the same order of magnitude as k_{-2} for $k_{cs} > k_c$ we will expect $K_{CS} > K_T$.

The consideration may be extended to more complex reaction sequences with qualitatively identical results.

2. In the presence of unstirred layers at the membrane faces, the apparent K_T will increase over the true one (cf. Winne, 1973; Wilson and Dietschy, 1974; Dugas et al., 1975). If $K_{T_{app}}$ is the concentration giving rise to half-maximum transport rate, we have

$$K_{T_{app}} = K_{T_{true}} + 0.5\ J_{max}\ \delta/D \tag{3}$$

where J_{max} is the maximum rate of transport (in $mol \cdot s^{-1} \cdot cm^{-2}$), δ, the thickness of the layer (in cm), and D, the diffusion coefficient in the unstirred layer (in $cm^2 \cdot s^{-1}$). Hence the difference between the true and the apparent half-saturation constants will be greater for transports with intrinsically high maximum rates and the smaller, the thinner the unstirred layer. For intensely stirred suspensions of cells the error will thus be diminished. It is thus imperative that values of K_T reported in the literature be accompanied by a statement on the effect of stirring on their magnitude.

Other, more biochemical, reasons for the discrepancy between K_T and K_{CS} may lie in the isolation procedure. (1) If two states of the binding protein of different affinities exist, these may be obscured in the process. (2) Alterations of the binding affinity may take place. (3) Cofactors (ions, lipids, etc.), which in the membrane bear on the binding constant may be lost.

However, my chief concern here is with causes of nonlinearity of reciprocal plots of rate vs. concentration that may influence the estimation of the apparent half-saturation constant. Deviation from linearity in the present context is equivalent to a deviation from simple hyperbolic dependence of the initial rate on substrate concentration, which, for obvious reasons, has a much lower diagnostic value. In the following, the Lineweaver-Burk plot ($J_s{}^{-1}:s^{-1}$) will be used for the demonstration of the effects, but other related plots ($s\,J_s^{-1}:s$ or $J_s:J_s s^{-1}$) as well as the linear plot of s and J_s in parametric space (Eisenthal and Cornish-Bowden, 1974) will show corresponding deviations either from linearity or from uniqueness of the intersection point, as the case may be.

Five causes of intrinsic nonlinearity will be considered: (1) the presence of more than one system of uptake for a given solute; (2) the presence of unstirred layers at the cell surface; (3) excess-substrate inhibition; (4) cooperativity between multiple binding sites; (5) obligatory dependence of transport on a limited supply of energy.

A. More Than One Simultaneous System of Uptake

The existence of multiple uptake systems has been demonstrated in various biological objects, in particular in microorganisms and especially with amino acids, nucleic acid bases, and sugars. The presence of additional transport systems is detected only under certain conditions, as computed for a 10% uncertainty of measurement by Gross et al. (1972) (cf. Fig. 1). It should be pointed out that to derive the true half-saturation constants (and maximum fluxes) from a plot like that shown in Figure 2 one should resolve the curve mathematically (a method for doing this was described by Neal, 1972). However, practical application shows that the accuracy of deriving the slopes and the intercepts is such that an error of several-fold for the numerically smaller component is easily incurred. For instance, Figure 2 was plotted for $K_{T_1} = 10^{-2}$, $K_{T_2} = 10^{-3}$, $J_{max_1} = 10$, $J_{max_2} = 1$. Values derived from the plot according to Neal are $K_{T_1} = 1.03 \cdot 10^{-2}$, $K_{T_2} = 1.76 \cdot 10^{-4}$(!), $J_{max_1} = 9.91$, $J_{max_2} = 1.2$. The discrepancy in K_T's is often greater than when reading the values directly from extrapolated linear segments of the graph when $K_{T_1} = 9.1 \cdot 10^{-3}$,

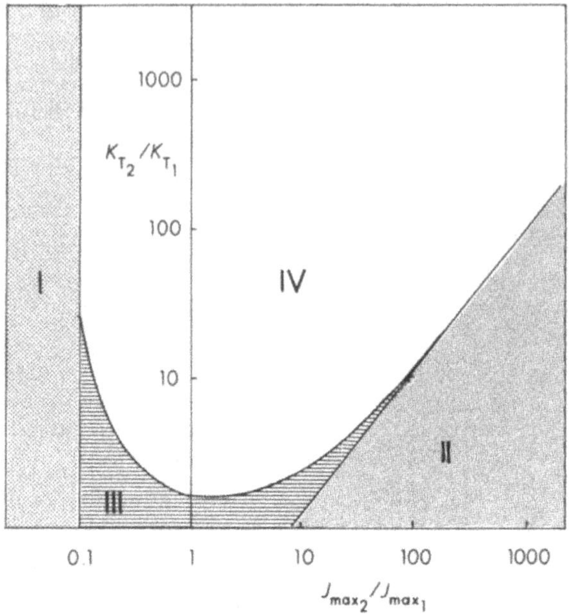

Fig. 1. Conditions under which two transport systems can be distinguished kinetically. Area I: only system I can be determined; area II: only system II can be determined; area III: presence of both systems is detectable but they cannot be separated; area IV: the two systems can be separated. (From Gross et al., 1972)

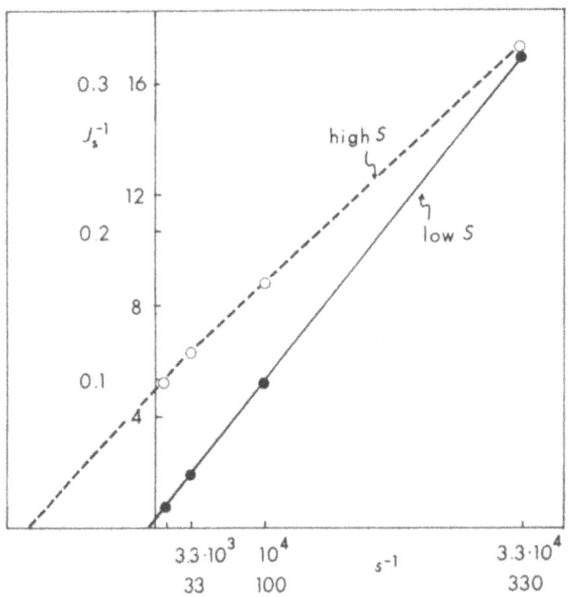

Fig. 2. A Lineweaver-Burk plot constructed for two-system transport, using $K_{T_1} = 0.01$, $K_{T_2} = 0.001$, $J_{max_1} = 10$, $J_{max_2} = 1$. Numerically smaller descriptions of axes refer to high substrate curve

$K_{T_2} = 1.2 \cdot 10^{-3}$, $J_{max_1} = 11.1$, $J_{max_2} = 2.8$. The only way to overcome this difficulty is to prepare mutants that lack one of the systems.

A situation resembling the presence of two or more saturable systems arises when there is a diffusion process accompanying the

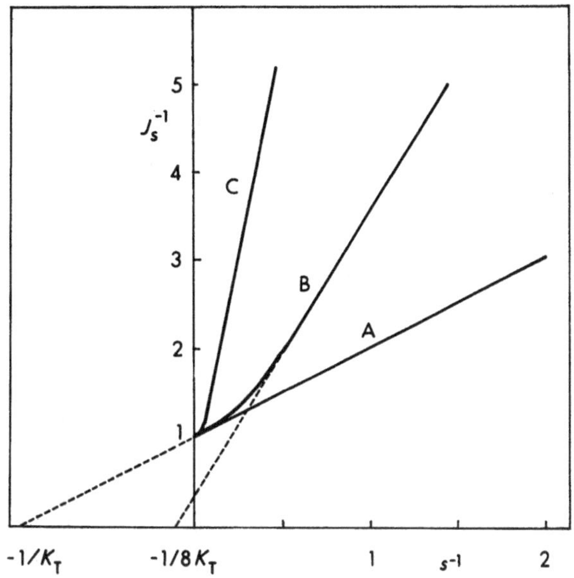

Fig. 3. Influence of an external unstirred layer on determination of K_T of carrier transport. Curve A: no unstirred layer; curve B: $K_T = 0.5\ J_{max}\ \delta/D$; curve C: $5K_T = 0.5\ J_{max}\ \delta/D$. Values used: $J_{max} = 1$ nmol $sec^{-1}cm^{-2}$, $K_T = 1$ mM, $D = 5 \cdot 10^{-6}\ cm^2\ sec^{-1}$, $\delta = 0.1$ mm (curve B) or 0.5 mm (curve C). (From Winne, 1973)

saturable one, with the Lineweaver-Burk plot converging on zero at high substrate concentrations.

B. Unstirred Layers at the Cell Surface

Besides the well-documented effect on the apparent K_T (see foregoing) there is a deviation from linearity at very high concentrations of substrate if the last term of Eq. (3) is large. The value of J_{max} read from the graph is correct, but the value of K_T can be assessed only if the segment of the plot near the J_s^{-1} axis can be linearly extrapolated whereupon the intercept on the abscissa yields the true value of $-1/K_T$ (Fig. 3).

C. Excess Substrate Inhibition

This is a phenomenon well known from enzymology, which, however, in one form or another, may occur in transport kinetics as well. In principle, the excess substrate inhibition can be caused by two mechanisms: (1) crowding of substrate molecules at the binding protein so that inefficient binding occurs; (2) binding to another, inhibitory, binding site on the protein. While the latter possibility is more likely, both result in a typical distortion of the reciprocal plots. The dissociation constant of the binary carrier substrate complex can be obtained from a Lineweaver-Burk plot; the dissociation constant of the ternary, abortive, substrate-carrier-substrate complex, can be read from a Dixon-type plot (Fig. 4).

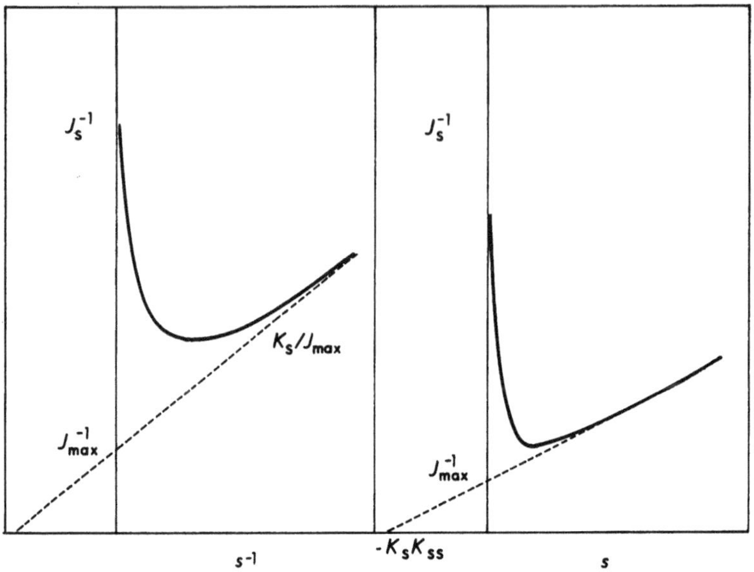

Fig. 4. Excess substrate inhibition as reflected in Lineweaver-Burk plot (left-hand panel) and in Dixon plot (right-hand panel)

D. Cooperativity Between Multiple Binding Sites

This is a situation well known from many regulatory enzymes and apparently not infrequent in transport kinetics (e.g., Borst-Pauwels et al., 1971). While negative cooperativity can be envisaged, positive cooperativity (for a single substrate at least) seems to predominate. The reciprocal plots (Fig. 5) resemble superficially those in which unstirred layers play an important role but can be distinguished from the latter by insensitivity to the rate of stirring and by more or less linear Hill plots, $\log[J_s/(J_{max} - J_s)]$ vs. log s, with slopes greater than unity.

Determination of the dissociation constants at the individual binding sites is beyond practical possibilities, and hence comparison with binding to an isolated protein is hardly meaningful. Since the saturation function (the fractional saturation with substrate) is linearly related to the relative rate of transport only under restricted conditions (only if the rate constant of transmembrane movement by either of the protein forms is much smaller than that of protein-substrate dissociation, and if the transmembrane movement by both protein forms proceeds at an equal rate), one cannot expect agreement between rate and binding in general. On the other hand, qualitative observations of cooperativity in a binding protein as well as in the suspected transport in vivo (plus possibly similar indexes of cooperativity) will be sufficient evidence for their identity.

Fig. 5. Positive cooperativity seen in Lineweaver-Burk plot, for different values of h, the Hill coefficient, using simplest phenomenologic equation $J_s = J_{max}s^h/(K + s^h)$

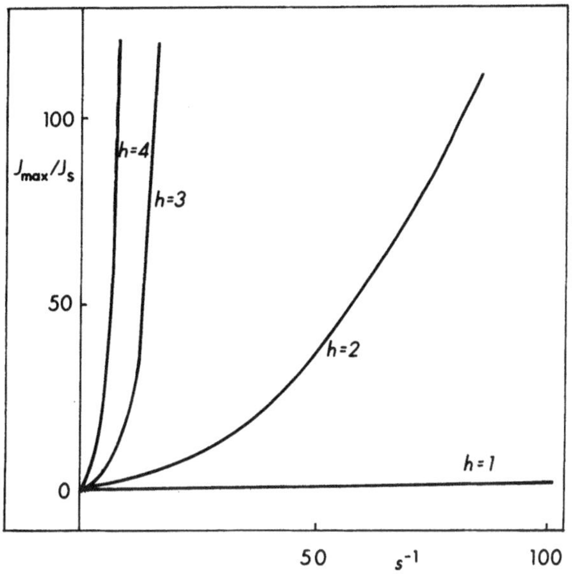

E. Obligatory Dependence on a Limited Supply of Energy

A hitherto unpublished source of nonlinearity of reciprocal plots in the case of active transport may lie in the dissipation of the energy source by the transport process. The observation has often been made that as concentration of transported substrate increases the steady state accumulation ratio decreases. It may either tend to unity or decrease below unity (Fig. 6). This last situation (e.g., Deák and Kotyk, 1968) has been described by some in terms of a unidirectional leak operating together with a unidirectional saturable uptake, but this is a physical impossibility unless sophisticated regulatory and energy-requiring processes are invoked.

A more realistic approach would be one in which the operation of a transport process is obligatorily linked to the interaction with a source of energy that is in limited supply. For the sake of simplicity, and in analogy to enzyme systems with relatively little substrate present, we may assume that the actually available amount of energy is the lower, the higher the amount of the energized carrier, thus $P_t = P + CS'$, where P is the free form of energy donor and CS' is the energized complex of carrier with substrate. We have designed several modifications of the carrier system in which P enters either the binding reaction I or the translocation II or the dissociation III. For illustration we prefer the translocation variant shown in the scheme:

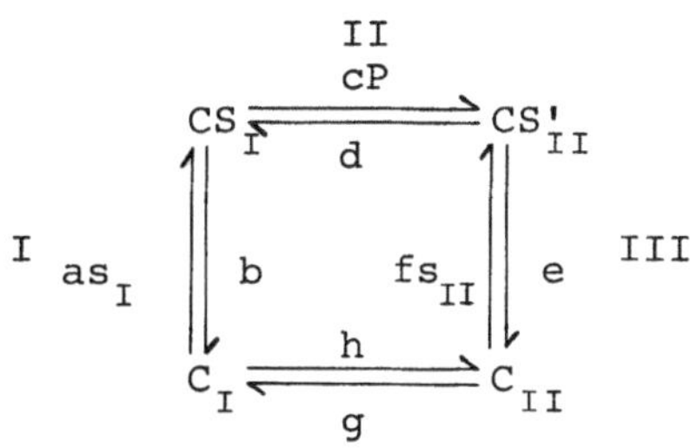

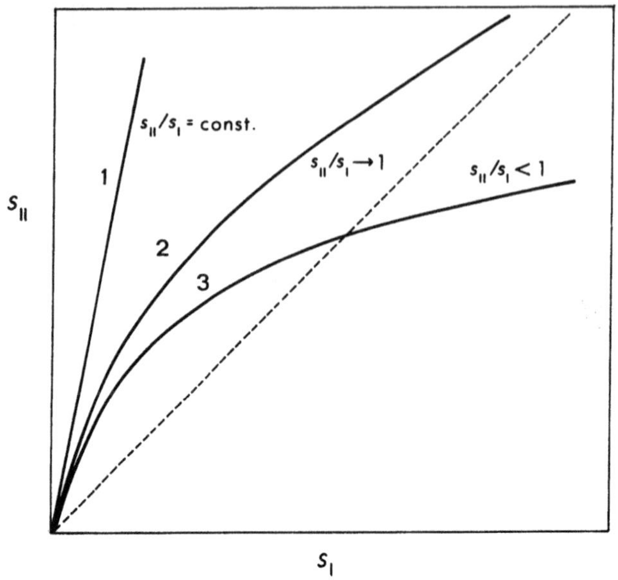

Fig. 6. Three possibilities of dependence of steady state accumulation ratio on concentration. Curve 1: with unlimited energy supply; curve 2: with limited energy supply but with nonobligatory coupling of movement to energy source; curve 3: with limited energy supply and with obligatory coupling of movement to energy source

The steady state ratio s_{II}/s_I can be readily shown to be equal to acegP/bdfh, and P can be computed by evaluating CS'/C_t (C_t being the total amount of carrier, equal to $C_I + CS_I + CS'_{II} + C_{II}$) and subtracting it from P_t (the total amount of energy supplied by the cell metabolism at a steady rate). The general character of the $s_{II}:s_I$ dependence under these conditions is shown in Figure 7, which may well reflect the situation in a living cell. (If CS_I could move without energization, the accumulation ratio s_{II}/s_I would tend to unity at high values of s_I).

A somewhat unexpected consequence of this treatment is the dependence of the initial rate of uptake on P, viz.

$$J_{s(0)} = C_t \frac{acegPs_I}{as_I[g(d+e) + cP(g+e)] + (g+h)[(d+e)b + ceP]}$$

$$= J_{max} \frac{s_I}{s_I + K_T} \qquad (4)$$

where

$$J_{max} = \frac{cegP}{g(d+e) + cP(g+e)} \qquad \text{and}$$

$$K_T = \frac{(g+h)[b(d+e) + ceP]}{a[g(d+e) + cP(g+e)]}$$

Fig. 7. Dependence of log s_{II} on log s_I computed from model shown in accompanying scheme using restrictions described in text. Numerical values employed were: a = 100, b = 10, c = 10, d = 0.001, e = 1, f = 100, g = 1, h = 1, C_t = 2, P_t = 1

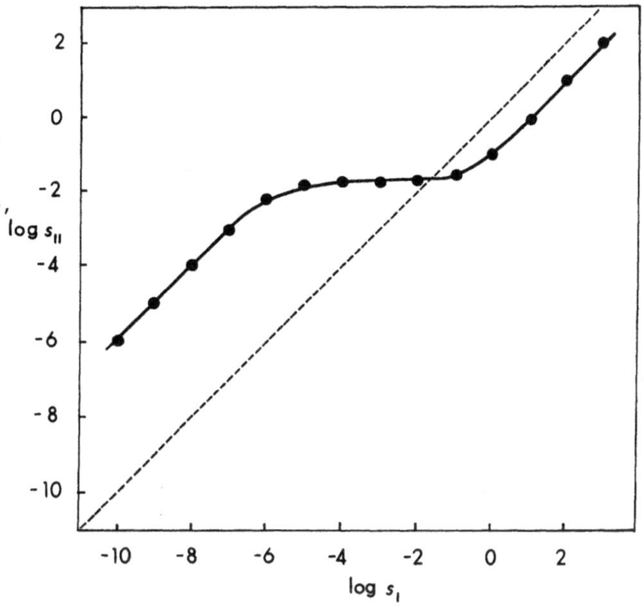

In this system the intrinsic J_{max} will decrease as s_I increases (since P decreases), while K_T will increase. In a plot like that in Figure 8 this is reflected in an opposite trend of the apparent maximum rate. Thus $J_{max(app)}$ increases with concentration (in this case from 0.33 at low to 0.40 at high values of s) while $K_{T(app)}$ changes from 0.013 at low to 0.022 at high values of s.

If the energization takes place at the association step (rate constant a), J_{max} will be unaffected while K_T will intrinsically decrease with increasing substrate concentration.

If the energization is associated with the dissociation step (rate constant e) intrinsic J_{max} will increase while K_T will also increase with increasing s_I.

It goes without saying that all these effects will be observable only if P_t is of the same order of magnitude or smaller than C_t.

A situation explainable by energization in step II (translocation) has actually been found for the transport of 6-deoxy-D-glucose in Candida parapsilosis and may be present in other cases not yet tested in detail.

In conclusion it should be stressed that linear reciprocal plots in transport kinetics, particularly in microorganisms, are an exception that should be viewed with caution.

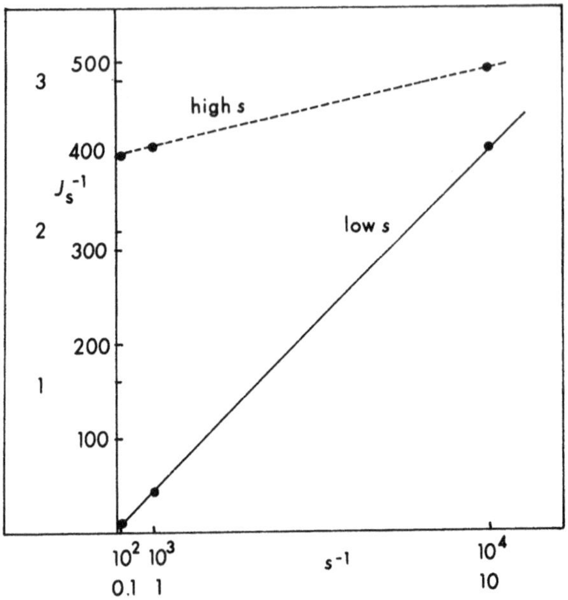

Fig. 8. Lineweaver-Burk plot constructed for initial rate of transport using same values of constants as in Figure 7. Numerically small descriptions of axes refer to high-substrate curve

References

Borst-Pauwels, G.W.F.H., Wolters, G.H.J., Henricks, J.J.G.: The interaction of 2,4-dinitrophenol with anaerobic Rb^+ transport across the yeast cell membrane. Biochim. Biophys. Acta 225, 269-276 (1971)

Deák, T., Kotyk, A.: Uphill transport of monosaccharides in Candida beverwijkii. Folia Microbiol. 13, 205-211 (1968)

Dugas, M.C., Ramaswamy, K., Crane, R.: An analysis of the D-glucose influx kinetics of in vitro hamster jejunum, based on considerations of the mass-transfer coefficient. Biochim. Biophys. Acta 382, 576-589 (1975)

Eisenthal, R., Cornish-Bowden, A.: The direct linear plot. A new graphical procedure for estimating enzyme kinetic parameters. Biochem. J. 139, 715-720 (1974)

Gross, W., Geck, P., Burkhardt, K.-L., Ring, K.: Kinetic analysis of two-compartment systems in transmembrane transport (multiple forms of transport systems). Biophysik 8, 271-279 (1972)

Kotyk, A., Janáček, K.: Cell Membrane Transport. Principles and Techniques. 2nd ed. New York-London: Plenum Press 1975, p. 77

Neal, J.L.: Analysis of Michaelis kinetics for two independent, saturable membrane transport functions. J. Theor. Biol. 35, 113-118 (1972)

Wilson, F.A., Dietschy, J.M.: The intestinal unstirred layer: its surface area and effect on active transport kinetics. Biochim. Biophys. Acta 363, 112-126 (1974)

Winne, D.: Unstirred layer, source of biased Michaelis constant in membrane transport. Biochim. Biophys. Acta 298, 27-31 (1973)

Transport in Epithelia and in Some Other Mammalian Cells

Hydrogen-Ion Dissociation as a Factor in Amino Acid Transport

H. N. CHRISTENSEN

Summary

The pH dependence of the entry of amino acids into the Ehrlich ascites tumor cell usually corresponds to simple expectations. As histidine is titrated from a neutral amino acid to a cation, its principal route of uptake changes from System L for neutral amino acids to System Ly$^+$ for cationic amino acids. The uptake of anionic amino acids (aspartic, glutamic, α-aminoadipic acids, and S-carboxymethylcysteine; also diiodotyrosine and two isomeric nitrotyrosines) increases by System L as these substances lose their side-chain charge by protonation. Comparisons within this group show that no specific recognition of the anionic group is necessary for this effect. In all these cases, a moderate shift is seen in the pH at which the acceleration is half-maximal, from the pK' value for titration in free solution at 37°C; but this shift can largely be accounted for by the selective binding of only one species of the amino acid.

The slow, difficult saturated uptake of glutamic acid by System A at pH 7.4 may prove an exception to the simple predictions. The loss of this transport as the pH is lowered is, however, intrinsic to System A and is not related to the protonation of the substrate. Exceptional behavior is also seen in the low pH transport of aspartic acid by System L, for which the usual stereospecificity is counteracted, apparently because the β-carbonyl group lies too close to the α-functional groups. The same shift in stereospecificity is seen in going from glutamine to asparagine.

The transport by Na$^+$-dependent System A and by Na$^+$-independent System L of several diamino acids occurs even though their pK_2' values may be high enough to make them 90% cationic in free solution. Furthermore, these uptake processes are unusually concentrative when pK_2' in free solution applies mainly to the α-amino group, i.e., when their dissociation should give mainly a nonsubstrate species. In contrast, structural modifications that specifically lower the pK' of the distal amino group do not lead to intense accumulation. This behavior shows that the membrane modifies both the course and the apparent pK' of the dissociation of some of these amino acids, with potentiating consequences for the uphill transport. On this basis we propose that the transport systems couple proton gradients generated within the membrane to amino acid movements through differential recognition of at least two different species of the substrate molecule. A possible role of amino acids in pairing to charge separations is also considered.

The three classes of amino acids represented by $H_3\overset{+}{N}\overset{R}{C}H - COO^-$, $H_3\overset{+}{N}\overset{R-NH_3^+}{C}H - COO^-$, and $H_3\overset{+}{N}\overset{RCOO^-}{C}H - COO^-$ present different problems for membrane transport, both for the binding step and for the subsequent translocation step. Long ago we all became familiar with the idea that three different kinds of transport systems handle these three different kinds of substrates. Suggestions that some transport systems might serve for two or for all three of these classes have been put forward from time to time, perhaps without an appropriate degree of concern as to how this could be managed, possibly without an appreciation of the physical problems introduced by the symbols - COOH or $-NH_2$, shown attached to the amino acid side-chain. If we wrote more correctly $-COO^-$ and NH_3^+ for these groups as above, this problem would be pressed on us more strongly. I do not mean to deny the existence of such systems, but to urge a proper respect for them if they do exist.

For example, Dent and Rose in 1951 proposed that cystine was similar enough in structure to lysine and arginine by reason of its second amino group to share a mechanism for renal tubular resorption with them, without concern as to how the predictable effects of its second anionic group might be nullified.

Questions of this kind make the amino acids such interesting and valuable solutes for the study of membrane transport - their structures are susceptible to so much variation without producing high physical permeation and without doing away with specific transport across biological membranes, provided certain requirements are met. I can illustrate this advantage by telling you that our inert buffer for those of our present transport experiments made below pH 5 was an amino acid, ε-aminocaproic acid, $pK_1' = 4.4$, at 20 to 40 mM.

A. Effects of Side-Chain Groups That Can Lose Their Charge Through pH Change, and Thereby Accommodate to a Transport System for Neutral Amino Acids

I. Histidine

Normalized pH profiles for the transport of histidine into the Ehrlich cell are shown in Figure 1 (Im and Christensen, 1976) at the right as a zwitterion and at the left as a cation, by the Na^+-independent System L for neutral amino acids and by the cationic amino acid transport system. These changes are logical consequences of the protonation of the imidazole group, although because of the energy of binding of only one molecular species by each transport system, the cross-over point does not fall exactly at the midpoint of the titration curve. The pK' for the imidazole group of histidine in free solution at 37°C is 5.9. Sodium-dependent System A also shows a curve resembling the one drawn here for System L, except that the corrections for the inhibition of the transport system itself by H^+ are in that case severe.

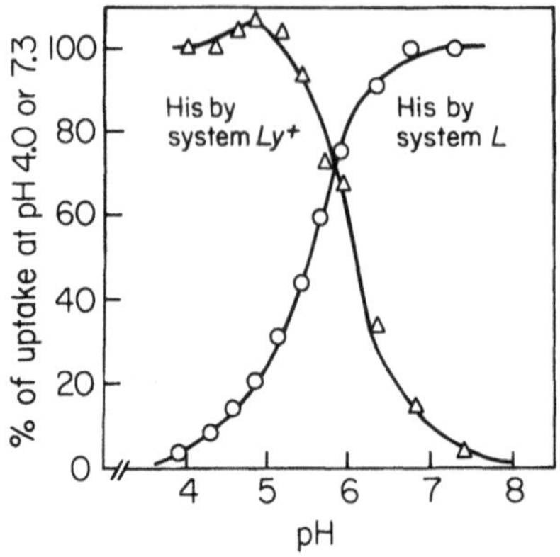

Fig. 1. Changes in rates of L-histidine uptake by Systems L and Ly^+ of Ehrlich cell as charge on molecule changes with pH. Actual rates observed for ^{14}C-histidine uptake were normalized for changes in operation of each transport system by reference to rates for model substrates, 2-aminoorbornane-2-carboxylic acid and homoarginine. Rates were then referred to rate observed at optimal pH values of 4 and 7.3 observed for histidine. Rate for System A declined somewhat more steeply than that for System L as pH was lowered. Results of Im and Christensen, 1976

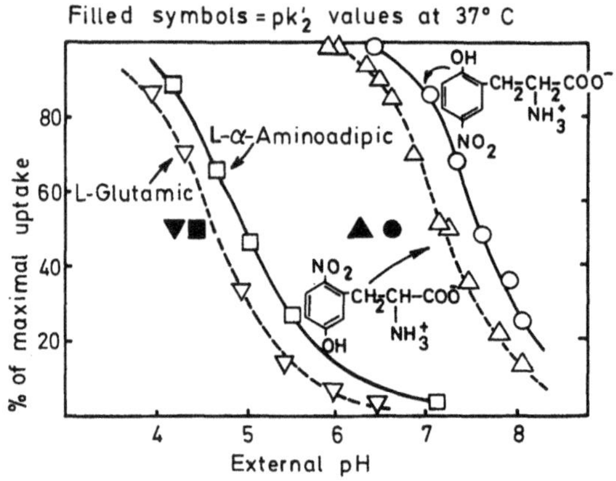

Fig. 2. Changes in rates of uptake of four anionic amino acid as pH is changed. Open circles, triangles and squares: rates of uptake at $37^{o}C$. Filled symbols: corresponding values of pK_2' for these amino acids at same temperature. Results of Garcia-Sancho et al., 1977a

II. System L Transport of Acidic Amino Acids

Another straightforward example of a similar kind is shown by glutamic acid uptake by System L (Fig. 2). As the glutamate anion is converted to the neutral zwitterion, glutamic acid, its uptake by that Na^+-independent system rises to a high rate (Garcia-Sancho et al., 1977a). Aspartic acid shows the same behavior, although its uptake is only about one-fifth as fast. The higher homologs of glutamic, α-aminoadipic, and S-carboxymethylcysteine each in turn show a higher rate of uptake, but cysteic acid shows neither mediated uptake nor inhibition of glutamic acid uptake. Since we can scarcely protonate the sulfonic acid group within physiological pH ranges, this amino acid remains an anion. If the Ehrlich cell had a transport system for anionic amino acids, both cysteic and glutamic acids might have been expected to be transported by it, even at neutral pH as is the case in some other cells and organelles. Cysteic and homocysteic acids accordingly deserve examination as model substrates for any transport system presumed to serve for anionic amino acids.

To identify the principal route of glutamic acid uptake at pH 4 to 5 with System L, we showed that this uptake was insensitive to in-

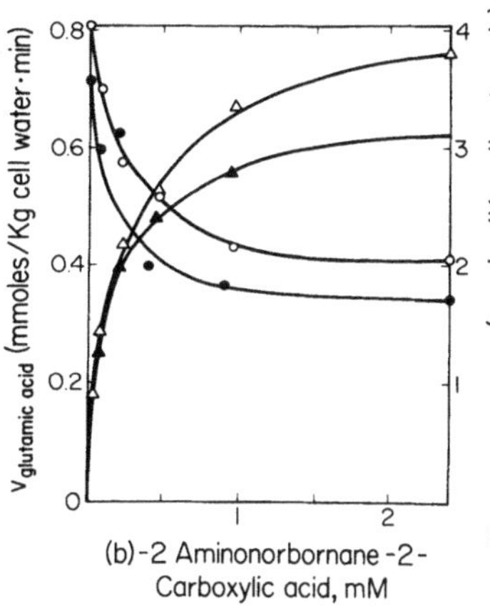

Fig. 3. Action of 2-aminonorbornane-2-carboxylic acid to inhibit uptake of glutamic acid (circles), compared with its action to stimulate exodus of previously accumulated ^{14}C-labeled norbornane amino acid from cell (triangles), both at pH 4.5. Concentrations producing half-maximal effects in both cases are indistinguishable. Results of Garcia-Sancho et al., 1977a

hibition by 2-(methylamino)-isobutyric acid (MeAIB), also almost entirely Na^+-independent, and fully sensitive to inhibition by the model substrate for System L, 2-aminonorbornane-2-carboxylic acid. Furthermore the kinetic parameters describing the interaction between glutamic acid and the norbornane amino acid are internally consistent, namely the concentrations of each half-saturating its uptake, and also those half-saturating their mutual trans-stimulation of uptake and exodus (see Fig. 3). We were driven to a thorough demonstration of the route of glutamic acid transport at low pH by the observation that 2-aminoisobutyric acid (AIB) becomes an effective inhibitor of glutamic acid uptake in this low pH range. This effect arises because System L becomes overwhelmingly the principal route of AIB uptake. Even the reactivity of glycine becomes measurable at pH 4.5. Therefore we found ourselves questioning the validity even of MeAIB and the norbornane amino acid as model substrates for Systems A and L over the full pH range 4 to 8. On reanalysis they proved, however to retain their validity throughout this range. Figure 4 shows the increasing rate with which AIB stimulates the exodus of the norbornane amino acid as the pH is lowered. Figure 5 shows the shifts in the roles of Systems A and L in AIB uptake as the pH is lowered.

Let me emphasize in passing that in this increase in the Na^+-dependent reactivity of AIB seen at low pH (seen earlier for its higher homolog, α,α-diethylglycine, and to some degree for a number of other amino acids; Christensen, 1973) we have a phenomenon not falling within our expectations. Although no group in the AIB molecule has been titrated, and although System L shows no other evidence of being itself titrated by the hydrogen ion in lowering the pH from 7.4 to 4.5, nevertheless that degree of acidification changes the specificity of System L and intensifies the reactivity of selected amino acids with it. What have we "titrated" in this case? We should like to find out.

We were also able to show that the uptake of glutamic acid by System L occasions the uptake of the hydrogen ion, as expected. For this purpose, Garcia-Sancho and Sanchez introduced the use of a new indicator for the hydrogen ion gradient across the membrane, namely salicylic acid. The principle is the same as that for the well-known DMO, except that the pK' of 3.0 for salicylic acid makes accessible a new pH range (Waddell and Butler, 1959).

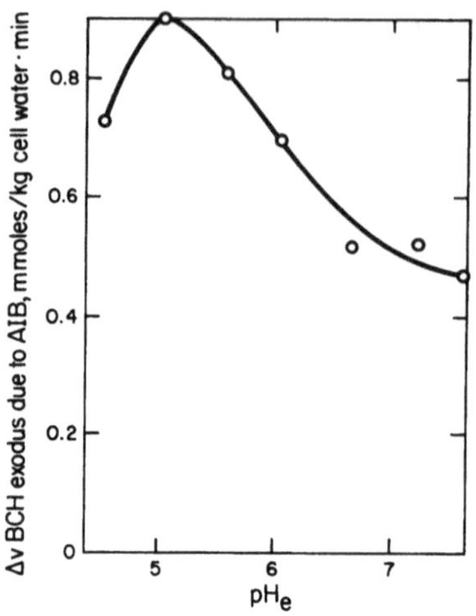

Fig. 4. Changing effectiveness of extracellular 2-aminoisobutyric acid (5 mM) in stimulating exodus of previously accumulated ^{14}C-b-2-aminonorbornane-2-carboxylic acid, as external pH is changed. Apparent internal concentration of norbornane amino acid, 1.7 mmol/kg cell water^{-1}. Exodus rate in absence of external amino acid ranged from 0.015 to 0.131 mmol/kg cell water^{-1} · min^{-1}. In agreement the rate of AIB uptake was doubled by lowering the pH. For sym. difluoro AIB, pK_2' 7.4, Na^+-independent uptake was accelerated 5-fold by the same pK_2 fall. Experiment by M.E. Handlogten

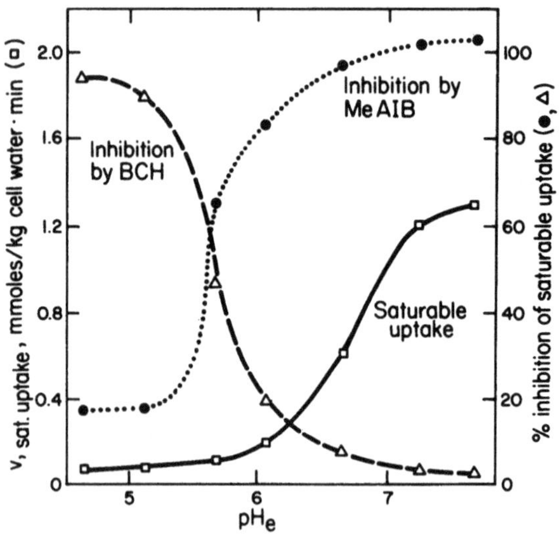

Fig. 5. Change with pH in sensitivity of uptake of 2-aminoisobutyric acid (0.2 mM) to two system-specific inhibitors. (.....): % inhibition of saturable uptake by 20 mM 2-(methylamino)-isobutyric acid. (-----): % inhibition by 20 mM 2-aminonorbornane-2-carboxylic acid (BCH). Note that rate of total saturable uptake is shown by solid line below

So far I have oversimplified my statement about the pH profiles showing the gain in transport rate by dicarboxylic acids. The mid-point for the increase in System L transport for each of the acidic amino acids lies at a pH about 0.4 units above the titration curve it shows in free solution (Fig. 2). The mid-point is also somewhat concentration-dependent. These departures can be largely accounted for by taking into account the strength of binding of the amino acids, which applies in this case only for the protonated species, which is thereby stabilized. With histidine we had a case in which the protonated species was the one bound and also a case in which the unprotonated species was the one stabilized.

This explanation does not exclude the possibility that the side-chain carboxyl group of the dicarboxylic amino acids is specifically recognized and accommodated by a special subsite at the receptor site. Perhaps the uptake of aspartic acid is so slow be-

Table 1. Reactivity of various amino acids with System L of the Ehrlich cell at pH 4.2, as shown by their K_i values in inhibiting uptake of norbornane amino acid and K_e values in stimulating its exodus from that cell. Results are expressed as percent of reactivity of glutamic acid. Results of Garcia-Sancho et al., 1977a

	L	DL	D	
S-carboxymethyl-L-cysteine	975			Ordinary stereospecificity
α-aminoadipic acid	407	190		
Glutamic acid	100		9	
Glutamine	90		5	
Aspartic acid	23		21	Counteracted stereospecificity
Asparagine	27		29	
Cysteic acid	1			
α-methyl-DL-glutamic acid		58		The β-carboxyl group is readily shifted to an even less favorable position in space[a]
α-methyl-DL-aspartic acid		0		
(β-alanine, iminodiacetic acid)		0		

[a]Also shown by contrast between cis and trans isomers of 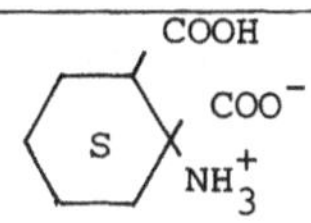

cause its short chain leads to difficulty in accommodating it to the position of such a possible subsite. Therefore, we tested some very different acidic amino acids whose side chains are anionic through the presence of a phenolic rather than a carboxylic group, namely diiodotyrosine, 2-hydroxy-5-nitrophenylalanine, and 5-hydroxy-2-nitrophenylalanine, whose pK_2' values range from 6.3 to 6.7. These amino acids also gained transport by System L as they were titrated to their neutral forms, even though their anionic groups have a quite different nature and take quite different relative positions in space (Fig. 2). Accordingly we conclude that the carboxyl group of the glutamic acid side chain receives no special recognition - it merely becomes acceptable as it loses its charge. In agreement its uptake and that of the higher homologs shows a degree of stereospecificity normal to System L (Table 1), namely ca. a 10:1 preference for the L isomer. Why then does L-aspartic acid show low reactivity for this system? We see in Table 1 that D-aspartic acid is fully as reactive with System L at pH 4.3 as L-aspartic acid. What we should say, I think, is that L-aspartic is as poorly transported as the D-isomer. We believe that the distal carboxyl group in this amino acid falls too close to the α-amino and the α-carboxyl groups, and that this nearness causes more difficulty for the L than for the D isomer, so that the normal preference for the L isomer is counteracted. This view receives support from the dramatic loss of transport reactivity seen when an α-methyl group is introduced into aspartic acid, a change that has little effect in the case of glutamic acid (see Table 1). The implication is that the orientation of the distal carboxyl group is critical in aspartic acid, a conclusion also supported by a comparison of the two isomeric 1-aminocyclohexane-1,2-dicarboxylic acids. Both of these aspartic acid analogs are weakly transported, but one of them is transported much more weakly than the other (Garcia-Sancho et al., 1977a).

We may then ask, what feature of the carboxyl group of L-aspartic acid interferes with its transport. Is it the tendency of the protonated oxygen to lose its hydrogen ion and then to be confounded with the corresponding portion of the α-carboxyl group? Or does the carbonyl group, given a true carbonyl character by the protonation, cause the difficulty? We know from experiments with amides and esters and with chloromethylketone analogs of amino acids that the carbonyl portion of the carboxyl group may sometimes be sufficient for reactivity with a transport system. We believe that we have answered that question in favor of the β-carbonyl group as the troublesome one in the case of aspartic acid. Asparagine uptake shows at pH 4.5 the same apparent absence of stereospecificity as does aspartic acid, whereas glutamine shows much the same stereospecificity as glutamic acid (Table 1). The $-NH_2$ group can thus substitute rather stably for the hydroxyl group of -COOH without loss of the effect on stereospecificity (Garcia-Sancho et al., 1977a).

An apparently unrelated aspect is that on lowering the pH below 4.5 System L itself tends to lose activity. But the pH at which this effect sets in is also determined by the structure of the amino acid substrate being tested. With falling pH, this effect occurs in the sequence, the norbornane amino acid, valine, and phenylalanine, the same sequence taken by their pK_1' values (2.8, 2.3, 1.8). An unfavorable protonation of the carboxylate group apparently is facilitated by the transport system.

III. System ASC and System A Transport of Glutamic Acid

Glutamine and asparagine are transported into various cells by another transport system known as ASC (for <u>alanine</u>, <u>serine</u>, and <u>cysteine</u> (Christensen et al., 1967; Koser and Christensen, 1971; Thomas and Christensen, 1971)), as well as by Systems A and L. Furthermore, system ASC is much less pH sensitive than System A. Hence it was no surprise to find that L-glutamic acid shows a small component of uptake at pH 4.5 inhibitable by methionine but not by MeAIB or by the norbornane amino acid. The kinetic plot, not presented, corresponds to a K_m of 12 mM and a V_{max} of 1.3 mmol per kg cell water·min for this component. We could observe no such component of uptake for D-glutamic acid. These features, including the high stereospecificity, are characteristic of System ASC.

At pH 7.4 the principal part of the slow uptake of glutamic acid occurs by System A, this component being both Na^+-dependent and inhibitable by MeAIB at the appropriate concentration. This uptake is rapidly lost as the pH is lowered, as is always the case for System A. This uptake could prove very interesting, because it might represent uptake of an anionic amino acid by a neutral system, since it occurs when the pH is over 3 pH units removed from the pK_2' of glutamic acid. On the other hand the uptake at pH 7.4 cannot readily be saturated with glutamate, so even in this case the real substrate may be the protonated species.

B. Effects on Transport of a Second, Distinctly Cationic Group on the Amino Acid Molecule

In the normal case in which the pK' of the side chain cationic group is 10 or higher (e.g., lysine or arginine), transport occurs by one or more separate systems that require the presence of a cationic group on the side chain. For the most commonly occurring system in cells of higher animals, all that is required for recognition, beyond the usual structures on the α carbon atom, is the cationic group on the side chain. Homoarginine, because of its high pK_3', or a metabolism-resisting arginine analog built on the piperidine ring (Christensen and Cullen, 1973) serves as a model substrate for this system. Cationic systems of another type specifically require the side chain of L-ornithine (seen in the hepatic mitochondrion, Gamble and Lehninger, 1973) or the guanidinium group of L-arginine (Rosen, 1971) and its homologs. We had the pleasure of identifying the extra structural feature of the arginine side chain, which permits such a system specifically to recognize that amino acid in *Salmonella typhimurium* (Quay and Christensen, 1974). This feature is the hydrogen atom on the secondary N in the guanidinium group. Our evidence comes from the specific inhibitory action of homoserine and 4-hydroxyproline on the arginine-specific system of this organism. This inhibitory action of the hydroxyamino acid is not accompanied by any measurable transport for it, presumably because of its lack of a side chain cationic group. Furthermore, only when hydroxyproline presents its hydroxyl group in the *trans* orientation does inhibition take place. In addition, replacing the hydrogen atom on the secondary N of homoarginine with a methyl group abolishes all its reactivity with this transport system (Quay and Christensen, 1974).

The picture is simpler for the more general transport system for cationic amino acids, such as is seen in the Ehrlich cell, than it is in *E. coli*. A quaternary nitrogenous group on the side chain will not, however, serve as the cationic structure. Therefore the nitrogenous group must have characteristics other than simple cationicity. Two principal peculiarities deserve further study:

1. A neutral (i.e., zwitterionic) amino acid plus Na^+ can mimic the cationic amino acid in reacting with this system in the Ehrlich cell (see Fig. 6 in Christensen et al., 1969)

2. Arginine analogs for which pK_2' (certainly in this case applying largely to the α-amino group) is unusually low, show especially strong accumulation into the Ehrlich cell via this system (Christensen and Handlogten, 1975).

C. Effects on Transport of a Second Cationic Group Where pK_2' Has a Rather Low Value

Lysine shows in the Ehrlich cell a small component of transport by System L, presumably assignable to the very small portion present in a zwitterionic rather than the predominant form. As we bring the two amino groups closer together in the sequence, lysine, ornithine, 2,4-diaminobutyric acid, and 2,3-diamino propionic acid, transport by System A comes to predominate and to generate relatively enormous gradients (see Fig. 3 in Christensen and Handlogten, 1975; also Christensen et al., 1952, 1974; Christensen and Liang, 1966; Christensen, 1973). These four homologs show values for pK_2' in the sequence, 9.5, 8.7, 8.4, and 6.8. It is logical enough that diaminopropionic acid, which we can calculate to take the cationic form to only the extent of 20% at pH 7.4, would be transported principally as a neutral amino acid; but diaminobutyric acid, still 90% cationic at pH 7.4, shows by far the highest V_{max} and ultimately the highest degree of accumulation by System A. Clearly we are not dealing here with a situation as straightforward as the one presented by glutamic acid. Furthermore, pK_2' was shown by proton magnetic resonance to refer in free solution, as one would expect, to the α-amino group of 2,4-diaminobutyric acid, so that our calculation of 10% zwitterionic for diaminobutyric acid in free solution actually refers mainly to the α,γ-zwitterion, and a much smaller proportion of the molecule in free solution will take the α,α-zwitterionic form, which is the form ordinarily accepted for transport by neutral systems. Clearly the transport system strongly modifies the sequence and extent of dissociation of these amino acids, and in coincidence a remarkably intense accumulation takes place. Some twenty-four years ago we showed that the accumulation of the four-carbon diamino acid was so intense as to cause a four-fold osmotic swelling of the Ehrlich cell, followed by osmotic cytolysis (Christensen et al., 1952).

Similar transport activity has been seen for every diamino acid we have been able to prepare or obtain in which pK_2' lies between 6.7 and 8.4. Examples are S-aminoethylcysteine (thialysine), pK_2' = 8.4, azaleucine, pK_2' = 6.8, and 4-amino-4-carboxyl-l-methylpiperidine, pK_2' = 7.2 (Christensen, 1973). The behavior applies to System L as well as to System A in these cases. Note also that unusual accumulation is shown in System Ly^+ for cationic amino acids when pK_2' is quite low (Fig. 5 in Christensen and Handlogten, 1975). That result seems especially significant, because for these arginine analogs there seems little likelihood that pK_2' refers to any group but the α-amino group, even in a modified environment within the transport system.

For the ordinary, neutral substrates of System A and L, one achieves only a transport disadvantage by lowering the pK' value of the α-amino group. Only in the presence of a second amino group does this lowering become an advantage (Christensen and Handlogten, 1975). The advantage in their ready dissociation cannot lie in the ability of these amino acids to take the α,α-zwitterionic form in a static way, nor in their ability to take statically the α,ω-zwitterionic form. More likely then, these forms are taken in sequence. We know from the pH profile that these amino acids show for uptake (Fig. 6 in Christensen, 1973) that the reactant is a zwitterion, and we infer from the failure of β- and γ-amino acids to react appreciably with this system that the zwitterion must be of the α,α-type at the reception stage for transport. But the ability of the electrical charges to take an α,ω-distribution, presumably at a later stage, proves somehow to be a strong advantage to uphill transport. Our proposal for "a cycle of deprotonation and reprotonation during transport" has that basis (Christensen and Handlogten, 1975). We have called these selected diamino acids "gradient-sensing transport substrates" to call attention to the problem (Christensen et al., 1974).

In the diamino acids selected so far for study of this transport behavior, the structural factor that lowers pK_2' has unfortunately

Fig. 6. Time course of uptake of N^5-methoxyornithine, 3 mM, by Ehrlich cell at 37°C, pH 7.4. Upper curve from Krebs-Ringer bicarbonate medium. Lower curve: from similar medium in which Na^+ had been replaced by choline

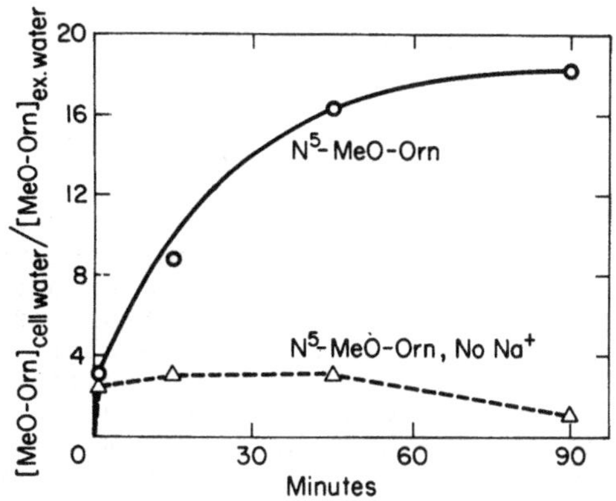

also lowered pK_3', which applies largely to the distal amino group. Examples are the closeness of the amino groups to each other, or their closeness to a sulfur atom or an unsaturated link lying between them in the chain. To avoid ambiguity, suppose we were instead to lower systematically the pK' of the distal amino group so that we could be sure that it would represent the second dissociation, pK_2', and in the meanwhile keep the pK' of the α-amino group approximately in its normal position. For example, histidine has these properties, but is not accumulated to an exceptional degree. We have accomplished this purpose in several ways, but most clearly perhaps by attaching a trifluoroethyl group to the ε-amino group of lysine (Christensen et al., 1976), and an N-methoxy group to the δ-amino group of ornithine. The pK_2' values obtained then were 5.7 and 5.2, respectively, and the pK_3' values 9.7 and 9.5, respectively. These amino acids showed degrees of accumulation typical for the systems involved, as illustrated in Figure 2 of Christensen et al. (1976) and in Figure 6 of the present paper. Note particularly how weak accumulation is by System L. Therefore, the unusual degrees of accumulation of diamino acids by neutral systems does not depend simply on the loss of the charge on the side chain - there is no obvious reason why the removal of that charge would make a "supersubstrate" of a diamino acid.

D. Proposals as to Why Diamino Acids With pK_2' in the Range 6.8 to 8.5 Show Strong Accumulation by Neutral Transport Systems

Let me summarize the proposals we have already made in efforts to explain the foregoing paradox:

Once we get beyond the first recognition stage in neutral amino acid transport, we propose that the amino acid molecule enters an environment in which the molecule will respond vectorially (1) by reason of its ability to resume a positive charge, to a charge gradient, or (2) by reason of its ability to lose and regain a proton, to a proton gradient (or by both) to a protonmotive force.

We are impressed by the idea that the fundamental event in transport is the production in the membrane of a charge separation. This charge separation probably involves the separation of a proton from an electron. This fundamental event could then be capable of driving all other transports as cotransports or countertransports, given either that the transported substrate is a charged species, or that it is capable of complexing with a charged species within the membrane. At least for the Na^+-dependent System A, cotransport with Na^+ appears to us not to be an adequate and universal mode of energization for amino acid transport (Christensen et al., 1975). Hence I think we need to be alert to indications of reactivities of the amino acids with H^+ during transport. The evidence just considered suggests that the inherent reactivities of the amino acids with H^+ have been harnessed within the membrane to couple amino acid movements so that they will respond to H^+ movements and vice versa.

Peter Mitchell set us all to thinking especially about proton gradients that extend all the way across the biological membrane, an idea that seems very useful and promising for various bacterial species as well as for the mitochondrion and the chloroplast. But most of the cells of the higher animals are much more permeable than these bodies to the hydrogen ion, and the electrical gradients generated across them are generally rather small, if we except nerve and muscle. Should we suppose that the same general principle of a fundamental charge separation applying to transport into the cells and organelles able to maintain a high gradient of H^+ and a high electrical potential across the membrane, has been adapted during evolution for cells unable to do so? Or should we suppose instead that a totally different principle drives active transport in the latter case? I urge the first choice. A charge separation is a microscopic event, and it is not implicit in this event how far the charges will come to be separated or how many charged particles will be stored in each phase. That is, we do not know a priori how much energy storage the process will occasion, whether through accumulation of the separated charges in large bulk phases or only in tiny amounts in "microphases". Let us therefore suppose that within the general principle of the Mitchell hypothesis we have the possibility, developed especially in cells of the higher animal, of maintaining the separation of the initial charged species within the substance of a membrane, perhaps even within an oligomeric protein molecule deeply embedded in the membrane. The receptor sites present on the immediately adjacent structures, and their functional behavior, will determine then whether the gradient that is transmitted to the bulk phases outside the membrane will be a gradient of the originally separated species, or whether instead gradients of other species such as Na^+, K^+, and amino acids will be substituted. These gradients would then have developed at the expense of the initial gradients through a kind of cotransport in which the original gradient is largely confined to the microenvironment in which it is generated. The alkali-ions come then to substitute in many cells for the proton gradient as forms of energy storage. This concept seems to me compatible with the irregularly observed evidence of structure-dependent cotransport of H^+ with neutral amino acids, for example with α,α-diethylglycine (Christensen, 1973). This concept can include the idea that the usual functional groups of the amino acid, e.g., the α-amino group, may through a cycle of deprotonation and reprotonation serve to carry the hydrogen ion.

E. Concluding Discussion

Blondin and Green (1975) recently suggested that the fundamental charge separation becomes energetically feasible only when it is paired with an adjacent separation of two opposite charges one from the other. I believe these authors may have misapprehended the calculations made by Guggenheim (1949) of the energy needed to produce an isolated charge separation. It is when we try to produce energy storage by repeated charge separations between the same two finite phases, so that high potential differences gradually are generated, that the energy cost rises strongly over that needed to produce transport upon reversal of the charge separation. Furthermore, Mitchell has pointed out that significant energy storage can be produced by a continuing process of charge separation, if it takes place across an interface with a relatively enormous area (Mitchell, 1969). He discusses these ideas under the heading, "the fiction of electroneutrality".

If a pairing of charge separations with the fundamental endergonic separation of H^+ from e- should prove to be essential, it is hard to imagine a more suitable charge separation that could readily be coupled to proton movements within the membrane than this one, which is strongly exergonic from right to left in aqueous solutions (Christensen, 1976):

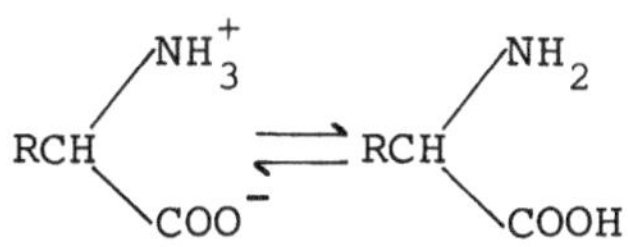

If we accept such a role for the amino and carboxyl groups, either of a free amino acid or perhaps of an ω-peptide such as glutathione, we could readily project an adaptation by which the amino acid molecule itself would fall under the vectorial influence of the basic charge separation. Furthermore an osmotic barrier could readily be traversed by the partner shown at the right of the equation.

In this connection, we are very interested in a strong transport stimulation arising when a 2-hydroxyl group is introduced into 1-aminocyclohexanecarboxylic acid. In one of the isomers, obtained by the Bucherer-Libe synthesis, pK_2' and pK_1' are separated by the ordinary value of (9.6 - 2.1) = 7.5 pH units. In the other isomer (by the Strecker synthesis), 2.6 times as rapidly transported by System L (as measured by V_{max}/K_m), the difference $pK_2' - pK_1'$, is only (9.0 - 2.7) or 6.3 pH units. This result might imply that a significant transport advantage arises when the energy required to transfer a proton from the amino to the carboxyl group is lowered. A control test in which a 2-methyl group was introduced in place of the 2-hydroxy group produced a 6.7-fold difference in transport rate between the two isomers, however, and only a 0.1 unit difference in the separation of the pK' values. Hence steric effects, rather than the change in the readiness of charge transfer, appear to account for the transport differences.

If we examine again my first figure, we can divide the changes in amino acid transport produced by pH changes into two groups:

1. Those that are logical consequences of the titration of the substrate to a different state of charge

2. Those that are not seen as straightforward results of the titration of the substrate.

The latter group contains the effects that we think are likely to be revealing as to how the amino acid molecule is given a directional thrust within the plasma membrane.

Acknowledgments. Support from the Institute of Child Health and Human Development (Grant HD01233) of the National Institutes of Health, U.S.P.H.S., is gratefully acknowledged.

References

Blondin, G.A., Green, D.E.: A unifying model of bioenergetics. Chemical and Engineering News, 26-42, Nov. (1975)

Christensen, H.N.: On the meaning of effects of substrate structure on biological transport. J. Bioenergetics 4, 31-61 (1973)

Christensen, H.N.: Towards a sharper definition of energetic coupling through integration of membrane transport into bioenergetics. J. Theoret. Biol. 57, 419-431 (1976)

Christensen, H.N., Cullen, A.M.: Synthesis of metabolism-resistant substrates for the transport system for cationic amino acids; their stimulation of the release of insulin and glucagon and of the urinary loss of amino acids related to cystinuria. Biochim. Biophys. Acta 298, 932-950 (1973)

Christensen, H.N., Handlogten, M.E.: A cycle of deprotonation and reprotonation energizing amino acid transport? Proc. Natl. Acad. Sci. U.S. 72, 23-27 (1975)

Christensen, H.N., Liang, M.: Transport of diamino acids into the Ehrlich cell. J. Biol. Chem. 241, 5542-5551 (1966)

Christensen, H.N., de Cespedes, C., Handlogten, M.E., Ronquist, G.: Modified transport substrates as probes for intramembrane gradients. Ann. N.Y. Acad. Sci. 227, 355-379 (1974)

Christensen, H.N., de Cespedes, C., Handlogten, M.E., Ronquist, G.: Energization of amino acid transport, studied for the Ehrlich ascites tumor cell. Biomembrane Reviews, Biochim. Biophys. Acta 300, 487-522 (1975)

Christensen, H.N., Handlogten, M.E., Garcia-Sancho, J., Sanchez, A.: Protonations and deprotonations in amino acid transport. Possible relations to energy transfer. In: Amino Acid Transport and Uric Acid Transport. Silbernagl, S., Lang, F., Greger, R. (eds.). Stuttgart: Georg Thieme 1976, pp. 5-12

Christensen, H.N., Handlogten, M.E., Thomas, E.L.: Na^+-facilitated reactions of neutral amino acids with a cationic amino acid transport system. Proc. Natl. Acad. Sci. U.S. 63, 948-955 (1969)

Christensen, H.N., Liang, M., Archer, E.G.: A distinct, Na^+-requiring transport system for alanine, serine, cysteine and similar amino acids. J. Biol. Chem. 242, 5237-5246 (1967)

Christensen, H.N., Riggs, T.R., Fischer, H., Palatine, I.M.: Intense concentrative uptake of α,γ-diaminobutyric acid by cells. J. Biol. Chem. 198, 17-22 (1952)

Dent, C.E., Rose, G.A.: Amino acid metabolism in cystinuria. Quart. J. Med. N.S. 20, 205-219 (1951)

Gamble, J.G., Lehninger, A.L.: Transport of ornithine and citrulline across the mitochondrial membrane. J. Biol. Chem. 248, 610-618 (1973)

Garcia-Sancho, J., Sanchez, A., Christensen, H.N.: Role of H^+-dissociation in the transport of acidic amino acids by the Ehrlich ascites-tumor cell. Biochim. Biophys. Acta, in press (1977a)

Guggenheim, E.A.: Thermodynamics. Amsterdam: North Holland 1949

Im, Wha Bin, Christensen, H.N.: Heterogeneity of histidine transport systems in the Ehrlich cell. Biochim. Biophys. Acta 455, 144-151 (1976)

Koser, B.H., Christensen, H.N.: Effect of substrate structure on coupling ratio for Na^+-dependent transport of amino acids. Biochim. Biophys. Acta 241, 9-19 (1971)

Mitchell, P.: Chemiosmotic coupling and energy transduction. In: Theoretical and Experimental Biophys., Vol. 2. Cole, A. (ed.). New York: Marcel Dekker 1969, pp. 177-179

Quay, S., Christensen, H.N.: Basis of transport discrimination of arginine from other basic amino acids in Salmonella typhimurium. J. Biol. Chem. 249, 7011-7017 (1974)

Rosen, B.P.: Basic amino acid transport in Escherichia coli. J. Biol. Chem. 246, 3653-3662 (1971)

Thomas, E.L., Christensen, H.N.: Nature of the cosubstrate action of Na^+ and neutral amino acids in a transport system. J. Biol. Chem. 246, 1682-1688 (1971)

Waddell, W.J., Butler, J.C.: Calculation of intracellular pH from the distribution of 5,5-dimethyl-2,4-oxazolidinedione (DMO). Application to skeletal muscle of the dog. J. Clin. Invest. 38, 720-729 (1959)

Electrogenic Ion Pump as an Energy Source for Active Amino Acid Transport in Ehrlich Cells

E. HEINZ, P. GECK, C. PIETRZYK, and B. PFEIFFER

Summary

K^+-depleted Na^+-enriched Ehrlich cells have a powerful electrogenic Na^+ pump that can be maximally stimulated by raising the extracellular K^+ to about 10 mM. At this $[K^+]$, both the Na^+ output rate and the distribution ratio of the lipid-soluble cation tetraphenylphosphonium (TPP^+) used here as an indicator of the electrical potential difference, are maximal. Also the rate of uptake of the amino acid aminoisobutyrate is maximal under the same conditions, as should be expected if the electrogenic pump contributes to the driving force of this transport. Since Na^+ output as well as potential generation and amino acid transport can be strongly inhibited by ouabain, all effects are probably due to the ATPase-operated Na^+-K^+ pump. The ions Cl^- and K^+, however, show a deviating behavior that becomes increasingly more pronounced with increasing extracellular $[K^+]$: Cl^- ions tend to enter the cell against their electrochemical potential gradient, suggesting a second pathway of penetration besides free diffusion, which appears to be independent of ouabain and of the electrical potential difference. Under the same conditions the K^+ ions also move into the cell in an amount excess to that accounted for by the Na^+-K^+ exchange pump; and this extra influx is also insensitive to ouabain inhibition. Evidence has been produced that the second pathway of Cl^- is due to an electrically silent Cl^--base exchange, driven by an outwardly directed, ouabain insensitive proton pump. This proton pump apparently requires an exchange for K^+ ions to maintain electroneutrality. The action of this proton pump thus amounts to a net transport of K^+Cl^- into the cell, which is unrelated to the ATPase-operated Na^+-K^+ pump. Since the exchange portion of the Cl^- movement is controlled by the pH difference rather than by an electrical potential difference (pd), it is unsuitable to indicate the electrical potential.

The gradient hypothesis (Crane, 1962) postulates that the active transport of neutral amino acids in Ehrlich cells is energized by the electrochemical potential gradient of Na^+ ions across the plasma membrane (Eddy, 1968). Under normal conditions of electrolyte distribution between cells and medium, it could be shown that both the Na^+ electrochemical potential gradient, in the following called simply gradient, and the degree of coupling (q) between the fluxes of Na^+ and amino acids account for adequate energy transfer, to drive the amino acid transport actually observed (Pietrzyk and Heinz, 1974). It is in seeming contradiction, however, to the gradient hypothesis that also with inverted Na^+ and K^+ gradients, some active transport of aminoisobutyrate takes place (Schafer and Heinz, 1971). This "paradox" transport could be accounted for by the Na^+ gradient only if the electrical pd ($\Delta\psi$) across the cell membrane were under these conditions at least 40 mV higher than had been estimated for normal cells.

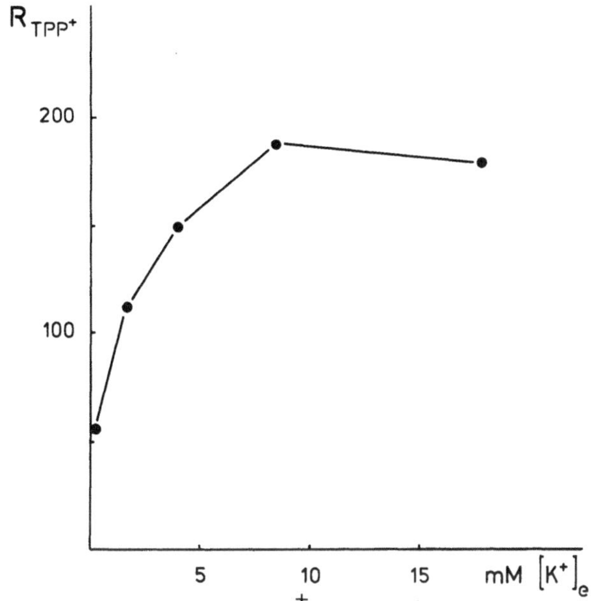

TPP^+ accumulation

Fig. 1. Distribution ratio (cell:medium) of TTP^+ is plotted as function of extracellular $[K^+]$. Cells were depleted of K^+ by preincubation in K^+-free Ringer solution at 1^oC. Cellular K^+ drops to about 10 to 20 mM by this procedure whereas Na^+ rises accordingly. Main incubation was at 37^oC in phosphate Ringer with varying $[K^+]$ (as indicated). $[Na^+]$ was adjusted to make $[Na^+] + [K^+] = 150$ mM

Such a high potential could be plausibly attributed only to the activation of an electrogenic pump. An electrogenic Na^+ pump was postulated previously (Levinson and Hempling, 1967; Gibb and Eddy, 1972) without an estimate, however, as to the magnitude of its contribution to the electrical pd. More recently, we found strong evidence of an electrogenic pump, stimulatable by inversion of Na^+ and K^+ distribution, which makes a considerable, probably adequate contribution to the electrical potential to drive the paradox active amino acid transport in these cells (Heinz et al., 1975). Still, some problems remained, especially with respect to the behavior of K^+ and Cl^-, which does not easily agree with the foregoing picture.

It has been puzzling that the potential attributed to the electrogenic Na^+ pump is not evident in the Cl^- distribution. Since the net movement of Cl^- ions has been shown to be very slow under certain conditions, it was thought that the chloride movement could not adjust rapidly enough to any short-term rise of the electrical potential. Meanwhile, some doubts have arisen as to this explanation: the Cl^- movement, even if it is slow, should at least show a tendency in the right direction to equilibrate with the electrical potential, which could not be demonstrated. Furthermore, under certain conditions the chloride appeared to move faster, certainly fast enough to equilibrate with the potential during the time of observation. It had to be shown, therefore, that under such conditions the chloride movement does not follow, or does not follow exclusively, an electrical potential difference.

It could indeed be demonstrated experimentally that under certain conditions Cl^- ions enter the cell even though its electrochemical potential difference is clearly directed in the outward direction. The experiments were carried out with K^+-depleted and Na^+-enriched cells, in which an electrogenic pump could be shown to be activated

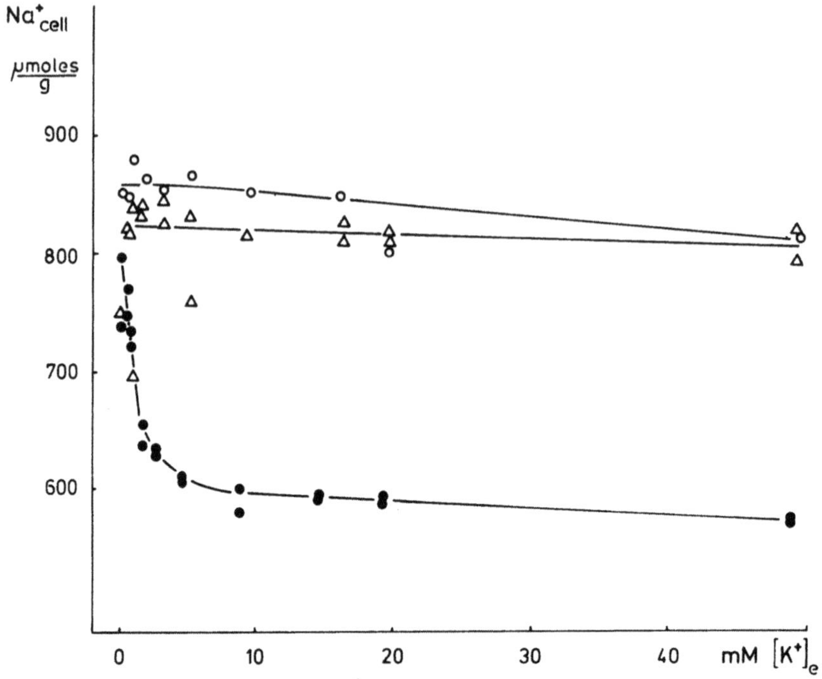

Na^{+} output rate

Fig. 2. Na^+ content after 0 min (Δ) and after 10 min (•) is plotted as function of extracellular [K^+]. Since the Na^+ output is almost linear with time within this range, ordinate values give approximately the output rate. Cells had been depleted of K^+ as indicated in Fig. 1. The Na^+ content after 10 min in presence of 1 mM ouabain is also given (o). Extracellular solution as in Figure 1, except that temperature was 26°C. Corresponding results obtained at 37°C (not shown) were similar in principle

by raising the extracellular K^+. This is illustrated in Figure 1, which shows that the electrical pd, as derived from the distribution of the lipid soluble cation tetraphenylphosphonium (TPP^+), rises if the extracellular K^+ is raised to about 10 mM and remains high upon further increasing the K^+ level. In close agreement, the initial 10 min Na^+ output also rises sharply to a plateau under the same conditions (Fig. 2). Obviously, the electrogenic Na^+ output is activated by extracellular K^+ but reaches saturation speed at 10 mM K^+. It is also seen that ouabain completely inhibits the K^+-activated Na^+ output. Since the electrical pd is also depressed by ouabain (Heinz et al., 1975), it may be concluded that the electrogenic Na^+ output is related to the ATPase-operated Na^+-K^+-pump.

Figure 3 shows that also the amino acid uptake rises almost parallel with pd and Na^+ output upon increase of extracellular K^+, as expected if the electrogenic Na^+ output adds to the driving force for cotransport.

Quite unexpected was the behavior of the Cl^- and K^+ ions under the same experimental conditions (Fig. 4). Only up to a K^+ concentration of about 2 to 4 mM, the Cl^- output appears to follow the Na^+

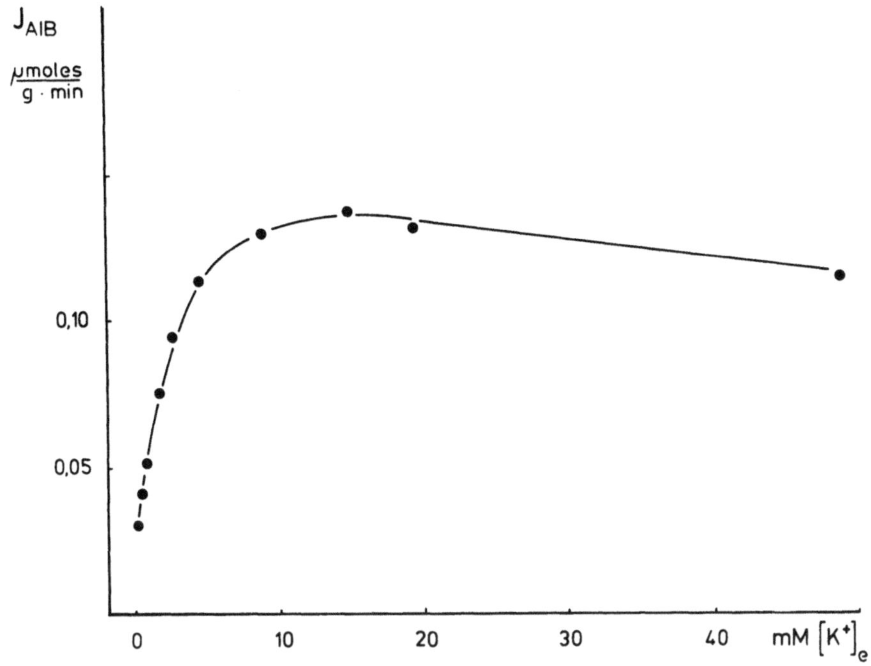

Fig. 3. Uptake of aminoisobutyrate (AIB) is plotted as function of extracellular $[K^+]$. Experimental conditions were the same as in Figure 2

extrusion, according to expectation in order to maintain electroneutrality. At higher concentrations of K^+, however, the Cl^- output becomes smaller, disappears almost completely, and at still higher K^+ concentrations in the medium, even chloride moves into the cell, obviously despite the electrical potential difference, which is still high and outside positive. Clearly Cl^- follows the electrical potential difference only in the lower range of $[K^+]_e$, whereas at higher concentrations it becomes independent of it, and at still higher K^+ concentrations it clearly migrates against its own electrical potential difference. In the presence of ouabain, the outward movement of Cl^- disappears completely; it migrates into the cell at very low K^+ concentrations (Fig. 5). It looks as if there were two components of chloride movement overlapping each other, one which depends on the electrical potential difference and which is inhibitable by ouabain, and another fraction rising with increasing extracellular K^+, which is independent of the electrical potential and not inhibitable by ouabain.

The movement of K^+ is closely related to that of Cl^-. It appears to meet throughout the requirement of charge transfer to maintain electroneutrality under the shifting conditions (Fig. 6). In the range of low extracellular K^+, the K^+ influx rises toward a plateau lower than would be equivalent to that of the concomitant Na^+ output, as could be expected if the electrogenic extra output of Na^+ is largely covered by chloride exit. In contrast to Na^+ extrusion, however, the K^+ uptake rate rises still further beyond this

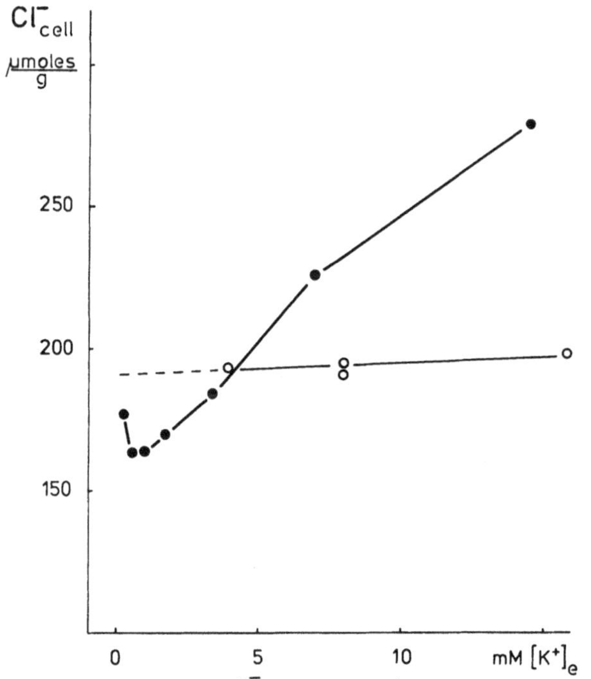

Cl^-- movement

Fig. 4. Movement of Cl^- during 10 min is plotted as function of extracellular $[K^+]$. Experimental conditions were the same as in Figure 2; o: Cl^- content after 0 min; •: Cl^- content after 10 min

plateau with increasing extracellular $[K^+]$, and eventually even exceeds the concomitant but constant Na^+ output. At first sight one may interpret this extra uptake of K^+ as representing the flow of free K^+ ions driven by the electrical pd generated by the electrogenic Na^+ pump. Provided that the K^+ permeability of the cellular membrane is sufficiently high, one could imagine that with increasing extracellular K^+ more of this ion species becomes available to replace Cl^- in maintaining electroneutrality. This interpretation, however, does not explain why at higher extracellular K^+ the chloride movement is reverted and the K^+ uptake rises above the concomitant Na^+ output. It is clearly incompatible with the observation that ouabain, while inhibiting Na^+ output completely, leaves the extra K^+ uptake apparently unchanged. It rather seems that also the K^+ ions enter, besides via the Na^+-K^+ pump, via a second pathway that is insensitive to ouabain inhibition.

If the K^+ and Cl^- movements are corrected for their ouabain-insensitive portions, it looks indeed as if above an extracellular $[K^+]$ of 10 mM Na^+ and K^+ are transported at a constant rate, the electrogenic part of Na^+ output being neutralized by an equivalent Cl^- outflow (Fig. 7). For better understanding, the ion movements at the three characteristic K^+ concentrations are visualized schematically in Figure 8.

Two questions now arise: which is the nature of the two "second" pathways for Cl^- and K^+, respectively, and which is the force to drive these two ions in probably equivalent amounts along these pathways?

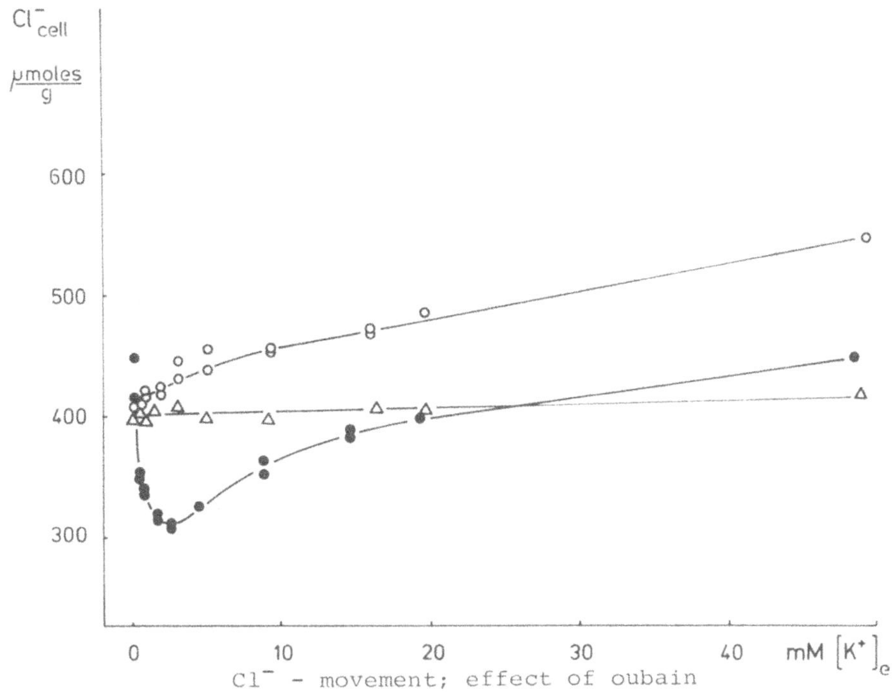

Fig. 5. Same as Fig. 4 with another batch of cells with a higher water content, which shifts intersection between chloride curve and abscissa to the right. Effect of 1 mM ouabain (o) is also given. Symbols are as in Fig. 2

As to the second pathway of chloride movement, its indifference toward the electrical potential difference suggests that it is not electrogenic but rather due to an electrically silent exchange for another anion, presumably a base. It was found that the exit of chloride is associated with an acidification of the medium, which rises with the rate of Cl^- exit (Fig. 9). Conversely, the Cl^- exit rises with increasing pH of the extracellular solution (Fig. 10). The entry can be strongly stimulated by alkalinization of the cell interior brought about by the addition of NH_4Cl (Fig. 11). All these observations can be taken as evidence that the second pathway of chloride movement represents an electrically silent exchange mechanism between Cl^- and a metabolic base, presumably OH^- or HCO_3^-. Since this exchange of chloride is likely to be faster than the electrogenic net movement of this ion, the chloride movement appears to be controlled primarily by a pH difference and only to a lesser degree by an electrical potential difference. Accordingly, a close relation between chloride distribution and electrical potential cannot be expected.

The Cl^- uptake by exchange would easily be accounted for by an outward proton pump, which be the generation of residual base inside the cell, indirectly promotes this uptake. It has been observed before that Ehrlich cells aerobically secrete protons at a rate of 8 to 10 µmol min^{-1} g^{-1} dry wt (Selby-Smith and Heinz, 1975). This H^+ output is insensitive to ouabain; it stops complete-

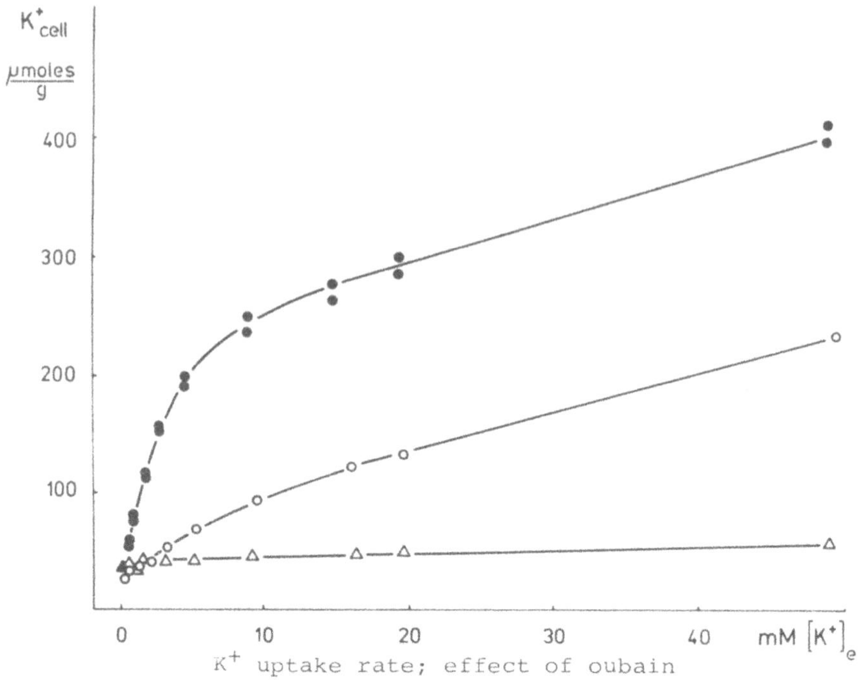

Fig. 6. K^+ content as function of extracellular K^+. Experimental conditions and symbols are the same as in Figs. 2 and 5

Table 1. Proton transport in Ehrlich cells

Addition	Gassing	H^+-extrusion (μmol/g min)
None	O_2	7.9
None	N_2	<0.1
Ouabain 1 mM	O_2	8.2
Ouabain 1 mM	N_2	<0.1

ly if O_2 supply is discontinued and it recovers completely if O_2 is readmitted (Table 1). It should be pointed out, however, that this rate need not represent the real acid-secreting capacity, since in the steady state the rate of H^+ production is limited by the rate at which the evolving base, OH^-, is being removed from the cell. This can be done effectively only by the net acid production of cellular metabolism, e.g., by cellular metabolic CO_2, which will limit the secretion of H^+ ions. Accordingly, the rate of H^+ production measured under these conditions gives us only the rate of net acid production of the cell metabolism, whereas

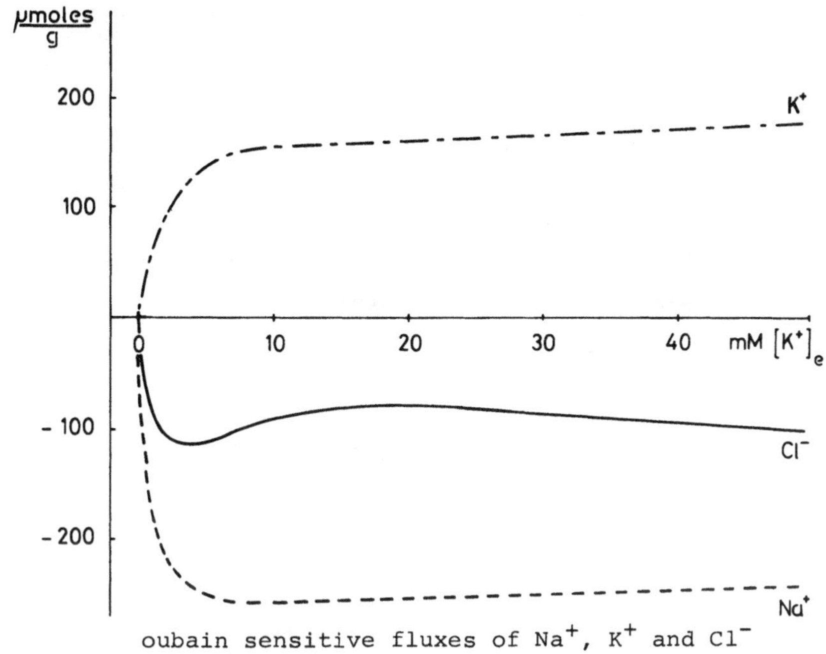

Fig. 7. Uptake rate of Na^+, K^+, and Cl^- (10 min), as obtained from Figs. 2,5 and 6 after correction for ouabain-insensitive movement of K^+ and Cl^-, respectively, is plotted as function of extracellular $[K^+]$

the true H^+-secreting capacity may in reality be much higher. In any case, however, in the presence of a rapid, presumably electrically silent, Cl^- base exchange mechanism, such an H^+ ion pump should affect Cl^- movements. The extrusion of H^+ ions by the H^+ pump necessarily leads to accumulation of residual basic ions inside the cell, which could then rapidly equilibrate with outside chloride, thus collapsing the pH gradient and thereby allowing further H^+ secretion until a new steady state is reached in which a pH difference equivalent to the chloride distribution can be maintained.

Since the output of H^+ ions, for reasons of electroneutrality, must be associated with the movement of another ion, presumably the inward movement of K^+ ions, it becomes limited by the extracellular K^+ level. Accordingly it has been shown experimentally that by increasing the extracellular K^+, the steady state distribution of both K^+ and Cl^- is shifted toward higher intracellular concentrations of these ions, corresponding to the "second" pathway postulated. Hence an electrogenic proton pump together with a Cl^--base exchange mechanism would indeed account for the so far unexplained observations, in particular for the ouabain-insensitive K^+ and Cl^- movement into the cell.

At present, it cannot be said whether the proton pump is electrogenic or whether it is due to an electrically silent H^+-K^+ exchange. Either alternative would be compatible with the present

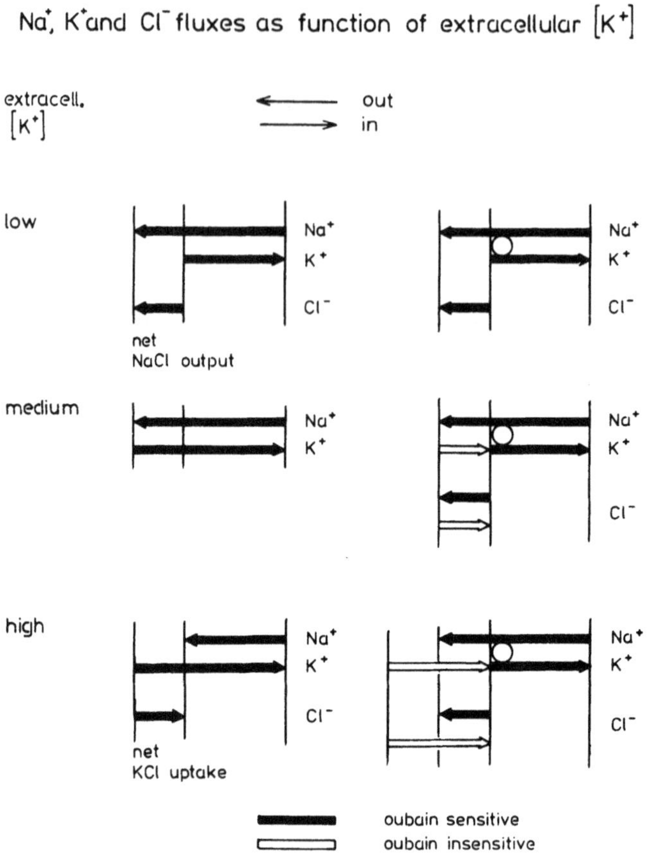

Fig. 8. Schematic presentation of direction of fluxes of Na^+, K^+, and Cl^-, respectively, for three representative ranges of extracellular $[K^+]$, low $[K^+]$ (2-5 mM), medium $[K^+]$ (5-15 mM), and high $[K^+]$ (above 15-20 mM)

results. A K^+-H^+ exchange pump would fit better if it could be shown that the ouabain-insensitive K^+ uptake could occur against the electrochemical potential gradient of K^+ ion, which has not been done yet.

Attempts to monitor intracellular pH changes associated with the various ion movements were not satisfactory, because the determination of intracellular pH changes by the usual basic or acidic probes is misleading in the presence of pH differences within the cellular space. It can be predicted that in a completely homogeneous intracellular phase, the distribution ratios of the appropriate basic and acidic probes are reciprocal: in other words, the distribution ratio of the acid multiplied by that of the base under the same conditions should give unity. If there is, however, compartmentalization with different pH, the product of the two distribution ratios may greatly exceed unity. Hence all determinations of intracellular pH by such probes should include both acidic and basic ones, and a clear statement about intracellular pH can be given only if the indicated product is unity. In our cells we applied dimethyloxazolidindione (DMO) as an acidic, and methylammonium as a basic indicator; we found that the product of

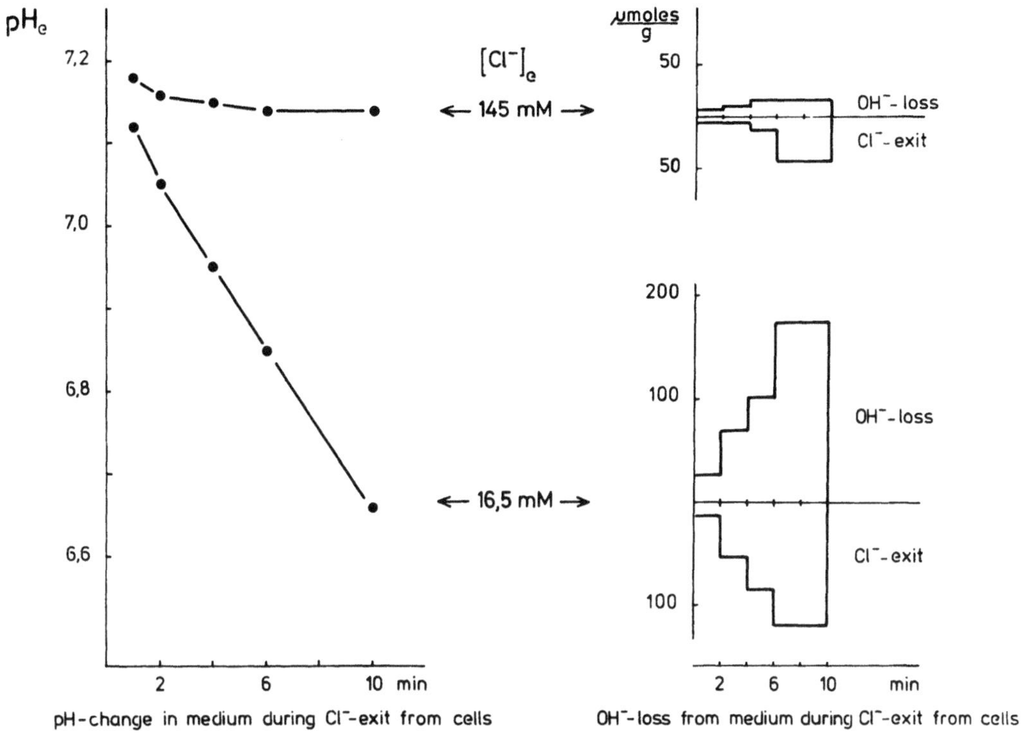

Fig. 9. Left: Change in extracellular pH during Cl^- exit from normal cells into low and high Cl^--medium, respectively, as indicated. Right: Actual Cl^- exit is compared with concomitant loss of OH^- ions in medium, as determined by titration with NaOH to original OH. Intracellular $[Cl^-]$ at beginning of experiment was about 80 mM. Medium was phosphate, Ringer with low buffer capacity (1 mM phosphate). Temperature 37°C

Table 2. Distribution ratio (R) of acidic (A) and basic pH probes between cell and medium

Inhibitor	R_A (DMO)	R_B (MA)	$R_A \times R_B$
None	1.0 (ΔpH = 0)	11.0 (ΔpH = -1.04)	11.0
Antimycin A	0.47 (ΔpH = 0.33)	2.28 (ΔpH = 0.36)	1.07

DMO = Dimethyloxazolidin-dione (Acidic)
MA = Methylamine (Basic)

the distribution ratio of DMO times that of methylamine (MA) greatly exceeded unity in metabolizing cells, but approached unity upon metabolic inhibition (Table 2). Unless the base applied is actively accumulated inside the cell, the outcome of these experiments appears to indicate that intracellular regions of higher acidity

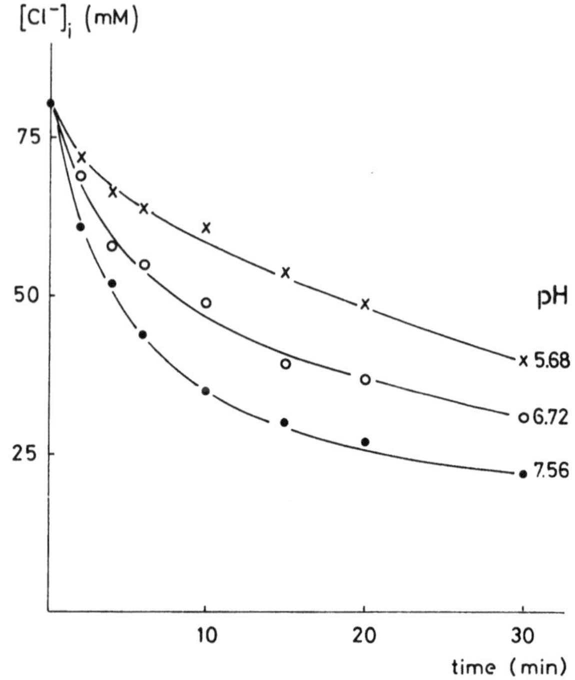

Cl^-- exit at varying pH

Fig. 10. Exit of Cl^- from normal cells with initial intracellular $[CL^-]$ of about 70 mM with time at varying extracellular pH. Normal KRB with adjusted buffer. Temperature 37°C

are created during metabolism and vanish during metabolic inhibition. Accordingly, we are at the moment not in a position to give clear information about the behavior of intracellular pH. It may be pointed out en passant that in many cases of alleged active accumulation of basic material in biological compartments, this accumulation may in reality be due to intracellular pH gradients.

Figure 12 is a summary of our experimental observations to the present. It appears that there is an electrogenic pump that is strongly stimulated by decreasing intracellular $[K^+]$ and increasing intracellular Na^+, and that can be activated by raising the extracellular $[K^+]$ up to about 10 mM. This electrogenic pump is inhibitable by ouabain and therefore probably due to the ATPase-operated Na^+-K^+ pump with a stoichiometry of Na^+ to K^+ greater than 1. This ratio appears to be about 3:1 at low $[K^+]$ but seems to tend toward 3:2 with increasing $[K^+]$ in the medium (Fig. 13). The extra Na^+ output by this pump appears to be associated with an electrogenic Cl^- extrusion, driven by the electrical potential difference. This Cl^- output is slow, in agreement with the previously determined rate coefficient of electrogenic Cl^- net movement, and may therefore by obscured by the faster Cl^- base exchange which is related to the ouabain-insensitive, probably non-electrogenic, second K^+ uptake mechanism. It seems that most of the K^+ movement across the cellular membrane is due to electrically silent mechanisms whereas the electrogenic movement of K^+ ions may also be very slow. The electrogenic movements of both Cl^- and K^+ being slow, the contribution of the electrogenic pump to the

Fig. 11. Uptake of chloride in normal cells with time at various extracellular concentrations of NH_4CL to raise intracellular pH

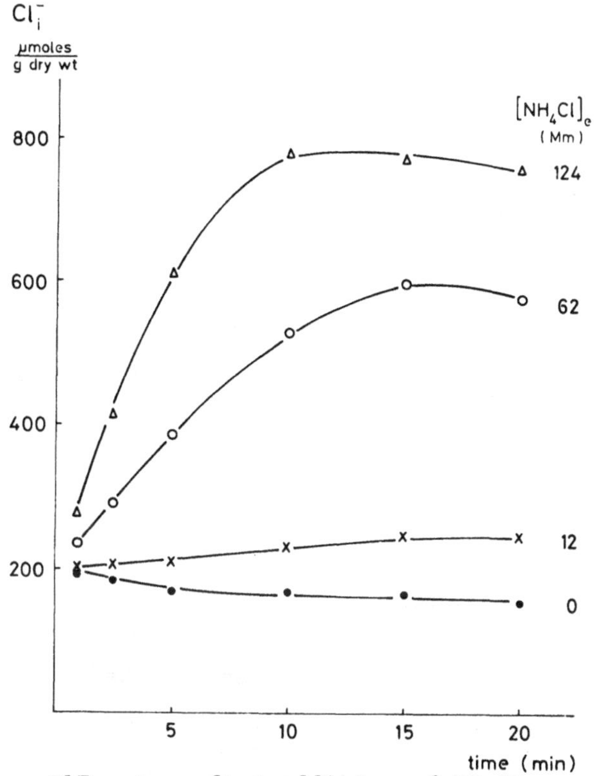

Cl^- entry after addition of NH_4Cl

Transport mechanisms in Ehrlich cells

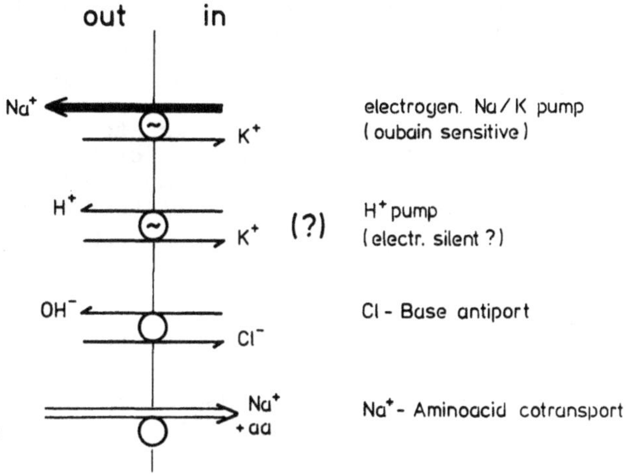

Fig. 12. For explanation see text

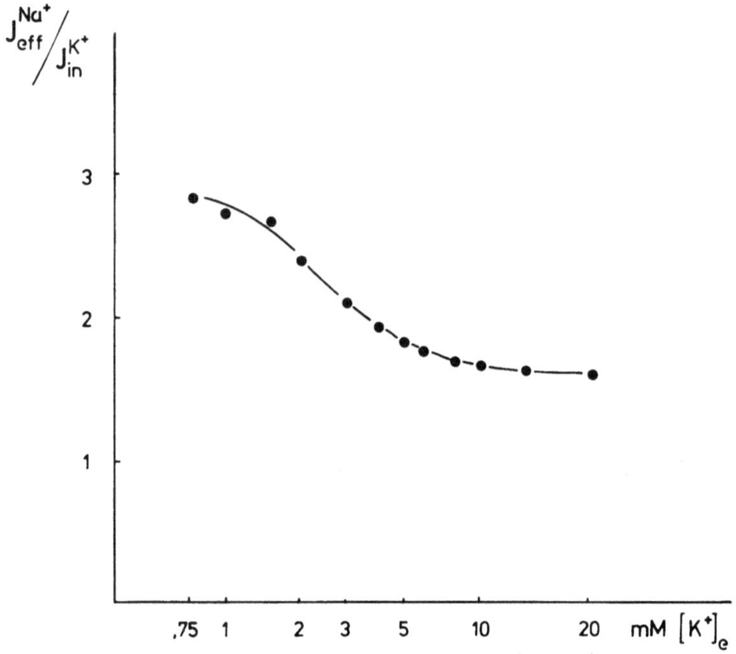

Stoichiometric ratio of Na^+ - K^+ pump

Fig. 13. Stoichiometric ratio of Na^+ output vs. K^+ uptake of Na^+-K^+ pump is plotted as function of extracellular $[K^+]$. Both Na^+ output and K^+ uptake were corrected for ouabain-independent parts of overall fluxes. Temperature 26°C

membrane potential may at times be considerable, such as to raise the potential high enough to account alone for the uptake of amino acids with inverted Na^+-K^+ gradients. Of the chloride movement only the slower part follows the electrical potential difference, whereas the faster exchange part appears to follow the pH difference. Accordingly, the overall Cl^- distribution, especially if the latter exchange mechanism dominates, is poorly related to the electrical potential difference. This explains why the contribution of the electrogenic pump to the electrical membrane potential appears in the TPP^+ distribution but not in the Cl^- distribution.

Acknowledgments. The authors wish to thank Mrs. E. Heyne, Miss G. Werner, and Miss C. Henschel for valuable technical assistance and Mrs. E. Kemsley for preparing the manuscript. The work has been supported by Grant No. 102/14 from Deutsche Forschungsgemeinschaft.

References

Crane, R.K.: Hypothesis of mechanism of intestinal active transport of sugars. Federation Proc. 21, 891-895 (1962)

Eddy, A.A.: The effects of varying the cellular and extracellular concentrations of sodium and potassium ions on the uptake of glycine by mouse ascites tumour cells in the presence and absence of sodium cyanide. Biochem. J. 108, 489-498 (1968)

Gibb, L.E., Eddy, A.A.: An electrogenic sodium pump as a possible factor leading to the concentration of amino acids by mouse ascites tumour cells with reversed sodium ion concentration gradients. Biochem. J. 129, 979-981 (1972)

Levinson, C., Hempling, H.G.: The role of ion transport in the regulation of respiration in the Ehrlich mouse ascites tumour cell. Biochim. Biophys. Acta 135, 306-318 (1967)

Heinz, E., Geck, P., Pietrzyk, C.: Driving forces of amino acid transport in animal cells. Ann. N.Y. Acad. Sci. 264, 428-441 (1975)

Pietrzyk, C., Heinz, E.: The sequestration of Na^+, K^+ and Cl^- in the cellular nucleus and its energetic consequences for the gradient hypothesis of amino acid transport in Ehrlich cells. Biochim. Biophys. Acta 352, 397-411 (1974)

Schafer, J.A., Heinz, E.: The effect of reversal of Na^+ and K^+ electrochemical potential gradients on the active transport of amino acids in Ehrlich Ascites tumor cells. Biochim. Biophys. Acta 249, 15-33 (1971)

Selby-Smith, J.D., Heinz, E.: unpublished (1975)

Some Common Aspects of Active Solute Transport in Yeast and Mouse Ascites Tumour Cells

A. A. EDDY, R. PHILO, P. EARNSHAW, and R. BROCKLEHURST

There is growing evidence that the concentration of specific amino acids, carbohydrates, and other solutes, such as phosphate, by certain eukaryotic transport systems is not energized directly by ATP or some similar compound, but instead is driven by the spontaneous flow of cosubstrate ions (H^+ or Na^+) across the plasmalemma (Mitchell, 1970; Harold, 1972; Reid et al., 1974; Slayman and Slayman, 1974; Hamilton, 1975; Seaston et al., 1976). Similar ideas may also apply to the transport systems found both in bacteria and in mitochondria (Mitchell, 1970; Harold, 1972). A further category of bacterial systems, transporting specific amino acids and sugars, appears to require ATP as such, however, to function optimally (Berger, 1974; Harold and Spitz, 1975).

The cosubstrate hypothesis, a term synonymous with the gradient hypothesis, is illustrated in Figure 1 (a,b). The solute S crosses the plasmalemma by symport mechanisms involving either H^+ or Na^+ as respective cosubstrates. The latter subsequently re-cycle through the appropriate ionic pumps and leak pathways. Little is known about the molecular basis of these processes in the two systems under discussion, namely yeast and mouse ascites tumour cells. However, the ouabain-sensitive pump of the latter cells superficially resembles the classical sodium pump based on a functional Na^+/K^+ ATPase. The activity of the proton pump of yeast, original-

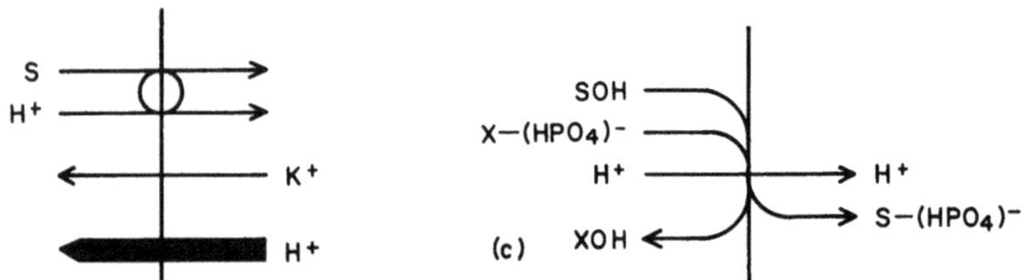

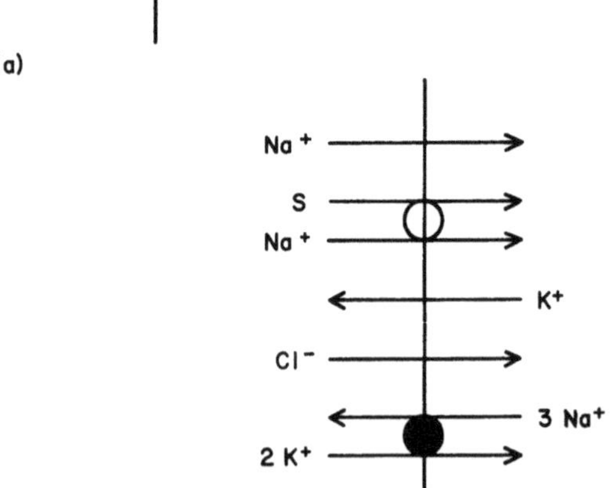

Fig. 1a–c. (a) Proton symport in yeast, with substrate S neutralized either by efflux of K^+ or ejection of H^+ through proton pump. Absorption of K^+ would tend to occur when the proton pump hyperpolarized the cell membrane. (b) Sodium ion symport in mouse ascites-tumour cells. Sodium pumping is represented as electrogenic process, passive fluxes of Na^+, Cl^-, and K^+ also being shown. (c) One form of sugar phosphorylation mechanism proposed by Van Steveninck (1970). Uptake of S is accompanied by H^+

Table 1. Number (n) of equivalents of cosubstrate ions absorbed with each equivalent of substrate in various systems. (Observations are taken from the literature and from a series of new assays based on procedures described by Seaston et al., 1973.)

	Substrate	n	Reference
Mouse ascites tumour cells	glycine	0.90 ± 0.11 (5)	Eddy, 1968
Na^+ absorption	leucine	0.72 ± 0.12 (6)	Morville et al., 1973
Yeasts			
H^+ absorption			
Sac. cerevisiae	glycine	2.00 ± 0.11 (62)	Seaston et al., 1976
	lysine	2.13 ± 0.26 (8)	Seaston et al., 1973
	glutamate	2.91 ± 0.10 (19)	Cockburn et al., 1975
	phosphate	3.16 ± 0.07 (15)	Cockburn et al., 1975
	maltose	0.95 ± 0.06 (29)	
Candida utilis	glycine	1.20 ± 0.10 (3)	
	arginine	1.04 ± 0.10 (6)	
	glutamate[a]	2.10 ± 0.17 (7)	
	glutamate[b]	1.10 ± 0.08 (6)	

[a] 18-h culture of strain N.C.Y.C. No. 193 from glucose-mineral salts medium containing 2.5 mM glutamate as sole source of N

[b] 42-h culture in above solution

ly studied by Conway and O'Malley (1946) and by Rothstein and Enns (1946), may depend on an ATPase like that postulated in Neurospora (Slayman and Slayman, 1974).

Support for the cosubstrate hypothesis has come from the study of the coupling between the respective flows of substrate and cosubstrate into cell preparations containing metabolic inhibitors that inhibit energy metabolism and, therefore, the re-cycling of the cosubstrate ions through the ionic pumps.

A. Cosubstrate Stoicheiometry (n)

Some examples studied in these laboratories are illustrated in Table 1 and in Figures 2, 3, and 4. The fact that the ratios are roughly integers suggests that the different carrier systems involved bind a specific number of cosubstrate ions per solute binding site. Clearly n, the number in question, may exceed unity, a possibility to which inadequate attention has been paid in the literature. For instance, Hamilton (1975) has suggested that cationic amino acids are expected to be absorbed without a proton cosubstrate, responding simply to the electrical potential difference across the membrane, whereas neutral and acidic amino acids would be accompanied by one proton equivalent. Accordingly, the influx of neutral amino acids would respond to the gradient of

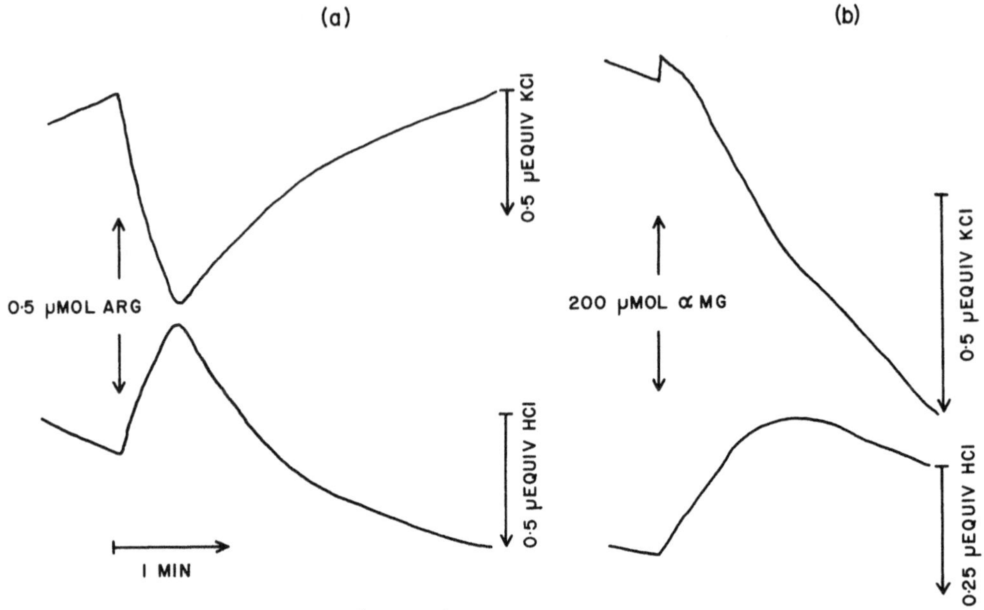

Fig. 2a,b. Net movements of H^+ and K^+ caused by L-arginine and α-methylglucoside, respectively. (a) Uptake of protons and their subsequent ejection from yeast (*Candida utilis*, N.C.Y.C. No. 193) following addition of 0.5 μmol of arginine to 5 ml of yeast (50 mg dry wt) suspension at pH 4.8, 30°C. No metabolic inhibitors were present to inhibit proton ejection, otherwise conditions were those described by Seaston et al. (1973). (b) Uptake of protons and efflux of K^+ caused by addition of 200 μmol of α-methylglucoside to yeast (50 mg dry wt) at pH 5. Strain N.C.Y.C. No. 240 of *Sac. cerevisiae* was employed and was grown in yeast-extract, peptone medium containing maltose as principal carbon source

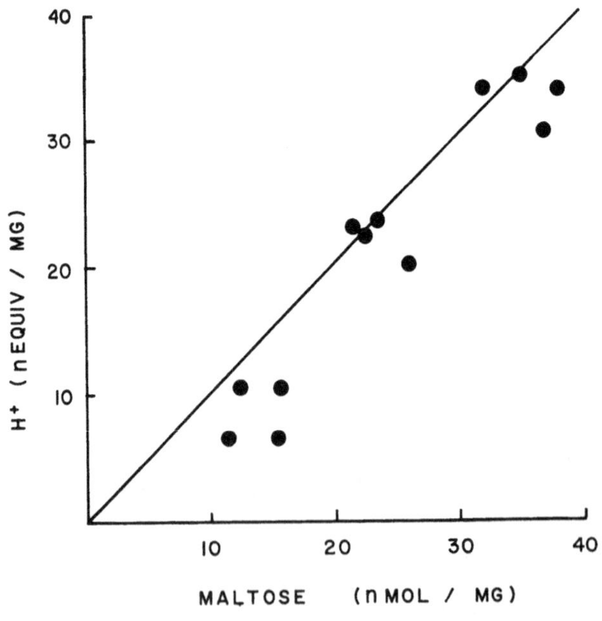

Fig. 3. Absorption of H^+ with maltose. Seaston et al. (1973) observed stoicheiometry of about 2 H^+ per maltose equivalent absorbed, but underestimated uptake of maltose, some of which was hydrolysed and released into solution as glucose. New observations shown here have been corrected for this effect

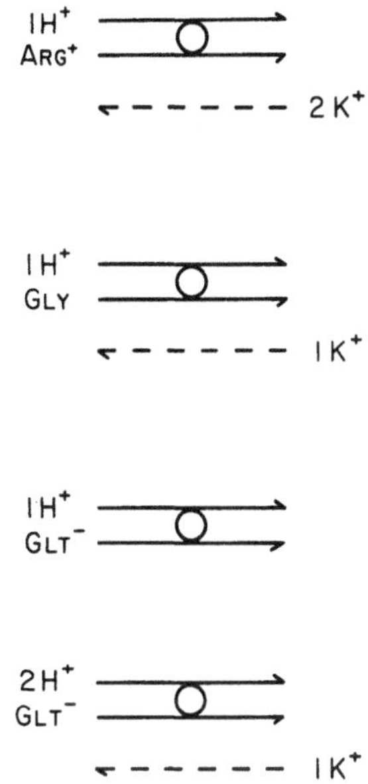

Fig. 4. Proton symport mechanisms absorbing glycine, arginine, and glutamate in Candida utilis. See also Table 1. Efflux of K^+ represents facultative process presumed to be governed by prevailing value of membrane potential (Seaston et al., 1973)

electrochemical potential of H^+, whereas the acidic amino acids would respond simply to the pH gradient across the plasmalemma. The observations summarized in Table 1 demonstrate that Hamilton's proposals represent only one, however, of several possible modes of absorption of a given substrate species. The behavior of Candida utilis towards glutamate shows that n is not indeed constant for a given organism and substrate. A possible explanation is that distinct carriers with different proton stoicheiometries are used in different physiological conditions, rather than that the value of n for a given carrier system varies.

In terms of the cosubstrate hypothesis, the magnitude of n is one factor governing the energy input into the carrier system. The behaviour of the anionic substrates is interesting in this connexion. Both phosphate and glutamate appear, in certain instances, to be absorbed as positively charged complexes, a circumstance that would permit relatively large concentration gradients of the anions to form between the yeast cells and their environment (Cockburn et al., 1975).

When the yeasts first absorbed protons with maltose, α-methylglucoside, and the other substrates listed in Table 1, the cells ejected roughly the amount of K^+ required to maintain electroneutrality (Fig. 2). Van Steveninck (1970) has suggested that α-methylglucoside and certain other carbohydrates are absorbed not by means of proton symports but by a phosphorylation mechanism (Fig. 1c) during which the methylglucoside itself becomes phosphorylated. His proposal is that a phosphate group is transferred from certain polyphosphate molecules, situated on the outer surface of the yeast cells, to the methylglucose molecule undergoing absorption. That proposal is inconsistent, however, with the observed exchange of H^+ for an equivalent of K^+ (Fig. 2b), predicting rather that the transferred phosphate would neutralize the ingoing H^+. Another reason for rejecting the group-transfer model is that the analogue of α-methylglucoside, α-thioethylglucoside, causes a similar exchange of H^+ for K^+ and yet appears not to be metabolized by the yeast. Our work further indicates that α-methylglucoside absorption and sucrose absorption by these yeast strains are closely related processes, both being governed by the disaccharide fermentation gene ssf described by Kahn et al. (1973).

B. Solute Gradients Formed Without the Intervention of Energy Metabolism During the Influx of H^+ or Na^+

A crucial test of the schemes shown in Figure 1a,b is that the solute would accumulate in the cellular phase even when energy metabolism had stopped, provided that the electrochemical gradient of the respective cosubstrate ions across the plasmalemma was manipulated so as to cause the spontaneous flow of the cosubstrate ions into the cellular interior. There is now clear evidence that this phenomenon occurs in both systems under discussion.

I. Mouse Ascites Tumour Cells

Reid et al. (1974) studied the accumulation of L-methionine in mouse ascites tumour cells depleted of ATP. The extracellular concentration of Na^+ was kept at 80 mEqu./l and $[Na^+]$ in the cells varied in different cellular preparations from about 130 mEqu./l to 60 mEqu./l. Under these conditions the membrane potential was varied systematically by varying the ratio of the cellular to extracellular potassium ion concentrations ($[K^+]_i/[K^+]_o$ in the presence of valinomycin. Striking changes in the extent to which the cells accumulated methionine were found to take place. Thus when $[K^+]_i/[K^+]_o$ varied systematically from 1 to a factor of 30, the ratio of the cellular methionine concentration to the extracellular methionine concentration increased proportionately from about 1-up to about 17-fold. Because the process was insensitive to ouabain, the sodium pump appeared not to be involved. These observations, and others involving nigericin, gramicidin, and specific proton conductors, showed that methionine accumulation was indeed driven by the spontaneous flow of Na^+ into the tumour cells down its gradient of electrochemical potential. The maximum accumulation of methionine observed was similar to the ratio of about 20-fold observed during energy metabolism. The symport mechanism thus seems to be a relatively efficient process, despite earlier indications to the contrary (Christensen et al., 1973).

II. Yeast

Somewhat analogous observations have been made by Seaston et al. (1976) with preparations of a strain of *Saccharomyces cerevisiae* depleted of ATP. Two contrasting situations were studied. (1) The yeast preparation was kept at pH 4.5 with extracellular $[K^+]$ less than 1 mEqu./l. The pH gradient (ΔpH) across the plasmalemma was then about 2 units and the ratio $[K^+]_i/[K^+]_o$ was at least 300-fold. (2) The pH was raised to 7.2 in the presence of 100 mM KCl. ΔpH would then approach zero, as would $\Delta\psi$. The maximum ratio of the

Fig. 5a,b. Alternative carrier models with different formal charges. (a) Protonated carrier bears unit positive charge. Kinetic parameters of glycine absorption are compatible with such a model, but not with (b) the case in which the protonated carrier lacks a net charge (Seaston et al., 1976)

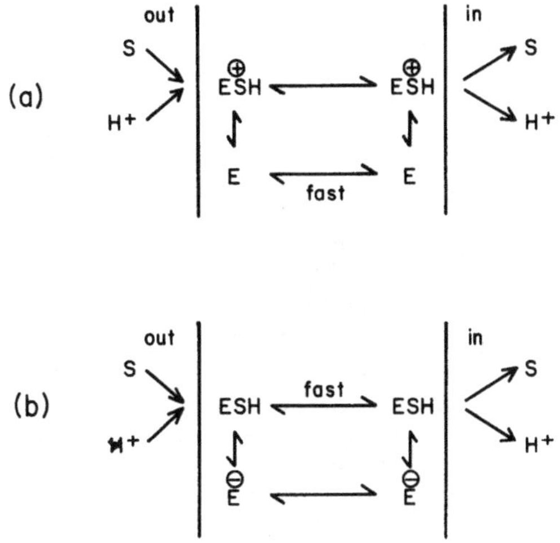

for Glycine

ΔpH	Δψ	Km	Vmax
2	∼ 120 mv	60 μM	10 nmol / mg / min
0	0 mv	80 μM	0·2 nmol / mg / min

cellular glycine concentration to the extracellular concentration was about 2 × 10^4 in condition (1) and about 10 in condition (2). That the electrochemical gradient of H^+ was coupled to the absorption of glycine was indicated by the fact that the proton conductor dinitrophenol markedly lowered the former ratio. Such coupling is energetically feasible because these preparations absorbed glycine with two equivalents of H^+.

The kinetic parameters defining the absorption of glycine under these two contrasting conditions varied in an interesting way with the magnitude of the proton gradient. The K_m with respect to glycine was roughly constant, whereas V_{max} varied by a factor of about 50. Study of the two kinetic models shown in Figure 5 showed that only the system bearing a formal positive charge after protonation would be expected to exhibit such behaviour. Seaston et al. (1976) have suggested that the fall in V_{max} with the magnitude of the proton gradient acting on the glycine carrier system may account for the failure of the yeast cells to release absorbed amino acids in the presence of dinitrophenol and other inhibitors of energy metabolism, a property that is in striking contrast with the corresponding behaviour of the β-galactoside permease of Escherichia coli (Schuldiner et al., 1975). Based on that view, the irreversibility of the process of amino acid absorption by the yeast is only apparent, arising from the difficulty of actually inverting the proton gradient rather than simply lowering its magnitude.

C. Charge Displacements During Solute Absorption: Studies with Fluorescent Probes

The absorption of S with H^+ or Na^+ would tend to depolarize the plasmalemma, that is, make $\Delta\psi$ less negative. The subsequent ejection of the cosubstrate ions through the proton pump or sodium pump, respectively, would tend to hyperpolarize the plasmalemma (Fig. 1). This sequence of events has been clearly demonstrated in *Neurospora mycelia* by means of recordings from micro-electrodes (Slayman and Slayman, 1974), a technique that is not feasible with yeast and that has proved difficult to interpret when applied to the relatively small mouse ascites tumour cells (Lassen et al., 1971; Smith et al., 1972). We therefore looked for the expected changes in membrane potential by an indirect method, based on the fluorescent carboycanine dyes studied by Hoffman and Laris (1974). These workers showed that hyperpolarization of red blood cells in the presence of 3,3'-dihexyl-2,2'-oxacarbocyanine, which bears a screened net positive charge, led to a decrease in fluorescence, whereas their depolarization led to an increase in fluorescence. Hyperpolarization caused the cells to take up the dye from the surrounding medium, and depolarization had the reverse effect. The redistribution of the dye is probably a relatively slow process in comparison with the changes in membrane potential that govern the redistribution.

I. Yeast

The effects of lactose and of its non-metabolized analogue, β-thiomethylgalactoside on the fluorescence of 3,3'-dipropyl-2,2'-thiadicarbocyanine is illustrated in Figure 6. Both substrates initially caused a rapid proton absorption into the preparations of *S. fragilis* used (Seaston et al., 1973). During that period the membrane potential was expected to be relatively low. A net proton efflux occurred after about 1 min, when the membrane potential was expected to rise. The observations shown in Figure 6 thus support, in a general way, the notion that the absorption of the substrate caused the successive depolarization and repolarization of the plasmalemma. The proton-dependent absorption of various α-glucosides, such as maltose, by *Sac. cerevisiae* was associated with similar fluorescence changes.

II. The Mouse Ascites Tumour Cells

Work in these laboratories with 3,3'-dipropyl-2,2'-oxadicarbocyanine (Philo and Eddy, 1975) and elsewhere with 3,3'-dipropyl-2,2'-thiadicarbocyanine (Laris et al., 1976) has shown that glycine and L-methionine each cause a Na^+-dependent increase in fluorescence in the presence of the tumour cells. This was attributed to the partial depolarization of the cells. The interpretation of the foregoing experiments is complicated by the possibility, seemingly ignored by Laris et al. (1976), that the fluorescent probe might respond not only to (1) the electric field across the plasmalemma, but also to (2) the electric field across the membranes of cellular organelles such as the mitochondria. Indeed isolated mitochondria

Fig. 6. Fluorescence changes following addition of lactose or β-thiomethylgalactoside to preparations of *Sac. fragilis* (Seaston et al., 1973). Yeast (2 mg dry wt) was suspended at pH 5 in 4 ml of buffer containing 2 μg of 3,3'-dipropyl-2,2'-thiadicarbocyanine iodide. Fluorescence was excited at 577 and 579 nm and recorded at 670 nm. Direction of increase in fluorescence is indicated by scale. Arrows: lactose (25 μmole) or β-thiomethylgalactoside (50 μmole) were added

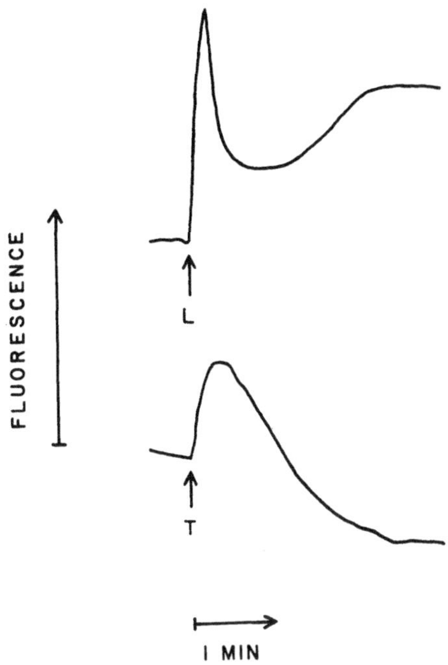

Fig. 7. Increases in fluorescence signal from tumour cells caused by metabolic inhibitors. Fluorescence was excited at 577 and 579 nm and recorded at 620 nm. Tumour cells (2 mg dry wt) were suspended in 4 ml of standard Ringer containing 4 μg of 3,3'-dipropyl-2,2'-oxadicarbocyanine at 37°C. Glucose was also present at concentration of 2 mg/ml. Following additions were made: A, 5 μg of antimycin; B, 50 ng of oligomycin; C, 1.6 nmol of carbonyl cyanide p-trifluoromethoxyphenylhydrazone. Scale of fluorescence is an arbitrary one

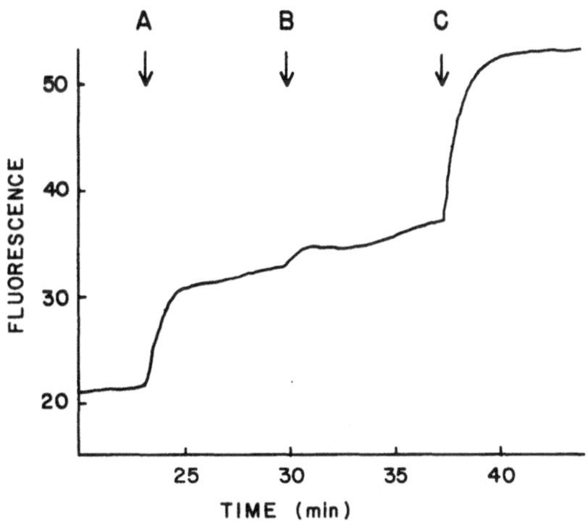

accumulate these dyes, apparently in response to the potential difference across the inner mitochondrial membrane (Laris et al., 1975). Figure 7 shows some observations with the intact tumour cells where such a problem of interpretation arises. We have therefore attempted to distinguish between fluorescence changes due to (1) above and those due to (2) above. Changes in the latter category might be caused by, for instance, the utilization of ATP by transport processes at the plasmalemma that were not themselves electrogenic, but that indirectly affected mitochondrial metabolism.

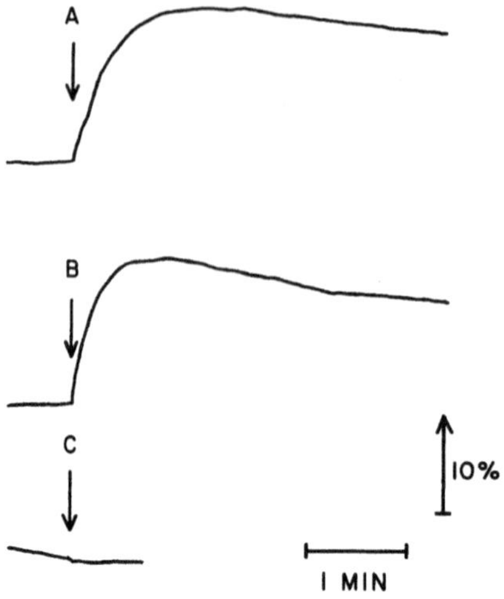

Fig. 8. Change in fluorescence signal caused by addition of amino acids. Tumour cells were suspended in 4 ml of Ringer solution in presence of dye (4 µg) and amino acid was added as indicated. Other conditions were as described in legend to Fig. 7. A, 7.5 µmol of glycine was added to standard Ringer containing 50 ng of oligomycin, 2 µg of antimycin, and 250 nmol of 2,4-dinitrophenol. B, as above except that 5 µmol of L-methionine was used. C, as for A except that Ringer solution contained K^+ instead of Na^+

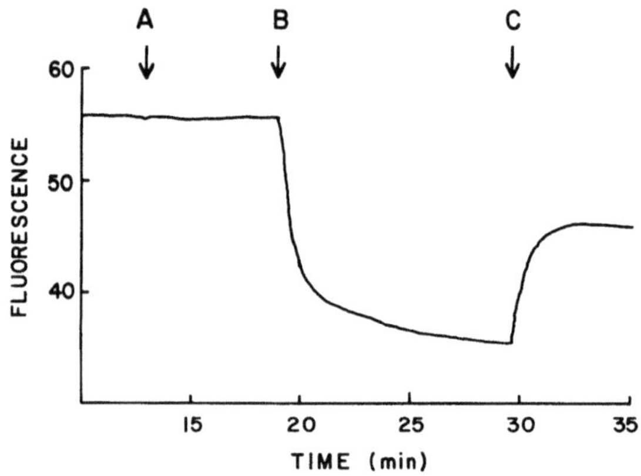

Fig. 9. Fluorescence changes associated with activity of sodium pump. Tumour cells were suspended in presence of metabolic inhibitors in sodium Ringer (4 ml) lacking K^+ ions under conditions like those described in legend to Fig. 8. Following additions were made: A: 10 mg of glucose; B: 20 µmol of KCl; C: 1.2 µmol of ouabain

Our work suggests that, when glucose is added to initiate glycolysis, the problem of possible mitochondrial involvement in the fluorescence signals can be overcome by the addition of a mixture of appropriately selected amounts of oligomycin, antimycin, and 2,4-dinitrophenol. The intention was to abolish the membrane potential in the mitochondria of the intact tumour cells, without affecting transport at the plasmalemma. The choice of inhibitors depends on the circumstance (Reid et al., 1974) that a concentration of dinitrophenol that is a weak proton conductor at the plasmalemma may nevertheless efficiently uncouple oxidative phosphorylation. Under these specific conditions, the addition of either methionine or glycine caused a relatively fast increase in fluorescence provided Na^+ was present (Fig. 8). Thus the absorption of Na^+ with the amino acid (Table 1) appeared to depolarize the plasmalemma. Figure 9 shows that the addition of K^+ in the presence of glucose, initiated a marked decrease in fluorescence, partially reversed by ouabain. We attribute this behaviour to the hyperpolarization of the plasmalemma when the addition of K^+ caused the electrogenic sodium pump to start. The existence of the preceding phenomena, in conditions where the mitochondria were prob-

Table 2. Comparison of the fluorescence (F) of the tumour cell suspension and the magnitude of the membrane potential ($\Delta\psi$) in equilibrium with the potassium ion gradient

Cellular history	F (arbitrary units)	$-\Delta\psi$ (mv)
1. 0.2% glucose for 10 min	26	88
2. 0.2% glucose for 30 min	43	38
3. 0.2% glucose and 1 mM glycine for 10 min	37	63
4. 0.2% glucose + 0.3 mM ouabain for 9 min	45	56
5. 5 mM deoxyglucose for 10 min	50	36

As outlined in text, $\Delta\psi$ was determined by method of Hoffman and Laris (1974). Tumour cells (1 vol) were suspended in Ringer solution (400 vol) of required K^+ concentration containing 60 µM 2,4-dinitrophenol, 12.5 ng of oligomycin/ml, 500 ng of antimycin/ml and incubated as shown below, at 37°C. Dye (3,3'-dipropyl-2,2'-oxadicarbocyanine iodide) was added to concentration of 1 µg/ml of cell suspension, 4 ml of which was placed in the cuvette. When trace had reached steady level (about 5 min), 10 µg valinomycin was added. Intracellular concentration of K^+ was also determined in remainder of cell suspension as described previously (Reid et al., 1974). Correlation coefficient of 0.946 was found between 12 pairs of values of F and $\Delta\psi$

ably not involved, is thus evidence in favour of the scheme shown in Figure 1b.

There is some evidence (Table 2) that when the experimental conditions are strictly standardized, the fluorescence (F) of the tumour cell suspension in the presence of the probe is linearly related to the potential difference across the plasmalemma ($\Delta\psi$) as assayed by the technique of Hoffman and Laris (1974). Their procedure involves an estimate, for a given cell preparation, of the ratio [cellular K^+]/[extracellular K^+] that is in equilibrium with $\Delta\psi$. The criterion of equilibrium used is that the addition of valinomycin causes no change in F. Thus the method measures $\Delta\psi$ in the ionic conditions prevailing at the null point. These are not necessarily the conditions of physiological interest. However, it seems possible that the same standard relationship between F and $\Delta\psi$, as shown in Table 2, holds when $\Delta\psi$ is not equal to the equilibrium potential for K^+ ions. The magnitude of $\Delta\psi$ for the cellular preparations kept with glucose in the standard Ringer was estimated to be about -38 mV on that basis. We suggest that a similar procedure might be used to estimate $\Delta\psi$ in other conditions where the null-point method is not applicable. For instance, Gibb and Eddy (1972) and Heinz et al. (1975) have defined circumstances in which (1) the concentration gradients of both Na^+ and K^+ across the plasmalemma are small and (2), owing to the operation of the sodium pump, the membrane potential of the tumour cells appears to approach -60 mV, if it is indeed correct to explain the accumulation of L-methionine and 2-aminoisobutyrate in terms of the gradient hypothesis. We have made preliminary observations that show that the fluorescence of the probe in the presence of such cell preparations is consistent with $\Delta\psi$ being of that magnitude.

D. Conclusions

The preceding discussion concerns three aspects of the cosubstrate hypothesis: (1) the stoicheiometry of the substrate and cosubstrate absorption; (2) the accumulation of the amino acids at the expense of the gradient of cosubstrate ions; (3) the electrogenicity of the two partial processes of solute absorption and ion ejection. It will be apparent that, in general terms, the transport systems of the yeasts and of the mouse tumour cells are broadly similar in these three respects and that their behaviour is consistent with the models shown in Figure 1a and b.

Acknowledgement. This work was supported in part by grant B/RG/42357 from the Science Research Council, U.K.

References

Berger, E.A.: Proc. Nat. Acad. Sci. U.S. 70, 1514-1518 (1974)
Christensen, H.N., De Cespedes, C., Handlogten, M.E., Ronquist, C.: Biochim. Biophys. Acta 300, 487-522 (1973)
Cockburn, M., Earnshaw, P., Eddy, A.A.: Biochem. J. 146, 705-712 (1975)
Conway, E.J., O'Malley, E.: Biochem. J. 40, 59-67 (1946)
Eddy, A.A.: Biochem. J. 108, 195-206 (1968)
Gibb, L.E., Eddy, A.A.: Biochem. J. 129, 979-981 (1972)
Hamilton, W.A.: Adv. Microb. Physiol. 12, 1-53 (1975)
Harold, F.M.: Bacteriol. Rev. 36, 172-230 (1972)
Harold, F.M., Spitz, E.: J. Bacteriol. 122, 266-277 (1975)
Heinz, E., Geck, P., Pietrzyk, C.: Ann. N.Y. Acad. Sci. 264, 428-441 (1975)
Hoffman, J., Laris, P.C.: J. Physiol. 239, 519-552 (1974)
Khan, N.A., Zimmermann, F.K., Eaton, N.R.: Mol. Gen. Genet. 123, 43-50 (1973)
Laris, P.C., Bahr, D.P., Chafee, R.R.J.: Biochim. Biophys. Acta 376, 415-425 (1975)
Laris, P.C., Pershadsingh, H.A., Johnstone, R.M.: Biochim. Biophys. Acta 436, 475-488 (1976)
Lassen, U.V., Nielsen, A.M.T., Pape, L., Simonsen, L.D.: J. Membrane Biol. 6, 269-288 (1971)
Mitchell, P.: Symp. Soc. Gen. Microbiol. 20, 121-166 (1970)
Morville, M., Reid, M., Eddy, A.A.: Biochem. J. 134, 11-26 (1973)
Philo, R.D., Eddy, A.A.: Biochem. Soc. Trans. 3, 886-888 (1975)
Reid, M., Gibb, L.E., Eddy, A.A.: Biochem. J. 140, 383-393 (1974)
Rothstein, A., Enns, L.H.: J. Cell. Comp. Physiol. 28, 231-252 (1946)
Schuldiner, S., Kerwar, G.K., Kaback, R., Weil, R.: J. Biol. Chem. 250, 1361-1370 (1975)
Seaston, A., Inkson, C., Eddy, A.A.: Biochem. J. 134, 1031-1043 (1973)
Seaston, A., Carr, G., Eddy, A.A.: Biochem. J. 154, 669-676 (1976)
Slayman, C.L., Slayman, C.W.: Proc. Natl. Acad. Sci. U.S. 71, 1935-1939 (1974)
Smith, T.C., Mikiten, T.M., Levinson, C.: J. Cell Physiol. 79, 117-126 (1972)
Van Steveninck, J.: Biochim. Biophys. Acta 203, 376-384 (1970)
Zimmermann, F.K., Khan, N.A., Eaton, N.R.: Mol. Gen. Genet. 123, 29-41 (1973)

Reconstitution of Na^+ Gradient-Coupled Carrier Functions of Brush Border Membranes of Intestine and Kidney in Sonicated Liposomes

R. K. CRANE, P. MALATHI, and H. PREISER

A. Introduction

Some years ago, Crane and his colleagues (1960; Crane et al., 1961) explicitly formulated noncovalent coupling to the Na^+ circuit as the driving force for organic solute accumulation across the brush border membrane of the mammalian intestinal epithelial cell. As finally expressed (Crane, 1962, 1965; Crane et al., 1961), the formulation consisted of the coupling being made possible by the formation of a ternary complex of ion, solute, and membrane transport carrier. It was assumed that the ternary complex was able to respond to the transmembrane electrochemical potential gradient; thus one of the terms used to identify the concept is the gradient hypothesis, and the transport carriers have been called gradient-coupled. Over the years, many studies have confirmed the essential features and predictions of the gradient hypothesis not only with intact cells and tissues but also with vesiculated membrane preparations; not only in mammals but also in many other living forms including microorganisms (Crane, 1976). The Na^+-dependent D-glucose transport modalities of the intestinal and kidney brush border membranes have now been reconstituted in sonicated liposomes (Crane et al., 1976a,b), and reconstitution is the subject of this paper.

To provide an appropriate background against which to view the experiments on reconstitution, the results of studies with vesiculated brush border membrane preparations will be briefly reviewed. The use of vesiculated membrane preparations was introduced by Kaback (1960; Kaback and Stadtman, 1966), and their special value has been to demonstrate properties of the plasma membrane substantially free of complications from the presence of the usual intracellular biochemical machinery.

B. Intestinal Brush Border Membrane Vesicles

Starting with a preparation of rat small intestine brush border membranes (Forstner et al., 1968) like that produced earlier in our laboratory from the hamster (Miller and Crane, 1961; Eichholz and Crane, 1965), Hopfer et al. (1973) produced a membrane preparation having an osmotically responsive intravesicular space. The entry of D-glucose into this space was compared to that of L-glucose, which can be used as a control sugar despite being a substrate for the D-glucose carrier (Caspary and Crane, 1968) because its affinity as compared to D-glucose is very low. In these experiments as well as later ones, the characteristics of D-glucose entry were qualitatively the same as the characteristics of entry into energy-depleted intact cells (Bihler et al., 1962). In the presence of Na^+, and not other cations, D-glucose entered more rapidly than L-glucose, and the enhanced D-glucose rate was completely inhibited by an equal concentration of the competitive inhibitor phlorizin (Alvarado and Crane, 1962). The entry of Na^+ along with D-glucose was electrogenic (Murer and Hopfer, 1974) as predicted by the findings with intact cells of Bihler and Crane (1962) with impermeant anions and confirmed by the electrical measurements of Schultz and Zalusky (1964) and Rose and Schultz (1971). This means that Na^+ enters on the carrier without an attached counterion. With intact cells, entering Na^+ may be removed by the Na^+ pump or it may exchange with intracellular K^+. With vesicles that lack the Na^+ pump and do not exchange cations readily, rapid entry of Na^+ requires the addition of a pathway for the simultaneous rapid entry of an anion or for the simultaneous rapid exit of a cation. In the experiments of Murer and Hopfer (1974), an anion pathway was provided by use of the lipophilic anion, SCN^-, added outside. A cation pathway was provided by the K^+-specific ionophore, valinomycin, or the H^+-specific ionophore, carbonyl cyanide p-trifluoromethoxy phenylhydrazone (CF-CCP) incorporated into the membranes. When the ionophores were used the vesicles were prepared with either high K^+ or high H^+ concentrations inside. In all three instances, the addition of a transmembrane pathway for a counterion enhanced D-glucose entry, and an overshoot (above equilibrium value) in the early time periods was observed that decayed eventually to the common equilibrium value also achieved by L-glucose. In tests comparing the entry of L-alanine to that of D-alanine (Sigrist-Nelson et al., 1975), a relatively more rapid entry of L-alanine was observed, and a similar overshoot phenomenon was obtained under the appropriate experimental conditions.

C. Kidney Brush Border Membrane Vesicles

Studies with a variety of in vivo and in vitro techniques have indicated that glucose uptake at the brush border membrane of the epithelial cells of the proximal convoluted tubules of the mammalian kidney is in all important respects the same as in the intestine (Ullrich, 1976). Kidney brush border membranes were first isolated by Thuneberg and Rostgaard (1968), and the methods were improved by Kinne and Kinne-Saffran (1969). A vesiculated membrane preparation was produced by Busse et al. (1972). Studies with vesiculated kidney membrane preparations have now produced results that are strikingly similar to those described previously for intestinal membrane preparations. Aronson and Sacktor (1974, 1975) and Beck and Sacktor (1975) have studied D-glucose uptake. Evers et al. (1976) have studied L-phenylalanine uptake. These studies confirm the presence of the respective Na^+ gradient-coupled carriers in the kidney tubular brush border membranes.

D. Attempts to Isolate the Glucose Carrier

From the first (Fox and Kennedy, 1965) attempts to isolate carriers have been hampered by the need to use chemical rather than functional tests of identity. Phlorizin-sensitive D-glucose binding proteins have been separated from both kidney (Thomas, 1973) and intestinal (Smith et al., 1975) brush border membranes, labeled with radioactivity for identification in the manner introduced by Fox and Kennedy (1965). Functional tests of these isolated entities have not been reported, presumably due to their inactivation by the reagents employed and to the lack of a suitable assay system for transport function. The report on reconstitution to follow is equally a report that a suitable assay system has been achieved.

E. Reconstitution

I. Materials and Methods

The details of materials and methods are given in the original papers (Crane et al., 1976a,b). Briefly, however, D[U-^{14}C] glucose and L-[1-^{3}H(N)] glucose were used as substrates. D-[U-^{14}C] mannitol and D-[1-^{3}H(N)] mannitol were used to measure extravesicular space. Soybean phospholipids, acetone-extracted as described by Kagawa and Racker (1971), were used to prepare liposomes. Purified brush border membranes were prepared according to Schmitz et al. (1973) and extracted with 1% Triton X-100, which was later removed on a column of Biobeads (Holloway, 1973). The method described by Racker (1973) was used with some modifications for reconstitution. Sugar transport was measured by adding an aliquot (50 μl) of suspended liposomes into buffers containing the labeled sugar at a final volume of 350 μl. Aliquots were removed at intervals and diluted into 1 ml of cold stopping solution (0.154 M NaCl + 0.5 mM phlorizin at 4°C) containing tracer amounts of radioactive mannitol. The mixture was filtered through 0.3 μm millipore filters and washed with 5 ml of cold stopping solution. The radioactivity remaining on the filters was counted.

II. Results With Extracts From Intestinal Brush Border Membranes

Some of our results are shown in Figure 1.

Liposomes reconstituted with extracted proteins in 0.1 M KCl buffer showed time-dependent nonspecific uptake of both D- and L-glucose from a 0.1 M KCl medium. Protein-free liposomes showed a similar nonspecific but greatly reduced (25-30%) uptake characteristic. Extracted proteins subjected to the procedures involved in reconstitution but without phospholipids showed no uptake.

When inward Na^+ movement was induced by addition of the lipophilic anion, SCN^-, substantial specific entry of D-glucose but not of L-glucose occurred. Phlorizin and galactose both inhibited the enhanced D-glucose entry.

In these experiments, the excess entry of D-glucose in the presence of Na^+ is maintained over the time tested. We have not tested whether the same equilibrium is finally reached with D-glucose and L-glucose in the presence and absence of sodium. Movement of sugar against a gradient has not yet been achieved, and this may suggest that there are volume changes in the liposomes induced by D-glucose entry. We suspect that the use of a heterogeneous protein extract has probably reconstituted a number of other trans-

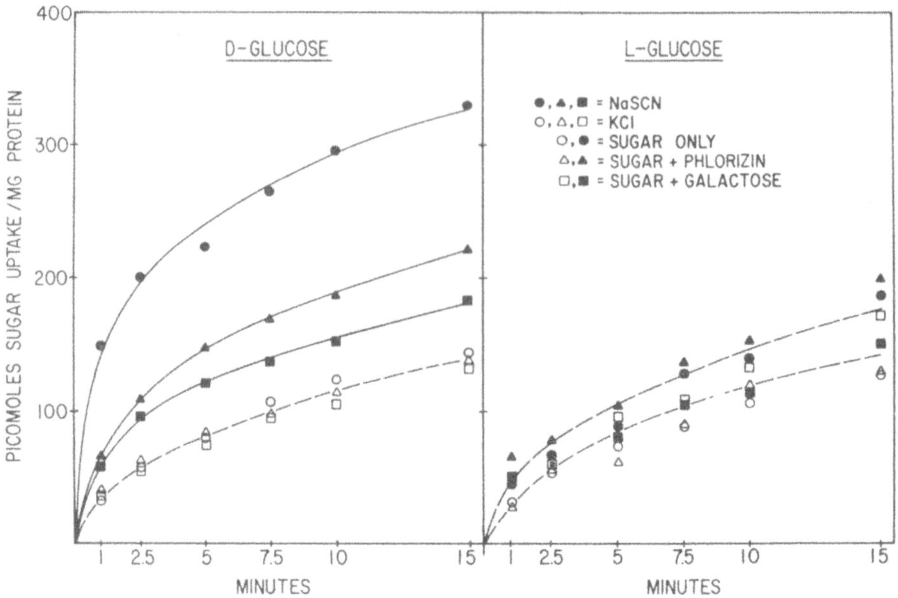

Fig. 1. Reconstitution with intestinal brush border membrane proteins. Triton X-100-extracted membrane proteins were reconstituted into soybean phospholipid liposomes in 0.1 M Kcl - 5 mM Hepes-Tris buffer containing 1 mM dithiothreital and 0.1 mM $MgSO_4$, as described in text. Incubation was at room temperature (25°C) in total volume of 350 µl containing 0.1 mM D-[U-^{14}C] glucose or 0.1 mM L-[1-^{3}H(N)]-glucose, 0.1 m KCl or NaSCN, 5 mM Hepes-Tris (pH 7.5) and 0.1 mM $MgSO_4$. Reaction was started by addition of 50 µl of reconstituted liposomes. When present, phlorizin was 5 mM and D-galactose, 10 mM

port systems, not yet tested for. In fact, our preliminary data indicate a Na^+-dependent uptake of L-alanine by these reconstituted liposomes.

Some of the brush border hydrolases are incorporated into the liposomes along with the D-glucose transport activity. Consequently, the question may be raised that D-glucose enters by binding to a hydrolase and undergoing translocation in a form of hydrolase related transport (HRT) (Ramaswamy et al., 1974, 1976; Semenza, 1976). This is highly improbable for the following reasons. First, experiments on HRT indicate that it is not a significant route for the entry of free glucose. Second, HRT is not dependent on the presence or the coflux of Na^+. Third, phlorizin has no appreciable effect on hydrolase activity or on HRT. Fourth, $Tris^+$ is a potent, competitive inhibitor of several disaccharidases (Larner and Gillespie, 1956), and the concentration used for buffering in these experiments is high enough (3.3 mM) to cause substantial inhibition of glucose (0.1 mM) binding to the active sites of these enzymes.

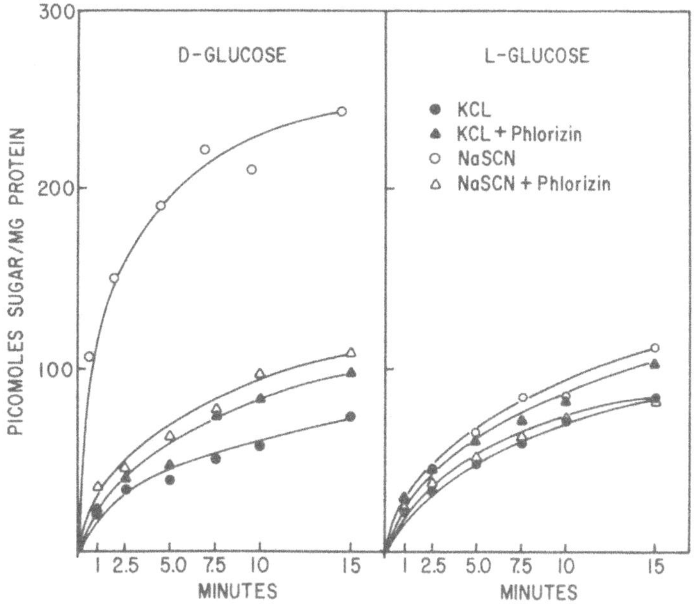

Fig. 2. Reconstitution with kidney brush border membrane proteins: Effects of NaSCN and phlorizin. Triton X-100-extractable membrane proteins were reconstituted with soybean phospholipid vesicles in 0.15 M KCl 5 mM Hepes-Tris 1 mM DTT 0.1 mM $MgSO_4$. Incubation was at room temperature (≈ 25°C) in total volume of 350 µl. Incubation medium contained 0.1 mM D-[U-^{14}C] glucose or 0.1 mM L-[1-^{3}H(N)]-glucose, 0.15 M KCl, or NaSCN 5 mM Hepes-Tris (pH 7.5) and 0.1 mM $MgSO_4$. Reaction was initiated by addition of 50 µl reconstituted liposomes. Phlorizin was added at concentration of 5 mM

F. Results With Kidney Brush Border Membranes

Our results with kidney brush border membranes shown in Figures 2 and 3 are comparable to those with intestinal membranes except perhaps even more clear.

There is a time-dependent nonspecific uptake of D- and L-glucose from 0.1 M KCl. With NaSCN there was a marked enhancement of D-glucose uptake whereas there were no discernable effects on L-glucose. Phlorizin completely inhibited NaSCN-enhanced D-glucose uptake and was without effect on L-glucose.

Sugars that share the same transport mechanism as D-glucose - viz. D-galactose, 6-deoxyglucose, and α-methylglucoside (Ullrich, 1976) also substantially inhibited NaSCN-enhanced D-glucose uptake. L-glucose had no effect. Cl^- is a usable, though not ideal, counter-ion for Na^+ in this liposome system. Replacing KCl with NaCl enhanced D-glucose uptake although to a markedly lesser extent as compared to NaSCN.

To the extent thus far studied our results are consistent with those obtained with vesiculated membrane preparations. They also extend to gradient-coupled carriers the successful reconstitution

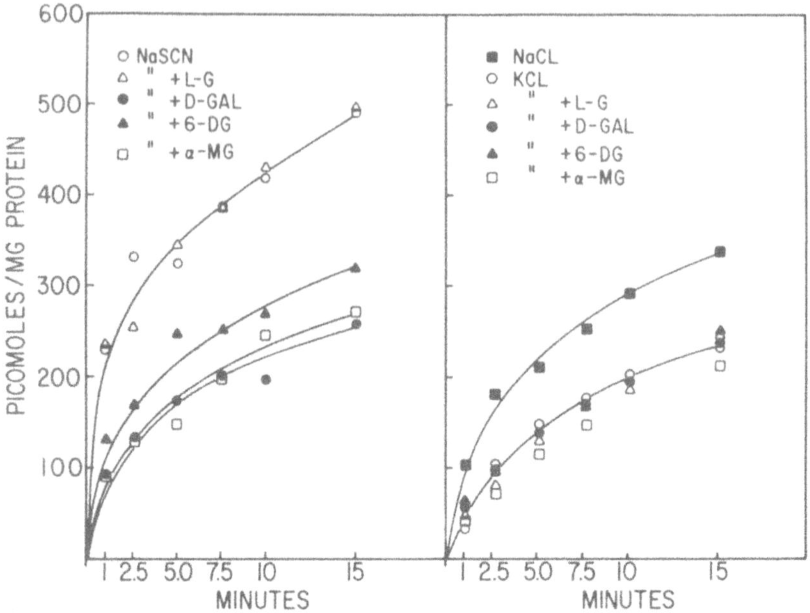

Fig. 3. Reconstitution with kidney brush border membrane proteins: effects of other substrates. Experimental details same as in Figure 1. NaCl (0.15 M) replaced KCl where indicated. L-glucose, D-galactose, 6-deoxyglucose, and α-methylglucoside were used at final concentration of 5 mM

of the red cell facilitated diffusion sugar carrier by Kasahara and Hinkle (1976).

G. Conclusions

In our experiments we have naturally followed in the footsteps of the pioneers of reconstitution - Racker, Hokin, Rothstein, Semenza, Hinkle, and others. However, with our experiments the ring seems to have been closed. Previous reports have demonstrated reconstitution of ATP-dependent ion transport, oxidation-dependent and light-dependent proton transport, carrier-mediated ion transport, carrier-mediated sugar transport, and hydrolase-related transport. With reconstitution of gradient-coupled transport, all the known varieties of systems seem to be in hand for studies that hopefully will lead us out of phenomenology into the real molecular world.

Acknowledgements. We wish to thank Dr. K. Ramaswamy for his participation in our early attempts at reconstitution, Mr. F. Durando for his participation in the early work on reconstitution from kidney brush border membranes, and Mr. B. Miller for excellent technical assistance. This work was supported by a research grant from the National Institute of Arthritis, Metabolism and Digestive Diseases.

References

Alvarado, F., Crane, R.K.: Phlorizin as a competitive inhibitor of the active transport of sugars by hamster small intestine, in vitro. Biochim. Biophys. Acta 56, 170-172 (1962)

Aronson, P.S., Sacktor, B.: Transport of D-glucose by brush border membranes isolated from the renal cortex. Biochim. Biophys. Acta 356, 231-243 (1974)

Aronson, P.S., Sacktor, B.: The Na^+ gradient-dependent transport of D-glucose in renal brush border membranes. J. Biol. Chem. 250, 6032-6039 (1975)

Beck, J.C., Sacktor, B.: Energetics of the Na^+-dependent transport of D-glucose in renal brush border membrane vesicles. J. Biol. Chem. 250, 8674-8680 (1975)

Bihler, I., Crane, R.K.: Studies on the mechanism of intestinal absorption of sugars. V. The influence of several cations and anions on the active transport of sugars, in vitro, by various preparations of hamster small intestine. Biochim. Biophys. Acta 59, 78-93 (1962)

Bihler, I., Hawkins, K.A., Crane, R.K.: Studies on the mechanism of intestinal absorption of sugars. VI. The specificity and other properties of Na^+-dependent entrance of sugars into intestinal tissue under anaerobic conditions, in vitro. Biochim. Biophys. Acta 59, 94-102 (1962)

Busse, D., Elsas, L.J., Rosenberg, L.E.: Uptake of D-glucose by renal tubule membranes. J. Biol. Chem. 247, 1188-1193

Caspary, W.F., Crane, R.K.: Inclusion of L-glucose within the specificity limits of the active sugar transport system of hamster small intestine. Biochim. Biophys. Acta 163, 395-400 (1968)

Crane, R.K.: Intestinal absorption of sugars. Physiol. Rev. 40, 789-825 (1960)

Crane, R.K.: Hypothesis of mechanism of intestinal active transport of sugars. Federation Proc. 21, 891-895 (1962)

Crane, R.K.: Na^+-dependent transport in the intestine and other animal tissues. Federation Proc. 24, 1000-1006 (1965)

Crane, R.K.: The gradient hypothesis and other models of carrier-mediated active transport. Rev. Physiol. Biochem. Pharmacol. In press, 1976, Springer-Verlag, Heidelberg

Crane, R.K., Malathi, P., Preiser, H.: Reconstitution of specific Na^+-dependent D-glucose transport in liposomes by Triton X-100 extracted proteins from purified brush border membranes of hamster small intestine. Biochem. Biophys. Res. Commun. 71, 1010-1015 (1976a)

Crane, R.K., Malathi, P., Preiser, H.: Reconstitution of specific Na^+-dependent D-glucose transport in liposomes by Triton X-100 extracted proteins from purified brush border membranes of rabbit kidney cortex. FEBS Lett. 67, 214-216 (1976b)

Crane, R.K., Miller, D., Bihler, I.: The restrictions on the possible mechanism of intestinal active transport of sugars. In: Membrane Transport and Metabolism. Kleinzeller, A., Kotyk, A. (eds.). Prague: Czech. Acad. Sci. Press 1961, pp. 439-449

Eichholz, A., Crane, R.K.: Studies on the organization of the brush border in intestinal epithelial cells. I. Tris disruption of isolated hamster brush borders and density gradient separation of fractions. J. Cell Biol. 26, 687-691 (1965)

Evers, J., Murer, H., Kinne, R.: Phenylalanine uptake in isolated renal brush border vesicles. Biochim. Biophys. Acta 426, 598-615 (1976)

Forstner, G.G., Sabesin, S.M., Isselbacher, K.J.: Rat intestinal microvillus membranes. Purification and biochemical characterization. Biochem. J. 106, 381-390 (1968)

Fox, C.F., Kennedy, E.P.: Specific labeling and partial purification of the M protein, a component of the β-galactoside transport system of Escherichia coli. Proc. Natl. Acad. Sci. U.S. 54, 891-899 (1965)

Holloway, P.W.: A simple procedure for removal of Triton X-100 from protein samples. Anal. Biochem. 53, 304-308 (1973)

Hopfer, U., Nelson, K., Perrotto, J., Isselbacher, K.J.: Glucose transport in isolated brush border membrane from rat small intestine. J. Biol. Chem. 248, 25-32 (1973)

Kaback, H.R.: Uptake of amino acids by "Ghosts" of mutant strains of E. Coli. Federation Proc. 19, 130 (1960)

Kaback, H.R., Stadtman, E.R.: Proline uptake by an isolated plasma membrane preparation of E. Coli. Proc. Natl. Acad. Sci. U.S. 55, 920-927 (1966)

Kagawa, Y., Racker, E.: Partial resolution of the enzymes catalyzing oxidative phosphorylation. J. Biol. Chem. 246, 5477-5487 (1971)

Kasahara, M., Hinkle, P.C.: Reconstitution of D-glucose transport by a protein fraction from human erythrocytes in sonicated liposomes. Proc. Natl. Acad. Sci. U.S. 73, 396-400 (1976)

Kinne, R., Kinne-Saffran, E.: Isolierung und enzymatische Charakterisierung einer Burstensaumfraktion der Rattenniere. Pflügers Arch. 308, 1-15 (1969)

Larner, J., Gillespie, R.E.: Gastrointestinal digestion of starch. II. Properties of the intestinal carbohydrases. J. Biol. Chem. 223, 709-726 (1956)

Miller, D., Crane, R.K.: A procedure for the isolation of the epithelial brush border membrane of hamster small intestine. Anal. Biochem. 2, 284-286 (1961)

Murer, H., Hopfer, U.: Demonstration of electrogenic Na^+-dependent D-glucose transport in intestinal brush border membranes. Proc. Natl. Acad. Sci. U.S. 71, 484-488 (1974)

Racker, E.: A new procedure for the reconstitution of biologically active phospholipid vesicles. Biochem. Biophys. Res. Commun. 55, 224-230 (1973)

Ramaswamy, K., Malathi, P., Caspary, W.F., Crane, R.K.: Studies on the transport of glucose from disaccharides by hamster small intestine in vitro. II. Characteristics of the disaccharidase-related transport system. Biochim. Biophys. Acta 345, 39-48 (1974)

Ramaswamy, K., Malathi, P., Crane, R.K.: Demonstration of hydrolase-related glucose transport in brush border membrane vesicles prepared from guinea pig small intestine. Biochem. Biophys. Res. Commun. 68, 162-168 (1976)

Rose, R.C., Schultz, S.G.: Studies on the electrical potential profile across rabbit ileum. J. Gen. Physiol. 57, 639-663 (1971)

Schmitz, J., Preiser, H., Maestracci, D., Ghosh, B.K., Cerda, J.J., Crane, R.K.: Purification of the human intestinal brush border membrane. Biochim. Biophys. Acta 323, 98-112 (1973)

Schultz, S.G., Zalusky, R.: Ion transport in isolated rabbit ileum. II. The interaction between active sodium and active sugar transport. J. Gen. Physiol. 47, 1043-1059 (1964)

Semenza, G.: Small intestinal disaccharidases: their properties and role as sugar translocators across natural and artificial membranes. In: The Enzymes of Biological Membranes, Vol. 3. Martonosi, A. (ed.). New York: Plenum Press 1976, pp. 349-382

Sigrist-Nelson, K., Murer, H., Hopfer, U.: Active transport in isolated brush border membranes. J. Biol. Chem. 250, 5674-5680 (1975)

Smith, M.W., Ferguson, D.R., Burton, K.A.: Glucose and phlorizin protected thiol groups in pig intestinal brush-border membranes. Biochem. J. 147, 617-619 (1975)

Thomas, L.: Isolation of N-ethylmaleimide-labelled phlorizin-sensitive d-glucose binding protein of brush border membrane from rat kidney cortex. Biochim. Biophys. Acta 291, 454-464 (1973)

Thuneberg, L., Rostgaard, J.: Isolation of brush border fragments from homogenates of rat and rabbit kidney cortex. Exp. Cell Res. 51, 123-140 (1968)

Ullrich, K.J.: Renal tubular mechanisms of organic solute transport. Kidney Intern. 9, 134-148 (1976)

A Possible Common Activation Site in Na^+-Driven Transport Systems. Partial Purification of the Na-Dependent D-Glucose Transport System From Small Intestinal Brush Borders

G. SEMENZA, C. TANNENBAUM, M. KESSLER, G. TOGGENBURGER, and L. WAHLGREN

A. Some Kinetic Considerations

In a number of laboratories attempts are presently underway to purify proteins involved in transmembrane transport. One of the criteria that will have to be employed in the identification of such component(s) is the comparsion of substrate binding with parameters derived from transport kinetics. Thus a discussion of the complex significance of the macroscopic kinetic constants, as obtained from the operation of the original transport systems, seems appropriate at this time. A similar, and perhaps more refined, treatment was given by Heinz et al. (1972), who also point out the fundamental unity of cation-driven co-transport systems. When expressed in velocity constants, the ratio between C_t (total carrier concentration) and $J^{max}_{A^{II}=0}$ (the unidirectional substrate flux at zero trans substrate concentration and infinitely large cis substrate concentration) is given by:

$$\frac{C_t}{J^{max}_{A^{II}=0}} = \frac{1}{k_4} + \frac{1}{k_3} + \frac{k_{-2}}{k_2 k_3} + \frac{1}{k_2} = \frac{1}{k_4} + \frac{1}{k_3} + \frac{1}{k_2}\left(1 + \frac{k_{-2}}{k_3}\right)$$

(For definition of symbols, see scheme 1; see also Regen and Morgan, 1964; Geck, 1971; Segel, 1975; Kotyk, 1973, 1976).

The expression for the K_m of initial uptake velocity is

$$K_{mA^I} = \frac{(k_4 + k_{-4})\ (k_{-1}\ k_3 + k_{-1}\ k_{-2} + k_2\ k_3)}{k_1 (k_2\ k_3 + k_2\ k_4 + k_{-2}\ k_4 + k_3\ k_4}$$

It is immediately apparent that $K_{mA I}$ can be identical with, smaller than, or larger than K_{sI} value (k_{-1}/k_1). In addition, the K_m value at the I side is even more remotely related to the K_s value at the II side (k_{-3}/k_3).

These considerations are rather trivial, but sometimes are apparently neglected.

The situation becomes even more complicated in the case of cotransport. The complete scheme 1A would lead to equations of forbidding and unmanageable complexity.

Yet it is possible and appropriate to consider, at least qualitatively, how the presence of a cotransported cation in the presence of a $\Delta\psi$ (negative inside) may change macroscopic kinetic parameters, such as K_m or V_{max}. In addition to the reason mentioned

---

This paper is dedicated to Walther Wilbrandt on the occasion of his 70th birthday.

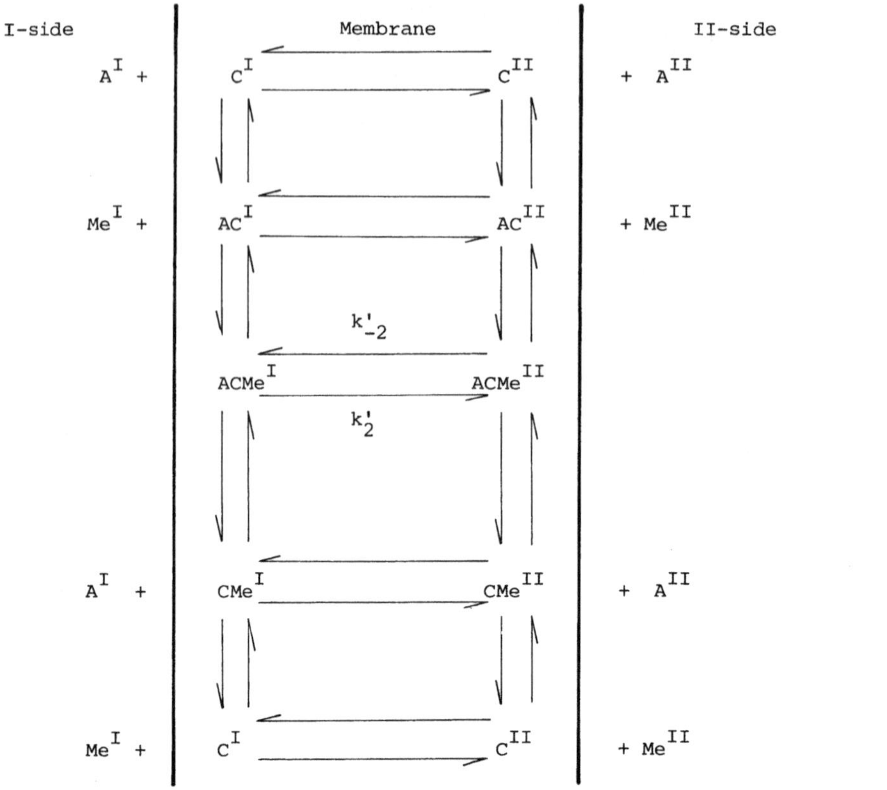

Scheme 1A. C: carrier; A: noncharged solute; Me: monovalent cotransported cation (Na^+ or H^+, for example)

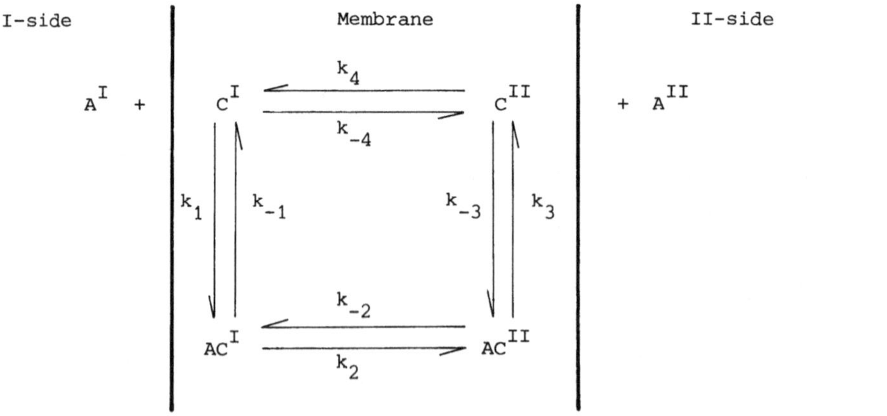

Scheme 1B

previously, there is also the following justification. The cotransport mechanism in Na^+- or H^+-activated systems (Crane, 1962, 1965; Mitchell, 1967) can now be regarded as universally accepted, particularly after the demonstration of their electrogenic character (Gibb and Eddy, 1972; Murer and Hopfer, 1974; West and Mitchell, 1972; Harold and Papineau, 1972a,b; Komor and Tanner, 1974; Deshusses and Reber, 1976; Seaston et al., 1976; Eddy et al., 1976; Heinz et al., 1976).

However, it is intellectually unsatisfactory that the apparent mode of activation of solute transport by Na^+ seems to follow different types of kinetics (although all imply the formation of a tertiary or quaternary (Vidaver, 1964) carrier-substrate-Na^+ complex). For example, Na^+ activation of sugar uptake in the small intestine of the hamster (Crane et al., 1965; Semenza, 1967) takes place via a decrease in the apparent K_m with no detectable effect on the V_{max}; whereas in the rabbit it takes place via an increase in V_{max}, with little effect on the apparent K_m (Goldner et al., 1969; Semenza, 1967). This difference is not due to the different techniques used or to unstirred layer effects, since data from different laboratories using different techniques agree very well and since the same laboratory has obtained different types of Na^+ activation for different systems in the same species and with the same techniques (amino acids vs. sugars in the rabbit, Curran et al., 1967; Goldner et al., 1969). It seems a priori unlikely that Na^+ should affect the rate of membrane translocation in some systems while affecting the apparent "affinity" for the substrate in others (in so far as $\frac{1}{K_m}$ is to be taken as a measure of this affinity).

Similarly, it is puzzling that there should be different kinetic types of heterologous inhibition of influx among Na^+-dependent systems. For example the mutual heterologous inhibition of amino acid and sugar uptake in hamster small intestine is due to an increase in K_m values ("partial" competitive inhibition - reviewed by Semenza, 1971), whereas the inhibition of inositol uptake by sugars in the same tissue is "noncompetitive" (Caspary and Crane, 1970).

It is indeed more likely that Na^+ together with $\Delta\tilde{\mu}_{Na^+}$ affects the same few kinetic parameters in all transport systems, and that the apparent mode of activation (or of inhibition, in the case of heterologous interactions) actually ensuing is related to the complex kinetic significance of "transport K_m" and "transport V_{max}".

A detailed discussion of this problem will be presented elsewhere. The following considerations may suffice here. Since the driving force of cation-cotransporting systems has been identified with the $\Delta\tilde{\mu}_{Na^+}$ (or $\Delta\tilde{\mu}_{H^+}$, respectively) across the membrane, we may start by the reasonable assumption that in all cation-cotransporting systems the rate of translocation of the substrate in the presence of the cotransported cation (whatever the "carrier" model may be) is larger than that of the substrate-loaded carrier, i.e., in all these cotransport systems, $k_2 \cdot AC + k_2' \cdot ACMe^+ > k_2 \cdot AC$ (where k_2' depends on $\Delta\tilde{\mu}Me^+$, and $ACMe^+$ is a nonlinear function of $[Me^+]$ and [A] in water phase I). The complex dependence of k_2' on $\Delta\tilde{\mu}_{Me^+}$ and on the charge of the tertiary complex was studied recently by Geck and Heinz (1976).

Another effect of the cation that may (or may not) be common to all cation-cotransporting systems is a decrease in K_s (k_{-1}/k_1) for the substrate: the dissociation constant for high affinity binding of phlorizin in renal proximal tubuli has indeed been shown to depend on Na^+ concentration (Frasch et al., 1970). It is conceivable, e.g., that neutralization of a charge at the substrate binding site may in some way increase its affinity for the substrate. Other possibilities include the effect of the high electric field density on the structure of the transport protein, etc.

We now ask: can the same effect of $Me^+ + \Delta\tilde{\mu}_{Me^+}$ (be it solely an increase in translocation rate or be it associated with an increase in affinity for the substrate) be common to all cation transporting systems, although resulting in different macroscopic kinetic effects?

The answer is yes and is based on the following (admittedly arbitrarily selected) examples.

For the k_2 being the rate-limiting step (with $k_2 << k_3, k_4, k_{-1}$) and $k_{-2} << k_3$, the maximum initial unidirectional flux is given by:

$$J^{max}_{A^{II}=0} = k_2 C_t$$

and the K_m by:

$$K_{mA^I} = \frac{k_{-1}}{k_1}\left(1 + \frac{k_{-4}}{k_4}\right)$$

In this example an increase in the translocation step (from $k_2 \cdot AC$ to $k_2' \cdot ACMe^+ + k_2 \cdot AC$) produces a pure "V-effect".

In contrast, in a system in which the rate limiting step is at k_3 with $k_3 << k_2, k_4$; $k_{-1}k_{-2} >> k_2k_3$ and $k_{-2} << k_2$, the maximum velocity is given by $J^{max}_{A^{II}=0} = k_3C_t$ and the K_m by

$$K_{m_A I} = \frac{k_{-2}}{k_2}\,\frac{k_{-1}}{k_1}\left(1 + \frac{k_{-4}}{k_4}\right)$$

In this example (and in others, not mentioned here) an increase in the translocation step produces a pure "K effect" with no effect on V_{max}.

As pointed out previously, these are rather arbitrary examples. They only illustrate the point that a fundamental unity is likely to underlie the variety of macroscopic kinetic effects produced by the cotransported cation. The fundamental unity can be a stimulation step alone (as in the foregoing example) or, more likely, a combination of stimulation of substrate translocation step plus a decrease in K_s (for a more detailed discussion, see Semenza, 1977). In addition, the model used here is an obvious oversimplification (see e.g., Kotyk, 1976; Wilbrandt, 1976). Its justification is only that of illustrating the point just made.

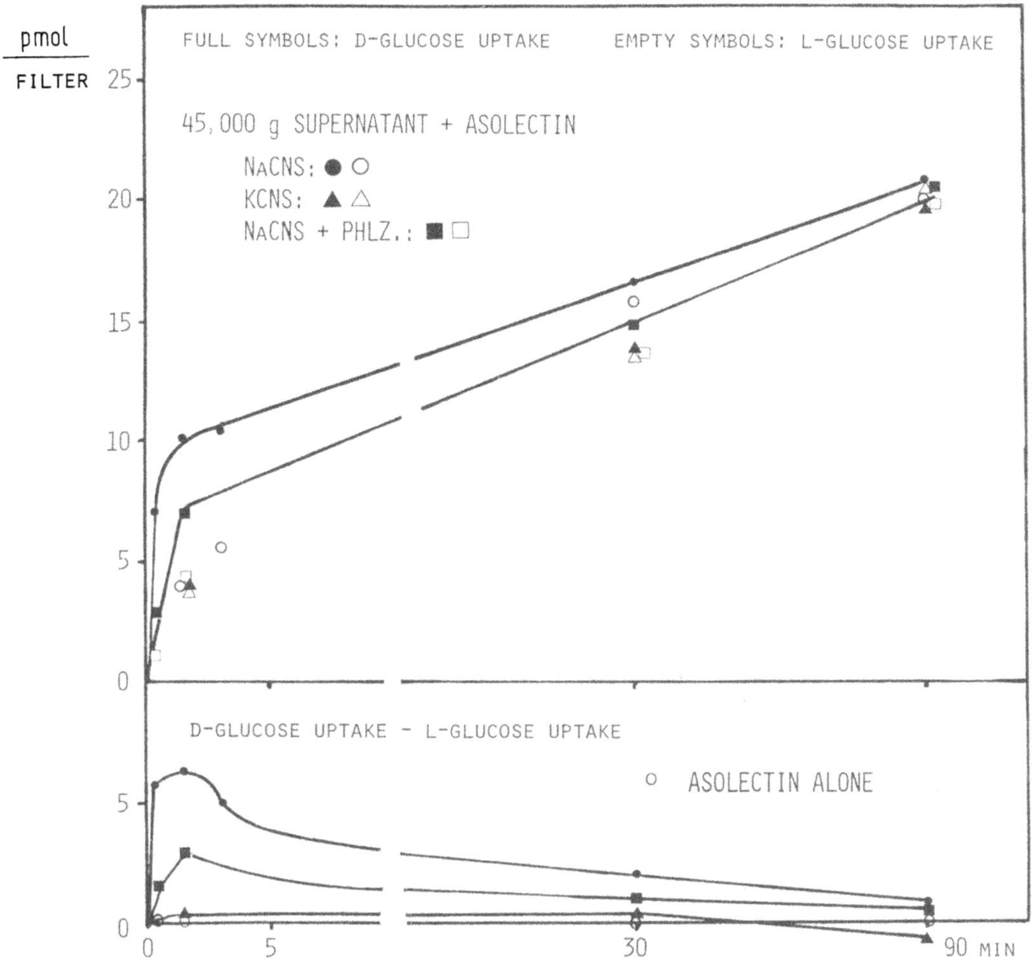

Fig. 1. Uptake of D- and L-glucose by 45,000 g "solubilizate" from small intestinal brush border membranes with 1% cholate, after removal of detergent and addition of asolectin. For details, see text

B. Cholate Solubilization and Reconstitution of the Na-Dependent Glucose Transport System From the Small Intestine?

As an attempt toward purifying the Na-dependent glucose transport system of small intestine, brush border vesicles prepared from rabbit intestine by the Ca^{2+}-precipitation method (Storelli et al., 1977) were solubilized with 1% cholate. After centrifugation at 45,000 or 100,000 g for one h, the supernatant was freed of cholate by percolation through Sephadex G-25 and spun down again at 140,000 g for 2 h. The pellet was mixed with sonicated asolectin and sonicated for three minutes.

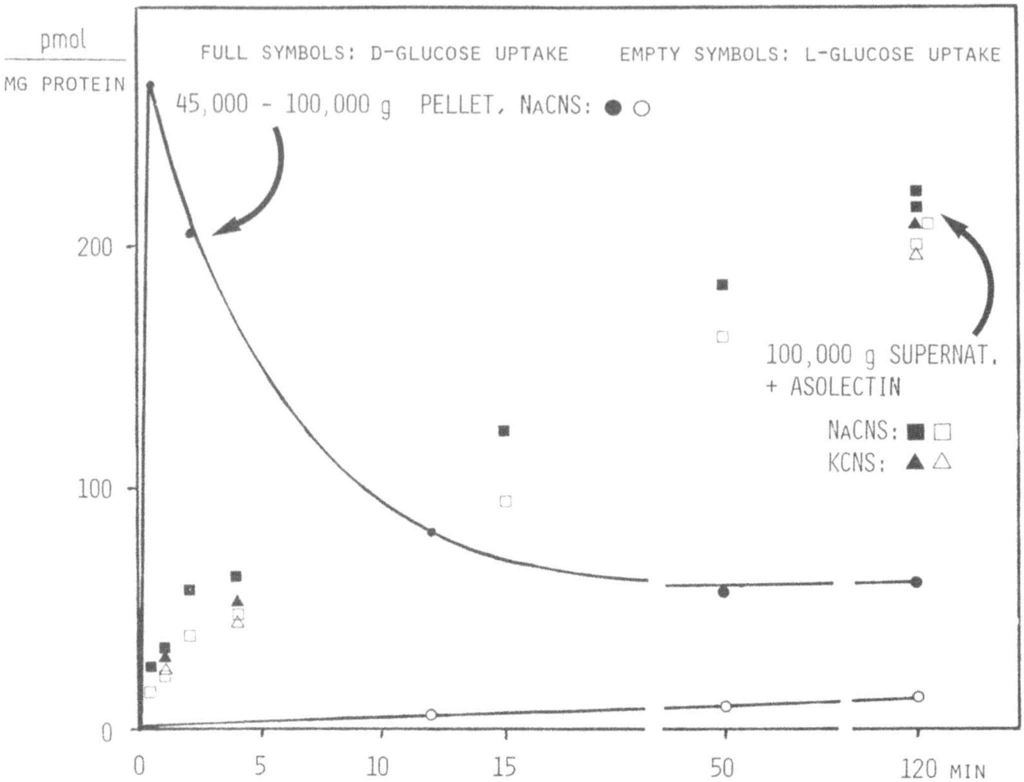

Fig. 2. Uptake of D- and L-glucose by pellet sedimenting between 45,000 and 100,000 g after cholate "solubilization" (o, •) and uptake by 100,000 g supernatant after removal of cholate and addition of asolectin

Uptake of D- and L-glucose was measured in the presence of a NaSCN gradient. The results are reported in Figures 1 and 2. It is apparent that the original 45,000 g supernatant, when mixed with asolectin, while not producing an overshoot (as the original brush border vesicles did), could discriminate between D- and L-glucose during at least the first 15 min, although asolectin alone showed a large, time-dependent nonspecific uptake (or binding), the extent of which was almost identical to that observed with protein-containing asolectin in KCNS.

The initial discrimination between D- and L-glucose was dependent on the presence of Na^+ (compare with the data in KSCN) and was inhibited by 0.5 mM phlorizin, indicating an analogy with the original Na^+-dependent glucose transport.

As a second step, the 45,000 g supernatant after cholate solubilization was separated into two fractions: the pellet sedimenting at 100,000 g and its supernatant. The 45,000-100,000 pellet was used as such (i.e., without either further cholate removal or asolectin addition) for uptake studies (Fig. 2, o and •). It showed excellent discrimination between D- and L-glucose and a very high overshoot in the uptake of D-glucose. It clearly consisted of brush border membrane vesicles that had resisted "solubilization" by 1% cholate under the conditions employed.

The 100,000 g supernatant (Fig. 2) behaved in a manner similar to that of the 45,000 g supernatant (Fig. 1), although it discriminated less well between the two glucose isomers. Thus, at present, at least two interpretations are possible:

1. True solubilization had been achieved by cholate treatment, and the results in Figure 1 show true reincorporation of a protein fraction, although still crude into artificial membranes.

2. Cholate had produced fractionation of brush border membranes into small vesicles of different sizes (see e.g., Hatefi et al., 1962), some of which are large enough to sediment at 45,000 g, others of which may need a stronger centrifugational field. Asolectin would only provide a support, allowing the Millipore technique to be used for measuring uptake.

This alternative - which is not merely semantic in nature - i.e., dealing only with the operational definitions of solubilization and reconstitution - is presently the object of further studies, although we regard explanation 2 as the more likely one.

C. Partial Purification of the Na-Dependent D-Glucose Transport System by Selective Extraction of Membrane Proteins Not Involved in This Transport

An alternative route toward the goal of purifying - even if partially - a transport system is the removal of proteins other than those involved in the transport proper. Controlled papain digestion of the membranes of intestinal brush borders has proved of value in solubilizing a number of "intrinsic" (or rather "fibrous", in Bretscher's (1976) nomenclature) membrane proteins. On the other hand, papain treatment of brush border from kidney proximal tubuli did not significantly decrease the number of high affinity phlorizin binding sites to these structures (Thomas, 1973). Thus, controlled papain digestion of small intestinal brush border membranes was the obvious first choice. Membrane vesicles prepared from rabbit small intestine by Ca^{2+} were used (Storelli et al., 1977).

Figure 3 contains the results of a representative experiment. In A, the data are presented as pmol uptake/mg protein where the protein content is that actually measured in each incubation medium; in B, the results are normalized to the amount of protein present in the vesicles prior to papain solubilization. It is immediately apparent (Fig. 3B) that papain treatment under the conditions used did not reduce the function of the Na-dependent D-glucose transport system - as shown by the initial velocity of uptake - nor did it make the vesicles leaky - as shown by the equilibrium concentration of the substrate in the vesicles. Thus, the recovery of the transport activity after papain treatment was quantitative or nearly so.

Whereas the total transport activity was unaffected, some 50% of the proteins - apparently not associated with this transport sys-

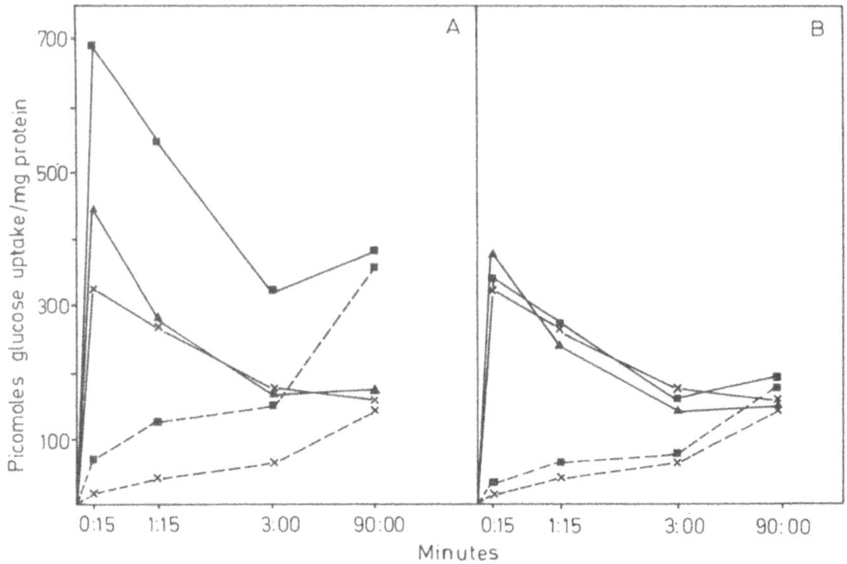

Fig. 3A,B. Uptake of D- and L-glucose by papain-(+ EDTA) treated brush border vesicles from rabbit small intestine. (A) data are given in terms of amount of protein remaining in membrane vesicles after papain treatment; (B) in terms of amount of protein present in original vesicle preparation prior to papain treatment. (----), L-glucose uptake; (——), D-glucose uptake; (x-x), control; (■-■), vesicles treated with 0.8 U papain mg^{-1} protein; (▲-▲), vesicles treated with 0.4 U papain mg^{-1} protein

Table 1. Effect of papain treatment on glucose uptake by vesicles of brush border membranes (Ca^{2+}-method)

	Uptake[a]		Sucrase[b]
	D	L	
Control	215	11	1.08
After EDTA treatment (1 mM)	295	34	1.11
After EDTA + papain (0.8 U/mg protein)	460	44	<0.05

[a]Uptake (of D- or L-glucose, respectively) in pmol · mg^{-1} protein · 10 sec^{-1} at room temp

[b]Sucrase activity left in vesicles in units · mg^{-1} protein

tem - had been removed from the vesicles by papain. This resulted in a two-fold increase of specific transport activity (Fig. 3A and Table 1). In addition, the treated vesicles retain the ability to discriminate between D- and L-glucose that is present in the untreated vesicles. There is no indication that we are seeing the exposing of cryptic transport sites or that we have destroyed the ability of the vesicles to accumulate glucose.

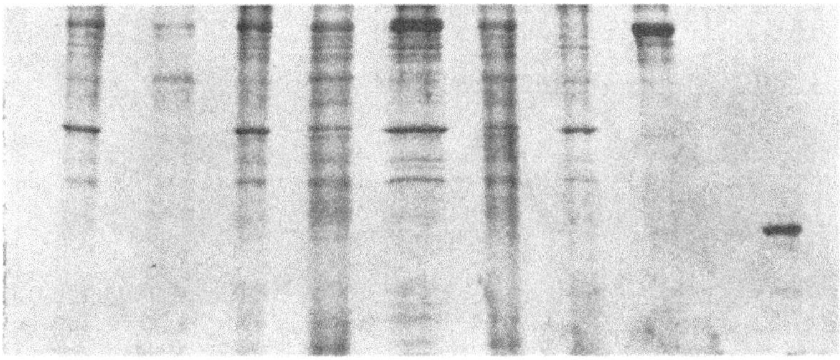

Fig. 4. SDS-PAGE (start from top) of small intestinal rabbit brush border membranes (BBM) with no treatment (left), after EDTA treatment (center), or after EDTA + papain treatment (right). After each treatment, vesicles were spun down (P: pellet; S: supernatant)

The vesicles were assayed for sucrase and lactase activity before and after treatment with papain. Treatment of the vesicles with EDTA resulted in the liberation of only about 4% of the sucrase activity; after papain treatment less than 2% of the sucrase activity was found in the vesicles (see Table 1). In both cases the total sucrase activity (vesicles plus supernatant) was not different from that of the untreated control. No significant amount of sucrase activity was destroyed by papain treatment. Papain treatment also caused almost complete solubilization of lactase activity.

If the sodium dodecyl sulfate polyacrylamide gel electrophoresis patterns of the control vesicles, the treated vesicles, and the vesicle-treating solution are compared, then we can see that the proteins that have been removed from the vesicles by papain treatment appear to be in the supernatant (Fig. 4). There appear to be only two new bands present in either gel pattern. None of the bands found in the control is absent from the patterns obtained after papain treatment, and there is no increase in bands found in the low molecular weight region of the gel. Thus, there is no indication of extensive degradation of membrane proteins.

The objective of this line of investigation was to remove as much of the nonessential vesicle protein as possible without severely affecting the transport capacity of the vesicles. At best, we hoped to remove some of the loosely associated external proteins and possibly some of the "fibrous" proteins. Our results indicate that we have achieved our goal and obtained a quantitative recovery of transport activity with a two-fold purification. Whether the resultant vesicles will be susceptable to further selective extraction is the object of future experimentation.

Acknowledgments. This work was partially supported by the Swiss National Science Foundation, Berne (grant No. 3.2370.74) and by Nestlé Alimentana, Vevey. Thanks are also due to Mrs. B. O'Neill for her competent assistance.

References

Bretscher, M.S.: Membranes at Embo. Nature (London) 261, 366-367 (1976)

Caspary, W.F., Crane, R.K.: Active transport of myo-inositol and its relation to the sugar transport system in hamster small intestine. Biochim. Biophys. Acta 203, 308-316 (1970)

Crane, R.K.: Hypothesis for mechanism of intestinal active transport of sugars. Federation Proc. 21, 891-895 (1962)

Crane, R.K.: Na^+-dependent transport in the intestine and other animal tissues. Federation Proc. 24, 1000-1005 (1965)

Crane, R.K., Forstner, G., Eichholz, A.: Studies on the mechanism of the intestinal transport of sugars. X. An effect of Na^+ concentration on the apparent Michaelis constants for intestinal sugar transport in vitro. Biochim. Biophys. Acta 109, 467-477 (1965)

Crane, R.K., Malathi, P., Preiser, H.: Reconstitution of specific Na^+-dependent D-glucose transport in liposomes by Triton X-100 extracted proteins from purified brush border membranes of hamster small intestine. Biochem. Biophys. Res. Commun. 71, 1010-1015 (1976)

Curran, P.F., Schultz, S.G., Chez, R.A., Fuisz, R.E.: Kinetic relations of the Na-amino acid interaction at the mucosal border of intestine. J. Gen. Physiol. 50, 1261-1286 (1967)

Deshusses, J., Reber, G.: Asymmetry of the myo-inositol transport system in Klebsiella aerogenes. Energy is necessary to create the asymmetry of the transport system. Eur. J. Biochem. In press (1977)

Eddy, A.A., Philo, R., Earnshaw, P., Brocklehurst, R.: Some common aspects of active solute transport in yeast and mouse ascites tumour cells. This Symposium (1976)

Frasch, W., Frohnert, P.P., Bode, F., Baumann, K., Kinne, R.: Competitive inhibition of phlorizin binding by D-glucose and the influence of sodium: a study on isolated brush border membrane of rat kidney. Pflügers Arch. 320, 265-284 (1970)

Geck, P.: Eigenschaften eines asymmetrischen Carrier-Modells für den Zuckertransport am menschlichen Erythrozyten. Biochim. Biophys. Acta 241, 462-472 (1971)

Geck, P., Heinz, E.: Coupling in secondary transport. Effect of electric potentials on the kinetic of ion-linked cotransport. Biochim. Biophys. Acta 443, 49-53 (1976)

Gibb, L.E., Eddy, A.A.: An electrogenic sodium pump as a possible factor leading to the concentration of amino acids by mouse ascites-tumor cells with reversed sodium ion concentrations. Biochem. J. 129, 979-981 (1972)

Goldner, A.M., Schultz, S.G., Curran, P.F.: Sodium and sugar fluxes across the mucosal border of rabbit ileum. J. Gen. Physiol. 53, 362-383 (1969)

Harold, F.M., Papineau, D.: Cation transport and electrogenesis by Streptococcus faecalis. I. The membrane potential. J. Membrane Biol. 8, 27-44 (1972a)

Harold, F.M., Papineau, D.: Cation transport and electrogenesis by Streptococcus faecalis. II. Proton and sodium extrusion. J. Membrane Biol. 8, 45-62 (1972b)

Hatefi, Y., Haavik, A.G., Fowler, L.R., Griffiths, D.E.: Studies on the electron transfer system. XLII. Reconstitution of the electron transfer system. J. Biol. Chem. 237, 2661-2669 (1962)

Heinz, E., Geck, P., Pietrzyk, C., Pfeiffer, B.: Electrogenic ion pump as an energy source for active amino acid transport in Ehrlich cells. This Symposium (1976)

Heinz, E., Geck, P., Wilbrandt, W.: Coupling in secondary active transport. Activation of transport by co-transport and/or counter transport with the fluxes of other solutes. Biochim. Biophys. Acta 225, 442-461 (1972)

Komor, E., Tanner, W.: The hexose proton symport system of Chlorella vulgaris: specificity, stoichiometry and energetics of sugar induced proton uptake. Europ. J. Biochem. 44, 219-233 (1974)

Kotyk, A.: Mechanismus of nonelectrolyte transport. Biochim. Biophys. Acta 300, 183-210 (1973)

Kotyk, A.: Nonlinearity of reciprocal plots in transport kinetics. This Symposium (1976)

Mitchell, P.: Active transport and ion accumulation. In: Comprehensive Biochemistry. Florkin, M., Stotz, E.M. (eds.). Amsterdam: Elsevier 22, 167-197 (1967)

Murer, H., Hopfer, U.: Demonstration of the electrogenic Na^+-dependent D-glucose transport in intestinal brush border membranes. Proc. Natl. Acad. Sci. U.S. 71, 484-488 (1974)

Regen, D.M., Morgan, H.E.: Studies on the glucose-transport system in the rabbit erythrocyte. Biochim. Biophys. Acta 79, 151-166 (1964)

Seaston, A., Carr, G., Eddy, A.A.: The concentration of glycine by preparations of the yeast Saccharomyces carlsbergiensis depleted of adenosine triphosphate. Biochem. J. 154, 669-676 (1976)

Segel, I.H.: Enzyme Kinetics. New York: Wiley 1975

Semenza, G.: Sucrase and sugar transport in the small intestine. In: Protides of the Biological Fluids. Amsterdam: Elsevier 15, 201-208 (1967)

Semenza, G.: On the mechanism of mutual inhibition among sodium-dependent transport systems in the small intestine. A hypothesis. Biochim. Biophys. Acta 241, 637-649 (1971)

Semenza, G.: On the mechanism of H^+- and Na^+-driven cotransport systems: could the variety of macroscopic kinetic effects be given a single explanation? To be submitted (1977)

Storelli, C., Kessler, M., Murer, H., Müller, M., Semenza, G.: A simple and efficient procedure for the preparation of transporting vesicles from small intestinal brush border membranes. To be submitted (1977)

Thomas, L.: Isolation of N-ethylmaleimide-labelled phlorizin-sensitive D-glucose binding protein of brush border membrane from rat kidney cortex. Biochim. Biophys. Acta 291, 454-464 (1973)

Vidaver, G.A.: Some tests of the hypothesis that the sodium ion gradient furnishes the energy for glycine active transport by pigeon red cells. Biochemistry 3, 803-808 (1964)

West, I.C., Mitchell, P.: Proton-coupled β-galactoside translocation in non-metabolizing Escherichia coli. J. Bioenerg. 3, 445 (1972)

Wilbrandt, W.: The asymmetry of sugar transport in the red cell membrane. This Symposium (1976)

Amino Acid Transport and Binding Activity from Membrane Preparations of Animal Cells

D. L. Oxender, G. Cecchini, M. Lee, and P. Moore

A. Introduction

The transport of neutral amino acids in animal tissues is carried out by several distinct transport systems. This conclusion has resulted from our early studies on the detailed interactions for transport in Ehrlich ascites cells among a large group of natural and synthetic amino acids (Oxender and Christensen, 1963). We determined that the affinities of the neutral amino acids clustered into two groups, indicating the presence of at least two distinct heavily overlapping mediated systems identified as the A and the L transport systems. System A (alanine-preferring) serves mainly for such amino acids as alanine, glycine, and serine while System L (leucine-preferring) shows a preference for the branched chain and aromatic amino acids. Straight chain amino acids such as methionine have a high affinity for both systems. These two systems were shown to differ in their sensitivity to metabolic inhibitors, pH changes, and external sodium ions with System A being more sensitive to the changes in each case.

Christensen (1969, 1975b) examined the homogeneity of the systems in a variety of tissues and found that, in addition to System A and System L, some tissues contain additional transport systems such as the ASC system and the imino amino acid system for the neutral amino acids.

We recently examined the transport systems for neutral amino acids in animal cells grown in tissue culture. For these studies we used 3T3 mouse cells, baby hamster kidney cells, and Chinese hamster ovary cells. We have shown the presence of Systems A and L in each of these cell types. In general the interactions of the various amino acids with the transport systems cultured cells grown in vitro are similar to those previously observed in the Ehrlich ascites tumor cells that are grown intraperitoneally in mice.

We have observed large changes in the transport activities for several of the amino acids as the growth conditions of the cells are varied. These regulatory changes could result from a repression-depression mechanism or from kinetic changes as a consequence of the direct effect of certain nutrients on the transport system. The first type of regulatory mechanism has been described by Guidotti et al. (1975). This mechanism appears to be a time-dependent repression of transport activity at the transcriptional level and is controlled by cellular amino acid levels. We have studied the second type of regulation involving the trans-inhibition or trans-stimulation phenomena, which result from a direct feedback regulation of the transport process by accumulated amino acids.

In addition, we have prepared plasma membrane vesicles from Chinese hamster ovary cells for studying amino acid uptake in vesicle preparations. Sodium-dependent System A uptake can be observed in the vesicle preparations using α-aminoisobutyric acid as a substrate. We next attempted to isolate transport components from the animal cells. In order to get sufficient quantities of cells to attempt to isolate membrane transport components we used Ehrlich ascites tumor cells. Plasma membrane fractions were prepared from the Ehrlich cells and treated with the detergent, Triton X-100. We were able to obtain solubilized leucine-binding activity from the plasma membrane fraction. The binding activity has a similar specificity to that of the L system in Ehrlich cells. This communication describes our studies on the transport systems in tissue culture cells and the leucine binding activity from membrane preparations of the Ehrlich tumor cells.

B. Methods

I. Cells and Culture Methods

The Balb/3T3 cells were obtained from Dr. G.J. Todaro of the National Cancer Institute. We determined that the cells were free of mycoplasma infection using the uridine phosphorylase assay for mycoplasma (Levine, 1972). The Chinese hamster ovary cells were obtained from Dr. L.H. Thompson of Livermore, California. The baby hamster kidney cells were obtained from Dr. W. Folk of this Department. Ehrlich ascites tumor cells were obtained from Dr. H.N. Christensen of this Department.

The tissue culture cells were grown in Dulbecco's modified Eagle's medium (Eagle, 1965) adjusted so that a pH of 7.4 was obtained when the medium was equilibrated in a 5% CO_2 atmosphere. In addition, the growth medium contained penicillin (120 U/ml), streptomycin (120 μg/ml), and 10% calf serum obtained from Gibco.

II. Uptake Methods

Amino acid transport was measured on glass coverslips (11 × 22 mm) by techniques similar to that described by Foster and Pardee (1969). The cultured cells were rinsed three times in phosphate-buffered saline (PBS, pH 7.4) with 0.1% glucose and 0.01% $CaCl_2$ and $MgCl_2$ (PBSG).

After this treatment the coverslips were incubated for a specified period of time in PBSG plus labeled amino acid and then rinsed rapidly by dipping in PBSG at 4°C. Each coverslip was placed in 0.6 ml of 0.2 M sodium hydroxide to dissolve the cells. A sample was removed for determining the radioactivity, and another sample was assayed for protein by the method of Lowry et al. (1951).

III. Analysis of Amino Acid Pools

Cells were cultured in 9-cm tissue culture dishes as described previously for the uptake measurements. For the analysis of the cellular amino acids the media was aspirated from the culture and the cells were rapidly rinsed once with PBS. One ml of 3% 5-sulfosalicylic acid (wt/vol) was added to the dish, and the cells were scraped from the plate with a rubber-tipped spatula and transferred to a test tube. The dish was rinsed with an additional 1 ml of 3% sulfosalicylic acid, and this was combined with the original material. The extracted cells were heated in a boiling water bath for 5 min. After cooling the extract was centrifuged at 20,000 g for 10 min. The pellet was analyzed for protein, and a sample of the supernatant fluid was placed directly on the resin column of a Beckman 120 c amino acid analyzer.

IV. Membrane Vesicle Preparations

1. Chinese Hamster Ovary Cells

These cells were used for making membrane vesicles. These cells were grown in one liter roller bottles. We could obtain about 0.5 g of cells per bottle. Using 12 to 15 roller bottles we obtained 4-6 g of cells. Methods for preparing membranes from cultured animal cells have been extensively reviewed by Hochstadt et al. (1975). The cells are removed from the roller bottles by scraping with a rubber-tipped glass rod. The cells are washed three times in 50 vol of 0.5 M sucrose, 10 mM Tris-HCl, pH 8.0, and 0.1 mM EDTA. For the washing procedure the cells were centrifuged at 800 g for 10 min. Washed cells are resuspended in medium and homogenized using a Dounce glass homogenizer. The homogenization is continuously monitored by phase contrast microscopy to determine how many strokes are required to gently rupture the plasma membranes. The nuclear material is removed by centrifugation at 1000 g for 5 min. The supernatant fluid contains the membrane vesicles. The membrane vesicles can be sedimented at 16,000 g for 15 min. The pellet of membranes is resuspended in 4 to 6 vol of medium and subjected to a linear gradient of 5 to 25% ficoll. Centrifugation is carried out for 5 h at 56,000 g in a SW27 rotor. The membrane vesicles form two bands in the ficoll gradient. The bands are removed and the membranes centrifuged at 16,000 g for 20 min. The pellet is resuspended in 0.25 M sucrose, 10 mM Tris-HCl, pH 8.0. They can be stored frozen at -70°C.

For transport studies a suspension of washed membrane vesicles is warmed to 37°C and combined with labeled amino acids. At the end of the incubation period the suspension is diluted, filtered, and washed using a cellulose acetate filter with a pore size of 0.2 μ.

2. Ehrlich Ascites Cells

Plasma membrane fractions from Ehrlich cells were prepared by a modification of the method of Forte et al. (1973). We harvested 8-10 g of cells from the tumor-bearing Swiss white mice for each preparation. Washed cells were suspended in 15 mM sodium phosphate buffer, pH 7.4, containing 1 mM $MgCl_2$. The cells were homogenized using a Dounce glass homogenizer.

The homogenized cell preparation was further purified by differential centrifugation. The enriched plasma membrane-containing fraction was then placed on a discontinuous glycerol gradient. A band at the 60-67% glycerol interface contained the majority of the plasma membrane and was enriched 25-30-fold, based on Na^+-K^+ ATPase activity.

V. Solubilization of Membranes with Triton X-100

Leucine-binding activity was extracted from the plasma membrane fraction using 0.5% Triton X-100 in 15 mM $NaPO_4$, pH 7.4. The membranes were incubated for 30 min at 37°C and then centrifuged at 30,000 g for 30 min. The binding activity remains in the supernatant fluid under these conditions. The leucine-binding activity is further purified using a concanavalin A-Sepharose 4B affinity column. The binding activity is not tightly bound, but trails the large amount of protein that is eluted as the column is washed with buffer. The binding activity produces from 4 to 7 bands on SDS gels that are stained with Coomassie blue.

C. Results and Discussion

I. Characterization of Amino Acid Transport in 3T3 Mouse Cells

We measured the uptake of glycine, alanine, leucine, and phenylalanine in 3T3 cells and Chinese hamster cells. We also use non-metabolizable amino acid analogues. For System A we use α-aminoisobutyric acid and its N-methylated derivative, and for System L we use a bicycloheptyl amino acid, 2 aminonorbornane-2-carboxylic acid (BCH). These analogues have been developed and studied by Christensen (1969, 1975b).

In general the results we have obtained using various tissue culture lines are remarkably similar to those obtained using the Ehrlich cell (Oxender and Christensen, 1963). The presence of System A and System L was confirmed in the 3T3 mouse cells, Chinese hamster ovary cells, and the baby hamster kidney cells.

Figure 1 presents the time course of the uptake of several amino acids into 3T3 mouse cells grown in a monolayer on glass coverslips. See the methods section for details. Amino acids that primarily use System L such as leucine and phenylalanine show a rapid initial rate of entry and reach a steady state level in the cells within 2 to 4 min. Amino acids such as glycine, alanine, and α-aminoisobutyric acid, which primarily use System A, are accumulated to much higher levels. The latter group of amino acids reach steady state levels in the cells only after 20 to 60 min. These uptake measurements were made on rapidly growing cells at less than 50% of confluent growth on the coverslips. The effect of confluency on uptake is discussed below.

We found that alanine, glycine, and α-aminoisobutyric acid uptake was sodium dependent and sensitive to pH changes in the 3T3 mouse cells. When the 3T3 cells were preloaded with System L amino acids, the initial rates of uptake of these amino acids were accelerated, a phenomenon often referred to as trans-stimulation (Christensen, 1975a; Stein, 1967).

When 3T3 cells were preloaded with System A amino acids, the uptake of amino acids such as α-aminoisobutyric acid was inhibited, suggesting System A may be subject to trans-inhibition by high levels of cellular amino acids.

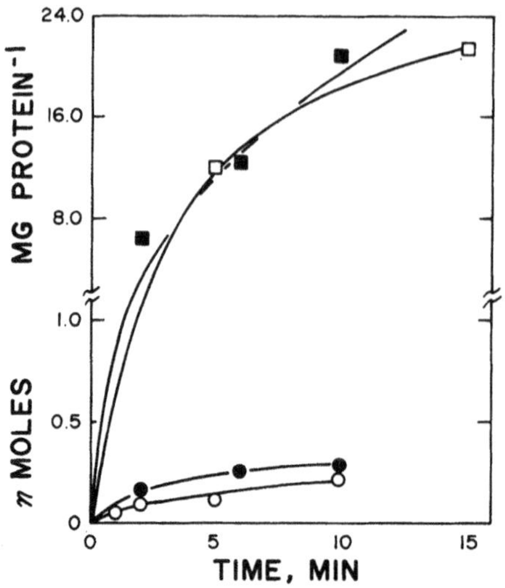

Fig. 1. Time course for uptake of amino acids by Balb/3T3 cells on glass coverslips. Cells were approximately 50% confluent. Concentration of amino acid in incubation medium (PBSG) was 0.1 mM. Each point represents average of duplicate determinations. Leucine: (●); phenylalanine: (o); α-aminoisobutyric acid: (□); alanine: (■)

Table 1. Kinetic constants for amino acid transport

Amino acid	3T3 mouse cells		Chinese hamster ovary cells	
	K_m	V_{max}	K_m	V_{max}
Alanine	300	5.3	200	7.5
Glycine	5000	13.3	1000-2000	3.6
α-aminoisobutyric acid	1000-3000	2.5	1000	10.7
Leucine	10	1.3	10	3.6
	90 - 100	10.0	70-100	12.2
Phenylalanine	20	13.3	20	3.7
	n.d.[a]	n.d.[a]	50-100	18.3

Apparent K_m and V_{max} values are expressed as μM and nmoles mg protein^{-1} min^{-1}, respectively

[a]n.d. = not determined

The rate of uptake increases with external concentration, and kinetic parameters were determined for some of the amino acids using the 3T3 mouse cells.

Table 1 presents the K_m and V_{max} values obtained from double reciprocal plots of the initial rates of uptake for alanine, glycine, α-aminoisobutyric acid, leucine, and phenylalanine in 3T3 mouse cells and Chinese hamster ovary cells. These K_m values are similar to those obtained for Ehrlich cells (Oxender and Christensen, 1963) with the exception of the two K_m values obtained for leucine and phenylalanine. Double reciprocal plots of uptake of the latter two amino acids were distinctly biphasic, and two K_m values were obtained using the method described by Neal (1972) for analysis of

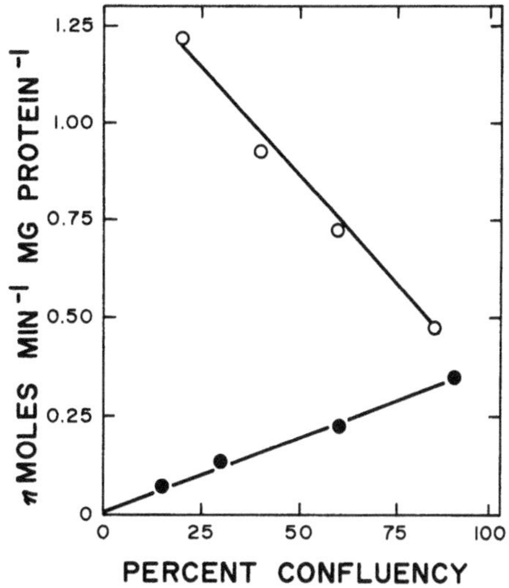

Fig. 2. Effect of cell density on transport in Balb/3T3 cells. Cells were inoculated onto glass coverslips at different cell densities, when visual observation indicated most dense cells were confluent (24-30 h); transport assays were done on each culture and the most dense culture was assigned a value of 100% confluency. Cultures designated 100% confluent had 51-60 μg of protein per coverslip; other cultures had proportionately less protein per coverslip as indicated by their degree of confluency expressed as percent. Glycine: (o); leucine: (●)

biphasic plots. The high affinity component of leucine-transport could also be observed in Ehrlich cells when the appropriate range of external concentrations of leucine were used. The high affinity of this component of uptake prompted us to examine membrane preparations for leucine-binding activity using equilibrium dialysis techniques. These results are reported below.

In order to establish the specificity of the transport systems, we measured the inhibition of various amino acids on the uptake of glycine and leucine in the tissue culture cells. The results of this study showed that all amino acids tested produced significant inhibition of both System A and System L test substrates indicating considerable overlap in the use of transport systems.

II. Changes in Transport Activity with Different Cell Growth Conditions

Altering the growth conditions of growth of cells grown in vitro produced significant changes in transport activity. It became necessary to carefully control the growth conditions in order to compare the transport activity of one cell culture with another.

1. Degree of Confluency

The 3T3 mouse cells show density dependent growth when grown in monolayers on glass coverslips. The initial rate of glycine uptake into 3T3 cells decreases several fold as cells approach confluency and stop growing. Under these same conditions leucine transport activity increases several-fold. These results are presented in Figure 2. Since we had previously shown that cells preloaded with leucine showed a stimulation of System L amino acids in the Ehrlich cells (Oxender and Christensen, 1963), we measured the cellular amino acid changes in 3T3 cells at different states of confluency. We found that the levels of most of the amino acids in-

Table 2. Ratio of amino acid pools during various growth conditions

Amino acid	Ratio: 80% Confluent cells[a] / 20% Confluent cells	Quiescent cells[b] / Growing cells	Cycloheximide-treated cells[c] / Growing cells
Aspartic acid	5.3	2.2	2.8
Glutamic acid	2.4	1.5	1.6
Threonine	2.0	2.0	1.7
Serine	1.1	2.0	1.7
Glycine	1.8	1.7	1.2
Alanine	3.8	1.6	0.9
Valine	1.2	2.2	1.6
Isoleucine	1.3	2.1	1.6
Leucine	1.3	2.1	1.6
Tyrosine	1.5	2.1	1.6
Phenylalanine	1.8	2.5	1.3

[a]Ratio of cellular amino acids in 80% confluent to those in 20% confluent cells

[b]Quiescent cells were quiescent for 24 h after removal of calf serum and are compared to growing cells

[c]Cycloheximide (100 μg/ml) cells were treated for 2 h with cycloheximide and are compared to growing cells

crease around two-fold as the cell density changed from 20% confluency to 80% confluency. These results are presented in Table 2. These results show that when cell growth decreases as cells approach confluency, the cellular amino acid levels rise significantly and the System A transport activity decreases while System L transport activity increases.

2. Transport and Amino Acid Levels in Quiescent Cells

The 3T3 mouse cells require serum in the growth medium. If serum is removed, the cell growth is arrested in the G_1 or G_0 phase of growth within approximately 12 h and the cells become quiescent. The transport activity for alanine and leucine was measured in actively growing cells and compared to that of quiescent cells. Figure 3 presents these results in graphic form. System A amino acids show decreased transport in quiescent cells and System L amino acids show increased transport activities. When calf serum was added back to the quiescent cells, rapid growth resumed within 24 h and the transport activity returned to that found previously for these cells under rapid growth conditions.

We measured the cellular amino acid levels in rapidly growing 3T3 cells and in quiescent 3T3 cells. These results are presented in Table 2 as the ratio of the level of amino acid in quiescent cells to that in growing cells. As the table shows all the cellular amino

Fig. 3. Effect of quiescence on transport in Balb/3T3 cells. Actively growing cells at about 30% confluency were measured for transport activity (open bars). Cultures then had their growth arrested by replacing calf serum in media with bovine serum albumin (250 μg/ml) and were incubated under these conditions for 24 h in CO_2 incubator (slashed bars). Calf serum (10%) was then added back to cultures, and growth was allowed to resume for 24 h and transport then measured (solid bars)

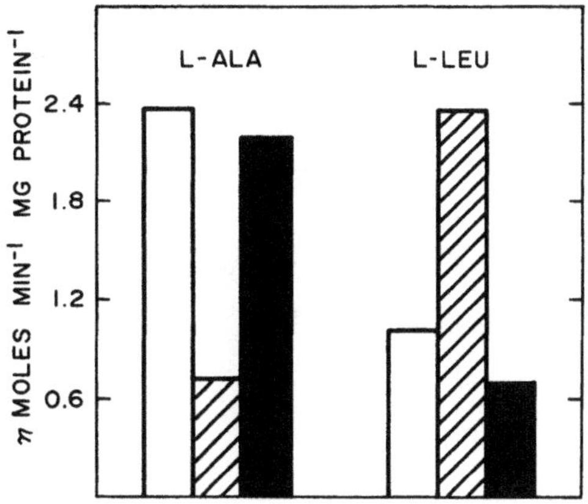

acids increase about twofold as cell growth is arrested by serum removal and the cells become quiescent. When 10% calf serum was added back to the quiescent cells, cell growth resumed within 24 h and the cellular amino acids decreased to levels characteristic for rapidly growing 3T3 mouse cells.

3. Effect of Protein Synthesis Inhibitors

Arresting the growth of cells by the inhibition of protein synthesis also results in changes in cellular amino acid levels and in transport activities. We added 50 or 100 μg per ml of cycloheximide to actively growing Balb/3T3 mouse cells that were about 30% confluent. These levels of cycloheximide arrest cell growth very rapidly. The transport activities for leucine, alanine, and α-aminoisobutyric acid were measured before treatment and 2-5 h after cycloheximide addition. The results are presented in Table 3. System A substrates such as alanine and α-aminoisobutyric acid show decreases in transport activity while leucine transport activity goes up several-fold. The cellular level of amino acids was determined before and again two hours after addition of 100 μg per ml cycloheximide. The results are presented in Table 2 as the ratio of amino acid level in cycloheximide-treated cells to that of growing cells. Most amino acids show a significant rise as protein synthesis is inhibited. These increases could result from continued degradation of proteins or increased accumulation.

Hypertonicity also inhibits protein synthesis in animal cells. The addition of 100 mM sodium chloride to the growth medium has been shown to inhibit protein synthesis (Saborio et al., 1974; England et al., 1975). We found that after two hours in 100 mM sodium chloride, leucine transport increased 3 to 4-fold in 3T3 cells. When the cells were returned to isotonic conditions, where protein synthesis could resume, leucine transport returned to that of rapidly growing cells within 45 min.

The preceding results show that whenever conditions are established that arrest cell growth, the cellular amino acids rise and transport activities are altered. As noted above, System A activity is decreased and System L activity is increased. Since the L system is subject to trans-stimulation by exchange of external labeled amino acid with unlabeled cellular amino acids, the transport ac-

Table 3. Effect of cycloheximide on transport activity

Conditions	Time (h)	Leucine	α-aminoisobutyric acid	Alanine
		(nmoles min^{-1} mg $protein^{-1}$)		
Control	0	1.0	3.7	1.5
A. 50 μg/ml	2	2.2	n.d.[a]	n.d.
cycloheximide	5	3.0	n.d.	n.d.
B. 100 μg/ml	2	3.2	n.d.	n.d.
cycloheximide	4	n.d.	0.9	0.8

[a] n.d. = not determined

tivity changes observed above would be expected on the basis of the exchange properties of System L and the cellular amino acid level changes that were observed. On the other hand the transport activity decreases for System A during the approach to confluency or quiescence are greater than would be predictable on the basis of the trans-inhibition. This has been observed previously in a variety of tissues (Ring et al., 1970; Kadner, 1975; Benko et al., 1969; Crabeel and Grenson, 1970).

The regulation of System A must be mediated via additional mechanisms. Presumably the derepression-repression type of regulation described by Guidotti et al. (1975) is also operating under the experimental conditions used in the studies reported here. The mechanism of System A regulation is currently under study in our laboratory as well as in a number of other laboratories.

III. Transport Activity of Membrane Vesicle Preparations

Membrane vesicles offer the important advantage of looking at the transport event apart from other cellular activities. We are currently making membrane vesicles from Ehrlich ascites tumor cells as well as from Chinese hamster ovary cells. The procedures used are described in the Methods section. We have used the membrane preparations from the Ehrlich cells in preliminary studies using α-aminoisobutyric acid as System A test solute. Since α-aminoisobutyric acid uptake can be coupled to a sodium ion gradient, we add 90 mM sodium chloride to the vesicles at the same time as the labeled amino acid is added. Tris buffer was added in place of NaCl to control vesicle preparations. The results are presented in Table 4. As shown in the table, there is a transient sodium-dependent accumulation of AIB that decreases to the level of uptake in the control cells with time, presumably because the sodium ion gradient initially present is rapidly dissipated. This active accumulation of α-aminoisobutyric acid produced by the sodium ion gradient in membrane vesicle preparations is similar to that observed in a number of other laboratories.

Table 4. Transport of α-aminoisobutyric acid in membrane vesicles of Ehrlich ascites cells

Time (min)	pmol uptake mg protein	
	$-Na^{+}$ [a]	$+Na^{+}$ [b]
1	n.d. [c]	60
2	40	70
4	22	
7	0	40
9	0	42

[a] Tris buffer pH 7.4 was added to control vesicles

[b] 90 mM NaCl was added at same time that labeled amino acid was added

[c] n.d. = not determined

IV. Leucine-Binding Activity from Ehrlich Tumor Cells

In order to isolate membrane components from mammalian cells we need relatively large amounts of cells. We therefore decided to use Ehrlich ascites tumor cells, which can be obtained from mice in large quantities. A large number of techniques were initially used to obtain suitable membrane preparations for measuring leucine-binding activity by equilibrium dialysis. As mentioned earlier in the section on kinetic studies, we detected a high-affinity component of leucine uptake in 3T3 cells and subsequently in Ehrlich cells, which encouraged us to use the equilibrium dialysis technique to look for leucine-binding activity in membrane preparations.

Details of the procedures for isolating the plasma membranes from Ehrlich cells and treating them with the detergent, Triton X-100, are presented in the section on methods. The leucine-binding activity of the partially purified and solubilized preparation is presented in Table 5. As shown in the table, the leucine-binding activity is saturable since the addition of 100 μM unlabeled leucine decreases the ratio of counts inside the dialysis bag to that found outside. In addition, unlabeled phenylalanine inhibits the binding activity. These properties of the solubilized membrane preparation would be expected from the kinetic parameters found for System L. The leucine-binding activity is being further purified using additional chromatographic procedures.

D. Summary

We have examined neutral amino acid transport in various animal cells using Ehrlich ascites cells, 3T3 mouse cells, and Chinese

Table 5. Leucine-binding activity from Ehrlich cell plasma membrane solubilized with Triton X-100

	Ratio[a] cpm inside/cpm outside
1 nM leucine	1.42
100 µm leucine	1.02
50 µm phenylalanine with 1 µm leucine	1.01

[a]Binding activity was measured by equilibrium dialysis at 4°C in 15 µM $NaPO_4$, pH 7.4. Activity is expressed as ratio of cpm inside dialysis bag (solubilized membrane fraction) to cpm outside. Sodium azide (0.03%) was included in all dialysis solutions to prevent bacterial contamination

hamster ovary cells. We found that all of these cultured cells contain at least two neutral amino acid transport systems, previously identified as Systems A and L. The A system serves for amino acids such as alanine and glycine and is dependent on sodium ions, whereas the L system, which serves for aromatic and branched chain amino acids, is relatively sodium independent. The A system transport activity decreases as cells approach confluency and L system activity increases. Similar changes are observed when cell growth is arrested by removal of serum or by the addition of protein synthesis inhibitors such as cycloheximide or subjecting the cells to hypertonic conditions.

When cell growth is arrested, the cellular levels of most amino acids increase, producing a trans-stimulation of System L by exchange and trans-inhibition of System A. System A is subject to additional regulation by a derepression-repression phenomenon.

The kinetics of amino acid transport were examined. Leucine and phenylalanine produced biphasic kinetic plots that indicated two K_m values of around 10 µM and 0.1 mM.

Plasma membrane vesicle preparations were prepared from Ehrlich ascites cells and used for transport studies. Sodium-dependent uptake of α-aminoisobutyric acid was observed in the preparations when sodium was added simultaneously with the labeled amino acid. Plasma membrane fractions from the Ehrlich cells were treated with 0.5% Triton X-100 to obtain leucine-binding activity using equilibrium dialysis techniques. The leucine-binding activity was purified using a combination of density gradient centrifugation and concanavalin A-Sepharose affinity chromatography. The solubilized binding activity shows 4 to 7 bands on SDS gel electrophoresis. The binding activity has a specificity similar to that of System L in Ehrlich cells.

Acknowledgments. The expert technical assistance of Gregory Payne is greatly appreciated. This research was supported by the National Institutes of Health, grants GM20737 (D.O.) and GM1517 (G.C.).

References

Benko, P.V., Wood, T.C., Segal, I.H.: Multiplicity and regulation of amino acid transport in Penicillium chrysogenum. Arch. Biochem. Biophys. 129, 498-508 (1969)

Christensen, H.N.: Some special kinetic problems of transport. Advan. Enzymol. 32, 1-20 (1969)

Christensen, H.N.: Biological Transport. 2nd Ed. Reading, Mass.: W.A. Benjamin 1975a, p. 144

Christensen, H.N.: Recognition sites for material transport and information transfer. In: Current Topics in Membranes and Transport, Vol. 6. Bonner, F., Kleinzeller, A. (eds.). New York: Academic Press 1975b, pp. 227-258

Crabeel, M., Grenson, M.: Regulation of histidine uptake by specific feedback inhibition of two histidine permeases in Saccharomyces cerevisiae. Europ. J. Biochem. 14, 197-204 (1970)

Eagle, H.: Metabolic controls in cultured mammalian cells. Science 148, 42-51 (1965)

England, J.M., Howett, M.K., Tan, K.B.: Effect of hypertonic conditions on protein synthesis in cells productively infected with simian virus 40. J. Virol. 16, 1101-1107 (1975)

Forte, J.G., Forte, T.M., Heinz, E.: Isolation of plasma membranes from Ehrlich ascites tumor cells. Biochem. Biophys. Acta 298, 827-841 (1973)

Foster, D.O., Pardee, A.B.: Transport of amino acids by confluent and non-confluent 3T3 and polyoma virus-transformed 3T3 cells growing on glass cover slips. J. Biol. Chem. 244, 2675-2681 (1969)

Guidotti, G.G., Gazzola, G.C., Borghetti, A.F., Franchi-Gazzola, R.: Adaptive regulation of amino acid transport across the cell membrane in avian and mammalian tissues. Biochem. Biophys. Acta 406, 264-279 (1975)

Hochstadt, J., Quinlan, D.C., Rader, R.L., Li, C.C., Dowd, D.: Use of isolated membrane vesicles in transport studies. In: Methods in Membrane Biology, Vol. 5. Korn, E. (ed.). New York: Plenum Press 1975, pp. 117-162

Kadner, R.J.: Regulation of methionine transport activity in Escherichia coli. J. Bacteriol. 122, 110-119 (1975)

Levine, E.M.: Mycoplasma contamination of animal cell cultures: A simple rapid detection method. Exp. Cell Res. 74, 99-109 (1972)

Lowry, O.H., Rosebrough, N.J., Farr, A.L., Randall, R.J.: Protein Measurement with the folin phenol reagent. J. Biol. Chem. 193, 265-275 (1951)

Neal, J.L.: Analysis of Michaelis kinetics for two independent saturable membrane transport functions. J. Theoret. Biol. 35, 113-118 (1972)

Oxender, D.L., Christensen, H.N.: Distinct mediating systems for the transport of neutral amino acids by the Ehrlich cell. J. Biol. Chem. 238, 3686-3699 (1963)

Ring, K., Gross, W., Heinz, E.: Negative feedback regulation of amino acid transport in Streptomyces hydrogenans. Arch. Biochem. Biophys. 137, 243-253 (1970)

Saborio, J.L., Pong, S.S., Koch, G.: Selective and reversible inhibition of initiation of protein synthesis in mammalian cells. J. Mol. Biol. 85, 195-211 (1974)

Stein, W.D.: The Movement of Molecules Across Cell Membranes. New York: Academic Press 1967, p. 369

Sidedness and Coupling of Transport Processes in Small Intestinal and Renal Epithelia

H. MURER, and R. KINNE

Epithelial cells separate phases of different ionic composition and by transepithelial transport maintain or establish the asymmetric distribution of ions. The epithelia of the small intestine and the proximal tubule are additionally specialized in the (re)absorption of a variety of solutes, such as sugars, amino acids, and phosphate. Such vectorial transports can only be performed if a sidedness of the epithelial cell in respect to the transport properties of the luminal and contraluminal membrane exists. Maintenance of asymmetric distribution of solutes involves active transport processes that are energized by cellular metabolism. The coupling of transepithelial transport to energy can be direct or indirect. In primary active transport the movement of a solute across the cell membrane is directly coupled to an energy-transducing chemical reaction, such as the coupling of sodium transport via the Na^+-K^+-ATPase to ATP hydrolysis. In secondary active transport the movement of the solute across the membrane is coupled to a flux of a primary actively transported substrate.

This contribution will deal with studies on the transport properties of isolated luminal and contraluminal membrane vesicles derived from small intestinal and renal proximal tubular epithelial cells as revealed by rapid filtration techniques and will consider mainly the functional polarity of the epithelial cell and the coupling phenomena in sodium-dependent transport.

The results demonstrate that in the epithelial cell a delicate array of sodium-solute cotransport systems, sodium-proton exchange, and a primary active sodium transport system exists that enables the cell to perform secondary active, sodium-dependent transepithelial solute transport in small intestine and proximal tubule.

A. Sidedness of Small Intestinal and Renal Proximal Tubular Epithelial Cells

In addition to morphologic polarity and asymmetric distribution of membrane-bound enzymes, recent studies on the transport properties of plasma membrane vesicles isolated from the luminal (brush border membranes) and from the contraluminal (basal-lateral membranes) cell face revealed a sidedness of transport functions in epithelial cells (Murer et al., 1974; Kinne et al., 1975; Kinne and Murer, 1976; Murer and Hopfer, 1976). As shown in Figure 1 for the rat small intestine, the transport of D-glucose and L-valine into brush border membrane vesicles is stimulated by a sodium gradient as compared to a potassium gradient. In contrast glucose and amino acid transport into basal-lateral membrane vesicles is insensitive to sodium. Similar results were obtained with renal plasma membrane

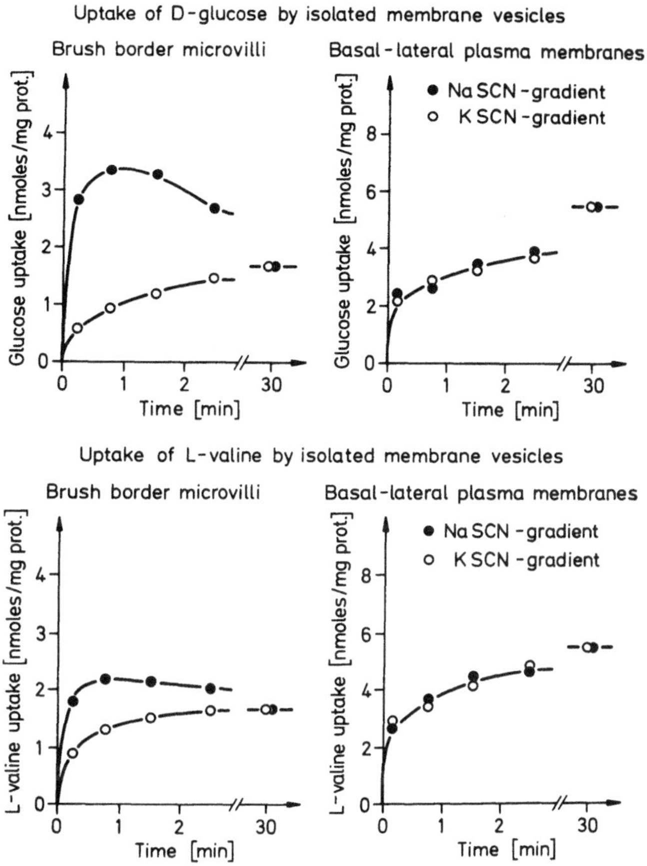

Fig. 1. Sidedness of transport processes in small intestinal epithelium. Uptake of D-glucose and L-valine by isolated intestinal brush border and basal-lateral membrane vesicles. Incubation medium contained: 100 mM D-mannitol, 1 mM HEPES-Tris, pH 7.4, 1 mM ^{3}H-labeled substrate and 100 mM salt as indicated in figure (HEPES = 2-(N-2-hydroxyethylpiperazine-N'-yl) ethanesulphonic acid)

vesicles. The sugar transport systems in the brush border membrane and in the basal-lateral membranes not only differ in sodium sensitivity but also show different substrate and inhibitor specificities. Thus in basal-lateral membrane vesicles, D-glucose transport is strongly inhibited by D-mannose and 2-deoxy-D-glucose, whereas D-glucose transport in the brush border membrane is not inhibited. Further the D-glucose transport system in the contraluminal membranes is more sensitive to phloretin and cytochalasin B than the D-glucose transport system in the luminal membrane. In contrast D-glucose uptake by luminal membranes is more affected by phlorizin than is the uptake by contraluminal membranes (Kinne et al., 1975; Hopfer et al., 1975).

Figure 2 shows a similar sidedness of the proximal tubular cell with respect to inorganic phosphate (P_i) transport and p-aminohippuric acid (PAH) transport. Phosphate transport in the luminal membranes is stimulated more strongly by sodium than in the contraluminal membranes (Hoffmann et al., 1976). On the other hand a probenecid-sensitive transport system for organic anions such as PAH seems to be present in the contraluminal membranes (Berner and Kinne, 1976).

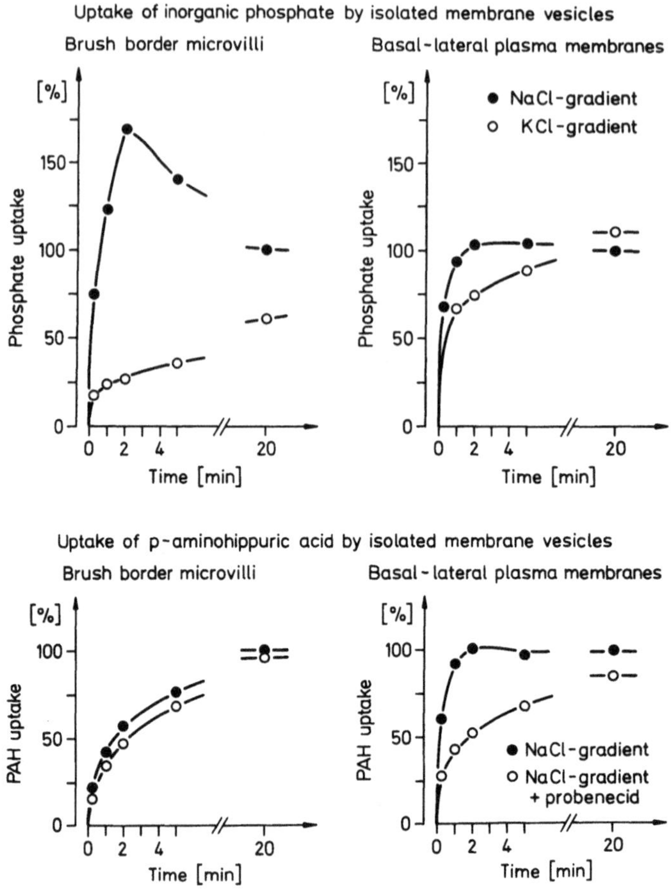

Fig. 2. Sidedness of transport processes in renal proximal tubular epithelium. Uptake of P_i and PAH by isolated renal brush border and basal-lateral membrane vesicles. Incubation medium contained: 100 mM D-mannitol, 20 mM HEPES-Tris, pH 7.4, and 100 mM salt as indicated in figure. Phosphate concentration was 1 mM, PAH concentration was 0.01 mM, and probenecid concentration was 0.5 mM

B. Coupling Between Sodium and Nonelectrolyte Transport Across Plasma Membrane Vesicles

In Figure 3A the effect of sodium on D-glucose uptake by intestinal brush border membrane vesicles under gradient ($Na_i < Na_o$) and under nongradient conditions ($Na_i = Na_o$) is compared. Under nongradient conditions for sodium the initial rate of D-glucose uptake is increased twofold as compared to the uptake in the presence of potassium. In the presence of a gradient for sodium, the stimulation of initial uptake is fourfold, and a transient accumulation of glucose inside the vesicles is observed. These findings suggest that a glucose-sodium cotransported system is operating in the brush border membrane and that the sodium gradient - via flux coupling - can provide a driving force for intravesicular accumulation of D-glucose. Such coupling of glucose and sodium flux can also be demonstrated in experiments in which the movement of sodium across the brush border membrane in the absence and in the presence of a glucose gradient is investigated. As shown in Figure 3B, addition of

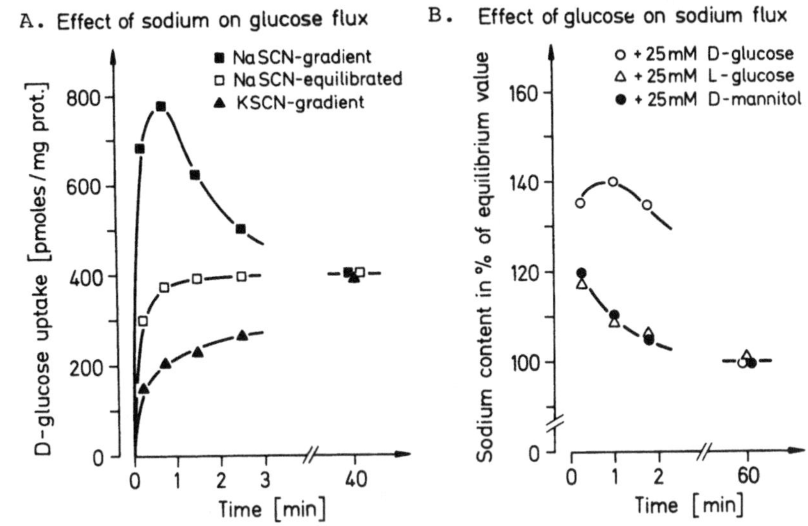

Fig. 3. Coupling between glucose and sodium flux across brush border membranes (intestine). Incubation media for glucose uptake contained: 1 mM D-(U-^{14}C) glucose, 100 mM D-mannitol, and 1 mM HEPES-Tris, pH 7.4, with either 100 mM NaSCN or 100 mM KSCN. NaSCN gradient was abolished by preincubation of membranes with 100 mM NaSCN for 10 min at 25°C. Glucose-induced Na^+ accumulation was studied as follows. Isolated brush border microvilli were incubated for 1 h at 25°C in a solution containing 100 mM D-mannitol, 5 mM HEPES-Tris, pH 7.4, 95 mM KSCN, 5 mM ^{22}NaSCN. At zero time, 30 µl containing 5 mM HEPES-Tris, 75 mM KSCN and 150 mM nonelectrolyte (indicated in figure) were added to 150 µl of membrane suspension. Values are expressed as percent of new equilibrium reached after 60 min of incubation

D-glucose (but not of D-mannitol or L-glucose) to vesicles equilibrated with ^{22}Na (95 mM KSCN, 5 mM NaSCN) leads to a transient accumulation of sodium inside the brush border membrane vesicles. This adds further support for a coupling of the glucose flux and the sodium flux via a cotransport system.

C. Coupling Between Sodium and Anion Transport in Plasma Membrane Vesicles

Stimulation of anion flux by sodium does not necessarily mean a flux coupling via a cotransport system. As represented schematically in Figure 4, coupling between the fluxes of cations and anions can also be due to an "electrical coupling"[1] via a diffusion potential, which might be generated across the vesicle membrane in the presence of a sodium gradient. A distinction between these two possibilities can be made by the following criteria. Coupling via a cotransport system would be indicated by a stimulation specific for

[1]The term "electrical coupling" is used for a situation in which no sodium-anion cotransport system is present in the membrane and in which the movement of anions and cations following the law of electroneutrality is governed only by the passive permeabilities of the membrane

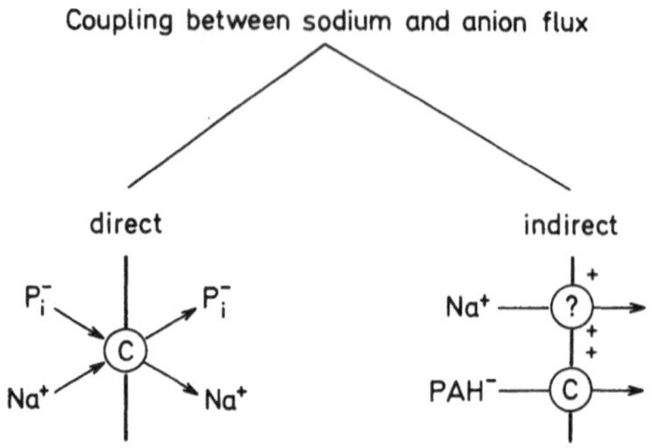

Fig. 4. Schematic representation of possibilities of coupling between sodium and anion flux. Ⓒ = specific transport system: "carrier"

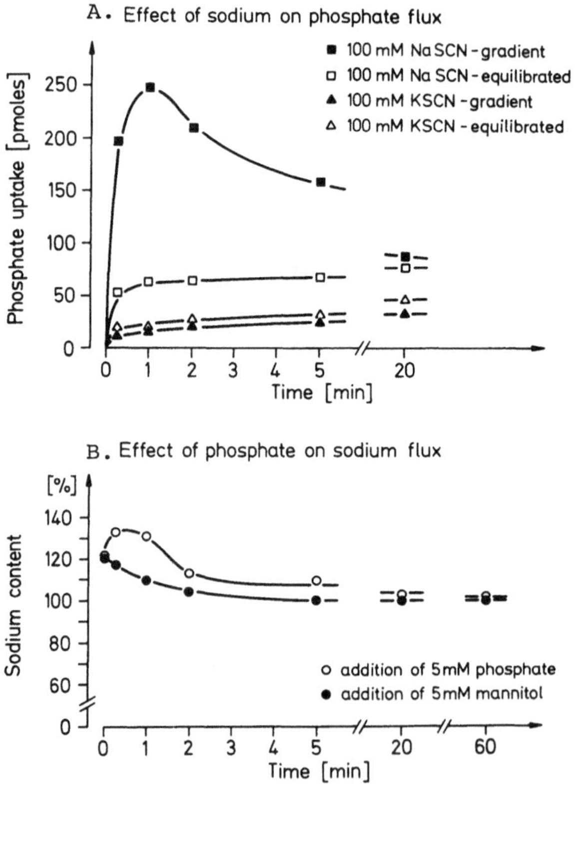

Fig. 5. Coupling between phosphate and sodium flux across renal brush border membranes (cotransport). Incubation media for phosphate uptake contained: 1 mM labeled phosphate, 100 mM D-mannitol, 20 mM HEPES-Tris, and 100 mM salt as indicated in figure. Salt gradient was abolished by preincubation of membranes with 100 mM salt for 1 h at 25°C. Phosphate-induced Na^+ accumulation was studied as follows: Membranes were incubated for 1 h at 25°C in solution containing 100 mM D-mannitol, 5 mM HEPES-Tris, 95 mM KSCN, 5 mM ^{22}NaSCN. At zero time 30 µl of solution containing 5 mM HEPES-Tris, 95 mM KSCN, and 5 mM inorganic phosphate or 5 mM D-mannitol was added to 150 µl of membrane suspension. Values are expressed as percent of new equilibrium reached after 60 min of incubation

sodium under gradient and nongradient conditions, by saturability of the sodium stimulation, and by coupling of sodium flux and anion flux under conditions where the probability for the generation of a membrane potential is minimal. Sodium-dependent phosphate uptake by renal brush border membranes fulfils all the foregoing criteria (Hoffmann et al., 1976). As shown in Figure 5A, for example, sodium stimulates the phosphate uptake under gradient as well as under nongradient conditions. Furthermore it should be noted that the stimulation of phosphate uptake by the sodium gra-

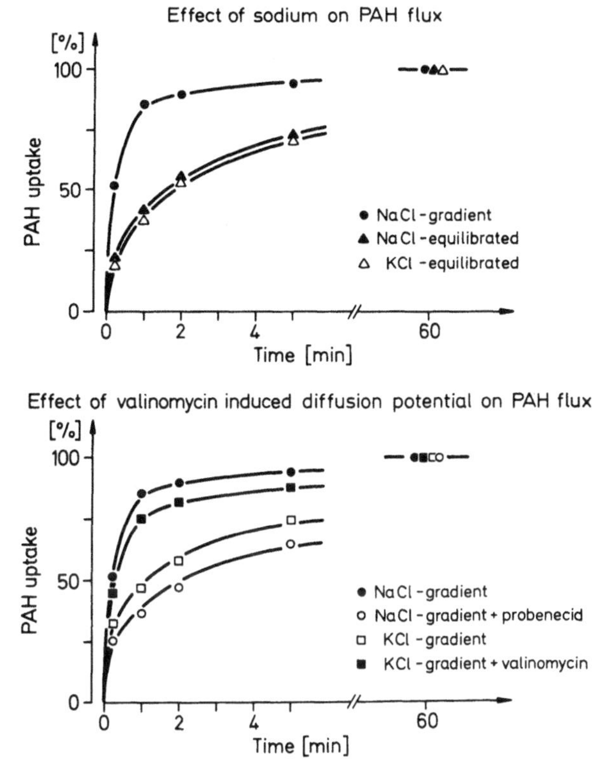

Fig. 6. "Electrical coupling" between p-aminohippuric acid flux and sodium flux across renal basal-lateral plasma membranes. Experiment was carried out as described in Fig. 2. Salt gradient was abolished by preincubation of membranes with 100 mM salt for 1 h at 25°C. Valinomycin, 30 µg per mg of membrane protein, was added

dient is observed in the presence of thiocyanate. The thiocyanate anion is lipophilic and therefore would electrically compensate diffusion potentials that might build up during the movement of sodium ions across the membranes. From the stimulation by inorganic phosphate of sodium flux (Fig. 5B) the coupling of phosphate and sodium via a cotransport system is also evident.

In contrast, a solely potential-mediated stimulation of anion flux by sodium should only be observed in the presence of a salt gradient and not under nongradient conditions. Similarly, the degree of stimulation should be a function of the relative permeabilities of the membrane to cations and anions. Thus, for example, in the presence of a potassium concentration higher in the medium than in the vesicles, a valinomycin-induced increase in potassium permeability of the membrane should stimulate the anion uptake. As shown in Figure 6, PAH transport seems to belong to this category. Sodium exerts its stimulatory effect only under gradient conditions. In the presence of a KCl gradient, valinomycin increases the rate of PAH uptake (Berner and Kinne, 1976).

The small stimulation of phosphate uptake by a NaCl gradient in the basal-lateral membranes might also be due to such a membrane potential-mediated stimulation.

D. Coupling at the Membrane Level: Mutual Interaction Between Sodium-Dependent Cotransport Systems

The presence of several sodium-dependent cotransport systems in one membrane (i.e., the brush border membrane) has several implications that can be delineated from the phenomenological description of sodium-solute cotransport. Using the terminology of irreversible thermodynamics, the coupling between the flux J of a substrate i and the cosubstrate Na^+ can be described as follows (Frömter et al., 1973):

$$J_i = L_{ii}\Delta\eta_i + L_{iNa^+}\Delta\eta_{Na^+} \tag{1}$$

The equation relates the net flux of a substrate $i(J_i)$ to the two driving forces derived from the electrochemical potential difference of the substrate ($\Delta\eta_i$) and from the electrochemical potential difference of sodium ($\Delta\eta_{Na^+}$). L_{ii} is a straight coefficient and L_{iNa^+} is the cross coefficient that expresses the coupling of the substrate flux to the cosubstrate (sodium) flux.

According to Eq. (2),

$$\Delta\eta_{Na^+} = RT\,\frac{\Delta c_{Na^+}}{\bar{c}_{Na^+}} + z_i F\Delta\varphi \tag{2}$$

the driving force experted by $\Delta\eta_{Na^+}$ is composed of two components, one related to the chemical concentration difference of sodium, Δc_{Na^+}, divided by the mean concentration of sodium $\bar{c}_{Na^+}$ (Frömter et al., 1973) and the other related to the electrical potential difference, $\Delta\varphi$.

Cotransport of sodium with a nonelectrolyte is an "electrogenic", potential-sensitive process, unless it is coupled to a stoichiometric movement of an anion in the same direction or of a cation in the opposite direction. Cotransport of sodium with anions, however, can be electroneutral or "electrogenic", depending on the charge of the transported substrate and the coupling ratio to the sodium flux.

To analyze an eventual "electrogenicity" of sodium-dependent substrate transport experimentally, approaches were developed to manipulate the membrane potential (Murer and Hopfer, 1974).

As can be seen in Figure 7 for intestinal brush border membrane transport processes, manipulation of membrane potential affects only sodium-dependent nonelectrolyte transport. The transport rates for L-alanine and D-glucose are drastically increased by an inside negative diffusion potential (relative to the NaCl condition), whereas transport of inorganic phosphate remains unchanged (Murer and Hopfer, 1974; Sigrist-Nelson et al., 1975). This and other similar experiments strongly suggest that sodium-dependent sugar and amino acid transport are indeed "electrogenic". On the other hand, although coupled in a carrier complex, sodium-dependent phosphate movement across the membrane is electroneutral (Berner et al., 1976). Here, the positive charge(s) of the sodium ion(s) seems to be compensated by the negative charge(s) of the anion phosphate.

Salt gradient (medium > vesicles)	Membrane potential (relative to NaCl-gradient)	Phosphate uptake	L-alanine uptake	D-glucose uptake
		in % of uptake in the presence of NaCl		
NaSCN (100 mM)	inside negative	97	235	370
NaCl (100 mM)	0	100	100	100
Na_2SO_4 (50 mM)	inside positive	106	-	65

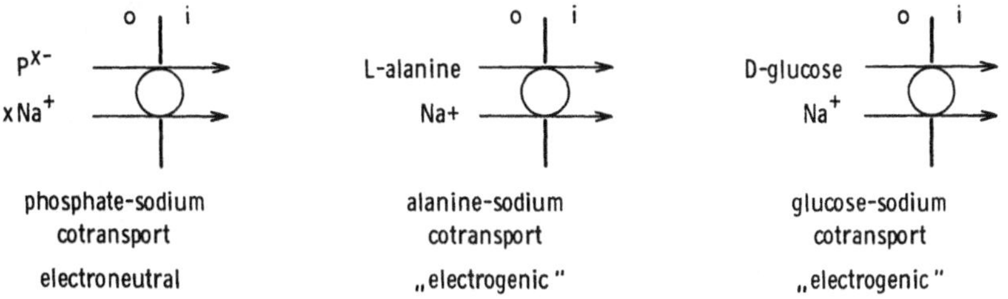

Fig. 7. Influence of membrane potential on sodium-dependent phosphate, alanine, and glucose uptake by isolated intestinal brush border membrane vesicles. Incubation medium contained: 100 mM D-mannitol, 20 mM HEPES-Tris, pH 7.4, and 100 mM salt as indicated in figure. Phosphate concentration: 0.1 mM; glucose concentration: 1 mM; alanine concentration: 1 mM

The relative contribution of the two components of $\Delta\eta_{Na^+}$ to the driving force thus differs depending on the electrical nature of the cotransports. If a charge transfer occurs (glucose sodium cotransport), both Δc_{Na^+} and $\Delta\varphi$ influence the rate of transfer across the membrane. If, however, the transfer is electroneutral, mainly Δc_{Na^+} contributes to the driving force (phosphate sodium cotransport).

Accordingly interaction of electroneutral cotransport processes with other sodium-dependent transport processes located in the same membrane will be mediated via changes in the concentration difference (Δc_{Na^+}) for sodium across the membrane, whereas interdependence of "electrogenic" transport processes will be mediated by changes in the electrochemical potential difference ($\Delta\eta_{Na^+}$) as a whole.

An example of the mutual interaction of electrogenic sodium-dependent cotransport systems in isolated brush border vesicles is given in Figure 8. Here the membrane vesicles were equilibrated with sodium sulfate - this means $\Delta\eta_{Na^+}$ is zero - and the uptake of D-glucose was measured. Addition of L-alanine inhibits D-glucose uptake by the brush border membrane vesicles (Murer et al., 1975). This can be explained by the assumption that L-alanine promotes electrogenic sodium entry into the vesicles via the sodium-alanine cotransport system - similar to the experiments shown in Figure 3. The electrogenic movement of sodium with alanine induces, in the presence of impermeable anions, a membrane potential (vesicle interior positive) that in turn inhibits the entry of glucose into

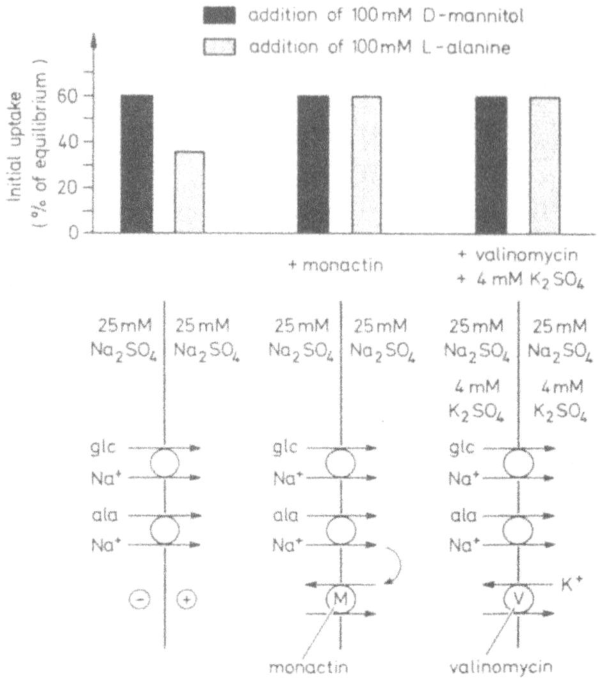

Fig. 8. Inhibition of D-glucose transport by alanine. Intestinal brush border membranes were preincubated in 100 mM D-mannitol, 1 mM HEPES-Tris, pH 7.4, and 25 mM Na_2SO_4 for 30 min at 25°C. Uptake of D-glucose was initiated by addition of D-(1-^{3}H) glucose (final concentration: 1 mM) plus L-alanine (final concentration: 100 mM). In control experiments mannitol instead of L-alanine was added. Other additions were as indicated: monactin (10 µg mg protein^{-1}); valinomycin (10 µg mg protein^{-1}); K_2SO_4 (4 mM). Results from Murer et al. (1975)

the vesicles via the electrogenic sodium-glucose cotransport system (see also Fig. 8). This assumption can be supported by experiments in which the generation of a membrane potential is prevented. Thus for example addition of monactin, which allows a rapid reequilibration of sodium across the brush border membrane, abolishes the inhibition of D-glucose flux by alanine. That the coupling between the two electrogenic fluxes of glucose and alanine involves mainly $\Delta\varphi$ is also evident from experiments with valinomycin in the presence of potassium. Here charge-compensating movements of potassium out of the vesicles via the valinomycin-potassium complex neutralize the membrane potential-generating movements of the sodium ions into the vesicles via the cotransport system. Accordingly no inhibition of D-glucose uptake by alanine is observed in the presence of valinomycin and potassium.

E. Interdependence Between Sodium-Dependent Cotransport Systems and Cotransport-Independent Sodium Permeability

There exists in addition an interrelationship between sodium-dependent cotransport systems and the substrate-independent sodium permeability of the membrane. The common denominator is again the electrochemical potential difference of sodium. Interestingly the substrate-independent sodium movement across the brush border membrane involves mainly an electroneutral proton-sodium exchange. This system could be detected in renal and intestinal brush border

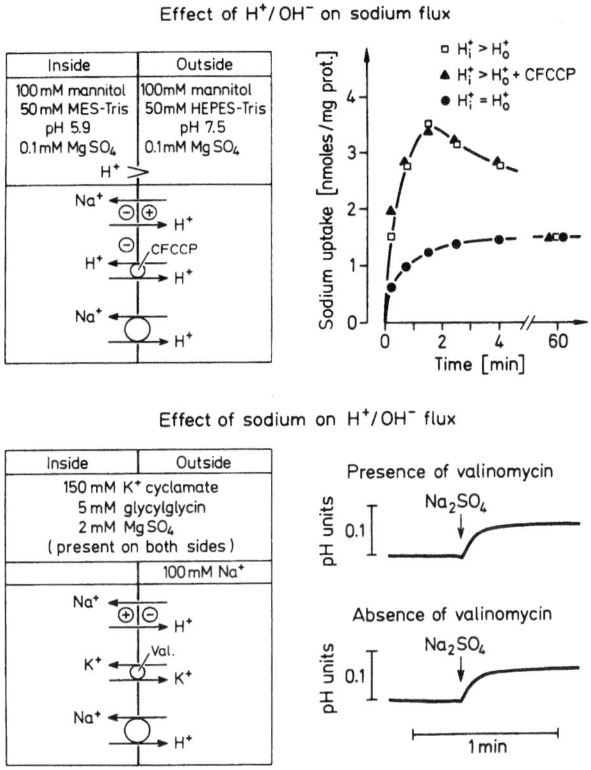

Fig. 9. Coupling between sodium and H^+/OH^+ flux across brush border membranes (kidney). Experimental conditions are given in scheme. CFCCP (= carbonyl cyanide p-trifluoromethoxyphenylhydrazone) and valinomycin were added to incubation medium as ethanolic solutions (final concentration: 0.5% ethanol; 10 µg CFCCP mg protein^{-1}; 10 µg valinomycin mg protein^{-1}). Sodium concentration in H^+ efflux experiments was 100 mM. Sodium concentration in sodium uptake experiments was 1 mM. Results from Murer et al. (1976)

vesicles (Murer et al., 1976) by measuring acidification of incubation medium during sodium influx into the vesicles and by measuring sodium uptake during proton efflux out of the vesicles. As shown in Figure 9 for renal brush border membrane vesicles, a proton gradient is able to accomplish a transient overshoot in the uptake of sodium. The overshoot is not observed under nongradient conditions at either pH 7.4 or pH 5.9. A coupling of proton and sodium flux can also be demonstrated if proton transfer across the brush border membrane is monitored by pH measurements. If a sodium gradient is established across the brush border membrane by addition of a concentrated salt solution to membrane vesicles suspended in a medium with low buffer capacity, an acidification of the incubation medium is observed. The increased H^+ concentration in the medium can either be due to replacement of protons from protonated groups in the membrane surface or represent translocation of protons from the intra- to the extravesicular space. We have been able to rule out the former possibility by several control experiments. Translocation of protons can be caused by the two different processes represented schematically in Figure 9. Firstly, the two ions could use different transport systems, but their fluxes would influence each other via the membrane potential. Secondly, a molecular coupling by counterflow via a transport system could occur. This "antiporter" would accomplish an electroneutral exchange of sodium and protons, and the rate of transport therefore should not be dependent on the membrane potential. If Na^+ influx and H^+ efflux would be coupled via the membrane potential, the addition of valinomycin in the presence of equal potassium concentrations at both sides of the membrane should reduce the membrane potential by charge-compensating potassium movements and thereby the acidification. As shown in Figure 9, valinomycin does not alter the response of the brush border membrane vesicles to sodium. These observations indicate that a sodium-proton exchange system is operating in the brush border membrane, which facilitates an electroneutral entry of sodium into the epithelial cell across the luminal membrane.

In view of the magnitude of the cotransport-independent sodium movement (Schultz and Curran, 1968), this electroneutral entry of sodium into the cell seems to be favorable for the coupling of electrogenic sodium-solute cotransport to the electrochemical potential difference of sodium across the membranes, since an electrogenic transport of sodium would be expected to lead to mutual interaction as discussed previously.

F. Coupling at the Cellular Level

The foregoing results demonstrate that sodium-dependent cotransport systems for D-glucose, neutral amino acids, and phosphate are present in the luminal (brush border) membrane of intestinal and renal epithelial cells, whereas sodium-independent transport systems are concentrated in the contraluminal (basal-lateral) membranes. This sidedness of the epithelial cell with respect to the distribution of transport systems is depicted in Figure 10. The sodium solute cotransport systems provide a mechanism by which $\Delta\eta_{Na^+}$, which is present across the brush border membrane, can be utilized by flux coupling for the intracellular accumulation and active transepithelial transport of solutes.

This coupling of the cotransport systems to $\Delta\eta_{Na^+}$ is improved by the presence of an electroneutral Na^+/H^+ exchange mechanism at the luminal membrane. The charge of the sodium ions that enter the epithelial cell during transcellular sodium transport - the maximal sodium transport capacity is much higher than the transport capacity of the cotransport systems (Schultz and Curran, 1968) - is thereby compensated.

$\Delta\eta_{Na^+}$ is maintained at the cellular level by the action of the Na^+K^+-ATPase, which maintains the low intracellular sodium concentration and - by its electrogenic action - also generates the electrical membrane potential. This coupling between the Na^+-K^+-ATPase and the sodium-dependent cotransport systems is obvious from the inhibition of sodium-dependent transport processes in the epithelium by ouabain.

In addition, it has to be considered that the transport systems of the brush border membrane and of the contraluminal membrane are interrelated by the intracellular concentration of the compound that is to be actively transported. Finally, in the most general sense, all active transepithelial transport processes - primary active and secondary active - are coupled, since they are energized by the cellular metabolism and therefore under conditions of limited energy production may interfere with each other.

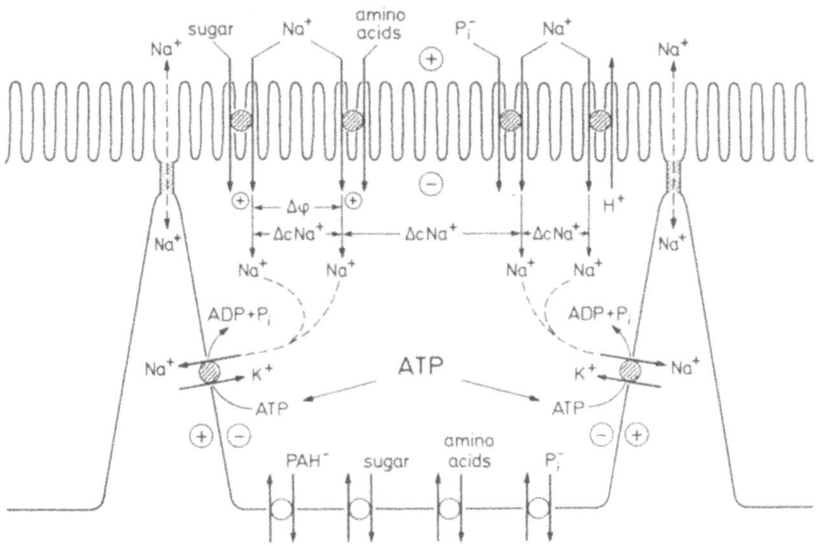

Fig. 10. Sidedness and coupling of transport processes in renal and small intestinal epithelia

References

Berner, W., Kinne, R.: Transport of p-aminohippuric acid by plasma membrane vesicles isolated from rat kidney cortex. Pflügers Arch. 361, 269-277 (1976)

Berner, W., Kinne, R., Murer, H.: Phosphate transport into brush border membrane vesicles isolated from rat small intestine. Biochem. J., in press (1976)

Frömter, E., Rumrich, G., Ullrich, K.J.: Phenomenologic description of Na^+, Cl^- and HCO_3^- absorption from proximal tubules of the rat kidney. Pflügers Arch. 343, 189-220 (1973)

Hoffmann, N., Thees, M., Kinne, R.: Phosphate transport by isolated renal brush border vesicles. Pflügers Arch. 362, 147-156 (1976)

Hopfer, U., Sigrist-Nelson, K., Murer, H.: Intestinal sugar transport. Studies with isolated plasma membranes. Ann. N.Y. Acad. Sci. 264, 414-426 (1975)

Kinne, R., Murer, H.: Polarity of epithelial cells in relation to transepithelial transport in kidney and intestine. In: Intestinal Ion Transport. Robinson, J.W.L. (ed.). Lancaster: MTP Press Ltd., 1976, pp. 79-95

Kinne, R., Murer, H., Kinne-Saffran, E., Thees, M., Sachs, G.: Sugar transport by renal plasma membrane vesicles. Characterization of the systems in the brush-border microvilli and basal-lateral plasma membranes. J. Membrane Biol. 21, 375-395 (1975)

Murer, H., Hopfer, U.: Demonstration of electrogenic Na^+ dependent D-glucose transport in intestinal brush border membranes. Proc. Natl. Acad. Sci. U.S. 71, 484-488 (1974)

Murer, H., Hopfer, U.: The functional polarity of the intestinal epithelial cell. Studies with isolated plasma membrane vesicles. In: Intestinal Permeation. Lauterbach, F., Kramer, M. (eds.). Amsterdam, Excerpta Medica Monographs. In press (1976)

Murer, H., Hopfer, U., Kinne, R.: Sodium/proton antiport in brush-border-membrane vesicles isolated from rat small intestine. Biochem. J. 154, 597-604 (1976)

Murer, H., Hopfer, U., Kinne-Saffran, E., Kinne, R.: Glucose transport in isolated brush-border and lateral-basal plasma-membrane vesicles from intestinal epithelial cells. Biochim. Biophys. Acta 345, 170-179 (1974)

Murer, H., Sigrist-Nelson, K., Hopfer, U.: On the mechanism of sugar and amino acid interaction in intestinal transport. J. Biol. Chem. 250, 7392-7396 (1975)

Schultz, S.G., Curran, P.F.: Intestinal absorption of sodium chloride and water. In: Handbook of Physiology, Vol. III, Section 6. Code, C.F. (ed.). Washington: American Physiological Society 1968, pp. 1245-1275

Sigrist-Nelson, K., Murer, H., Hopfer, U.: Active alanine transport in isolated brush border membranes. J. Biol. Chem. 250, 5674-5680 (1975)

Transport in Erythrocyte Membranes

The Action of 2-(4'-Amino Phenyl)-6-Methyl Benzene Thiazol-3,7'-Disulfonic Acid (APMB) on Anion Transport and the Protein in Band 3 of the Red Blood Cell Membrane

H. Passow, M. Pring, B. Legrum-Schuhmann, and L. Zaki

Summary

The disulfonic acid APMB produces maximally about 85% inhibition of sulfate equilibrium exchange. Inhibition is half maximal at about 0.4 mM APMB. The reversibly binding APMB shares binding sites with the irreversibly binding inhibitors DNFB and H_2DIDS on the protein in band 3 and on the anion transport system. This was demonstrated by observing the inhibitory effect of increasing concentrations of APMB on the rate of dinitrophenylation of common binding sites of DNFB and H_2DIDS on band 3 and on the anion transport system. A mathematical description of the effect of APMB on dinitrophenylation of the anion transport system was consistent with the assumption that APMB reacts with 85% of the DNFB binding sites with a K_m value which is close to the K_I value observed in studies of the inhibitory effect of APMB.

An investigation of the effects of increasing sulfate concentration on sulfate transport showed, in accordance with previous work of Schnell (1975), that the sulfate flux passes through a maximum which suggests the existence of an inhibitory modifier site in addition to a transfer site. Inhibition by APMB decreases slightly over the sulfate concentration range 30 to 300 mM, with half maximal inhibition at APMB concentrations of 0.9 and 0.4 mM, respectively, at the upper and lower limits of this concentration range. The inferred low affinity of SO_4 for modifier and transfer sites and the variations of ionic strength associated with variations of sulfate concentration made it impossible to decide whether sulfate reduces inhibition by a specific reaction with one or both of the two sites, or by other interactions not involving specific binding sites.

A. Introduction

The electrically neutral anion exchange across the red blood cell membrane is facilitated by a specific transport system. Although the lipid composition of the membrane has a modifying influence on its rate (Gruber and Deuticke, 1973) the transport process is mediated by membrane proteins (Passow, 1971; Obaid et al., 1972). Using radioactively labeled inhibitors it was shown that of the

---

Abbreviations. APMB = 2-(4'-amino phenyl)-6-methyl benzene thiazol-3,7'disulfonic acid; DNFB = 1-fluoro-2,4-dinotrobenzene; H_2DIDS = 4,4'-diisothiocyanato dihydro stilbene-2,2'-disulfonic acid

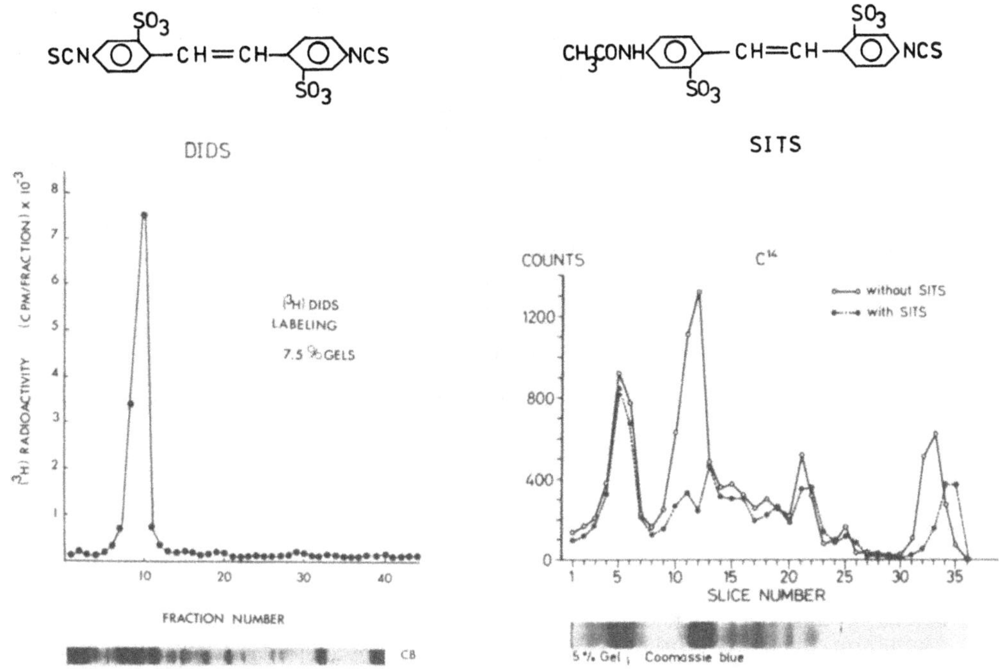

Fig. 1. Binding of inhibitors of anion transport to the protein in band 3. Left: Binding of tritiated DIDS (according to Cabantchik and Rothstein, 1974). Right: Protection of band 3 against dinitrophenylation by treatment of red cells with non-radioactive SITS prior to the exposure to ^{14}C DNFB. Shows that band 3 carries common binding sites for two different inhibitors of anion transport (according to Passow et al., 1975)

"major" membrane proteins (i.e., the more abundant proteins revealed by Coomassie blue staining of SDS polyacrylamide gel electropherograms), the protein in band 3 is the sole possible candidate for some role. Only this protein carries binding sites for a number of reversibly and irreversibly binding inhibitors of anion transport (Cabantchik and Rothstein, 1974; Passow et al., 1975). Figure 1 illustrates this statement and shows that the protein in band 3 is labeled by a strongly inhibitory isothiocyanate and that such isothiocyanates have common binding sites with DNFB, another potent inhibitor, only on band 3 but not on other membrane proteins.

The transport system can be purified by the incorporation of trypsin into resealable erythrocyte ghosts. The intracellular enzyme digests most membrane proteins. However, three, sometimes four, peptides are found to survive. Two of these peptides with molecular weights of 58,000 and 48,000 daltons carry the binding sites for the isothiocyanate with which the protein in band 3 can be labeled. If enzymatically "cleaned" membranes are prepared without labeling with an inhibitor, the ghosts continue to perform anion transport at a somewhat reduced rate, which is still susceptible to a variety of representative inhibitors (Passow et al., 1975; Lepke and Passow, 1976). This indicated that essential components of the anion transport mechanism survived the enzymatic treatment and suggested that one or both of the peptides derived from the protein in band 3 are involved in mediating anion transport.

Labeling with covalently binding inhibitors represented the initial step in the identification of the membrane constituent involved in anion transport. Delineation of the function of the sites identified calls for the use of reversibly binding inhibitors and the investigation of their competition with the ion to be transported. The present paper shows that the disulfonic acid APMB produces a reversible inhibition and has common binding sites with H_2DIDS and DNFB on the protein in band 3. Like previous observations on the action of certain reversibly binding stilbene disulfonic acids (Passow et al., 1975; Zaki et al., 1975) this finding indicates the participation of the protein in band 3 in anion transport. A study of the effects of sulfate concentration on the inhibition by APMB suggested in addition a slight competition. However, the enormous changes of ionic strength associated with the variations of sulfate concentration and other factors which will be dealt with below made it impossible to decide whether this competition takes place at a modifier site or a transfer site.

B. Reversible Inhibition of Anion Transport by APMB

Figure 2 demonstrates the inhibitory effect of APMB on the rate of sulfate equilibrium exchange. At maximal inhibition the rate of exchange remains about 15% of the control value. With the sulfate concentration used, the half-maximal effect is obtained at an APMB concentration of 0.54 ± 0.13 mM[1] (pooled data of a number of similar experiments, n = 18). It should be noted that at maximally inhibitory concentrations H_2DIDS is capable of reducing the rate of sulfate equilibrium exchange to a much smaller value than is APMB[2]. There are at least two alternative explanations for this striking difference: (1) The population of H_2DIDS binding sites is nonuniform, only 85% of them being susceptible to APMB. This could reflect differences in location, chemical reactivity, or both; it might be related to the existence of functionally different types of binding sites, but does not, of course, prove the existence of such sites. (2) All sites are common binding sites for APMB and H_2DIDS and have identical functions. However, in contrast to H_2DIDS, at complete occupancy of the sites, APMB does not produce complete inhibition.

C. Common Binding Sites for APMB, DNFB and H_2DIDS on the Protein in Band 3

The observation that the reversibly binding APMB shares the binding sites on band 3 with the irreversibly binding DNFB and H_2DIDS is illustrated in Figure 3. It is shown that dinitrophenylation

---

[1]All ranges are quoted as maximum likelihood value ± 95% confidence limits

[2]See Passow and Zaki (1976). Furthermore we have found that, if 15 μM H_2DIDS is added to whole erythrocytes under standard conditions in the presence of APMB concentrations of 2.5, 5.0 and 10.0 mM, sulfate exchange is further reduced from 21.8% to 0.21%, from 16.9% to 0.15%, and from 12.1% to 0.21% respectively of the control value in the absence of either inhibitor

Fig. 2. Inhibition of sulfate equilibrium exchange by APMB. The flux measurements were carried out at 30°C or 37°C with red cell ghosts equilibrated with media containing 5 mM Na_2SO_4, 20 mM Na-phosphate, 122.5 mM NaCl, pH 7.4. Ordinate: sulfate flux as a % of control. Abscissa: concentration of APMB in the external medium. Circles: measured fluxes; Bars: range of values obtained in several experiments, of which the mean was taken. Solid line: function $1 - a[APMB]/(K_I + [APMB])$; with a = 0.847 ± 0.034 and K_I = 0.54 ± 0.13 mM (95% confidence limits)

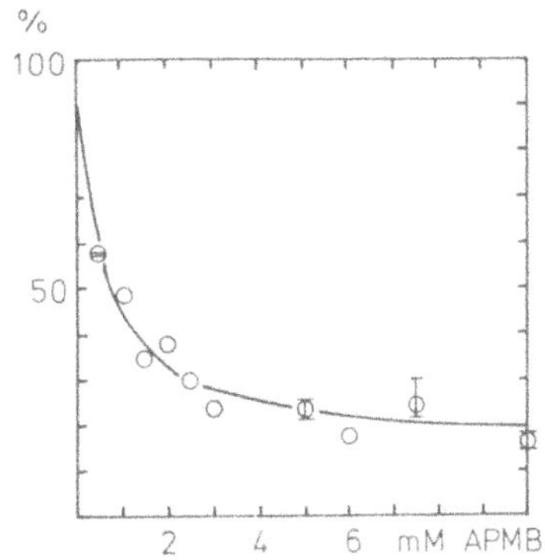

prior to the exposure to H_2DIDS leads to a sharp decrease of H_2DIDS binding. This decrease is largely prevented if dinitrophenylation is carried out in the presence of APMB. Control experiments showed no reaction between DNFB and APMB under these conditions.

D. Common Binding Sites for APMB, DNFB and H_2DIDS on the Anion Transport System

APMB has common binding sites with DNFB and H_2DIDS on the sulfate transport system. This is suggested by the results presented in Figure 4. They were obtained in an experiment in which red cell ghosts had been exposed to DNFB in the presence of APMB. After removal of external APMB and unreacted DNFB by washing, the ghosts were hemolyzed for a second time in order to remove traces of APMB which had slowly penetrated into them. After resealing, SO_4 equilibrium exchange was measured in the absence of added inhibitors. A comparison of Figures 4a, b and c shows that inhibition of SO_4 transport is much smaller in ghosts which had been dinitrophenylated in the presence rather than the absence of APMB. It can also be shown that the SO_4 exchange observed after removal of APMB and DNFB is still inhibitable by H_2DIDS.

The results of a more systematic study of the protection afforded by APMB against dinitrophenylation of the sulfate transport system are depicted in Figure 5. The experimental procedure was the same as in the experiment in Figure 4. With increasing APMB concentration the protection increases until about 85% of the control rate is reached. The remaining 15% inhibition could reflect inaccuracies of the rather involved experimental procedure. Moreover, APMB is a reversibly acting inhibitor while DNFB acts irreversibly. Thus, APMB should not be able to afford a time-independent protection against dinitrophenylation. The concomitant slow irreversible modification could contribute to the impression that complete protection cannot be achieved. However, we are tempted to believe that 85% represents a real limit for the protection by APMB against dinitrophenylation, and that the remaining 15% demonstrate the existence of DNFB binding sites on the transport system which do not react with APMB and hence cannot be protected.

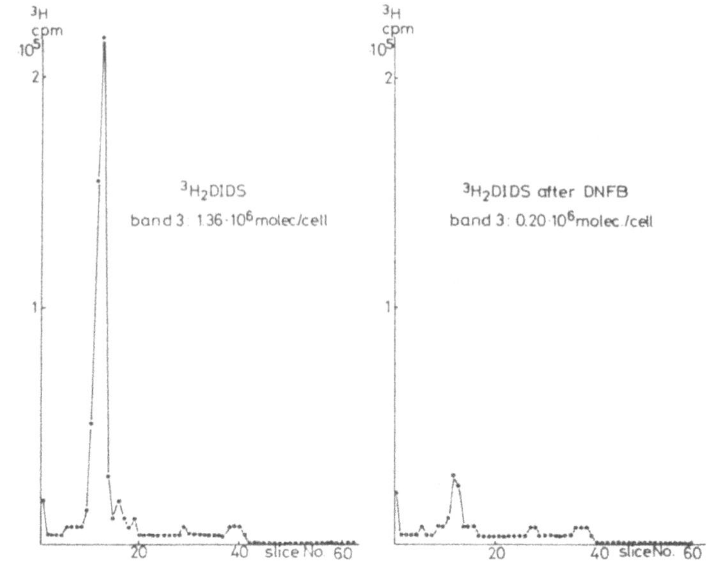

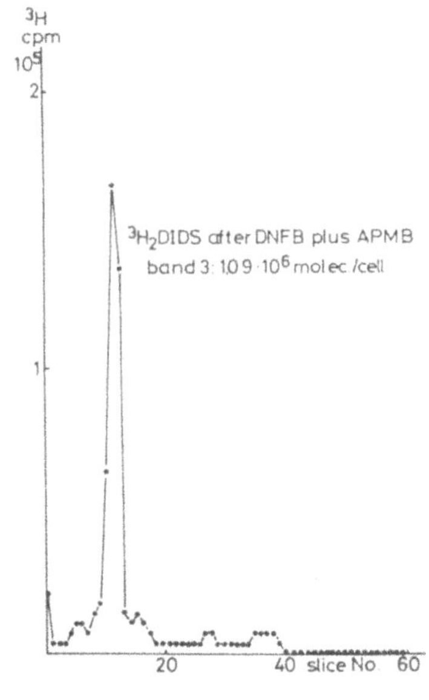

Fig. 3a-c. Distribution of radioactivity on SDS polyacrylamide gel electropherograms of red cell membranes derived from resealed ghosts which had been exposed to 25 µM $^3H_2$DIDS for 90 min at 37°C. Exposure took place in cells which had received the following pretreatment (a) none (b) dinitrophenylation in presence of 600 µM DNFB for 30 min at 37°C, (c) dinitrophenylation in presence of 600 µM DNFB plus 25 mM APMB for 30 min at 37°C. Location of peaks at slices 12-13 corresponds to location of band 3 in Coomassie blue-stained gels. Numbers on figures: binding of $^3H_2$DIDS to band 3 in molecules/cell

It is instructive to consider the mechanism of the protection by APMB against dinitrophenylation in some more detail. The reversibly binding APMB reduces the time average of the number of unoccupied sites S on the transport system and leads to a decrease of the probability for a dinitrophenylation of S:

$$S + APMB \underset{}{\overset{K_I}{\rightleftharpoons}} S\text{-}APMB$$

$$S + DNFB \xrightarrow{k} S\text{-}DNFB.$$

Fig. 4. Effects of inhibitors of anion transport on sulfate equilibrium exchange in red cell ghosts. Prior to the flux measurements, the ghosts received the following treatment: a: none, (= control); b: dinitrophenylation in the presence of 200 µM DNFB for 30 min at 37°C; c: dinitrophenylation in the presence of 200 µM DNFB plus 35 mM APMB for 30 min at 37°C. Fluxes were measured in the absence of inhibitors after removal of APMB and unreacted DNFB. External and internal media had the composition described in the legend to Figure 2. Ordinate: Rate constant for SO_4 efflux

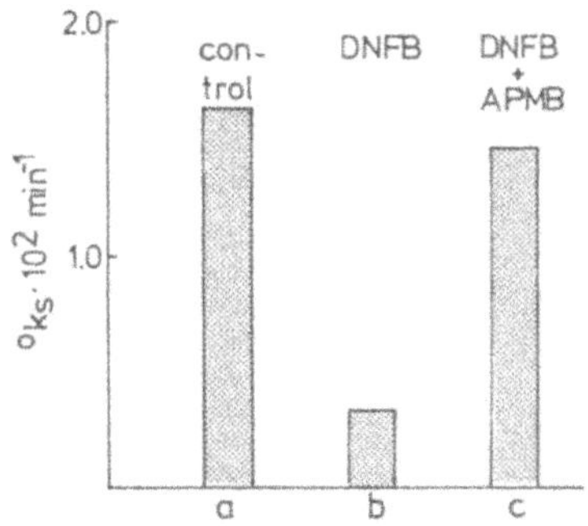

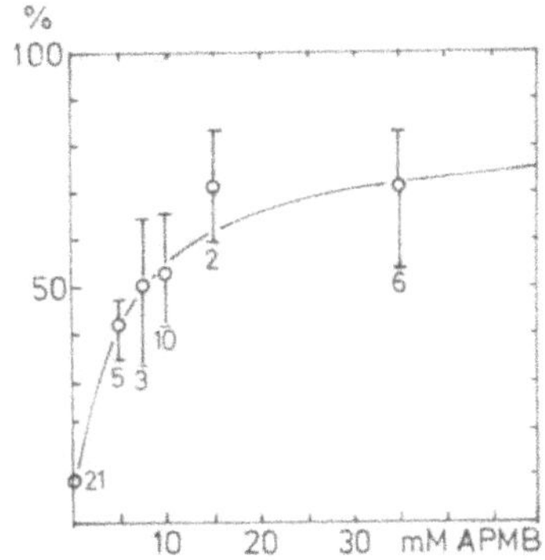

Fig. 5. Protection by APMB against dinitrophenylation of the sulfate transport system. Red cell ghosts were dinitrophenylated at 200 µM DNFB in the presence of APBM at the concentrations indicated on abscissa for 30 min at 37°C. After removal of APMB and unreacted DNFB by washing and rehemolysis (see text), SO_4 fluxes were measured in twice-sealed ghosts at 37°C. Ordinate: sulfate flux as a % of a control consisting of unmodified twice-sealed ghosts. Abscissa: APMB concentration in the medium during exposure to DNFB. Figures, circles and bars: respectively the number, mean and range of measurements obtained in several experiments. Solid line: fitted to Eq. (1), using a = 0.829 ± 0.112, K_I = 2.29 ± 1.08 mM, k[DNFB]t = 2.18 ± 0.96

If one assumes that (1) the rate of sulfate transport is proportional to the number of unreacted sites S, (2) only a fraction, α, of S reacts with APMB, (3) the reaction between DNFB and S follows pseudo-first order kinetics, and (4) the equilibration between S and APMB is rapid compared to the irreversible reaction with DNFB; the effect of varying the concentration of APMB and DNFB on sulfate transport is described by:

$$\frac{{}^{o}k_{S\ DNFB}}{{}^{o}k_{S}} = \alpha\, e^{-\left(\frac{K_I\, k\ [DNFB]\ t}{K_I + [APMB]}\right)} \tag{1}$$

In this equation ${}^{o}k_{S\ DNFB}$ represents the rate constant for sulfate equilibrium exchange as measured in the modified cells after removal of APMB and unreacted DNFB. ${}^{o}k_{S}$ refers to the rate of ex-

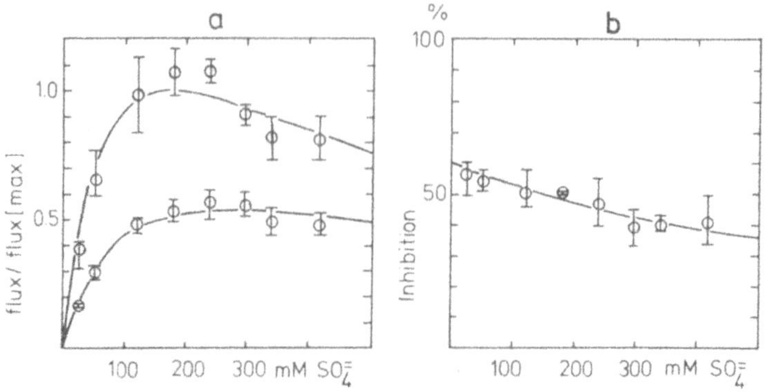

Fig. 6a,b. (a) Effect of increasing sulfate concentrations on sulfate flux in absence (upper curve) and presence (lower curve) of 1 mM APMB. Ordinate: sulfate flux as a fraction of the maximal flux ($2.2\text{-}3.1 \cdot 10^{-9}$ mM/cm^2/sec, see text). Abscissa: sulfate concentration in mM. Circles and bars: respectively mean and range of measurements obtained in 3 or 4 experiments. Solid line: calculated as described in the text (30°C); (b) Inhibition of sulfate flux by 1 mM APMB as function of sulfate concentration in cells and medium. Ordinate: inhibition, I, as percent reduction of control flux. Abscissa: as in Figure 6a. Same data as in Figure 6a. Solid line: fitted to the expression

$$I = a_2 \frac{[APMB]}{K_I} / \left(1 + \frac{[SO_4^{2-}]}{K_2} + \frac{[APMB]}{K_I}\right).$$

The parameter values are listed in the text.

In both sets of experiments, as well as in the experiments represented in Figure 7, prior to the flux measurements cells and media had been equilibrated at the various sulfate concentrations by means of the technique of Cass and Dalmark (1973)

change in a control of untreated twice-resealed ghosts. According to this equation the ratio of the rates observed in modified and untreated cells depends on the magnitude of the rate constant k for the dinitrophenylation of S, and on the equilibrium constant K_I for the binding of APMB to these sites. The solid line drawn in Figure 5 was obtained by optimizing the equation to the experimental data with respect to α, K_I and k[DNFB]t. The values obtained for the various constants were 0.171 ± 0.113, 2.29 ± 1.08, 2.18 ± 0.96, respectively. It is obvious that this simple theoretical treatment is compatible with the data. It is also clear, however, that further work is required to test the assumptions of this treatment more adequately.

E. Effects of Sulfate on Inhibition by APMB

On the basis of the previously presented information there is little doubt that APMB, DNFB, and H_2DIDS have common binding sites on the transport system. We thought it possible, therefore, to study the function of these common sites by means of APMB and tried to investigate competition between APMB and SO_4, the anion to be transported. For this purpose we varied the sulfate concentration at constant APMB concentration, and the APMB concentration

Fig. 7. Effect of increasing concentrations of APMB on sulfate efflux as measured at sulfate concentrations of 30 mM (lower curve) and 300 mM (upper curve). Ordinate: sulfate flux as fraction of maximal flux defined in text; Abscissa: APMB concentration in medium; Circles: experimental measurements; Solid line: calculated as described in text

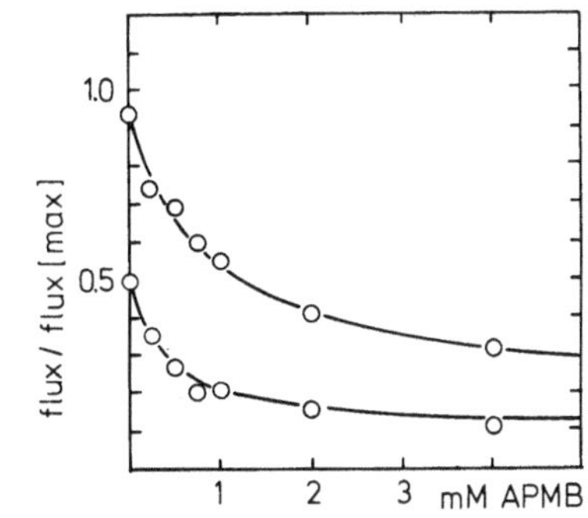

at 2 different and constant sulfate concentrations. Since the variations of concentration over a very wide range are associated with considerable changes of tonicity, the intact cells used in these experiments were equilibrated in the various media prior to the flux measurements by the method of Cass and Dalmark (1973). The results are represented in Figures 6 and 7.

The upper curve in Figure 6 demonstrates the dependence of SO_4 equilibrium exchange on SO_4 concentration. It confirms previous results of Schnell (1975) and shows that the sulfate flux passes through a maximum. If measured in the presence of a fixed concentration of APMB, the curve retains its characteristic features. Over the concentration range 25 to 300 mM inhibition decreases by about 30%. The APMB concentration of half-maximal inhibition increases from about 0.4 mM to about 0.9 mM.

In contrast to previous results with red cell ghosts in which a very much smaller sulfate concentration range was studied (Passow and Zaki, 1976) the results suggest a slight competition between APMB and SO_4.

The curve relating sulfate flux to concentration passes through a maximum. This indicates the existence of two counteracting effects. The activating effect is obviously related to the saturation of the transport system with sulfate. The inhibitory effect could represent the combination of additional sulfate ions with a modifier site, the existence of which is also suggested by the pronounced maximum of the pH dependence of sulfate transport (Schnell, 1975).

If one assumes that the SO_4 concentration at which half maximal inhibition of SO_4 transport is observed (about 30 mM) represents an equilibrium constant for the reaction between binding site S and sulfate ion, then our results are clearly not compatible with the assumption of a competition between APMB and SO_4 for a transfer site: at a K_I of 0.4 and a K_M of 30 mM one would expect that an increase of the sulfate concentration from 25 to 400 mM would reduce the inhibitory effect of APMB much more than actually observed.

A more detailed analysis of the data on the basis of competition between APMB and SO_4 for a common modifier site shows, however, the ambiguities involved in this conclusion. It is possible to fit the data to Eq. (2) which represents sulfate flux as a function of the concentrations of SO_4 and APMB, assuming that only a proportion, a_2, of the carrier is susceptible to APMB:

$$\text{Flux} = \frac{\dfrac{a_1[SO_4^{2-}]}{K_1}}{\left(1 + \dfrac{[SO_4^{2-}]}{K_1}\right)} \left[\frac{(1-a_2)}{\left(1 + \dfrac{[SO_4^{2-}]}{K_2}\right)} + \frac{a_2}{\left(1 + \dfrac{[SO_4^{2-}]}{K_2} + \dfrac{[APMB]}{K_I}\right)} \right] \tag{2}$$

To obtain the lines drawn in Figures 6 and 7 all experiments were fitted together imposing the same values for a_2, K_1, K_2 and K_I but allowing each experiment to have a different value of a_1. The data and calculated curves were then normalized for plotting by dividing by the maximal flux derived from the above equation with $[APMB] = 0$, $SO_4^{2-} = \sqrt{K_1 \cdot K_2}$. The following parameter values were obtained: $a_1 = 8.7 - 12.3 \cdot 10^{-9}$ mM/cm²/sec; $a_2 = 0.815 \pm 0.074$; $K_1 = 170 \pm 63$ mM; $K_2 = 178 \pm 72$ mM; $K_I = 0.343 \pm 0.156$ mM.

The large values of the equilibrium constant obtained for the transfer and modifier sites explain why the effect of SO_4 on APMB inhibition remains small even though both substances are assumed to compete for a common site. Since modifier and transfer sites have similar binding constants for SO_4 it would seem impossible to discriminate on the basis of the experiments described between effects on either of the two sites.

The difficulties associated with the interpretation of the data on the basis of Eq. (2) are further augmented by a problem which is inherent in all studies involving variations of sulfate concentration over a wide range. In the derivation of Eq. (2) activity coefficients were assumed to be 1.0 and independent of sulfate concentration. However, these coefficients must be expected to vary considerably and in an unknown fashion with the enormous variations of ionic strength in the experiments described. Moreover, the changes of ionic strength could also produce changes in the exposed portions of the transport protein, affecting the sulfate flux without involving sulfate binding to a specific modifier site.

Nevertheless, it is of interest to note that under the three conditions reported - inhibition, protection, and competition - the values obtained for the proportion of the transport system that can be inhibited or protected by APMB, 0.847 ± 0.034, 0.829 ± 0.112, and 0.815 ± 0.074, respectively, are quite consistent. The corresponding values obtained for the apparent K_I at 30 mM SO_4 were 0.54 ± 0.13 mM, 2.29 ± 1.08 mM, and 0.40 ± 0.18 mM. Though somewhat more scattered, these are not inconsistent when viewed in the light of the fact that they represent different experimental systems (ghosts, twice sealed ghosts, and intact erythrocytes), and that the most deviant was the least well defined, and also exhibited large positive covariance with another loosely bounded parameter, k[DNFB]t. This overall consistency confirms that these phenomena are manifestations of a common mechanism, but the detailed interpretation of their complexities must await more definitive studies.

Acknowledgments. We thank Mr. Wolfgang Ruffing, Ms. Sigrid Lepke and Mrs. Eva-Maria Gärtner for their collaboration, and Professor H. Fasold for the supply of $^3H_2DIDS$ and much valuable advice.

References

Cabantchik, Z.I., Rothstein, A.: Membrane proteins related to anion permeability of human red blood cells. I. Localization of disulfonic stilbene binding sites in proteins involved in permeation. J. Membrane Biol. 15, 207-226 (1974)

Cass, A., Dalmark, M.: Equilibrium dialysis of ions in nystatin-treated cells: Nature (New Biol.) 244, 47-49 (1973)

Gruber, W., Deuticke, B.: Comparative aspects of phosphate transfer across mammalian erythrocyte membrane. J. Membrane Biol. 13, 19-36 (1973)

Lepke, S., Passow, H.: The effect of incorporated trypsin on membrane proteins and anion transport in human red blood cell ghosts. Biochim. Biophys. Acta 455, 353-370 (1976)

Obaid, A.L., Rega, A.F., Garrahan, P.: The effects of maleic anhydride on the ionic permeability of red cells. J. Membrane Biol. 9, 385-401 (1972)

Passow, H.: Effects of pronase on passive ion permeability of the human red blood cell. J. Membrane Biol. 6, 233-258 (1971)

Passow, H., Fasold, H., Zaki, L., Schuhmann, B., Lepke, S.: Membrane proteins and anion exchange in human erythrocytes. In: Biomembranes: Structure and Function. Proc. 9th Meet. FEBS, Budapest 1974. Gardôs, G., Szâsz, I. (eds.). 1975, pp. 197-214

Passow, H., Zaki, L.: Studies on the molecular mechanism of anion transport across the red blood cell membrane. In: 30th Ann. Symp. Biophys. Lab. Harvard Medical School. Molecular Specialization and Asymmetry in Membrane Function. Boston: Harvard Univ. 1976

Rothstein, A., Cabantchik, Z.I., Knauf, P.: Mechanism of anion transport in red blood cells: role of membrane proteins. Federation Proc. 35, 3-10 (1976)

Schnell, K.F.: Untersuchungen zum Mechanismus des Sulfattransportes durch die Erythrozytenmembran. Regensburg, Habilitationsschrift 1975

Zaki, L., Fasold, H., Schuhmann, B., Passow, H.: Chemical modification of membrane proteins in relation to inhibition of anion exchange in human red blood cells. J. Cell. Physiol. 86, 471-494 (1975)

NAP-Taurine, a Photoaffinity Probe for the Anion Transport System of the Red Blood Cell

A. ROTHSTEIN, P. A. KNAUF, and Z. I. CABANTCHIK

The membrane components involved in anion permeation have been tentatively identified through the use of chemical probes that are potent irreversible inhibitors. These include the disulfonic stilbenes (Cabantchik, 1972; Cabantchik and Rothstein, 1974a; Lepke et al., 1976) and 1-isothiocyanate-4-benzenesulfonate (ITA), a derivative of sulfanilic acid (Ho and Guidotti, 1975). The predominant protein labeled by these agents is known as band 3 (based on its position in SDS-polyacrylamide gels after electrophoresis (Fairbanks et al., 1971). The anion transport system behaves kinetically as though the anions must bind to a specific carrier site in order for the translocation to occur (Passow, 1969; Gunn et al., 1973; Dalmark, 1976). It is not clear, however, that the irreversible inhibitors are labeling the presumptive transport sites. The disulfonic stilbenes and sulfanilic acid derivatives are themselves anions and they might be expected to react with an anion binding site. On the other hand, a number of neutral compounds are also inhibitors of anion transport (Passow, 1969; Knauf and Rothstein, 1971) and one, fluorodinitrobenzene (FDNB), interacts with the same sites in band 3 as the disulfonic stilbenes (Zaki et al., 1975). Furthermore, it is possible that the agents may be inhibiting by an indirect mechanism involving sites at a distance from the transport system. In the case of sulfanilic acid, a non-reactive analog of ITA, a case can be made for interaction occurring with the transport site inasmuch as phosphate, a substrate for the anion transport system, affords protection against inhibition (Ho and Guidotti, 1975).

In attempting to designate the site of anion transport, the best probe would be a transport substrate that could, under appropriate conditions, bind irreversibly to the transport site so that it could be isolated and identified. This requirement for a probe that can bind to the site either reversibly or irreversibly has been met in two cases. Pyridoxal phosphate (PDP) inhibits anion transport reversibly, but after addition of the reducing agent $NaBH_4$ inhibition becomes irreversible (Cabantchik et al., 1975). PDP interacts to form a stable but reversible Schiff's base with amino groups, but reduction results in formation of an irreversible bond. The extent of the irreversible inhibition for a given concentration of PDP is the same as the reversible inhibition, suggesting that the inhibitory sites marked by the agent are those with which it was associated before addition of the reducing agent.

Although PDP seems to be a substrate for the anion transport system in addition to being a potential irreversible marker, it suffers certain disadvantages as a probe. In the first place, although it is highly specific for amino groups, it is not as specific for membrane components as the disulfonic stilbene, DIDS (4,4'-diisothiocyano-2,2'-stilbene disulfonate (Cabantchik and Rothstein, 1974a). At low concentrations, it reacts predominantly with band 3 at sites that also bind DIDS, but at higher concentrations it also reacts with several other surface proteins (Cabantchik et al., 1975). Secondly, although it penetrates the membrane by a mechanism inhibitable by DIDS (which is a highly specific inhibitor of the anion transport system), its flux is extremely low so that it is difficult to determine the kinetic parameters of its transport.

N-(-4-azido-2-nitrophenyl)-2-aminoethylsulfonate (NAP-taurine), a photoreactive compound used as a non-penetrating surface probe at low temperature (Staros and Richards, 1974), can penetrate the membrane at high temperature in the dark (Staros et al., 1975; Rothstein et al., 1976; Cabantchik et al., 1976). On exposure to light, its azido group is converted to a highly reactive nitrene, which can form an irreversible bond with almost any ligand, even displacing H from C. If it binds reversibly to anion transport sites in the dark, it should react irreversibly on exposure to light, regardless of the particular ligands that may be present in the local environment.

In order to use NAP-taurine as a probe for the anion transport site, it should be demonstrated that: (1) in the dark, it permeates the membrane by the normal anion system; (2) that it is a reversible competitive inhibitor of inorganic anion transport; (3) that when exposed to light it reacts irreversibly with the same sites with which it was reversibly associated; and (4) that its irreversible binding to specific membrane components is inhibited by transportable inorganic anions.

Some data bearing on each of these points will be presented, as well as data on the relationship of NAP-taurine and DIDS binding sites.

Fluxes of NAP-Taurine

The efflux of NAP-taurine from cells preloaded with the agent and kept in the dark is demonstrated in Figure 1. The half-time of the efflux is about 20 min. In cells treated with DIDS, or to which dipyridamole is added, the flux is reduced to about 10% of the control rate (Cabantchik et al., 1976). These two agents are also potent inhibitors of inorganic anion transport, one (DIDS) being irreversible (Cabantchik and Rothstein, 1974a) and the other reversible (Deuticke, 1970). Because they are chemically quite dissimilar and presumably act by different mechanisms, their common action on inorganic anion and NAP-taurine transport suggests that inorganic anions and NAP-taurine are transported by the same system.

NAP-Taurine as a Reversible Inhibitor of Anion Transport

In the dark, NAP-taurine is not only transported, but also acts as an effective reversible inhibitor of sulfate or Cl transport (Cabantchik et al., 1976; Rothstein et al., 1976) with 50% inhibition being produced by a concentration of about 50 $\mu\underline{M}$. The exact degree of inhibition is reduced when the Cl^- or SO_4^{--} concentrations are raised (Table 1). The effect is not due simply to

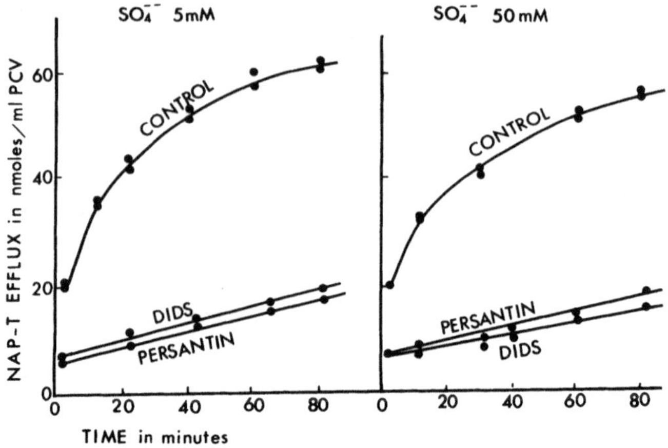

Fig. 1. Effects of DIDS and of dipyridamole on NAP-taurine efflux from the red blood cell. Efflux of ^{35}S NAP-taurine measured in the dark from cells preloaded by exposure to 5 mM NAP-taurine for 10 h at 37°C. One portion of cells was pretreated with 25 μM DIDS before flux measurements. Dipyridamole (Persantin) (25 μM) was present in another portion during flux measurements. (From Cabantchik et al., 1976)

Table 1. Inhibition of sulfate efflux by NAP-taurine in the presence of various anions

	Expt. A	Expt. B	Expt. C
Low SO_4 buffer (S)	62.1	66.3	
S + 20 mM NaCl		61.9	56.6
S + 60 mM NaCl	33.8	33.6	
S + 134 mM NaCl	30.6	36.0	35.6
S + 134 mM $NaNO_3$	25.4		
S + 85 mM Na_2SO_4		44.3	
S + 85 mM Na_3 citrate	65.8		61.0

Values represent % inhibition. Low sulfate buffer (S) contained 5 mM Na_2SO_4 and 25 mM Tris base titrated to pH 7.4 at room temperature with H_2SO_4. Sucrose was added to make the solution isotonic. The various anions were added by isosmotic replacement of sucrose. Cells were equilibrated in the various media and loaded with ^{35}S-sulfate prior to exposure to NAP-taurine. In Expts. A and B, the temperature was 37°C; in Expt. C it was 28°C. Inhibition is expressed as the % decrease in ^{35}S efflux produced by 70 μM NAP-taurine relative to a control in same medium without NAP-taurine. (From Rothstein et al., 1976)

changes in external ionic strength, since citrate which is not transported does not reduce the inhibition. A more quantitative assessment of the Cl^- effect is shown in Figure 2. The reciprocal of the fractional inhibition ($^1/_i$), determined by using SO_4^{--} as a test anion for flux measurements, is plotted against the reciprocal of the NAP-taurine concentration ($^1/I$) (Webb, 1963). Straight lines for the data are consistent with Michaelis-Menten kinetics, with the inhibitor reacting 1 to 1 with a single population of sites and producing an inhibition in proportion to the

Fig. 2. Kinetics of inhibition of sulfate fluxes by NAP-taurine in the presence of different concentrations of CL^- or NO_3^-. Cells were loaded with ^{35}S-sulfate and fluxes were measured in the dark with NAP-taurine present as described in Table 1, except that temperature was 27°C. Media used were either S + 20 mM NaCL (●), S + 134 mM NaCL (■) or S + 134 mM $NaNO_3$ (▲) . Reciprocal of fractional inhibition (1/i) is plotted on ordinate against reciprocal of NAP-taurine concentration in mM on abscissa

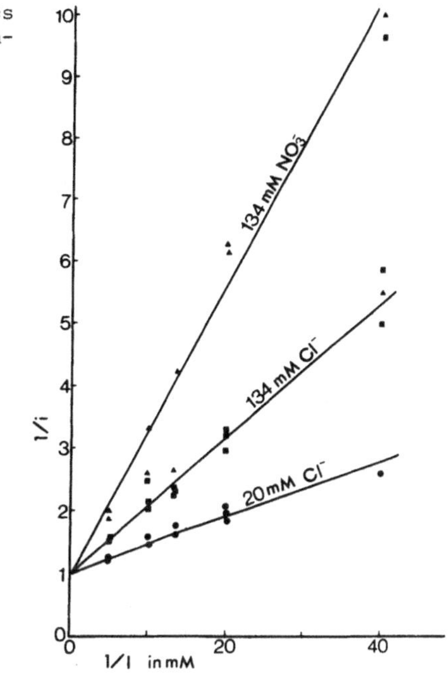

fraction of sites occupied. An increase in slope with higher concentrations of Cl^- is inconsistent with a "non-competitive inhibition" and is consistent with either a "competitive inhibition" or a "mixed inhibition" (in which the inhibitor does not react with the substrate binding site, but does interact with another site that inhibits the reaction and also reduces the affinity for the substrate).

Changing the external anion concentrations could influence the inhibitory effects of NAP-taurine because of competition for a common site, or alteration of the membrane potential. Because the red cell membrane is considerably more permeable to anions than to cations, the gradient of anions from inside to outside determines the potential according to the Nernst equation (Sachs et al., 1974). A reduction in external Cl^- would increase the potential, with the outside becoming more negative. This could increase the concentration of the hypothetical positively charged anion carrier at the outside surface, resulting in increased binding of NAP-taurine. It seems unlikely that the effects of potential would mimic the Michaelis-Menten kinetics shown in Figure 2, but a more direct assessment of the possible contribution of the potential is also illustrated in Figure 2. The kinetics of inhibition are compared in situations in which the external anion concentrations were the same, but 134 mM Cl^- in one case was replaced by NO_3^- in the other. After equilibration, the potentials would not be expected to differ substantially. Nevertheless, the effectiveness of NO_3^- in reversing the inhibition is substantially greater than that of Cl^-. This difference can be attributed to the greater affinity of NO_3^- than Cl^- for a common binding site for these anions and NAP-taurine. In a following section, it will be demonstrated that HCO_3^- also has a greater capacity than Cl^- to prevent the irreversible

Table 2. The protective effect of reversibly bound NAP-taurine on the inhibition of sulfate efflux by DIDS

	Inhibition in % of control	
	8 μM DIDS 50 μM NAP-taurine	4 μ DIDS 100 μM NAP-taurine
DIDS only	88, 92	72, 83
NAP-T (dark) + DIDS, then washed	63, 58	22, 25
NAP-T (dark), then washed + DIDS	95 -	66, 78

Cells (10% hematocrit) preloaded with ^{35}S sulfate (5 mM) were exposed to DIDS for 5 min at $10^{\circ}C$ in presence or absence of NAP-taurine. They were then washed twice with albumin (0.5%) containing buffer, and twice with buffer. In some aliquots of cells, NAP-taurine, 50 or 100 μM, was present during DIDS-treatment; in others it was added and then washed out prior to DIDS treatment. Control samples of red blood cells were subjected to same washing procedures. Fluxes were measured at $37^{\circ}C$ at 5% hematocrit after washing procedure. All procedures were carried out in semi-darkness to avoid photolysis of NAP-taurine. (From Cabantchik et al., 1976)

NAP-taurine binding to an inhibitory site. Again, this difference cannot be attributed to the potential, but most reflect different affinities of anions for a common site.

In addition to competing with inorganic anions, NAP-taurine also competes with DIDS for inhibitory sites (Cabantchik et al., 1976; Rothstein et al., 1976) (Table 2). DIDS binds irreversibly to membrane proteins producing an irreversible inhibition of anion transport (Cabantchik and Rothstein, 1972, 1974a). If NAP-taurine (in the dark) is present during the DIDS treatment, the binding and inhibitory effect of DIDS are reduced. This "protective effect" of NAP-taurine is reversible and can therefore be removed by washing. These data indicate that NAP-taurine interacts with the sites responsible for the DIDS inhibition and that the two agents may interfere with anion transport by a common mechanism.

The Irreversible Inhibition of Anion Transport by NAP-Taurine after Exposure to Light

On exposure to light, the azido group of NAP-taurine is converted to a nitrene that reacts irreversibly with membrane components (Staros and Richards, 1974). The inhibition of anion transport also becomes irreversible (Cabantchik et al., 1976), so that the agent can potentially be used to identify the inhibitory sites. It is essential, however, to know that the sites that are "marked" by the irreversible binding are the same as those with which the agent was associated in the dark (i.e., that it acts as a photoaffinity probe). Under these circumstances its irreversible binding can be unequivocally related to its reversible interaction

Table 3. Relative sulfate effluxes in the presence of NAP-taurine and p-amino benzoic acid (PABA) in the dark and after exposure to light

	Dark	Light
NAP-taurine	54 ± 6	-
NAP-taurine washed (2x)	102 ± 10	43 ± 1
Prephotolyzed NAP-taurine	98 ± 1	114 ± 10
NAP-taurine	54 ± 6	43 ± 1
NAP-taurine + PABA	76 ± 2	68 ± 3

In the "dark" experiments, NAP-taurine was present during the flux except in the case indicated (washed 2x); in the "light" experiments, cells at 5% hematocrit were irradiated for 20 min in presence of NAP-taurine with light from a General Electric DX8 Photospot at a distance of 10 cm, then washed at least two times in buffer containing 0.5% albumin prior to flux measurements. The concentration of NAP-taurine was 50 μM, that of PABA was 100 μM.

Data for NAP-taurine + PABA express relative rates compared to PABA alone, which causes considerable reversible inhibition of sulfate exchange. In parallel, PABA decreases ability of NAP-taurine to reversibly inhibit sulfate exchange. To assess possible effect of PABA as a scavenger, it is necessary to compare the reversible (dark) and irreversible (light) effects of NAP-taurine with and without PABA. PABA did not inhibit conversion of reversible to irreversible inhibition, suggesting that the nitrene reacts with its binding site before it can be removed by scavengers in the external solution

with the transport system. If, during the exposure to light, additional sites become labeled, or occupied sites are not labeled, the potential of the agent as a transport probe is reduced.

By a number of criteria (Ruoho et al., 1973), NAP-taurine behaves as a photoaffinity probe:

1. In the dark it is a reversible inhibitor, but after exposure to light its effect is irreversible (Table 3).

2. The extent of reversible and irreversible inhibition is about the same for a wide range of NAP-taurine concentrations (Fig. 3).

3. Pre-photolyzed NAP-taurine produces no irreversible inhibition (Table 3).

4. The irreversible inhibition on exposure to light is not reduced by the presence of the scavenger, p-amino benzoic acid (Ruoho et al., 1973) (Table 3). These data suggest that on exposure to light, the photoactivated molecules present in the bulk solution react so rapidly with water that few of them reach membrane sites related to the inhibition. On the other hand, those molecules already associated with membrane sites become fixed by an irreversible reaction. Included will be those sites responsible for transport of NAP-taurine, for its inhibitory effect, and for the inhibitory effect of DIDS.

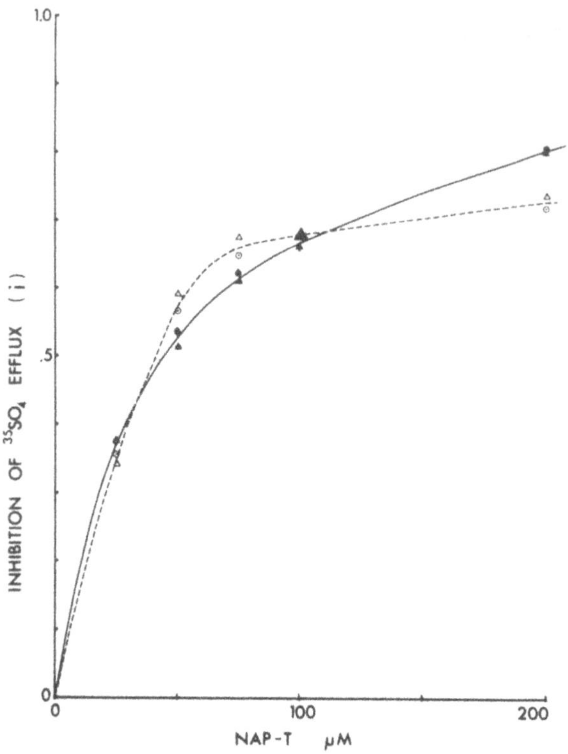

Fig. 3. Comparison of reversible inhibition by NAP-taurine in the dark and irreversible inhibition after exposure to light. Cells were loaded with ^{35}S-sulfate and fluxes measured at 27°C in medium S + 20 mM NaCl as described in Fig. 2. NAP-taurine was present during sulfate efflux in "dark" experiments (closed symbols). In "light" experiments (open symbols) cells were exposed to light and washed as described in Table 3. Circles and triangles: data from two separate experiments

The Irreversible Binding of NAP-Taurine

When NAP-taurine is present in the medium, its binding is primarily to band 3, although other minor binding components are also observed (Staros and Richards, 1974; Cabantchik et al., 1976). After pronase treatment of the cells, most of the NAP-taurine is found in the large pronase-resistant segment of 65,000 daltons (Cabantchik et al., 1976). In these respects, NAP-taurine binding resembles that of DIDS (Cabantchik and Rothstein, 1974b). For this reason, and because of the demonstrated protective effect of NAP-taurine against DIDS binding (Table 2), it is not surprising that both agents largely bind to common sites in band 3. In cells pretreated with DIDS, most of the binding of NAP-taurine to band 3 is eliminated (Cabantchik et al., 1975; Rothstein et al., 1976), and in cells pretreated with NAP-taurine, the binding of dihydro DIDS is reduced (Table 2).

In Figure 4, the distribution of NAP-taurine in SDS-polyacrylamide gels from normal cells and pronase-treated cells is illustrated. In the normal cell, most of the binding is to band 3, but in the pronase-treated cell, much of the agent is in band 3' (the 65,000 dalton segment of band 3). A small fraction is resistant to pronase and remains at the band 3 location (95,000 daltons), and a small fraction is found in a 35,000 dalton segment (also derived from band 3). In the upper panel of Figure 4, the Cl^- concentration was relatively low (20 mM) and the inhibitory effect of NAP-

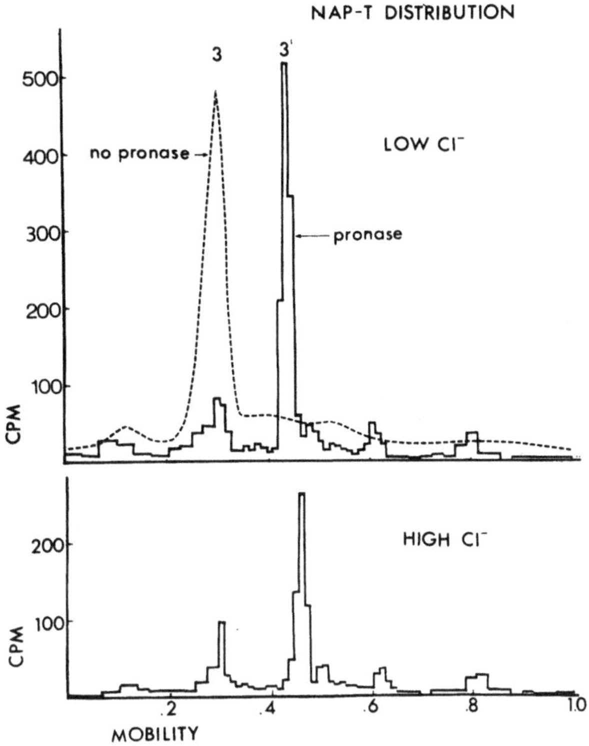

Fig. 4. Effect of Cl^- on distribution of NAP-taurine in membrane proteins of pronase-treated cells. Ghosts prepared from cells exposed to light in presence of 30 µM ^{35}S-NAP-taurine at 0^oC in S + 20 mM NaCl medium as described in Table 3 were solubilized in sodium dodecyl sulfate and electrophoresed on 5.6% polyacrylamide gels as described by Fairbanks et al., 1971. In upper panel, broken line indicates that most of NAP-taurine is bound to 95,000 dalton protein, band 3. When intact cells are treated with pronase (0.1 mg/ml for 10 min at 32^oC), the result shown in upper histogram is obtained. NAP-taurine labeling is shifted to 65,000 dalton region parallel to conversion of band 3 to band 3'. In lower panel, cells were suspended in S + 60 mM NaCl during photolysis and then treated with pronase

taurine was substantial (about 35%). In the lower panel, the Cl^- was raised to 60 mM and there was no significant inhibition. The binding to the membrane was reduced by about 50%, with the predominant effect on band 3'. These data suggest that Cl^- not only reverses the inhibition of anion transport, but that it causes a reduction of NAP-taurine binding to band 3'. They suggest that Cl^- and NAP-taurine compete for sites in band 3' that are involved in the inhibition of anion transport.

It was pointed out earlier that when the Cl^- concentration of the medium is reduced, the potential across the membrane is altered. It seems unlikely, however, that the potential could account for the effect of Cl^-, because in experiments carried out using the Nystatin technique (Dalmark, 1975) to maintain constant potentials, high Cl^- also reduced the effectiveness of NAP-taurine as an inhibitor of chloride exchange.

The sites of irreversible inhibition by NAP-taurine can be protected more effectively by HCO_3^- than by Cl^- (Table 4). In these experiments, the total diffusible anion concentrations (Cl^- + HCO_3^-) were about equal, so that the potentials were approximately the same. The greater protection by HCO_3^- must, therefore, be attributed to a greater affinity of HCO_3^- than Cl^- for the NAP binding sites.

The total binding of NAP-taurine to achieve complete inhibition of anion transport is about 1,400,000 molecules per cell (determined by extrapolation from data at lower levels of inhibition), of which about 1,100,000 are in band 3 (Cabantchik et al., 1976). The data of Figure 4 indicate that some binding occurs even when

Table 4. Effects of Cl^- and HCO_3^- on NAP-taurine inhibition of sulfate efflux

	Cl^-	HCO_3
Control	.12	.12
NAP	.058	.076
Inhibition	53	37

Inhibitions are in % and rate constants for SO_4 efflux in min^{-1}. All values are averages of 3 experiments. Cells were loaded with $^{35}SO_4$ as described in Table 1 in a Cl medium containing 20 mM HEPES, pH 7.4 at room temperature, 5 mM Na_2SO_4, 192 mM sucrose and 25 mM NaCl. The cells were then divided into 2 portions, one of which (Cl) was suspended in the Cl medium and the other of which was suspended in a similar medium in which NaCl was replaced by $NaHCO_3$. The cells in bicarbonate medium were gassed with 5% CO_2 at 0-4°C, pH 7.6. After 2 washes in these media, the cells were irradiated in presence of 30 μM NAP-taurine as described in Table 3 and then washed in Cl medium containing 0.5% albumin and incubated for 0.5 h in Cl medium, after which the flux was measured in Cl medium at 27°C

inhibition is almost zero. This binding, that is not reversed by Cl^-, cannot be concerned with inhibition. Thus the Cl^--reversible inhibition-related sites are about 50% of the total in band 3, or about 500,000 to 600,000 per cell (compared to 1,000,000 monomers of band 3 per cell (Fairbanks et al., 1971)).

Discussion

A number of chemical probes, including the disulfonic stilbenes (Cabantchik and Rothstein, 1972, 1974a), pyridoxal phosphate (Cabantchik et al., 1975), NAP-taurine (Cabantchik et al., 1976; Rothstein et al., 1976), a sulfanilic acid derivative (Ho and Guidotti, 1975), and fluorodinitrobenzene (Zaki et al., 1975) are effective inhibitors of anion transport and also interact substantially with band 3 protein. The disulfonic stilbenes are the most potent, with inhibitory concentrations in the case of DIDS being less than 10 micromolar. The effect with this probe is highly specific, with no evidence that other functions are disturbed and with up to 95% of the agent localized in band 3 (Cabantchik and Rothstein, 1974a, 1974b), or in segments derived from it by pronase treatment. The DIDS sites in band 3 are largely in common with those that react with the other probes (Cabantchik et al., 1975, 1976; Zaki et al., 1975; Rothstein et al., 1976), so it can be suggested that all of the probes may have a common site of action.

It is not absolutely certain that band 3 is the site of inhibition. Two other proteins, the sialoglycoprotein (glycophorin) and the "satellite" band (Lepke et al., 1976), may each contain a small percentage of the DIDS. Involvement of glycophorin in anion exchange is highly unlikely, however, since the DIDS bound to it can be removed by trypsin treatment without changing the anion permeability or the effect of DIDS (Cabantchik and Rothstein, 1974b). In view of limitations in the ability of present methods to detect the existence of minor DIDS-binding components, the conclusion that band 3 is the site of inhibition must be considered tentative, although highly probable.

From inhibition studies above, although it is quite possible that the probes react directly with the transport site, it can only be concluded that band 3 contains an inhibitory site. With pyridoxal phosphate (Cabantchik et al., 1975) and NAP-taurine (Cabantchik et al., 1976; Rothstein et al., 1976) on the other hand, because they behave as substrates, it is possible that they can be used to identify the transport system directly. The studies are most complete in the case of NAP-taurine, whose transport is inhibited (Fig. 1) by DIDS and dipyridamole (Cabantchik et al., 1976), agents also effective against inorganic anion transport (Deuticke, 1970; Cabantchik and Rothstein, 1974a). The conclusion that NAP-taurine and the inorganic anions utilize a common transport pathway is reinforced by the finding that the kinetics of inhibition of inorganic anion fluxes by NAP-taurine (Fig. 2) are consistent with competition for a common site. On the other hand, the kinetics are also consistent with a mixed inhibition, in which the NAP-taurine would react to a second site but thereby alter the affinity of the transport site. The reversal of NAP-taurine inhibition by inorganic anions (Table 1, Fig. 2) indicates that the NAP-taurine binding site is also a Cl^- binding site. This information does not, however, resolve the question of inhibition at the transport site versus inhibition at a second site, because Cl^- has been shown by a kinetic analysis to react not only with the transport site, but also with a second, inhibitory site (Dalmark, 1976). Thus Cl^- in high concentrations is self-inhibitory. The data of Table 4, showing that HCO_3^- is more effective than Cl^- in protecting against NAP-taurine inhibition, is consistent with direct interaction at the transport site, for which HCO_3^- has a relatively high affinity compared to Cl^- (Dalmark, 1976). HCO_3^- shows little affinity for the second site.

In view of the complex nature of the anion system, it is difficult at this time to define the interactions of NAP-taurine in an unequivocal manner. It seems likely, nevertheless, that NAP-taurine is transported in large part by the inorganic anion system and that it inhibits that system by interaction with the transport site or other closely associated sites. This would imply that the covalent probes such as DIDS also interact directly with components of the anion transport system.

Because NAP-taurine behaves as a photoaffinity probe (Table 3, Fig. 3), its irreversible binding sites, after exposure to light, directly reflect its interaction sites in the dark, including its associations with the anion transport system. NAP-taurine does not exclusively bind to band 3, or its products of hydrolysis by pronase (Fig. 4), but the largest number of sites are in the 65,000 dalton segment (band 3') derived from band 3. This component is also the site of localization of most of the DIDS (Cabantchik and Rothstein, 1974b) and protein fractions enriched in this component (90% of the total protein) can be used to reconstitute the anion transport activity (Rothstein et al., 1975). Furthermore,

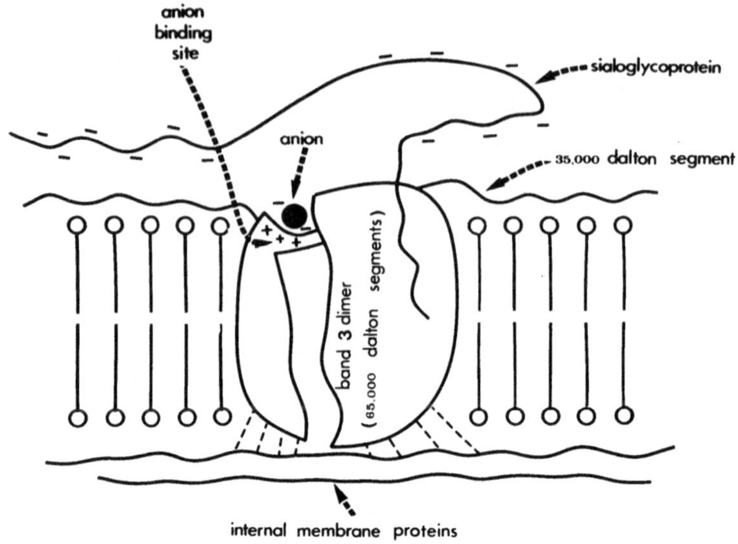

Fig. 5. Model of arrangement of band 3 in membrane with respect to anion transport. (From Rothstein et al., 1976)

when high Cl^- is used to reverse the NAP-taurine inhibition, its binding to this segment is substantially reduced (Fig. 4), so that it must contain Cl^- binding sites. It is the best candidate, at the present time, for the location of the anion transport system.

Band 3 is a reasonable candidate for the anion transport protein on structural grounds (reviewed by Rothstein et al., 1976). It is a transmembrane protein, of sufficient abundance, that might provide a largely aqueous pathway for anion transport through the lipid bilayer (Fig. 5). It has been proposed that the transport may be accomplished by an anion binding site that can exist in two conformations, one facing outward and one inward. The movement of the site from one to the other allows the exchange of anions to proceed.

Acknowledgment. This work was supported in part by grants from the Medical Research Council (Canada) MT 4665 and MA 5149.

References

Cabantchik, Z.I., Balshin, M., Breuer, W., Rothstein, A.: Pyridoxal phosphate: An anionic probe for protein amino groups exposed on the outer and inner surfaces of intact human red blood cells. J. Biol. Chem. 250, 5130-5136 (1975)

Cabantchik, Z.I., Knauf, P.A., Ostwald, T., Markus, H., Davidson, L., Breuer, W., Rothstein, A.: The interaction of an anionic photoreactive probe with the anion transport system of the human red blood cell. Biochim. Biophys. Acta 455, 526-532 (1976)

Cabantchik, Z.I., Rothstein, A.: The nature of the membrane sites controlling anion permeability of human red blood cells as determined by studies with disulfonic stilbene derivatives. J. Membrane Biol. 10, 311-330 (1972)

Cabantchik, Z.I., Rothstein, A.: Membrane proteins related to anion permeability of human red blood cells. I. Localization of disulfonic stilbene binding sites in proteins involved in permeation. J. Membrane Biol. 15, 207-226 (1974a)

Cabantchik, Z.I., Rothstein, A.: Membrane proteins related to anion permeability of human red blood cells. II. Effects of proteolytic enzymes on disulfonic stilbene sites of surface proteins. J. Membrane Biol. 15, 227-248 (1974b)

Dalmark, M.: Chloride and water distribution in human red cells. J. Physiol. 250, 65-84 (1975)

Dalmark, M.: Effects of halides and bicarbonate on chloride transport in human red blood cells. J. Gen. Physiol. 67, 223-234 (1976)

Deuticke, B.: Anion permeability of the red blood cell. Naturwissenschaften 57, 172-179 (1970)

Fairbanks, G., Steck, T.L., Wallach, D.F.H.: Electrophoretic analysis of the major polypeptides of the erythrocyte membrane. Biochemistry 10, 2606-2617 (1971)

Gunn, R.B., Dalmark, M., Tosteson, D.C., Wieth, J.O.: Characteristics of chloride transport in human red blood cells. J. Gen. Physiol. 61, 185-206 (1973)

Ho, M.K., Guidotti, G.: A membrane protein from human erythrocytes involved in anion exchange. J. Biol. Chem. 250, 675-683 (1975)

Knauf, P.A., Rothstein, A.: Chemical modification of membranes. I. Effects of sulfhydryl and amino reactive reagents on anion and cation permeability of the human red blood cell. J. Gen. Physiol. 58, 190-210 (1971)

Lepke, S., Fasold, H., Pring, M., Passow, H.: A study of the relationship between inhibition of anion exchange and binding to the red blood cell membrane of 4,4'-diisothiocyano stilbene-2,2'-disulfonic acid (DIDS) and its dihydro derivative (H_2DIDS). J. Membrane Biol. 29, 147-178 (1976)

Passow, H.: Passive ion permeability of the erythrocyte membrane. In: Progress in Biophysics. Butler, J.A.V., Noble, D. (eds.). New York: Pergamon 1969, Vol. XIX, pp. 425-467

Rothstein, A., Cabantchik, Z.I., Balshin, M., Juliano, R.: Enhancement of anion permeability in lecithin vesicles by hydrophobic proteins extracted from red blood cell membranes. Biochem. Biophys. Res. Commun. 64, 144-150 (1975)

Rothstein, A., Cabantchik, Z.I., Knauf, P.: Mechanism of anion transport in red blood cells: role of membrane proteins. Federation Proc. 35, 3-10 (1976)

Ruoho, A.E., Kiefer, H., Roeder, P.E., Singer, S.J.: The mechanism of photoaffinity labeling. Proc. Natl. Acad. Sci. U.S. 70, 2567-2571 (1973)

Sachs, J.R., Knauf, P.A., Dunham, P.B.: Transport through red cell membranes. In: The Red Blood Cell, 2nd ed. Surgenor, D.M. (ed.). New York: Academic Press 1974, Vol. II, pp. 613-705

Staros, J.W., Richards, F.M.: Photochemical labelling of the surface proteins of human erythrocytes. Biochemistry 13, 2720-2726 (1974)

Staros, J.W., Richards, F.M., Haley, B.E.: Photochemical labelling of the cytoplasmic surface of the membranes of intact human erythrocytes. J. Biol. Chem. 250, 8178 (1975)

Webb, J.: Enzyme and Metabolic Inhibitors. New York: Academic Press 1963, Vol. I, pp. 149-164

Zaki, L., Fasold, H., Schuhmann, B., Passow, H.: Chemical modification of membrane proteins in relation to inhibition of anion exchange in human red blood cells. J. Cell Physiol. 86, 471-494 (1975)

Structural and Functional Properties of the Anion Transport System Isolated from Human Erythrocyte Membranes

Z. I. CABANTCHIK, J. M. WOLOSIN, H. GINSBURG, and O. ZEMEL

The red blood cell membrane is endowed with a high capacity system for the specific and rapid exchange of anions (Sachs et al., 1975; Rothstein et al., 1976). This system is involved in the cellular-for-extracellular Cl^--HCO_3^- exchange, a property which is efficiently carried out while blood cells flow through lung or tissue capillaries. From electrical as well as from kinetic measurements it was recently recognized that a predominant component of Cl^- exchange is electrically silent and is indicative of a mechanism which displays an obligatory one-for-one anion exchange (Harris and Pressman, 1967; Hunter, 1971; Sachs et al., 1975). Conversely, for anions such as sulfate, it is thought that the predominant path of exchange permeation is probably conductive in nature (Knauf et al., 1977; Passow, 1976) and thus sensitive to the membrane potential. However, the exact relationship between the exchange and the conductive pathways and the possibility that both pathways may share a common underlying mechanism remains to be firmly established.

In general, most attempts to identify transport components involve serious difficulties, since they can be characterized only after membrane disruption, solubilization and fractionation. Unless the isolated membrane components display some features identifiable with the transport function, no progress can be made. In transport systems where no obvious built-in markers such as an enzyme activity are present, the identity of the functional membrane component has to be revealed by highly specific probes (i.e. inhibitors) which can also serve to label the system. In the case of anion transport, the solution of some aspects of the problem has been expedited by the use of chemical modifiers which are themselves anions, some of which are transported as such, so that they are expected to interact directly with components of the transport system (Knauf and Rothstein, 1971; Cabantchik and Rothstein, 1972, 1974a; Staros and Richards, 1974; Cabantchik et al., 1975, 1976; Rothstein et al., 1976). These compounds (or closely related analogs) were originally designed to probe the outer surface of the membrane (Fig. 1) but most of them, especially the anionic ones, have been found to be specific and potent inhibitors of the anion transport system as well, and instrumental in studying its chemical properties.

A. Nature of the Inhibitory Site and its Location in Membrane Components

The chemical architecture of the anion recognition site has been explored primarily with a series of structurally related analogs of disulfonic stilbene (Fig. 2) which are demonstrably impermeable agents. All of these analogs, whether or not capable of covalent reaction, inhibit anion transport in red blood cells. However, the

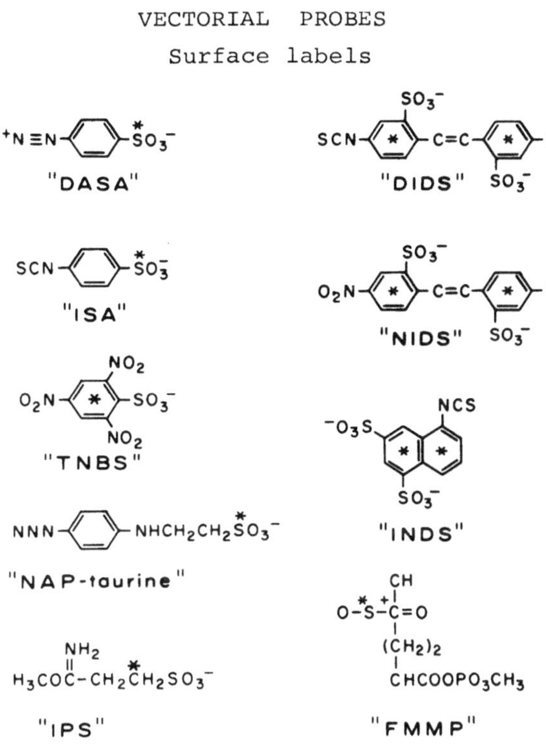

Fig. 1. Chemical structure of various surface labels. DIDS: 4,4'-diisothiocyano-2,2'-stilbene disulfonate; NIDS: 4-nitro-4'-isothicyano-2,2'-stilbene disulfonate; NAP-taurine: N-(4-azido-2-nitrophenyl)-2-aminoethylsulfonate; ISA: p-isothiocyanobenzene-sulfonate; DASA: p-diazobenzenesulfonate; FMMP: formylmethionylsulfone methylphosphate; INDS: 8-isothiocyano-naphthalene-2,4-disulfonate; IPS: imido-methoxypropyl-sulfonate (i.e. isothionyl acetiimidate); Asterisk: commonly used radioactive label

concentrations required for 50% of maximal effect vary over a 5000-fold range (Cabantchik and Rothstein, 1972). From the dose-response curves and from evaluation of their binding modes (Figs 2 and 3) it was proposed (Cabantchik and Rothstein, 1972; Rothstein et al., 1976) that the binding of these and other anionic probes occurs as schematically portrayed in Figure 4. The initial binding is electrostatic in nature and occurs within a microenvironment composed of positively charged groups (i.e. amino and probably guanidino), admirably suited to binding of anions. An adjacent hydrophobic center accounts for the increased effectiveness of certain analogs (Fig. 2). The presence of a covalent binding moiety reactive towards electrophilic groups (i.e. -NCS moiety) allows subsequent irreversible fixation, preferentially at the anionic sites to which the probe was originally bound. This was most clearly demonstrated with the photoreactive probe NAP-taurine (Fig. 1), which interacts reversibly with the site in the dark, but is fixed irreversibly upon in situ photoactivation (Cabantchik et al., 1976). Both the reversible and irreversible binding are competitively reduced

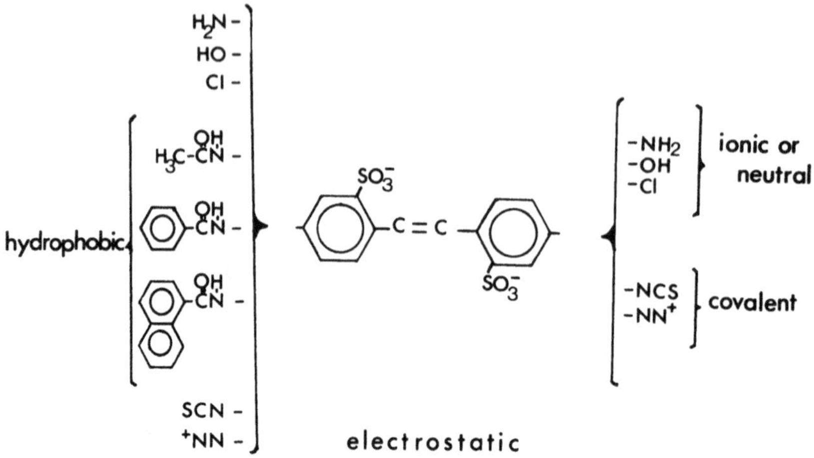

Fig. 2. Structures of several mono- and bifunctional derivatives of disulfonic stilbene. Possible contribution of different substituents on 4,4' positions of disulfonic stilbene core to binding modes of probes. Effects on anion transport are presented in Figure 3

by anions such as CL^- or $SO_4^=$, indicating that the photoaffinity probe interacts directly with the anion transport system (Cabantchik et al., 1976; Rothstein et al., this volume).

With radioactively labeled covalent reagents (Fig. 1) it has been possible to determine which membrane components contain the inhibitory site (Cabantchik and Rothstein, 1972, 1974a; Cabantchik et al., 1975, 1976; Zaki et al., 1976). The most specific of these reagents, DIDS, blocks the function almost completely at concentrations where it is over 95% associated with polypeptides of apparent molecular weight 95,000 daltons (Fig. 5) (Cabantchik and Rothstein, 1974a). These polypeptides are known also as band 3, from its position in SDS-polyacrylamide gel electrophoretograms (Fairbanks et al., 1971). Other agents that inhibit anion transport and have their effect localized in band 3 include (see Fig. 1): DASA (Bender et al., 1971; Rothstein et al., 1976), ISA (Ho and Guidotti, 1975), TNBS (Arrotti and Garvin, 1975; Rothstein et al., 1976), PDP-$NaBH_4$ (Cabantchik et al., 1975), NAP-taurine (Rothstein et al., 1977, this volume; Zaki et al., 1976), and FDNB (Zaki et al., 1976). The binding sites for the latter three agents in band 3 seem to be largely the same as those for the stilbenes, inasmuch as their interactions with polypeptides of band 3 are specifically reduced in cells pretreated with stilbene analogs (Cabantchik et al., 1975; Cabantchik et al., 1976; Zaki et al., 1976). After a mild pronase treatment which does not affect anion fluxes in intact cells, the polypeptides of band 3 are split into two parts containing the respective segments of 65,000 and 35,000 daltons molecular weight (Fig. 5). Most of the stilbene, however, remains associated with the 65,000 dalton segment (Cabantchik and Rothstein, 1974b). Moreover, the 65,000 dalton segment shows a direct relationship between DIDS binding and inhibition of anion transport (Cabantchik and Rothstein, 1974b). On this basis it has been suggested that the pronase-resistant (in the intact cell) 65,000 dalton segment is the probable location of the stilbene inhibitory site (Cabantchik and Rothstein, 1974b; Rothstein et al., 1976). This segment is also the predominant site of binding of

COMPOUND	STRUCTURE	BINDING	INHIBITORY EFFECT
4,4'-DIAMINO- -AS	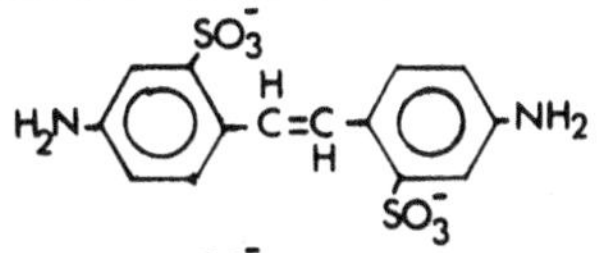	ELECTROSTATIC	(+) READILY REVERSIBLE
4-AMINO-4'- -BENZOAMIDO-AS	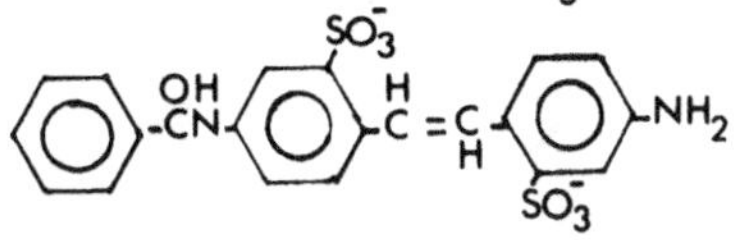	ELECTROSTATIC HYDROPHOBIC	(++) REVERSIBLE
4,4'-DIISOTHIOCYANO- -AS	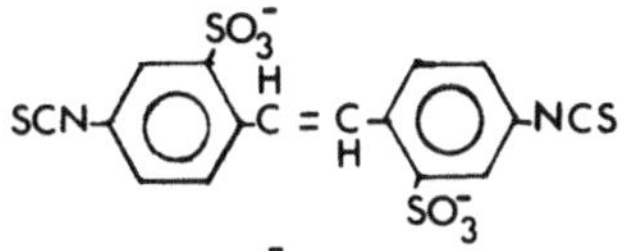	ELECTROSTATIC COVALENT	(++++) IRREVERSIBLE
4-ISOTHIOCYANO- -4'-BENZOAMIDO- -AS	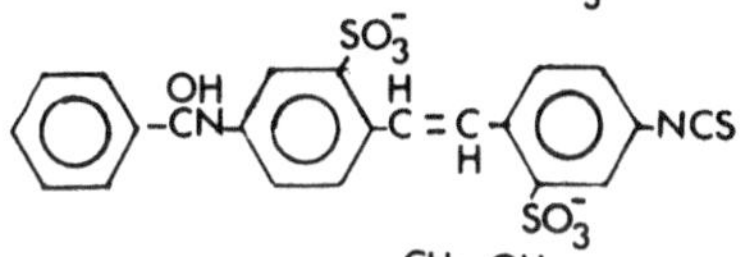	ELECTROSTATIC HYDROPHOBIC COVALENT	(+++) PARTLY REVERSIBLE
PYRIDOXAL-5-PHOSPHATE	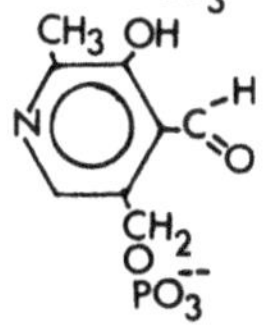	ELECTROSTATIC COVALENT	(+++) PARTLY REVERSIBLE

AS: 2,2'-STILBENE DISULFONATE

Fig. 3. Interactions between stilbene disulfonates and anion permeation system. Relative effects of the various probes on anion transport, binding affinities to cell membrane and their reversibility properties are compared. Data based primarily on results shown in Cabantchik and Rothstein, 1972

pyridoxal phosphate (Cabantchik et al., 1975) and of NAP-taurine (Cabantchik et al., 1976; Zaki et al., 1976) in inhibitory concentrations of reagents. It also constitutes the hydrophobic moiety of the 95,000 dalton polypeptide which is embedded in the lipid matrix of the membrane.

Concurrently with the incision of band 3, the carbohydrate-containing portions of the sialoglycoprotein are excised by proteolytic treatment, as visualized in Figure 5 (right).

B. Isolation and Reconstitution of the Anion Transport System

The studies discussed above suggest that the irreversible inhibitors of anion transport elicit their effect by interacting with the 65,000 dalton, pronase-resistant segment of the 95,000 dalton polypeptides. This has been corroborated by other attempts to identify the putative functional component of transport, all of which have led to the suggestion that the 95,000 dalton protein,

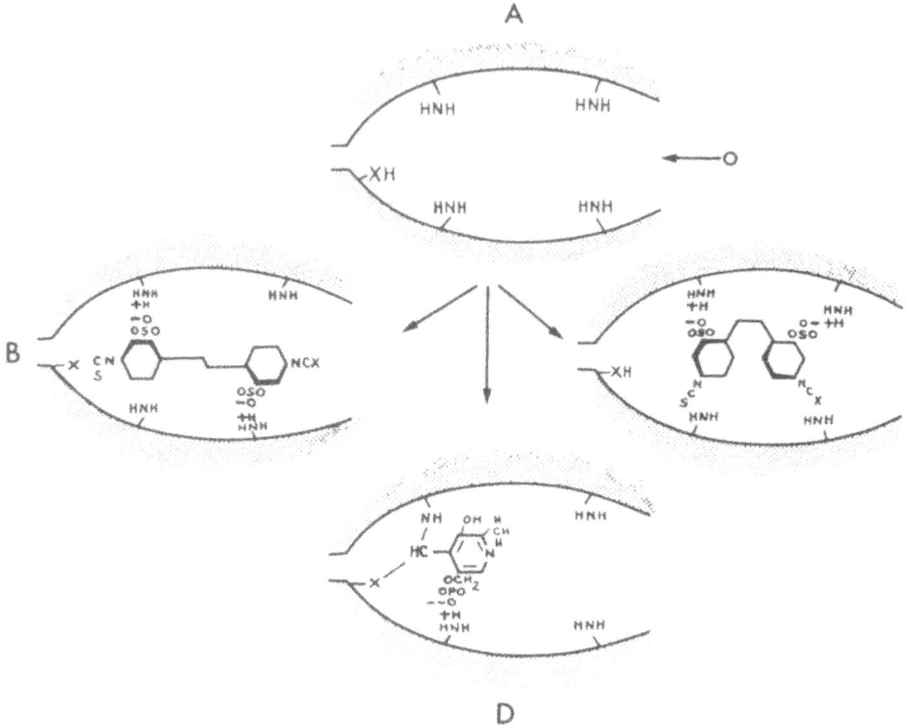

Fig. 4. Schematic representation of an idealized "inhibitory site". Site A is composed of a cluster of positively ionizable groupds (amino and/or guanidino) and a nucleophilic group X. Stilbene disulfonates interact with the system in a reversible fashion (B,C) but can also react with X or NH_2 if an NCS group is present in ring (Cabantchik and Rothstein, 1972, 1974a). Another probe, PDP (Fig. 1) can form aldimino bonds (D) or a Schiff base which can be subsequently reduced by $NaBH_4$ (Cabantchik et al., 1975)

the major polypeptide component of the red blood cell membrane, is an essential component of the anion permeation system. However, direct evidence for the functional involvement of these polypeptides has only recently been presented (Rothstein et al., 1975). Both the original 95,000 dalton polypeptide and its derivatized 65,000 dalton segment were extracted from erythrocyte membranes, with some specificity, by the nonionic detergent Triton X-100 (Fig. 5). In either case, the extracted proteins, when reconstituted in lecithin vesicles, enhanced the anion permeability (Fig.6) (Rothstein et al., 1975). Conversely, the same polypeptides from cells pretreated with DIDS, when extracted and reconstituted in parallel, did not enhance the anion fluxes. Addition of DIDS to reconstituted vesicles inhibited the function only partially, suggesting that not all the potentially reactive sites were available (i.e. accessible) to the chemical modifier in these vesicles (Fig. 6). Persantine, a noncovalent inhibitor of anion fluxes, also inhibited the function (Rothstein et al., 1975). Evidently all these effects could not be unequivocally assigned to the polypeptides derived from band 3, because the detergent had also extracted a fraction of sialoglycoprotein and traces of other minor proteins as well (Fig. 5). However, in the case of pronase-treated cells, at least 85% of the protein and 98% of the DIDS in the

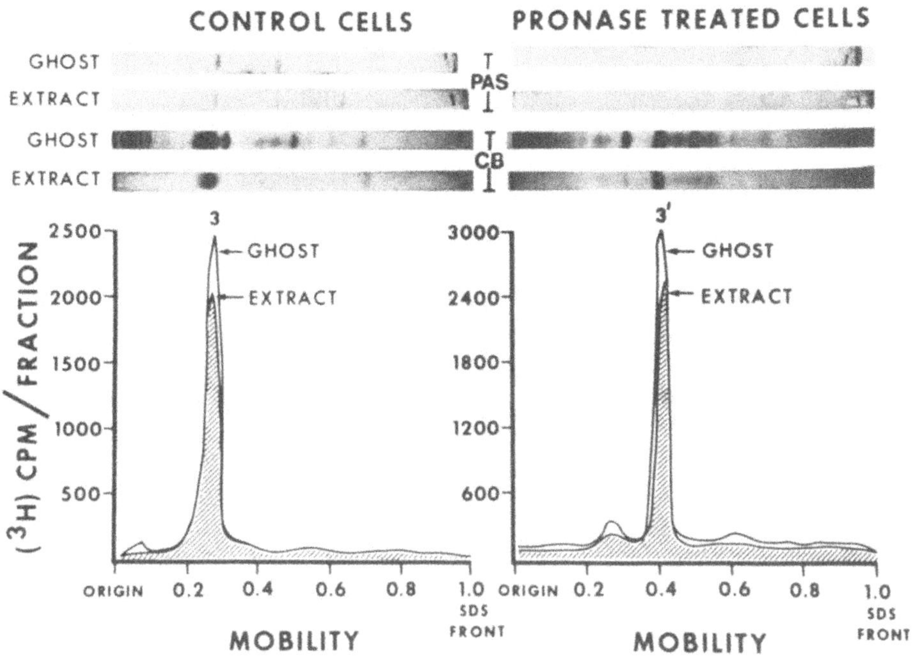

Fig. 5. Extraction of (^{3}H) DIDS-labeled proteins from normal and pronase treated cells by Triton X-100. Upper sections: photographs of sodium dodecylsulfate polyacrylamide gels of proteins derived from ghosts or from Triton extracts of ghosts. The two upper gels were stained for carbohydrate with periodic acid-Schiff base technique (PAS) and the two lower gels were stained for proteins with Coomassie Blue (CB). Lower sections: radioactivity profiles from parallel gels. In control cells most of the (^{3}H) DIDS is located at position of band 3 (95,000 mol wt) and in pronase treated cells at position of band 3' (65,000 mol wt). Reproduced with permission from Rothstein et al., 1975

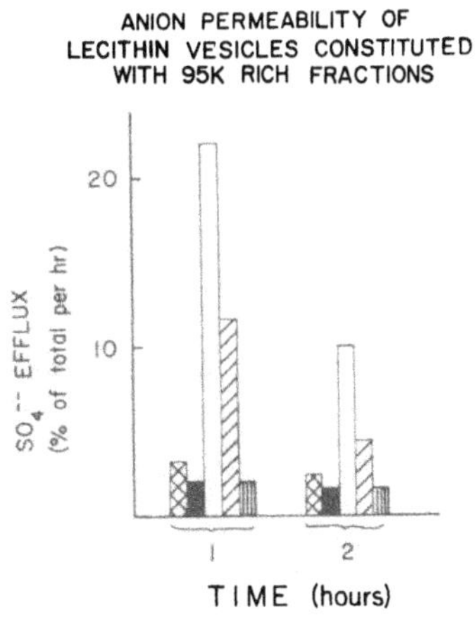

Fig. 6. Anion exchange fluxes in lecithin vesicles containing Triton-extracted fractions. Extracts enriched in 95,000 dalton polypeptides (Fig. 5) were incorporated in lecithin vesicles by procedures described by Rothstein et al., 1975. Sulfate fluxes were measured in parallel with protein from normal ▭ or DIDS-treated cells ▥. In ▨, the protein-reconstituted vesicles were treated with DIDS. In ▩ and ■ sulfate fluxes were measured in lecithin vesicles with or without Triton X-100 at same levels as those reconstituted with protein fractions. Data adapted from Rothstein et al., 1975

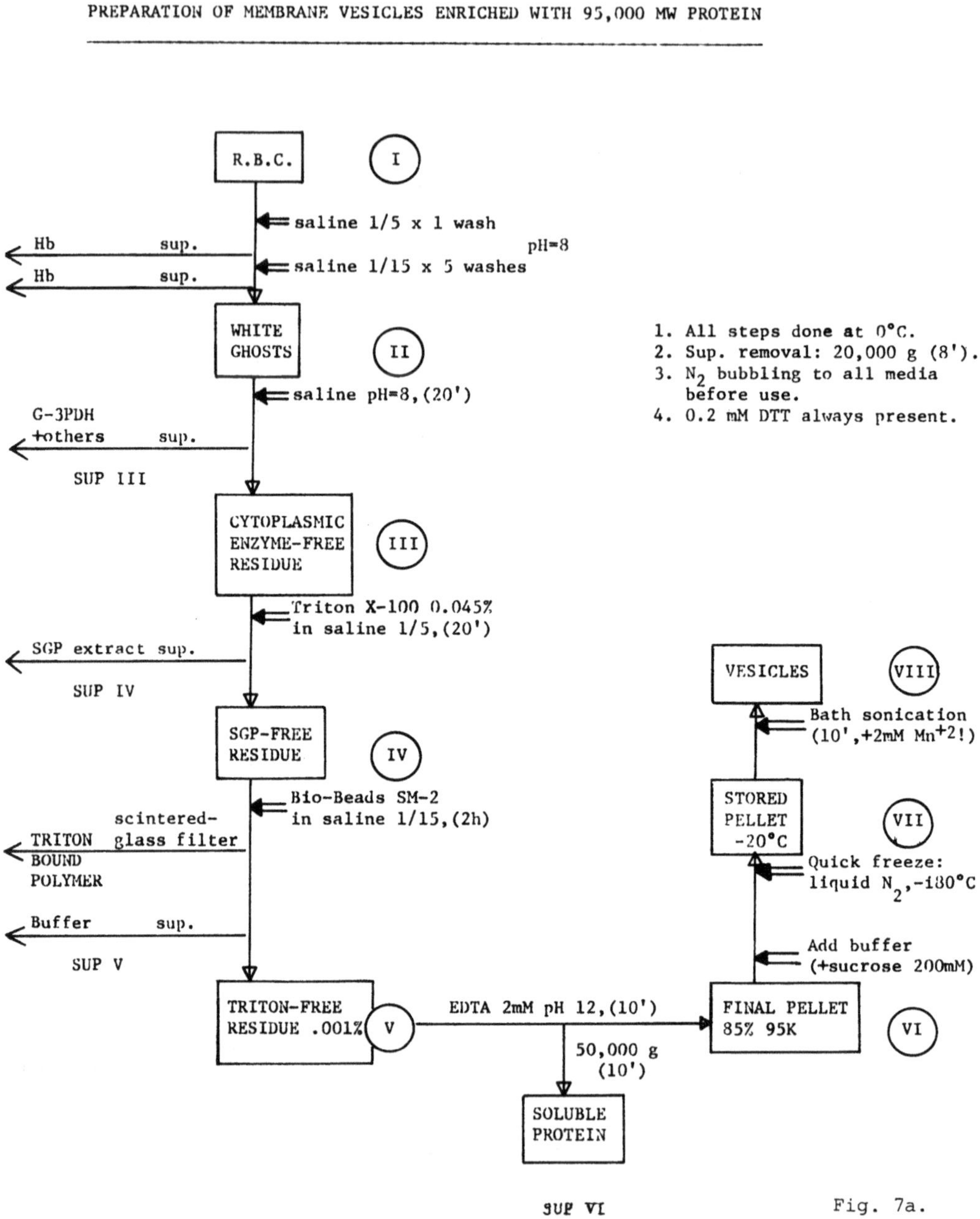

Fig. 7a.

extract could be accounted for by the 65,000 dalton segment. The possible contribution of the sialoglycoproteins to the aforementioned effects can be eliminated not only on the basis of the results of DIDS labeling in conjunction with proteolytic treatments, but also in view of the fact that human red cells lacking the sialoglycoproteins have normal anion permeability properties which are susceptible to DIDS (Tanner et al., 1976).

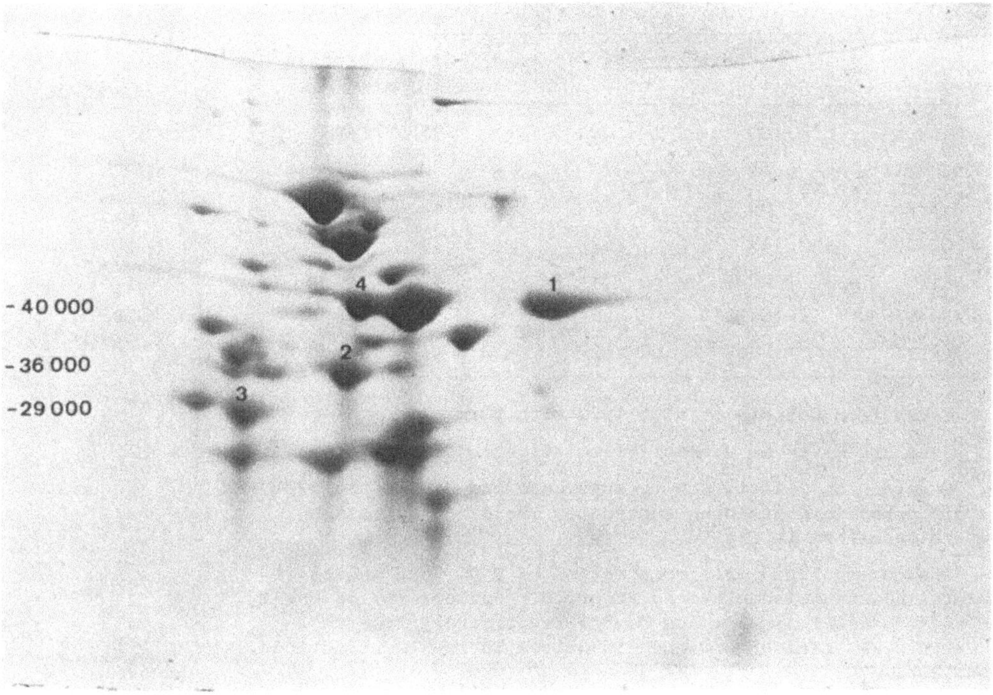

b

Fig. 7. Procedure for isolation of membrane vesicles enriched in 95,000 dalton polypeptides. All steps were carried out at 0°C. Solutions were bubbled through with N_2 before use

Thus, in order to resolve the identity of the putative functional components of anion transport, a procedure had to be devised to isolate a highly purified system in a state where all anion transport properties found in intact cells could also be preserved (Wolosin et al., 1977). For the latter, an improved methodology to measure anion fluxes in membrane vesicles accurately had to be developed as well (Wolosin et al., 1977).

I. Chemical and Structural Characterization of Isolated Membranes and Vesicles

A more selective preparation of the 95,000 dalton polypeptides from erythrocyte ghosts was achieved by combining the use of mild detergent treatment to deplete selectively the membranes of sialoglycoproteins and their subsequent exposure to alkaline EDTA to remove all but the very intrinsic components of the membrane (Wolosin et al., 1977). All these steps were so devised that minimal damage would be caused to the functional entities as a result of the fact that the polypeptides remain in their original lipid environment throughout all the extraction steps. The procedure is schematized in Figure 7 and the results are depicted in Table 1 and Figure 8. It is shown that the 95,000 dalton polypeptides could be obtained reasonably pure (more than 90%) and in association with lipids which have a composition identical to that of the native membrane.

Table 1. Chemical composition of membrane fractions[a]

Step[c]	Protein[b] (%)	Sialic acid[b] (%)	Sialic/protein (mM/mg)
Ghosts-II	100	100	95
SUP-III	11	0	0
P-III	89	100	112
SGP-IV	19	62	320
P-IV	63	31	47
SUP-VI	55.5	29	52
P-VII[d]	7.5	2[e]	31

[a] Taken from Wolosin et al., 1977 with permission

[b] Data are given as yields (%) relative to Ghosts-II

[c] P refers to pellet, SUP to supernate and SGP to sialoglycoprotein-containing supernate. The different steps are described in Fig. 8

[d] The phospholipid (PL) composition of P-VII and Ghosts-II are indistinguishable; the PL/protein ratio (w/w) of P-VII and Ghosts-II are 2/1 and 0.65/1 respectively; the cholesterol/PL ratios (w/w) are identical in the two fractions, 0.25/1

[e] Glycolipids

These components are organized in membranous structures resembling those of intact cell membranes. This is based primarily on freeze fracture electron microscopy observation in which the aforementioned polypeptides are shown to be associated with intralipid particles (Fig. 9, lower level). Upon mild sonication the membranes yield vesicles which are apparently homogeneous in a linear Dextran gradient showing a specific gravity of 1.030. These vesicles show radii of 0.1-0.15 microns and are apparently impermeable to phosphotungstic acid, as revealed by negative staining electron microscopy (Fig. 9, upper level). The vesicles were made tightly closed to small ions by addition of divalent ions such as Mn^{2+} (Wolosin et al., 1977).

II. Functional Characterization of Vesicles

Most flux measurements were performed under equilibrium exchange conditions in media where normally sulfate and occasionally phosphate were the only permeable anions. The methodological aspects are schematized in Figure 10. The time course of sulfate egress in vesicles loaded with $^{35}SO_4^{=}$ is depicted in Figure 11. In the particular experimental conditions (25°C) the half-time ($t_{1/2}$) of efflux and influx (not shown) were virtually identical (3.3 and 3.4 min, respectively). At 0°C the corresponding $t_{1/2}$ of efflux was not less than 2.5 h, a property used to stop fluxes by quick cooling of small aliquots to 0°C as shown in Figure 12. In order to evaluate to what extent the transport features of the isolated

Fig. 8. Sodium dodecylsulfate-polyacrylamide gel electrophoresis of membrane fractions (Table 1). Membrane fractions P-II, S-VI (=SUP-VI) and P-VI were solubilized, electrophoresed (3% acrylamide) and stained with Coomassie Blue according to Wolosin et al., 1977. S: supernate, P: pellet. Amount of protein loaded on gels was 70 µg for Ghosts-II and 30 µg for other fractions. Nomenclature of bands taken from Fairbanks et al., 1971

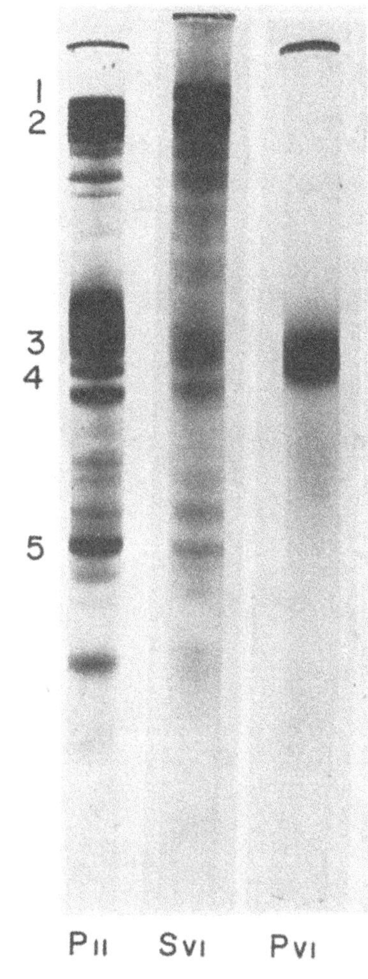

system resemble those displayed by the native system, six characteristic properties of anion transport were examined.

1. Anion exchange in vesicles shows an unusually high temperature dependence (Fig. 12), similar to that observed in intact cells. The activation energy obtained in vesicles (26 kcal/mol) is comparable to that obtained in intact cells (28-33 kcal/mol) (Schnell, 1972; Schwoch et al., 1974).

2. The pH profile of sulfate exchange in vesicles (Fig. 13) shows the general trend found in intact cells (Passow, 1976). However, this profile is less pronounced in vesicles, indicating probably that a large component of the fluxes could be attributed to the less pH-dependent flux found under conditions of net sulfate transport (Passow, 1976).

3. The isolated system does not only transport divalent anions as sulfate, but monovalent anions as well. Since the experimental setup was not amenable to measuring Cl^- fluxes, the monovalent phos-

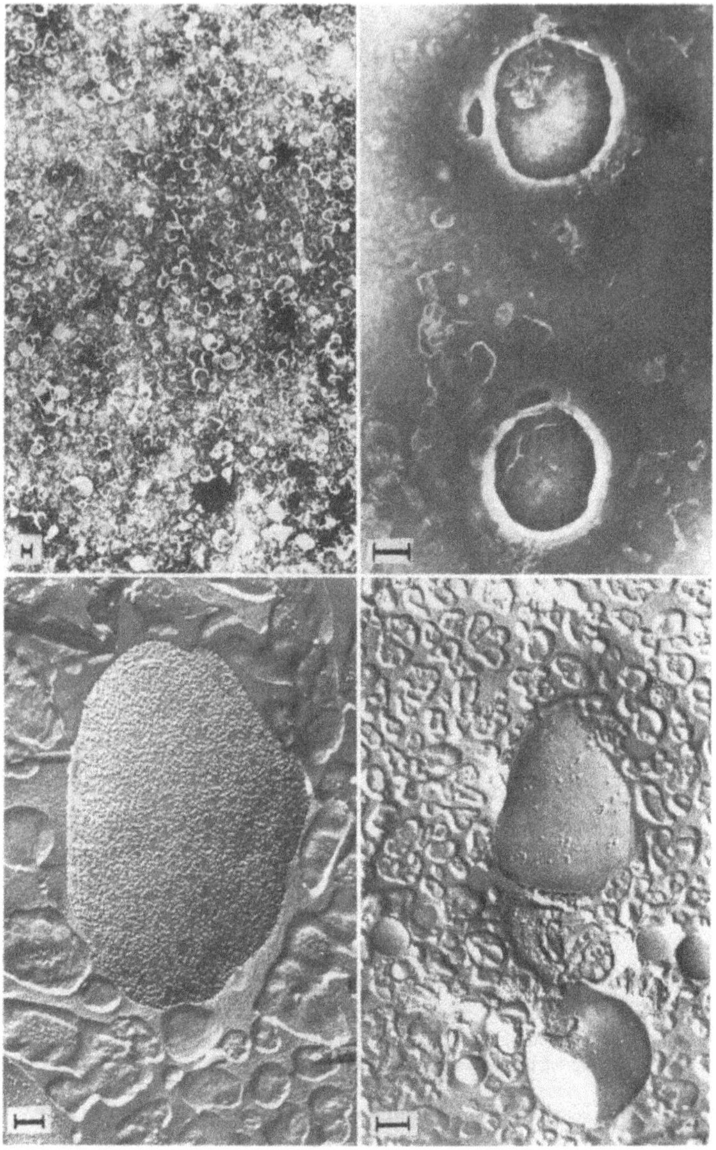

Fig. 9. Structural features of isolated vesicles. Upper figures: negative staining pictures of isolated vesicles. Bars: 0.1 μm length. Lower figures: freeze fracture replicas obtained from Ghosts-II (left) and from pellet of fractions VI (right). Bars: 0.1 μm length. For details see Wolosin et al., 1977

phate was chosen as representative of the class (Fig. 14). The specificity ratio $k_{P_I^-}/k_{SO_4^=}$ (k = rate constant) obtained in vesicles is comparable with that calculated for the intact system (4.7 and 3.5 respectively) in the particular experimental conditions of Fig. 13 and those of Ho and Guidotti, 1975. We conclude therefore that the selectivity properties of the intact system are largely retained in the isolated membrane vesicles.

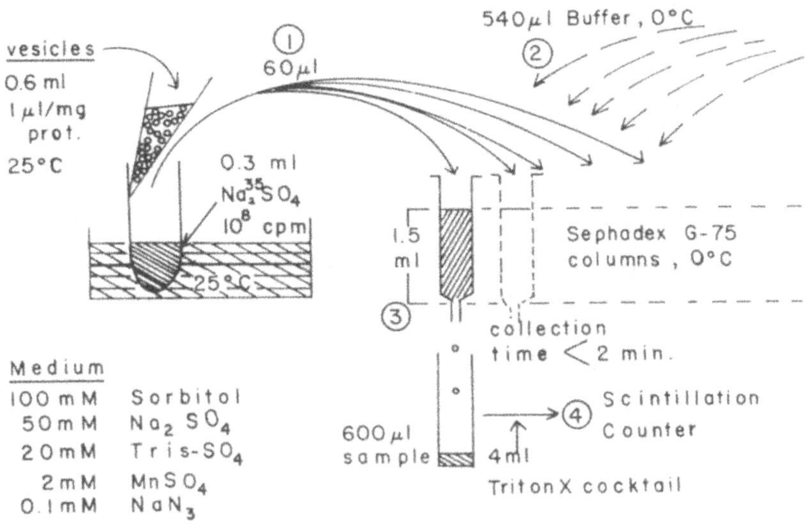

Fig. 10. Schematic representation of methodology of flux measurements in isolated vesicles. Method shown is for influx studies. For efflux studies, a similar procedure was used except that sonicated membrane vesicles were first loaded for 20 min at 30°C with solution containing $^{35}SO_4^=$ and extravesicular $^{35}SO_4^=$ removed by Sephadex G-75 gel filtration at 0°C. For details see Wolosin et al., 1977

4. Transport of monovalent and divalent anions in vesicles is susceptible to the inhibitory effect of certain reagents as in intact cells. The covalent reactive agent DIDS and the noncovalent probe persantine are typical examples of such compounds (Figs. 14 and 15 for the former and Fig. 16 for the latter) (Cabantchik and Rothstein, 1972; Schnell, 1972). Similar inhibitions could be observed in vesicles obtained from DIDS-treated cells (Fig. 15). All these observations suggest that the integrity of the functional component in terms of susceptibility to specific probes has largely been preserved in the isolated vesicles. This may also imply that the actual disposition of the polypeptides in the membrane matrix may have retained its "sidedness" inasmuch as the impenetrable DIDS has access to all the functional sites. These sites are known to be asymmetrically disposed in the intact cell, virtually inaccessible to stilbene disulfonates acting from the interior of the cell (Zaki et al., 1976).

Evidently these results differ from those obtained with the partially reconstituted system where lipids are externally added and the disposition of the polypeptides either randomized or internalized upon sonication (Fig. 7). In this procedure not more than 40% inhibition could be obtained with externally added DIDS.

5. The anion transport properties of the vesicles were also studied under net transfer conditions (zero-trans efflux) by replacing sulfate on the trans side of the membrane (exterior) with the impermeable anion TES (N-tris (hydroxymethyl) methyl-2-aminoethane

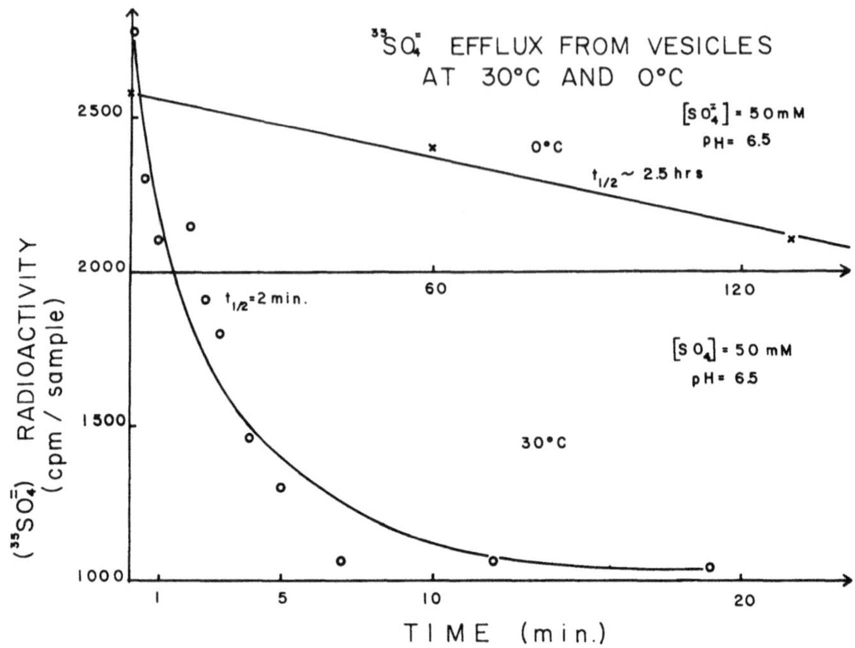

Fig. 11. Equilibrium exchange efflux of $^{35}SO_4^=$ in isolated vesicles. Vesicles were incubated in 50 mM $SO_4^=$, 78 mM Na^+, 20 mM Tris, 2 mM Mn^{2+}, 100 mM sorbitol and 0.1 mM N_3^-. Egress of $^{35}SO_4^=$ was measured as a function of time at 25°C as detailed. Efflux also measured at 0°C. Extremely low rate constant obtained at 0°C constitutes basis for stopping technique employed in present work (Fig. 10)

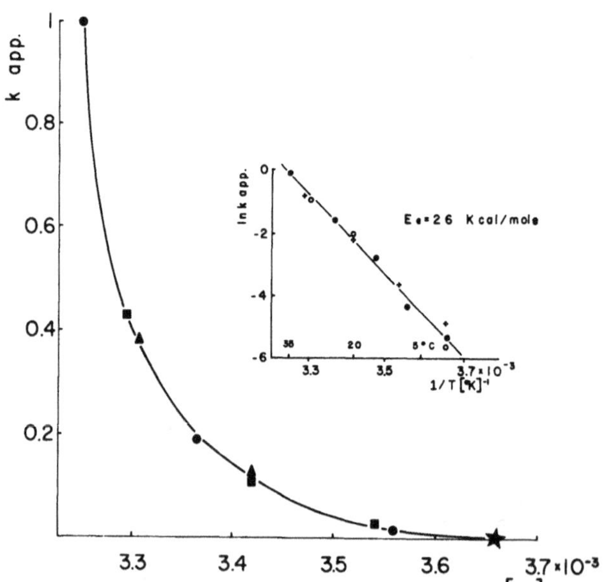

Fig. 12. Temperature dependence of $^{35}SO_4^=$ exchange efflux in vesicles. Vesicles loaded with $^{35}SO_4^=$ and free of extracellular medium were brought to desired temperature by addition of medium (as in Fig. 11) and sampled at different time intervals. Logarithm of rate constant ($k_{app} = \ln 2/t_{1/2}$) obtained for every temperature is plotted versus reciprocal of absolute temperature (inset). From slope of straight line an energy of activation of 26 kcal/mol is calculated. Different symbols represent 3 independent experiments

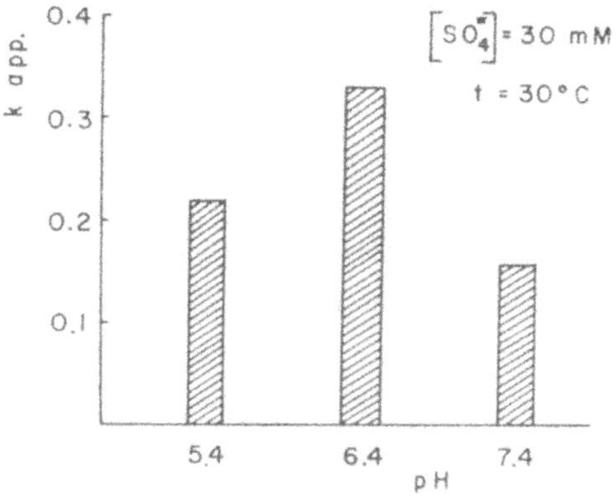

Fig. 13. pH dependence of $^{35}SO_4^=$ exchange in vesicles. Prior to sonication, vesicles were suspended in a medium of the following composition: 100 mM sorbitol, 60 mM Na^+, 2 mM Mn^+, 30 mM Tris base, 0.1 mM N_3^-, 12 mM $SO_4^=$, and citrate up to desired pH. Egress of $^{35}SO_4^=$ was measured as detailed in Fig. 11. Fractional radioactivity remaining in vesicles is plotted against time. Maximal exchange rate observed at pH 6.4

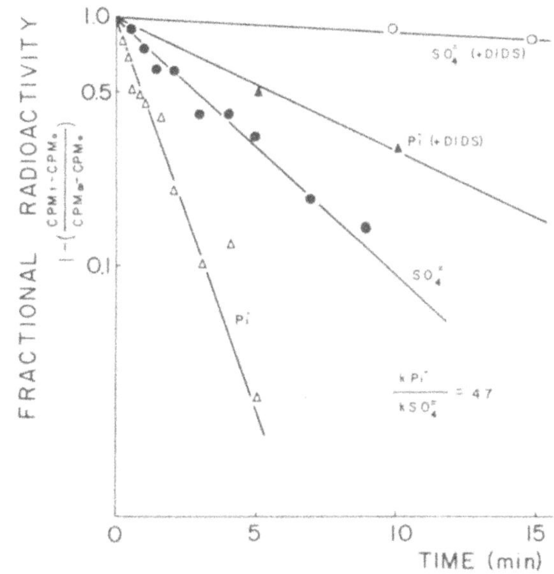

Fig. 14. Uptake of $^{35}SO_i^=$ and P_i^- in isolated vesicles. Influx of either anion in untreated or DIDS-treated vesicles was measured as described in Fig. 10. For P_i^- the uptake medium consisted of 100 mM Na phosphate, 100 mM sorbitol, 2 mM $MnSO_4$ and 0.1 mM NaN_3. For sulfate the uptake medium consisted of 50 mM $SO_4^=$, 78 mM Na^+, 20 mM Tris, 2 mM Mn^{2+}, 100 mM sorbitol and 0.1 mM NaN_3. Uptake under equilibrium exchange conditions is given as [1-fractional radioactivity] appearing in vesicles as function of time

sulfonic acid), used here also for its buffering capacity. The cation present in the interior was K^+. Under these conditions, the K^+ flux is demonstrably rate-limiting, so no detectable efflux of $^{35}SO_4^=$ was observed (Fig. 17). This and other observations prove not only that the vesicles are tightly sealed but also that in the absence of an exchangeable anion no net salt movement occurs, as in intact cells. However, if K^+ permeability is markedly increased by saturating concentrations of valinomycin, the resulting sulfate efflux reaches values virtually indistinguishable from those obtained under equilibrium exchange conditions. DIDS inhibits the zero-trans and the exchange fluxes of sulfate to approximately the same extent (Fig. 17). Since all these properties were also observed in intact cells (Knauf et al., 1977), it is strongly suggested that some mechanistic features of the native system are preserved in the isolated vesicles. The observation that in vesicles sulfate exchange fluxes are commensurate with net fluxes could have some implications for understanding the pH

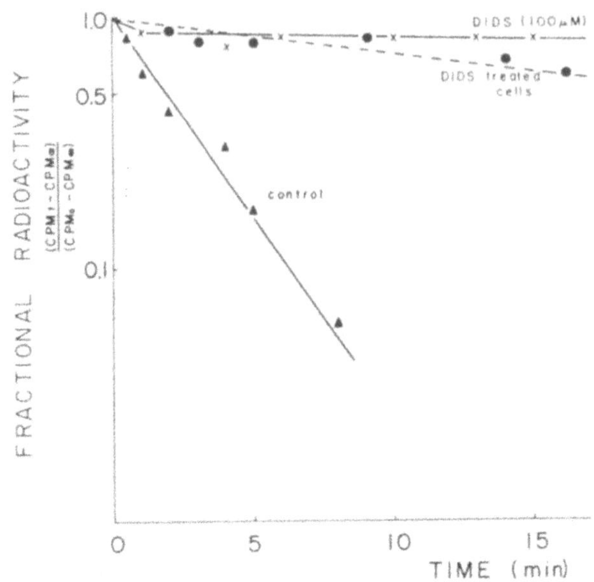

Fig. 15. DIDS inhibition of $^{35}SO_4^=$ efflux in isolated vesicles. ▲: Control vesicles, as in Fig. 11 but at 30°C. X: Vesicles loaded with $^{35}SO_4^=$ were treated with 0.1 mM DIDS at 0°C for 20 min. Extravesicular ^{35}S and DIDS were removed by gel filtration and fluxes measured as in control. ●: Vesicles derived from DIDS treated cells were equilibrated with $^{35}SO_4^=$ overnight at 5°C and made free of extracellular label as described above. Fluxes measured as in control

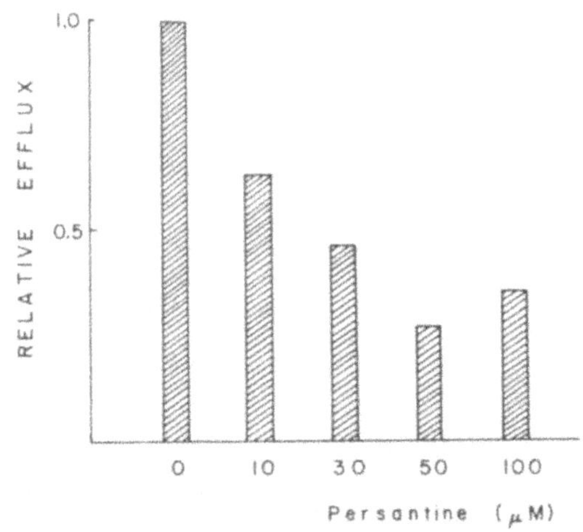

Fig. 16. $^{35}SO_4^=$ efflux in vesicles treated with various concentrations of persantine. Vesicles equilibrated with $^{35}SO_4^=$ were loaded on a gel column containing shown concentration of persantine at 0°C, and made free of extracellular radioactivity. Fluxes were measured as in control but in presence of appropriate concentration of persantine in efflux medium

effect on anion transport. In intact cells the pH profile of sulfate exchange fluxes differs considerably from that obtained for net fluxes (Passow, 1976). In fact, the low pH dependence obtained with vesicles under equilibrium exchange conditions (Fig. 13) resembles somewhat the pH profile of net flux rather than the exchange fluxes of intact cells. It is therefore conceivable that the relative contribution of the conductive (i.e. net) path to the overall flux which is observed under exchange conditions is larger in vesicles than in intact cells (Knauf et al., 1977; Passow, 1976). Alternatively, or additionally, some subtle structural alterations

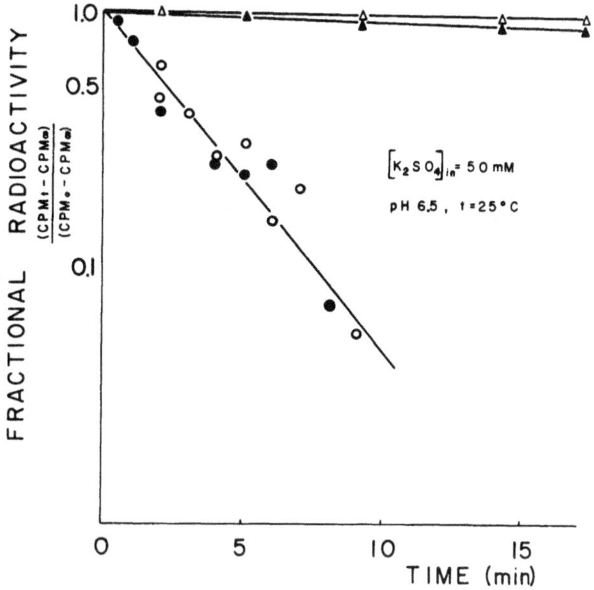

Fig. 17. Net and exchange fluxes of $^{35}SO_4^=$ in isolated vesicles. Vesicles were equilibrated in a medium containing 30 mM $K_2\ ^{35}SO_4$, 53 mM Na-TES, 133 mM sorbitol, 2mM Mn citrate and 0.1 mM NaN_3. Extravesicular radioactivity removed by gel filtration of Sephadex G-75 columns preequilibrated with appropriate efflux medium as indicated below. Tonicity of medium kept at 284 mOsm by addition of Na-TES and sorbitol. Efflux performed into the following media: X: no sulfate; ●; no sulfate, 1μM valinomycin; o: 30 mM Na_2SO_4; ▲: no sulfate, 1 μM valinomycin, but vesicles were reacted with DIDS (0.1 mM) as in Figure 15

of the polypeptides caused by the various factors to which the system was subjected could have contributed to functional alterations as well.

6. The most critical test for evaluating the functional resemblance of the isolated and the native system is based on a direct comparison of the rate constants of exchange (k_v and k_g, where v is for vesicles and g for ghosts). Since these values are directly affected by the surface/volume ratio (A/V) of the systems, it was expected that the quotient $Q = k_v/k_g$ would be equivalent to the ratio $R = ((A/V)_v/(A/V)_g)$. This expectation relies on the assumption that the number of transport sites per unit surface area is the same in vesicles and in ghosts (Wolosin et al., 1977). For Q we have calculated a value of 17/1 ($t_{1/2}$ in vesicles is 3.3 min at 25°C, Figs. 11 and 12, and 58 min in ghosts (Schwoch et al., 1974). For R we have estimated a value of 22/1 (for vesicles we assumed an average radius of 0.15 μm, Fig. 9; for ghosts data are taken from Seeman, 1968). We note that a 30% difference exists between vesicles and ghosts. Granted that the estimations carry experimental errors, the correspondence of numbers is remarkable. Of course, several factors could affect this number, particularly the fact that some partial loss of 95,000 daltons polypeptide does in fact occur, so that the assumption of equal protein density may not be entirely justified. It is also conceivable that throughout the various treatments all the polypeptides have been partially altered or that some polypeptides have been denatured so that the average transport efficiency has been somewhat reduced. Nevertheless, the close similarity between the two values is highly indicative of the functional preservation of the anion transport system in the isolated purified vesicles.

In conclusion, the early indication that the 95,000 dalton polypeptides were the target of specific inhibitors of anion transport and were therefore implicated in the function (Cabantchik and Rothstein, 1972, 1974a; Ho and Guidotti, 1975) can now be substantiated by the functional isolation and reconstitution of the system (Rothstein et al., 1975; Wolosin et al., 1977 respectively). This system (Wolosin et al., 1977) shows structural and functional properties which are commensurate with those of intact cell membranes and it is therefore amenable to molecular approaches aimed at elucidating the intricacies of the mechanism of transport.

Acknowledgments. This work was supported by the Israel National Academy of Sciences and Humanities. Z.I. Cabantchik was also supported by the Lady Davis Research Foundation. Parts of this work have been taken from the doctoral dissertation of J.M. Wolosin, which was done in partial fulfilment of the requirement for Ph.D. at the Hebrew University.

References

Arrotti, J.J., Garvin, J.E.: Selective labelling of human erythrocyte components with tritiated trinitrobenzenesulfonic acid and picryl chloride. Biochem. Biophys. Res. Commun. 49, 205-211 (1975)

Bender, W.W., Garan, H., Berg, H.C.: Proteins of the human erythrocyte membrane as modified by pronase. J. Mol. Biol. 58, 783-797 (1971)

Cabantchik, Z.I., Rothstein, A.: The nature of the membrane sites controlling anion permeability of human red blood cells as determined by studies with disulfonic stilbene derivatives. J. Membrane Biol. 10, 311-330 (1972)

Cabantchik, Z.I., Rothstein, A.: Membrane proteins related to permeability of human red blood cells. I. Localization of disulfonic stilbene binding sites in proteins involved in permeation. J. Membrane Biol. 15, 207-226 (1974a)

Cabantchik, Z.I., Rothstein, A.: Membrane proteins related to anion permeability of human red blood cells. II. Effects of proteolytic enzymes on disulfonic stilbene sites of surface proteins. J. Membrane Biol. 15, 227-248 (1974b)

Cabantchik, Z.I., Balshin, M., Breuer, W., Rothstein, A.: Pyridoxal phosphate: an anionic probe for protein amino groups exposed on the outer and inner surfaces of intact human red blood cells. J. Biol. Chem. 250, 5130-5136 (1975)

Cabantchik, Z.I., Ostwald, T., Knauf, P.A., Markus, H., Rothstein, A.: The interaction of a photoreactive label with the anion transport system in red blood cells. Biochim. Biophys. Acta. In press (1976)

Fairbanks, G., Steck, T.L., Wallach, D.F.H.: Electrophoretic analysis of the major polypeptides of the human erythrocyte membrane. Biochemistry 10, 2606-2617 (1971)

Harris, E.J., Pressman, B.C.: Obligate cation exchanges in red cells. Nature (London) 216, 918-920 (1967)

Ho, M.K., Guidotti, G.: A membrane protein from human erythrocytes involved in anion exchange. J. Biol. Chem. 250, 675-683 (1975)

Hunter, M.J.: A quantitative estimate of the non-exchange-restricted chloride permeability of the human red cell. J. Physiol. 218, 49-50 (1971)

Knauf, P.A., Fuhrman, G.F., Rothstein, A.: The relationship between anion exchange and net anion flow across the human red blood cell membrane. J. Gen. Physiol. (in press, 1977)

Knauf, P.A., Rothstein, A.: Chemical modification of membranes. I. Effects of sulfhydryl and amino reactive reagents on anion and cation permeability of the human red blood cell. J. Gen. Physiol. 58, 190-210 (1971)

Passow, H.: Sulfate/sulfate exchange and K_2SO_4 net movements in valinomycin-treated red cells and red cell ghosts. J. Gen. Biol. In press (1977)

Rothstein, A., Cabantchik, Z.I., Balshin, M., Juliano, R.L.: Enhancement of anion permeability in lecithin vesicles by hydrophobic proteins extracted from red blood cell membranes. Biochem. Biophys. Res. Commun. 64, 144-149 (1975)

Rothstein, A., Cabantchik, Z.I., Knauf, P.A.: Mechanism of anion transport in red blood cells: role of membrane proteins. Federation Proc. 35, 3-10 (1976)

Rothstein, A., Knauf, P.A., Cabantchik, Z.I.: this vol., Chap. 23

Sachs, J.R., Knauf, P.A., Dunham, P.B.: Transport through red cell membranes. In: The Red Blood Cell, 2nd ed. Surgenor, D.M. (ed.). New York: Academic Press 1975, Vol. II, pp. 613-703

Schnell, K.F.: On the mechanism of inhibition of the sulfate transfer across the human erythrocyte membrane. Biochim. Biophys. Acta 282, 265-276 (1972)

Schwoch, G., Rudloff, V., Wood-Guth, I., Passow, H.: Effect of temperature on sulfate movement across chemically or enzymatically modified membranes of human red blood cells. Biochim. Biophys. Acta 339, 126-138 (1974)

Seeman, P.: Membrane stabilization and expansion by drugs. In: Metabolism and Membrane Permeability of Erythrocytes and Thrombozytes. Deutsch, E., Gerlach, E., Moser, K. (eds.). Stuttgart: Georg Thieme 1968, pp. 384-391

Staros, J.W., Richards, F.M.: Photochemical labelling of the surface proteins of human erythrocytes. Biochemistry 13, 2720-2726 (1974)

Tanner, M.J.A., Jenkins, R.E., Anstee, D.J., Clamp, J.R.: Abnormal carbohydrate composition of the major penetrating membrane protein of En(a-) human erythrocytes. Biochem. J. 155, 701-703 (1976)

Wolosin, J.M., Ginsburg, H., Cabantchik, Z.I.: Functional characterization of the anion transport system isolated from human erythrocytes. J. Biol. Chem. (in press, 1977)

Zaki, H., Fasold, H., Schuhman, B., Passow, H.: Chemical modification of membrane proteins in relation to inhibition of anion exchange in human red blood cells. J. Cell Physiol. 86, 471-494 (1976)

Reconstitution and Purification of the D-Glucose Transport Protein from Human Erythrocytes

M. KASAHARA and P. C. HINKLE

The D-glucose facilitated diffusion system of human erythrocytes has been studied extensively in whole cells and ghosts (see Jung, 1975) but has largely resisted attempts to identify the protein involved. Several workers have used binding of D-glucose or inhibitors to approach the problem. Studies with the affinity label, D-glucosylisothiocyanate (Taverna and Langdon, 1973) indicated that D-glucose protected against the labeling of two proteins of molecular weight 100,000 and 75,000. Another approach using the enhancement by D-glucose of labeling by 1-fluoro-2,4 dinitrobenzene indicated a protein of molecular weight 180,000 (Jung and Carlson, 1975). Noncovalent binding of cytochalasin B (Lin and Spudich, 1974) or D-glucose (Kahlenberg, 1976) to ghosts treated with 2,3-dimethylmaleic anhydride or NaI indicated that the membranes contained the glucose transport protein. These membrane residues contained bands 3,4.5,7 and glycoproteins PAS 1-4 (nomenclature according to Steck, 1974).

We have found conditions for reconstitution of a Triton X-100 extract of erythrocyte ghosts into liposomes which show specific permeability to D-glucose (Kasahara and Hinkle, 1976). In this paper we report an improved method for reconstitution and preliminary results on identification of the protein.

Methods

Triton X-100 Extract

Ghosts washed with EDTA and NaCl were extracted with 0.5% Triton-X-100 as described (Kasahara and Hinkle, 1976). Treatment of Triton X-100 to remove peroxides was found unnecessary. Triton X-100 was removed by incubation at 4°C overnight with Bio-Beads SM-2.

Reconstitution of D-Glucose Transport

The Triton extract or purified protein were added to sonicated liposomes made from acetone-washed soybean phospholipids in 10 mM Tris HCl and the mixture frozen quickly in dry ice-ethanol and thawed. The resulting flocculant suspension was sonicated for less than 30 sec in a bath type sonicator. The addition of a freeze-thaw step improved the original method (Kasahara and Hinkle, 1976) by increasing the amount of D-glucose uptake and making the procedure more reproducible.

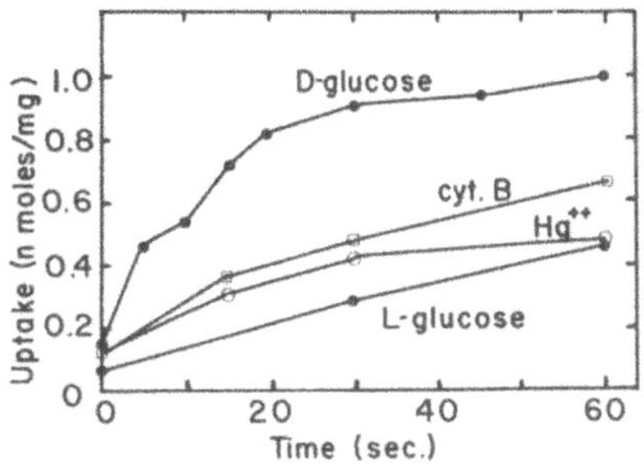

Fig. 1. Time course of D-glucose uptake. Reconstituted liposomes (0.67 mg ml^{-1} Triton X-100 extract, 15 mg phospholipids ml^{-1}) were prepared as described in Methods. Uptake was initiated by addition of radioactive D-glucose or L-glucose. At indicated times, it was stopped by cold stopping solution and followed by filtration. 2 μM cytochalasin B or 0.1 mM $HgCl_2$ was added, where indicated, to the assay mixture and D-glucose uptake was measured

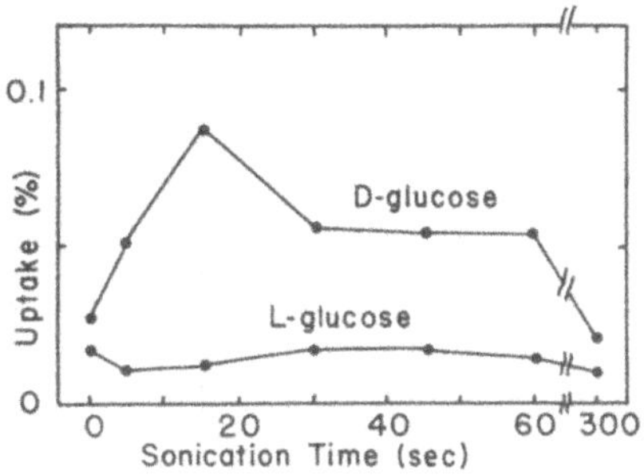

Fig. 2. Effect of sonication time on reconstitution. Mixture of liposomes (15 mg ml^{-1}) and Triton X-100 extract (0.67 mg ml^{-1}) was freeze-thawed followed by sonication for indicated time at 30°C. A portion of 53 μl was used for a 15 sec uptake assay

Transport Assay

Reconstituted liposomes (0.7-1.0 mg phospholipid) were incubated in a medium of 2 mM $MgSO_4$, 10 mM Tris HCl, pH 7.4; total volume 0.2 ml. Glucose uptake was initiated by addition of D-[^{14}C]glucose or L-[^{14}C]glucose, (0.4 μCi, final concentration 0.2 mM). The uptake was stopped by addition of 5 ml of cold reaction medium containing 0.5 mM $HgCl_2$. The mixture was passed through a membrane filter (Millipore, 0.22 μM, Type GSWP) and the filter washed with 10 ml of cold stopping solution. Radioactive glucose trapped on the filter was counted in Bray's solution.

Results and Discussion

An assay of D-gucose uptake by liposomes reconstituted with Triton X-100 extract is shown in Figure 1. The rapid uptake was specific for D-glucose, the low level of L-glucose uptake being similar to that observed with liposomes alone. Mercuric ions and cytochalasin B inhibited the rapid uptake of D-glucose. Under our conditions phloretin and stilbestrol did not inhibit the uptake. Heat denaturation of the Triton extract for 5 min at 70°C caused a 50% loss of activity and 90°C caused complete loss of activity.

There was a sharp optimum time for sonication of the freeze-thaw mixture (Fig. 2). This was presumably a reflection of an optimum size of the liposomes for the transport assay. With no sonication the phospholipids appeared by optical microscopy to be large string-like aggregates which broke up on short sonication to single membrane vesicles of about 500-3000 Å diameter as shown by thin section electron microscopy.

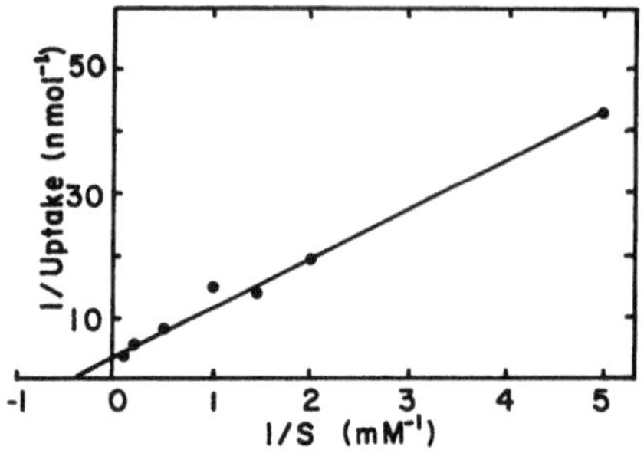

Fig. 3. Concentration dependence of D-glucose uptake. 53 µl of reconstituted vesicles (0.67 mg ml^{-1} of Triton X-100 extract and 15 mg ml^{-1} of phospholipid) was used for a 15-sec uptake. D-glucose-specific uptake was obtained by subtracting L-glucose uptake from D-glucose uptake

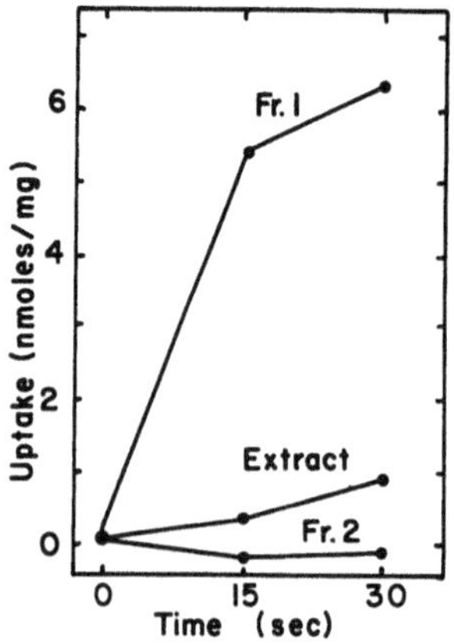

Fig. 4. D-glucose uptake activities of DEAE cellulose column fractions. Fraction I (0.31 mg ml^{-1}) or Fraction II (0.46 mg/ml) of DEAE cellulose column or Triton X-100 extract (0.46 mg ml^{-1}) was reconstituted with phospholipids (15 mg ml^{-1}). A portion of 53 µl was used for each uptake assay. D-glucose-specific activity (nmol·mg $protein^{-1}$) was obtained by subtracting L-glucose uptake from D-glucose uptake

Measurement of the difference between D-glucose and L-glucose for 15 sec at different concentrations allowed estimation of the K_m for D-glucose uptake by the reconstituted liposomes. The value of 2.2 mM obtained (Fig. 3) is comparable to that for intact erythrocyte. In previous studies it was also shown that the sugar specificity of the reconstituted system corresponded well with that for intact erythrocytes. Furthermore, the addition of nonradioactive D-glucose caused a rapid transient loss of D-[^{14}C]glucose followed by recovery, which indicates transport rather than binding in the reconstituted vesicles (Kasahara and Hinkle, 1976).

Purification of the glucose transport protein from the Triton extract is in progress. Preliminary results have been obtained by passing the Triton extract through a DEAE-cellulose column, equilibrated with 0.5% Triton X-100 and 50 mM Tris.HCl, pH 7.4. The initial eluate was collected and combined to give "Fraction I". The salt concentration was then increased to 1 M NaCl and a second protein fraction collected (Fraction II). Fraction I, which was not retarded by the column, contained about 12% of the protein and was very high in phospholipids (11 µmol P_i · mg $protein^{-1}$). The uptake of D-glucose by liposomes reconstituted with these fractions is shown in Figure 4. The specific activity of Fraction I was 15 times higher than the original Triton extract, and Fraction II was inactive. The proteins of the Triton extract and Fraction I were analyzed by polyacrylamide gel electrophoresis in the presence of SDS (Steck and Yu, 1973) (Fig. 5). Fraction I contained predominantly one protein band which was rather broad with an apparent

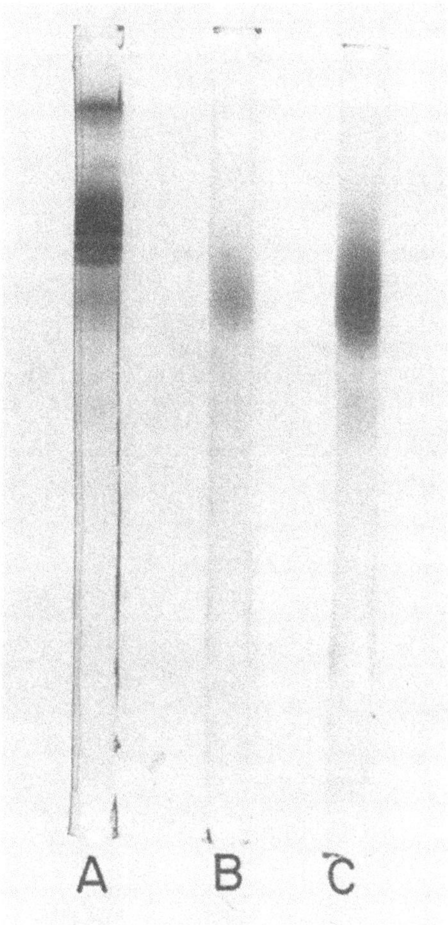

Fig. 5. SDS-polyacrylamide gel electrophoresis of: A: Triton X-100 extract (50 µg); B: Fraction I (20 µg); C: Fraction I (50 µg), stained with Coomassie Blue G

molecular weight of 50,000 to 60,000. It corresponds to a component of band 4.5 and is clearly different from band 3 or other proteins which have been suggested to be the glucose transport protein. The band seems to be inherently broad, since coelectrophoresis of other proteins with Fraction I showed no effect on the sharpness of other bands and partial removal of the lipids from Fraction I by washing with ether did not change the appearance of the band. When high amounts of the protein were loaded, low levels of bands 3 and 7 were also observed (Fig. 5C). Periodic acid-Schiff staining (PAS) revealed only the major broad band and glycolipids, suggesting that the protein is a glycoprotein.

The yield of protein in Fraction I from the DEAE cellulose column was 2 mg which came originally from 100 mg of ghosts. If the molecular weight is 60,000 this would represent 1.6×10^5 molecules per cell. This value is of the same order of magnitude as recent estimates of the number of glucose transport molecules made by glucose (Kahlenberg, 1976) or inhibitor binding methods (Jung and Carlson, 1975; Lin and Spudich, 1974).

Acknowledgments. We are grateful to Dr. L.A. Heppel for many discussions and for encouragement. We wish to thank Mrs. Yen-sheng L. Tu for excellent technical assistance. This study was supported by National Institutes of Health Grants CA-14454 and HL 14483.

References

Jung, C.Y.: In: The Red Blood Cell. Surgenor, D.M. (ed.). New York: Academic Press 1975, Vol. II, p. 705

Jung, C.Y., Carlson, L.M.: J. Biol. Chem. 250, 3217 (1975)

Kahlenberg, A.: J. Biol. Chem. 251, 1582 (1976)

Kasahara, M., Hinkle, P.C.: Proc. Natl. Acad. Sci. U.S. 73, 396 (1976)

Lin, S., Spudich, J.A.: Biochem. Biophys. Res. Commun. 61, 1471 (1974)

Steck, T.L.: J. Cell Biol. 62, 1 (1974)

Steck, T.L., Yu, J.: J. Supramol. Struct. 1, 220 (1973)

Taverna, R.D., Langdon, R.G.: Biochem. Biophys. Res. Commun. 54, 593 (1973)

Transport ATPases

Titration of Sodium Against Potassium by Their Action on (Na^+, K^+) Transport ATPase

R. L. Post

In this paper the features of the sodium and potassium ion pump are selected and organized in the hope of finding a conceptual framework useful for understanding past experiments and for designing future ones. I apologize for incompleteness and neglect of details. I assume that the pump has much the same features in all experimental preparations.

A. Rationale of Ligand Binding

In a system as complicated as the sodium and potassium ion pump, simple relationships are important. I am trying this approach. Bind ligands at equilibrium one at a time and then in pairs, looking for interactions of ligand binding. Consider next equilibria of enzyme-ligand complexes in partial exchange reactions. Finally, fit these components to steady state kinetics.

As ligands for titration, Na^+ and K^+ contrast great chemical similarity against great differences in the responses of the enzyme. Furthermore, of all the ligands, Na^+ is unique in all its actions on this enzyme; in producing its contrasting responses, K^+ has as congeners the other alkali metal cations together with NH_4^+ and Tl^+.

This approach grew out of my preoccupation with the idea of a strong antagonism between the binding of sodium and the covalent binding of phosphate to the pump enzyme. I thought the antagonism was powerful since binding of Na^+ raised the energy level of a phosphate group on the enzyme. First, phosphoenzyme was formed from inorganic phosphate and then second, Na^+ was added at a high concentration together with ADP. The low-energy phosphate group on the enzyme was transferred to ADP making a high-energy group in ATP (Post et al., 1975; Taniguchi and Post, 1975).

B. Past Experience and Concepts

Seven interesting review articles are available (De Weer, 1975; Glynn and Karlish, 1975; Jørgensen, 1975a,b; Skou, 1975; Whittam and Chipperfield, 1975; Albers, 1976; Garrahan and Garay, 1976).

I. Pump Transport in Erythrocytes

Per cycle 3 intracellular Na^+ go outward through the membrane, 2 extracellular K^+ go inward, and the terminal phosphate bond of one

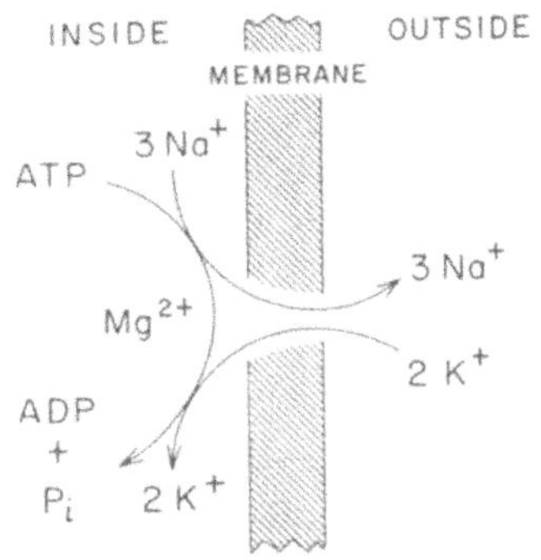

Fig. 1. Stoichiometry and sidedness of the sodium and potassium ion pump of plasma membranes. (From Post et al., 1975)

intracellular ATP molecule is hydrolyzed with intracellular Mg^{2+} as a catalyst. Intracellular ADP and P_i are products (Fig. 1). Thus, intracellular Na^+ and extracellular K^+ can be considered as substrates and extracellular Na^+ and intracellular K^+ can be corresponding products of translocations. According to this view, there are four substrates and four products. This classification includes water as a substrate. The implied inequality of charge transfer is real; the pump is electrogenic. The stoichiometry is a little flexible. In the absence of extracellular K^+, uncoupled outward transport of Na^+ may still occur at a rate up to 10% of capacity (Glynn and Karlish, 1976). For net outward transport, Na^+ is unique; for net inward transport, K^+ has congeners.

In preparations of broken membranes, the implied $(Na^+ + K^+)$-dependent ATPase activity persists, although concentration gradients of Na^+ and K^+ cannot accumulate. Persistence of the transport function is shown by reconstitution experiments of Hokin (Chap. 28) and of other investigators.

In addition to net transport, the transport kinetics shows in $\rightleftharpoons$ out exchange reactions. These are particularly well characterized in human erythrocytes. For Na^+, in $\rightleftharpoons$ out exchange requires: (1) absence of extracellular K^+ and (2) presence of intracellular ATP and ADP. The ATP is not hydrolyzed. In broken membranes, there is a corresponding Na^+-dependent ATP $\rightleftharpoons$ ADP exchange reaction. For K^+, in $\rightleftharpoons$ out exchange requires: (1) absence of intracellular Na^+ and (2) presence of intracellular ATP and inorganic phosphate, P_i. In this case also, the ATP is not hydrolyzed. In broken membranes, there is a corresponding K^+-dependent $P_i \rightleftharpoons$ HOH exchange. These translocation exchanges show that translocation of each ion can proceed independently of translocation of the other.

Another feature of the pump is sidedness asymmetry with respect to affinities for Na^+ and K^+. With respect to intracellular ions, Na^+ is favored; with respect to extracellular ions, K^+ is favored (Table 1). For estimation of affinities, in $\rightleftharpoons$ out exchanges are more useful than net transport since the hydrolysis of ATP does not drive the reaction in the forward direction and modify equilibrium between liganded and unliganded species. Actually, intracellular Na^+ and extracellular K^+ show the same affinities in the exchange reactions as in their roles as substrates for net transport. It is in their roles as products that intracellular K^+ and extracellular Na^+ produce second-order complexities. These complexities may reflect hybrid occupancy of a three-site translocation center (Skou, 1975; Glynn and Karlish, 1976).

Table 1. Semiquantitive affinities for in ⇌ out exchange of monovalent cations through the (Na,K) pump in human erythrocytes corrected for competitive inhibition by the other cation and expressed as concentrations for a half-maximal effect in mM. Second-order complexities are ignored (Garay and Garrahan, 1973; Simons, 1974)

Cation	Inside	Outside
Na^+	1	100
K^+	10	0.1

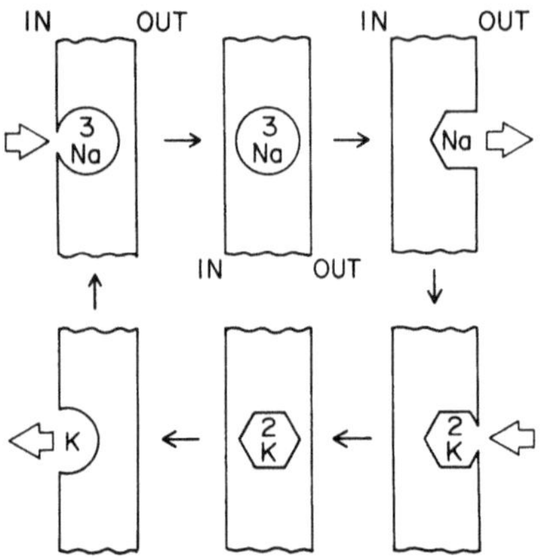

Fig. 2. Hypothesis for: (1) alternating occupancy (2) alternating translocation, and (3) linkage of carrier affinity to sidedness in the mechanism of the (Na,K)-pump. Upper row: three successive conformations (encounter, transition, and farewell) of a carrier center in the enzyme during translocation of Na^+ outward. Lower row: corresponding conformations for translocation of K^+ inward. Change in affinity is indicated by change in the shape of center from circular to hexagonal. (Modified from Post et al., 1973). This scheme leaves unspecified the transition between Na_3- and K_2-forms. Some sort of hybrid occupancy is likely (Glynn and Karlish, 1976)

In summary, the transport kinetics emphasizes (1) non-equivalent linked transport of Na^+ outward and K^+ inward, (2) possible separation of partial reactions involving in ⇌ out exchange of Na^+ and ADP from those involving in ⇌ out exchange of K^+ and inorganic phosphate, and (3) marked asymmetry of preference of the active transport center for Na^+ over K^+ with respect to the intracellular solution and for K^+ over Na^+ with respect to the extracellular solution (Fig. 2). If the model in Figure 2 were equilibrated with Na^+ and K^+ in the absence of other ligands, the conformations in the upper right and lower left would be disfavored and titration would take place between the conformations in the upper left and lower right, including possible contributions from the occluded, transition conformations.

II. Sodium, Potassium Adenosine Triphosphatase

The richest sources of this enzyme are tissues specialized for transport of salt solutions across epithelia, such as kidney, or for electrical activity, such as brain. The purest enzyme has been prepared from the electric organ of the electric eel, from the salt glands of the dogfish or of the duck, and from the outer medulla of mammalian kidney.

The composition of the enzyme is partly known. It is an intrinsic enzyme of plasma membranes and contains at least two peptides and

Fig. 3. Hypothesis of control of transphosphorylation pathway according to occupancy of carrier center by Na^+ or K^+. E1 and E2: conformations of the transphosphorylation center shaped for transfer of a high-energy or low-energy phosphate group, respectively. Active site of phosphorylation is the beta carboxyl group of same aspartyl residue in each case. (From Post et al., 1973)

PHOSPHORYLATION PATH

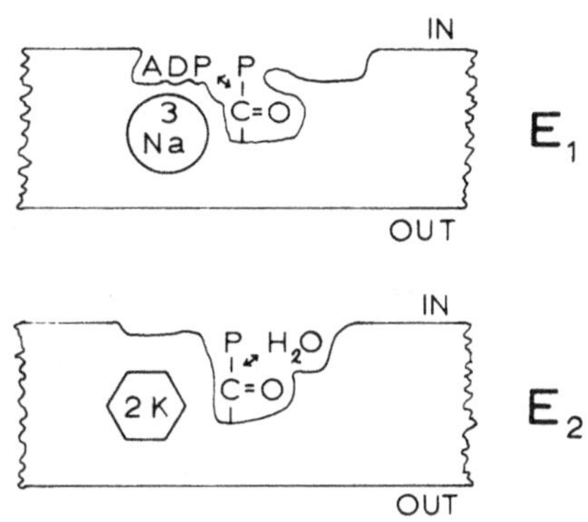

CONTROL BY CATION BINDING

phospholipid. The molecular weight of one peptide is about 100,000 daltons; this peptide carries an active site for phosphorylation and an active site for binding a cardioactive steroid inhibitor such as ouabain. The other peptide has a molecular weight of about 50,000 daltons and is a glycopeptide.

In the presence of Na^+ and Mg^{2+}, the enzyme accepts the terminal phosphate group of ATP onto a beta carboxyl group of an aspartyl residue forming a mixed anhydride bond. In the presence of K^+ and Mg^{2+}, it accepts a phosphate group rapidly and reversibly from inorganic phosphate at the same active site. Since it equilibrates with inorganic phosphate, this group is functionally a low-energy phosphate group. The mixed anhydride bond by itself does not provide sufficient stability for a low-energy group. Consequently, it is natural to imagine that the low-energy phosphate group is also held on the enzyme by other, non-covalent, bonds.

The phosphoenzyme from ATP changes its reactivity as the concentration of Na^+ is varied. The reactivity can be estimated from the rate of dephosphorylation following addition of a chelator of Mg^{2+} or a chase of unlabeled ATP. (The phosphoenzyme is estimated as the [^{32}P]form.) At 16 mM Na^+, inclusion of 0.1 mM K^+ with the chase produces immediate, complete dephosphorylation, whereas inclusion of ADP has no effect. At 400 mM Na^+, inclusion of saturating concentrations of K^+ or ADP produces almost the same partial effect for each ligand in recent experiments in our laboratory. At 3 M Na^+, inclusion of K^+ has no effect and ADP produces immediate and complete dephosphorylation. At 16 mM Na^+, the phosphoenzyme from ATP equilibrates slowly with P_i. These and other experiments (Post et al., 1975) support a hypothesis of control of the pathway of phosphorylation by the binding of Na^+ or K^+ to the enzyme (Fig. 3). Since the active site of phosphorylation is the same in both cases, the change in reactivity implies a change in the conformation of the active center surrounding the active site. A reaction sequence for phosphorylation and dephosphorylation is shown in Figure 4.

Jørgensen (1975a,b) has produced the most direct evidence for a conformational change induced by the binding of Na^+ or K^+. He digested the purified enzyme with trypsin in the presence of NaCl or KCl and obtained different kinetics for the rate of loss of

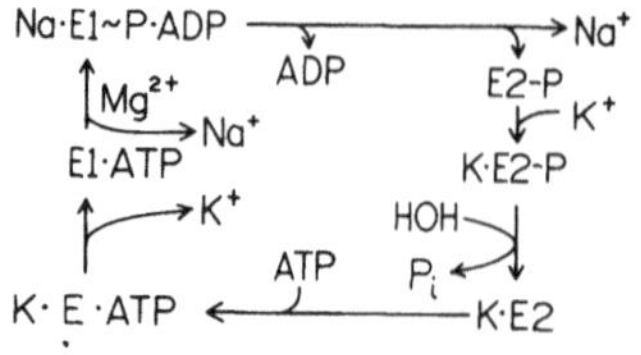

Fig. 4. Reaction sequence for phosphorylation and dephosphorylation of (Na,K)-transport ATPase. E: enzyme. The forms of the enzyme are arranged according to their patterns of tryptic digestion (cf. Fig. 5). One set is on left, the other is on right. Ligands inside the loop of the reaction sequence are intracellular and those outside are extracellular. Sidedness assignments for Na^+ and K^+ were made by comparison of their half-maximal concentrations for phosphorylation and dephosphorylation with those for translocation (cf. Table 1). The Mg^{2+} requirement for phosphorylation from P_i is not shown

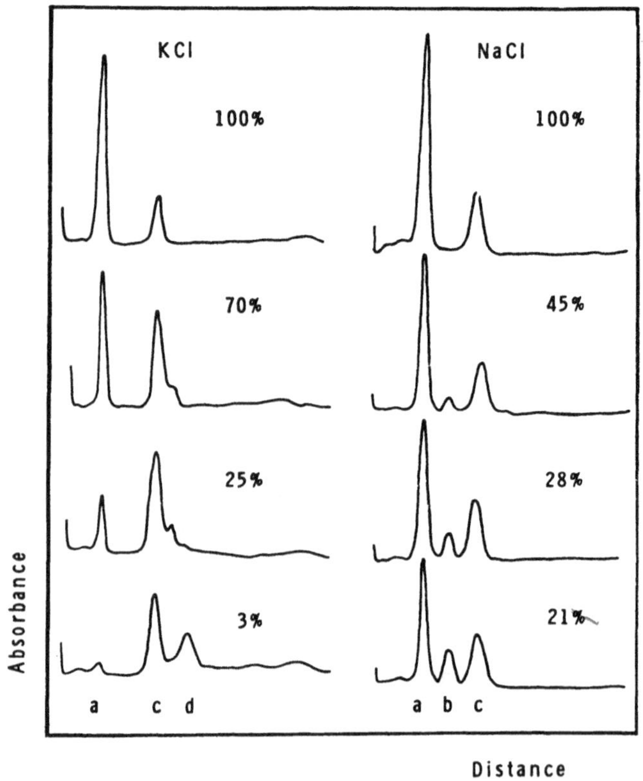

Fig. 5. Electrophoretograms of the peptides of purified (Na,K)-ATPase during digestion with trypsin in the presence of KCl (left) or of NaCl (right). Each sample was applied to a polyacrylamide gel in the presence of sodium dodecyl sulfate and then an electric potential moved the peptides from left to right. The larger peptides moved more slowly than the smaller. From above downward, the duration of tryptic digestion was 0, 5, 20, or 40 min. The % of activity remaining is shown with each scan. Native peptides: a and c; products of digestion: b and d. (From Jørgensen, 1975b)

activity. With Na^+, the loss was diphasic; with K^+, it was monophasic. Furthermore, the visible product peptide was larger than the glycoprotein with Na^+ and was smaller than the glycoprotein with K^+ (Fig. 5). He named the conformations of the native enzyme "t(Na)-form" and "t(K)-form". These forms were identified in the absence of other ligands of the enzyme. When other ligands were added, they modified the forms in ways consistent with the scheme in Figure 4. In particular, addition of ATP to t(K) converted it to t(Na) in the absence of Na. Addition of Mg^{2+} to t(Na) in the presence of both Na and ATP produced t(K) in the absence of K^+.

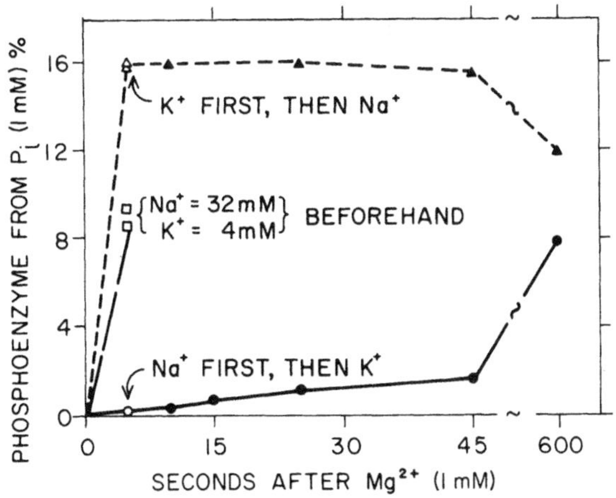

Fig. 6. Slow equilibration of free (Na,K)ATPase with Na^+ and K^+. In 0.9 ml at 0°C and pH 7.0, 2.4 mg of washed membrane protein from guinea pig kidney was incubated with 15 µmol of imidazole, 20 µmol of 2-(N-morpholino)propane sulfonic acid, 1 µmol of $^{32}P_i$ and 32 µmol of Na acetate (o) or 4 µmol of KCl (Δ) or both (□). The reaction was started with 1 µmol of $MgCl_2$ in 0.1 ml. At 5 sec the mixture was completed with 4 µmol of KCl (●) or 32 µmol of Na acetate (▲). At the times indicated, the reaction was stopped with acid, the precipitate was digested with pepsin and active site phosphopeptides were isolated by paper electrophoresis and counted. Similar incubation for 15 min with 0.25 µmol of ouabain in place of Na^+ and K^+ salts estimated 100% phosphorylation. Further technical details are given in Taniguchi et al., 1975

C. New Experiments

These experiments estimate the distribution of the enzyme between sodium and potassium forms at saturating concentrations of sodium and potassium and in the absence of other ligands. The enzyme was equilibrated with mixtures of sodium and potassium salts and then the occupancy of the enzyme was estimated by its response to an addition of inorganic phosphate and magnesium. Potassium permits phosphorylation from this source and sodium prevents it.

I. Slow Equilibration with Sodium and Potassium Ions

Sodium and potassium ions equilibrated relatively slowly at zero degrees in the absence of nucleotides. If potassium was added to the enzyme first and then magnesium and P_i, a maximal level of phosphorylation was obtained. This level was maximal relative to the capacity of the potassium enzyme complex to accept inorganic phosphate for formation of a covalent bond. The complex of the enzyme with the inhibitor ouabain accepts phosphate to a five- or six-fold higher level. The equilibrium between covalently bound phosphate and noncovalently bound phosphate is probably different in the two cases. The time constant for dephosphorylation is 76 sec for the ouabain enzyme and less than 1 sec for the potassium

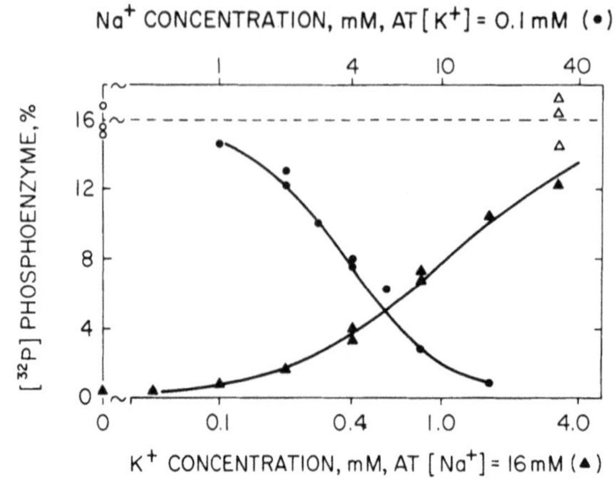

Fig. 7. Titration of Na^+ against K^+. Two experiments are shown, one with variable Na^+ and constant K^+, the other vice versa. The procedure was that of the samples of Figure 6 in which Na^+ and K^+ were incubated with the enzyme before addition of Mg^{2+}. The reaction was stopped after 5 sec. Curved lines fit Hill equations. For $[K^+]$ = 0.1 mM, the $K_{0.5}(Na^+)$ = 3.7 mM, and n_H = -1.96. For $[Na^+]$ = 16 mM, the $K_{0.5}(K^+)$ = 1.1 mM, and n_H = 1.27

enzyme. In any case, once potassium phosphoenzyme had been formed, subsequent addition of sodium had little effect on the level of phosphoenzyme for the next 20 sec (Fig. 6). Similarly, if sodium was added to the enzyme first and then magnesium and inorganic phosphate, no phosphoenzyme appeared. Then if potassium was added, it had little effect for the next 20 sec. If potassium and sodium were added together and then magnesium and inorganic phosphate, an intermediate level of phosphoenzyme appeared (Fig. 6). Other experiments have shown that such a level was stable for at least 4 min. Thus relative occupancy of the enzyme by sodium or potassium could be estimated provided that these ions were incubated with the enzyme before phosphorylation from inorganic phosphate.

II. Titration of Sodium Against Potassium

When the enzyme was incubated with a constant concentration of sodium ion and an increasing concentration of potassium ion, a progressive increase in the level of phosphoenzyme was seen. When the results were fitted to a Hill equation, the Hill coefficient was 1.3 (Fig. 7). When the concentration of potassium was held constant and the concentration of sodium increased, the level of phosphoenzyme decreased. In this case, the Hill coefficient was larger, namely -2.0 (Fig. 7). The clear difference between the Hill coefficients in the two cases shows that competition between sodium and potassium ions was not on a 1-for-1 basis.

Further experiments were done also at pH 7.0 and zero degrees. Useful data were obtained at potassium concentrations from 0.1 to 5 mM and at sodium concentrations from 2 to 128 mM. Data at 23 pairs of concentrations were fitted to a modified Hill equation and the following result was obtained as the best least squares fit. Where "f" is the ratio of phosphoenzyme to the maximal level of potassium-complexed phosphoenzyme, then

$$f/(1-f) = ([K^+]/0.1\ \text{mM})^{1.38}/([Na^+]/3.81\ \text{mM})^{2.13} \qquad (1)$$

In fitting this equation to the data, there is only one independent coefficient of the ion concentrations. The value of 0.1 mM was taken

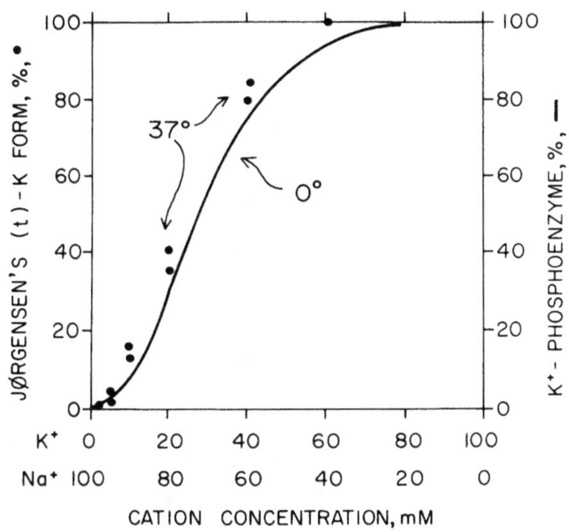

Fig. 8. Comparison of titration of Na^+ against K^+ at pH 7.5 according to tryptic digestion (●) or according to phosphorylation from inorganic phosphate (——). Data for tryptic digestion are from Figure 2 of Jørgensen (1975b). The line for phosphorylation from P_i fits the equation $[K \cdot EP]/[K \cdot EP_{max}] = f$, where $f/(1-f) = ([K^+]/0.10 \text{ mM})^{1.62}/([Na^+]/2.68 \text{ mM})^{2.78}$. $K \cdot EP_{max}$ was 16% of the total amount of enzyme estimated with ouabain, Mg^{2+}, and $^{32}P_i$. This equation was fitted to 30 data points at concentrations of K^+ from 0.125 to 8 mM and concentrations of Na^+ from 4 to 64 mM in experiments like that in Figure 7. The choice of 0.1 mM as a constant was taken arbitrarily from transport kinetics (cf. Table 1)

arbitrarily as the value for the coefficient of the $[K^+]$ from the data on transport in erythrocytes (Table 1). The differences between observed and calculated values were squared and summed. The square root was taken and divided by the number of data points. The estimated discrepancy was 0.8% of the maximal amount of potassium-complexed phosphoenzyme.

The ratio of the exponents of the cation concentrations is 1.54. This means that for titration of $[Na^+]$ against $[K^+]$ at any constant level of partial phosphorylation, an increase in the concentration of Na^+ by a factor, R, must be matched by an increase in the concentration of K^+ by a larger factor, $R^{3/2}$, since $[K^+]$ is less powerful than $[Na^+]$ in changing the level of the phosphoenzyme. This ratio of exponents is consistent with an equilibrium like the following:

$$3\ Na^+ + K_2 \cdot E \rightleftharpoons Na_3 \cdot E + 2\ K^+ \qquad (2)$$

which is consistent with the stoichiometry of transport (Fig. 1). The relative affinities for Na^+ and K^+ with a ratio of 25-40 : 1 (see also Fig. 8) are in fair agreement with those of the inside and outside sites, respectively. Gache et al. (1976) obtained fairly similar values for activation of (Na,K)ATPase: for Na^+, $K_{0.5}$ = 3 mM or less and Hill coefficient = 1.9; for K^+, $K_{0.5}$ = 0.1 mM or less and Hill coefficient = 1.1. It is possible to believe that these experiments titrated the affinity of the inside site for Na^+ against the affinity of the outside site for K^+.

Comparison of data from these experiments with those of Jørgensen (1975b) shows a good correspondence (Fig. 8). It is likely that conformational changes estimated by tryptic digestion correspond to those estimated by changes in the pathway of phosphorylation. The small discrepancy may be due to the difference in temperature. Incidentally, comparison of Figure 8 with Equation (1) above shows that an increase in pH increased the affinity for Na^+ relative to that for K^+.

III. Discussion of Alternations

The properties of the sodium enzyme contrast with those of the potassium enzyme not only with respect to phosphorylation pathway and pattern of tryptic digestion but also with respect to affinity for ATP and inhibition by Be^{2+}, F^-, or dicyclohexylcarbodiimide. A major change in conformation is induced simply by exchanging Na^+ for K^+ (and vice versa) in the absence of other ligands. One conformation must prefer Na^+ and the other must prefer K^+. These preferences correspond fairly well to the relative affinities for Na^+ and K^+ as substrates on the inside and outside of the cell, respectively. It is therefore attractive to include in the alteration of properties an alteration of the sidedness of access of monovalent cations to a translocation center. According to this line of thought, State One of the enzyme would be stabilized by binding nucleotides or Na^+ which exchanges into the solution in contact with the intracellular face of the membrane. In contrast, State Two would be stabilized by binding K^+ which exchanges into the solution in contact with the extracellular face of the membrane and by covalently bound phosphate and by Mg^{2+}. In State Two, ouabain could substitute for K^+ to produce an inhibited complex (Giotta, 1976). This bimodal classification of ligands suggests a bistable alternation of conformations as a significant feature of the mechanism. A bistable alternation is a flip-flop mechanism as follows. State One stabilized by intracellular Na^+ and ATP is attacked by Mg^{2+}, a State Two ligand. It transphosphorylates, loses its nucleotide, translocates and releases its Na^+ into the extracellular solution. It now flips into State Two and accepts extracellular K^+ easily. In the presence of a high concentration of inorganic phosphate and a low concentration of nucleotide, State Two is stable (Karlish et al., 1976). However, a high concentration of ATP displaces the bound phosphate and causes the enzyme to flop into State One which translocates and releases K^+ into the intracellular solution (Fig. 4).

D. Afterword

I. Simultaneous Transport

This alternation hypothesis faces apparent incompatibilities with available data. A long-standing incompatibility is the observed kinetics of Na^+ and K^+ transport (Garrahan and Garay, 1976). This kinetics appears to deny the simplest alternation, a ping-pong mechanism, in which inward and outward conformations of the free

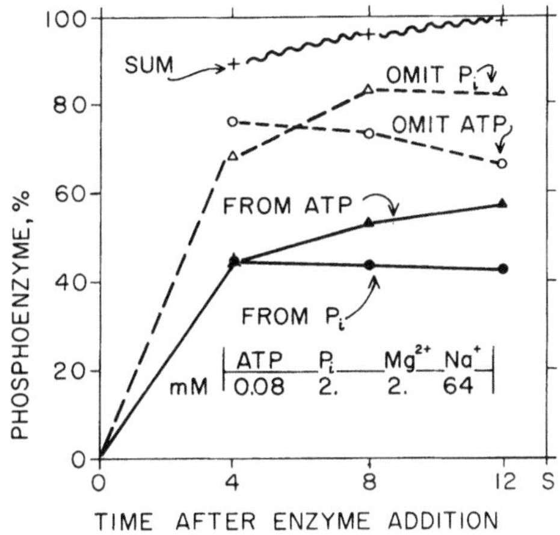

Fig. 9. Simultaneous phosphorylation from ATP and inorganic phosphate. Experiment conducted as in Figure 6. The enzyme was washed free of all reactive ligands and incubated in 0.8 µl of buffer. Reaction was started by addition of 0.2 ml containing ATP, $(Tris)_3$phosphate, $MgCl_2$, and Na acetate to produce the concentrations shown. ^{32}P was in the gamma phosphate group of ATP (▲,Δ) or in P_i (●,o). The plus mark (+) shows the sum of the separate values (▲ + ●). The 100% value was estimated with ouabain (without Na^+ or ATP) in triplicate with a standard deviation of ± 2.1%

enzyme are intermediate between a Na-form and a K-form (compare Fig. 2). For net transport a ping pong mechanism fits the observed kinetics approximately provided that (1) Na^+ or K^+ saturates the enzyme on at least one side of the membrane and (2) the rate constants for dissociation of Na^+ into the intracellular and extracellular solutions are equal to the corresponding rate constants for dissociation of K^+. However, for in ⇌ out exchange of labeled Na^+ or K^+ this mechanism does not fit the data.

To resolve the incompatibility Stein et al. (1973) and Repke et al. (1974) proposed tetramer or dimer models with half-of-the-sites reactivity. Perhaps it will be enough to find new cation binding sites intermediate between the aqueous solutions and the sites controlling phosphorylation pathways and conformations.

II. Simultaneous Phosphorylation from ATP and Inorganic Phosphate

In order to look for rigid linkage between dimers in a half-of-the-sites system, we phosphorylated the enzyme simultaneously from ATP and P_i. The idea was that a rigid linkage would permit the enzyme to be phosphorylated from either donor at only half of the sites. In this case, phosphorylation from both donors would produce twice the level of phosphorylation that could be obtained from either donor alone. This result did not appear (Fig. 9). The sum of phosphoenzyme from both donors acting simultaneously was almost equal to the maximal level obtainable from inorganic phosphate alone in the presence of Mg^{2+} and ouabain. If this is a half-of-the-sites dimer enzyme, the linkage between the oligomers is not rigid.

Acknowledgments. The author is indebted to Jean Sorrells and Phyllis Hodges for attentive technical assistance. This work was supported by a grant from the National Heart, Lung, and Blood Institute, 5R01 HL-01974, and from the National Institute of Arthritis and Metabolic Diseases, 5P01 AM-07462, of the National Institutes of Health.

References

Albers, R.W.: The (Sodium plus Potassium)-transport ATPase. In: The Enzymes of Biological Membranes, Vol. III. Martonosi, A.N. (ed.). New York: Plenum Publ. Corp. 1976, pp. 283-301

De Weer, P.: Aspects of the recovery processes in nerve. In: MTP Intern. Rev. Sci. Physiology, Series One, Vol. III. Hunt, C.C. (ed.). Baltimore: University Park Press 1975, pp. 231-278

Gache, C., Rossi, B., Lazdunski, M.: (Na^+,K^+)-Activated adenosinetriphosphatase of axonal membranes, cooperativity and control. Europ. J. Biochem. 65, 293-306 (1976)

Garay, R.P., Garrahan, P.J.: The interaction of sodium and potassium with the sodium pump in red cells. J. Physiol. 231, 297-325 (1973)

Garrahan, P.J., Garay, R.P.: The distinction between sequential and simultaneous models for sodium and potassium transport. In: Current Topics In Membranes and Transport. Bronner, F., Kleinzeller, A. (eds.). New York: Academic Press 1976, Vol. VIII, pp. 29-97

Giotta, G.J.: Native (Na^+ + K^+)-dependent adenosine triphosphatase has two trypsin-sensitive sites. J. Biol. Chem. 250, 5159-5164 (1975)

Glynn, I.M., Karlish, S.J.D.: The sodium pump. Ann. Rev. Physiol. 37, 13-55 (1975)

Glynn, I.M., Karlish, S.J.D.: ATP hydrolysis associated with an uncoupled sodium flux through the sodium pump: Evidence for allosteric effects of intracellular ATP and extracellular sodium. J. Physiol. 256, 465-496 (1976)

Jørgensen, P.L.: Isolation and characterization of the components of the sodium pump. Quart. Rev. Biophys. 7, 239-274 (1975a)

Jørgensen, P.L.: Purification and characterization of (Na^+, K^+)-ATPase. V. Conformational changes in the enzyme. Transitions between the Na-form and the K-form studied with tryptic digestion as a tool. Biochim. Biophys. Acta 401, 399-415 (1975b)

Karlish, S.J.D., Yates, D.W., Glynn, I.M.: Transient kinetics of (Na^+ + K^+)-ATPase studied with a fluorescent substrate. Nature (London) 263, 251-253 (1976)

Post, R.L., Kume, S., Rogers, F.N.: Alternating paths of phosphorylation of the sodium and potassium ion pump of plasma membranes. In: Mechanisms in Bioenergetics. Azzone, G.F., Ernster, L., Papa, S., Quagliariello, E., Siliprandi, N. (eds.). New York, London: Academic Press 1973, pp. 203-218

Post, R.L., Toda, G., Kume, S., Taniguchi, K.: Synthesis of adenosine triphosphate by way of potassium sensitive phosphoenzyme of sodium, potassium adenosine triphosphatase. J. Supramol. Struct. 3, 479-497 (1975)

Repke, K.R.H., Schön, R., Henke, W., Schönfeld, W., Streckenbach, R., Dittrich, F.: Experimental and theoretical examination of the flip-flop model of (Na,K)-ATPase function. In: Properties and Functions of (Na^+ + K^+)-Activated Adenosinetriphosphatase. Askari, A. (ed.). N.Y. Acad. Sci. 1974, pp. 203-219

Simons, T.J.B.: Potassium:potassium exchange catalyzed by the sodium pump in human red cells. J. Physiol. 237, 123-155 (1974)

Skou, J.C.: The (Na^+ + K^+) activated enzyme system and its relationship to transport of sodium and potassium. Quart. Rev. Biophys. 7, 401-434 (1975)

Stein, W.D., Lieb, W.R., Karlish, S.J.D., Eilam, Y.: A model for active transport of sodium and potassium ions as mediated by a tetrameric enzyme. Proc. Natl. Acad. Sci. U.S. 70, 275-278 (1973)

Taniguchi, K., Post, R.L.: Synthesis of adenosine triphosphate and exchange between inorganic phosphate and adenosine triphosphate in sodium and potassium ion transport adenosine triphosphatase. J. Biol. Chem. 250, 3010-3018 (1975)

Whittam, R., Chipperfield, A.R.: The reaction mechanism of the sodium pump. Biochimica Biophysica Acta 415, 149-171 (1975)

Concept on the Link Between Chemical and Electro-Chemical Free Energies in (NaK)-ATPase Transport Function

K. R. H. REPKE

A. Aim and Scope

When reviewing the more than 3000 papers published in the last 15 years on (NaK)-ATPase, I came to the conclusion that at the present stage the development of new concepts on the key problems can become even more important for fostering progress than the production of new data on special questions. As stated by H.A. Krebs (1966): "Unifying concepts are needed more urgently now than ever before because new facts are brought to light at a breath-taking tempo. They are needed because they may tell us what the observed facts mean; they guide us in the search for new relevant facts and guard us against accumulating useless facts." Thus, I decided to focus my contribution to this Symposium on the conceptual developments initiated in my group during the last three years (Repke and Schön, 1973; Dittrich et al., 1974; Schön et al., 1974; Repke et al., 1974, 1975). Specifically, I wish to tackle the following fundamental problems: first, the nature of the free-energy releasing chemical reaction in ATP cleavage utilized for transport work; second, the mode of utilization of chemical free energy in transport work; third, the source of reversibility and high thermodynamical efficiency of the (NaK)-ATPase system. The concepts to be presented are based on a large body of special experimental data and general theoretic arguments that cannot be detailed here but will be reviewed elsewhere (Repke, in preparation).

B. Nature of Free-Energy Releasing Chemical Reaction in ATP Cleavage Utilized for Transport Work

In ATP cleavage, the length of the bond to be broken between the terminal phosphorus atom and the bridge oxygen amounts to 1.68 Å corresponding to a bond energy of 86 kcal (Matheja and Degens, 1971). Hence, in the enzymatic cleavage of this bond, a special mechanism must exist for the change of the strong bond into a weak bond so that in bond splitting not too much energy is required and net free energy can be released.

Molecular orbital calculations revealed that considerable free energy is stored in ATP by the change from the tetrahedral arrangement of oxygen ligands in the orthophosphate ion to their unsymmetrical arrangement in the polyphosphate chain (Alving and Laki, 1972). X-ray analysis showed that the oxygen ligands of the terminal and the middle phosphorus atom are completely staggered (Kennard et al., 1970). These "high energy rearrangements", the result of mutually repulsive electrostatic charges, are thought to be a form of potential mechanical energy (Alving and Laki, 1972).

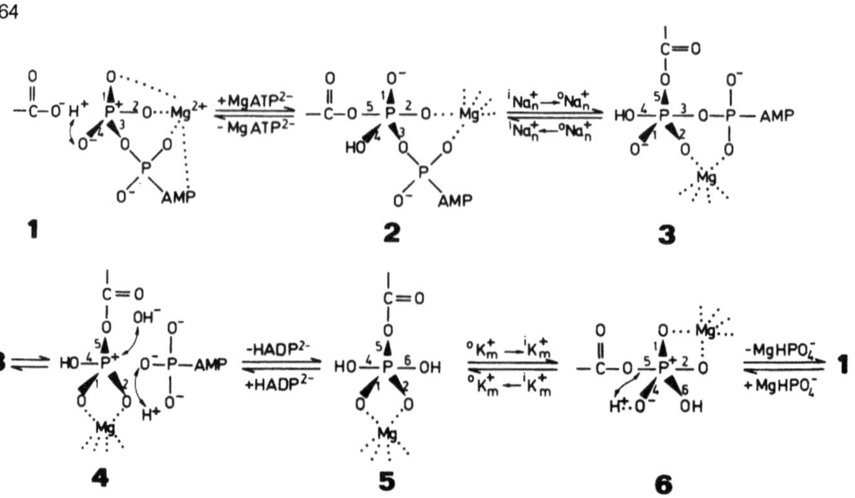

Fig. 1. Hypothetical associative phosphorane mechanism of phosphoryl transfer reactions involved in (NaK)-ATPase functions. In transport function, reaction sequences are: phosphoryl transfer from ATP to enzyme (reactions 1 → 2 → 3 → 4) proceeding through pseudorotation of ligands of ATP terminal phosphorus (reaction 2 → 3), and phosphoryl transfer from formed enzyme carboxyl phosphate to medium (reactions 4 → 5 → 6 → 1) proceeding through pseudorotation of ligands of enzyme carboxyl phosphate phosphorus (reaction 5 → 6). Extended version of earlier published reaction schemes (Dittrich et al., 1974; Repke et al., 1975)

This information was the starting point for the design of the mechanism of ATP cleavage by (NaK)-ATPase shown in Figure 1. This hypothetical, associative phosphorane mechanism is based on the current theories of exchange reactions on phosphorane intermediates (Boyd, 1969; Mislow, 1970; Gillespie et al., 1971; Marquarding et al., 1973), their application to different enzymatic phosphoryl transfer reactions (Mildvan, 1970; Benkovic and Schray, 1973; Mildvan and Grisham, 1974), and of course the known data on ATP cleavage and ATP synthesis effected by (NaK)-ATPase. The carboxyl residue shown in the reaction scheme represents the β-carboxyl of an aspartyl group in the catalytic center of the enzyme. This carboxylate anion is the phosphate acceptor in the intermediary transphosphorylation reaction (Degani and Boyer, 1973; Post and Kume, 1973; Nishigaki et al., 1974). The reactions 1 → 2 → 3 → 4 comprise the phosphorylation of the enzyme from ATP and the reactions 4 → 5 → 6 → 1 the dephosphorylation of the phosphoenzyme formed.

The available chemical, physical and kinetic evidence favors the occurrence of the intermediates shown in Figure 1, namely the noncovalent enzyme-ATP association complex (state 1), the covalent enzyme-ATP intermediates prior to and after phosphorus ligands pseudorotation (states 2 and 3 respectively), the phosphoenzyme with attached ADP (state 4), and the enzyme carboxyl phosphate intermediates prior to and after phosphorus ligands pseudorotation (states 5 and 6 respectively). The reaction scheme is in compliance with and supported by the results of studies on the source of the C-O-P linkage formed in phosphoryl transfer from ATP and orthophosphate and on the oxygen exchange effected in the same transfer reactions (Skworzewitsch et al., 1972; Dahms and Boyer, 1973; Dahms et al., 1973).

Concerning the connection between the described overall chemical reactions and the transport of Na^+ and K^+, a careful analysis of the available literature data supports the following conclusion. The phosphoryl transfer from ATP to enzyme, catalyzed by internal Na^+, is coupled with the outward transport of Na^+, and the phosphoryl transfer from the formed phosphoenzyme to medium, being catalyzed by external K^+, is coupled with the inward transport of K^+. Clearly, these statements leave unanswered the key question as to the proper nature of the free-energy releasing reactions utilized to drive the transport processes. The described associative phosphorane mechanism of the phosphoryl transfer reactions offers the answer that the free-energy releasing processes are the pseudorotations of the phosphorus ligands (reactions 2 → 3 and 5 → 6). These processes will therefore be considered in greater detail.

Due to the stereochemical nonrigidity of the phosphorane intermediates, their ligands can undergo a nondissociative intramolecular exchange of positions. More specifically, the Berry pseudorotation, shown in the figure, is a pairwise exchange of apical and equatorial ligands taking place in a concerted fashion. The phosphorus ligands pseudorotation can be accompanied by high free-energy changes as calculations on comparable model compounds have shown (Gillespie et al., 1971). The ligands pseudorotations occurring in reactions 2 → 3 and 5 → 6 include the shift of the ADP residue and of the enzyme carboxyl residue, i.e. of rather bulky ligands, from constrained equatorial positions to energetically favored axial positions. Hence, the pseudorotations of these ligands are suggested to release in two packets the free energy stored in ATP.

Moreover, the phosphorus ligands pseudorotations allow an understanding of the transduction of chemical free energy, as the following arguments show. The ligand shifts occur without development of a turning momentum of the residual molecule and, of course, in close interaction with the polypeptide chains forming the catalytic center. Thus, the ligands pseudorotations necessarily enforce a structural reorganization of the catalytic center. The emerging concept then is that the pseudorotations of the phosphorus ligands constitute the basic processes of the conversion of skalar chemical free energy into vectorial mechanical free energy of polypeptide chains able to be transported from the catalytic center to the ionophoric centers of the enzyme and to produce there the vectorial and geometric rearrangements required for the outward transport of Na^+ and the inward transport of K^+.

C. Mode of Utilization of Chemical Free Energy in Transport Work

A thorough analysis of the published data on the Na^+ and K^+ affinities of the (NaK)-ATPase system shows that in the various modes the system can be made to operate (Glynn and Karlish, 1975), the affinities of the ionophoric centers are always much the same, namely in their inward exposure high for Na^+ and low for K^+, and vice versa in their outward exposure, low for Na^+ and high for K^+. The simplest deduction to be made is that one and the same type of ionophoric center provides the basic structure for complexing with Na^+ and K^+, and that the preference given for either Na^+ or K^+ rests at any rate solely on its location-determined, different geometry. Similar to the inward- or outward-exposed ionophoric

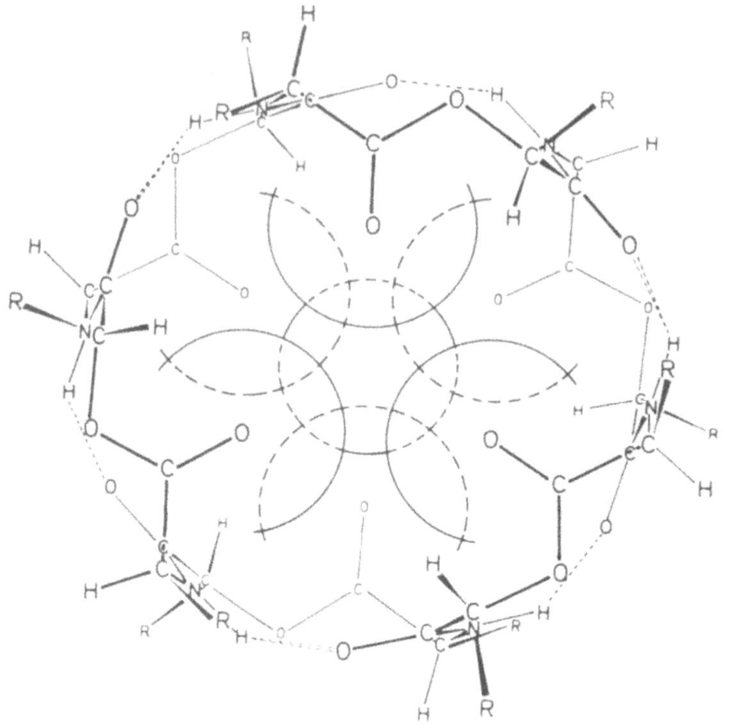

Fig. 2. Three-dimensional structure of valinomycin-cation complex, determined by nuclear magnetic resonance and conformational energy studies. Size of polar core determined by orientation of ester C=O moiety. Redrawn from Urry, 1972

centers of (NaK)-ATPase, the neutral cyclic peptide compounds antamanide and valinomycin energetically highly prefer complexation with Na^+ and K^+ respectively. Obviously, similar although not necessarily cyclic structures could be found in proteins (Winkler and Eigen, 1972) and so in (NaK)-ATPase (Ovchinnikov, 1974). Thus, these ionophores offer suitable models for the understanding of the geometries and energetics of the ionophoric center of (NaK)-ATPase preferentially attracting Na^+ or K^+.

Concerning antamanide, the binding constant for Na^+ is an order of magnitude greater than that of K^+ (Wieland et al., 1972) so that the conformational energy of the cation-complexed ionophore is with Na^+ much lower than with K^+. On the conformational energy aspect to cation selectivity, more information is available for valinomycin (Urry, 1972), preferentially attracting K^+. The ionophore holds the cation in a polar core formed by six acyl oxygens of ester moieties. Figure 2 shows that the size of the polar core is determined by the orientation of the ester C=O moieties. If the oxygen atoms tilt inward toward the center of the molecule, a smaller core is formed. Similarly, the polar centers enlarged when the C=O moieties are rotated outward. Accordingly, the energetics of this process would bear on the size of the cation that is favored by the structure. Figure 3 gives a plot of conformational energy versus core size. On the curve points are noted at which the size of the polar core would correspond to the size of Na^+, K^+, Rb^+ and Cs^+. K^+ and Na^+ thus form with valinomycin complexes which differ in having either low or high conformational free energies. In other words, according to the Boltzmann distribution, the low or high free energy conformations found with the cation complexes have with K^+ a high and with Na^+ a low energetic probability to exist.

Fig. 3. Conformational free energy of complexes of valinomycin with Na^+, K^+, Rb^+ or Cs^+. Redrawn from Urry, 1972

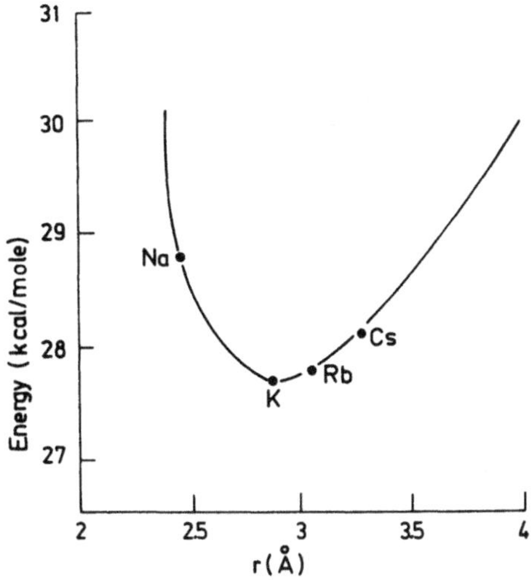

The reported data on the cation complexation features of antamanide and valinomycin provide the experimental basis of a key idea for the understanding of the differences in cation affinities of the ionophoric center of (NaK)-ATPase when facing either the intracellular space with high Na^+ and low K^+ affinity or when facing the extracellular space with high K^+ and low Na^+ affinity. As deduced before, the preference given for either Na^+ or K^+ rests solely on the location-determined difference in the geometry of the ionophoric center.

The energetic consequences are best illustrated by Figure 4, which shows a hypothetical plot of conformational free energy versus core size when the ionophoric centers have either the antamanide-like, inward-exposed geometry (I) or the valinomycin-like, outward-exposed geometry (II). On the curves points are noted at which the size of the ionophoric core corresponds to the size of either Na^+ or K^+. In the inward-exposed geometry (I), Na^+ complexation is associated with relaxed core size and so with low conformational energy, whereas K^+ complexation is associated with constrained core size and so with high conformational energy. Corresponding to the conformational energies, this geometry complexes with Na^+ at low concentration, but with K^+ at high concentration only. Vice versa, in the outward-exposed geometry (II), Na^+ complexation is associated with constrained core size and so with high conformational energy, whereas K^+ complexation is associated with relaxed core size and so with low conformational energy. Again corresponding to the conformational energies, this geometry complexes with Na^+ at a high concentration only, but with K^+ already at low concentration. Clearly, the conversion of the Na^+- or K^+-complexed ionophoric centers from the geometry with low conformational energy into the geometry with high conformational energy requires free energy. This energy is delivered by the mechanical free energy that is produced by the phosphorus ligands pseudorotations as deduced in Section B above.

The interplay between the chemical and the conformational subsystem is illustrated in Figure 5. The mechanical free energy pro-

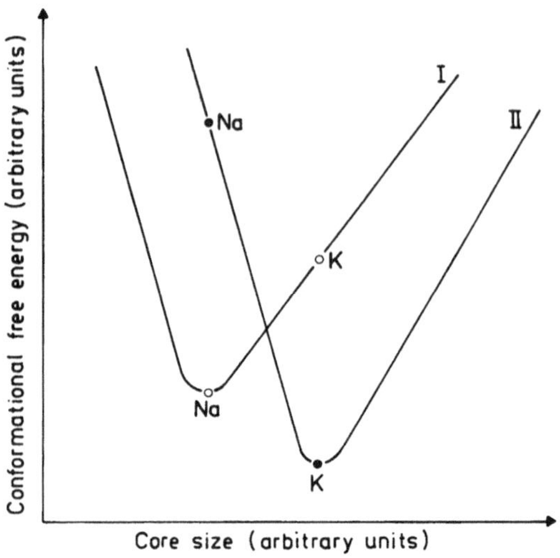

Fig. 4. Hypothetical plot of conformational free energy versus core size for ionophoric center of (NaK)-ATPase in inward-exposed, antamanide-like geometry (I) and in outward-exposed, valinomycin-like geometry (II)

duced by the Na^+-catalyzed pseudorotation of the ligands of ATP terminal phosphorus is supposed to transform the Na^+-complexed ionophoric center from the inward-exposed, antamanide-like geometry (I) with low conformational free energy into the outward-exposed, valinomycin-like geometry (II) with high conformational free energy. Since this constrained conformation has a low energetical probability to exist, Na^+ will be released so that K^+ can combine, resulting in a relaxed, energetically favored conformation. The mechanical free energy produced by the K^+-catalyzed pseudorotation of the ligands of the enzyme carboxyl phosphate phosphorus is supposed to transform the K^+-complexed ionophoric center from outward-exposed, valinomycin-like geometry (II) with low conformational free energy into the inward-exposed, antamanide-like geometry (I) with high conformational free energy. Since this constrained conformation has a low energetical probability to exist, K^+ will be released so that Na^+ can combine, starting the next cycle.

D. Source of Reversibility and High Thermodynamic Efficiency

The designed concept on the tight coupling between free-energy releasing chemical reactions and free-energy consuming cation transports is an application of the more general theory that in all highly efficient energy interconverting processes the exergonic and endergonic reactions proceed simultaneously, although separated in space (Fig. 6). Due to coupling between the exergonic and endergonic centers during energy transduction, the system as a whole is in local thermodynamic equilibrium and is already at local minimum of free energy (Kemeny, 1974). Of course, this conservative mechanism ensures maximum thermodynamic efficiency.

For simplicity, (NaK)-ATPase has been treated up to now as if it were a monomeric enzyme. However, the enzyme is a tetramer with two functional subunits (Fig. 7). Hence, the requirement of coupling between the exergonic and endergonic centers must also apply

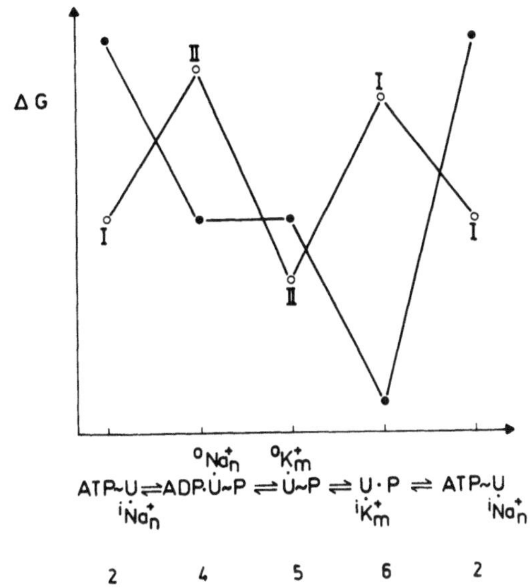

Fig. 5. Hypothetical plot of free energies of conformational subsystem (o - o) and chemical subsystem (● - ●) versus coupled processes of cation movements and chemical reactions occurring during transport work (reaction coordinate from left to right) and ATP synthesis (from right to left). The figure illustrates the basic assumption that for transport work the chemical free energies of stepwise ATP cleavage, released through pseudorotations of the ligands of the pentacoordinate phosphorane intermediates, pull the conformational free energies of the cation-complexed ionophoric center uphill as they themselves go downhill, and that for ATP synthesis the high free-energy conformations of the cation-complexed ionophoric centers pull, through pseudorotations of the ligands, the free energies of the pentacoordinate phosphorane intermediates uphill as they themselves go downhill. It is arbitrarily assumed that the free energies transduced to the enzyme during the pseudorotations of the ligands of the two pentacoordinate phosphorane intermediates are similar and that the efficiency of the interconversion of chemical and conformational free energy equals 100%. Basically, this representation is an application of the Gibbs-Duhem relation saying that the thermodynamic potentials of a system are interdependent and that the sum of the potential changes equals zero. For clarity, free energies of activation and interplay between the 2 functional subunits of enzyme (cf. Fig. 7) are neglected. Roman numerals on points characterizing the conformational subsystem refer to inward-exposed, antamanide-like geometry (I) and outward-exposed, valinomycin-like geometry (II) of ionophoric center, whose properties are illustrated in Fig. 4. Arab numerals on reaction equations refer to Fig. 1. U stands for subunit. Superscripts o and i: outward or inward exposure of cation-complexed ionophoric centers

to the functional dimer to account for the reversibility of the (NaK)-ATPase system and the high thermodynamic efficiency of its transport work, reported to be 80%. As deduced in earlier papers (Repke and Schön, 1973; Dittrich et al., 1974; Schön et al., 1974; Repke et al., 1974, 1975), (NaK)-ATPase adds to the rapidly growing class of enzymes with kinetic anticooperativity. In fact, the available evidence suggests the anticooperative coupling of the following events: binding of substrate versus release of products, transphosphorylation versus dephosphorylation, translocation of the Na^+-complexed ionophoric center from facing inside

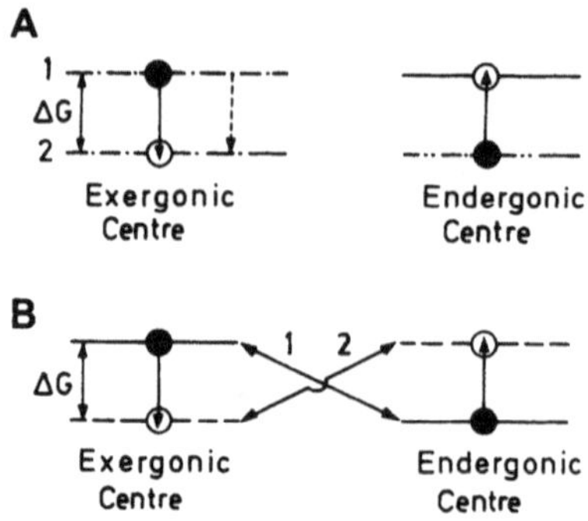

Fig. 6. Schematic representation of 2 extreme possibilities of free-energy release: A, dissipation of free energy due to absence of coupling; or B, conservation of free energy due to presence of coupling between exergonic and endergonic centers (redrawn from Kemeny, 1974)

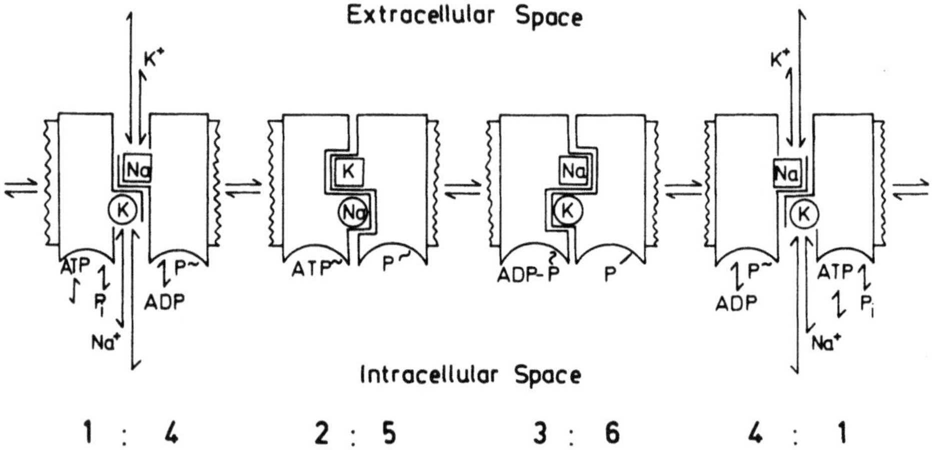

Fig. 7. Hypothetical scheme of arrangement of functional subunits of (NaK)-ATPase in membrane and of concerted interplay of subunits coupling cation movements and chemical reactions. The functional subunits (the glycoprotein subunits are omitted) span the membrane and form down the central axis of symmetry relating both monomers a discontinuous channel. In the middle of the membrane plane, the channel is interrupted by the polypeptide chain arrangements which form the ionophoric centers. In the transport function, three major steps are assumed to occur: (1) closing of the half-channels through the concomitant processes of Na^+ and K^+ complexation of ionophoric centers and the formation of the covalent bond between ATP and enzyme; (2) conversion of the Na^+-complexed ionophoric center from inward-exposed, antamanide-like geometry into outward-exposed, valinomycin-like geometry parallel to the conversion of the K^+-complexed ionophoric center from outward-exposed, valinomycin-like geometry to inward-exposed, antamanide-like geometry; (3) opening of the half-channels through the concomitant processes of ATP binding and orthophosphate plus ADP release followed by cation exchanges. Arab numerals refer to intermediary reaction states shown in Fig. 1

to facing outside versus translocation of the K^+-complexed ionophoric center from facing outside to facing inside, K^+ release versus Na^+ binding and Na^+ release versus K^+ binding. The common basis of the described diverse anticooperative actions of the various effectors is that their chemical free energy, via interaction

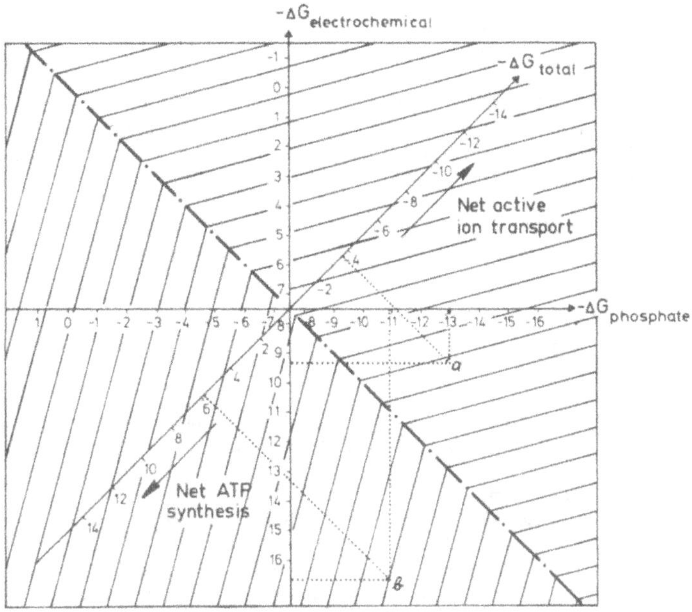

Fig. 8. Nomogram on relationship between total free enthalpy (ΔG_{total}), net active cation transport and net ATP synthesis both effected by (NaK)-ATPase.

$$\Delta G_{total} = -RT \ln \left(\frac{[ADP] \cdot [PO_4]}{[ATP]} + \frac{[^{o}Na^+]^3}{[^{o}K^+]^2} \cdot \frac{[^{i}K^+]^2}{[^{i}Na^+]^3} \right)$$

The nomogram is calculated on basis of experimental data determined under nearly physiological conditions (a) or under reversal conditions (b) (experimental data from Garrahan and Glynn, 1967)

with their centers on the enzyme, is converted into mechanical free energy of a special polypeptide chain. In this form, the free energy is transported from the signal-producing center to the signal-realizing center in compliance with the focused programming of conformation change potentials of the enzyme protein (cf. Lumry and Biltonen, 1969; Lumry, 1971, 1974; Byers and Koshland, 1975). The resulting geometric rearrangement of the signal-realizing center then becomes manifest as anticooperative effect.

The outlined kinetic anticooperativity of the coupled chemical and transport events appears to account for the fact that the (NaK)-ATPase system can work near thermodynamic equilibrium (Fig. 8). The nomogram shows the relationship between the total free enthalpy, net active transport and net ATP synthesis. The nomogram is calculated on the basis of experimental data determined under nearly physiological or under reversal conditions. Thus, under physiological conditions, the total free enthalpy value is not far from zero, so that even under physiological conditions microscopically ATP synthesis must also occur, although the net process, of course, is cation transport driven by ATP cleavage.

E. Concluding Remarks

In principle, the concepts presented may have bearing on the understanding of the mechanism of other membrane transport processes and of other energy interconverting proteins such as myosin ATPase and ATP synthetase of oxidative and light phosphorylation. A first attempt to apply the concept of kinetic anticooperativity to the mechanism of energy interconversion by ATP synthetase has already been published (Repke and Schön, 1974).

References

Alving, R.E., Laki, K.: J. Theor. Biol. 34, 199-214 (1972)

Benkovic, S.J., Schray, K.J.: The Enzymes. 3rd ed. Boyer, P.D. (ed.). New York: Academic Press 1973, Vol. VIII, pp. 201-237

Boyd, D.B.: J. Am. Chem. Soc. 91, 1200-1205 (1969)

Byers, L.D., Koshland Jr., D.E.: Biochemistry 14, 3661-3669 (1975)

Dahms, A.S., Boyer, P.D.: J. Biol. Chem. 248, 3155-3162 (1973)

Dahms, A.S., Kanazawa, T., Boyer, P.D.: J. Biol. Chem. 248, 6592-6595 (1973)

Degani, C., Boyer, P.D.: J. Biol. Chem. 248, 8222-8226 (1973)

Dittrich, F., Schön, R., Repke, K.R.H.: Acta Biol. Med. Germ. 33, K 17- K 25 (1974)

Garrahan, P.J., Glynn, I.M.: J. Physiol. 192, 237-256 (1967)

Gillespie, P., Hoffmann, P., Klusacek, H., Marquarding, D., Pfohl, S., Ramirez, F., Tsolis, E.A., Ugi, J.: Angew. Chem. 83, 691-721 (1971)

Glynn, I.M., Karlish, S.J.D.: Ann. Rev. Physiol. 37, 13-55 (1975)

Kemeny, G.: Proc. Natl. Acad. Sci. U.S. 71, 3669-3671 (1974)

Kennard, O., Isaacs, N.W., Coppola, J.C., Kirby, A.J., Warren, S., Motherwell, W.D.S., Watson, D.G., Wampler, D.L., Chenery, D.H., Larson, A.C., Kerr, K.A., Riva di Sanseverino, L.: Nature (London) 225, 333-336 (1970)

Krebs, H.A.: In: Current Aspects of Biochemical Energetics. Kaplan, N.O., Kennedy, E.P. (eds.). New York: Academic Press 1966, pp. 83-92

Lumry, R., Biltonen, R.: In: Biological Macromolecules. Timasheff, S.N., Fasman, G.D. (eds.). New York: Marcel Dekker 1969, Vol. II, pp. 65-212

Lumry, R.: In: Electron and Coupled Energy Transfer in Biological Systems. King, T.E., Klingenberg, M. (eds.). New York: Marcel Dekker 1971, Vol. I, Part A, pp. 1-116

Lumry, R.: Ann. N.Y. Acad. Sci. 227, 46-76 (1974)

Marquarding, D., Ramirez, F., Ugi, J., Gillespie, P.: Angew. Chem. 85, 99-127 (1973)

Matheja, J., Degens, E.T.: Structural Molecular Biology of Phosphates. Stuttgart: Gustav Fischer 1971, pp. 1-180

Mildvan, A.S.: In: The Enzymes, 3rd ed. Boyer, P.D. (ed.). New York: Academic Press 1970, Vol. II, pp. 445-536

Mildvan, A.S., Grisham, C.M.: In: Structure and Bonding. Berlin-Heidelberg-New York: Springer 1974, Vol. XX, II, pp. 1-21

Mislow, K.: Acc. Chem. Res. 3, 321-331 (1970)

Nishigaki, J., Chen, F.T., Hokin, L.E.: J. Biol. Chem. 249, 4911-4916 (1974)

Ovchinnikov, Yu.A.: FEBS Lett. 44, 1-21 (1974)

Post, R.L., Kume, S.: J. Biol. Chem. 248, 6993-7000 (1973)

Repke, K.R.H., Schön, R.: Acta Biol. Med. Germ. 31, K 19 - K 30 (1973)

Repke, K.R.H., Schön, R.: Acta Biol. Med. Germ. 33, K 27 - K 38 (1974)

Repke, K.R.H., Schön, R., Henke, W., Schönfeld, W., Streckenbach, B., Dittrich, F.: Ann. N.Y. Acad. Sci. 242, 203-219 (1974)

Repke, K.R.H., Schön, R., Dittrich, F.: In: Proceedings of IXth FEBS Meeting. Gårdos, G., Szasz, J. (eds.). Budapest: Akadêmiai Kiadô 1975, Vol. XXXV, pp. 241-243

Schön, R., Dittrich, F., Repke, K.R.H.: Acta Biol. Med. Germ. 33, K 9 - K 16 (1974)
Skworzewitsch, E.G., Pantelejewa, N.S., Pisarjewa, L.N.: Dokl. Akad. Nauk SSSR 206, 240-242 (1972)
Urry, D.W.: Biochim. Biophys. Acta 265, 115-168 (1972)
Wieland, T., Faulstich, H., Burgermeister, W.: Biochem. Biophys. Res. Commun. 47, 984-992 (1972)
Winkler, R., Eigen, M.: In: Molecular Bioenergetics and Macromolecular Biochemistry. Weber, H.H. (ed.). Berlin-Heidelberg-New York: Springer 1972, pp. 130-148

Purification and Properties of Na, K-ATPases from the Rectal Gland of *Squalus acanthias* and the Electric Organ of *Electrophorus electricus* and Reconstitution of the Na^+ - K^+Pump from the Purified Enzyme

L. E. HOKIN

Introduction

About twenty years ago Skou (1957) discovered an adenosine triphosphatase in the microsome fraction of the leg nerve of the shore crab which required Mg^+, Na^+ and K^+ for optimal activity. He suggested that this enzyme, which will henceforth be referred to as the Na,K-ATPase, might be involved in the coupled Na^+-K^+ pump (Baker, 1972), which serves many important physiological functions, such as maintaining osmotic equilibrium in all animal cells, maintaining the gradients of Na^+ and K^+ in excitable tissues such as nerve and muscle, which permits waves of excitation down their membranes, and effecting transepithelial transport of Na^+, and in some cases, K^+ in such structures as the intestinal epithelium, the renal tubules, and various glandular structures. It has been estimated that in the mammal about 30% of the ATP which is produced by resting respiration is hydrolyzed via the Na,K-ATPase (Whittam, 1965). Edelman and his coworkers (1974) have demonstrated that thyroid hormone elevates the level of Na,K-ATPase in thyroid-sensitive tissues, and he has suggested that the resultant increased hydrolysis of ATP could be responsible for the calorigenic effect of thyroid hormone.

Since Skou's discovery the Na,K-ATPase has been extensively studied, particularly in the erythrocyte membrane (see recent reviews by Dahl and Hokin, 1974; Glynn and Karlish, 1975; Schwartz et al., 1975; Whittam and Chipperfield, 1975; Jorgensen, 1975; Skou, 1975) and these studies have provided strong circumstantial evidence that the Na,K-ATPase might be the molecular machine for effecting Na^+ and K^+ transport. To prove this point, however, it is necessary to purify the Na,K-ATPase, insert it into a lipid bilayer membrane and reconstitute the coupled transports of Na^+ and K^+.

Purification of the Na,K-ATPase

The purification of membrane-bound enzymes such as the Na,K-ATPase to near homogeneity has proved to be very difficult. Many of these enzymes, particularly the Na,K-ATPase, are deeply embedded in the membrane bilayer, and because of their membranous, lipoprotein nature, classical enzyme purification procedures have been of limited value. In the case of the Na,K-ATPase all purification procedures have required the use of detergents. With this enzyme two strategies have been employed. One of these employs negative purification, i.e., impurities are extracted from the membrane, while the Na,K-ATPase remains membrane bound. A membrane skeleton is left behind, and this contains the Na,K-ATPase as its chief protein component. This approach has only been successful with the outer renal medulla of mammals. It is quite remarkable that this procedure is so effective, because, based on purely statistical grounds, one would not have predicted that of all the integral proteins in the membrane only the Na,K-ATPase would be left behind after detergent treatment. This procedure has required the use of anionic detergents such as deoxycholate or sodium dodecyl sulfate at critical concentrations. Our Laboratory has utilized positive purification, which is to solubilize the Na,K-ATPase with a nonionic detergent, Lubrol WX, and to purify the solubilized enzyme. Nonionic detergents, in contrast to the anionic detergents mentioned above, rather selectively solubilize integral proteins such as the Na,K-ATPase. Once solubilized, the Na-K-ATPase can be purified by methods which, although perhaps not conventional, are at least not wildly unconventional (see Hokin et al., 1973; Dixon and Hokin, 1974).

In our attempts to purify the Na,K-ATPase we have kept two goals in mind: (1) the purified enzyme should not be denatured due to the purification procedures and (2) the purified enzyme should be obtained in sufficient yield to carry out the necessary studies without difficulty. Working with beef brain, which is fairly rich in Na,K-ATPase, we worked out a basic purification scheme which gave high yields and minimal denaturation (Uesugi et al., 1971), but unfortunately the final membranous enzyme was only about 50% pure. Extensive efforts to resolubilize this membranous enzyme and to purify it further were unsuccessful.

It appeared that with essentially the same purification scheme but with a starting material which was considerably richer in Na,K-ATPase it might be possible to reach our goal of complete purification. Nature has provided us with at least two very rich sources of Na,K-ATPase - the salt glands of marine birds and elasmobranchs and the electric organs of electric eels and rays. Preliminary studies with salt glands of marine birds indicated that because of the very tough connective tissue in this structure and the difficulty of obtaining large amounts of these glands they would not be suitable for the goals enumerated above. However, the rectal salt gland of the spiny dogfish shark, *Squalus acanthias*, is rich in Na,K-ATPase (Bonting, 1966), it has the consistency of liver once the connective tissue capsule is removed, and dogfish are readily available in large numbers on most continental coasts. In addition, the gland can be frozen in dry ice immediately after its removal, shipped to the laboratory, and stored at -70°C for 1-2 years without any loss in Na,K-ATPase activity. The frozen glands can then be thawed, and the enzyme can be purified by the following simple steps: (1) A membrane fraction is isolated by differential centrifugation. (2) The membrane fraction is solubilized by the nonionic detergent, Lubrol WX. The ratio of Lubrol to protein is important for this step, and with starting material from different species or prepared in different ways, the optimum concentrations of protein and Lubrol must be worked out. In some cases the enzyme is also somewhat unstable in the presence of the concentrations of Lubrol required, so the free Lubrol should be removed as soon as possible, which is accomplished by the next purification step. The addition of 1 mM Na_2ATP at the Lubrol extraction stage also stabilizes the enzyme in some instances (Perrone et al., 1975). (3) The Lubrol extract is placed on a sucrose gradient and submitted to zonal centrifugation. The free Lubrol remains at the top of the gradient and the Na,K-ATPase moves ahead of a considerable amount of protein. Thus, two important objectives are achieved here. The free Lubrol is removed, and the enzyme is enriched about two-fold. The appropriate fractions from the zonal run are pooled

Table 1. Highly purified preparations of Na,K-ATPase reported in the literature

Source	Yield (mg protein)	Specific Activity (μmol Pi/mg protein/h)	Reference
Outer renal medulla of dog	1-2	600-800	Kyte (1971a,b)
Rectal gland of dogfish	20-30	1500	Hokin et al. (1973)
Outer renal medulla of dog	16-18	1500	Lane et al. (1973)
Pig brain	0.1-0.3	500-7000	Nakao et al. (1973a,b)
Outer renal medulla of rabbit or pig	4-5	1900-2200	Jorgensen (1974)
Electric organ of electric eel	90	1900	Dixon and Hokin (1974) Perrone et al. (1975)

and concentrated in a Diaflo ultrafiltration cell and washed several times to remove sucrose and other components from the sucrose gradient solution. (4) The greatest purification is achieved at the last stage. It is a novel ammonium sulfate fractionation procedure which consists of incubation of the zonal concentrate with 1 M ammonium sulfate at 30°C for 2-4 min followed by dilution of the ammonium sulfate solution to 0.4 M and further incubation at 30°C for 10 min. The initial incubation with ammonium sulfate precipitates down all of the protein, while the second incubation resolubilizes the impurities, leaving behind purified Na,K-ATPase.

The Na,K-ATPase from the electric organ of *Electrophorus electricus* has been purified by the same procedure (Dixon and Hokin, 1974; Perrone et al., 1975). The more recent procedure (Perrone et al., 1975), which is somewhat modified over the original procedure (Dixon and Hokin, 1974), has given specific activities as high as 32 μmol of Pi/mg protein/min and yields as high as 100 mg (Table 1). The enzyme in the microsomes or "pellet" (Perrone et al., 1975) of the electric organ of *Electrophorus electricus* is somewhat less stable than that from the rectal gland of *Squalus acanthias*, but it can be stabilized to a considerable extent by adding 1 mM Na_2ATP throughout the purification procedure. After purification the enzymes from both sources are quite stable without any additions, losing very little activity on storage at 0° for one to two weeks and are "indefinitely" stable on freezing at -70°C.

Four other highly active Na,K-ATPase preparations have been reported (Table 1). In three of the cases (Kyte, 1972; Lane et al., 1973; Jorgensen, 1974) the purification has been the alternative one mentioned in the Introduction, which is to extract all of the protein impurities from membrane fractions from the outer renal medulla, leaving the Na,K-ATPase membrane-bound. The specific activity of two of these preparations (Lane et al., 1973; Jorgensen, 1974) appears to be comparable and of the same magnitude as that obtained by the method involving initial extraction with Lubrol WX (Hokin et al., 1973; Dixon and Hokin, 1974; Perrone et al., 1975). The enzyme preparation of Kyte (1972) has a lower specific activity; based on gel electropherograms this is probably due primarily to denaturation during purification rather than inadequate purification. The preparation of Nakao and his associates (Nakao et al., 1973a,b), which involves Lubrol extraction of brain microsomes followed by purification over columns, is difficult to assess because the specific activities of the preparations ranged ten-fold, and the amounts of protein and the method of protein estimation raise some questions about the validity of these determinations. With the exception of this preparation the gel patterns of the remaining preparations appear to be essentially comparable, although one must be careful when interpreting gel patterns in assessing purity, since different gel electrophoresis techniques have quite different resolving powers. For example, Deupree and Hokin (unpublished observation) compared several published gel electrophoresis systems and found that the procedure of Laemmle (1970) showed remarkably greater power in resolving minor protein bands which would be missed by the other methods.

In addition to purity, what is equally important is that the biological function of the Na,K-ATPase be preserved on purification, i.e., that the coupled transports of Na^+ and K^+ can be reconstituted when the purified enzyme is reinserted into a lipid membrane. To date, only one purified preparation has met this criterion, i.e., the purified Na,K-ATPase from the rectal gland of *Squalus acanthias*. The reconstitution of this system and other systems will be discussed below.

Molecular Properties of the Na,K-ATPase

A variety of chemical studies has been carried out with the two purified fish enzymes (Perrone et al., 1975). Sodium dodecyl sulfate (SDS) polyacrylamide gel electrophoresis shows that 90-95% of the protein from the eel enzyme and the shark enzyme is accounted for by a 95,000 molecular weight protein band and a smaller glycoprotein band (55,000 daltons in the case of the shark enzyme and 47,000 daltons in the case of the eel enzyme). In our hands the ratio of large chain to small chain was 2:1 for both the eel and the shark enzyme. This gives a minimum molecular weight of 250,000 which is in agreement with radiation inactivation studies based on target theory (Kepner and Macey, 1968). Both enzymes also give 4,000-4,300 pmol per mg protein for ouabain binding or phosphorylation from [γ-^{32}P]ATP in the presence of Na^+ and Mg^+. This also gives a minimum molecular weight of 250,000, if one assumes "half-of-sites" reactivity, for which there is kinetic evidence (Stein et al., 1973; Repke et al., 1974). The data on phosphorylation, ouabain binding, subunit composition, and molecular weight based on radiation inactivation are thus all internally consistent for both the shark and eel enzyme. Pure catalytic subunits and glycoproteins were isolated from both enzymes by preparative SDS-gel electrophoresis and various chemical studies were carried out (Perrone et al., 1975). The amino acid composition of the catalytic subunit was different from that of the glycoprotein, but the amino acid composition of the catalytic subunit or the glycoprotein was essentially the same for both species. In fact, the amino acid composition of the large or small chain is essentially the same throughout evolution from the elasmobranchs to mammals, suggesting considerable conservation of structure throughout evolution (Hokin, 1974). However, the NH_2-terminal amino acid for the large chain differs through evolution, suggesting some variation in amino acid sequence (Hokin, 1974). The fish enzymes also show a somewhat lower content of sulfur-containing amino acids than the mammalian enzymes (Hokin, 1974). The carbohydrate compositions of the shark and eel enzymes were qualitatively the same but quantitatively different. The carbohydrates were glucosamine, sialic acid, fucose, galactose and mannose (Perrone et al., 1975). All of the sialic acid could be released by neuraminidase without affecting enzyme activity. The enzymes contained about 33% phospholipid by weight. The major phospholipids were phosphatidyl choline, phosphatidyl ethanolamine, phosphatidyl serine with lesser amounts of sphingomyelin and phosphatidylinositol. With the exception of phosphatidyl serine the content of a given phospholipid was quite different for the eel enzyme and the shark enzyme.

Since the large chain carries the acyl phosphate (aspartyl-β-phosphate) which is formed from ATP in the presence of Mg^+ and Na^+ (Alexander and Rodnight, 1970; Uesugi et al., 1971; Kyte, 1971; Avruch and Fairbanks, 1972; Collins and Albers, 1972; Hokin et al., 1973) it must carry the catalytic site. The large chain also appears to carry the ouabain binding site, since it reacts with diazoethylmalonyl cymarin on photoaffinity labeling (Ruoho and Kyte, 1974). These two observations are of interest since it is known from work with the squid axon and the red cell (Baker, 1972; Baker and Manil, 1968; Gardner and Conlon, 1972; Caldwell et al., 1960; Hoffman, 1960) that the ATP binding site is on the inside surface of the membrane and the ouabain binding site is on the outside surface of the membrane. This must mean that the large chain spans the membrane.

Evidence that the glycoprotein is an integral component of the Na,K-ATPase is based on the fact that it copurifies with the catalytic subunit, it is present as the only other major protein in all purified preparations with one possible exception (Nakao et al., 1973a,b) it crosslinks with the catalytic subunit with crosslinking reagents (Kyte, 1972) and antibody prepared against it inhibits the Na,K-ATPase activity (Rhee and Hokin, 1975). Its functional role is not understood, but one possibility is that it is involved in the orientation of the Na,K-ATPase in the membrane, since the oligosaccharides are likely to project from the outer surface of the membrane.

Reconstitution of Coupled Na^+ and K^+ Transport in Vesicles Containing the Purified Na,K-ATPase from the Rectal Gland of Squalus acanthias

Following a report by Jain et al. (1972) that a ouabain-inhibitable shortcircuit current could be demonstrated in black lipid membranes in contact with Na^+, K^+, and Mg^{2+} on both sides, and ATP and a rat brain particulate preparation containing crude Na,K-ATPase on one side, our laboratory attempted to reproduce without success these studies as well as studies with the purified rectal gland Na,K-ATPase. After finally abandoning these studies we turned our attention to the use of vesicles containing the purified rectal gland Na,K-ATPase in the hope that coupled Na^+ and K^+ transport could be reconstituted in this system. Considerable success had been achieved with vesicles containing purified proteins in the transport of other ions (Racker, 1972; Meissner and Fleischer, 1974; Hinkle et al., 1972; Racker and Stoekenius, 1974; Racker and Eytan, 1973; Racker, 1973). The two most commonly used techniques for preparation of vesicles for reconstitution studies are (1) sonication of the transport protein in the presence of phospholipid and (2) slow dialysis of a phospholipid-cholate mixture containing the transport protein. The first procedure is a classic method for preparing liposomes. The latter procedure was developed in the laboratory of Racker and has been used by him and others with considerable success in reconstituting several transport systems (Racker, 1972; Meissner and Fleischer, 1974; Hinkle et al., 1972; Racker and Stoekenius, 1974). Removal of the detergent by dialysis leads to spontaneous formation of vesicles containing both phospholipid and protein. We have not been successful to date in reconstituting coupled Na^+-K^+ transport by the sonication technique. We have had considerable success, however, in reconstituting coupled Na^+-K^+ transport by the dialysis technique.

Our method for preparing vesicles is a slight modification (Hilden et al., 1974; Hilden and Hokin, 1975) of the method of Kagawa and Racker (1971). Purified rectal gland Na,K-ATPase (1 mg/ml) is mixed with egg lecithin (20 mg/ml), cholate (10 or 20 mg/ml), NaCl (20 mM), KCl (100 mM), $MgCl_2$ (5 mM) and 30 mM imidazole buffer (pH 7.0). The mixture is dialyzed at 4°C for 2 days against a solution containing the same composition of salts and buffer but without enzyme, phospholipid or detergent. In the K^+ transport experiments half of the KCl in the dialyzing medium is replaced with choline chloride. Proof that the vesicles were formed was shown by dialyzing in the presence of either radioactive glucose or inulin and demonstrating radioactivity in the void volume with the protein after passage of the dialyzed material over Sephadex G-150. Similar results were obtained if vesicles were allowed to stand in the presence of $^{22}Na^+$, $^{86}Rb^+$, or $^{42}K^+$ and then chromatographed over Sephadex G-150. A peak consisting of the free ions or molecules eluted after the peak containing vesicles with trapped ions or molecules. As to be expected, the peak containing the free ions or molecules was much larger, since the vesicular volume was calculated to be only 1-2% of the total volume of the incubation medium. After one day's dialysis more than 90% of the cholate was dialyzed out; very little additional cholate dialyzed out on dialysis up to five days. The simplest explanation of these data is that on removal of the cholate the lecithin, endogenous phospholipid and the Na,K-ATPase self-assembled into phospholipid bilayer vesicles containing Na,K-ATPase. Bilayer vesicles were demonstrated by electron microscopy (Hilden, Perdue, and Hokin, unpublished observations). It would be expected that the Na,K-ATPase would be inserted into the vesicle randomly so that some enzyme molecules would have their substrate site facing inward and some would have their substrate site facing outward. The simplest way to orient the system was to add substrate to the outside of the vesicles by adding MgATP to the incubation system. Thus, only the Na,K-ATPase molecules with their substrate sites facing outward would be utilized and active transport of Na^+ would be from outside to inside the vesicles, and active transport of K^+ would be from inside to outside. This is opposite to the normal situation in which MgATP is inside the cell and Na^+ transport is from inside the cell to the outside and K^+ transport is from outside to inside. We could not duplicate the normal situation because we were not successful in incorporating MgATP or Mg^{2+} plus an ATP generating system inside the vesicles, presumably because of the limited vesicular volume and the very vigorous ATPase activity. Thus, in all of our studies the movements and the binding sites are opposite to those found in normal cells.

In the earlier studies only $^{22}Na^+$ influx was examined (Hilden et al., 1974). For these studies $^{22}Na^+$, Mg^{2+} and ATP were added at the beginning of incubation and the uptake of the isotope into the vesicles was followed. In later studies (Hilden and Hokin, 1975) $^{42}K^+$ or $^{86}Rb^+$ efflux was studied, often in conjunction with $^{22}Na^+$ influx. For these studies the vesicles were "preloaded" with radioactive ions by letting them stand at 4°C for at least 24 h which allowed equilibration of the radioactive ion between the inside and outside of the vesicles. Magnesium and ATP were then added to the system, it was warmed to 25°C and incubated.

Studies on the optimum temperature and pH were carried out with $^{22}Na^+$ and not with $^{42}K^+$ or $^{86}Rb^+$. The optimum temperature of ATP-dependent uptake was about 25°C. Transport fell sharply above this temperature, presumably as a result of increased lability of the vesicles. On the other hand, transport fell very little on lowering the temperature to 4°C, suggesting that the lipids in the membrane did not "freeze out" down to this temperature. In this connection it is of interest that a high percentage of the endogenous phosphatidyl serine and phosphatidyl ethanolamine crosslinks with the protein on treatment with difluorodinitrobenzene (Dixon and Hokin, unpublished observations) indicating that they may be "boundary lipids", and in the rectal gland Bergh et al. (1975) have shown that these two phospholipids have a high content of fatty acids with multiple double bonds (up to 6).

ATP-dependent sodium uptake rose very rapidly during the first 10 minutes of incubation, although the rate of uptake began to fall off after 10 minutes. After 15 min the $^{22}Na^+$ content had reached a steady state which remained fairly constant for 2 h before the $^{22}Na^+$ content began to fall slowly, presumably as a result of leakage or disruption of vesicles. No time points earlier than 10 min were measured. The pH optimum for Na,K-ATPase in the vesicles and the ATP-dependent $^{22}Na^+$ uptake was the same (pH 7.0).

When the same type of experiment was carried out in separate batches of vesicles, in one case $^{22}Na^+$ uptake being studied and in the other case $^{42}K^+$ or $^{86}Rb^+$ efflux from "preloaded" vesicles being studied, the process will be referred to as the coupled transports of Na^+ and K^+. In actuality, $^{86}Rb^+$ was used in place of $^{42}K^+$ for many of the experiments because of the much greater convenience of this longer lived isotope. $^{86}Rb^+$ gave identical results to $^{42}K^+$ (Hilden and Hokin, 1975). If MgATP was added to the vesicles there was a much greater uptake of $^{22}Na^+$ and a marked fall in the $^{42}K^+$ content of the vesicles (Fig. 1). If the cholate-phospholipid-Na,K-ATPase mixture was not dialyzed there was no significant $^{22}Na^+$ or $^{86}Rb^+$ in the void volume, indicating that no vesicles were formed. If Na,K-ATPase was added after dialysis of a cholate-lecithin mixture $^{22}Na^+$ and $^{86}Rb^+$ were found in the void volume, but there was no change in vesicular content of either $^{22}Na^+$ or $^{86}Rb^+$ if the system had been incubated with MgATP. This indicates that vesicles were formed, as to be expected, but that $^{22}Na^+$ and $^{86}Rb^+$ had entered the vesicles by diffusion only. Thus, for the Na,K-ATPase to be incorporated into the vesicles and for active transport to occur, the enzyme, phospholipid and detergent must be present before dialysis, and as the cholate is dialyzed out the system self-assembles.

Fig. 1. Time course for $^{22}Na^+$ entry and $^{42}K^+$ exit. Vesicles were preloaded with isotope for 2 days. During incubation at 25°C, 0.5 ml of vesicles was removed at indicated times and vesicular isotope contents measured in usual way. Reprinted by permission of J. Biol. Chem.

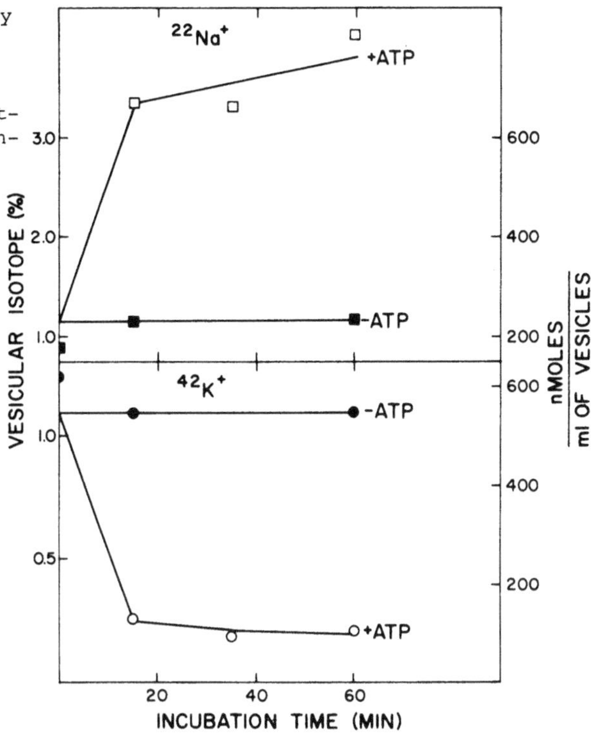

The coupled transports of Na^+ and K^+ were markedly reduced if Mg^{2+} was omitted. There was also a good parallelism between the nucleotide specificity for the Na,K-ATPase activity and the coupled transports of Na^+ and K^+. For example, CTP is a fair substrate for the Na,K-ATPase activity but UTP is not a substrate. CTP was fairly effective in driving the coupled transports but UTP was without effect.

The coupled transports of Na^+ and K^+ were not inhibited by external ouabain as would be predicted from the sidedness of ouabain inhibition, i.e. ouabain does not inhibit transport when it is only present on the inside of red cell ghosts or squid axons (Baker and Manil, 1968; Gardner and Conlin, 1972; Perrone and Blostein, 1973). In other words, ouabain does not inhibit on the side of the membrane containing the ATP site, which would be the external surface of the vesicles in our case. On the other hand, ouabain inhibits when applied to the external surface of cells. If vesicles were prepared by adding ouabain to the dialyzing medium so that ouabain was trapped inside the vesicles, the coupled transports were inhibited. Rather high concentrations of ouabain (10^{-4} M) in the dialyzing medium during preparation of the vesicles were required for complete or near complete inhibition (Hilden et al., 1974; Hilden and Hokin, 1975); however, calculations of the amounts of ouabain trapped in the vesicular volume and the nanomoles of Na,K-ATPase in the same vesicular volume indicated that at 10^{-4} M ouabain the amount of ouabain trapped inside the the vesicles was roughly equal to the nanomoles of Na,K-ATPase (Hilden and Hokin, 1975).

If vesicles were prepared in the absence of Na^+, and no Na^+ was added during incubation, no $^{86}Rb^+$ efflux was observed (Hilden and Hokin, 1975). It is well known that a variety of monovalent cations such as Rb^+, Cs^+, Li^+, NH_4^+ and Tl^+ can substitute for K^+ at the K^+ site, but the Na^+ site is absolutely specific for Na^+. As to be expected, when LiCl was substituted for NaCl no ATP-dependent $^{86}Rb^+$ efflux occurred. This experiment demonstrates the obligatory coupling of Rb^+ efflux to external Na^+. This experiment is also another example of the sidedness of the pump, since it is only necessary to add NaCl to the vesicles at the beginning of the incubation to demonstrate normal $^{86}Rb^+$ efflux.

After "preloading", the concentrations and specific activities of $^{42}K^+$ and $^{22}Na^+$ are the same on both sides of the vesicular membrane. Since the specific activities of the ions inside and outside the vesicles is known at the start of incubation one can calculate the concentration gradients established by the pumps at the steady state (see Hilden and Hokin, 1975). A typical concentration gradient for Na^+ at the steady state was 20 mM on the outside and 72 mM on the inside. A typical concentration gradient for K^+ at the steady state was 10 mM inside and 50 mM outside. Thus, the vesicles were capable of establishing considerable concentration gradients.

As discussed above, external ouabain did not inhibit coupled Na^+-K^+ transport. However, it did inhibit about 70% of the Na,K-ATPase. Thus, a maximum of only 30% of the Na,K-ATPase molecules was oriented in such a way as to support the influx of Na^+ and the efflux of K^+. If all of the Na,K-ATPase were oriented randomly and perpendicular to the plane of the membrane, and all of the vesicles were tightly sealed, one would expect to see 100% inhibition of the Na,K-ATPase, since the remainder of the Na,K-ATPase would be inaccessible to the external ATP and would be silent. It is possible that a small amount of externally added ATP enters "tight" vesicles, since if external ouabain was added the mean sodium accumulation was 25% higher than in the non-ouabain control (not statistically significant), suggesting that there may have been a small amount of diffusion of ATP into the vesicles which would drive Na^+ exit and thus oppose the normal Na^+ entry (see Hilden et al., 1974). However, it is probable that there was a significant number of leaky vesicles in which the Na,K-ATPase would presumably be fully susceptible to ouabain. To determine stoichiometries of ions transported to molecules of ATP hydrolyzed the experiments were carried out in the presence of 2×10^{-4} M external ouabain to inhibit Na,K-ATPase molecules which could not drive Na^+ entry, either because of the wrong orientation or because of leaky vesicles. The Na^+/ATP and the K^+/ATP ratios were 1.48 and 1.04, respectively. These compare to ratios of 3 and 2 in the red blood cell (Sen and Post, 1964; Whittam and Ager, 1965). Although these ratios are lower in vesicles than in red blood cells this difference is probably technical rather than fundamental. This contention is supported by the fact that the Na/ATP ratio increased from 0.3 in the earlier study (Hilden et al., 1974) to 1.5 in the later study (Hilden and Hokin, 1975). But what did remain constant and what did appear to remain a fundamental aspect of the pump was the ratio of Na^+/K^+ transport, which was 1.43. This is very close to 3 Na^+ transported in for 2 K^+ transported out. This stoichiometry is the same as that observed by others in the red blood cell (Hoffman, 1973; Whittam and Ager, 1965; Sen and Post, 1964; Glynn, 1962; Gardos, 1964; Garrahan and Glynn, 1967a). In any event, it is of considerable interest that the Na^+-K^+ pump reconstituted from a purified Na,K-ATPase should give the same stoichiometry as has been observed in intact cells or erythrocyte ghosts.

Role of Phospholipids in the Coupled Transports of Na^+ and K^+ in Vesicles

There has been much discussion about the role of phospholipids in the Na,K-ATPase (see review by Dahl and Hokin, 1974). Warren et al. (1974a,b) developed a technique by which the endogenous lipids in a membrane lipoprotein preparation can be displaced by a known phospholipid. Basically, the technique consists of mixing the membrane lipoprotein with a solution containing cholate and the phospholipid with which one wishes to substitute for the endogenous phospholipid. The mixture is placed on a discontinuous sucrose density gradient, and the material at the 15-50% sucrose interface (plus any material which had pelleted) is collected. If this procedure was repeated twice with the Na,K-ATPase and egg lecithin as the substituting agent, all of the endogenous phospholipid as measured by thin layer chromatography was substituted by lecithin (Hilden and Hokin, 1976). Vesicles were then prepared from the "lecithin" Na,K-ATPase, and the coupled transports of Na^+ and K^+ were measured. The lecithin Na,K-ATPase vesicles actually pumped Na^+ and K^+ about twice as effectively as the usual vesicles containing endogenous phospholipids (Hilden and Hokin, 1976). The transports were inhibited by ouabain. These experiments show that the "bulk" of the phospholipid in the vesicle can be lecithin and transport is as good or better than with vesicles containing endogenous phospholipids. Although thin layer chromatography is a sensitive indicator of the presence of phospholipids, these experiments do not rule out the presence of trace amounts of non-lecithin phospholipids which may be required for transport.

Exchange Diffusion of Na^+

With erythrocyte preparations exchange diffusion of Na^+ without net Na^+ transport can be demonstrated in the absence of K^+ and in the presence of ATP (Garrahan and Glynn, 1967b). This process is ouabain inhibitable. This exchange of one Na^+ on one side of the membrane for one Na^+ on the other side of the membrane appears to be a property of the Na,K-ATPase. Vesicles were prepared in the absence of KCl and in the presence of 120 mM NaCl (Hilden et al., 1974). If these vesicles were incubated as usual there was a much smaller uptake of $^{22}Na^+$ than under standard conditions. If this small ATP-dependent Na^+ uptake were due to exchange diffusion rather than active transport, the ATP-dependent uptake should disappear when the specific activity is the same inside and outside the vesicles. If vesicles were preloaded with $^{22}Na^+$ as described above, but in the absence of K^+ so as to equilibrate the $^{22}Na^+$ on both sides of the membrane, there was no ATP-dependent $^{22}Na^+$ uptake on incubation in the presence of MgATP. This shows that the small uptake of $^{22}Na^+$ in the non-preloaded vesicles in the absence of K^+ was due to exchange diffusion. If vesicles were prepared and preloaded with $^{22}Na^+$ in the presence of K^+ and incubated in the presence of K^+, an ATP-dependent $^{22}Na^+$ uptake could be demonstrated. This shows that active transport of Na^+ still occurred in the preloaded vesicles in the presence of K^+. The exchange diffusion of Na^+ in the absence of K^+ was inhibited by ouabain as has been observed in erythrocyte preparations (Garrahan and Glynn, 1967b). Exchange diffusion of Na^+ was inhibited only when ouabain was present inside the vesicles.

Exchange Diffusion of K^+

Exchange diffusion of K^+ can also be demonstrated in erythrocyte ghosts if they are prepared in the absence of Na^+ (Simons, 1974). This exchange diffusion of K^+ is ouabain sensitive and is dependent on ATP and Pi. If vesicles containing Na,K-ATPase were prepared in a medium not containing Na^+, K^+-K^+ exchange should result in a faster rate of equilibration of external $^{42}K^+$ with internal nonradioactive K^+ than that due to simple diffusion. This was in fact demonstrated by the observation that the movement of $^{42}K^+$ into the vesicles in the absence of Na^+ was much faster if ATP and Pi were added (Hilden and Hokin, 1975). If Na^+ was added to the vesicles that were also in contact with ATP and Pi, the active transport of $^{42}K^+$ outward was activated. Addition of Na^+ had no effect on the time course of equilibration in control vesicles incubated without ATP and Pi. These data demonstrate exchange diffusion of K^+ in the Na,K-ATPase vesicles. This exchange diffusion of K^+ was inhibited by internal ouabain but not by external ouabain, as might be predicted.

Other Studies on the Reconstitution of Na^+ and K^+ Transport

Goldin and Tong (1974) were able to reconstitute Na^+ transport but not coupled K^+ transport when they used vesicles prepared from the Na,K-ATPase purified from the outer renal medulla of the dog by the method of Kyte (1971a,b). Instead, Cl^- was co-transported with Na^+ to maintain electroneutrality. In a later note Sweadner and Goldin (1975) reconstituted coupled Na^+ and K^+ transport from a dog brain microsome preparation treated with deoxycholate and sodium iodide (Nakao et al., 1973a,b). Based on its specific acitivity this Na,K-ATPase preparation was only 5% pure. Using microsomes from the electroplax of *Electrophorus electricus* prepared by a modification of the method of Albers et al., (1963) and a sonication method to prepare vesicles, Racker and Fisher (1975) reconstituted Na^+ transport. Based on their reported specific activity their enzyme would be about one-third pure. No data on K^+ transport were presented. By measuring initial rates of Na^+ transport at 2 min Racker and Fisher (1975) found Na^+ translocation rates of 100-200 natoms/min/mg protein. They claimed that these rates were 10 to 20 times higher than in vesicles reconstituted by the cholate-dialysis procedure. However, in comparing their rates with those reported by others in some instances they did not calculate from initial rates but from steady states at 1 h or longer. For example, in the studies of Hilden and Hokin (1975) the earliest time studied was 10 min at which point the steady state was being approached (see Fig. 1). Even at this point an average of 1,000 nanomoles Na^+ had been accumulated per mg protein. This gives a rate of 100 nmol Na^+/mg protein/min, which is within the same range found by

Racker and Fisher (1975) who used 2 min to measure their initial rates.

Acknowledgments. This work was aided by grants from the National Institutes of Health (HL-16318) and the National Science Foundation (BMS 73-01506).

References

Albers, R.W., Fahn, S., Koval, G.J.: The role of sodium ions in the activation of electrophorus electric organ adenosine triphosphatase. Proc. Natl. Acad. Sci. U.S. 50, 474-481 (1963)

Alexander, D.R., Rodnight, R.: Separation at neutral pH of a ^{32}P-labelled membrane protein associated with the sodium-plus-magnesium ion-activated adenosine triphosphatase from ox brain. Biochem. 119, 44-45 (1970)

Avruch, J., Fairbanks, G.: Demonstration of a phosphopeptide intermediate in the Mg^{++}-dependent, Na^{+}- and K^{+}-stimulated adenosine triphosphatase reaction of the erythrocyte membrane. Proc. Natl. Acad. Sci. U.S. 69, 1216-1220 (1972)

Baker, P.F.: Metabolic Pathways. Vol. VI: Metabolic Transport: The Sodium Pump in Animal Tissues and its Role in the Control of Cellular Metabolism and Function. New York: Academic Press 1972, pp. 243-268

Baker, P.F., Manil, J.: The rates of action of K^{+} and ouabain on the sodium pump in squid axons. Biochim. Biophys. Acta 150, 328-330 (1968)

Bergh, C.H., Larson, G., Samuelson, B.E.: Fatty acid and aldehyde composition of major phospholipids in salt gland of marine birds and spiny dogfish. Lipids 10, 299-302 (1975)

Bonting, S.L.: Studies on sodium-potassium-activated adenosine triphosphatase. XV. The rectal gland of the elasmobranchs. Comp. Biochem. Physiol. 17, 953-966 (1966)

Caldwell, P.C., Hodgkin, A.L., Keynes, R.D., Shaw, T.I.: The effects of injecting "energy rich" phosphate compounds on the active transport of ions in the giant axons of LOLIGO. J. Physiol. 152, 561-590 (1960)

Collins, R.C., Albers, R.W.: The phosphoryl acceptor protein of Na-K-ATPase from various tissues. J. Neurochem. 19, 1209-1213 (1972)

Dahl, J., Hokin, L.E.: The sodium-potassium adenosine triphosphatase. Ann. Rev. Biochem. 43, 327-356 (1974)

Dixon, J.F., Hokin, L.E.: Studies on the characterization of the sodium-potassium transport adenosine triphosphatase. Arch. Biochem. Biophys. 163, 749-758 (1974)

Edelman, I.S.: Drugs and Transport Processes: Thyroid and Sodium Transport. London: Macmillan 1974, pp. 101-110

Gardner, J.D., Conlon, T.P.: The effects of sodium and potassium on ouabain binding by human erythrocytes. J. Gen. Physiol. 60, 609-629 (1972)

Gardos, G.: Connection between membrane adenosine triphosphatase activity and potassium transport in erythrocyte ghosts. Experientia 20, 387 (1964)

Garrahan, P.J., Glynn, I.M.: Factors affecting the relative magnitude of the sodium-potassium and sodium: sodium exchanges catalysed by the sodium pump. J. Physiol. 192, 189-216 (1967a)

Garrahan, P.J., Glynn, I.M.: The stoicheiometry of the sodium pump. J. Physiol. 192, 217-235 (1967b)

Glynn, I.M.: Activation of adenosine triphosphatase activity in a cell membrane by external potassium and internal sodium. J. Physiol. 160, 18 (1962)

Glynn, I.M., Karlish, S.J.D.: The sodium pump. Ann. Rev. Physiol. 37, 13-55 (1975)

Goldin, S.M., Tong, S.W.: Reconstitution of active transport catalyzed by the purified sodium and potassium ion-stimulated adenosine triphosphatase from canine renal medulla. J. Biol. Chem. 249, 5907-5915 (1974)

Hilden, S., Hokin, L.E.: Active potassium transport coupled to active sodium transport in vesicles reconstituted from purified sodium and potassium ion-activated adenosine triphosphatase from the rectal gland of Squalus acanthias. J. Biol. Chem. 250, 6296-6303 (1975)

Hilden, S., Hokin, L.E.: Coupled Na^+-K^+ transport in vesicles containing a purified (NaK)-ATPase and only phosphatidyl choline. Biochem. Biophys. Res. Commun. 69, 521-527 (1976)

Hilden, S., Rhee, H.M., Hokin, L.E.: Sodium transport by phospholipid vesicles containing purified sodium and potassium ion-activated adenosine triphosphatase. J. Biol. Chem. 249, 7432-7440 (1974)

Hinkle, P.C., Kim, J.J., Racker, E.: Ion transport and respiratory control in vesicles formed from cytochrome oxidase and phospholipids. J. Biol. Chem. 247, 1338-1339 (1972)

Hoffman, J.F.: The link between metabolism and the active transport of Na^+ in human red cell ghosts. Federation Proc. 19, 127 (1960)

Hoffman, J.F.: Organization of Energy-Transducing Membranes: The Molecular Aspects of Na^+,K^+-pump and Red Blood Cells. Baltimore: University Park Press 1973, pp. 9-21

Hokin, L.E.: Purification and properties of the (sodium + potassium)-activated adenosine triphosphatase and reconstitution of sodium transport. Ann. N.Y. Acad. Sci. 242, 12-23 (1974)

Hokin, L.E., Dahl, J.D., Deupree, J.D., Dixon, J.F., Hackney, J.F., Perdue, J.R: Studies on the characterization of the sodium-potassium transport adenosine triphosphatase. J. Biol. Chem. 248, 2593-2605 (1973)

Jain, M.K., White, F.P., Strickholm, A., Willians, E., Cordes, E.H.: Studies concerning the possible reconstitution of an active cation pump across an artificial membrane. J. Membrane Biol. 8, 363-388 (1972)

Jorgensen, P.L.: Purification and characterization of (Na^++K^+)-ATPase. III. Purification from the outer medulla of mammalian kidney after selective removal of membrane components by sodium dodecylsulphate. Biochim. Biophys. Acta 356, 36-52 (1974)

Jorgensen, P.L.: Isolation and characterization of the components of the sodium pump. Quart. Rev. Biophys. 7, 239-274 (1975)

Kagawa, Y., Racker, E.: Partial resolution of the enzymes catalyzing oxidative phosphorylation. XXV. Reconstitution of vesicles catalyzing ^{32}Pi-adenosine triphosphate exchange. J. Biol. Chem. 246, 5477-5487 (1971)

Kepner, G.R., Macy, R.I.: Membrane enzyme systems molecular size determinations by radiation inactivation. Biochim. Biophys. Acta 163, 188-203 (1968)

Kyte, J.: Phosphorylation of a purified ($Na^+ + K^+$) adenosine triphosphatase. Biochem. Biophys. Res. Commun. 43, 1259-1265 (1971a)

Kyte, J.: Purification of the sodium- and potassium-dependent adenosine triphosphatase from canine renal medulla. J. Biol. Chem. 246, 4156-4165 (1971b)

Kyte, J.: Properties of the two polypeptides of sodium- and potassium-dependent adenosine triphosphatase. J. Biol. Chem. 247, 7642-7649 (1972)

Laemmle, U.V.: Cleavage of structural proteins during the assembly of the head of bacteriophage T4. Nature (London) 227, 680-685 (1970)

Lane, K., Copenhaver, J.H., Lindenmayer, G.E., Schwartz, A.: Purification and characterization of and [^{3}H] ouabain binding to the transport adenosine triphosphatase from outer medulla of canine kidney. J. Biol. Chem. 248, 7197-7200 (1973)

Meissner, G., Fleischer, S.: Dissociation and reconstitution of functional sarcoplasmic reticulum vesicles. J. Biol. Chem. 249, 302-309 (1974)

Nakao, T., Nakao, M, Mizuno, N., Komatsu, Y., Fujita, M.: Purification and some properties of Na,K-ATPase. I. Solubilization and stability of Lubrol extracts. J. Biochem. 73, 609-619 (1973a)

Nakao, T., Nakao, M., Nagai, F., Kawai, K., Fujihara, U., Hara, Y., Fujita, M.: Purification and some properties of Na,K-transport ATPase. II. Preparations with high specific activity obtained using aminoethyl cellulose chromatography. J. Biochem. 73, 781-791 (1973b)

Perrone, J.R., Blostein, R.: Asymmetric interaction of inside-out and right-side-out erythrocyte membrane vesicles with ouabain. Biochim. Biophys. Acta 291, 680-689 (1973)

Perrone, J.R., Hackney, J.F., Dixon, J.D., Hokin, L.E.: Molecular properties of purified (sodium and potassium)-activated adenosine triphosphatase and their subunits from the rectal gland of Squalus acanthias and the electric organ of Electrophorus electricus. J. Biol. Chem. 250, 4178-4184 (1975)

Racker, E.: Reconstitution of a calcium pump with phospholipids and a purified Ca^{++}-adenosine triphosphatase from sarcoplasmic reticulum. J. Biol. Chem. 247, 8198-8220 (1972)

Racker, E.: A new procedure for the reconstitution of biologically active phospholipid vesicles. Biochem. Biophys. Res. Commun. 55, 224-230 (1973)

Racker, E., Eytan, E.: Reconstitution of an efficient calcium pump without detergents. Biochem. Biophys. Res. Commun. 55, 174-178 (1973)

Racker, E., Fisher, L.W.: Reconstitution of an ATP-dependent sodium pump with an ATPase from electric eel and pure phospholipids. Biochem. Biophys. Res. Commun. 67, 1144-1150 (1975)

Racker, E., Stoekenius, W.: Reconstitution of purple membrane vesicles catalyzing light-driven proton uptake and adenosine triphosphatase formation. J. Biol. Chem. 249, 662-663 (1974)

Repke, K.R.H., Schon, R., Henke, W., Schonfeld, W., Streckenbach, B., Dittrich, F.: Experimental and theoretical examination of the flip-flop model of (Na-K)-ATPase function. Ann. N.Y. Acad. Sci. 242, 203-219 (1974)

Rhee, H.M., Hokin, L.E.: Inhibition of the purified sodium-potassium activated adenosine triphosphatase from the rectal gland of Squalus acanthias by antibody against the glycoprotein subunit. Biochem. Biophys. Res. Commun. 63, 1139-1145 (1975)

Ruoho, A., Kyte, J.: Photoaffinity labeling of the ouabain-bending site on $(Na^+ + K^+)$ adenosine triphosphatase. Proc. Natl. Acad. Sci. U.S. 71, 2352-2356 (1974)

Schwartz, A., Lindenmayer, G.E., Allen, J.C.: The sodium-potassium adenosine triphosphatase: pharmacological, physiological and biochemical aspects. Pharmacol. Rev. 27, 3-134 (1975)

Sen, A.K., Post, R.L.: Stoichiometry and localization of adenosine triphosphatase-dependent sodium and potassium transport in the erythrocyte. J. Biol. Chem. 239, 345-352 (1964)

Simons, T.J.B.: Potassium:potassium exchange catalysed by the sodium pump in human red cells. J. Physiol. 237, 123-155 (1974)

Simpkins, H., Hokin, L.E.: Studies on the characterization of the sodium-potassium transport adenosine triphosphatase. XIII. On the organization and role of phospholipids in the purified enzyme. Arch. Biochem. Biophys. 159, 897-902 (1973)

Skou, J.C.: The influence of some cations on an adenosine triphosphatase from peripheral nerves. Biochim. Biophys. Acta 23, 394-401 (1957)

Skou, J.C.: The $(Na^+ + K^+)$ activated enzyme system and its relationship to transport of sodium and potassium. Quart. Rev. Biophys. 7, 401-434 (1975)

Stein, W.D., Lieb, W.R., Karlish, S.J.D., Eilam, Y.: A model for active transport of sodium and potassium ions as mediated by a tetrameric enzyme. Proc. Natl. Acad. Sci. U.S. 70, 275-278 (1973)

Sweadner, K.J., Goldin, S.M.: Reconstitution of active ion transport by the sodium and potassium ion-stimulated adenosine triphosphatase from canine brain. J. Biol. Chem. 250, 4022-4024 (1975)

Uesugi, S., Dulak, M.C., Dixon, J.F., Hexum, T.D., Dahl, J.D., Perdue, J.F., Hokin, L.E.: Studies on the characterization of the sodium-potassium transport adenosine triphosphatase. VI. Large scale partial purification and properties of a Lubrol-solubilized bovine brain enzyme. J. Biol. Chem. 246, 531-543 (1971)

Warren, G.B., Toon, P.A., Birdsall, N.J.M., Lee, A.G., Metcalfe, J.C.: Complete control of the lipid environment of membrane-bound proteins: Application to a calcium transport system. FEBS Lett. 41, 122-124 (1974a)

Warren, G.B., Toon, P.A., Birdsall, N.J.M., Lee, A.G., Metcalfe, J.C.: Reconstitution of a calcium pump using defined membrane components. Proc. Natl. Acad. Sci. U.S. 71, 622-626 (1974b)

Whittam, R., Ager, M.E.: The connexion between active cation transport and metabolism in erythrocytes. Biochem. J. 97, 214-227 (1965)
Whittam, R., Chipperfield, A.R.: The reaction mechanism of the sodium pump. Biochim. Biophys. Acta 415, 149-171 (1975)

Some Aspects of the Ca-Pump in Human Red Blood Cells

H. J. SCHATZMANN and B. ROELOFSEN

Characteristics of the Ca-Pump in Human Red Blood Cells (HRBC)

HRBCs are endowed with a membrane-bound system which extrudes Ca^{2+} ions against a large gradient from the cytosol to the medium (Schatzmann, 1966; Schatzmann and Vincenzi, 1969; Lee and Shin, 1969; Olson and Cazort, 1969; Ferreira and Lew, 1976; Sarkadi et al., 1976). The intracellular Ca^{2+} concentration is presumably not higher than 10^{-7} M (Simons, 1976b) and in consequence there is a gradient of the order of 10^4 across the membrane under normal conditions. The leak flux of Ca^{2+} down this gradient is 1 to 10 μmol/h · (l cells) (Lew, 1971; Lew, personal communication). Unless there were an outwardly directed Ca pump even this small Ca leak would allow, within a few hours, Ca^{2+} concentrations to build up inside the cell that interfere with the normal barrier function of the membrane for K (Simons, 1976a,b; Sarkadi et al., 1976) and with the normal shape and compliance of the cells (Sarkadi et al., 1976).

The Ca pump is directly dependent on ATP hydrolysis for its energy supply, it requires the presence of Mg^{2+} ions inside the cell and its affinity for Ca^{2+} is high on the internal surface of the membrane ($K_{Ca} \sim 10^{-6}$ M) (Schatzmann, 1973). It is insensitive to cardiac glycosides at concentrations which fully inhibit the Na-K pump. The specificity for ATP as compared to other nucleotides is very high (Watson et al., 1971). The system accepts Ca^{2+} or Sr^{2+} probably equally well but does not transport other divalent cations (Schatzmann and Vincenzi, 1969; Olson and Cazort, 1969).

The system may be looked upon as an ATPase requiring Mg^{2+} and Ca^{2+} ions. The (Ca^{2+} + Mg^{2+})-activated ATPase function of the membrane shows the characteristic asymmetry which one would expect from an uphill transport system: Ca^{2+} activates only if it is present on the internal membrane surface (Schatzmann and Vincenzi, 1969), ATP has access to the system only from inside the cell and inorganic phosphate (P_i) is liberated at the internal membrane surface. The latter point is illustrated in Table 1.

There is hardly any room for doubt that it is the "high affinity" (Ca + Mg)-ATPase, with a $K_{Ca} \sim 10^{-6}$ M, demonstrated in disrupted HRBC membranes, which reflects the activity of the Ca-transporting system. Wolf et al. (1976) have recently succeeded in partially purifying the corresponding protein by solubilizing HRBC membranes with Triton X-100 and transferring the protein to mixed micelles of phosphatidylcholine and Triton X-100. The protein was found to have a molecular weight of about 140,000 and its properties in the artificial lipid environment such as K_m, K_{Ca} and pH-profile after the purifying procedure did not significantly differ from those found in the native, membrane-bound enzyme.

Table 1. Liberation of inorganic phosphate (P_i) from ATP in resealed HRBCs during Ca-pumping activity (single experiment)

	P_i in cells		P_i in medium	
	Net concentration (µmol/ml cells)	Increment (µmol/ ml cells · 5 min)	Net concentration (µmol/ml medium)	Increment (µmol/ ml cells · 5 min)
0 min	1.26		0.087	
		0.64		0.099
5 min	1.90		0.195	
		0.56		0.10
10 min	2.46		0.304	

(Schatzmann, unpublished)

Fresh cells loaded with 5 mM Ca-EGTA buffer (Ca^{2+} conc. = 2 × 10^{-5}M), 2 mM Na-ATP, 4 mM $MgCl_2$. Resealing with KCl. Medium: (mM) 100 KCl, 50 imidazole-Cl, 4 $MgCl_2$, 0.1 $CaCl_2$, pH 6.8. Hematocrit 0.54. Notice that P_i leakage is sufficiently slow to allow the amount of P_i appearing inside cells to exceed that in medium by a factor of 6

Several authors convincingly demonstrated that Ca^{2+} stimulates the formation of a phosphorylated protein in HRBC membranes incubated with ATP. The incorporated phosphate undoubtedly is transferred from the γ-position of ATP to the protein and the phosphoprotein is in all likelihood an intermediate product in the cyclic performance of the Ca-pump, because this phosphate transfer was also found by Wolf et al. (1976) in the purified (Ca^{2+} + Mg^{2+})-activated ATPase. Rega and Garrahan (1975) propose that at non-limiting ATP concentrations Mg^{2+} is not required for phosphorylation, but that it is instrumental in transforming the original phosphorylated intermediate into a form which readily yields the phosphate group to water, which would explain the requirement for Mg^{2+} found in the ATPase reaction.

It has been shown that the (Ca^{2+} + Mg^{2+})-stimulated activity of the membrane ATPase is further enhanced by the presence of Na^+ or K^+ (also Rb^+ and NH^+_4) (Schatzmann and Rossi, 1971; Bond and Green, 1971; Wolf, 1973a). This property is preserved in the purified material and awaits an explanation, since evidence is lacking that Ca^{2+} transport requires the presence of any of these monovalent cations.

For the maximal rate of transport Lew (1976) found 7-12 µmol/(ml cells)·h at 37°C in intact red cells loaded with Ca by means of the ionophore A 23187, which is in accord with what we find in our more recent experiments with resealed cells (see Fig. 2) in which the initial rate at an internal Ca^{2+} concentration of 10^{-5} M was 9.6 µmol/(ml cells)·h. In earlier experiments we arrived at approximately the same figure at 28°C. This is a marked discrepancy, in view of the fact that the transport has a Q_{10} exceeding 3 (Schatzmann and Vincenzi, 1969; Lee and Shin, 1969), for which we have no explanation.

Both the transport and the (Ca + Mg)-stimulated ATPase activity are blocked by lanthanum^{III} (Quist and Roufogalis, 1975) or other trivalent lanthanides like holmium or praseodymium (Schatzmann and Tschabold, 1971) at about equal concentrations (10^{-3} to 10^{-4} M). The site attacked by lanthanides is probably located on the external surface of the membrane, because the effect is seen in resealed cells (Fig. 1) or in intact cells (Sarkadi et al., 1976) when lanthanides are only present in the medium.

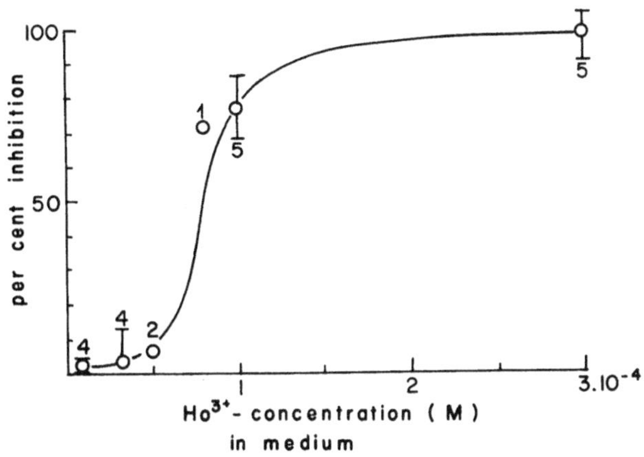

Fig. 1. Inhibitory action of external holmium^{III} on Ca^{2+} transport in resealed HRBC. Cells starved for 17 h at 37°C in glucose-free medium and loaded by haemolysis in (mM) 2 Na_2 ATP, 4 $MgCl_2$, 5 Ca-EGTA (Ca^{2+} = 2.10^{-5} M), 5 imidazole-Cl, pH 6.8. Resealing in KCl-imidazole solution. Total internal Ca conc. achieved: 1-3 mmol/l cells. Medium: (mM) KCl 100, imidazole-Cl 50 (pH 6.8), $CaCl_2$ 0.1, $MgCl_2$ 4, $HoCl_3$ as indicated. Incubation 10 min at 28°C. Ca conc. in medium and cells measured by atomic absorption flame photometry. Ordinate: Inhibition of Ca transport measured during 5 or 10 min. Abscissa: Ho^{3+} conc. in medium. Points: mean ± SEM (number of experiments near points). ATP-free cells in 0.5 mM $CaCl_2$ medium showed no inward leak for Ca^{2+}. Note: the inhibition is reversible upon washing of cells (not shown) (Schatzmann, unpublished)

Stoichiometry

In studying the relation between the number of Ca^{2+} ions transported and the number of ATP molecules hydrolyzed (Ca/ATP ratio) by the system, it is important to use resealed cells loaded with exogenous ATP and prepared from cells with thoroughly depleted energy stores, in order to avoid any interference from ATP regenerating processes. In our experiments depletion was achieved by starving the cells for 17 h at 37°C in a glucose-free medium prior to loading with Ca-EGTA and MgATP and resealing. If appropriate corrections for cells leaky to Ca^{2+} and for the basal Mg-activated ATPase are applied (see legend, Fig. 2) and if Na+K-activated ATPase is suppressed by keeping Na and K concentrations low, the Ca/ATP ratio appears to be 1 (Fig. 2). Quist and Roufogalis (1975) have published results suggesting that the La^{3+}-inhibited activity is characterized by a Ca/ATP ratio of 2. This result would be of considerable interest because it might mean that some ATPase activity (50%) remains when the transport of Ca^{2+} is blocked by La^{3+} from the outside surface of the membrane. The possibility that in our experiments, with an internal Ca^{2+} concentration of 10^{-5} M, spectrin contributes 50% of the ATPase activity is unlikely in view of its high K_{Ca} of 10^{-3} M (Rosenthal et al., 1970).

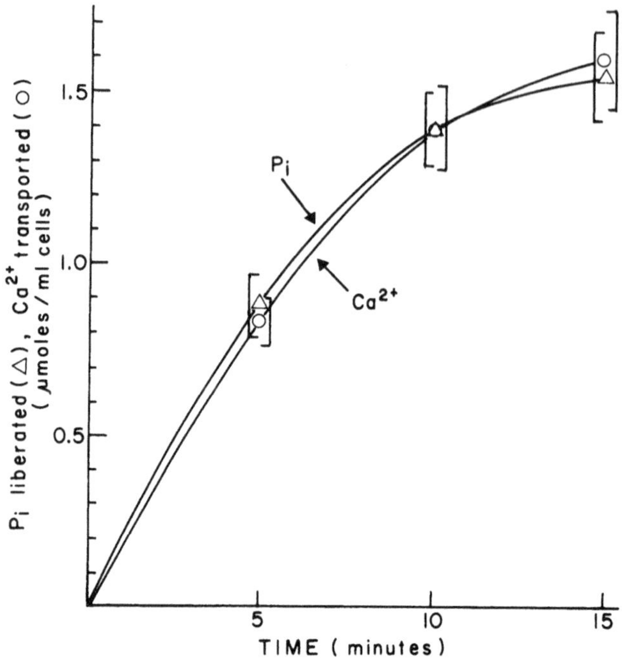

Fig. 2. Stoichiometry of Ca transport and ATPase reaction in resealed HRBC. Washed HRBCs were starved in (mM) NaCl 120, KCl 5, tris-Cl 30, $MgCl_2$ 2, pH 7.4 for 17 h at 37°C, and loaded by hemolysis in (mM) tris-EGTA-$CaCl_2$ buffer 5 (Ca^{2+} conc. = 10^{-5} M), $MgCl_2$ 2, MgATP 2, imidazole-Cl 10, pH 7.0. Isotonicity restored after 135 sec with choline-Cl + imidazole-Cl. Medium: (mM) choline-Cl 120, imidazole-Cl 30, $MgCl_2$ 2, MgATP 0.1, $CaCl_2$ 0.1 or $CaCl_2$ 0.1 + tris-EGTA 5; pH 7.0. Mean hematocrit: 0.204; mean K^+ conc.: 0.15 mM; mean Na^+ conc.: 0.068 mM in suspension. Incubation at 37°C for time indicated. Ca conc. in cells measured by atomic absorption flame photometry. P_i measured in suspension after precipitation with trichloroacetic acid according to Berenblum and Chain (1938). Leak for Ca^{2+} measured during 10 min in ^{45}Ca containing solution, expressed as % of leaky cells (average = 1.75%). 3 different blood samples. Points: mean of result in Ca^{2+} 0.1 mM and Ca^{2+}-free medium (not different) ± SEM (6 determinations). Correction for Mg-ATPase: After the experiment one sample was frozen-thawed, washed without losses and incubated in (mM) choline 120, imidazole 30, $MgCl_2$ 7, Mg-ATP 2, tris-EGTA 5, pH 7.0 for 10 min at 37°C (Schatzmann, unpublished)

Kinetics

Recently Lew (1976) has provided evidence to the effect that more than one Ca^{2+} ion is involved in the transport cycle. His results can be fitted by assuming two simultaneously occupied sites for Ca^{2+} of about equal affinity for Ca^{2+}. His argument does not touch upon the question of the number of Ca^{2+} ions transported per one ATP molecule hydrolyzed, but he explicitly states the opinion that both Ca^{2+} ions bound to the system are moved across the hydrophobic barrier. If the Ca/ATP ratio as defined above is in fact unity his scheme would require also two ATP molecules per cycle. However, the possibility remains that only one of the two Ca^{2+} ions is trans-

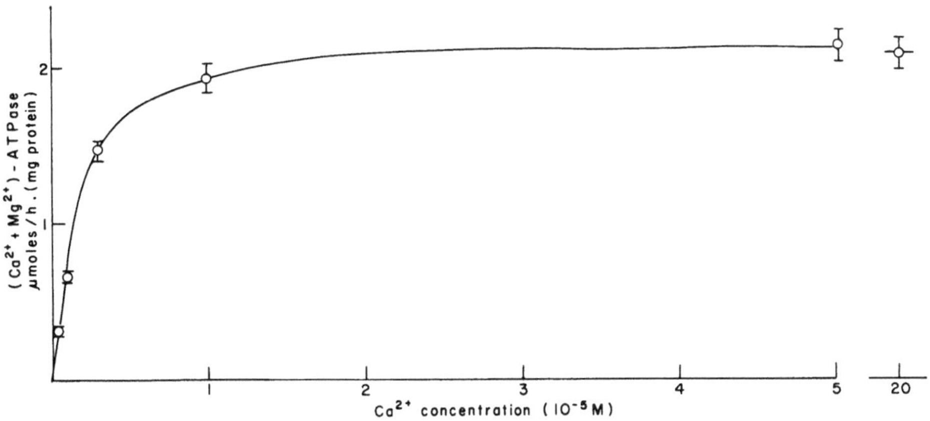

Fig. 3. Activation by Ca^{2+} of ATPase in isolated membranes prepared in absence of Ca complexing agents as described earlier (Schatzmann, 1973). Ordinate: (Ca + Mg)-activated ATPase. Points: mean ± SEM, 23 experiments. Medium: (mM) choline-Cl 120, imidazole-Cl 30, $MgCl_2$ 2, MgATP 2, EGTA-Ca buffer 0.5, ouabain 0.17; pH 7.0, 90 min, 37°C. Protein in sample 0.3-0.6 mg, vol. 2.5 ml. P_i liberated measured according to Berenblum and Chain (1938) (Roelofsen and Schatzmann, unpublished)

ported while the other, although necessary for activation, is not translocated. So far it has been held that the (Ca + Mg)-activated ATPase function of the system is related to the Ca^{2+} concentration by a simple saturation curve which, however, seems incompatible with Lew's observation on Ca transport. The kinetics of the ATPase are complicated by the appearance of a "low affinity" enzyme under certain isolation procedures designed to yield hemoglobin-free membranes. The method of preparing HRBC membranes which we adopted essentially from Wolf (see Schatzmann, 1973) and which consists of using low ionic strength, slightly elevated pH (7.7) and no Ca-chelating agents, circumvents this difficulty by yielding rather high activity of a homogenous enzyme. As can be seen from Figure 3, the curve for activation of ATPase by Ca^{2+} does not show any evidence for a dual nature of the enzyme. If, however, a large number of experiments are pooled a slight sigmoidicity in the activation curve is detected (Fig. 3), which suggests some cooperativity with respect to Ca^{2+}. As may be seen from Figure 4, the Hill coefficient is 1.3 and differs from 1 in a statistically significant way. In hoping that cooperative behavior of Ca^{2+} can be inferred from such experiments one ought to realize that the conclusion depends entirely on the reliability of EGTA as a buffer substance for Ca^{2+} concentration. Presently there is no way of checking experimentally whether the buffer system behaves as expected from the calculation (Wolf, 1973b) at all concentrations.

Effect of Manganese^{II} on the $(Ca^{2+} + Mg^{2+})$-ATPase

Many divalent cations other than Ca^{2+} and Sr^{2+} activate the ATPase function of isolated HRBC membranes in the presence of Mg^{2+} (Pfleger and Wolf, 1975). Among these is Mn^{2+}. The affinity for Mn^{2+} of the

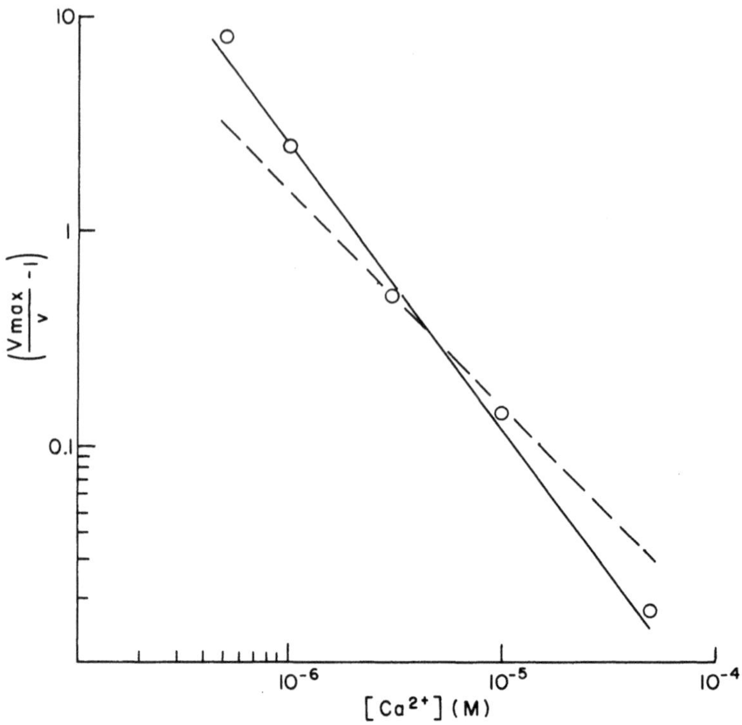

Fig. 4. Hill-plot of data of Fig. 3. Calculated regression line (solid) with a slope of 1.3 (P = 0.012 for deviation from slope 1 (dashed line))

system is higher than for Ca^{2+} ($K_{Mn} \leq 10^{-8}$ M) but the maximal rate obtained with Mn^{2+} is only about one half of that elicited by Ca^{2+}. However, Mn^{2+} is not transported in resealed cells (Fig. 5) and there is some evidence that the ATPase activation observed in isolated membranes is absent in resealed cells. For the latter reason the ATPase activation by Mn^{2+} in isolated membranes might be dismissed as an artifact. It has nonetheless some peculiarities which might shed light on the nature of the Ca transporting system. Figure 6 demonstrates that the Mn^{2+} activation is not additive with the effect of maximal stimulation by Ca^{2+}, which indicates that Mn^{2+} and Ca^{2+} act on the same system. Furthermore, it may be seen that there is no competition between the two ions of the sort one would expect if the action were on one and the same site. The inhibition of the Ca effect by Mn^{2+} which is observed does not tend to be complete at infinite Mn^{2+} concentration (Fig. 6, inset), a behavior which is usually taken to indicate a mechanism of allosteric inhibition. Mn^{2+} does not seem to stimulate the formation of a phosphorylated intermediate of the molecular weight observed when Ca^{2+} is used to activate the system (Fig. 7) but possibly some phosphorylation is induced by Mn^{2+} in proteins of about 200,000 molecular weight.

These observations suggest that Mn^{2+} activates the same (transport-) protein as Ca^{2+} but by occupying a position other than the transport site, which becomes accessible to Mn^{2+} during isolation of membranes, possibly due to increased mobility of membrane proteins after partial removal of spectrin. It might be speculated that this second site is accessible to Ca^{2+} even in the intact

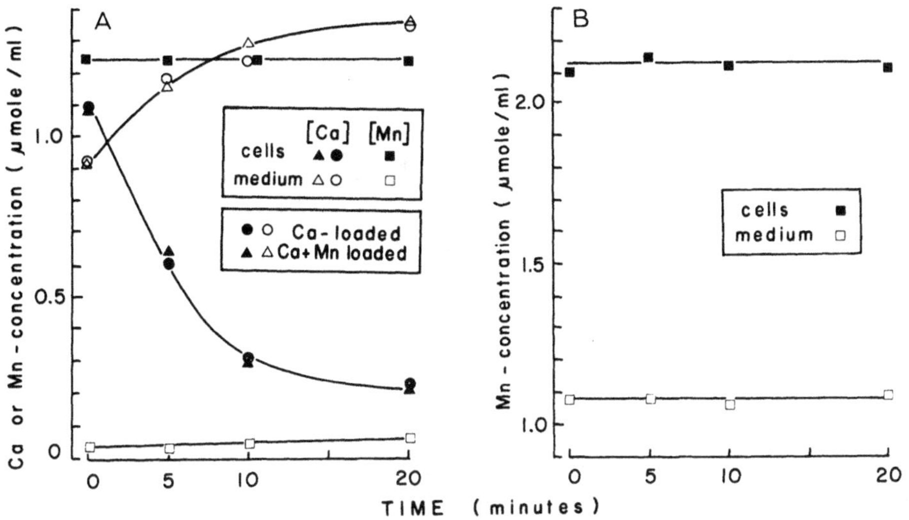

Fig. 5. Transport in resealed HRBCs. Cells hemolyzed in presence of 0.75 mM $CaCl_2$ (A) or $MnCl_2$ (B), or both (A). Isotonicity restored after 135 sec at room temp. with KCl. Other conditions as in Fig. 2. Medium: (mM) KCl 100, imidazole-Cl 50, $MgCl_2$ 4, $CaCl_2$ 1 (in A) and $MnCl_2$ 1 (in B); pH 7.0. Single experiment. Ca determined by atomic absorption flame photometry in cells and medium. Note that Mn^{2+} is not transported and does not interfere with Ca transport (Schatzmann, unpublished)

membrane of whole or resealed cells and must be occupied by Ca^{2+} in order to allow the actual transport site to operate.

Phospholipid Requirement

Several authors have pointed out both for the sarcoplasmic reticulum (Martonosi et al., 1971; The and Hasselbach, 1973; Nakamura and Ohnishi, 1975) and the red cell membrane (Coleman and Bramley, 1975) that phospholipids are necessary for the performance of the (Ca^{2+} + Mg^{2+})-activated system. The procedure which was successful in demonstrating the phospholipid requirement of the Na-K pump (Roelofsen and van Deenen, 1973) can also be applied to the Ca-transport system. It consists in applying pure phospholipases (devoid of any proteolytic or hemolytic activity) with different substrate specificities either to intact HRBC or to disrupted red cell membranes, which amounts to enzymatic degradation of phospholipids either present in the outer leaflet alone or in the total membrane. Table 2 shows that outer leaflet phospholipids can be degraded completely without affecting the (Ca + Mg)-ATPase activity. On the other hand the activity is reduced to zero by attack of disrupted membranes with B. cereus phospholipase C, degrading 100% of the total glycerophospholipids (phosphatidylcholine, phosphatidylethanolamine and phosphatidylserine), whereas complete degradation of sphingomyelin has no effect whatsoever (Roelofsen and Schatzmann, 1976). From Figure 8 it is evident that an excellent fit to a line indicating direct proportionality between residual activity and residual glycerophospholipid content is obtained when only

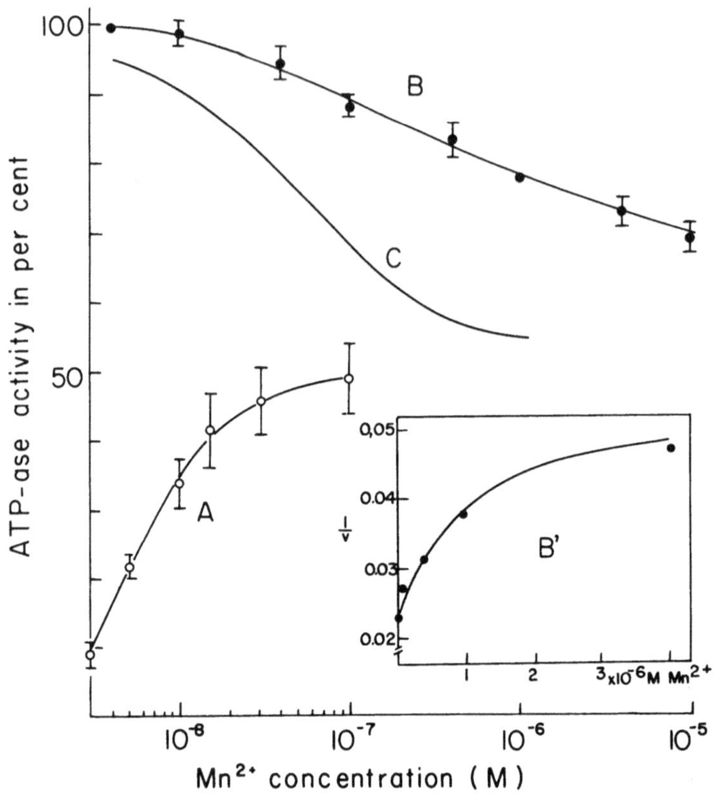

Fig. 6. Effect of Mn^{2+} and (Mn^{2+} + Ca^{2+}) on ATPase in isolated HRBC membranes (prepared as in Fig. 3). Medium: (mM) choline-Cl 110, imidazole-Cl 30 (pH 7.0), Ca-Mn-EGTA buffer or Mn-EGTA buffer 1, $MgCl_2$ 4, Na-ATP 2, ouabain 0.17. Controls: Ca-EGTA buffer 1 mM (Ca^{2+} = 10^{-5} M) and tris-EGTA 1 mM. ATPase activity measured by assay of P_i. Ordinate: Mn^{2+} or (Mn^{2+} + Ca^{2+}) induced ATPase obtained by subtracting the activity due to Mg^{2+} from the total. 100 = activity at 10^{-5} M Ca^{2+} (= 94% of v_{max}). Line A: Mn-activation; line B: combined effect of Ca^{2+} 10^{-5} M and variable Mn^{2+} conc.; Line C: expected behavior for Mn-Ca competition at one site. Inset: Reciprocal of difference between B and v_{max} of A plotted against Mn^{2+} conc. Line B': rectangular hyperbola $(1/v - 0.023) = 0.036\,[Mn^{2+}] / (9.7 \times 10^{-7} + [Mn^{2+}])$. Note that inhibition of Ca^{2+} effect by Mn^{2+} approaches finite value at infinite Mn^{2+} conc. Mean K_{Ca} = $1.02 \cdot 10^{-6}$ M, mean K_{Mn} = $0.52 \cdot 10^{-8}$ M, v_{max} for Mn activity = 53% of v_{max} for Ca. Points: mean of 4 experiments ± SEM. All data collected in same blood samples (Schatzmann, unpublished)

the inner leaflet glycerophospholipids are considered (Roelofsen and Schatzmann, 1976). Since the ratio of the three glycerophospholipids changes in favor of phosphatidylserine toward the lower end of the curve in Figure 8 (due to relative resistance of this phospholipid to phospholipase C from B. cereus and total resistance to phospholipase C from C. Welchii) it may be concluded that phosphatidylserine and (phosphatidylcholine + phosphatidylethanolamine) are equipotent in maintaining (Ca + Mg)-stimulated ATPase.

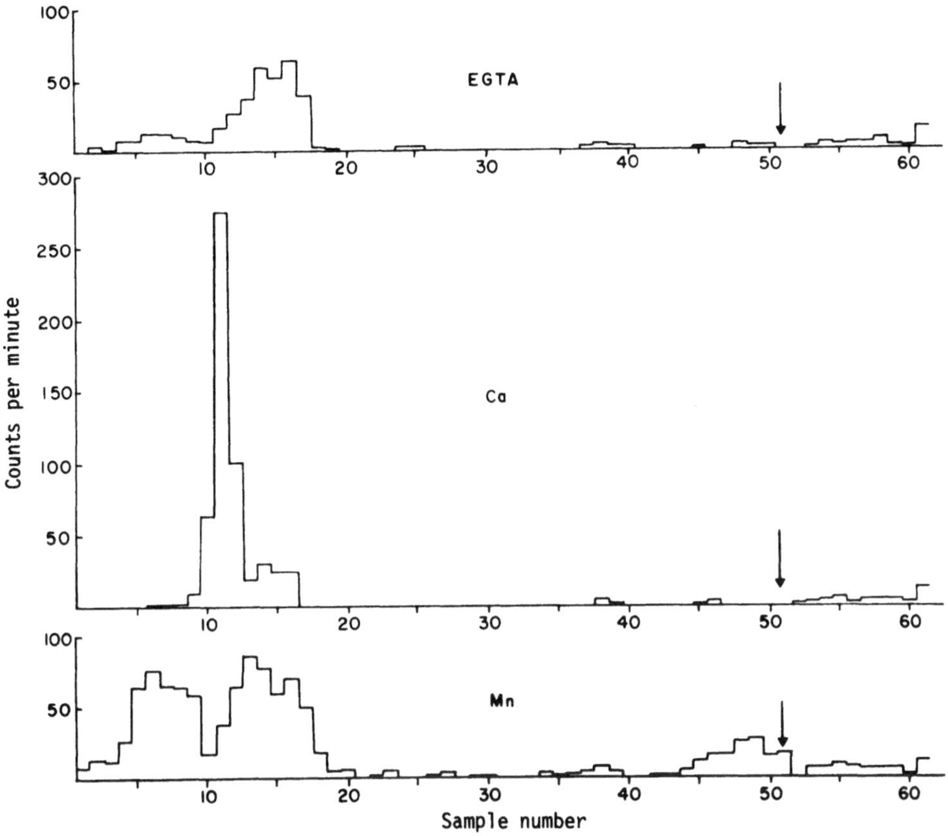

Fig. 7. SDS-gel electrophoresis of HRBC membranes prepared as in Figure 3 and washed 3 times in (mM) choline-Cl 120, imidazole-Cl 35; pH 7.0. Incubation for 30 sec at $0^{o}C$ with $\gamma^{32}P$-ATP. Polyacrylamide gel 7% as described by Fairbanks et al. (1971). Membrane pellet dissolved in 4% SDS, 5% sucrose, 10 mM tris-Cl (pH 8.0), 1 mM EDTA, 40 mM dithiothreitol, at approx. $15^{o}C$. Electrophoresis run at $15^{o}C$. 0.36 mg protein per gel. Origin at left. Arrow: tracking dye (methyl-green). Samples: slices of gel 1 mm wide, dissolved and counted by liquid scintillation. Top (control): Incubation in (mM) choline-Cl 120, imidazole-Cl 35; pH 7.0, $MgCl_2$ 0.06, tris-EGTA 0.05. Middle: Incubation in (mM) $CaCl_2$ 0.1, tris-EGTA 0.05, $MgCl_2$ 0.01 (+ choline-imidazole). Bottom: Incubation in (mM) Mn-EGTA buffer 0.5 (Mn^{2+} conc. = 3.10^{-8} M), $MgCl_2$ 0.03 (+ choline-imidazole). Volume of all samples 0.5 ml. The added $MgCl_2$ provides 0.01 mM Mg^{2+} in all samples. Reaction started by injecting $\gamma^{32}P$-ATP of 16.2 Ci/mM in H_2O to give approx. 0.3 μM ATP. Reaction stopped after 30 sec at $0^{o}C$ by injecting 6% trichloroacetic acid, 50 mM H_3PO_4, 0.5 mM tris-ATP; precipitate washed 5 times with 10 ml of the same solution. Radioactive ATP contained less than 2% $^{32}P_i$. Cold ATP contained 4% ADP. Note: Ca^{2+} induces a peak at mol wt 150,000 whereas Mn^{2+} does not. Blurred peak at approx. 200,000 mol wt in Mn^{2+} sample (Schatzmann, unpublished)

Table 2. Effect of degradation of the outer phospholipid layer in human erythrocytes on the $(Ca^{2+} + Mg^{2+})$-ATPase activity

Exp. No.	% degradation of phospholipids					$(Ca^{2+} + Mg^{2+})$-ATPase	
	Total phospholipid	Sphingomyelin	Phosphatidylcholine	Phosphatidylethanolamine	Phosphatidylserine	V_{max} (µmol Pi/h/mg ghost protein)	K_{Ca} ($M \times 10^{-6}$)
1. Control	0	0	0	0	0	1.67	4.1
treated	48	79	76	25	0	1.67	4.8
2. Control	0	0	0	0	0	1.70	1.8
treated	48	80	78	23	0	1.87	1.7
3. Control	0	0	0	0	0	1.62	5.6
treated	47	79	78	21	0	1.82	5.3

(Roelofsen and Schatzmann, 1976)

$(Ca^{2+} + Mg^{2+})$-ATPase activity was assayed in ghosts derived from intact erythrocytes, previously treated with phospholipase A_2 and sphingomyelinase C under isotonic conditions, which treatment has been shown to cause a selective degradation of the phospholipids present in the outer membrane layer (Verkleij et al., 1973). The maximal rate (V_{max}) and K_{Ca} values were found by plotting activation curves obtained with six different concentrations of Ca^{2+} between 5×10^{-7} and 2×10^{-4}M, according to Lineweaver-Burk

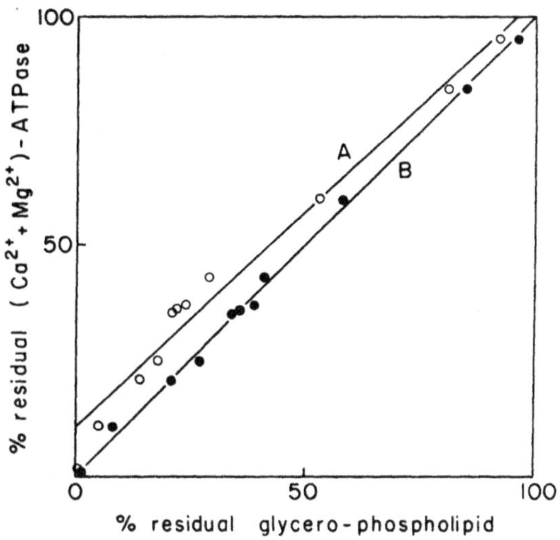

Fig. 8. Percent residual ATPase activity after phospholipase C (from B. cereus or C. Welchii) as function of % residual total glycero-phospholipid content of total membrane (A) and inner leaflet (B). o: experimental points from experiments with total membrane; •: values for inner leaflet calculated on basis of known distribution of phospholipids between inner and outer layer (Verkleij et al., 1973) and assuming all phospholipids are equally accessible to phospholipases. A ($r = 0.9899$) and B ($r = 0.9989$) are calculated regression lines. Slopes of A and B not statistically different from each other or from 1. Intercept of A with the ordinate is significant ($P < 0.001$), which demonstrates deviation from direct proportionality between ATPase activity and glycerophospholipid content of total membrane (Roelofsen and Schatzmann, 1976)

Reconstitution of (Ca + Mg)-ATPase in phospholipase C-treated ghosts by adding phospholipid suspensions confirmed the degradation experiments. Each of the glycerophospholipids is able to restore (Ca + Mg)-stimulated ATPase even after full inactivation. In addition, lysolecithin shares this property, whereas fatty acids do not reactivate and in fact inhibit the effect of lysolecithin to some extent. The reconstituted (Ca + Mg)-ATPase is indistinguishable from the native form with regard to Mg^{2+} requirement and affinity for Ca^{2+}.

The conclusions drawn from these experiments then are (1) that only the phospholipids of the inner leaflet are relevant for the (Ca + Mg)-ATPase, and (2) that any of the glycerophospholipids is apt to maintain (Ca + Mg)-ATPase activity and that in addition to phosphatidylserine one, if not both of the two other types of glycerophospholipids contribute to the functional state of the enzyme under natural conditions.

Since it has been shown that (Na + K)-ATPase of HRBC membranes has a specific requirement for phosphatidylserine (Roelofsen and van Deenen, 1973), conclusion (2) affords an additional indication in favor of the assumption that the (Na + K)-ATPase and the (Ca + Mg)-ATPase, and hence the Na-K- and the Ca-transport system are distinct entities. Other reasons for this contention are the strict insensitivity of the Ca system toward cardiac glycosides and the difference in molecular weight of the phosphorylated intermediates formed in the presence of Na^+ and Ca^{2+} (Knauf et al., 1974).

Acknowledgments. We are greatly indebted to Dr. H.U. Wolf (Dept. of Biochemistry, University of Mainz) for the computer program and the numerical calculations for the mixed metal ion buffers and to Dr. G.F. Fuhrmann (Dept. of Pharmacology, University of Bern) for helping us with the polyacrylamide gel electrophoresis. One of us (B.R.) was a recipient of a grant of Roche Research Foundation for Scientific Exchange and Biomedical Collaboration with Switzerland.

References

Berenblum, J., Chain, E.: Biochem. J. 32, 286-294 (1938)
Bond, G.H., Green, J.W.: Biochim. Biophys. Acta 241, 393-398 (1971)
Coleman, R., Bramley, T.A.: Biochim. Biophys. Acta 382, 565-575 (1975)
Fairbanks, G., Steck, T.L., Wallach, D.F.H.: Biochemistry 10, 2606-2617 (1971)
Ferreira, H.G., Lew, V.L.: Nature (London) 259, 47-49 (1976)
Knauf, P.A., Proverbio, F., Hoffman, J.: J. Gen. Physiol. 63, 324-336 (1974)
Lee, K.S., Shin, B.C.: J. Gen. Physiol. 54, 713-728 (1969)
Lew, V.L.: Biochim. Biophys. Acta 233, 827-829 (1971)
Martonosi, A., Donley, J.R., Pucell, A.G., Halpin, A.R.: Arch. Biochem. Biophys. 144, 529-540 (1971)
Nakamura, M., Ohnishi, S.: J. Biochem. 78, 1039-1045 (1975)
Olson, E.J., Cazort, R.J.: J. Gen. Physiol. 53, 311-322 (1969)
Pfleger, H., Wolf, H.U.: Biochem. J. 147, 359-361 (1975)
Quist, E.E., Roufogalis, B.D.: FEBS Lett. 50, 135-139 (1975)
Rega, A.F., Garrahan, P.J.: J. Membrane Biol. 22, 313-327 (1975)
Roelofsen, B., van Deenen, L.L.M.: Europ. J. Biochem. 40, 245-257 (1973)
Roelofsen, B., Schatzmann, H.J.: Biochim. Biophys. Acta (in press, 1976)
Rosenthal, A.S., Kregenow, F.M., Moses, H.L.: Biochim. Biophys. Acta 196, 254-262 (1970)
Sarkadi, B., Szasz, I., Gárdos, G.: J. Membrane Biol. 26, 357-370 (1976)
Schatzmann, H.J.: Experientia 22, 364-368 (1966)
Schatzmann, H.J.: J. Physiol. 235, 551-569 (1973)
Schatzmann, H.J., Rossi, G.L.: Biochim. Biophys. Acta 241, 379-392 (1971)
Schatzmann, H.J., Tschabold, M.: Experientia 27, 59-61 (1971)
Schatzmann, H.J., Vincenzi, F.F.: J. Physiol. 201, 369-395 (1969)
Simons, T.J.B.: J. Physiol. 256, 209-225 (1976a)
Simons, T.J.B.: J. Physiol. 256, 227-244 (1976b)
The, R., Hasselbach, W.: Europ. J. Biochem. 39, 63-68 (1973)
Verkleij, A.J., Zwaal, R.T.A., Roelofsen, B., Comfurius, P., Katelijn, B., van Deenen, L.L.M.: Biochim. Biophys. Acta 323, 178-193 (1973)
Watson, E.L., Vincenzi, F.F., Davis, P.W.: Life Sci. 10, 1399-1404 (1971)
Wolf, H.U.: Hab.schrift, Univ. Mainz, 1973a
Wolf, H.U.: Experientia 29, 241-249 (1973b)
Wolf, H.U., Dieckvoss, G., Lichtner, R.: Experientia 32, 776 (1976)

Protein-Protein Interactions and the Functional States of Sarcoplasmic Reticulum Membranes

A. MARTONOSI, H. NAKAMURA, R. L. JILKA, and J. H. VANDERKOOI

Summary

Rapid kinetic and electron microscope observations suggest that cooperative phenomena play an important role in the mechanism of Ca transport and Ca release by sarcoplasmic reticulum. The observations are consistent with the occurrence of tetramers of Ca transport ATPase in the membrane in equilibrium with monomers and dimers.

Fluorescence energy transfer between ATPase molecules labeled with N(iodoacetylamino)-1-naphthylamine-5-sulfonic acid (1,5-IAEDANS) and iodoacetamidofluorescein (IAF) support these conclusions.

A hypothesis is presented in which the physiological release of Ca from sarcoplasmic reticulum occurs through hydrophilic channels generated by the reversible formation of ATPase tetramers under the influence of action potential.

A. Introduction

In the resting muscle the [Ca^{2+}] concentration of the cytoplasm is of the order of 10^{-7} M - 10^{-8} M (Ebashi and Endo, 1968) maintained by the limited Ca permeability of the surface membrane and by Ca pumps located in the sarcolemma, mitochondria, and sarcoplasmic reticulum.

During activation the depolarization of sarcolemma and T-tubule membranes is transmitted at the triad to the sarcoplasmic reticulum, causing a rapid release of calcium and the activation of contractile material. The increased Ca influx through the depolarized surface membrane contributes only a small portion of activating calcium in skeletal muscle (Bianchi, 1961) but in cardiac and smooth muscle it may represent a significant element of the activation process (Winegrad, 1961).

The conduction of excitation through the T-tubules is an active process (Costantin, 1970; Peachey, 1972) but the concentration of Na channels in the T-tubule membrane may be only one-twentieth of that in the sarcolemma (Peachey, 1972) and T-tubules may contribute little to the Ca flux across the surface membrane (Van der Kloot, 1968) or to the chloride conductance of the muscle cell (Eisenberg and Gage, 1969).

During transmission of the excitation from the T system to the sarcoplasmic reticulum a voltage-dependent charge movement occurs presumably within the highly differentiated junctional region of the triad (Schneider and Chandler, 1973; Chandler et al., 1976)

which may be a key element in the eventual release of calcium (Ashley and Caldwell, 1974). The Ca release is accompanied by a potential change across the sarcoplasmic reticulum membrane as suggested by the potential sensitive fluorescence probe, Nile-Blue A (Bezanilla and Horowicz, 1975). Parallel with the potential change there is a change in the birefringence of muscle fibers (Baylor and Oetliker, 1975). These observations imply that the structural change of sarcoplasmic reticulum connected with Ca release extends over most of its surface. Earlier experiments (Winegrad, 1965a,b) suggested a functional differentiation within the sarcoplasmic reticulum with Ca uptake occurring primarily in the longitudinal tubules and Ca release through the terminal cisternae.

As this brief survey indicates, our knowledge of the mechanism of Ca release, so far, is derived almost entirely from physiological studies on living muscle fibers. The only hint that similar process may occur in isolated microsomes in vitro is the observed Ca release from skeletal muscle microsomes under conditions which were assumed to alter membrane potential (Kasai and Miyamoto, 1973). Even in this case, osmotic effects may have contributed to the observed Ca release. The reversal of the Ca pump which couples ATP synthesis to Ca release (Barlogie et al., 1971) is too slow to explain the rate of Ca release in activated muscle (Martonosi, 1972).

The main purpose of this report is to consider a reversible interaction between the principal protein components of sarcoplasmic reticulum as the molecular mechanism responsible for the changes in the permeability of sarcoplasmic reticulum membranes during excitation.

B. The Cooperativity of Ca Transport and Membrane Phosphorylation

The accumulation of 2-3 Ca for each mol of ATP hydrolyzed by sarcoplasmic reticulum (for review see Martonosi, 1972) long suggested that cooperative interaction within the Ca^{2+}-ATPase complex may be an essential feature of Ca transport by sarcoplasmic reticulum. This is also reflected in the dependence of the initial rate of membrane phosphorylation by [^{32}P]-ATP upon the ionized [Ca^{2+}] concentration of the medium (Fig. 1) which gives a Hill coefficient greater than 2. As similar Ca dependence is observed after tryptic cleavage of the Ca transport ATPase and in reconstituted ATPase vesicles, the mechanism does not depend upon other proteins of the sarcoplasmic reticulum membrane or upon a covalent interaction between the two polypeptide segments of the ATPase which are separated by tryptic cleavage. The observations are consistent with multiple binding sites for Ca within a single polypeptide chain or with interaction between several ATPase molecules within an oligomer, and so far we cannot distinguish between these two possibilities.

In either case the conformational change during Ca transport cannot involve a major rearrangement of the ATPase molecule within the membrane in view of the relative ineffectiveness of anti-ATPase antibodies to influence Ca transport or ATPase activity in the absence of complement (Martonosi and Fortier, 1974). In the light of these observations the rotary carrier model of Ca transport is considered unlikely. Among more acceptable models is a concerted

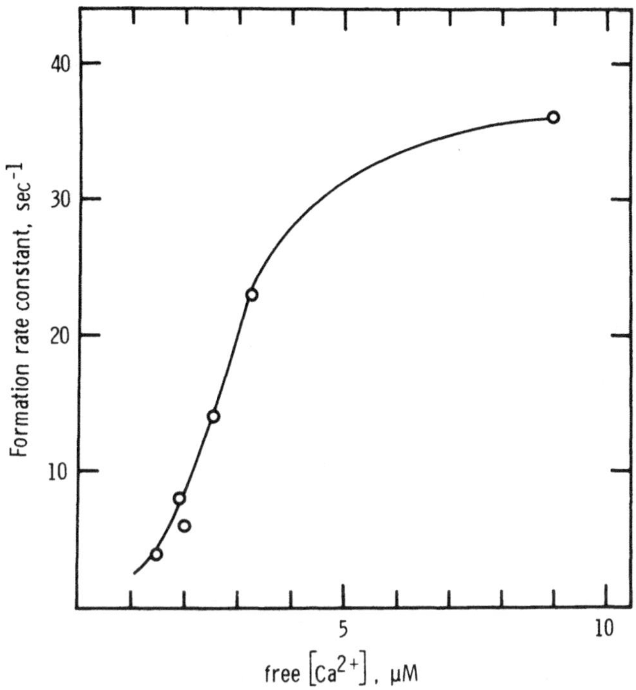

Fig. 1. Effect of $[Ca^{2+}]$ upon the rate of phosphoprotein formation. Experiments were performed in rapid mixing apparatus at $5^{o}C$ using native sarcoplasmic reticulum as enzyme, in a medium of 0.1 M KCl, 10 mM imidazole pH 7.0, 5 mM $MgCl_2$, 1 mM ^{32}P-ATP, 0.5 mM EGTA, and free Ca^{2+} concentrations indicated on abscissa

exchange of 2 Mg for 2 Ca during the cycle of ATP hydrolysis according to the general scheme of Stein (Stein et al., 1974). The basic feature of this scheme may also be represented as a row of binding sites with increasing affinities for Ca and decreasing affinities for Mg arranged in a channel across the membrane. A sequential change in the affinity of the sites produced by phosphorylation with ATP would propel Mg and Ca from site to site in opposite directions across the membrane. The relatively small free energy change between successive steps, in analog with the energy transfer during muscle contraction (Huxley and Simmons, 1972), constitutes an important element of the model.

C. The Mechanism of Ca Release. The Role of Ca Transport ATPase in the Regulation of Ca Permeability

The passive Ca^{2+} permeability of sarcoplasmic reticulum is orders of magnitude greater than that of bilayer membranes prepared from extracted lipids of sarcoplasmic reticulum (Boland et al., 1975). Incorporation of the purified Ca transport ATPase into bilayer lipid vesicles increases their passive Ca permeability to that of sarcoplasmic reticulum membranes (Jilka et al., 1975).

These observations indicate that the Ca transport ATPase, in addition to its role in active Ca accumulation, also contributes to the regulation of the passive Ca permeability of the membrane.

Some recent observations with electron microscopy, chemical cross-linking agents, and fluorescence techniques suggest that the permeability change may be connected with the association of ATPase molecules into tetramers surrounding a hydrophilic channel.

Table 1. The average densities of intramembranous and surface particles on sarcoplasmic reticulum and reconstituted ATPase vesicles

Vesicle preparation	Average densities of 75 Å intramembranous particles particle/μ^2			Average densities of 40 Å surface particles particles/μ^2
	Outer (concave) face	Inner (convex) face	Total	Surface count after background correction
ATPase reconstituted with microsomal lipids	1134	1391	2525	13,280
Fragmented sarcoplasmic reticulum (rabbit)	2238	517	2755	16,543

For details see Jilka et al., 1975

D. The Ultrastructure of Sarcoplasmic Reticulum. The Relationship Between the 85 Å Intramembranous and 40 Å Surface Particles

The concentration of 85 Å intramembranous particles seen by freeze-etch electron microscopy in native sarcoplasmic reticulum membranes and in reconstituted ATPase vesicles is about 4-5 times smaller (Table 1) than the density of 40 Å surface particles observed by negative staining (Jilka et al., 1975; Martonosi, 1975). Both sets of particles are associated with the Ca transport ATPase. In favorably oriented regions of the negatively stained membrane the 40 Å surface particles appeared in oligomeric clusters (Fig. 2) which may be identical with the 85 Å particles seen by freeze-etch electron microscopy (Jilka et al., 1975). Further analysis of the electron micrographs of negatively stained membranes by image enhancement techniques (Markham et al., 1963) revealed a pattern of four subunits within the clusters. The regular array of four subunits occurred only on portions of the surface and required some selection.

There are two possible interpretations of these observations: (1) The clusters of four subunits may represent tetramers of four ATPase molecules in equilibrium with a less orderly arrangement of monomers and dimers. This interpretation is supported by the observation of LeMaire et al. (1976) of the existence of 400,000 dalton oligomers in detergent solubilized sarcoplasmic reticulum preparations. The calculated concentration of ATPase molecules in the membrane may also be greater than the concentration of 85 Å intramembranous particles using available data on the surface area of the microsomes (Martonosi, 1964; Finean and Martonosi, 1965) and on the contribution of the ATPase to the protein content of the membrane (Martonosi and Halpin, 1971). However the combined error of these calculations is probably too great to permit firm conclusion. (2) Although perhaps less likely, the possibility must also be considered that the polypeptide chain of the ATPase is folded into several helical segments with their axis oriented perpendicularly to the plane of the membrane. In this interpretation the 40 Å subunits seen by negative staining may

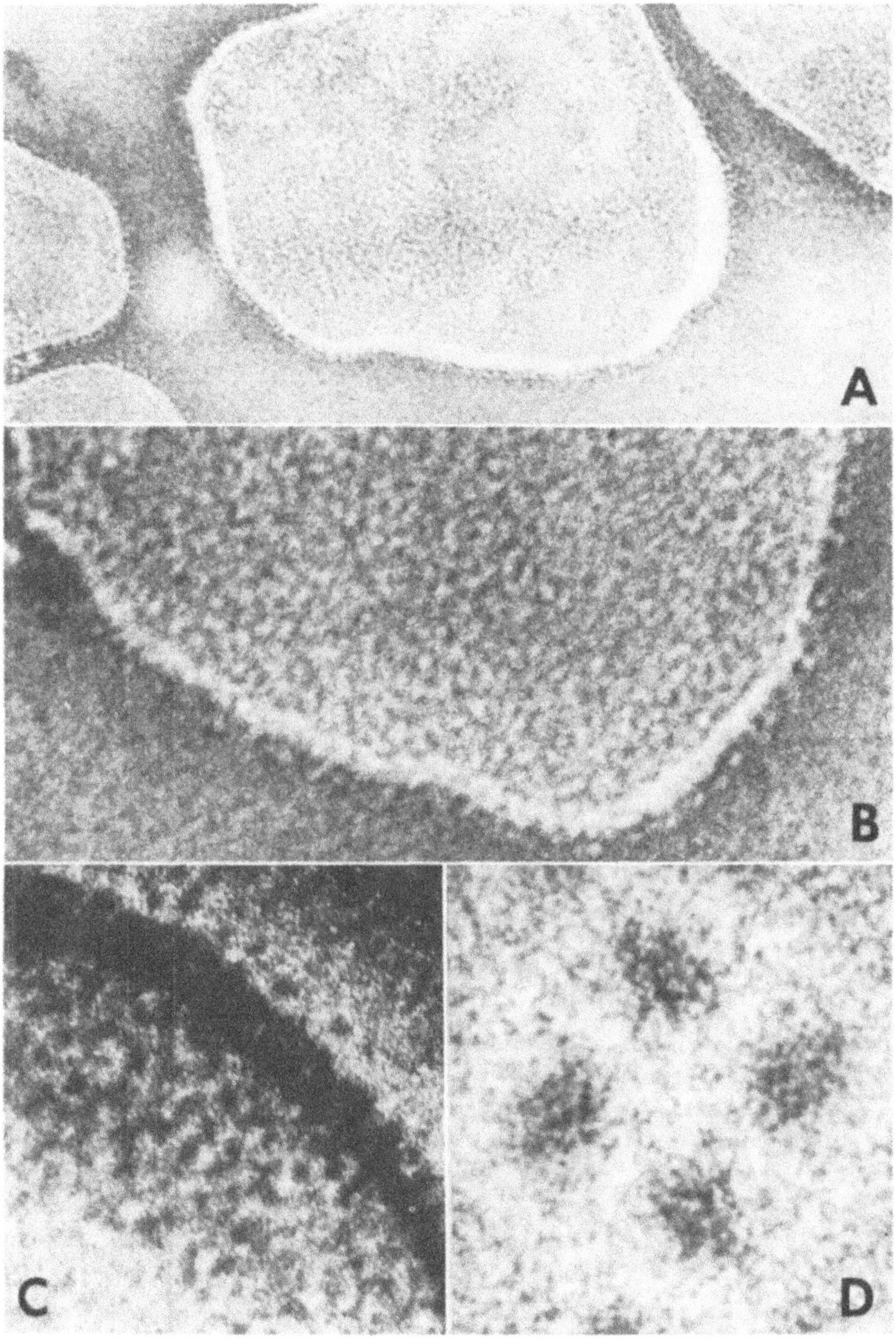

Fig. 2A-D. Electron microscopy of reconstituted ATPase vesicles. For technical details see Jilka et al., 1975. (A) x 23500; (B) x 510,000; (C) Inverted x ≃ 910,000; (D) Enlarged image of a tetramer obtained by the rotation by the rotation technique x ≃ 9,000,000

correspond to the nonhelical bends in the polypeptide chains which project into the water phase. Such arrangement may be similar to the structure of bacteriorhodopsin in the purple membrane described by Henderson and Unwin (1975). The nonhelical regions exposed on the surface may correspond to the areas where attack by trypsin produces the cleavage into the characteristic fragments observed in several laboratories (MacLennan and Holland, 1976).

Whatever is the structural basis of 40 Å particles, their widespread occurrence on negatively stained membranes isolated from a variety of sources suggests that they are a fairly general property of membrane proteins.

E. Chemical Crosslinking

Distinction between these alternatives should be possible by chemical crosslinking techniques which are expected to stabilize the oligomers into covalently bonded structures that can be separated from monomers by SDS-polyacrylamide gel electrophoresis.

In practice these methods rarely yield convincing results. The observed oligomers (Louis and Shooter, 1972; Ikemoto et al., 1975) show unexplained size differences and usually constitute only a small fraction of the total protein. Even in the presence of Cu-phenanthroline (Murphy, 1976) or toluene diisocyanate which produce oligomers most effectively, the bulk of the cross linked protein is present in the form of high molecular weight aggregates which do not enter the gel.

In fact many preparations of sarcoplasmic reticulum (see Fig. 1 in Martonosi and Halpin, 1971) contain an oligomer of 400,000 molecular weight without the use of crosslinking agents, presumably stabilized by disulfide bonds. Whether this component is an artifact or a functionally significant oligomer, remains to be decided.

F. Fluorescence Energy Transfer

The dynamic equilibrium between the hypothetical ATPase oligomers and monomers can be readily analyzed by measuring the efficiency of energy transfer between two populations of ATPase molecules, one labeled with a fluorescent energy donor (N(iodoacetylamino)-1-naphtylamine-5-sulfonic acid-IAEDANS) and the other with a fluorescent energy acceptor (iodoacetamido fluoresceine-IAF).

The emission spectrum of IAEDANS-labeled ATPase overlaps with the absorption spectrum of IAF-labeled ATPase permitting efficient energy transfer, provided the distance of separation between the donor and acceptor fluorophores is less than about 60 Å (Stryer and Hoagland, 1967). Since the average distance between ATPase molecules in reconstituted membranes is in excess of 100 Å, efficient energy transfer implies association with a lifetime sufficiently long for energy transfer to occur.

The technique produced useful results in analyzing proximity relationships in rhodopsin (Wu and Stryer, 1972), immunoglobulin (Bunting and Cathou, 1973), ribosomes (Huang et al., 1975), and

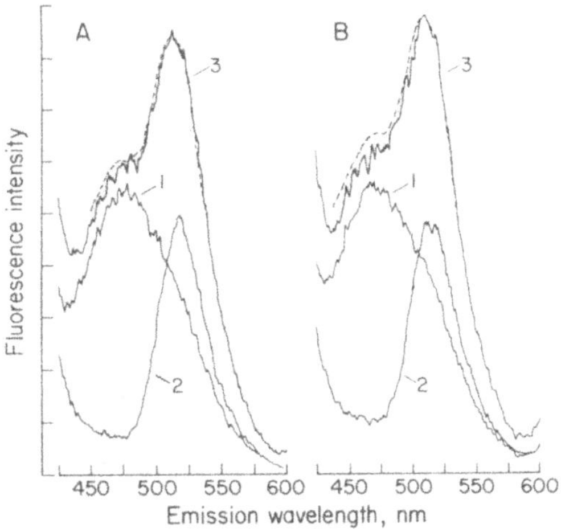

Fig. 3A,B. Fluorescence spectra of mixtures of IAEDANS-ATPase and IAF-ATPase in absence of deoxycholate. (A) Medium 0.1 M KCl, 10 mM Tris-maleate, pH 7.0; (B) Medium same as A + 5 mg egg lecithin liposomes per mg protein. 1: IAEDANS-ATPase; 2: IAF-ATPase; 3: IAEDANS-ATPase + IAF-ATPase. Broken lines: sum of fluorescence intensities of individual components

and in demonstrating the formation of gramicidin A dimers as conducting channels in artificial membranes (Veatch and Stryer, 1976).

Upon mixing IAEDANS-labeled ATPase vesicles with IAF-ATPase vesicles in the absence of detergent, the fluorescence spectrum is usually the sum of the spectra of the two components, indicating that no significant energy transfer occurs (Fig. 3).

When the donor and acceptor labeled vesicles are mixed in the presence of deoxycholate followed by dilution with 0.1 M KCl-10 mM Tris-maleate buffer, pH 7.0, one finds that the fluorescence intensity of the donor is smaller, while that of the acceptor is greater than the sum of the fluorescence intensities of separated components, suggesting energy transfer (Fig. 4).

The magnitude of these differences was essentially the same with deoxycholate concentrations ranging from 0.5 to 5.0 mg deoxycholate per mg protein, although the intensity of scattered exciting light diminished with increasing deoxycholate concentrations (Fig. 5).

A change in the intensity ratio of donor and acceptor fluorescence was also observed if the ATPase preparations were mixed with deoxycholate in the presence of 5 mg egg lecithin per mg protein, but in this case the required concentration of deoxycholate was usually higher than 1 mg deoxycholate per mg protein (Fig. 4-5).

Addition of this amount of egg lecithin should increase the lipid phase of the membrane about ten-fold and therefore increase the distance between ATPase molecules more than three-fold. If the observed energy transfer is the result of random collision between ATPase molecules in the membrane, one would expect a decrease in energy transfer after addition of egg lecithin. The absence of significant change in transfer efficiency supports the view that the energy transfer takes place in associated clusters of ATPase molecules.

In some preparations, energy transfer was observed even in the absence of deoxycholate but usually this required longer incubation times and may have been caused by residual deoxycholate in the ATPase preparations which promoted the fusion of the vesicles and the intermixing of their contents.

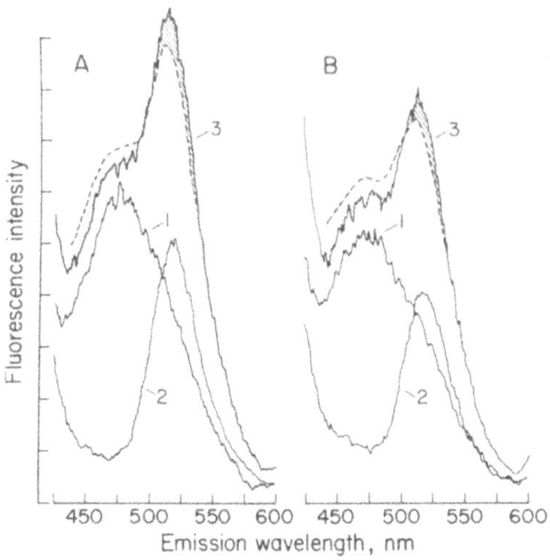

Fig. 4A,B. Fluorescence spectra of IAEDANS-ATPase and IAF-ATPase vesicles in the presence of deoxycholate. Donor and acceptor vesicles were mixed with 0.5 mg deoxycholate per mg protein (A) or 0.5 mg deoxycholate and 5 mg egg lecithin mg protein (B) and diluted with 0.1 M KCl, 10 mM Tris-maleate pH 7.0. Excitation at 320 nm. 1: IAEDANS-ATPase alone; 2: IAF-ATPase alone; 3: IAEDANS-ATPase + IAF-ATPase. Broken lines: sum of fluorescence intensities of individual components

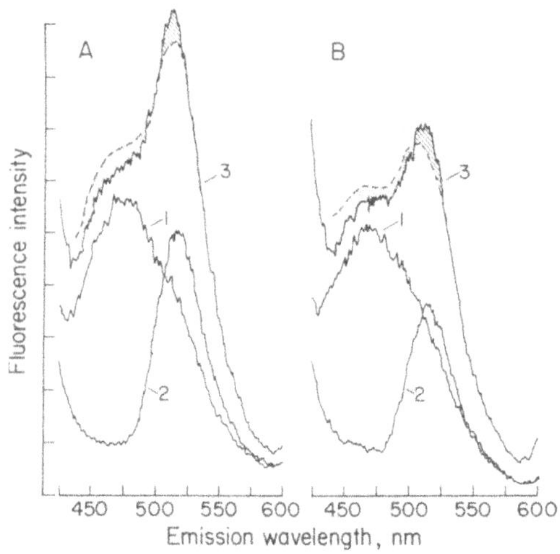

Fig. 5A,B. Fluorescence spectra of mixed IAEDANS-ATPase and IAF-ATPase vesicles in the presence of deoxycholate. Vesicles were mixed with 5 mg deoxycholate per mg protein (A) or with 5 mg deoxycholate and 5 mg egg lecithin per mg protein (B) and diluted with 0.1 M KCl, 10 mM Tris-maleate buffer, pH 7.0 before measurement. Excitation at 320 nm. 1: IAEDANS-ATPase; 2: IAF-ATPase; 3: IAEDANS-ATPase + IAF-ATPase

The Effects of ATP and Ultrasonic Treatment upon the Fluorescence of IAEDANS and IAF-labeled ATPase Preparations. The intensity of fluorescence in IAEDANS-labeled ATPase preparations decreases upon the addition of MgATP (Fig. 6A). The effect is saturated with 10^{-3} M ATP. Smaller changes were observed with Mg-ADP or with β-γ CH_2-ATP. Addition of EDTA (1 mM) diminished the effect of ATP on the fluorescence intensity. These observations suggest that the effect of ATP on fluorescence is connected with one of the steps that follow the binding of ATP to the enzyme. IAF-ATPase was generally less sensitive than IAEDANS-ATPase to ATP (Fig. 6B).

Although in dispersions of IAEDANS-ATPase and IAF-ATPase reconstituted with deoxycholate addition of 1.6 mM ATP decreases the ratio of donor and acceptor fluorescence intensities, a major part of this change may result from the effect of ATP upon the fluorescence intensities of individual components (Fig. 7). A similar conclusion applies to the effect of ultrasonic treatment on the intensity ratio of the fluorescence of donor and acceptor vesicles mixed in the absence of deoxycholate.

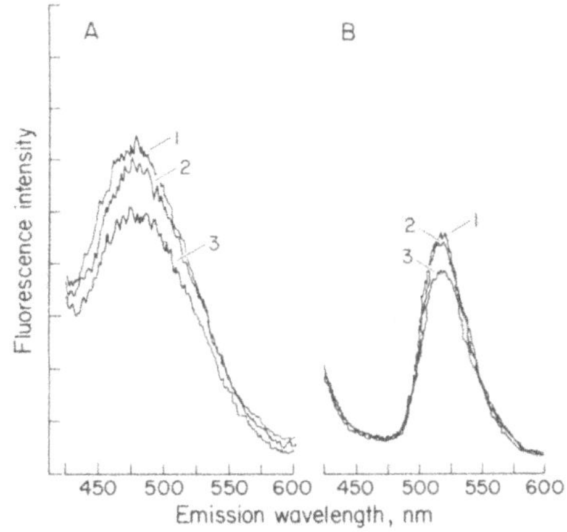

Fig. 6A,B. Effect of ATP upon fluorescence of IAEDANS-ATPase (A) and IAF-ATPase (B). Medium 0.1 M KCl, 10 mM Tris-maleate pH 7.0. Excitation: 320 nm. 1: Control; 2: with 1.6 mM $MgCl_2$: 3: with 1.6 mM Mg-ATP

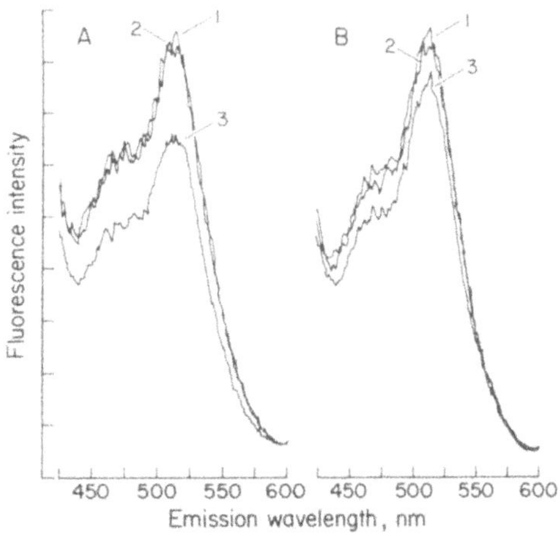

Fig. 7A,B. Effect of ATP upon fluorescence of mixed IAEDANS-ATPase and IAF-ATPase vesicles. Donor and acceptor vesicles were mixed in the presence of 0.5 mg DOC/mg protein followed by dilution with 0.1 M KCl, 10 mM Tris-maleate pH 7.0. Excitation at 320 nm. (A) 1: control; 2: with 1.6 mM $MgCl_2$; 3: with 1.6 mM Mg-ATP. (B) 1: control; 2: 1.6 mM EDTA; 3: 1.6 mM EDTA + 1.6 mM ATP

Nanosecond Fluorescence Decay in Reconstituted Vesicles Containing IAEDANS- and IAF-labeled ATPase Molecules. Independent evidence for energy transfer was obtained by measurement of the excited state lifetime (Weber, 1972) of IAEDANS-ATPase in the presence and absence of IAF-ATPase at a deoxycholate-protein weight ratio of 1.

The decay curve of IAEDANS-ATPase consists of two components with τ_1 = 2 nsec and τ_2 = 18.8 nsec respectively (Fig. 8).

The corresponding values in mixtures of IAEDANS-ATPase and IAF-ATPase are τ_1 = 0.4 nsec and τ_2 = 15.1 nsec respectively (Fig. 9). The observed difference in decay times is clearly significant and consistent with energy transfer.

The fluorescence lifetime was unchanged when the donor and acceptor preparations were mixed in the absence of deoxycholate.

After addition of 2-5 mg egg lecithin/mg protein in the presence of deoxycholate to IAEDANS-ATPase, the lifetime of the excited state remained close to 18 nsec; the corresponding value with a mixture of IAEDANS-ATPase and IAF-ATPase in the presence of egg

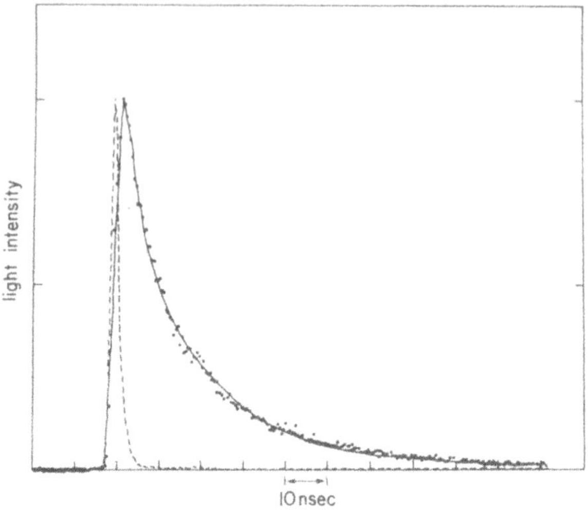

Fig. 8. Nanosecond fluorescence decay of IAEDANS-ATPase vesicles. The medium contained 0.1 M KCl, 10 mM Tris-maleate pH 7.0, and 0.5 mg deoxycholate per mg protein. Solid line over data points is computer fitted. Broken line: light-flash

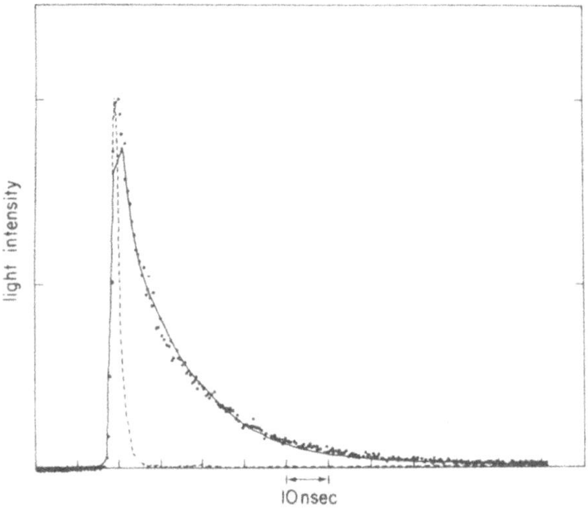

Fig. 9. Nanosecond fluorescence decay of IAEDANS-ATPase and IAF-ATPase vesicles. Vesicles were mixed in presence of 0.5 mg deoxycholate per mg protein and then diluted with 0.1 M KCl, 10 mM Tris-maleate buffer pH 7.0. Solid line over data points is computer fitted. Broken line: light-flash

lecithin and deoxycholate was close to 16 nsec. Therefore, addition of egg lecithin apparently left the energy transfer unaffected.

G. The Hypothetical Role of ATPase Oligomers in the Release of Ca from Sarcoplasmic Reticulum

The release of Ca from sarcoplasmic reticulum in Balanus muscle fibers is initiated a few milliseconds after stimulation and maximum Ca levels are reached in the cytoplasm in 50-100 msec at 10°C (Ashley and Caldwell, 1974). The underlying birefrigence and potential changes in the sarcoplasmic reticulum of frog semitendinous muscle are essentially complete in about 10 msec at 22°C (Baylor and Oetliker, 1975; Bezanilla and Horowicz, 1975).

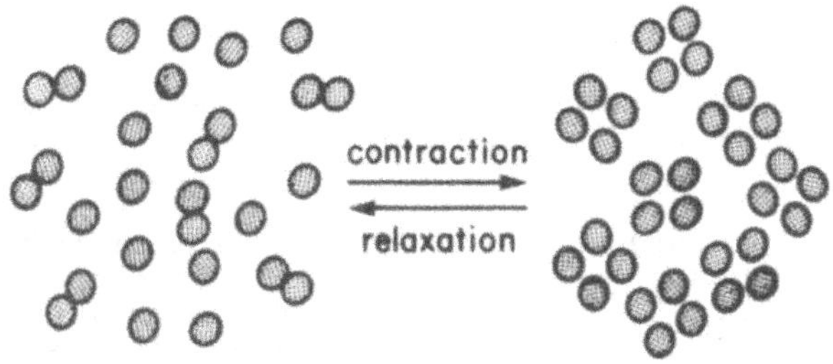

Fig. 10. Hypothetical arrangement of ATPase molecules in sarcoplasmic reticulum membranes during excitation and in rest

The observed rate of Ca release is several orders of magnitude greater than the passive permeability of sarcoplasmic reticulum membranes in vitro (Martonosi, 1971, 1972). We propose that the increase in the Ca permeability of sarcoplasmic reticulum upon excitation is mediated by a potential dependent shift in the equilibrium between ATPase monomers and tetramers in the membrane in favor of the tetramers (Fig. 10).

The hydrophilic channels formed by the tetramers are assumed to penetrate across the membrane, permitting the rapid release not only of Ca but also of other small molecular weight ions and metabolites at a rapid rate. During rest the equilibrium favors the monomers or dimers of Ca transport ATPase with a decrease in passive Ca permeability and an active uptake of Ca into sarcoplasmic reticulum.

Apart from the electron microscope and fluorescence energy transfer data indicating the existence of ATPase-ATPase interactions in the membrane, there is no definite evidence to support this scheme.

Nevertheless it represents a feasible and attractive model which can be tested experimentally.

The model associates both Ca uptake and Ca release with the Ca transport ATPase and therefore assumes, in agreement with fluorescence (Bezanilla and Horowicz, 1975) and birefringence observations (Baylor and Oetliker, 1975), that both processes occur essentially over the whole surface of sarcoplasmic reticulum. The proposed separation of Ca uptake and release functions in the longitudinal tubules and cisternae respectively (Winegrad, 1965a,b) conflicts with our hypothesis. As such separation is not based upon direct evidence but arose from the interpretation of complex autoradiographic data on the distribution of ^{45}Ca in resting and activated muscle, its reinvestigation may be necessary.

The formation of tetramers must be complete within about 10 msec in order to match the kinetics of Ca release during excitation, derived from aequorin fluorescence studies (Ashley and Caldwell, 1974).

The rate of lateral diffusion of intrinsic membrane proteins is relatively insensitive to molecular weight and the diffusion constant is generally close to 10^{-9} cm^2 sec^{-1} (Edidin, 1974; Elson and Webb, 1975). One can calculate the approximate diffusion rate in two dimensions (Saffman and Delbruck, 1975) using the simplified formula:

$$x^2 = D \cdot t$$

where X is the distance of movement by the particle in time t, with a diffusion constant $D = 10^{-9}$ cm^2 sec^{-1}.

The time required for the ATPase monomer to travel a linear distance of 200 Å in the membrane is of the order of 4 msec and even for a distance of 400 Å it is only 16 msec.

Therefore the rate of diffusion of ATPase monomers in sarcoplasmic reticulum membranes of fully developed muscles is not likely to retard the rate of tetramer formation beyond the physiologically set limit of Ca release rate. These approximate calculations also help to understand the fluorescence energy transfer observed in the presence of egg lecithin.

In preliminary experiments artificial iongradients across sarcoplasmic reticulum membranes had no effect upon the efficiency of energy transfer. This may suggest that the equilibrium between monomers and oligomers is not affected by the membrane potential but a potential dependent conformation change within the oligomers serves as a gating process for the regulation of Ca release.

The possibility must be raised that the energy transfer occurs during random collision between ATPase monomers dispersed in the lipid phase without their association into stable oligomers. An approximate quantitative evaluation of the contribution of such effects to the observed energy transfer is possible using the Stern-Volmer and Smoluchowski equations.

The concentration of Ca transport ATPase in the membrane of reconstituted vesicles is estimated to be about 5×10^{-3} M. Using the observed quenching of about 10% in the intensity of IAEDANS fluorescence in mixed vesicles or the similar change in the lifetime of excited state of IAEDANS fluorescence caused by the presence of IAF-ATPase, the diffusion limited bimolecular rate constant for the collision ATPase molecules is about 10^9 l/mol/sec.

The calculated value of rate constant for collision based on the Smoluchowski equation is 7.6×10^9 l/mol/sec, using 48 Å for R_o (Bunting and Cathou, 1973) and 10^{-9} cm^2/sec for the diffusion coefficient (Edidin, 1974).

The similarity between the values obtained by the two equations may imply highly efficient energy transfer upon random collision between two ATPase molecules. Alternatively, the efficiency of energy transfer in stable oligomeric structures may be limited by the distance of separation between ATPase molecules or by less than stoichiometric labeling with the donor or acceptor dyes. As the polarization of fluorescence of IAEDANS bound covalently to the ATPase was only 0.165-0.170, the orientation of the fluorophore is essentially random and therefore special orientation effects need not be considered.

Clear assessment of the contribution of the various alternatives may require energy transfer studies in mixtures of labeled and unlabeled ATPases at various concentrations.

The dependence of tetramer formation upon the concentration of ATPase monomer is unknown. Energy transfer studies on sarcoplasmic reticulum membranes obtained from early stages of embryonic development, where the concentration of Ca transport ATPase is 10-20 times smaller than in adult animals (Boland et al., 1974; Tillack et al., 1974), could provide interesting new information. One would expect, if our first hypothesis is correct, that the rate of formation of tetramers and the rate of Ca release would be a high-power function of the concentration of Ca transport ATPase in the membrane.

Acknowledgments. This work was supported at St. Louis University (A.M.) by research grants AM 18117 from the USPHS, PCM 7600707 from the National Science Foundation, and a grant-in-aid from the Missouri Heart Association. The nanosecond decay studies at the University of Pennsylvania (J.M.V.) were supported by GM 12202 from USPHS and by Career Development Award 5K04GM53. Dr. H. Nakamura was a postdoctoral research fellow of the Muscular Dystrophy Association of America Inc. on leave from the Faculty of Engineering Science, Osaka University, Osaka, Japan. We thank Dr. Carlos Gitler for interesting discussions.

References

Ashley, C.C., Caldwell, P.C.: Calcium movements in relationship to contraction. Biochem. Soc. Symp. 39, 29-50 (1974)

Barlogie, B., Hasselbach, W., Makinose, M.: Activation of calcium efflux by ADP and inorganic phosphate. FEBS Lett. 12, 267-268 (1971)

Baylor, S.M., Oetliker, H.: Birefringence experiments on isolated skeletal muscle fibers suggest a possible signal from the sarcoplasmic reticulum. Nature (London) 253, 97-101 (1975)

Bianchi, C.P.: Calcium movements in muscle. Circulation 24, No. 2, Part 2, 518-522 (1961)

Bezanilla, T., Horowicz, P.: Fluorescence intensity changes associated with contractile activation in frog muscle stained with Nile blue A. J. Physiol. 246, 709-735 (1975)

Boland, A.R., Jilka, R.L., Martonosi, A.: Passive Ca^{2+} permeability of phospholipid vesicles and sarcoplasmic reticulum membranes. J. Biol. Chem. 250, 7501-7510 (1975)

Boland, R., Martonosi, A., Tillack, T.W.: Developmental changes in the composition and function of sarcoplasmic reticulum. J. Biol. Chem. 249, 612-623 (1974)

Bunting, J.R., Cathou, R.E.: Energy transfer distance measurements in immunoglobulins. J. Mol. Biol. 77, 223-235 (1973)

Chandler, W.K., Rakowski, R.F., Schneider, M.I.: A non-linear voltage dependent charge movement in frog skeletal muscle. J. Physiol. 254, 245-283 (1976)

Costantin, L.L.: The role of sodium current in the radial spread of contraction in frog muscle fibers. J. Gen. Physiol. 55, 703-715 (1970)

Ebashi, S., Endo, M.: Calcium ion and muscle contraction. Progr. Biophys. Mol. Biol. 18, 123-183 (1968)

Edidin, M.: Rotational and translational diffusion in membranes. Ann. Rev. Biophys. Bioeng. 3, 179-201 (1974)

Eisenberg, R.S., Gage, P.W.: Ionic conductances of the surface and transverse tubular membranes of frog sartorius fibers. J. Gen. Physiol. 53, 279-297 (1969)

Elson, E.L., Webb, W.W.: Concentration correlation spectroscopy: a new biophysical probe based on occupation number fluctuation. Ann. Rev. Biophys. Bioeng. 4, 311-334 (1975)

Finean, J.B., Martonosi, A.: The action of phospholipase C on muscle microsomes: a correlation of electron microscopy and biochemical data. Biochim. Biophys. Acta 98, 547-553 (1965)

Henderson, R., Unwin, P.N.T.: Three-dimensional model of purple membrane obtained by electron microscopy. Nature (London) 257, 28-32 (1975)
Huang, K.H., Fairclough, R.H., Cantor, C.R.: Singlet energy transfer studies of the arrangement of proteins in the 30S Escherichia coli ribosome. J. Mol. Biol. 97, 443-470 (1975)
Huxley, A.F., Simmons, R.M.: Mechanical transients and the origin of muscular force. Cold Spring Harbor Symp. Quant. Biol. 37, 669-680 (1972)
Ikemoto, N., Garcia, A.M., O'Shea, P.A., Gergely, J.: New structural aspects of proteins (ATPase, calsequestrin) of sarcoplasmic reticulum. J. Cell Biol. 67, No. 2, Part 2, 187a (1975)
Jilka, R.L., Martonosi, A., Tillack, T.W.: Effect of the purified $[Mg^{2+} + Ca^{2+}]$-activated ATPase of sarcoplasmic reticulum upon the passive Ca^{2+} permeability and ultrastructure of phospholipid vesicles. J. Biol. Chem. 250, 7511-7524 (1975)
Kasai, M., Miyamoto, H.: Depolarization induced calcium release from sarcoplasmic reticulum membrane fragments by changing ionic environment. FEBS Lett. 34, 299-301 (1973)
LeMaire, M., Moller, J.V., Tanford, C.: Retention of enzyme activity by detergent-solubilized sarcoplasmic Ca^{2+}-ATPase. Biochemistry 15, 2336-2342 (1976)
Louis, C., Shooter, E.M.: The proteins of rabbit skeletal muscle sarcoplasmic reticulum. Arch. Biochem. Biophys. 153, 641-655 (1972)
MacLennan, D.H., Holland, P.C.: The calcium transport ATPase of sarcoplasmic reticulum. In: The Enzymes of Biological Membranes. Martonosi, A. (ed.). New York: Plenum 1976, Vol. III, pp. 221-259
Markham, R., Frey, S., Hills, G.J.: Methods for the enhancement of image detail and accentuation of structure in electron microscopy. Virology 20, 88-102 (1963)
Martonosi, A.: Role of phospholipids in ATPase activity and Ca transport of fragmented sarcoplasmic reticulum. Federation Proc. 23, 913-921 (1964)
Martonosi, A.: The structure and function of sarcoplasmic reticulum membranes. In: Biomembranes. Manson, L.A. (ed.). New York: Plenum Press 1971, Vol. I, pp. 191-256
Martonosi, A.: Biochemical and clinical aspects of sarcoplasmic reticulum function. In: Current Topics in Membranes and Transport. Bronner, F., Kleinzeller, A. (eds.). New York: Academic Press 1972, Vol. III, pp. 84-195
Martonosi, A.: The mechanism of Ca^{2+} transport in sarcoplasmic reticulum. In: Calcium Transport in Contraction and Secretion. Carafoli et al. (eds.). Amsterdam: North Holland Publ. Corp. 1975, pp. 313-327
Martonosi, A., Halpin, R.A.: Sarcoplasmic reticulum X. The protein composition of sarcoplasmic reticulum membranes. Arch. Biochem. Biophys. 144, 66-77 (1971)
Martonosi, A., Fortier, F.: The effect of anti-ATPase antibodies upon the Ca^{++} transport of sarcoplasmic reticulum. Biochem. Biophys. Res. Commun. 60, 382-389 (1974)
Murphy, A.J.: Crosslinking of the sarcoplasmic reticulum ATPase protein. Biochem. Biophys. Res. Commun. 70, 160-166 (1976)
Peachey, L.E.: Electrical events in the T-system of frog skeletal muscle. Cold Spring Harbor Symp. Quant. Biol. 37, 479-488 (1972)
Saffman, P.G., Delbruck, M.: Brownian motion in biological membranes. Proc. Natl. Acad. Sci. U.S. 72, 3111-3113 (1975)
Schneider, M.F., Chandler, W.K.: Voltage dependent charge movement in skeletal muscle: a possible step in excitation-contraction coupling. Nature (London) 242, 244-246 (1973)
Stein, W.D., Eilam, Y., Lieb, W.R.: Active transport of cations across biological membranes. Ann. N.Y. Acad. Sci. 227, 328-336 (1974)
Stryer, L., Hoagland, R.P.: Energy transfer: a spectroscopic ruler. Proc. Natl. Acad. Sci. U.S. 58, 719-726 (1967)
Tillack, T.W., Boland, R., Martonosi, A.: The ultrastructure of developing sarcoplasmic reticulum. J. Biol. Chem. 249, 624-633 (1974)
Van der Kloot, W.G.: The effect of disruption of the T-tubules on calcium efflux from frog skeletal muscle. Comp. Biochem. Physiol. 26, 377-379 (1968)

Veatch, W., Stryer, L.: The dimeric nature of the gramicidin A transmembrane channel: conductance and fluorescence energy transfer studies of hybrid channels. J. Mol. Biol. (1976, in press)
Weber, G.: Uses of fluorescence in biophysics: some recent developments. Ann. Rev. Biophys. Bioeng. 1, 553-569 (1972)
Winegrad, S.: The possible role of calcium in excitation-contraction coupling of heart muscle. Circulation 24, No. 2, Part 2, 523-529 (1961)
Winegrad, S.: Autoradiographic studies of intracellular calcium in frog skeletal muscle. J. Gen. Physiol. 48, 455-479 (1965a)
Winegrad, S.: The location of muscle calcium with respect to the myofibrils. J. Gen. Physiol. 48, 997-1002 (1965b)
Wu, C.W., Stryer, L.: Proximity relationships in Rhodopsin. Proc. Natl. Acad. Sci. U.S. 69, 1104-1108 (1972)

Energy Interconversion in Sarcoplasmic Membranes

W. HASSELBACH and F. U. BEIL

A. Introduction

The cytoplasmic calcium level changes more extensively and more rapidly in muscle than in any other tissue. That is because muscle contractile proteins need 0.05-0.50 μmol $Ca^{2+} \cdot ml^{-1}$ fiber water for activation (cf. Hasselbach, 1975a; Ebashi, 1976). In muscles composed of thick and fast contracting fibers, calcium ions cannot be supplied from the plasma membrane or from the extracellular space. Diffusion is too slow to account for the rapid inset of mechanical activity. Nature has circumvented this difficulty by the development of an elaborated network of internal membranes, the sarcoplasmic reticulum (SR). In 1961 and 1964 (Hasselbach and Makinose, 1961; Hasselbach, 1964) we showed that isolated vesicular fragments of the SR, as well as the SR membranes in situ, can store large amounts of calcium and that the vesicular membranes can thereby reduce the level of ionized calcium in the external medium below 10^{-7} M. This accumulation is brought by an ATP-driven calcium pump. Subsequently, this calcium pump became an essential element in the calcium concept of excitation contraction coupling. It is assumed:

1. That calcium is released from the cisternal elements of the SR into which it has been transported by the pump.

2. The calcium pump in the reticular membranes causes relaxation by lowering the calcium level and thereby removing calcium from the contractile proteins (Weber and Winicur, 1961). Storage and release of calcium seem to be the sole functions of the SR membranes.

Two obvious interests prompted the analysis of the properties of the SR membranes.

1. Interest may be focused on the function of the SR membranes in the living muscle. Calcium release and removal or the maintenance of the resting calcium level are some of the pertinent problems.

2. Interest may be focused on the function of the membranes as an active transport system. The extreme specialization makes the SR membranes most suitable objects for such studies. In return, the results of such transport studies may help to solve the problem connected with calcium movement in the living muscle. In the following, four aspects of the sarcoplasmic calcium transport will be discussed:

1. Structure and structural modification of the SR membranes
2. Energy transduction
3. Reconstitution problems
4. Transport intermediates.

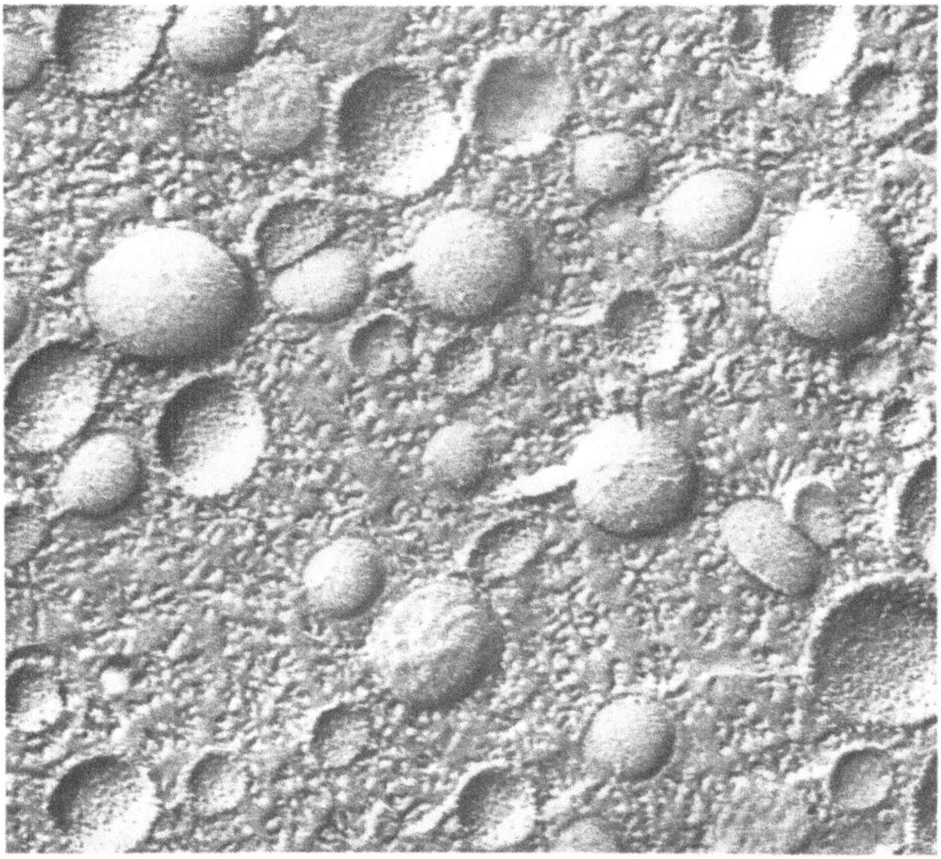

Fig. 1. Micrograph of a freeze-cleaved preparation of native vesicles. Sample was rapidly frozen in Freon 23, cooled by liquid N_2. Fracturing, etching (a min at -100°C) and replication with Pt-C (115 Å) and carbon (200 Å) were performed with a Balzers freeze-etch apparatus. The replicas were examined in a Siemens ELMISKOP 101 electron microscope (Hasselbach and Agostini, unpubl. results)

B. The Structural Component of the Sarcoplasmic Calcium Pump

A suitable procedure for isolation and purification of the vesicular fragments of the sarcoplasmic reticulum has been described in 1963 by Hasselbach and Makinose. All methods given later are small and immaterial variations of the original procedure. When an extremely pure preparation is desired, the vesicles can be loaded with calcium phosphate, and subsequently, the loaded fraction can be separated by centrifugation in a sucrose density gradient from the unloaded remainder. The isolated vesicles have a diameter of 800-3000 Å. The thickness of the membranes as observed in the electron microscope depends on the fixation procedure, 60-90 Å (Lengsfeld and Hasselbach, 1974). Small protein particles at the outer surface of the vesicles are visible (Fig. 1). SH groups at their tips can be decorated with mercuri-phenyl-azo-ferritin (Fig. 2a) (Hasselbach and Elfvin, 1967). These SH groups are functionally relevant. All membrane functions are blocked when the surfacial SH groups are substituted with NEM. The vesicles give rise to disclike structures when they are dehydrated under control conditions (Fig. 2b) (Hasselbach et al., 1975). From X-ray diffraction patterns of such preparations an asymmetric distribution of the membrane proteins has been deduced (Dupont et al., 1973). Like the discs in the rods of the eye, the highly ordered preparations are characterized by a negative form and a positive intrinsic birefringence with reference to the normal of the plane. Glycerol water mixtures were used to compensate the form birefringence. The ascending limb of the graph of Figures 3a and b represents mainly the contribution of the membrane lipids to intrinsic birefringence. When the same experiment is performed with lipid-deprived preparations, the positive intrinsic birefringence is maintained. This result proves the presence of highly ordered domains in the membrane proteins perpendicular to the plane of the membrane. Since 80% of the protein matrix of the SR membranes are occupied by one protein, we must assume that these highly ordered domains reside in it. Bile salts and Triton X-100 were used for isolation and separation of the protein of the SR membranes. The first procedure given by MacLennan (1970) can be considerably simplified. Low quantities of detergents are used in this procedure to remove the accessory proteins from the main protein and the lipid membrane matrix. Larger quantities of the detergents are required to release the main protein. It is much more firmly associated with the membrane lipid assembly than the other protein constituents. The lipid protein complex exhibits a high calcium-dependent ATPase activity. This activity identifies the complex as the functional unit of the SR calcium pump. If the lipids are stripped off from the protein by higher amounts of detergents, a complete and irreversible inactivation results (Hardwicke and Green, 1974). The most prominent minor protein component is a protein which can be precipitated by calcium. Its calcium affinity is relatively low. Its location in the SR membranes is a matter of conjecture (Ostwald and MacLennan, 1974; Hasselbach et al., 1975; Meissner, 1975).

C. Energy-Dependent Calcium Uptake

I. Energy Sources and Transport Capacity

The calcium accumulation by the sarcoplasmic membranes is energized not only by natural and artificial NTPs (Makinose and The, 1965), but also by acetyl phosphate (de Meis, 1969) or paranitrophenol-phosphate (Inesi, 1971). This is in contrast to the Na-K pump which translocates ions only when ATP is present as energy donor.

Under optimal conditions the uptake of calcium is strictly coupled to phosphate liberation from the various energy donors by a calcium activated splitting mechanism, the so-called extrasplitting.

Fig. 2a,b. Electron microscopic picture of SR membranes labeled with mercuri-phenyl-azo-ferritin. (a) Preparation of a vesicular pellet obtained by centrifugation at 127,000 g for 60 min. (b) Preparation of a flat vesicular pellet after dehydration in glycerol (Hasselbach and Stromer, Z. Naturforschung, in press) ▷

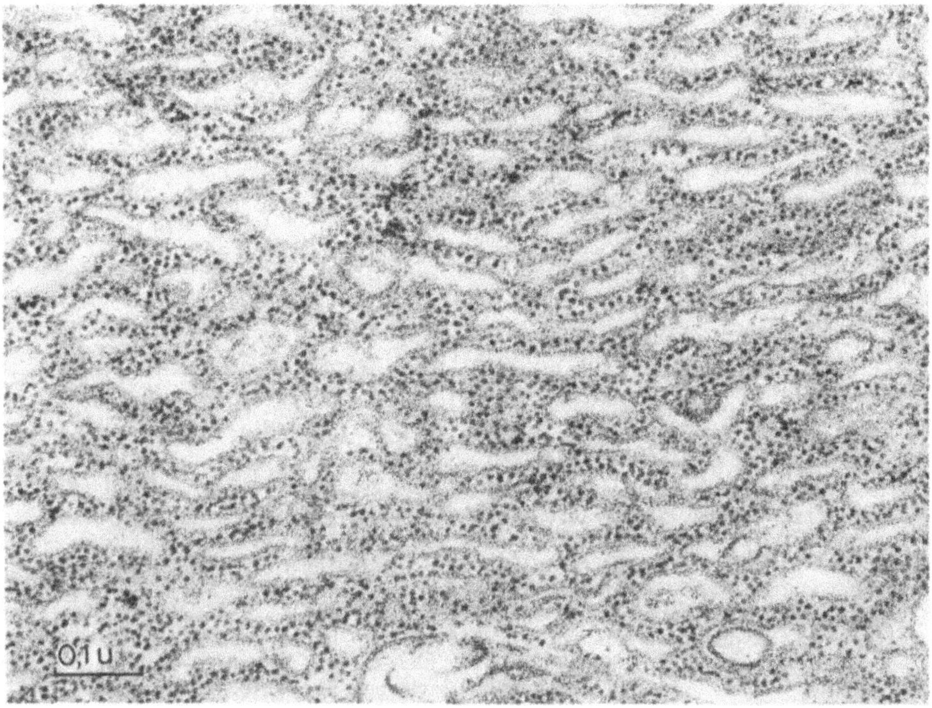

a

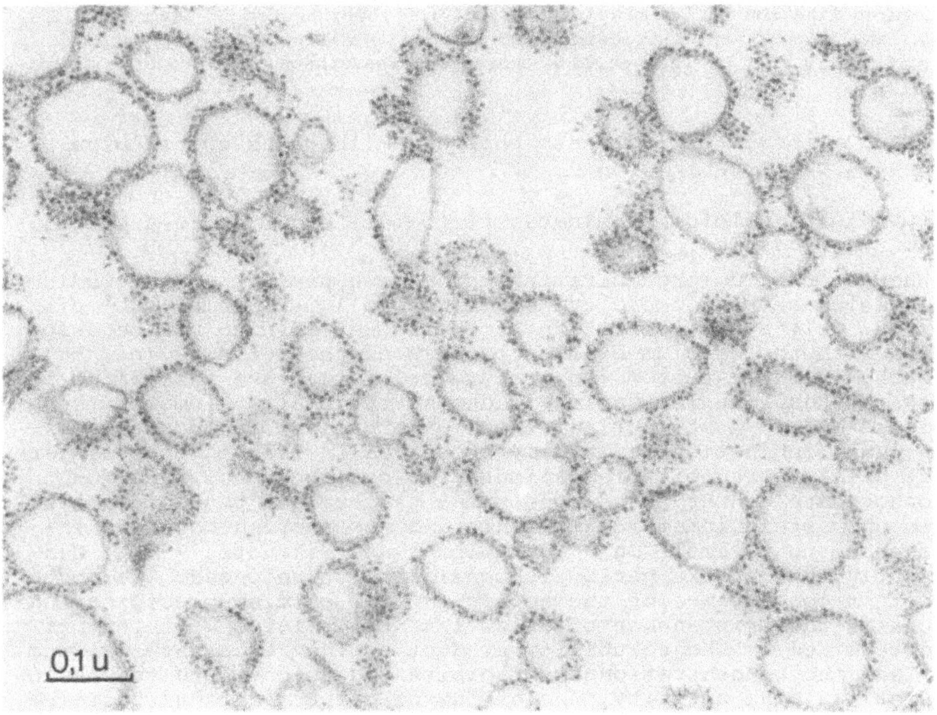

b

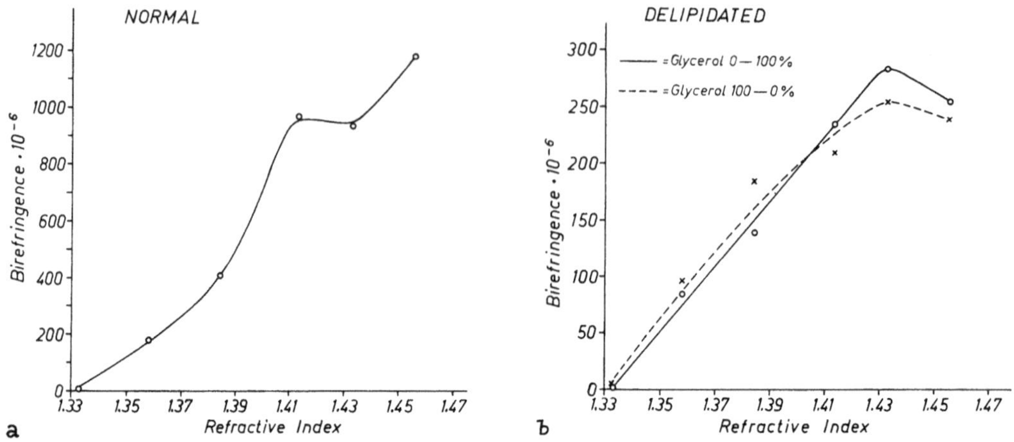

Fig. 3a,b. Birefringence of SR preparations. The vesicles were sedimented on a flat disc and subsequently dehydrated in the presence of glycerol until a protein content of 30-40% is reached. The disc-like protein pellet was cut and viewed perpendicular to the normal of the pellet surface in the polarizing microscope. (a) Native vesicles; (b) delipidated vesicles. The form birefringence is compensated by increasing concentrations of glycerol (abscissa). When the form birefringence is compensated, an intrinsic birefringence emerges which is positive in the direction of the normal to the vesicular surface. The step in the graph of the native preparation originates from the lipid bilayer, whose refractive index is lower than that of the membrane proteins. The intrinsic birefringence of delipidated preparations is smaller than that of native ones. On dehydration, delipidated vesicles produce elliptic profiles and not lamellar structures like the native preparations. If the lower degree of asymmetry is taken into account, the intrinsic birefringence of delipidated preparations reaches nearly that of the native preparation (Hasselbach and Stromer, unpubl. results)

For each molecule of inorganic phosphate liberated two calcium ions were taken up (Fig. 4).

The uptake of calcium terminates at two different degrees of filling of the vesicles:

1. When no calcium-precipitating anions are present in the medium, the calcium storing capacity is limited to 100-200 nmol·mg^{-1} of protein. Only a small fraction of the stored calcium can be associated with the calcium-binding protein of the vesicles (cf. Hasselbach, 1975a). At room temperature calcium uptake terminates in a few seconds. The limitation is due to the small volume of the vesicles and to a steep rise of the internal calcium concentration. This inhibition is largely prevented in the presence of permeant calcium-precipitating anions like oxalate, phosphate, or pyrophosphate. Under these conditions calcium and the anions are taken up in stoichiometrical amounts and the storing capacity increases up to 10 μmol · mg^{-1} (Hasselbach and Makinose, 1961). Consequently, the uptake period is considerably prolonged. The most important consequence of the application of calcium-precipitating anions is the maintenance of a low internal calcium activity. It is determined by the solubility product of the respective calcium salt and the concentration of precipitating anions. Due to the low internal calcium activity, the efflux of calcium during the initial phase of calcium uptake can be neglected. The initial rate of calcium uptake is therefore identical with the influx of cal-

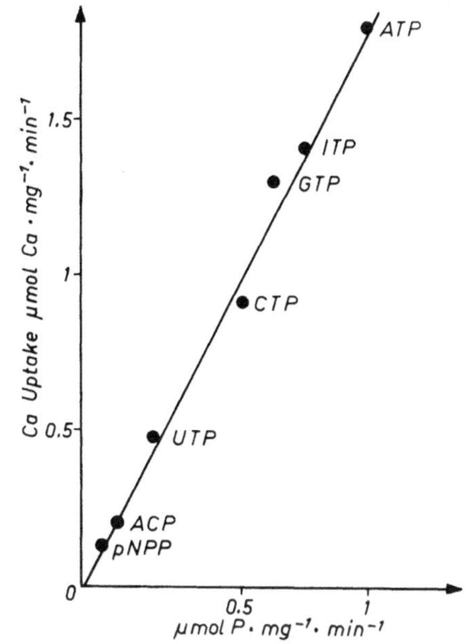

Fig. 4. The correlation between the rate of calcium uptake and the rate of calcium-activated phosphate liberation. The data for ATP, GTP, and UTP are taken from Makinose and The (1965). The transport ratio for acetyl-phosphate was determined at 30°C by measuring proton liberation and calcium uptake simultaneously. The transport ratio for para-nitrophenyl-phosphate is taken from Inesi (1971)

cium and can therefore be directly related to the calcium-stimulated phosphate liberation. These are the conditions under which the results described before were obtained. The low internal calcium activity does not only minimize passive calcium efflux, but also directly prevents the inhibition of energy transduction by high calcium activities. It has been shown that the occupation of low affinity calcium-binding sites at the internal surface of the membranes inhibits ATP hydrolysis (Makinose and Hasselbach, 1965; Weber, 1971; Ikemoto, 1974).

The application of calcium-precipitating anions makes it possible to determine the energy requirement of the calcium pump at every moment. This is because under these conditions the products of the activities of the respective calcium salts in the extra- and the intravesicular space can be obtained. In the extravesicular volume the activity can be determined experimentally. In the intravesicular space the activity product is identical with the solubility product of the precipitated calcium salt. Maximal activity ratios are reached when calcium transport terminates before the maximum calcium-storing capacity is reached. With ATP as energy donor, activity ratios between 3000 and 30,000 have been observed. They correspond to an expenditure of 4500-6000 cal per mol of transported calcium. Since for two calcium ions taken up one energy-rich is hydrolyzed, the efficiency of the pump approaches 100%. Beside ionized calcium the SR membranes can translocate its most closely related alcaline earth ion, strontium. Its affinity to the membranes is considerably lower, so that approximately a 100 times higher concentration of ionized strontium must be present to achieve optimal activation of the pump. The splitting of one molecule of ATP gives rise only to the uptake of one strontium ion. As a consequence, the concentration ratio for strontium which can be created by the pump can become much higher than that for calcium (Mermier and Hasselbach, 1976).

II. Energy-Transducing Calcium Release

The above-described high efficiency is difficult to reconcile with the concept that after the cessation of net calcium transport the pump must consume permanently energy to compensate passive calcium leakage. In fact, passive calcium leakage from calcium-loaded vesicles is small. Even when the external calcium concentration is kept low by the presence of high concentrations of EGTA, the rate of calcium release does not exceed 10 nmol $\cdot$ mg^{-1} $\cdot$ min^{-1} (25°C). The role of the membrane lipids as a barrier for calcium ions is revealed by the complete leakage which occurs when the lipid phase is chemically modified (Hasselbach et al., 1969), or when calcium ionophores are incorporated into the membrane bilayer (Scarpa et al., 1972). The role of the calcium transport protein as a gate for calcium clearly emerged when we found that a very fast calcium release could be induced by the addition of ADP to the phosphate and magnesium containing assay. 3 µM ADP and ~ 1 mM primary phosphate produce half-maximal activation of calcium efflux. Its rate increases when the internal calcium activity is raised. The stimulation of the calcium release is accompanied by an increase of the activation energy from 10,000 to 27,000 cal per mol (Hasselbach and Suko, 1974). An equally fast release of calcium is induced by arsenate of which 3 mM are required for half maximum activation. The conditions under which calcium efflux stimulation occurs suggest that the stimulated calcium efflux is the reversal of the ATP-dependent calcium influx. This suspicion had to be confirmed at once (Barlogie et al., 1971; Makinose and Hasselbach, 1971). The ADP-Pi-stimulated calcium efflux is stoichiometrically coupled to a net synthesis of ATP. As it is shown in Figure 5, the exit of two calcium ions leads to the formation of one molecule of ATP. These results were obtained in the presence of glucose and hexokinase, whereby the accumulation of ATP in the external medium is prevented by the transfer of its terminal phosphate to glucose. Consequently, the release of calcium and ATP synthesis continues until the calcium stores are empty. Hence, the stoichiometrical relationship between calcium release and ATP synthesis is observed under conditions which guarantee unidirectionality of the reactions. As described above, the stoichiometric relationship between calcium uptake and phosphate liberation was likewise established under conditions which provide unidirectionality.

D. Calcium and Phosphate Exchange at Equilibrium

In the following discussion the system near equilibrium will be considered. It can be reached either by calcium uptake or by calcium release. As mentioned before, when the forward reaction is allowed to continue, net calcium movement and phosphate liberation cease at a calcium concentration ratio of approximately 10,000. This ratio is in fair agreement with the figure calculated from the standard free energy of ATP hydrolysis and the concentrations of ATP, ADP, and Pi in the system assuming equilibrium conditions. The same relationship holds for the equilibrium reached by calcium release and ATP synthesis. The state of equilibrium is characterized by a slow exchange between external and internal calcium. As it is to be expected from the overall equation, the calcium exchange is paralleled by the incorporation of inorganic phosphate into the

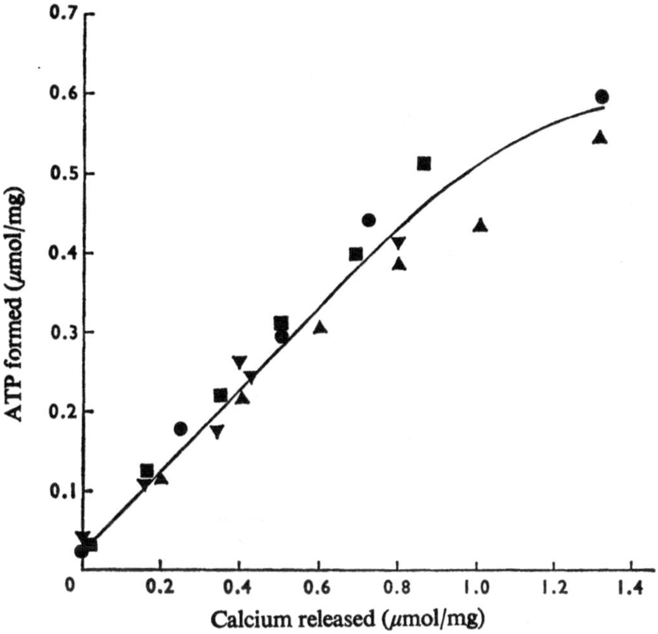

Fig. 5. Correlation between calcium release and ATP formation. The vesicles were loaded with calcium phosphate with acetylphosphate as energy donor. Calcium release and ATP formation were initiated by the addition of 2 mM ADP. The ATP concentration was kept low by using hexokinase to transfer the γ-phosphate of ATP to 1.0 M glucose for a different preparation

ATP fraction. The causal interrelationship between calcium movement and phosphate incorporation is stressed by two observations:

1. When the gradient is abolished, the phosphate incorporation is drastically reduced.

2. Agents which slow down calcium movement inhibit phosphate incorporation (Makinose, 1973).

E. Transport Intermediates

The nucleotide-dependent forward and backward calcium translocations share a common intermediate which can be isolated and stabilized as a phosphoprotein by acid quenching (Makinose, 1966; Yamamoto and Tonomura, 1967). In the forward reaction, the formation of this intermediate is preceded by the binding of calcium and ATP to sites at the external surface of the SR membranes. Both substrates are bound rapidly and with high affinity (Froehlich and Taylor, 1975). There is no indication of a mutual interaction between ATP and calcium-binding sites (Fiehn and Migala, 1971). ATP and calcium binding are followed by the transfer of the terminal phosphate of ATP to the protein moiety of the ATPase (Fig. 6). This phosphate transfer requires the presence of calcium and presumably also of magnesium ions. In the consecutive reactions the intermediate is relatively slowly hydrolyzed. However, if ADP is present in the assay, hydrolysis is suppressed and the phosphoryl group is predominantly transferred to ADP, thus giving rise to a phosphate exchange between ATP and ADP. We are dealing with a calcium-dependent nucleosiddiphosphokinase activity (Hasselbach and Makinose, 1962). This fast-occurring phosphoryl transfer re-

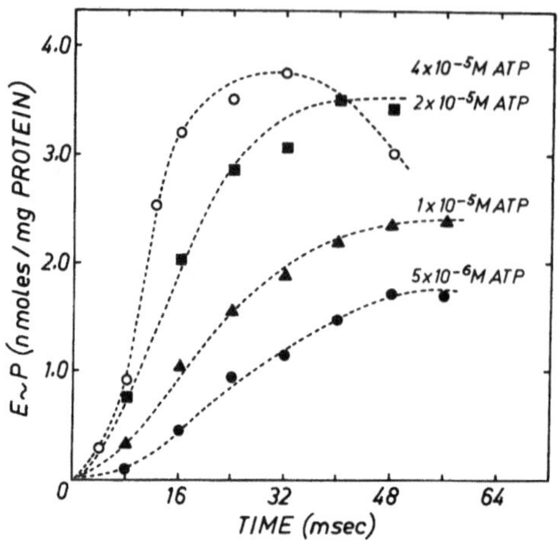

Fig. 6. The rate of phosphoprotein formation at different ATP concentrations (Hasselbach and Suko, 1974). ATP concentrations of 5 μM, 10 μM, 20 μM, and 40 μM. The assay media contain 1 mg of vesicular protein per ml, $MgCl_2$ 5 mM, histidine 20 mM, and KCl 40 mM at pH 7, 0.1 mM $CaCl_2$

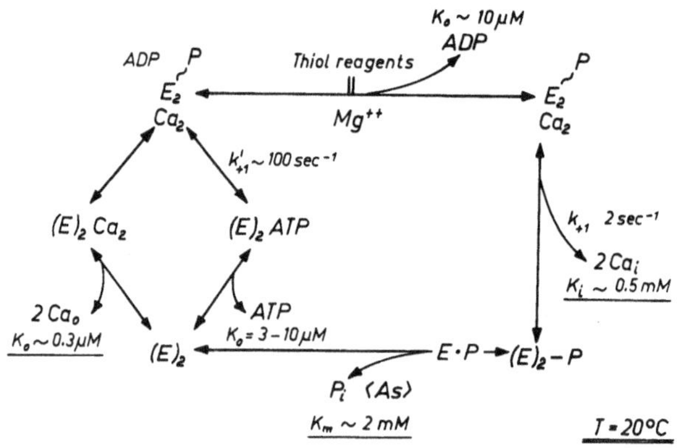

Fig. 7. Reaction scheme of SR calcium transport

action between ATP and ADP is a characteristic feature of the sarcoplasmic calcium transport.

The participation of the calcium-dependent nucleosiddiphosphokinase activity in the formation of ATP from ADP and inorganic phosphate during calcium release is a suggestive possibility, but has not been proven directly. The first step in this reaction sequence is the removal of calcium from its external high affinity binding sites. It is followed by the incorporation of inorganic phosphate into the calcium transport protein. Phosphorylation proceeds slowly and has a high temperature coefficient similar to the rate of calcium release. If ADP is added, the phosphoprotein level drops, ATP is formed, and calcium is released. This sequence continues until the internal calcium stores are depleted. Table 1 demonstrates that the incorporation of inorganic phosphate and the ADP-dependent calcium release have a number of features in common. We may conclude that the transfer of calcium from internal to external binding sites provides the energy required for the formation of the energy-rich phosphorylated intermediate.

In the last section two controversial problems will be discussed. Recently, both have received much attention, being namely the re-

Table 1. Conditions of phosphoprotein formation, ATP synthesis, and calcium efflux

	E-P	E ~ P	ATP synthesis	Ca release
Activation by P_i Km, mM	0.7 - 3	2 - 4	0.7 - 2	~ 2
Inhibition by Ca^{++} Ki, µM	0.3	0.3	0.3	0.3
pH Optimum	6.0	7.0	7.0	7.0
Dephosphorylation or activation by ADP Km, µM	none	0.3	0.3	0.3
Inhibition by KCl, M 50%	0.3	>> 0.3	>> 0.3	>> 0.3

E-P: phosphoprotein formed in the absence of a calcium gradient
E~P: phosphoprotein formed by vesicles loaded with calcium phosphate

construction of the sarcoplasmic calcium pump and ATP formation in the absence of a concentration gradient.

The reconstitution of structure and function of the calcium transport system must be judged according to the material from which reconstruction starts and the degree of reconstruction which is finally achieved.

1. Up to now, all attempts to reconstitute the calcium-dependent ATPase from its isolated purified membrane constituents, membrane proteins, and membrane lipids, have been unsuccessful. The calcium-dependent ATPase is irreversibly inactivated when it is removed from its natural lipid environment (Hardwicke and Green, 1974).

2. This conclusion is supported by the successful reconstruction of the calcium transport ATPase and some of the phosphoryl transfer reactions using enzymatically lipid-deprived SR vesicles. Their ATPase is rendered inactive by delipidation. However, the residual lipids preserve the structure, so that reactivation can be achieved by fatty acids and lysolecithin. For optimal ATPase activity the SR transport ATPase molecule must be surrounded by ~ 70 fatty acids. The fatty acids must be unsaturated and its carbon chain at least 16 carbon atoms long (The and Hasselbach, 1973). Various lysolecithins with appropriate acyl chains are as effective as the unsaturated fatty acids. Obviously the activating fatty acids must not be bound in diacylglycerophosphatidyl-cholin or -ethanolamin molecules. Up to now, attempts to reconstitute calcium accumulation starting from lipid-deprived transport protein preparations have been unsuccessful.

3. Reconstitution of calcium accumulation by inserting cholate-disaggregated sarcoplasmic membrane particles into liposomes must be considered to be the simplest attempt. The liposomes provide the space into which calcium can be transported by the intact calcium pump unit. In spite of the relatively mild procedure the reassembled preparations differ considerably from the native ones (Table 2).

Table 2. Properties of native and reconstituted vesicular preparations. Performance of native and reconstituted SR vesicles

	Native vesicles	Reconstituted vesicles
Rate of calcium uptake	20 nmol $\cdot$ sec^{-1} $\cdot$ mg^{-1}	5 nmol $\cdot$ sec^{-1} $\cdot$ mg^{-1}
Ca/ATP	2	0.4
Concentration ratio	3000-30000	0 0.1-0.01
$[Ca^{++}]_o$ $\frac{Vmax}{2}$	0.3 µM	25 µM
Anion uptake	Stoichiometrical amounts	none

(Data for the reconstructed preparations are taken from Knowles and Racker, 1975)

In addition to the phosphorylated intermediate mentioned earlier, two additional intermediates have recently been described. One intermediate has been found by Froehlich and Taylor (1975). It follows the known energy-rich phosphorylated intermediate in the reaction chain. It is acid-labile and gives rise to a burst of inorganic phosphate liberation during acid quenching. The appearance of the intermediate fits without difficulty into the reaction sequence. It is the immediate precursor of inorganic phosphate whose liberation terminates the forward reaction.

The second new intermediate has been found by Masuda and de Meis (1973), and its formation has been considered to be a very early step in the reverse reaction. The intermediate was detected when the following procedures were applied to both open and closed membrane preparations:

1. The preparations were incubated with inorganic phosphate and magnesium at pH 6.1.

2. A high affinity calcium-chelating agent was added.

3. The intermediate was isolated after acid quenching.

It is an important feature of this reaction that inorganic phosphate is incorporated into the protein, even if no calcium gradient exists across the membranes. In the acid-denaturated protein the phosphate group is bound to an acyl residue. In the native protein bound radioactive phosphate can rapidly be displaced by cold phosphate or other monovalent anions such as chloride. Therefore, it seems unlikely that the phosphate residue is covalently bound to the native protein. Furthermore, the phosphoryl group of this intermediate cannot be transferred to ADP (Beil and Hasselbach, unpublished experiments). This is in contrast to the properties of the corresponding intermediate of the Na-K pump. Post has shown that stoichiometrical amounts of ADP can be phosphorylated by inorganic phosphate when the intermediate was exposed to ADP and very high concentrations of sodium chloride. Fast sodium binding to low affinity binding sites should provide the energy to make the phosphate intermediate ADP acceptable. The calcium translocating SR membranes do not display a corresponding reactivity. The sudden exposure of membrane fragments phosphorylated by inorganic phosphate to ADP and high calcium does not give rise to the formation of ATP. The always present traces of myokinase in conjunction with a residual phosphate-ATP exchange can easily pretend ATP synthesis (Knowles and Racker, 1975). However, this negative result does not exclude that the formation of the discussed intermediate may be an early step in the reverse reaction. We assume that the ADP inacceptable form of the phosphorylated intermediate is interconverted into an ADP acceptable state when calcium is translocated across the membranes from inside to outside (Fig. 7). In this step the energy stored in the calcium gradient would become available for ATP synthesis. It is obvious that such a translocation of calcium can occur only when the internal binding sites remain occupied and the external binding sites are kept free. In this case, calcium can be released and phosphate can be accepted thus enabling continuous ATP formation.

Summary

The SR membrane is an ideal energy interconverter. Its isolated vesicular fragments allow to follow, simultaneously, calcium transport, its kinetics, and its energetics and, furthermore, to observe the formation of transport intermediates. For the function of the sarcoplasmic reticulum in the living muscle it is important, firstly, that the membranes can maintain high concentration gradient in equilibrium with the ATP potential, and that, secondly, the high density of transport sites in the SR membranes guarantees a fast elimination of calcium from the myoplasma. These properties are essential prerequisites for the economy of the muscle and for the control of its activity.

References

Barlogie, B., Hasselbach, W., Makinose, M.: FEBS Lett. 12, 267-268 (1971)
Dupont, Y., Harrison, S.C., Hasselbach, W.: Nature (London) 244, 555-558 (1973)
Ebashi, S.: Ann. Rev. Physiol. 38, 293-313 (1976)
Fiehn, W., Hasselbach, W.: Europ. J. Biochem. 13, 510-518 (1970)
Fiehn, W., Migala, A.: Europ. J. Biochem. 20, 245-248 (1971)
Froehlich, J.P., Taylor, E.W.: J. Biol. Chem. 250, 2013-2021 (1975)
Hardwicke, P.M.D., Green, M.N.: Europ. J. Biochem. 42, 183-193 (1974)
Hasselbach, W.: Fed. Proc. Fed. Am. Soc. Exp. Biol. 23, 309-312 (1964)
Hasselbach, W.: 26. Colloq. Mosbach. Berlin-Heidelberg-New York: Springer 1975a, pp. 81-92
Hasselbach, W., Elfvin, L.G.: J. Ultrastruct. Res. 17, 598-622 (1967)
Hasselbach, W., Fiehn, W., Makinose, M., Migala, A.: Molecular Basis of Membrane Function. New York: Prentice Hall 1969, pp. 299-316
Hasselbach, W., Makinose, M.: Biochem. Z. 333, 518-528 (1961)
Hasselbach, W., Makinose, M.: Biochem. Biophys. Res. Comm. 7, 132-136 (1962)
Hasselbach, W., Makinose, M.: Biochem. Z. 339, 94-111 (1963)
Hasselbach, W., Migala, A., Agostini, B.: Z. Naturforsch. 30c, 600-607 (1975)
Hasselbach, W., Suko, J.: Biochem. Soc. Spec. Publ. 4, 159-173 (1974)
Hasselbach, W., Suko, J., Stromer, M.H., Tke, R.: Ann. New York Acad. Sci. 264, 335-349 (1975)
Ikemoto, N.: J. Biol. Chem. 249, 649-651 (1974)
Inesi, G.: Science 171, 901-903 (1971)
Knowles, A.F., Racker, E.: J. Biol. Chem. 250, 3538-3544 (1975a)
Knowles, A.F., Racker, E.: J. Biol. Chem. 250, 1949-1951 (1975b)

Lengsfeld, A., Hasselbach, W.: Histochemistry 40, 113-127 (1974)
MacLennan, D.H.: J. Biol. Chem. 245, 4508-4518 (1970)
Makinose, M.: Proc. 2nd Intern. Biophys. Congr. No. 276, Verlag Wien. Med. Akad. Vienna
Makinose, M.: FEBS Lett. 12, 269-270 (1971)
Makinose, M.: FEBS Lett. 37, 140-143 (1973)
Makinose, M., Hasselbach, W.: Biochem. Z. 339, 94-111 (1963)
Makinose, M., Hasselbach, W.: Biochem. Z. 343, 360-382 (1965)
Makinose, M., The, R.: Biochem. Z. 343, 383-393 (1965)
Masuda, H., de Meis, L.: Biochemistry 12, 4581-4585 (1973)
de Meis, L.: J. Biol. Chem. 244, 3733-3739 (1969)
Meissner, G.: Biochim. Biophys. Acta 298, 906-926 (1973)
Meissner, G.: Biochim. Biophys. Acta 389, 51-58 (1975)
Meissner, G., Fleischer, S.: J. Biol. Chem. 249, 302-309 (1974)
Mermier, P., Hasselbach, W.: Europ. J. Biochem., in press (1976)
Ostwald, T.J., MacLennan, D.H.: J. Biol. Chem. 249, 974-979 (1974)
Racker, E., Eytan, E.: Biochem. Biophys. Res. Comm. 55, 174-178 (1975)
Scarpa, A., Baldassare, J., Inesi, G.: J. Gen. Physiol. 60, 735-749 (1972)
Tanigushi, K., Post, R.L.: J. Biol. Chem. 250, 3010-3018 (1975)
The, R., Hasselbach, W.: Europ. J. Biochem. 39, 63-68 (1973)
Warren, G.B., Toon, P.A., Birdsall, N.J., Lee, A.E., Metcalf, J.C.: Proc. Natl. Acad. Sci. U.S. 71, 622-626 (1974)
Weber, A.: J. Gen. Physiol. 57, 50-63 (1971)
Weber, A., Winicur, S.: J. Biol. Chem. 236, 3198-3202 (1961)
Yamamoto, T., Tonomura, Y.: J. Biochem. (Tokyo) 62, 558-575 (1967)

Transport of Ca^{2+} in Excitable and Other Cells

Calcium and the Control of Neurosecretion

P. F. Baker

A. Introduction

The process of neurosecretion involves calcium in at least three different kinds of membrane event: voltage-dependent changes in membrane permeability to Ca ions, Ca-mediated fusion of intracellular vesicles with the plasma membrane leading to transmembrane secretion (exocytosis) and activation of Ca pumping systems in the cell membrane that serve to terminate secretion by restoring the intracellular Ca to its initial level. Each of these membrane events will be discussed in turn.

B. Voltage-Dependent Changes in Membrane Permeability

One of the factors that contributes to the importance of Ca as a trigger substance is its low concentration inside cells; quite small changes in membrane permeability or intracellular binding serving to produce large changes in ionized calcium. Experiments in which reporters of ionized Ca, such as the Ca-sensitive photoprotein aequorin or the Ca-sensitive dye Arsenazo III, are injected into large nerve cells, such as the squid giant axon, have revealed that more than 99.9% of the total intracellular Ca is bound and the concentration of ionized Ca is less than 10^{-7} M (see Baker, 1972; Scarpa, Chap. 42, this vol.). As the interior of the cell is negative with respect to the exterior, it follows that the electrochemical gradient for Ca is inwardly directed and is very steep. At rest, the nerve membrane is relatively impermeable to Ca but depolarization results in a transient increase in permeability to Ca, allowing calcium ions to enter the cell and trigger many Ca-sensitive intracellular events including secretion.

Aequorin has proved useful in studying the properties of the channel through which extracellular calcium normally enters the axon (see Baker and Glitsch, 1975). It has been known for some time that massive stimulation of a squid axon results in an increased uptake of ^{45}Ca. In absolute terms the size of this extra flux is very small and no techniques were available to follow the changes in calcium entry during an action potential. Aequorin has the required sensitivity and experiments on aequorin-loaded axons have provided much new and interesting information on calcium channels in the nerve membrane.

It is well known that a reduction in potential, depolarization, of a nerve cell membrane leads to a rapid transient increase in the permeability to sodium ions and a slower maintained increase in the permeability to potassium ions. It is these changes in permeability, coupled with the existing ion gradients, that account for the characteristic shape of the action potential. It is of obvious

Table 1. Comparison of the effect of a variety of agents with transmitter release on the movements of Na, K and Ca that occur in response to a step depolarization of the squid axon membrane from -60 mV to zero potential

	Na current	K current	Ca entry: Early phase	Ca entry: Late phase	Transmitter release
Tetrodotoxin (TTX)	Blocks	No effect	Blocks	No effect	Little or no effect
Injected tetraethylammonium (TEA)	No effect	Blocks	No effect	No effect	No effect
Mg^{2+}, Mn^{2+}, Co^{2+}	Little effect	Little effect	Little effect	Blocks	Blocks
D-600 Iproveratril	Reduces	Little effect	Reduces	Blocks	Blocks
Inactivation rate	Rapid	Slow	Rapid	Slow	Slow in some systems. Unclear in others

interest to determine whether calcium enters the axon through either or both of these channels, or whether there is a special calcium channel. Experiments using aequorin revealed two phases of calcium entry following depolarization: an early phase that roughly parallels the rise in sodium conductance and a later phase that roughly parallels the rise in K conductance.

How can one find out whether or not the calcium is passing through the Na and/or K channels? Here one can make use of a variety of pharmacological agents that block these channels in a fairly specific manner. For instance, the puffer fish poison, tetrodotoxin (TTX), blocks the sodium channels very specifically and it also blocks the early phase of calcium entry: but substances such as tetraethylammonium ions which block the potassium channel have no effect on the late phase of calcium entry. This can, however, be blocked by a variety of substances including a number of divalent and polyvalent cations (Mg^{2+}, Co^{2+}, Mn^{2+}, La^{3+}) and organic Ca-antagonists, such as iproveratril and D600, which have little or no effect on the Na and K channels. For these, and a number of other reasons, it seems that the early phase of calcium entry results from Ca ions entering the axon through the sodium channels and the late phase results from calcium ions entering the axon through a channel that is pharmacologically distinct from the Na and K channels. This special calcium channel has been called "the late calcium channel".

Of particular interest to the present discussion are the striking number of features that the late calcium channel shares in common with the transmitter release mechanism (Table 1). Thus, both late channel and transmitter release persist in the presence of TTX and tetraethylammonium ions; but both are blocked by Mg^{2+}, Mn^{2+}, Co^{2+}, and La^{3+}. All the available evidence is consistent with the existence, both in the axon and at the nerve terminal, of a special voltage-sensitive permeability channel, the late calcium channel through which Ca ions can enter the axon and initiate transmitter release. So far, no naturally occurring highly specific toxins have been found which blocks the Na channel the way that tetrodotoxin blocks the Na channel.

One particularly interesting feature of the late Ca channel is that it only opens transiently in response to a maintained depolarization. In this respect it resembles the sodium channels; but whereas the Na channels close, or inactivate, within a few msecs inactivation of the late Ca channel takes many seconds. It follows that if Ca entry is essential for secretion and steady depolarization produces only a phasic entry of calcium, secretion elicited by a maintained depolarization should also be phasic. This is precisely what is observed in a number of secretory systems. For instance, exposure of the adrenal medulla to potassium-rich solutions only produces a transient secretion. In the past, failure to maintain secretion has been explained in terms of rapid depletion of a small releasable pool of transmitter. Recent experiments of Baker and Rink (1975) suggest that this explanation is unlikely, both because the rate of decline of secretion is largely independent of the amount of transmitter released and can even be detected in glands that have released no catecholamine, and also because in depolarized cells, once secretion has declined to a low level, it can be reactivated in a variety of ways that by-pass the normal route for calcium entry. A more likely explanation of the transient secretion in response to maintained depolarization is that the ability of Ca to enter the depolarized cell declines with time, probably by a process analogous to the inactivation of the late Ca channel seen in squid axons. The physiological significance of these observations is that small changes in the membrane potential of nerve terminals may alter the number of calcium channels available for activation by a nerve impulse and could have a profound influence on the amount of transmitter released by the nerve impulse. A mechanism of this kind may contribute to so-called presynaptic inhibition in the nervous system, where a small maintained depolarization of the presynaptic terminal causes a drastic reduction in transmitter release.

Some of the main properties of the late Ca channel are compared with the calcium dependency of transmitter release in Table 1. It should be stressed that, although there is a close parallel between the late Ca channel and Ca-dependent release of transmitter, calcium probably also enters nerve terminals through the sodium channels and this Ca presumably also participates in the activation of secretion, although its contribution is difficult to assess quantitatively. The available data suggest that the density of late Ca channels may be much higher at the presynaptic terminal than elsewhere along the axon and it may be that in regions specialized for transmitter release the ratio of late Ca channels to Na channels is much higher than in those regions of the axon specialized for conduction.

It may seem surprising that such a close parallel should exist between the properties of the late Ca channels studied in molluscan axons and transmitter release from mammalian nerve cells; but Ca channels very similar to the late Ca channel of squid have been found in most phyla from protozoa to mammals and in a very interesting recent paper Okamoto et al. (1976) have reported the existence of both calcium and sodium channels in unfertilized eggs of the tunicate <u>Halocynthia</u>.

C. Calcium-Mediated Exocytosis

The most direct evidence that a rise in intracellular Ca promotes exocytosis was obtained by Miledi (1973), who showed that microinjection of Ca into the presynaptic terminal of the squid giant synapse resulted in an increased rate of transmitter release, as measured by its effect on the electrical properties of the postsynaptic membrane. How calcium brings about these changes in exocytosis is still largely unknown; but it is instructive to examine in some detail the magnitude of the change precipitated by the arrival of a nerve impulse. For a brief period the rate of release of transmitter from frog neuromuscular junction may increase from a spontaneous level of 1 quantum sec^{-1} (presumed to reflect the release of the contents of one vesicle sec^{-1}) to a rate equivalent to 10,000 quanta sec^{-1}. Any proposed mechanism must take this into account.

One possibility is that the spontaneous release of transmitter takes place by a somewhat different mechanism from that involved in evoked release. This view was originally formed by Katz, largely on the basis of the differential sensitivity of evoked and spontaneous release to changes in external Mg and Ca. It now seems fairly certain that Ca is required inside the nerve terminal and on this basis it is possible to incorporate the properties of spontaneous and evoked release into a single mechanism of secretion. According to this interpretation, spontaneous release reflects activation of secretion by the resting level of ionized Ca inside the cell, whereas evoked release results from the entry of Ca into the cell from the external medium. As the intracellular Ca is strongly buffered, removal of external Ca will have little immediate effect on internal Ca and spontaneous release, but will completely block evoked release which is absolutely dependent on extracellular calcium. In support of this unified view of transmitter release are numerous findings that treatments which raise intracellular ionized Ca also increase the spontaneous rate of secretion and, in addition, provided the nerve terminal is not depolarized, evoked release is normally increased under conditions where spontaneous release is elevated, implying interaction between the pools of Ca involved in both processes. If it is accepted that spontaneous and evoked release involve two rates of activation of the same basic process, some mechanism must be found for the dramatic gear change effecting - for a short period - a 10^4 increase in the rate of exocytosis.

If secretion depends linearly on internal Ca, a 10^4 increase would require a rise in ionized Ca from 0.1 μM to 1 mM, which is greater than the total Ca inside the nerve terminal and far greater than the measured Ca entry. A way round this problem was first suggested by the observations of Jenkinson (1957) and Dodge and Rahamimoff (1967) working with the frog neuromuscular junction. They found that the dependence of transmitter release on external Ca is very nonlinear, increasing along a markedly sigmoidal curve. Subsequent experiments on other synapses revealed similar nonlinear dependences on external Ca. Analysis of the shape of this curve at the frog neuromuscular junction showed that it is quantitatively consistent with a requirement for 4 Ca ions to co-operate in effecting the release of a single quantum of transmitter. Provided this kind of analysis is justified - and this is open to debate - these results provide a simple explanation for the dramatic alteration in the rate of transmitter release brought about by quite small amounts of calcium. According to Dodge and Rahamimoff, transmitter release is proportional to $[Ca]^4$, which means that a ten-fold increase in internal Ca from 0.1 μM to 1.0 μM will increase transmitter release from $[0.1]^4$ to $[1.0]^4$, which is 10,000 times. Thus, the large alterations observed in the rate of transmitter release following arrival of a nerve impulse can be explained in terms of relatively small and entirely feasible changes in the absolute concentration of ionized Ca in the cytosol.

It is now time to return to the central question of what Ca does to stimulate exocytosis. At this point, hard experimental facts begin to run out and speculation prevails, but as we shall see there are a number of exciting pointers that this state of affairs may be about to change.

Examination of a presynaptic terminal in the electron microscope reveals a large number of vesicles clustering close to the plasma membrane and the appearance of many of the micrographs is suggestive of the existence of specialized presynaptic release sites. It is not hard to imagine that the arrival of a nerve impulse facilitates interaction between vesicle and release sites. The release process can be envisaged as occurring in a number of stages (1) mobilization of the vesicle, (2) interaction between vesicle and plasma membrane, (3) fusion of vesicle membrane with the plasma membrane, and (4) subsequent removal of the vesicle membrane so that the process can be repeated.

Within this framework Ca might increase the chances of a successful collision between vesicle and plasma membrane by facilitating any one of these four processes. For instance, the motion of the vesicle may be restricted by the gel-like properties of axoplasm and Ca may increase vesicle motion by promoting a change from gel to sol or, alternatively, if both vesicle membrane and plasma membrane are negatively charged, their interaction may be promoted by the presence of a divalent cation such as calcium.

These possibilities are difficult to examine experimentally. Control of vesicle motion by a calcium-mediated gel-sol conversion in axoplasm is supported by the observation that axoplasm freshly isolated from giant axons of both squid and the polychaete worm (Myxicola) is a very stiff gel which can be liquified by addition of calcium. Attractive as this possibility is, it is not supported on quantitative grounds. The basis of the Ca-dependent liquifaction seems to be a Ca-activated protease which cleaves the neurofilaments of axoplasm, causing an irreversible breakdown in structure (Gilbert et al., 1975). In order to activate protease activity, a Ca concentration in the millimolar range is required. Some activation may be brought about by physiological changes in Ca; but the extent of this has yet to be assessed. Another approach to this question is to examine vesicle motion in the intact cell. If the effective viscosity of the immediate environment of the vesicle alters following nerve stimulation, this might be detected by techniques sensitive to vesicle motion inside the cell. Such a technique is the scattering of laser light. A convenient preparation is the mammalian adrenal medulla which contains a large number of catecholamine-secreting cells. The secretory vesicles can be isolated and their motion characterized. A signal corresponding to vesicle motion can be detected from intact cells. The main finding is that the characteristics of vesicle motion in the intact adrenal medulla are similar to those of isolated granules immersed in isotonic saline - except that the amplitude of the signal from the intact cell is much smaller than from isolated vesicles. An essentially similar picture is found for adrenal vesicles trapped in agar. These observations suggest that the microenvironment of the presynaptic vesicle may have a viscosity similar to that of saline; but vesicle motion is restricted, perhaps simply because the vesicles are packed very close together or because the vesicles are trapped in a low viscosity environment within the interstices of the gel. It follows that, although a simple change in viscosity is ruled out, the possibility still exists that Ca may alter the gel in such a way that vesicles can move further. This might be detected in intact cells as an increase in the amplitude of the signal presumed to emanate from vesicles. As yet, no evidence has been found for such a change.

The possibility that calcium promotes vesicle fusion by charge neutralisation requires that calcium acts very specifically because there is a very large excess of ionized Mg inside cells (Baker and Crawford, 1972). Very large changes in transmitter release are brought about by micromolar concentrations of calcium in the presence of millimolar concentrations of magnesium. Although this degree of specificity is not impossible, examination of calcium binding to vesicles has failed to detect Ca-binding sites of high affinity. If vesicles are subjected to a longitudinal electric field, they move towards the anode, which is consistent with the existence of a net negative charge on the vesicle. In order to alter their mobility it is necessary to add Ca in millimolar concentrations and Mg is almost as effective. Although these observations suggest that large numbers of sites that bind Ca with high affinity are probably not present on the vesicle membrane, they do not exclude the existence of a few such sites - nor do they exclude the possibility that such sites may exist at the inner face of the plasma membrane.

Other possibilities are even more difficult to test. Thus, the fact that vesicles are often found in association with microtubules inside cells has led to the suggestion that a similar interaction may be involved in secretion. Although substances that dissociate microtubules such as colchicine fail to block secretion, this in no way excludes a role for tubulin. A similarly unsatisfactory situation applies to the possible involvement of actin and myosin. Here it is not even clear whether secretion requires ATP and the available results are open to more than one interpretation. Thus, in nerve cells, poisoning with dinitrophenol or cyanide results in an increase in the spontaneous release of transmitter - presumably secondary to a rise in ionized Ca as a result of release from intracellular binding sites; but in the mast cell the ionophore-mediated histamine release is blocked under conditions where cytoplasmic ATP is reduced to a low level.

It is clear from all that has been said that whole cells are too complex for a proper analysis of the Ca-dependent step in secretion and an enormous step forward would be to obtain a cell-free system capable of exocytosis. Recent reports suggest that something resembling exocytosis can be detected in a cell-free preparation containing vesicles and plasma membrane. Davis and Lazarus (1976) have reported that in the presence of calcium, isolated vesicles from β-cells of the pancreas will react with β-cell plasma membranes to release insulin. This reaction is inhibited by EGTA and fully activated by micromolar concentrations of calcium. It is specific in the sense that only β-cell membranes can promote release of β-cell granule insulin; liver membranes are, for instance, ineffective. Whitaker, working in the author's laboratory, has obtained essentially similar results with membranes and

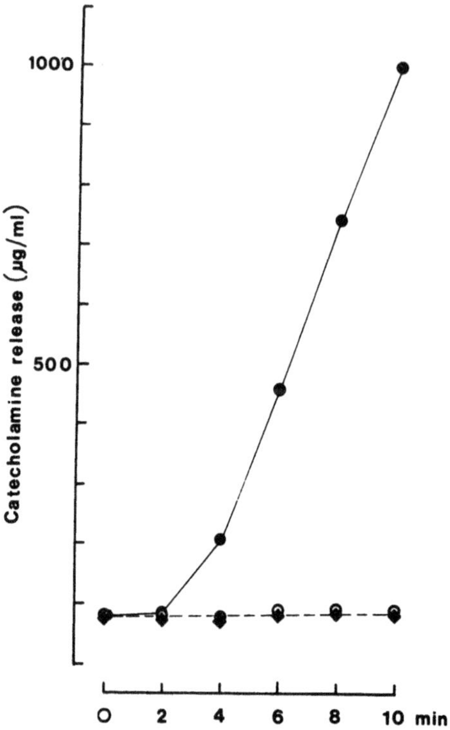

Fig. 1. Exocytosis in vitro? Release of catecholamines from bovine adrenal granules incubated with bovine adrenal plasma membranes: ●: in presence of 2 μM Ca; o: in presence of 1 mM EGTA. Incubation of adrenal granules with liver plasma membrane and 2 μM Ca is also shown (♦). Temp. 37°C. (Unpublished data of M. Whitaker)

vesicles obtained from the adrenal medulla (Fig. 1). As yet there is no conclusive proof that in these in vitro systems secretion is occurring by exocytosis, but it is very tempting to interpret the results in this way. If this proves to be justified, the stage should be set for major advances in our understanding of the mechanism of exocytosis.

D. Membrane Transport and the Cessation of Secretion

Once secretion has been started, what switches it off again? Withdrawal of the stimulus leaves the cell with more Ca than at the start and in view of the dependence of secretion on intracellular ionized Ca, it seems very likely that the cessation of secretion is associated with a return of the ionized Ca to its initial level. This process can be followed in motor nerve terminals by monitoring the spontaneous release of transmitter. After a period of repetitive stimulation, spontaneous release continues at a high rate but declines to its pre-stimulation level over a few minutes. Once again, the squid giant axon provides a convenient model system for studying the control of intracellular Ca in nervous tissue (Baker, 1972). Repetitive stimulation of an aequorin-loaded axon causes the light output to rise to a new steady level, from which it declines to its original level within a few minutes of stopping stimulation. Although part of this rapid recovery may reflect pumping of Ca out of the axon, the bulk seems to result from sequestration of Ca inside the cell and only in the longer term is the Ca completely removed from the axon by pumps located in the surface membrane. In the context of this meeting I shall only discuss transport of Ca across the plasma membrane.

Despite the fact that the axoplasmic binding systems can buffer appreciable changes in ionized Ca, if these were the only mechanisms involved in regulating intracellular Ca, the Ca content of axoplasm would rise steadily. The resting Ca influx into squid axons is about 0.1 p mol $cm^{-2} \cdot sec^{-1}$ and this would double the Ca content of the axon in 24 h. As the Ca content remains rather constant over much longer periods, mechanisms must exist for regulating the total amount of Ca available to the intracellular binding systems. These mechanisms are calcium pumps in the surface membrane.

By analogy with the Na-K exchange pump, Ca extrusion might be expected to depend for a source of energy on cellular ATP and poisoning the axon with cyanide should inhibit calcium extrusion. When this possibility was first examined by Rojas and Hidalgo (1968) and Blaustein and Hodgkin (1969) they found that metabolic inhibitors produced not the expected fall but a large rise in Ca efflux. Subsequent experiments with aequorin showed that the rise in efflux occurs at a time when the intracellular ionized Ca is elevated because of release of Ca from intracellular binding sites. The observation that poisoning did not inhibit Ca efflux led to a search for an alternative energy source to ATP. One such source is to couple Ca extrusion to the downhill movement of another ion or ions. In this way the energy available in one ion gradient can be used to generate a gradient of another ion. Examination of the ionic dependence of the Ca efflux from poisoned axons revealed that about half of the efflux is dependent on external Ca and half on external Na. It seemed reasonable to assume that the Ca-dependent component of Ca efflux reflects Ca-Ca exchange and attention was focused on the Na-dependent component as a possible basis for an ATP-independent Na-Ca counter transport system (see Baker, 1972). The concept of Ca extrusion by Na-Ca counter transport was strengthened by the knowledge that part of the Ca efflux from unpoisoned axons is also dependent on external Na and when the direction of the Na gradient is reversed it is possible to detect a Na efflux in exchange for Ca influx. Thus, it seemed possible that the direction of Ca movement might be governed solely by the magnitude and direction of the Na ion gradient.

Although there is no doubt that Na ions play an important part in the transport of Ca, a simple ATP-independent exchange of Na for Ca no longer seems very attractive. The main reasons for this statement are summarized below.

I. Stoichiometry and Sensitivity to Membrane Potential

In order for a simple Na-Ca counter transport to maintain the observed Ca gradient, at least 3, and possibly 4, Na ions must exchange with each calcium. Unless electroneutrality is maintained by involvement of ions other than Na, an exchange of 1 Ca for 4 Na should be sensitive to the membrane potential. This has been examined by a number of workers (Blaustein et al., 1974; Mullins and Brinley, 1975; Baker and McNaughton, 1976a,b). In poisoned axons and axons dialysed in the absence of ATP, the Na-dependent Ca efflux is increased by hyperpolarization which is consistent with a nonelectroneutral exchange of Na for Ca. However, in unpoisoned axons the Ca efflux is rather insensitive to membrane potential and this may be related to the small size of the Na-dependent Ca efflux under these conditions. The presumed reversal of Na_o-dependent Ca efflux, Ca_o-dependent Na efflux which is also seen in unpoisoned axons is strongly inhibited by hyperpolarization, suggesting that this exchange is electrogenic.

In summary, Ca transport is sensitive to membrane potential only under certain conditions. Ca efflux from poisoned axons is increased by hyperpolarization and Ca influx into unpoisoned axons is inhibited by hyperpolarization, but changes in membrane potential have rather little effect on Ca extrusion from unpoisoned axons suggesting that under physiological conditions Ca extrusion may be able to derive energy from a source other than the ion gradients.

II. Kinetics of Activation by Na

A dependence of Ca extrusion on more than one sodium ion might be detectable as a sigmoidal relation between external Na and Na-dependent Ca efflux. This has not been observed in unpoisoned axons where the dependence on external Na approximates very closely to a section of a rectangular hyperbola; but in fully-poisoned axons and axons dialysed in the nominal absence of ATP, the activation curve is markedly sigmoidal (Baker and Glitsch, 1973; Dipolo, 1974; Baker and McNaughton, 1976b). Although the failure to detect a sigmoidal activation curve in unpoisoned axons does not exclude a dependence on more than one external Na ion, the marked change in activation kinetics following poisoning provides further evidence that the Na-dependent extrusion of Ca is sensitive to the metabolic state of the cell. It should be noted that these changes in kinetics are not secondary to alterations in intracellular ionized Ca or membrane potential.

III. Kinetics of Activation by Ca_o

In both unpoisoned and poisoned axons there is a sizeable component of the Ca efflux that is sensitive to external calcium. The apparent dependence of this flux on external Ca is also very sensitive to the metabolic state of the cell (Baker and McNaughton, 1976a,b). In unpoisoned axons Ca_o-dependent Ca efflux requires only 2 μM Ca_o for half-maximal activation, whereas in fully-poisoned axons the apparent K_m^{Ca} is 3-8 mM. It is obviously of importance to know whether this Ca_o-dependent Ca efflux reflects exchange of internal Ca for external Ca under both sets of conditions. This can be easily tested by measuring Ca influx. The results are striking. In unpoisoned axons Ca influx increases linearly from 2 μM to 100 mM, but at 100 μM Ca_o, which is enough to saturate Ca-dependent Ca efflux, Ca influx is about 1% of Ca efflux. Under these conditions Ca is still being pumped out of the axon against an adverse electrochemical gradient. In fully poisoned axons Ca-dependent Ca efflux and Ca influx were similar at all calcium concentrations examined.

These observations show that poisoning not only changes the kinetics of activation of Ca efflux by external Ca, but also the mode of operation of the efflux. In unpoisoned axons Ca extrusion seems to be activated by, but not in exchange for, external Ca, whereas in poisoned axons the Ca-dependent Ca efflux reflects a genuine exchange of internal Ca for external calcium.

IV. The Residual Efflux

About half of the Ca efflux from unpoisoned axons persists in nominally Ca and Na-free sea waters, whereas in fully-poisoned axons the residual efflux in the nominal absence of Na and Ca is a very small fraction of the total efflux and in absolute terms is less than the residual efflux from unpoisoned axons. One possible explanation for the residual efflux from unpoisoned axons is that it reflects activation by traces of Ca present immediately external to the axolemma. According to this view the Ca-dependent Ca efflux and residual efflux have a common origin. Experiments with low concentrations of lanthanum provide some support for this possibility (Baker and McNaughton, 1976b). In the presence of 10 μM Ca, application of 300 μM La^{3+} inhibits both the Ca-dependent Ca efflux and residual efflux but does not abolish the Na-dependent component of Ca efflux, although the shape of the Na-activation curve becomes sigmoidal and is eventually inhibited by higher concentrations of lanthanum (Baker and McNaughton, 1976b; Van Breemen and De Weer, 1970). These experiments suggest that the Ca-dependent Ca efflux and residual Ca efflux from unpoisoned axons may reflect the same basic mode of operation of the Ca efflux system. If this is correct, about 2/3 to 3/4 of the total efflux reflects Ca extrusion activated by, but not in exchange for, external calcium. This component of the flux shows no obvious sensitivity to other external cations and it seems likely that this flux may be directly dependent on metabolic energy, a suggestion that is consistent with the virtual disappearance of the residual efflux in fully-poisoned axons.

V. Dependence on Metabolism

In view of the evidence that application of metabolic inhibitors leads to a rise in internal ionized Ca, conditions were sought where the effects of metabolic inhibitors could be studied in the absence of changes in ionized Ca. Two such conditions are (1) injection of apyrase which destroys ATP but leaves mitochondrial Ca accumulation unaffected and (2) application of cyanide to axons that have been pre-injected with enough Ca-EGTA buffer to clamp the intracellular ionized Ca at the physiological level. The results are quite clear-cut (Baker and Glitsch, 1973). When Ca_i is kept constant, a reduction in cellular ATP produces a large fall in Ca efflux. The Ca efflux can be reactivated by injection of ATP but not by injection of AMP or cAMP. The nonhydrolysable analogues of ATP, AMP-PCP and AMP-PNP either effect inhibition or a very variable amount of reactivation and it seems likely that any reactivation may be due either to contamination with ATP, or to a rise in endogenous ATP. Immediately after injection of ATP the kinetics of activation by external Na are no longer obviously sigmoidal and resemble quite closely those found in unpoisoned axons.

These effects of metabolism ATP are not associated with any detectable changes in either Na or K gradients or membrane potential.

Taken together these data provide compelling evidence for a central role of ATP in effecting Ca extrusion from unpoisoned squid axons. It is not yet know whether this ATP is hydrolysed during Ca transport; but it seems very likely that the Ca_o-dependent net extrusion of Ca from unpoisoned axons probably consumes ATP and is driven by a Ca-activated ATPase similar to that of red blood cells (see Schatzmann, 1975). A definitive answer must await isolation and characterisation of such an enzyme from axon membranes.

In the absence of specific inhibitors it is not possible to decide whether the various components of Ca-transport that can be detected in squid axons reflect different systems or different modes of operation of the same basic system. This is clearly a very important point, especially as it relates to the possible role of ATP hydrolysis in Na-dependent counter-transport of calcium. Three Na-dependent transport systems have now been examined in some detail in squid axons: Na-dependent counter transport of Mg (Baker and Crawford, 1972), Na-dependent co-transport of glutamate (Baker and Potashner, 1974) and the Na-dependent counter transport of Ca just described. In all three cases, at constant membrane potential and constant internal Na,K and Ca concentrations, application of metabolic inhibitors leads to a marked reduction in the rate of operating of the Na-dependent transport process. In the case of Ca and glutamate this results from an alteration in the kinetics of substrate binding consistent with an intimate involvement of ATP or some derivative of it in the transport process.

These changes in kinetics can be accounted for quantitatively by the assumption that ATP or some derivatives of it must first bind to the membrane before one of the activating substrates. Thus, for the Na-dependent activation of Ca extrusion, neglecting the involvement of more than one Na-ion and designating the membrane as M, one can write the following scheme:

$$M + ATP \underset{}{\overset{K_m^{ATP}}{\rightleftharpoons}} M \cdot ATP \overset{K_m^{Na_o}}{\rightleftharpoons} M \cdot ATP \cdot Na_o \longrightarrow \text{Ca extrusion}$$

from which

$$\frac{v}{V_{max}} = \frac{1}{1 + \frac{K_m^{Na_o}}{[Na_o]}\left(1 + \frac{K_m^{ATP}}{[ATP]}\right)}$$

and

$$K_{m\ apparent}^{Na_o} = K_m^{Na_o}\left(1 + \frac{K_m^{ATP}}{[ATP]}\right)$$

It follows that as the ATP concentration falls, $K_{m\ apparent}^{Na_o}$ increases, as is observed. This analysis does not make any statement about whether or not the ATP is hydrolysed and at least three possibilities are consistent with the available data:

1. The binding of ATP or some derivative of it facilitates transport, i.e. ATP acts as an allosteric effector.

2. Phosphorylation of the membrane by ATP or some derivative of it facilitates transport.

3. Transport involves the stoichiometric hydrolysis of ATP.

A clear distinction between these possibilities cannot at present be made; but in the first two, ATP merely serves to "gate" the Na-dependent transport system - the energy from transport being derived from the ion gradients - whereas in the third some of the energy for transport is derived directly from the hydrolysis of ATP.

E. Conclusion

In conclusion, the biochemical analysis of the mechanism of secretion is still very primitive. Further progress is likely to be directed towards an understanding of the chemistry of the voltage-sensitive calcium channel, an understanding of the part played by ATP in sodium-dependent extrusion of calcium and the detailed characterisation of the mechanism of exocytosis, with particular reference to the part played by calcium ions.

References

Baker, P.F.: Prog. Biophys. Mol. Biol. 24, 177 (1972)
Baker, P.F., Crawford, A.C.: J. Physiol. 227, 855 (1972)
Baker, P.F., Glitsch, H.G.: J. Physiol. 233, 44P (1973)
Baker, P.F., Glitsch, H.G.: Phil. Trans. Roy. Soc. B27, 270, 389 (1975)
Baker, P.F., McNaughton, P.A.: J. Physiol. 258, 97P (1976a)
Baker, P.F., McNaughton, P.A.: J. Physiol. Lond. 259, 103 (1976b)
Baker, P.F., Potashner, S.J.: Biochim. Biophys. Acta 318, 123 (1973)
Baker, P.F., Rink, T.J.: J. Physiol. 253, 593 (1975)
Blaustein, M.P.: Rev. Physiol. Biochem. Exp. Pharmacol. 70, 33 (1974)
Blaustein, M.P., Hodgkin, A.L.: J. Physiol. 200, 497 (1969)
Blaustein, M.P., Russell, J.M., De Weer, P.: J. Supramolec. Struct. 2, 558 (1974)
Davis, B., Lazarus, N.R.: J. Physiol. 256, 709 (1976)
Dipolo, R.: J. Gen. Physiol. 64, 503 (1974)
Dodge, F., Rahamimoff, R.: J. Physiol. 193, 419 (1967)
Gilbert, D.S., Newby, B., Anderton, B.: Nature (London) 256, 586 (1975)
Jenkinson, D.H.: J. Physiol. 138, 434 (1957)
Miledi, R.: Proc. Roy. Soc. B183, 421 (1973)
Mullins, L.J., Brinley, F.J.: J. Gen. Physiol. 65, 135 (1975)
Okamoto, H., Takahashi, K., Yoshii, M.: J. Physiol. 255, 527 (1976)
Rojas, E., Hidalgo, C.: Biochim. Biophys. Acta 163, 550 (1968)
Schatzmann, H.J.: Curr. Topics Membrane Transp. 6, 125 (1975)
Van Breemen, C., De Weer, P.: Nature (Lond.) 226, 760 (1970)

The Molecular Organisation of the Sodium Channels in Nerve

R. D. KEYNES

As was established by Hodgkin and Huxley (1952), the sequence of changes in the ionic permeability of electrically excitable membranes underlying the conduction of impulses depends on the presence of charged systems which open and close with appropriate timing two separate sets of ionic channels, selective respectively for Na^+ and K^+ ions. But although the well-known Hodgkin-Huxley equations describe with a fair degree of precision the time-course and voltage-dependence of both the sodium and the potassium conductance, they do not by themselves provide specific information about the molecular nature of the gating mechanism, for there are a number of different physical systems that would obey such first-order differential equations. In order to choose between alternative models, independent evidence is needed to show how the ionic gates are controlled by the membrane potential. It has recently proved possible to record the very small electrical currents that accompany the opening and closing of the sodium gates in the squid giant axon (Armstrong and Bezanilla, 1974; Keynes and Rojas, 1974, 1976; Meves, 1974) and at the node of Ranvier in myelinated frog nerve (Nonner et al., 1975). The object of this article is to discuss what can be learnt from gating current measurements about the organisation of the sodium channels.

The phenomenon is basically a displacement current, that is to say a current that flows across a capacitance when the potential is changed step-wise by ΔV. However, whereas for a simple capacitance the total transfer of charge varies linearly with ΔV, the integral of the gating current has decidedly nonlinear characteristics. The magnitude of the displacement current is in either case small compared with that of the ionic current which flows through the channels once they are open. Hence there are two problems to be overcome technically in making records of the gating current - the first, to block the ionic currents so that they do not obscure the displacement currents, and the second, to eliminate the linear capacity transient. Fortunately, it turns out that blockage of the sodium channels with tetrodotoxin (TTX) does not interfere with the operation of the voltage-dependent gate. The measurements are therefore made in the absence of sodium and in the presence of 300 nM-TTX in order to cut out all the sodium current, and the potassium channels are blocked with the aid of caesium or tetraethylammonium ions. A recording of the gating current in a squid giant axon is shown in Figure 1. The membrane was initially held at -100 mV to ensure that all the gates were in the resting configuration, and thus closed. Application of a hyperpolarizing voltage-clamp pulse gave, as expected, a fast capacity transient with a time constant no greater than 15 μsec even at the foot. A depolarizing pulse of the same size gave a much slower transient, both at the start and at the finish of the pulse, owing to the addition of a slow component arising from the charge transfer involved in opening and closing the gates. On summing the two current records, the symmetrical capacity transient cancels out, and the asymmetrical gating current is seen to be an exponential with

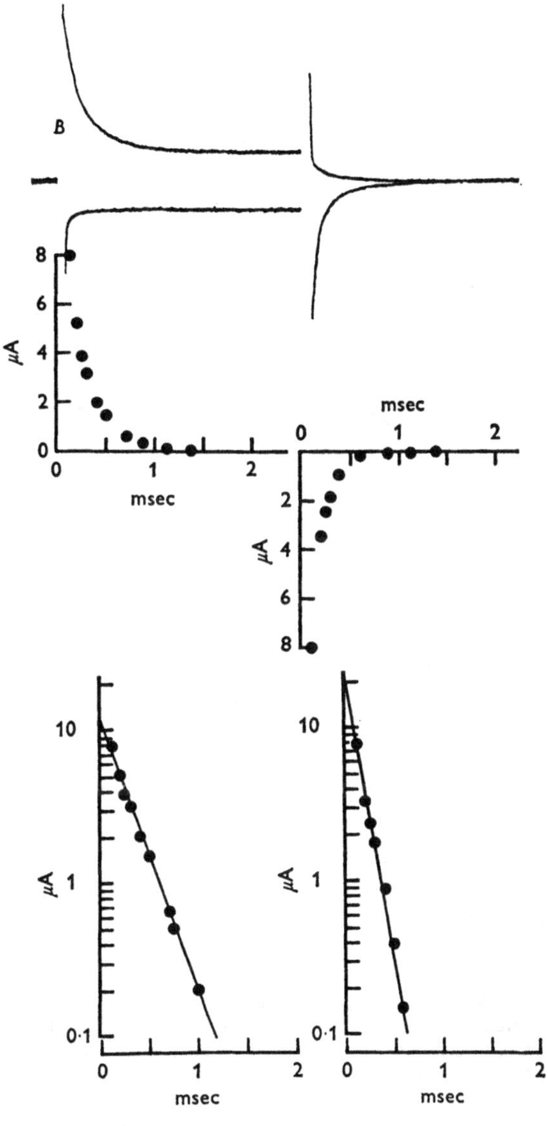

Fig. 1. Asymmetry of displacement current on application of equal and opposite voltage-clamp pulses to a squid axon perfused with 300 mM-CsF and bathed in Na- and K-free saline containing 1000 nM-TTX. Top traces: single-sweep records of the membrane current for ±120 mV pulses, applied at a holding potential of -100 mV. Difference between them is plotted beneath on linear and logarithmic scales. (From Keynes and Rojas, 1974)

a time constant of a few hundred microseconds, somewhat faster at "off" than at "on". Such graphical processing of the current records is far too laborious for normal use, and the recording technique originally employed was to add together with a signal averager the current transients for a series of equal and opposite voltage-clamp pulses, thus automatically eliminating the capacity transient and at the same time improving the signal-noise ratio. A typical family of records for pulses of different sizes is shown in Figure 2. This procedure has subsequently turned out to suffer from several serious limitations, and has now been superseded by the use of an on-line computer and other refinements to permit the gating current records to be corrected more accurately for leakage currents and the linear capacity transient (Keynes and Kimura, 1976).

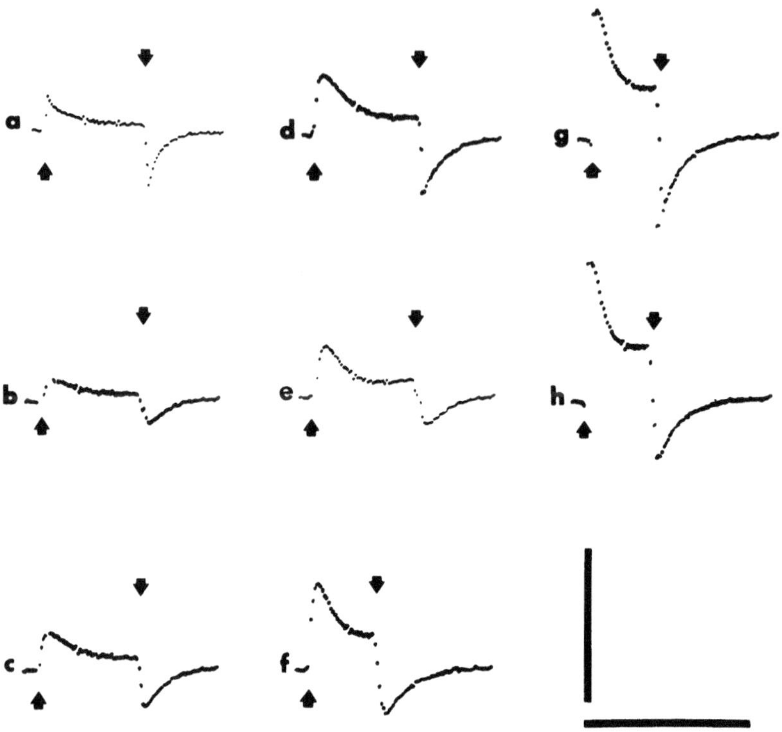

Fig. 2. Family of gating current records obtained by summation with a signal averager of membrane current for 60 positive and 60 negative voltage-clamp pulses, which started and finished at arrows. Axon was perfused with 55 mM-CsF and bathed in Na- and K-free saline containing 300 nM-saxitoxin. Pulse amplitude, a-h, 40-110 mV. Holding potential, -70 mV. Vertical bar, 5.56 μA. Horizontal bar: 2500 μsec. Membrane area, 0.06 cm^2. Temperature, 7^oC. (From Keynes and Rojas, 1974)

In considering the justification for identifying the asymmetrical displacement currents with charge movements involved in the operation of the sodium gating mechanism, both steady-state characteristics and kinetics have to be taken into account. Thus one may start by asking how far the voltage-dependence of the steady-state distribution of the mobile charges fits with the voltage-dependence of the sodium permeability. If the rise in permeability and the amount of charge transferred on altering the membrane potential from -100 mV to a less negative value are plotted together on a logarithmic scale against potential, it is at once seen that the permeability rises much faster than the charge transfer, the ratio of the limiting slopes being close to 3:1 (Keynes and Rojas, 1976). The inference is that each sodium channel behaves as if it were controlled by a group of three gating particles, this conclusion being satisfactorily consistent with Hodgkin and Huxley's (1952) quite different line of reasoning based on the inflexion of the time course for the initial rise in sodium conductance. Another permissible conclusion is that the effective valency of the individual particles is 1.3, although the evidence does not allow us to decide how the charge is distributed within the particles, nor on its polarity. Taking the sodium conductance or permeability as proportional to the cube of the total charge transferred, one may next compare the position on the voltage axis of the S-shaped curve for steady-state charge distribution with that of the similar curve for the cube root of the sodium conductance, the m_∞ curve in Hodgkin and Huxley's (1952) terminology. It turns out (Keynes and Rojas, 1976) that the midpoints of both curves lie between -30 and -25 mV, and although more accurate measurements may eventually place the charge distribution curve on the positive side of the m_∞ curve rather than cc inciding with it, the displacement is unlikely to be more than a few millivolts.

Up to a point, there is also good agreement between τ_m, the sodium conductance time constant as defined by Hodgkin and Huxley (1952), and the gating current time constant $\tau(V)$, as may be seen from Figures 6 and 13 of Keynes and Rojas (1976). However, there are certain differences that may prove to be rather significant in deciding between alternative molecular models for the gating system. One still unresolved issue concerns the ratio of the time constants for the gating current and for shutting off the sodium conductance at the end of a voltage-clamp pulse. If the three gating particles or sub-units return independently to the resting configuration on repolarization of the membrane, and if the movement of any one of them suffices to close the channel, then the time constant ratio should be 3. This is what was found by Keynes and Rojas (1976) in axons from <u>Loligo forbesi</u>, provided that large muscles were applied and the holding potential was around -100 mV; but according to Bezanilla and Armstrong (1975), the ratio in axons from <u>Loligo pealii</u> was never greater than 2. The discrepancy may perhaps arise from a species difference in the degree of interaction between the individual gating particles, for if they tended to stick together in the channel conducting configuration (C_3 in Fig. 3), the ratio would fall below 3. Further studies on this point may provide useful information about interactions within each channel.

Our analysis of the properties of the gating particles in the squid giant axon (Keynes and Rojas, 1974, 1976) indicates that in most respects the system obeys first-order kinetics with reasonable fidelity, but there is one important exception. In a first-order reaction the time constant for arrival at a new state is determined solely by the parameters for that state, and is independent of the previous history of the system. The gating current time constant at the end of a pulse should therefore be a function only of the holding potential to which the membrane returns. But as has been clear from the start (Keynes and Rojas, 1974; Meves, 1974), this expectation does not hold good in practice, and the "tail" time constant is appreciably greater after a large pulse that has maintained the membrane at a positive potential for some time, than it is following a smaller and briefer pulse. The time constant for shutting off the sodium conductance is, on the other hand, nearly independent of pulse size (see Table 4 of Keynes and Rojas, 1976). These facts can most easily be explained (Keynes, 1975) by the type of multi-state hypothesis illustrated in Figure 3, which supposes that the opening of the sodium channel involves two steps - the first, a voltage-dependent transition of the gating particles from the resting configuration R to the activated configuration A, and the second, a relatively rapid and mainly voltage-independent interaction between the particles that can take place only at channels where all three are in the appropriate state. It is further proposed that for those particles which have interacted to reach configuration C_3, the reverse transition to R* when the membrane potential is restored to the holding level is somewhat slowed, i.e. $\tau_{C_3 \to R^*}$ is about 50% greater than $\tau_{A \to R}$. Another experimental

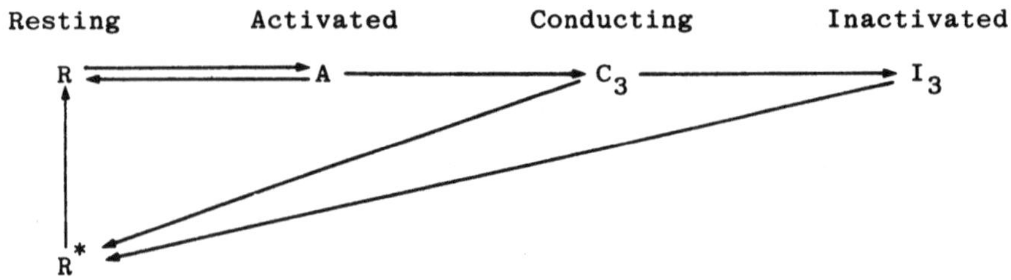

Fig. 3. Multi-state hypothesis to account for activation and inactivation of sodium channels

finding that can be explained plausibly by such a scheme is that replacement of H_2O by D_2O prolongs the sodium conductance time constant, but has relatively little effect on the gating current time constant (Meves, 1974). This might indicate that the configurational change R→A that takes place primarily under the influence of the electric field involves less hydrogen bond formation than the subsequent transformation $A \rightarrow C_3$.

At the node of Ranvier in frog nerve fibres, the behavior of the gating system departs from strict first-order kinetics in a manner that has been interpreted by Neumcke et al. (1976) in terms of a regime involving electro-diffusion of charged gating particles within the membrane. Their multi-state model thus differs from mine in proposing that the individual gating particles encounter a series of energy barriers in making the transition between two discrete states of the sodium channel, resting and conducting, whereas according to Keynes and Rojas (1976) there is only a single energy barrier between the resting and activated states of the gating particles, but the channel is to be regarded as passing through a sequence of three different configurations. The experimental evidence is as yet inadequate to distinguish decisively between the two hypotheses, and it may well turn out that both are valid, i.e. that the several states of Figure 3 are each separated by multiple rather than single energy barriers. Whichever theory is correct, one has still to account for the inactivation of the channel that occurs during prolonged depolarization of the membrane.

Although over the range of membrane potentials examined by Keynes and Rojas (1976) the sizes of τ_m and of $\tau(V)$ were much the same, suggesting that the time constant of the second step, $\tau_{A \rightarrow C3}$, was small throughout the range, there were indications that for very small depolarizing pulses this equality might not apply. As may be seen in Table 2 of Hodgkin and Huxley (1952), τ_m reaches a peak in the neighbourhood of -40 mV at which it may appreciably exceed the gating current time constant. This is an interesting point for further study, because if $\tau_m >> \tau(V)$ the postulated interaction between the activated gating particles would appear to be greatly slowed near threshold, when only a few of them are available to participate; and this conclusion carries important implications for the type of physical model compatible with the results.

The facts that I have recounted so far show that the parameter m in Hodgkin and Huxley's (1952) equation

$$g_{Na} = \bar{g}_{Na}\, m^3\, h$$

can, with certain reservations, be roughly equated with the fraction of gating particles in the activated state. Is it possible in a similar way to identify the inactivation parameter h as representing the fraction of a second set of more slowly moving particles that is not yet in the blocking position? From a knowledge of the size of the inactivation time constant τ_h, and of the steepness of the h_∞ curve relating the degree of inactivation in the steady state to membrane potential, the probable magnitude of the displacement current that would be generated by the movements of an independent set of inactivating particles can be calculated. In an experiment like that of Figure 1, the initial amplitude of the predicted "h" gating current would be about quarter of that of the "m" current, and its time constant would be about four times greater. Clearly no such slowly declining current was present, and indeed no component of the gating current with kinetics appropriate to the "h" system has yet been reported by anyone. It seems reasonable to conclude that the notion of a separate and independent "h" particle responsible for inactivation must be discarded.

An alternative proposed by Bezanilla and Armstrong (1975) is that steady-state inactivation involves a permanent immobilization of the gating particles, and hence an abolition of the gating current. However, this idea is contradicted by records like those of Meves (1974) and Keynes et al. (1974) which show that, although maintained depolarization of the membrane results in an inversion of the gating current, and probably some reduction in its amplitude, it certainly does not disappear altogether. Instead, a simple hypothesis to account for inactivation would be to suppose that after interacting to open the sodium channels, the gating particles may then undergo a further configurational change which inexorably closes the channels again. If the step $C_3 \rightarrow I_3$ in Figure 3 were time- rather than voltage-dependent, no gating current would accompany it. As Goldman (1977) has argued, a multi-state system of this general type can readily yield Hodgkin-Huxley kinetics for the ionic conductance. But although there is thus no conceptual difficulty in suggesting a mechanism for inactivation, we are still a long way from being able to produce a properly quantitative explanation for the bulk of the experimental observations. There are many complications to be considered, such as the occurrence of slow changes both in the peak sodium conductance and in the size of the gating current, and shifts in certain parameters of the system on lowering of the holding potential, which as yet have no place in the model. Nevertheless, there is good reason to hope that a systematic investigation in parallel of the sodium conductance changes on the one hand, and the gating currents on the other, will soon lead to a better understanding of the gating mechanism.

As was shown by Keynes and Rojas (1974), the gating current is not obviously affected by blocking the outward-facing mouth of the sodium channel with TTX. The part of the channel responsible for its high selectivity towards Na^+ ions, whose mode of action has been described by Hille (1971), can thus be regarded as functionally independent of the voltage-sensitive gate. We hence arrive at a picture of the sodium channel as consisting of a selectivity filter located at the outside of the membrane, arranged in series with three subunits, the gating particles, which straddle the low dielectric constant region. In the resting configuration, ions are effectively excluded from the membrane by the group of subunits, but when the electric field alters appropriately, they undergo a cooperative reorganisation that creates an aqueous channel between them and allows ions to pass. Closure is brought about either by a reversal of the sequence of events on restoration of the resting field, or, if the depolarization is prolonged, by a further slow conformational change that finally obliterates the aqueous channel. The sodium channel is an amphipathic membrane protein whose molecular weight is around 230,000 (Levinson and Ellory, 1973). However, little progress has yet been made towards extraction and purification of the sodium channel, and there is no information about its tertiary structure.

References

Armstrong, C.M., Bezanilla, F.: Charge movement associated with the opening and closing of the activation gates of the Na channels. J. Gen. Physiol. 63, 533-552 (1974)

Bezanilla, F., Armstrong, C.M.: Kinetic properties and inactivation of the gating currents of sodium channels in squid axon. Phil. Trans. Roy. Soc. B 270, 449-458 (1975)

Goldman, L.: Kinetics of channel gating in excitable membranes. Quart. Rev. Biophys. 9, 492-526 (1976)

Hille, B.: The permeability of the sodium channel to organic cations in myelinated nerve. J. Gen. Physiol. 58, 599-619 (1971)

Hodgkin, A.L., Huxley, A.F.: A quantitative description of membrane current and its application to conduction and excitation in nerve. J. Physiol. (London) 117, 500-544 (1952)

Keynes, R.D.: Organization of the ionic channels in nerve membranes. In: The Nervous System. Vol. 1. The Basic Neurosciences. Tower, D.B. (ed.). New York: Raven Press 1975, pp. 165-175

Keynes, R.D., Kimura, J.E.: Digital processing of ionic and gating current records. J. Physiol. (London) 263, 112 P. (1976)

Keynes, R.D., Rojas, E.: Kinetics and steady-state properties of the charged system controlling sodium conductance in the squid giant axon. J. Physiol. (London) 239, 393-434 (1974)

Keynes, R.D., Rojas, E.: The temporal and steady-state relationships between activation of the sodium conductance and movement of the gating particles in the squid giant axon. J. Physiol. (London) 255, 157-189 (1976)

Keynes, R.D., Rojas, E., Rudy, B.: Demonstration of a first-order voltage-dependent transition of the sodium activation gates. J. Physiol. (London) 239, 100-101P (1974)

Levinson, S.R., Ellory, J.C.: Molecular size of the tetrodotoxin binding site estimated by irradiation inactivation. Nature New Biol. 245, 122-123 (1973)

Meves, H.: The effect of holding potential on the asymmetry currents in squid giant axons. J. Physiol. (London) 243, 847-867 (1974)

Neumcke, B., Nonner, W., Stämpfli, R.: Asymmetrical displacement current and its relation with the activation of sodium current in the membrane of frog myelinated nerve. Pflügers Arch. 363, 193-203 (1976)

Nonner, W., Rojas, E., Stämpfli, R.: Displacement currents in the node of Ranvier. Voltage and time dependence. Pflügers Arch. 354, 1-18 (1975)

Voltage Clamp Analysis of Catecholamine-Induced Calcium Conductance Changes in Mammalian Cardiac Muscle

H. Reuter

In cardiac muscle a considerable fraction of the total membrane current (I_m) during excitation is carried by calcium ions. Voltage clamp analysis of I_m revealed a voltage- and time-dependent conductance change of the membrane which depends to a large extent on the external Ca ion concentration. The corresponding Ca-dependent inward current is of primary importance for the typical plateau phase of the cardiac action potential (Fig. 1). Ca ions flow down their electrochemical gradient into the cell, presumably through specific pores in the membrane, often referred to as "ionic conductance channels" (Hille, 1970; Armstrong, 1975). Although the conductance channels in question are about hundred times more selective for Ca ions than for Na or K ions, some fraction of the current is carried by these monovalent cations, because they are present at much higher concentrations than Ca ions in the extracellular fluid and in the myoplasm (Reuter and Scholz, 1977a). The Ca-dependent inward current is distinguished from the rapid excitatory Na current (I_{Na}), common to many excitable tissues, (1) by its different ionic dependence, (2) by its much slower kinetics and smaller current density, and (3) by its different sensitivity to pharmacological influences (Reuter, 1973). Therefore, it has been called slow or secondary inward current, I_s.

Catecholamines greatly increase the permeability of cardiac cell membranes to Ca ions. This has been demonstrated in a variety of cardiac tissues by measuring ^{45}Ca fluxes (Reuter, 1965; Meinertz et al., 1973) as well as by electrophysiological techniques (Reuter, 1967; Carmeliet and Vereecke, 1969; Pappano, 1970; Brown et al., 1975; Watanabe and Besch, 1975). Voltage clamp analysis of this catecholamine effect indicates an increase in I_s and a corresponding shift of the plateau height of the action potential towards more positive potentials (Figs. 1, 2).

Recently, we have analyzed this catecholamine effect in greater detail (Reuter, 1974; Reuter and Scholz, 1977b). The basic assumption is that the conductance channels of I_s can exist in the states closed or open (Hille, 1970; Armstrong, 1975). The channels open during depolarization and close upon repolarization with voltage-dependent rate constants α and β. Hence, the following reaction scheme can be applied:

$$\text{closed} \underset{\beta_d}{\overset{\alpha_d}{\rightleftharpoons}} \text{open} \tag{1}$$

The subscript d designates the voltage-dependent gating process determining the state of the channels. The fraction of channels in the open state is d, and in the closed state 1-d. A second conductance variable determining the state of the channels, the inactivation variable f (Reuter, 1973; Reuter and Scholz, 1977a), is neglected since its time constants are long compared to the short voltage clamp steps required for the present analysis; hence

This paper is dedicated to Professor Walther Wilbrandt on the occasion of his 70th birthday.

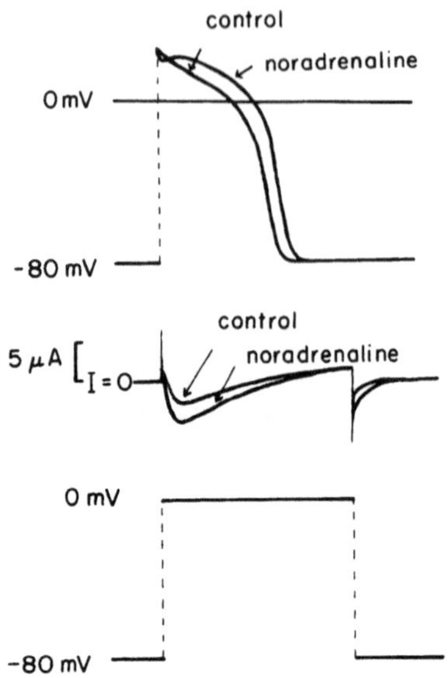

Fig. 1. Two superimposed action potentials (top traces) recorded from a calf ventricular trabecula in normal Tyrode's solution (control) and in the presence of noradrenaline (10^{-6} M), and two corresponding membrane currents (middle traces) without and with noradrenaline measured in the same experiment during voltage-clamp steps (bottom traces) from a holding potential -80 mV to 0 mV. The rapid initial sodium current was eliminated in the voltage clamp experiment by application of tetrodotoxin (5×10^{-5} M). Note increase in the slow inward current, I_s (downward deflection from I = 0 in current traces) and corresponding increase in plateau height in presence of noradrenaline

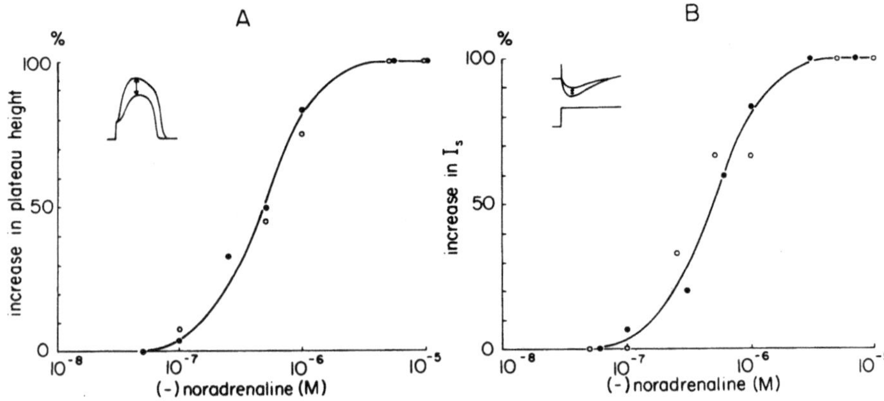

Fig. 2A,B. Log concentration-response curves of the effect of noradrenaline on plateau height of action potentials (A) recorded from two cat papillary muscles in tetrodotoxin containing (5×10^{-5} M) Tyrode's solution, and on the slow inward current, I_s (B) measured under voltage clamp conditions in two cat papillary muscles. Solid lines: concentration-response curves fitted to data to give a slope of 2 in the Hill plot. (Redrawn from Reuter, 1974)

f is close to 1. Therefore, the experimentally observed slow inward current is given by

$$I_s = \bar{g}_s \cdot (E_m - E_R) \cdot d \qquad (2)$$

where $\bar{g}_s$, the limiting conductance of the membrane, is determined by the total number of channels opening during strong depolarization, E_m and E_R are membrane potential and zero current potential (reversal potential) of I_s, respectively, and in the steady state

$$d_\infty = \frac{\alpha_d}{\alpha_d + \beta_d} \ . \qquad (3)$$

Since

$$\tau_d = \frac{1}{\alpha_d + \beta_d} \ , \qquad (4)$$

we obtain from Equations (3) and (4)

$$\alpha_d = \frac{d_\infty}{\tau_d} \qquad (5)$$

and

$$\beta_d = \frac{1-d_\infty}{\tau_d} \ . \qquad (6)$$

With these theoretical considerations in mind, which follow essentially the Hodgkin-Huxley hypothesis (Hodgkin and Huxley, 1952), the following questions pertinent to the action of adrenergic transmitters on I_s may be asked:

1. Are the kinetic properties of I_s, as defined by Equations (3-6), affected by catecholamines? This would be the case if the rate of opening or closing of the conductance channels is changed.

2. Is the selectivity of the conductance channels altered by catecholamines resulting in a shift of E_R and hence in a change of the driving force $(E_m - E_R)$ (Eq. (1))? This is to be expected if the conformation of the conductance channels is changed in such a way that they conduct more ions per unit time.

3. Is the total number of conductance channels increased as a result of the catecholamine-receptor interaction? This would be reflected by an increase in $\bar{g}_s$ without a change of E_R (Eq. (2)), and of the kinetics.

Methods

The experiments were performed on ventricular trabeculae of cow and cat hearts. The single sucrose gap voltage-clamp technique was applied, which has been described in detail elsewhere (Beeler and Reuter, 1970; Reuter and Scholz, 1977a). Briefly, the method permits rapid changes of the membrane potential. The current flowing through the membrane as a result of the stepwise potential changes is measured by an operational amplifier which holds the bath containing Tyrode's solution at virtual ground. (-) Adrenaline bitartrate or (-) noradrenaline bitartrate were added at different concentrations to the bathing solution. In some experiments tetrodotoxin (TTX) 5×10^{-5} M was used in order to eliminate I_{Na}.

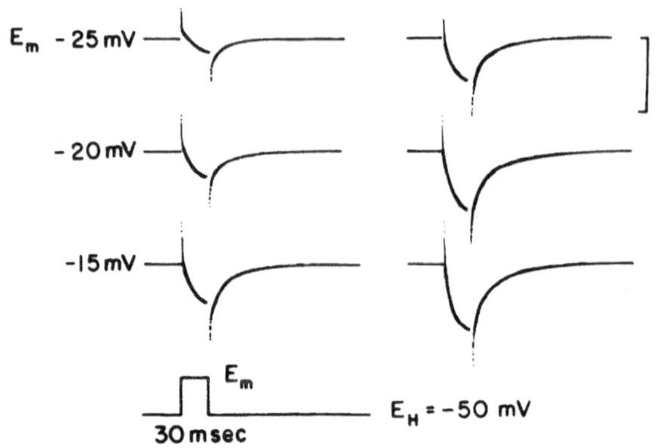

Fig. 3. Voltage clamp currents recorded from a cow ventricular trabecula during and after step changes of membrane potential from holding potential E_H = -50 to E_m -25, -20 and -15 mV. Note exponential increase in slow inward current, I_s, during depolarizing clamp steps (30 msec) and its exponential decay upon repolarization to -50 mV. Measurements were made without (left control) amd with (right) adrenaline (5 x 10^{-7} M)

Results

Membrane currents measured during and after 30 msec voltage clamp steps from a holding potential, E_H, of -50 mV to different membrane potentials (-25, -20 and -15 mV) are shown in Figure 3. E_H was adjusted to -50 mV in order to eliminate I_{Na} which is completely inactivated at this potential. Time-dependent outward current is negligible, and time-independent outward current shows very little voltage-dependence in the range -25 to -15 mV (Beeler and Reuter, 1970). Therefore, the inward current (downward deflection of current records from baseline current) shown in Figure 3 represents I_s which is largely carried by Ca ions (Reuter, 1973; Reuter and Scholz , 1977a).

The amplitude of I_s is considerably increased by adrenaline. While the rates of the developing inward currents during stepwise depolarization reflect the rate constants of opening of the conductance channels (α_d), the relaxation of I_s after repolarization gives a direct measure of β_d. Analysis of the current traces in Figure 3 reveals that (after the capacitative spikes at the beginning and the end of the clamp steps have subsided with a time constant < 1 msec) the currents relax exponentially to new steady states during and after the depolarizing clamp steps. Despite the considerable increase in current amplitude by adrenaline, the time constant τ_d of the relaxing current at each potential step is the same (10 msec) in the absence and presence of the drug. Since d_∞ is 0.55, 0.73 and 0.85 (see below) at -25, -20 and -15 mV respectively, α_d is 55, 73 and 85 sec^{-1} (Eq. (5)). Also the time constants of the relaxing currents after clamping back to E_H are identical (18 msec), although the amplitude of these tail currents are very different. Since d_∞ = 0 at -50 mV, β_d is 56 sec^{-1} (Eq. (6)). These results indicate that the rate constants of opening and closing of the conductance channels of I_s are not affected by adrenaline. Similar results were obtained in six other experiments.

The degree of steady-state activation (d_∞) of I_s during the voltage step can be obtained from the amplitude of the tail currents immediately after repolarization. Since the tail currents were always measured at -50 mV, the driving force (E_m - E_R) for I_s was constant (see below), and hence their amplitude is directly pro-

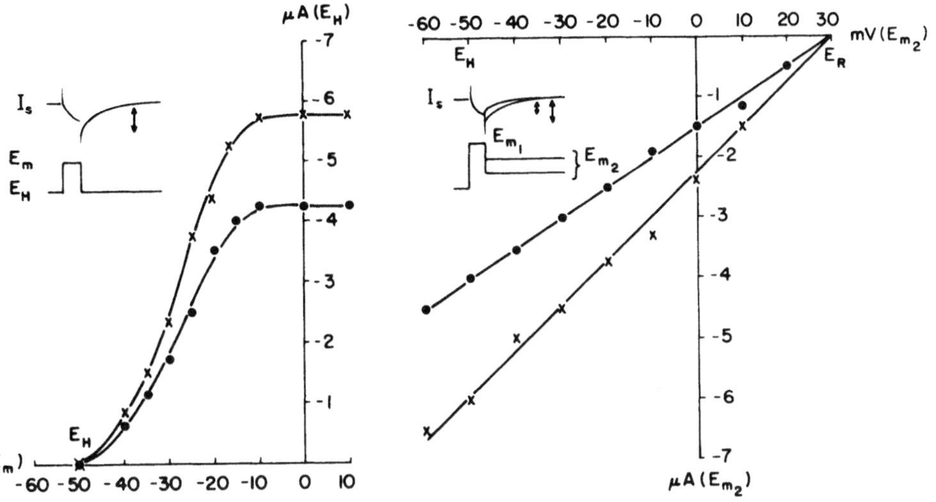

A B

Fig. 4A,B. (A) Activation curves of I_s measured in absence (●) and presence (×) of adrenaline (10^{-6} M) in cow ventricular trabecula. Plotted are maximal tail currents (ordinate) recorded at E_H and activated at E_m (abscissa) during short (20 msec) voltage clamp steps (inset); maximal tail currents were obtained from extrapolation to time zero of semilog plots of the current traces. (B): Same experiment as in (A); plotted are maximal tail currents recorded at E_{m_2} and activated by short (20 msec) depolarizing clamp steps to E_{m_1} = +10 mV (inset); abscissa: membrane potential E_{m_2}, where tail currents were measured in absence (●) and presence (×) of adrenaline (10^{-6} M). Note that clamp step E_{m_1} to +10 mV was sufficient to activate I_s fully (see (A))

portional to $\bar{g}_s \cdot d_\infty$ (Eq. (2)). In Figure 4A the amplitude of the tail currents at E_H is plotted as a function of the membrane potential (E_m) during the preceding depolarizing clamp steps. The conductance, $\bar{g}_s$, was maximal at potentials positive to -10 mV. The S-shaped relation indicates the degree of activation of I_s at each E_m, and is determined by the dimensionless steady-state activation variable d_∞ which can be obtained from normalization of the data in Figure 4A. Thus, d_∞ is zero at -50 mV and 1 at -10 mV; d_∞ is not shifted along the voltage axis by adrenaline, and its steepness is the same with and without the transmitter.

The maximal conductance, $\bar{g}_s$, is increased by catecholamines. This can be seen from Figure 4A, where the maximal amplitude of the tail currents, activated at E_m > -10 mV and measured at -50 mV, is increased (cf. also Fig. 3). Since $\bar{g}_s = \bar{I}_s/(E_m - E_R)$, the maximal amplitude of the tail currents, $\bar{I}_s$, measures $\bar{g}_s$ directly, provided ($E_m - E_R$) is constant (see below). Without adrenaline (filled circles in Fig. 4A) $\bar{g}_s$ is 5.25 × 10^{-5} Ω^{-1} and with adrenaline (5 × 10^{-7} M; × in Fig. 4A) $\bar{g}_s$ increased by 38% to 7.25 × 10^{-5} Ω^{-1}.

The effect of adrenaline on the reversal potential, E_R, which is the potential where I_s reverses its sign from inward current to outward current, was measured by instantaneous current voltage relations depicted in Figure 4B. As shown in the inset, the membrane was depolarized in a first step (E_{m_1}) to +10 mV for 20 msec. $\bar{g}_s$

was fully activated at this potential and during this time (Fig. 4A). Thereafter the potential was changed in a second step, E_{m_2}, to different levels and I_s relaxed exponentially to new steady states typical for the respective E_{m_2}. The maximal current amplitudes of I_s attained instantaneously after clamping to E_{m_2} (i.e. at time zero after subtraction of the capacity current) were plotted against the membrane potentials E_{m_2}. This yields linear current-voltage relations with a common intercept at +30 mV. The intercept is the reversal potential E_R of I_s. E_R can be used to determine the selectivity of the conductance channels of I_s (Hille, 1975; Reuter and Scholz, 1977a). For the present purpose it is enough to say that E_R, and hence the selectivity of I_s, is not changed by adrenaline. This also implies that the driving force ($E_m - E_R$) was constant in the experiment in Figure 4A, while it varied directly with E_{m_2} in the experiment illustrated in Figure 4B. Since $\bar{g}_s = \bar{I}_s/(E_m - E_R)$ the slopes of the instantaneous current-voltage relations in Figure 4B measure again $\bar{g}_s$. In the controls (filled circles), $\bar{g}_s$ is 5.2 × 10^{-5} Ω^{-1} and in the presence of adrenaline (×s) it is 7.3 × 10^{-5} Ω^{-1}. These values are identical to ones obtained from the experiment in Figure 4A.

This kinetic analysis of the effect of adrenaline on the Ca-dependent slow inward current, I_s, leads to the conclusion that the only factor which is changed by the neurotransmitter is the maximal conductance, $\bar{g}_s$.

Discussion

The present experiments show clearly that the membrane Ca conductance, measured as g_s, of mammalian cardiac muscle is increased by adrenergic neurotransmitters. This result agrees with ^{45}Ca influx measurements which also show an increase in the Ca permeability by catecholamines (Reuter, 1965; Meinertz et al., 1973). The most important result of this analysis is the fact that neither the kinetics of I_s nor the ion selectivity of the underlying conductance channels are changed by adrenaline. Therefore, I_s is increased only because the maximal conductance, $\bar{g}_s$, becomes larger. Since the conductance channels of I_s are about one hundred times more permeable to Ca ions than to Na or K ions (Reuter and Scholz, 1977a), the increase in Ca permeability measured by tracer fluxes during excitation is most likely the result of the increase in $\bar{g}_s$.

How can one explain the increase in $\bar{g}_s$ by catecholamines without any change of the forward and backward rate constants of reaction 1 or of the ion selectivity of the conductance channels? The most likely possibility is that more of the voltage-dependent conductance channels of the same type are opened in the presence of catecholamines. The other possibility, an increase in the ionic conductance of individual conductance channels, is less likely, because it is difficult to see how a conductance channel could become more permeable without any change in its ion selectivity. The narrowest part of a channel seems to be the selectivity filter (Hille, 1975) which must become wider, and therefore less selective if the permeability of the individual conductance channel becomes greater.

If one favors the idea of the increase in the number of functional Ca conductance channels by adrenergic transmitters one may ask how such a regulation could be achieved. There is considerable evidence that this effect is not a direct consequence of the drug-receptor reaction. Several other drugs, unrelated to catecholamines (methylxanthines, papaverine, histamine) have similar effects on I_s. All these drugs have in common that they increase the intracellular cyclic AMP level (Watanabe and Besch, 1975). Dibutyryl and monobutyryl cyclic AMP also have similar effects on I_s as catecholamines (Reuter, 1974). Therefore, it is tempting to speculate that cyclic AMP is involved in the catecholamine effect on $\bar{g}_s$. Such a membrane effect could result from a cyclic AMP-dependent phosphorylation of a membrane protein related to the Ca conductance channels. In this case, the number of channels opening during depolarization would be dependent on the steady-state cyclic AMP level in the cell. It is interesting to note in this context that another neurotransmitter, acetylcholine, reduces $\bar{g}_s$ in cardiac muscle (Giles and Tsien, 1975; Ikemoto and Goto, 1975; Ten Eick et al., 1976), and also causes a decrease of the intracellular cyclic AMP level.

To put this hypothesis in a more quantitative form, the dose-response curves in Figure 2 may give some hint. These dose-response curves are steeper than expected from a simple one-to-one binding reaction between catecholamines (C) and receptors (R). The adrenergic β-receptors located at the outside of the membrane (Reuter, 1974), however, seem to bind catecholamines in a one-to-one fashion (Harden et al., 1976; Lefkowitz et al., 1976). Therefore, suppose that the increase in $\bar{g}_s$ is not the direct consequence of the drug-receptor reaction. For example, this reaction may activate the adenylate cyclase in the membrane which causes an increase in the intracellular cyclic AMP level, which in turn produces these membrane effects.

Hence,

$$C + R \rightleftharpoons CR \longrightarrow \Delta cAMP \longrightarrow \Delta\bar{g}_s. \qquad (7)$$

However, rather than changing $\bar{g}_s$ directly, cAMP may cause phosphorylation of a membrane protein directly related to the conductance channel via a proteinkinase, P, i.e.

$$n\ cAMP + P \rightleftharpoons cAMP_n\ P \longrightarrow \text{phosphorylation}, \qquad (8)$$

where n cAMP denotes the number of cyclic AMP molecules bound to the enzyme. Another possibility is that P is an allosteric enzyme built from more than one, say m, subunits, In either case

$$\bar{g}_s \propto (cAMP_nP)^m, \qquad (9)$$

or

$$\bar{g}_s = \frac{1}{1 + (\frac{K}{cAMP_nP})^m}, \qquad (10)$$

where K is a complex constant. This is the form required for the dose response curves in Figure 2, where m must be greater than 1. A similar hypothesis has been applied to another catecholamine-induced conductance change by Tsien (1974). It is of interest to note, however, that different conductance systems in cardiac cell membranes are affected by catecholamines in very different ways (Reuter, 1974; Tsien, 1974).

Summary

Catecholamines increase the voltage- and time-dependent Ca conductance ($\bar{g}_s$) of cardiac cell membranes. In order to analyse the mechanism of this conductance change, the corresponding Ca-dependent slow inward current (I_s) was recorded from voltage-clamped cardiac muscle fibers in the absence and presence of catecholamines. Step changes in membrane potential caused an exponential increase in I_s during depolarization, and an exponential relaxation of I_s after repolarization. The amplitude of I_s was greatly increased by catecholamines, while its time constants were not affected. The zero current potential, E_R, which is a measure of the ion selectivity of the conductance channels was also unaffected. Since catecholamines increase I_s without changing the kinetics and the selectivity of the corresponding conductance channels, it is concluded that catecholamines increase $\bar{g}_s$ by increasing the number of channels which open during depolarization.

Cyclic AMP derivatives and phosphodiesterase inhibitors mimic the action of catecholamines on g_s. Therefore, the catecholamine action on the availability of Ca conductance channels in the membrane may not be a direct consequence of the drug-receptor interaction, but rather of the increased intracellular cyclic AMP level.

Acknowledgments. I am grateful to Professor H. Scholz (Hannover) for participating in some of the experiments, and to Dr. J. Stucki for stimulating discussions. This paper was supported by the Swiss National Science Foundation (grants 3.734.72 and 3.598-0.75).

References

Armstrong, C.M.: Ionic pores, gates, and gating currents. Quart. Rev. Biophys. 7, 179-210 (1975)

Beeler, G.W., Jr., Reuter, H.: Voltage clamp experiments on ventricular myocardium fibres. J. Physiol. 207, 165-190 (1970)

Brown, H.F., McNaughton, P.A., Noble, D., Noble, S.J.: Adrenergic control of pacemaker currents. Phil. Trans. Roy. Soc. Lond. 270B, 527-537 (1975)

Carmeliet, E., Vereecke, J.: Adrenaline and the plateau phase of the cardiac action potential. Importance of Ca^{++}, Na^+ and K^+ conductance. Pflügers Arch. ges. Physiol. 315, 300-315 (1969)

Giles, W., Tsien, R.W.: Effects of acetylcholine on membrane currents in frog atrial muscle. J. Physiol. 246, 64-66P (1975)

Harden, K.T., Wolfe, B.B., Molinoff, P.B.: Binding of iodinated beta adrenergic antagonists to proteins derived from rat heart. Mol. Pharmacol. 12, 1-15 (1976)

Hille, B.: Ionic channels in nerve membranes. Progr. Biophys. Mol. Biol. 21, 1-32 (1970)

Hille, B.: Ion selectivity, saturation, and block in sodium channels. A four-barrier model. J. Gen. Physiol. 66, 535-560 (1975)

Hodgkin, A.L., Huxley, A.F.: A quantitative description of membrane current and its application to conduction and excitation in nerve. J. Physiol. 117, 500-544 (1952)

Ikemoto, Y., Goto, M.: Nature of the negative inotropic effect of acetylcholine on the myocardium. Proc. Japan.Acad. 51, 501-505 (1975)

Lefkowitz, R.J., Limbird, L.E., Mukherjee, C., Caron, M.G.: The β-adrenergic receptor and adenylate cyclase. Biochim. Biophys. Acta 457, 1-39 (1976)

Meinertz, T., Nawrath, H., Scholz, H.: Stimulatory effects of db-c-AMP and adrenaline on myocardial contraction and ^{45}Ca exchange. Experiments at reduced calcium concentration and low frequencies of stimulation. Naunyn-Schmiedeberg's Arch. Pharmacol. 279, 327-338 (1973)

Pappano, A.J.: Calcium-dependent action potentials produced by catecholamines in guinea pig atrial muscle fibers depolarized by potassium. Circulation Res. 27, 379-390 (1970)

Reuter, H.: Über die Wirkung von Adrenalin auf den cellulären Ca-Umsatz des Meerschweinchenvorhofs. Naunyn-Schmiedeberg's Arch. Pharmacol. 251, 401-412 (1965)

Reuter, H.: The dependence of slow inward current in Purkinje fibres on the extracellular calcium-concentration. J. Physiol. 192, 479-492 (1967)

Reuter, H.: Divalent cations as charge carriers in excitable membranes. Progr. Biophys. Mol. Biol. 26, 1-43 (1973)

Reuter, H.: Localization of beta adrenergic receptors and effects of noradrenaline and cyclic nucleotides on action potentials, ionic currents and tension in mammalian cardiac muscle. J. Physiol. 242, 429-451 (1974)

Reuter, H., Scholz, H.: A study of the ion selectivity and the kinetic properties of the calcium-dependent slow inward current in mammalian cardica muscle. J. Physiol. In press (1977a)

Reuter, H., Scholz, H.: The regulation of the Ca conductance of cardiac muscle by adrenaline. J. Physiol. In press (1977b)

Ten Eick, R., Nawrath, H., Trautwein, W.: On the mechanism of the negative inotropic effect of acetylcholine. Pflügers Arch. ges. Physiol. 361, 207-213 (1976)

Tsien, R.W.: Mode of action of chronotropic agents in cardiac Purkinje fibers. Does epinephrine act by directly modifying the external surface charge? J. Gen. Physiol. 64, 320-342 (1974)

Vassort, G., Rougier, O., Garnier, D., Sauviat, M.P., Coraboeuf, E., Gargouil, Y.M.: Effects of adrenaline on membrane inward currents during the cardiac action potential. Pflügers Arch. ges. Physiol. 309, 70-81 (1969)

Watanabe, A.M., Besch, H.R.: The relationship between adenosine 3',5'-monophosphate levels and systolic transmembrane calcium flux. In: Recent Advances in Studies on Cardiac Structure and Metabolism. Vol. V. Fleckenstein, A., Dhalla, N.S. (eds.). Baltimore, London, Tokyo: University Park Press 1975, pp. 95-102

Regulation of Transport Processes

Effect of Cyclic-Adenosine-Monophosphate on Chloride Transport Across some Mammalian Epithelia

S. G. SCHULTZ AND R. A. FRIZZELL

Introduction

In recent years 3',5'-cyclic adenosine monophosphate (cAMP) has been shown to affect ion transport processes across a variety of vertebrate and invertebrate epithelia. In some instances cAMP is the direct or indirect mediator of some well-established hormonal action; in others, the physiologic significance of the action of this ubiquitous substance is unclear. In this communication we will briefly describe the effects of cAMP on Cl, and in some instances Na, transport by three gasto-intestinal epithelia from the rabbit. We will try to show that two distinct patterns of action emerge and that rather complex responses may reflect combinations of these two primary patterns.

Rabbit Ileum

When segments of rabbit ileum, stripped of much of the subepithelial muscle and connective tissues, are mounted in the short-circuit apparatus illustrated in Figure 1 and bathed on both surfaces with a mammalian bicarbonate-Ringer solution, the serosal solution is found to be electrically positive with respect to the mucosal solution. Under these conditions, Na and Cl are actively transported from the mucosal solution to the serosal solution and HCO_3 is secreted into the mucosal solution (Schultz and Curran, 1974). Typical values for the unidirectional and net fluxes of Na and Cl under control conditions are given in Table 1 where J^i_{ms} designates the unidirectional flux of i from the mucosal solution to the serosal solution, J^i_{sm} is the unidirectional flux in the opposite direction, and $J^i_{net} = J^i_{ms} - J^i_{sm}$; clearly, J^i_{net} is positive when i is actively absorbed and negative when it is actively secreted. The rate of HCO_3 secretion under these conditions is 1-2 $\mu Equ \cdot cm^{-2} \cdot h^{-1}$ (Dietz and Field, 1973). The short-circuit current across the tissue (I_{sc}) generally parallels the rate of active Na absorption suggesting that Cl absorption and HCO_3 secretion are the results of electrically silent, neutral transport processes (Schultz and Curran, 1974).

In 1968, Field and collaborators (Field et al., 1968; Field, 1971a,b) demonstrated that cAMP abolishes active Na and Cl absorption by rabbit ileum and elicits active Cl secretion. These investigators subsequently demonstrated that an enterotoxin elaborated by the *Cholera* vibrio mimics the effects of cAMP on Na and Cl transport by rabbit ileum and that exposure of this tissue to *Cholera* toxin elevates intracellular cAMP levels (Kimberg et al., 1971; Field et al., 1972). These findings have now been confirmed by numerous investigators and represent a major advance in our understanding of the pathogenesis of intestinal secretion elicited by some bacterial enterotoxins (Field, 1971, 1974; Schultz et al., 1974).

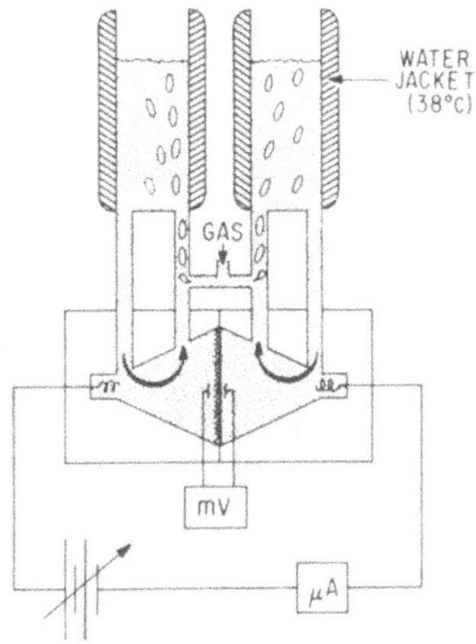

Fig. 1. Short-circuit apparatus. Both surfaces of tissue are bathed by aerated solutions at 38°C having identical compositions. Bidirectional fluxes of Na and Cl are determined using conventional isotope techniques. Provision is made for monitoring the transepithelial electrical potential difference and for passing sufficient current across the tissue to abolish this PD

Table 1. Effect of cAMP on Na and Cl fluxes across short-circuited rabbit ileum

	J^{Na}_{ms}	J^{Na}_{sm}	J^{Na}_{net}	J^{Cl}_{ms}	J^{Cl}_{sm}	J^{Cl}_{net}	I_{sc}
Control	16	12	4	11	8	3	3
+ cAMP	10	10	0	6	10	-4	5

Data from Field (1971). All values are in μEqu. cm^{-2} h^{-1} and have been rounded off without sacrificing statistical significance. The concentration of cAMP was 7.5 mM

The typical effect of cAMP on Na and Cl fluxes across rabbit ileum is given in Table 1; similar results are observed in the presence of theophylline or Cholera toxin and may be summarized as follows:

1. In general, net Na absorption is abolished due to a decrease in the unidirectional flux from mucosa-to-serosa (J^{Na}_{ms}) with little or no change in the serosa-to-mucosa flux (J^{Na}_{sm}). Although some investigators (Powell et al., 1973) have reported a small Na secretion (less than 0.5 μEqu · cm^{-2} · h^{-1}) it is difficult to attach much significance to this value since it represents a very small difference between two much larger unidirectional fluxes. Thus, given the errors involved in accurately short-circuiting and in determining unidirectional tracer fluxes across this very "leaky" epithelium, it is fair to say that, for all intents and purposes, active Na absorption is simply abolished.

2. Active Cl absorption is abolished and replaced by active Cl secretion. This is the result of a decrease in J^{Cl}_{ms} and an increase in J^{Cl}_{sm}. In addition, active Cl secretion is accompanied by an increase in the transepithelial potential difference (the serosal solution becomes electrically more positive) and I_{sc}. These observations have been repeatedly confirmed (Field, 1974).

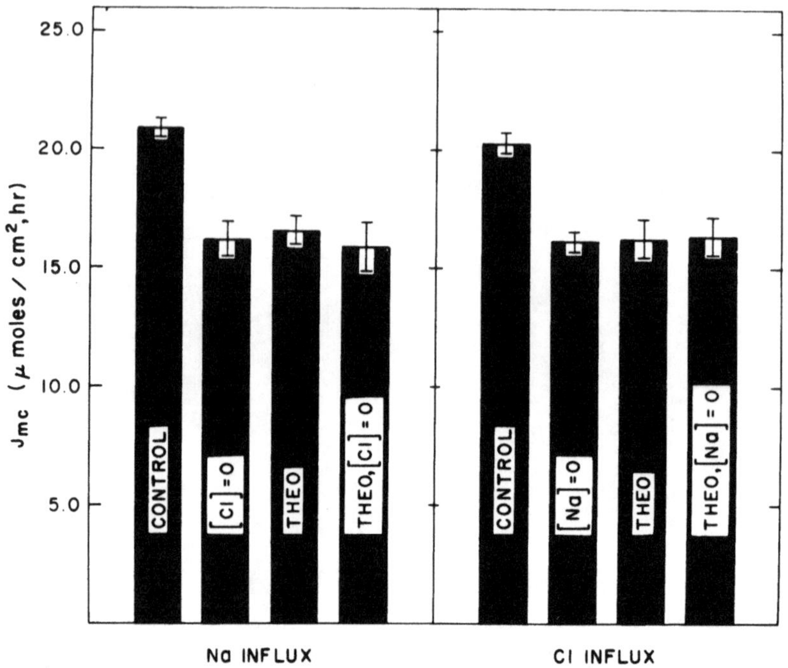

Fig. 2. Effects of Na-free of Cl-free media and theophylline on Na and Cl influxes across the mucosal membrane of rabbit ileum. (From Nellans et al., 1973)

A possible explanation for the finding that cAMP (or theophylline) brings about simultaneous decreases in J_{ms}^{Na} and J_{ms}^{Cl} emerged from the finding by Nellans et al. (1973) of a cAMP-sensitive process that mediates the neutral coupled unidirectional influx of Na and Cl from the mucosal solution into the rabbit ileal epithelium. Evidence for this process is illustrated in Figure 2. We see that approximately 20% of Na influx is inhibited by theophylline or by replacement of Cl in the mucosal solution with SO_4. Similarly, approximately 20% of Cl influx is inhibited by theophylline or by replacement of Na in the mucosal solution with choline; the inhibitions of Na and Cl influxes by theophylline or by removal of the other ion are essentially equal. Finally, theophylline has no effect on Na or Cl influx in the absence of one or the other of these ions. The kinetics of the neutral NaCl influx process at the brush border of rabbit ileum have been described in considerable detail by Nellans et al. (1973).

These findings together with the observations that (1) active Cl secretion by rabbit ileum in response to cAMP is abolished in the absence of Na and (2) in the absence of Cl, Na absorption persists in the presence of theophylline (Nellans et al., 1974) prompted the models shown in Figure 3.

Accordingly, under control conditions, active Na absorption is the result of a Cl-independent Na influx process and the coupled NaCl influx process. In the absence of Cl, Na absorption is reduced but not abolished. In the presence of theophylline, the coupled influx process is abolished and this unmasks a coupled NaCl efflux process. Na which enters the cell by means of the Cl-independent route simply

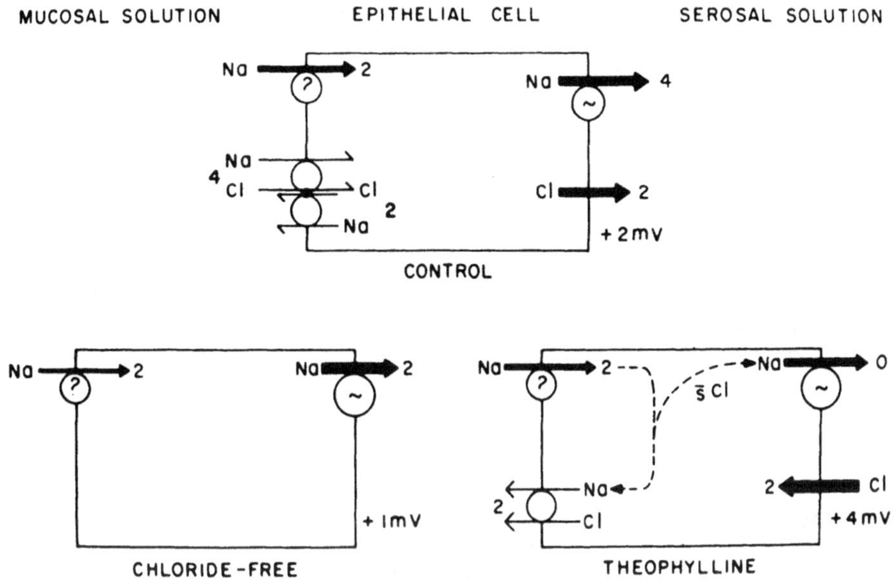

Fig. 3. Models of rabbit ileum relating brush border processes to transepithelial transport under short-circuit conditions. (From Nellans et al., 1974)

recycles across the mucosal membrane so that net Na absorption is abolished. Clearly, in the absence of Na, secretion of Cl is abolished; and, in the absence of Cl, absorption of Na would persist in the presence of theophylline.

In short, the model shown in Figure 3 could account for many of the observations of Nellans et al. (1974) and others (Powell et al., 1973) on the behavior of rabbit ileum under control conditions as well as in the presence of cAMP, theophylline or Cholera toxin. The explanation offered, however, focuses only on events at the mucosal membrane and provides no clue as to how cAMP reverses the direction of net Cl movements across the baso-lateral membranes.

Rabitt Gallbladder

It is well established that rabbit gallbladder actively absorbs Na and Cl but that, unlike many other Na-transporting epithelia, the transepithelial electrical potential difference is essentially zero. This prompted Diamond (1964) to propose that NaCl transport is the result of a neutral, coupled process. The finding of a neutral, coupled NaCl influx process in rabbit ileum prompted us to examine the properties of the mucosal membrane of rabbit gallbladder, and Frizzell et al. (1975) demonstrated a neutral coupled NaCl influx process at the mucosal membrane which could account entirely for the neutral transepithelial NaCl absorption by this tissue. In short, Na entry into the cell from the mucosal solution is abolished when Cl is replaced with SO_4 and Cl entry into the cell is abolished when Na is replaced with choline. Further, the inhibitions of Na or Cl transport when one or the other of these ions is removed from the mucosal solution do not differ significantly, indicating a one-for-one coupled entry.

Table 2. Effect of cAMP on Na and Cl fluxes across rabbit gallbladder

	J_{ms}^{Na}	J_{sm}^{Na}	J_{net}^{Na}	J_{ms}^{Cl}	J_{sm}^{Cl}	J_{net}^{Cl}	ψ_{ms}
Control	46	30	16	38	19	19	0
+ cAMP	37	32	5	28	20	8	0
Δ	-9	+2	-11	-10	+1	-11	0

Data from Frizzell et al. (1975). All fluxes are in μEqu. cm^{-2} h^{-1} and ψ_{ms} is in mV; all values have been rounded off to the nearest whole number. The concentration of cAMP was 7.5 mM. J_{net}^{Na} and J_{net}^{Cl} under control conditions and in the presence of cAMP do not differ significantly

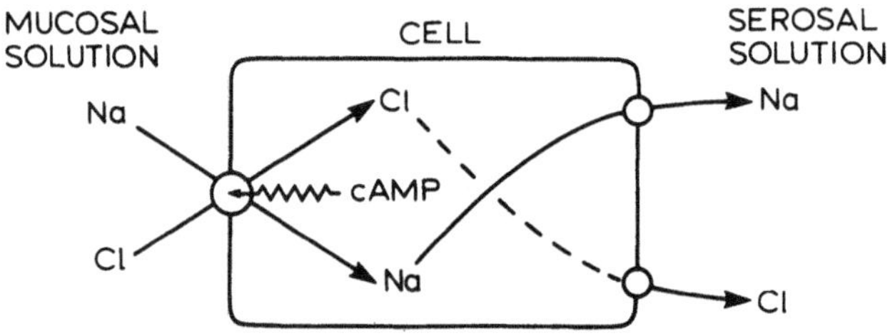

Fig. 4. Model illustrating the cAMP-sensitive coupled NaCl influx process at mucosal membrane of rabbit gallbladder. After entry, Na is extruded from cell against an electrochemical potential difference by a "pump" mechanism at the baso-lateral membranes. The mechanism responsible for Cl exit from cell is unknown

These findings have, for the most part, recently been confirmed (Cremaschi and Henin, 1975; Henin and Cremaschi, 1975).

The effect of cAMP on NaCl absorption by rabbit gallbladder is given in Table 2. We see that under control conditions Na and Cl absorption do not differ significantly and that in the presence of cAMP both are markedly inhibited to the same extent. This inhibition is due entirely to decreases in J_{ms}^{Na} and J_{ms}^{Cl}; the serosa-to-mucosa fluxes of these ions are not significantly affected. Thus, the effect of cAMP can be attributed entirely to inhibition of a neutral coupled NaCl influx process at the mucosal membrane similar to that identified in rabbit ileum (Nellans et al., 1973). However, it is important to note that Cl secretion is not elicited by cAMP; higher concentrations of cAMP simply reduced J_{net}^{Na} and J_{net}^{Cl} further, but at no time was Cl secretion or a change in transepithelial electrical potential difference (ψ_{ms}) observed! Thus, we can illustrate the effect of cAMP on NaCl transport by rabbit gallbladder as shown in Figure 4.

Table 3. Effect of cAMP on Na and Cl fluxes across short-circuited rabbit colon

	J^{Na}_{ms}	J^{Na}_{sm}	J^{Na}_{net}	J^{Cl}_{ms}	J^{Cl}_{sm}	J^{Cl}_{net}	I_{sc}
Control	3.1	1.5	1.6	5.5	4.3	1.2	1.6
+ cAMP	4.2	2.3	1.9	5.0	6.4[a]	-1.4[a]	3.8[a]
Δ	1.1	0.8	0.3	-0.5	2.1	-2.6	2.2

Data from Frizzell et al. (1976). All values are in μEqu. cm^{-2} h^{-1}. Errors have been omitted for clarity; values significantly different from control at $p < 0.01$ are designated by ([a]). The concentration of cAMP was 7.5 mM

Rabbit Colon

Recent studies by Frizzell et al. (1976) have demonstrated that rabbit descending colon actively absorbs Na and Cl but that the transepithelial electrical potential difference (ψ_{ms}) and short-circuit current (I_{sc}) are entirely attributable to active Na absorption. Cl absorption is electrically silent and appears to be coupled one-for-one to active HCO_3 secretion. Further, there appears to be no relation between active Na absorption and active Cl absorption inasmuch as (1) replacement of Na with choline does not affect Cl transport, (2) complete inhibition of Na transport with amiloride does not affect Cl transport, and (3) replacement of Cl with SO_4 or isethionate does not inhibit Na absorption.

The effect of cAMP on ion transport across rabbit colon is given in Table 3. Under control conditions there is excellent agreement between J^{Na}_{net} and the I_{sc} meaning that J^{Cl}_{net} must be balanced by a "residual" current" which, as noted above, appears to be attributable to HCO_3 secretion. In the presence of cAMP, Na absorption <u>is not</u> affected but Cl absorption is abolished and instead we observe active Cl secretion. This is due entirely to an increase in J^{Cl}_{sm}; the small change in J^{Cl}_{ms} is not statistically significant. Further, in the presence of cAMP, the I_{sc} does not differ significantly from the algebraic sum of J^{Na}_{net} and J^{Cl}_{net} meaning that the "residual current" (which under control conditions is almost certainly attributable to HCO_3 secretion) is reduced or abolished.

Given these data we may attempt initial "modeling" as a way of illustrating what has been experimentally observed, as well as guiding future investigations. Tentative models of Na, Cl and HCO_3 transport by rabbit colon under control conditions and in the presence of cAMP are illustrated in Figure 5. Under control conditions (Fig. 5a), Na and Cl are actively absorbed and HCO_3 is secreted. The HCO_3 that is secreted is probably derived from endogenous oxidative metabolism aided by the presence of a high level of carbonic anhydrase activity within the colonic mucosa; it is not dependent upon the presence of HCO_3 in the serosal solution (Carlinsky and Lew, 1970; Frizzell et al., 1976) and is inhibited by acetazolamide (Phillips and Schmalz, 1970). Clearly, under these conditions, the I_{sc} will be determined entirely by J^{Na}_{net}.

As shown in Figure 5b, in the presence of cAMP, HCO_3 secretion is abolished and Cl secretion now replaces Cl absorption; clearly the I_{sc} is now the sum of the Na and Cl currents. Comparison of Figure 5a

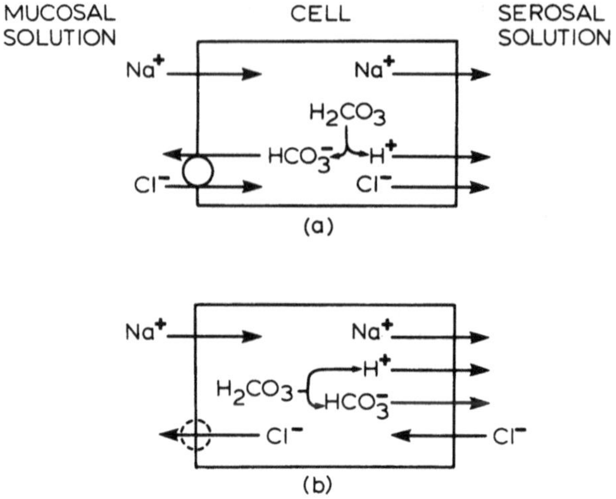

Fig. 5a,b. Models summarizing the movements of Na, Cl and HCO_3 across rabbit colon under control conditions (a) and in the presence of cAMP (b)

with Figure 5b suggests that the effect(s) of cAMP may be either (1) to abolish the Cl-HCO_3 exchange mechanism completely and elicit a *de novo* Cl secretory process, or (2) to alter the nature of the Cl-HCO_3 (anion) exchange process, so that it now extrudes Cl from the cell across the mucosal membranes rather than extruding HCO_3 in exchange for Cl. Since the effect of cAMP is rapid (the increase in I_{sc} is observed within a few seconds after the addition of cAMP to the serosal solution) it seems most reasonable to attempt to explain its effects on ion transport in terms of changes in the properties of pre-existing transport processes, rather than postulating the development (synthesis) of *de novo* transport processes. Thus, the possibility that cAMP somehow converts the "normal" Cl-HCO_3 exchange into a Cl-secretory process is most attractive; however, it should be stressed that this prejudice is based entirely on "intellectual appeal" and not on compelling, experimental evidence.

Finally, we have recently found that some of the effects of cAMP on rabbit colon can be mimicked by the addition of the Ca ionophore, A23187, to the mucosal solution. The results of these preliminary studies are summarized in Table 4. We see that A23187 has no effect on active Na absorption but markedly increases J_{sm}^{Cl}, thereby abolishing active Cl absorption and eliciting active Cl secretion; in these respects the actions of A23187 mimic those of cAMP. However, we also see that in the presence of the ionophore the sum of the currents attributable to J_{net}^{Na} and J_{net}^{Cl} differs from the I_{sc} and that the "residual current", which is presumably attributable to HCO_3 secretion, is the same under control conditions and in the presence of A23187. Thus, whereas cAMP appears to abolish HCO_3 secretion, A23187 appears to have no effect on this process. Further study is necessary, however, to identify the ion flows that are responsible for the residual current under both conditions. The findings that A23187 does not bring about a statistically significant increase in cell cAMP content and is not effective in the absence of extracellular Ca, suggest that its action on Cl transport, and perhaps the action of cAMP, is the result of a change in cell Ca content or distribution. This notion is entirely consistent with the hypothesis postulated by Rasmussen (1970) but further study is clearly necessary to define the rela-

Table 4. Effects of Ca Ionophore (A23187) on Na and Cl fluxes across rabbit colon

	J_{ms}^{Na}	J_{sm}^{Na}	J_{net}^{Na}	J_{ms}^{Cl}	J_{sm}^{Cl}	J_{net}^{Cl}	I_{sc}
Control	3.4	1.5	1.9	6.3	4.9	1.4	2.2
+ A23178	3.4	1.6	1.8	6.1	7.5[a]	-1.4[a]	4.6[a]
Δ	0	0.1	-0.1	0.2	2.6	-2.8	2.4

All values are in μEqu. cm^{-2} h^{-1}. Errors have been omitted for clarity; values significantly different from control at $p < 0.01$ are designated by ([a]). Concentration of A23187 was 0.1 μgm ml^{-1}

Table 5. Relation between presence of coupled NaCl transport mechanisms and effects of cyclic AMP in various epithelia

Epithelium	Rabbit gallbladder	Rabbit ileum	Rat colon	Rabbit colon	Cornea	Seminal vesicle
Increased I_{sc}	0	+	+	+	+	+
Cl secretion elicited	0	+	+	+	+	+
Na absorption decreased	+	+	+	0	0	0
Presence of coupled NaCl transport	+	+	+	0	0	0

tions among the effects of cAMP, A23187, and Ca on Cl transport by rabbit colon.

Summary

The effects of cAMP on Na and Cl transport by several mammalian epithelia are summarized in Table 5. Two, apparently distinct, patterns emerge.

1. In rabbit gallbladder (Frizzell et al., 1975), cAMP decreases both Na and Cl absorption equally; Cl secretion is not elicited and the transmural electrical potential difference is not affected. As discussed above, the reduction in NaCl absorption is due entirely to decreases in J_{ms}^{Na} and J_{ms}^{Cl} resulting from inhibition of a neutral, coupled NaCl influx mechanism.

2. In rabbit colon (Frizzell et al., 1976), which does not possess a neutral NaCl influx process, cAMP has no effect on Na absorption but abolishes Cl absorption and elicits Cl secretion. The latter is due to a marked increase in J_{sm}^{Cl} alone and is accompanied by an increase in the transepithelial electrical potential difference. The same response to cAMP is observed in rabbit cornea (Zadunaisky et al., 1973) and guinea pig seminal vesicle (Levine et al., 1975).

3. In rabbit ileum (Nellans et al., 1974), cAMP abolishes active Na absorption due to a decrease in J_{ms}^{Na} alone. At the same time active Cl absorption is abolished and active Cl secretion is elicited due to a decrease in J_{ms}^{Cl} and in increase in J_{sm}^{Cl}; this is associated with an increase in the transepithelial electrical potential difference. The same pattern of responses is observed in rat colon which also appears to possess a neutral NaCl absorptive mechanism (Binder and Rawlins, 1973a,b).

Clearly, the responses of rabbit ileum and rat colon to cAMP appear to be a combination, or superposition, of the two distinct patterns described above, i.e. the inhibition of a neutral NaCl absorptive process and the stimulation of an electrogenic Cl secretory process. This could be the result of the presence in rabbit ileum (and rat colon) of different cell types; i.e. some cells that resemble those found in rabbit gallbladder and others that resemble those found in rabbit colon. Alternatively, the "hybrid-like" responses of these tissues to cAMP could be the result of mechanisms that reside in the same cells. In either event, it appears that cAMP is capable of affecting Cl transport across mammalian epithelia through two different modes of action, and that apparently complex responses may simply represent the superposition of these actions. If so, attempts to explain these complex responses to cAMP at the cellular level by invoking a single mechanism of action are futile.

Acknowledgments. Research from the authors' laboratories cited in this paper was supported by grants from the USPHS National Institutes of Health (NIAMDD) and the American Heart Association.

References

Binder, H.J., Rawlins, C.L.: Effect of conjugated dihydroxy bile salts on electrolyte transport in rat colon. J. Clin. Invest. 52, 1460-1466 (1973a)

Binder, H.J., Rawlins, C.L.: Electrolyte transport across isolated large intestinal mucosa. Am. J. Physiol. 225, 1232-1239 (1973b)

Carlinsky, N.J., Lew, V.L.: Bicarbonate secretion and non-Na component of the short-circuit current in the isolated colonic mucosa of Bufo arenarum. J. Physiol. 206, 529-541 (1970)

Cremaschi, D., Henin, S.: Na and Cl transepithelial routes in rabbit gallbladder. Tracer analysis of the transports. Pflügers Arch. 361, 33-41 (1975)

Diamond, J.M.: Transport of salt and water in rabbit and guinea pig gallbladder. J. Gen. Physiol. 48, 1-14 (1964)

Dietz, J., Field, M.: Ion transport in rabbit ileal mucosa. IV. Bicarbonate secretion. Am. J. Physiol. 225, 658-861 (1973)

Field, M.: Ion transport in rabbit ileal mucosa. II. Effects of cyclic 3',5'-AMP. Am. J. Physiol. 221, 992-997 (1971a)

Field, M.: Intestinal secretion: Effect of cyclic AMP and its role in cholera. New Eng. J. Med. 284, 1137-1144 (1971b)

Field, M.: Intestinal secretion. Gastroent. 66, 1063-1084 (1974)

Field, M., Fromm, D., Al-Awqati, Q., Greenough, W.B. III: Effect of cholera enterotoxin on ion transport across isolated ileal mucosa. J. Clin. Invest. 51, 796-804 (1972)

Field, M., Plotkin, G.R., Silen, W.: Effects of vasopressin, theophylline and cyclic adenosine monophosphate on short-circuit current across isolated rabbit ileal mucosa. Nature (London) 217, 469-470 (1968)

Frizzell, R.A., Dugas, M.C., Schultz, S.G.: Sodium chloride transport by rabbit gallbladder: Direct evidence for a coupled NaCl influx process. J. Gen. Physiol. 65, 769-794 (1975)

Frizzell, R.A., Koch, M.J., Schultz, S.G.: Ion transport by rabbit colon. I. Active and passive components. J. Membrane Biol. 27, 297-316 (1976)

Henin, S., Cremaschi, D.: Transcellular ion route in rabbit gallbladder. Electric properties of the epithelial cells. Pflügers Arch. 355, 125-139 (1975)

Kimberg, D.V., Field, M., Johnson, J., Henderson, A., Gershon, E.: Stimulation of intestinal mucosal adenyl cyclase by cholera enterotoxin and prostaglandins. J. Clin. Invest. 50, 1218-1230 (1971)

Levine, N., Rinaldo, J.E., Schultz, S.G.: Ion transport by guinea pig seminal vesicle. J. Physiol. 246, 197-211 (1975)

Nellans, H.N., Frizzell, R.A., Schultz, S.G.: Coupled sodium-chloride influx across the brush border of rabbit ileum. Am. J. Physiol. 225, 467-475 (1973)

Nellans, H.N., Frizzell, R.A., Schultz, S.G.: Brush border processes and transepithelial Na and Cl transport by rabbit ileum. Am. J. Physiol. 226, 1131-1141 (1974)

Phillips, S.F., Schmalz, P.F.: Bicarbonate secretion by the rat colon: Effect of intraluminal chloride and acetazolamide. Proc. Soc. Exper. Biol. Med. 135, 116-122 (1970)

Powell, D.W., Binder, H.J., Curran, P.F.: Active electrolyte secretion stimulated by choleragen in rabbit ileum in vitro. Am. J. Physiol. 225, 781-787 (1973)

Rasmussen, H.: Cell communication, calcium ion, and cyclic adenosine monophosphate. Science 170, 404-412 (1970)

Schultz, S.G., Curran, P.F.: Sodium and chloride transport across isolated rabbit ileum. In: Current Topics in Membrane Transport, Vol. V. Bronner, F., Kleinzeller, A. (eds.). New York: Academic Press 1975, pp. 225-281

Schultz, S.G., Frizzell, R.A., Nellans, H.N.: Ion transport by mammalian small intestine. Ann. Rev. Physiol. 36, 51-91 (1974)

Zadunaisky, J.A., Lande, M.A., Chalfie, M., Neufeld, A.H.: Ion pumps in the cornea and their stimulation by epinephrine and cyclic AMP. Exp. Eye Res. 15, 577-584 (1973)

Steroid Hormone-Induced Alterations in Membrane Lipids – the Basis for Altered Ion Transport

H. Rasmussen, D. P. B. Goodman, and E. Max

In the fluid mosaic model of biological membrane structure, it is proposed that proteins float in a sea of lipids (Singer and Nicolson, 1972). A corollary of this model is that it is the interactions between the different proteins within this sea that control the functional properties of the membrane. Nonetheless, there is considerable evidence that alterations in membrane lipid structure alter the functional properties of membrane proteins (Smith and Kemp, 1969; Linden et al., 1973a,b; Wisnieski et al., 1973; Horowitz et al., 1974; Bretcher and Raff, 1975; Linden and Fox, 1975). However, nearly all of these studies have been conducted in vitro and the changes in lipid composition employed to alter membrane protein function have been quite marked. Thus, although artificially induced changes in membrane lipid composition can alter the function of membrane proteins, it remains to be demonstrated that, under physiological circumstances, control of the function of membrane proteins can be achieved by alterations in the surrounding lipid structure.

A. The Concept of the Phospholipid Annulus

Two approaches to exploring this possibility are: (1) to analyze the specific phospholipids closely associated with a functional membrane protein and determine whether their composition differs from the bulk lipid phase of the membrane, and (2) to examine whether in systems in which a change in membrane function, e.g. a change in Na^+ or Ca^{2+} transport, there is an associated change in membrane lipids.

A number of studies have suggested that the lipids immediately in contact with a membrane protein have a unique structure, and may differ in content from the bulk phase lipids (Jost et al., 1973; Dehlinger et al., 1974; Grant and McConnell, 1974; Lee et al., 1974). A specific example of this phenomenon are the recent studies of Metcalfe et al. (1976), in which it has been shown that a specific phospholipid annulus surrounds the Ca^{2+}-activated ATPase of sarcoplasmic reticulum. The composition of this annulus differs considerably from the composition of the bulk of the membrane lipids, is tightly bound to the enzyme, and, if altered by artificially induced phospholipid exchange in vitro, it alters the activity of this enzyme. The physiological implications of this concept are two-fold: (1) it raises the possibility that rather small but specific changes in membrane lipid composition could be a means of controlling the function of membrane proteins, and (2) it makes it difficult to interpret the physiological significance of a drug or hormone-induced change in membrane lipid content because such a change might reflect one of two processes. On the one hand, a change in lipid composition could be a primary event and a means by which the hormone or drug altered the function of existing proteins, or the introduction of new proteins into the membrane. On the other hand, if the hormone causes an increase in the synthesis and content of a specific membrane protein, one might anticipate that incorporation of new protein into the membrane would require the simultaneous incorporation of new lipid to form the annulus for this new protein. This alternative possibility makes it difficult to define the role of physiologically induced alterations in lipid structure in the control of membrane structure. Nonetheless, this is a role worth exploring.

B. Membrane Lipid Turnover

From an historical point of view, one can identify three successive steps in the biochemical examination of the possibility that membrane lipid turnover is involved in the control of cellular metabolism. In the first phase, studies concentrated on an examination of the turnover of phospholipid phosphorous. In spite of evidence that changes in cell metabolism are often associated with changes in phospholipid phosphorous turnover, no well-established concepts as to their functional significance of such a change has emerged, although Jones and Michel (1975) have argued that the specific turnover of phosphoinositide phosphorus is the basis for changes in the calcium permeability of the cell membrane.

In the second step this problem was approached by examining the possibility that changes in the relative content of specific classes of phospholipid in the membrane might alter membrane function. In spite of considerable effort, this approach has not been particularly rewarding, and the present authors are not aware of any well-characterized system in which physiologically induced changes in the relative content of the different classes of phospholipids is thought to play a key role in a physiologically induced change in membrane function.

The third approach has been that of analyzing the fatty acid content of either total or specific phospholipid classes within the membrane. Using this approach changes in the fatty acid composition of either total or specific phospholipids have been found in a number of cases in which an associated change in membrane function is seen (de Gier et al., 1968; Smith and Kemp, 1969; Goodman et al., 1971; Moore et al., 1975). Along with this evidence has come the recognition that a unique feature of naturally occurring phospholipids is the asymmetric distribution of their constitient fatty acids. This feature is that the long-chain polyunsaturated fatty acids are found predominantly in the 2- or B-position of the α-glycerol phosphate backbone. Furthermore, once a phospholipid molecule becomes a component of the membrane, the unsaturated fatty acids in this position turn over much more rapidly than either the glycerol backbone, or the more saturated fatty acids in the 1- or α-position of the glycerol. This cycle of deacylation and reacylation is a characteristic feature of the metabolism of many biological membranes, and each complete turn of the cycle requires the equivalent of one ATP in terms of energy expenditure. Clearly, such an energetically expensive process that has survived evolutionary pressure must be of biological value. Nevertheless the question of its specific function remains largely unanswered. It is possible that one of its functions is a control function. It is this possibility which we have been examining in relationship to the effect of aldosterone upon Na^+ transport in the toad bladder, and the effect of $1,25(OH)_2D_3$ in Ca^{2+} transport in intestinal mucosal cells.

C. Aldosterone, Lipid Metabolism, and Na^+ Transport

Transepithelial transport of Na^+ across the amphibian urinary bladder is stimulated by both vasopression (Leaf et al., 1958) and aldosterone (Crabbe, 1961). The two act differently. The action of vasopressin is immediate, is not blocked by inhibitors of RNA or protein synthesis, and is associated with a marked increase in H_2O and urea permeability of the epithelial cell layer. The action of aldosterone is delayed for 45-90 min, is blocked by inhibitors or RNA and protein synthesis, and causes no increase in H_2O permeability of the tissue. However, aldosterone pretreatment does lead to a change in three other membrane-related processes (Goodman et al., 1975). It increases (1) the effect of vasopression on H_2O and urea permeability (Goodman et al., 1969; Handler et al., 1969), (2) the sensitivity of sodium transport to inhibition by the cardiac glycoside, ouabain (Goodman et al., 1969), and (3) the sensitivity of sodium transport to inhibition by increased partial pressures of oxygen (Allen and Rasmussen, 1971; Allen et al., 1973).

In studies designed to elucidate the mechanism by which aldosterone treatment altered the response of the amphibian bladder to hyperbaric oxygen (Allen et al., 1973) we were led to the conclusion that aldosterone treatment altered the sensitivity of some membrane component to oxidative inactivation by high partial pressures of O_2. This conclusion was based upon the fact that pretreatment of this tissue with aldosterone altered the response to high partial pressures of O_2 (HPO). The rate of Na^+ transport was unaltered by HPO in untreated bladders, but was profoundly and reversibly inhibited by HPO in aldosterone-treated bladders. When the HPO-induced inhibition of Na^+ transport was maximal, there was an associated increase in tissue ATP content, an increase in the rates of glucose and pyruvate decarboxylations, and an increase in the NAD/NADH and NADPH ratios. These data indicated that the reduction in Na^+ transport could not be due simply to a reduction in the supply of energy for the Na^+ pump, but was probably due to oxidative alteration of some membrane component, possibly some of the membrane lipid. If the latter were the site of oxygen action, a logical explanation for the ability of aldosterone to alter the sensitivity of these cells to HPO was that it altered lipid metabolism and membrane lipid composition.

Based on this reasoning we undertook a systematic study of the effects of aldosterone upon lipid metabolism in this tissue (Goodman et al., 1971, 1975; Lien et al., 1975, 1976; Rasmussen et al., 1976). Presentation of the results of these studies falls naturally into two parts: (1) the metabolic effects of aldosterone seen in the 45-90 min latent period, i.e. before any aldosterone-induced change in Na^+ transport, and (2) the metabolic effects of aldosterone seen 4-6 h after hormone addition when the rate of Na^+ transport has become maximal (Table 1).

The results of metabolic studies carried out within the first 30 min after aldosterone addition to toad bladders show (Table 1) that the hormone causes an increase in linolenate oxidation, but

Table 1. Metabolic effects of aldosterone

Early (0-1 h)		Late (4-6 h)
increased	glucose [U-C^{14}] into DG	no change
increased	glucose [1-C^{14}] decarboxylation	decreased
no change	palmitate oxidation	increased
increased	linolenate oxidation	increased
no change	stearate oxidation	increased
no change	oleate oxidation	increased
no change	stearate [1-C^{14}] into PLFA	no change
increased	oleate [1-C^{14}] into PLFA	increased
no change	linolenate into PLFA	no change
-	oleate [9,10-^{3}H] into PLFA	increased
increased	deacylation of PLFA	-
no change	wt percent of LCPUFA	increased
increased	acetate [1-C^{14}] into PLFA	decreased
no change	Na^+ transport	increased

no change in the rate of oxidation of other fatty acids, an increase incorporation of carbon from uniformly labeled glucose into diglycerides, a stimulation of the deacylation of membrane phospholipids, a specific stimulation of [1-C^{14}]-labeled glucose, and a stimulation of acetate and [1-C^{14}] oleate incorporation into phospholipid fatty acids. All of these changes occur in the latent period before there is any change in Na^+ transport, or altered sensitivity to ouabain, hyperbaric oxygen or vasopressin.

We interpret these results to mean that aldosterone has nearly immediate and multiple effects upon tissue metabolism. It stimulates fatty acid synthesis, fatty acid chain elongation and desaturation, diglyceride synthesis, a specific cytosolic decarboxylation of oleic acid, and deacylation of membrane phospholipid. All of these changes represent a coordinated response aimed at eventually altering the profile of fatty acids in the membrane phospholipids.

When the same parameters are examined 4-6 h after hormone addition, the major changes are an increased rate of oxidation of all fatty acids probably secondary to the increased rate of ATP utilization and turnover secondary to the increased Na^+ transport; a continued stimulation of chain elongation and desaturation, and the incorporation of longer chained, more polyunsaturated fatty acids into the 2-position of membrane phospholipids, as determined either by the use of isotopically labeled precursors or by measurement of the weight percentage of the individual fatty acids contained in the phospholipid fraction.

These results indicated to us that aldosterone stimulates the deacylation and reacylation of membrane phospholipids, and that by this mechanism produces an alteration of membrane lipid structure consisting of an increase in the content of several longer chained, more polyunsaturated fatty acids.

The next question of major importance was that of determining the physiological significance of these hormone-induced changes in membrane lipid structure. Three possibilities were considered: (1) the change in lipid structure was a consequence rather than a determinant of the change in Na^+ transport; (2) the change in lipid structure was a necessary consequence of the synthesis of a new membrane protein in the aldosterone-treated tissue, or (3) the change in lipid structure was of primary importance, and led to an alteration of the catalytic activity of one or more membrane proteins involved in transcellular Na^+ transport.

The first possibility was ruled out by the use of a specific inhibitor of mucosal Na^+ entry, amiloride. When this agent is added to the mucosal medium of an isolated bladder, it totally blocks transcellular Na^+ transport in the absence or presence of aldosterone. However, if an amiloride-inhibited tissue is treated with aldosterone, the hormone exerts all of its usual effects upon lipid metabolism. If some 4-6 h after hormone addition, the amiloride is removed, there is an immediate increase in Na^+ transport to rates expected in a noninhibited, aldosterone-treated bladder. Thus, there is no latent period, implying that amiloride does not in any way interfere with the biochemical effects of aldosterone which are eventually responsible for the rise in Na^+ transport.

A choice between the second and third alternatives listed above is not easy. However, there are three facts which suggest that the changes in lipid metabolism and membrane lipid structure may be of primary importance in determining the change in transcellular Na^+ transport.

The first of these relates to the observation that the Na^+ carriers and the Na^+/K^+ ATPase of the cell membranes represent a small percentage of the total membrane protein. If the insertion of new protein molecules related directly to the Na^+ transport function required specific lipids to form their appropriate phospholipid annulus, the magnitude of the lipid change would not have to be large, and could probably not be detected by such insensitive methods as measurement of phospholipid fatty acid weight percentage. Thus, the magnitude of the lipid change is out of proportion to the presumed need.

The second fact is that an inhibitor of chain elongation and desaturation blocks the effect of aldosterone on lipid metabolism and on Na^+ transport without influencing bulk RNA or protein synthesis, and without altering the response of the tissue to vasopressin. This compound, 2-methyl-2-p-(1,2,3,4-tetrahydro-1-napthyl)-phenoxy-propionic acid (TPIA), in the absence of aldosterone, inhibits the incorporation of acetate carbon into all long chain polyunsaturated fatty acids, and blocks the effect of aldosterone on lipid synthesis, fatty acid incorporation into membrane phospholipids, and on the fatty acid composition of the membrane phospholipid as determined by an analysis of the weight percent of these compounds.

The third fact is that inhibitors of RNA and protein synthesis that are known to block the aldosterone-mediated increase in transcellular Na^+ transport (Chu and Edelman, 1972; Lahav et al., 1973) also block all of the effects of aldosterone on lipid metabolism and membrane lipid structure. These results cast a different light on the possible role of protein synthesis in the aldosterone-mediated response. They raise the possibility that the key proteins may not be components of the Na^+ transport system at all, but may be involved in some stage of lipid turnover.

The alteration in membrane lipid structure may be the basis for the potentiation by aldosterone of vasoprotein action on both Na^+ transport and water permeability in the toad bladder. This is made more likely by two observations. Stoff et al. (1972) have shown that when vasopressin acts upon bladders pretreated with aldosterone, there is a greater rise in intracellular cAMP in these bladders than in control bladders. One interpretation of their data is that aldosterone alters the nature of the coupling between the vasopressin receptor and the adenylate cyclase enzyme in the cell membrane. If this interpretation is correct then the recent observations of Houslay et al. (1976) are of considerable interest because they have shown that the lipid environment of the glucagen receptor in rat liver plasma membranes regulates the coupling between receptor interaction and adenylate cyclase activation.

Viewed in the perspective of these results, a single hormonally induced change in the membrane lipid composition can account for the multiple effects of aldosterone on the function and response of the toad bladder. This change can alter several membrane functions and thereby modify the long-term response of this tissue to other drugs and hormones.

An attractive hypothesis to account for these observations is that the aldosterone-mediated change in the fatty acid composition of the membrane phospholipids accounts for the effects of aldosterone on: (1) transcellular Na^+ transport; and (2) the sensitivity of response of this tissue to : (a) vasopressin, (b) ouabain, and (c) hyperbasic oxygen. It can be envisioned that by causing a change in the lipid environment, and eventually the composition of the phospholipid annulus of specific membrane proteins involved in transcellular Na^+ transport, and the binding of vasopressin, aldosterone alters the catalytic or transducing functions of these proteins. Proof of the hypothesis will require showing by some independent means that specific changes in fatty acid composition can alter the physiological functions in this tissue.

D. Membrane Lipids and Calcium Transport

An obvious question that arises as a result of this hypothesis is that of whether hormonally induced changes in cation transport in other systems are associated with changes in membrane lipid structure and turnover. Although data on this point are not extensive, there are two other membranes in which a change in calcium flux across the membrane is associated with an alteration in membrane lipid composition.

The first of these systems is the energy-dependent calcium sequestration by rat liver microsomes. Moore et al. (1975) have shown that these organelles accumulate calcium by an ATP-dependent mechanism similar to that seen in the endoplasmic reticulum in cardiac and skeletal muscle. Most germaine to the present discussion is their observation that the rate of calcium uptake is slower in microsomes from female as compared to male animals; and that this difference is associated with a difference in the fatty acid content of the phospholipids of this membrane. The membranes from the male animals have a higher percentage of palmitic and linoleic acids and lower percentage of stearic acid than membranes from female animals. These authors suggest that the change in lipid composition is a factor in the difference in the calcium pump activity of the membranes isolated from male and female animals.

More recent studies in our own laboratory (Max et al., in press) have been concerned with the possible role of changes in lipid structure in the action of $1,25(OH)_2D_3$ on transcellular calcium transport across the intestinal mucosal cell. Much previous work had indicated that at least one cellular site of action of the active forms of vitamin D in the intestine was the brush border of the mucosal cells (Martin and DeLuca, 1969; Adams et al., 1970; Melancon and DeLuca, 1970; Youmaszai et al., 1973; Kowarski and Schachter, 1975; Moriuchi and DeLuca, 1976).

Because our understanding of another function of the intestinal brush border - of sugar and amino acid transport - has been greatly clarified by investigations employing membranes vesicles derived from the organelle (Hopfer et al., 1973, 1975), we developed a method for preparing such membranes from the intestinal mucosal cells of the D-deficient chick (Max et al., in preparation). This membrane was found to have a lipid composition similar to that of brush border membranes prepared from rat or mouse (Forstner et al., 1968; Kawai et al., 1974). It is quite different from that of ordinary plasma membrane and is characterized by a high cholesterol to phospholipid ratio, and a predominance of phosphatidylethanolamine to phosphatidylcholine.

The two observations most germaine to the present discussion relate to the effects of prior treatment of the D-deficient chicks with $1\alpha(OH)D_3$ upon the structure and function of these isolated membranes. This metabolite, $1\alpha(OH)D_3$, is converted to $1\alpha,25(OH)_2D_3$ in vivo (Holick et al., 1976) and has a similar potency and time course of action when given by mouth to D-deficient animals (Pechet and Hesse, 1974). When given to D-deficient chicks, it causes a significant increase in intestinal calcium transport within 4-6 hours and produces a maximal effect within 12-18 hours. When membranes were isolated from the brush borders of animals treated with $1\alpha(OH)D_3$ 18 hours previously and compared to those prepared from suitable control animals, the two major differences seen are in the ability of the membranes to bind calcium, and in the lipid composition of the membrane.

Prior treatment with $1\alpha(OH)D_3$ leads to a significant increase in the ability of the brush border membrane to accumulate calcium. This accumulation takes place in the absence of ATP, and is made up both of binding to, and uptake within the membrane vesicles.

The two major changes in lipid structure seen after $1\alpha(OH)D_3$ therapy are: (1) an increase in lipid P(6.1 ± 0.5 to 6.9 ± 0.4 μg/mg protein) without a significant change in the relative distribution of the various phospholipid classes, and (2) a change in the fatty acid composition of phosphatidylcholine, characterized by an increase in 18:2 and 20:4 and a decrease in 22:0 and another unidentified minor fatty acid component.

Table 2.

	Aldosterone (Na^+ Toad bladder)	Testesterone (Ca^{2+}-liver ER)	$1,25(OH)_2D_3$ (Ca^{2+}-brush b.m.[a])
16:0	↓	↑	-
16:1	-	-	-
18:0	↓	↓	-
18:1	-	↑	-
18:2	-	↑	↑
18:3	↑	-	-
22:0	-	-	↓
20:4	-	-	↑
20:5	-	-	-
22:5	↑	-	-
22:6	↑	↓	-
24:0	↑	-	-
24:1	-	-	-

[a]border membrane

E. Conclusion

When the data on steroid-induced changes in membrane phospholipid fatty acid compositions are summarized (Table 2) two facts are immediately apparent: first, in all three changes the change would be expected to increase the fluidity of the membrane, and second, the alteration induced by aldosterone and presumably related to, or the cause of, the change in Na^+ transport, is quite different from the change seen in the other two systems.

The present discussion has pointed out that in three different systems in which a steroid or sterol hormone alters the cation transport properties of a cellular membrane, there is simultaneous change in the lipid composition of that membrane, and in particular in the fatty acid composition of one or more classes of membrane phospholipids. Furthermore, in the one case - aldosterone action on the toad bladder - there is good evidence, from studies with metabolic inhibitors, that the change in lipid structure is a necessary event in the change in transport function. Further work is clearly necessary in order to determine how commonly hormone-induced changes in membrane lipid structure occur, and whether, in fact, they are the primary basis of a change in membrane function. At present, it is only possible to propose that control of cell function, and in particular, membrane transport function can be regulated by altering the lipid environment of functional membrane proteins.

Acknowledgments. Our work was supported by Grants (AM-09650 and 7P01-AM-19813) from the National Institute of Health, United States Public Health Service.

References

Adams, T.H., Wong, R.G., Norman, A.W.: Studies on the mechanism of action of cholecalciferol II. Effects of the polyene antibiotic filipin on vitamin D mediated calcium transport. J. Biol. Chem. 245, 4432-4442 (1970)

Allen, J.E., Goodman, D.B.P., Besarb, A., Rasmussen, H.: Studies on the biochemical basis of oxygen toxicity. Biochim. Biophys. Acta 320, 708-728 (1973)

Allen, J.E., Rasmussen, H.: The effects of oxygen on cellular metabolism. Intern. Zert. Klin. Pharmab. Ther. Toxikol. 5, 26-33 (1971)

Bretcher, M.S., Raff, M.C.: Mammalian plasma membrane. Nature (London) 258, 43-49 (1975)

Chu, L.L.H., Edelman, J.S.: Cordycepin and α-Amanitin: Inhibitors of transcription as probes of aldosterone action. J. Membrane Biol. 10, 291-310 (1972)

Crabbe, J.: Stimulation of active sodium transport by the isolated toad bladder with aldosterone in vitro. J. Clin. Invest. 40, 2103-2110 (1961)

Cronan, J.R.: Physical properties of membrane lipids: Biological relevance and regulation. Bact. Rev. 39, 232-256 (1975)

de Gier, J., Mandersloot, J.G., Van Deenen, L.L.M.: Lipid composition and permeability of liposomes. Biochim. Biophys. Acta 150, 666-675 (1968)

Dehlinger, P.J., Jost, P.C., Griffith, O.H.: Lipid binding to the amphipathic membrane protein cytochrome b_5. Proc. Natl. Acad. Sci. U.S. 71, 2280-2284 (1974)

Forstner, G.G., Tanaka, K., Isselbacher, K.J.: Lipid Composition of the Isolated Rat Intestinal Microvillus Membrane. Biochem. J. 109, 51-59 (1968)

Goodman, D.B.P., Allen, J.E., Rasmussen, H.: On the mode of action of aldosterone. Proc. Natl. Acad. Sci. U.S. 64, 330-337 (1969)

Goodman, D.B.P., Allen, J.E., Rasmussen, H.: Studies on the mechanism of action of aldosterone: Hormone-induced changes in lipid metabolism. Biochemistry 10, 3825-3831 (1971)

Goodman, D.B.P., Wong, M., Rasmussen, H.: Aldosterone-induced membrane phospholipid metabolism in the toad urinary bladder. Biochemistry 143, 2803-2809 (1975)

Grant, C.W.M., McConnell, H.M.: Glycophorin in lipid bilayers. Proc. Natl. Acad. Sci. U.S. 71, 4653-4657 (1974)

Handler, J.S., Preston, A.S., Orloff, J.: Effect of adrenal steroid hormones on the response of the toad's urinary bladder to vasopressin. J. Clin. Invest. 48, 823-833 (1969)

Holick, M.F., Tavela, T.E., Holick, S.A., Schnoes, H.K., DeLuca, H.F., Gallagher, B.M.: Synthesis of 1α-hydroxy-6-^{3}H-vitamin D_3 and its metabolism to 1α,25-dihydroxy-6^{3}H vitamin D_3 in the rat. J. Biol. Chem. 251, 1020-1028 (1976)

Hopfer, V., Nelson, K., Perrotto, J., Isselbacher, K.J.: Glucose transport in isolated brush border membranes from rat small intestine. J. Biol. Chem. 248, 25-32 (1973)

Hopfer, V., Sigrist-Nelson, K., Murer, H.: Intestinal sugar transport: studies with isolated plasma membrane. Ann. N.Y. Acad. Sci. 264, 414-427 (1975)

Horowith, A.F., Hatten, M.E., Burger, M.M.: Membrane fatty acid replacement and their effect on growth and lectin-induced agglutinability. Proc. Natl. Acad. Sci. U.S. 71, 3115-3119 (1974)

Houslay, M.D., Hesketh, T.R., Smith, G.A., Warren, G.B., Metcalfe, J.C.: The lipid environment of the glucagon receptor regulates adenylate cyclase activity. Biochem. Biophys. Acta 436, 495-504 (1976)

Jones, L.M., Michel, R.H.: The relationship of calcium to phosphotidylenosital turnover. Biochem. J. 148, 479-485 (1975)

Jost, P., Griffith, O.H., Capaldi, R.A., Vanderkooi, G.: Identification and extent of fluid bilayer regions in membrane cytochrome oxidase. Biochim. Biophys. Acta 311, 141-152 (1973)

Kawai, K., Fujita, M., Nakao, M.: Lipid Composition of Two Different Regions of an Intestinal Epithelial Cell Membrane of Mouse. Biochim. Biophys. Acta 369, 222-233 (1974)

Kowarski, S., Schachter, D.: Vitamin D dependent particulate calcium binding activity and intestinal calcium transport. Am. J. Physiol. 229, 1198-1204 (1975)

Lahav, M., Dietz, T., Edelman, I.S.: The action of aldosterone on sodium transport: further studies with inhibitors of RNA and protein synthesis. Endocrinology 92, 1685-1699 (1973)

Leaf, A., Anderson, J., Page, L.B.: Active sodium transport by the isolated toad bladder. J. Gen. Physiol. 41, 657-668 (1958)

Lee, A.G., Budsall, N.J.M., Metcalfe, J.C., Toon, P.A., Warren, G.B.: Clusters in lipid bilayers and the interpretation of thermal effects in biological membranes. Biochemistry 13, 3699-3705 (1974)

Lien, E.L., Goodman, D.B.P., Rasmussen, H.: Effects of an acetyl-coenzyme A carboxylase inhibitor and a sodium-sparing diurectic on aldosterone-stimulated sodium transport, lipid metabolism and phospholipid fatty acid composition in the toad urinary bladder. Biochemistry 14, 2749-2754 (1975)

Lien, E.L., Goodman, D.B.P., Rasmussen, H.: Effects of inhibitors of protein and RNA synthesis on aldosterone stimulated changes in phospholipid fatty acid metabolism in the toad bladder. Biochim. Biophys. Acta 421, 210-217 (1976)

Linden, C.D., Fox, C.F.: Membrane physical state and function. Acc. Chem. Res. 8, 321-327 (1975)

Linden, C.D., Keith, A.D., Fox, C.F.: Correlations between fatty acid distribution in phospholipids and the temperature-dependence of the membrane physical state. J. Supra. Struct. 1, 523-534 (1973a)

Linden, C.D., Wright, K.L., McConnell, H.M., Fox, C.F.: Lateral phase separation in membrane lipids and the mechanism of sugar transport in E. coli. Proc. Natl. Acad. Sci. U.S. 70, 2271-2275 (1973b)

Martin, D.L., DeLuca, H.F.: Calcium transport and the role of vitamin D. Arch. Biochem. Biophys. 134, 139-148 (1969)

Max, E., Goodman, D.B.P., Rasmussen, H.: The effect of $1\alpha(OH)D_3$ treatment in D-deficient chicks on the calcium transport properties and structure of the microvillar membrane of the intestinal brush border. Submitted

Melancon, M.J., DeLuca, H.F.: Vitamin D stimulation of calcium-dependent adenosine triphosphotase in chicks intestinal brush borders. Biochemistry 9, 1658-1664 (1970)

Metcalfe, J.C., Bennett, J.P., Hesketh, T.R., Houslay, M.D., Smith, G.A., Warren, G.B.: The lateral organization of lipids around a calcium transport protein: Evidence for a phospholipid annulus that modulates function. In: The Structural Basis of Membrane Function. Hatefi, Y., Djavadi-Ohaniance, L. (eds.). New York: Academic Press 1976, pp. 57-67

Moore, L., Cohen, T., Knapp, H.R., Jr., Landon, E.J.: Energy-dependent calcium sequestration activity in rat liver microsomes. J. Biol. Chem. 250, 4562-4568 (1975)

Moriuchi, S., DeLuca, H.F.: The effect of vitamin D_3 metabolites on membrane proteins of chicks duodenal brush border. Arch. Biochem. Biophys. 174, 367-372 (1976)

Pechet, M.M., Hesse, R.H.: Metabolic and Clinical effects of pure crystalline 1α-hydroxyvitamin D_3 and 1α,25 dehydroxyvitamin D_3. Am. J. Med. 57, 13-20 (1974)

Rasmussen, H., Goodman, D.B.P., Lien, E.: Aldosterone-mediated changes in lipid metabolism in relationship to Na^+ transport in the amphibian urinary bladder. In: The Structural Basis of Membrane Function. Hatefi, Y., Djavadi-Ohaniance, L. (eds.). New York: Academic Press 1976, pp. 259-273

Singer, S.J., Nicolson, G.L.: The fluid mosaic model of the structure of cell membranes. Science 175, 720-731 (1972)

Smith, M.W., Kemp, P.: Phospholipase C-induced changes in intestinal adenosine triphosphatase prepared from goldfish acclimatized to different temperatures. Biochem. J. 114, 659-661 (1969)

Stoff, J.S., Handler, J.S., Orloff, J.: The effect of aldosterone on the accumulation of adenosine $3^1:5^1$ - cyclic monophosphate in toad bladder epithelial cells in response to vasopressin and theophylline. Proc. Natl. Acad. Sci. U.S. 69, 805-808 (1972)

Wisnieski, B.J., Williams, R.E., Fox, C.F.: Manipulation of fatty acid composition in animal cells grown in culture. Proc. Natl. Acad. Sci. U.S. 10, 3669-3673 (1973)

Youmaszai, M.K., Urban, E., Schedl, H.P.: Vitamin D and intestinal calcium fluxes in vivo in the rat. Am. J. Physiol. 225, 287-292 (1973)

Calcium, Glucose Transport and Insulin Action

T. CLAUSEN

A. Introduction

In insulin-sensitive cells, the glucose transport system serves the purpose of controlling the access of glucose to the cytoplasm. Like the other rate-limiting processes along the cellular metabolic pathways, the transport of glucose seems to be stimulated by either endogenous or exogenous factors.

For instance, we know that when the need for metabolizable substrate is increased during anoxia or contractile activity, the uptake of glucose is increased. It seems likely that this is the result of an activation of the glucose transport system brought about by endogenous factors, i.e. agents generated inside the cell following the metabolic perturbation.

The very same agents may be of importance for the activation induced by exogenous factors like insulin or other hormones.

If this is the case, it might even be possible to identify one common denominator for all the many conditions and factors which are known to stimulate glucose transport - a compound which in ideally all of these instances is made available at the plasma membrane and which can be shown to induce direct activation of the glucose transport system.

Ten years ago, Holloszy and Narahara (1965) proposed that the Ca^{2+} ion might be an important activator for the glucose transport system. This possibility may be tested by examining the consequences of exposing the glucose transport system to calcium.

B. The Distribution of Calcium Among Cellular Pools

Figure 1 shows a diagram of a typical cell. Calcium is mainly concentrated in the mitochondria, in the sarcoplasmic reticulum [and in fat cells, possibly also in the endoplasmic reticulum (Hales et al., 1974)], and on the surfaces of the plasma membrane. In addition, calcium is bound as complexes or salts, and the free Ca^{2+} ions in the cytoplasm probably constitutes the smallest single pool with a concentration around 10,000 times lower than that prevailing in the extracellular phase. This low level is ultimately the result of active transport across the plasma membrane, but due to the extensive surface of the calcium-accumulating cellular organelles, the short-term regulation of cytoplasmic Ca^{2+} is mainly dependent on the activity of mitochondria and sarcoplasmic reticulum. Indeed, changes in the transport across the plasma membrane are often only secondary manifestations of shifts in the cytosolic level of Ca^{2+} ions made available for transport from intracellular pools.

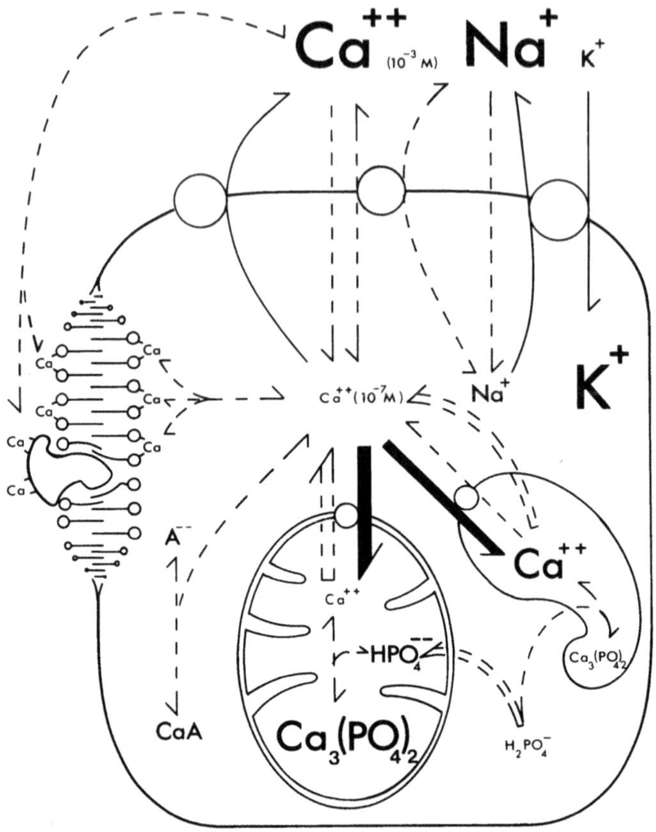

Fig. 1. The heterogeneity of cellular calcium distribution
Calcium is primarily concentrated as heavily soluble salts in the mitochondria, the sarcoplasmic reticulum and possibly the endoplasmic reticulum. Outer surface of plasma membrane contains sialoproteins and lipids which bind calcium with relatively low affinity, whereas some of the lipids at the inner surface may bind calcium with high affinity. Major part of the calcium in the cytoplasm is bound as salts or complexes with proteins and a variety of low-molecular weight compounds. The free Ca^{2+} ions in the cytoplasm are in equilibrium with all these pools, and in addition, calcium is transported across the plasma membrane via an ATP-dependent transport system and/or an Na-sensitive mechanism

Fig. 2. Effect of electrical stimulation on tension and the release of ^{45}Ca and 3-O-^{14}C-methylglucose from rat soleus muscle ▷
Isolated rat soleus muscles were loaded by incubation for 60 min in Krebs-Ringer bicarbonate buffer containing either 1 mM of 3-O-^{14}C-methylglucose (1 µCi.ml^{-1}) or 1.27 mM of ^{45}Ca (2 µCi.ml^{-1}). They were then transferred through a series of tubes containing 3 ml of unlabeled buffer, and at the end of washout, the radioactivity retained in the tissue and the amount released during each period were determined. The fraction of radioactivity released during each interval was calculated as described elsewhere (Clausen, 1969a). Each curve: mean of 3 observations with bars indicating S.E.M. where this exceeds the size of the symbols. Lowest curve: maximum isometric tension as recorded in parallel experiments with muscles suspended vertically in plastic holders. At the time indicated by arrow, the muscles were stimulated directly via platinum electrodes at a frequency of 10 impulses per sec (Kohn and Clausen, 1971). (Reprinted, with permission, from Clausen et al., 1975)

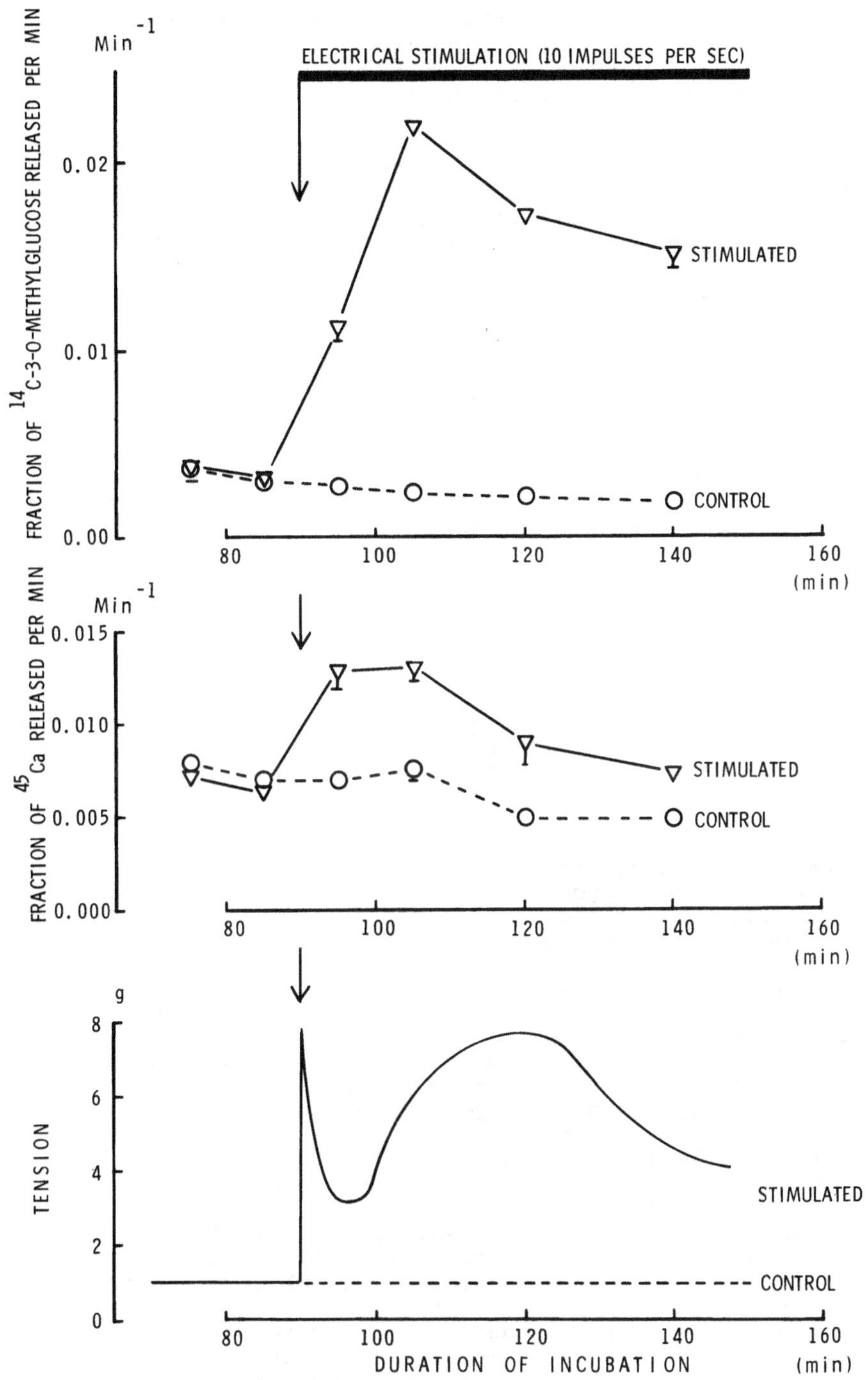

Min⁻¹
FRACTION OF 14C-3-O-METHYLGLUCOSE RELEASED PER MIN
ELECTRICAL STIMULATION (10 IMPULSES PER SEC)
0.02
0.01
0.00
STIMULATED
CONTROL
80
100
120
140
160
(min)
Min⁻¹
FRACTION OF 45Ca RELEASED PER MIN
0.015
0.010
0.005
0.000
STIMULATED
CONTROL
80
100
120
140
160
(min)
g
TENSION
8
6
4
2
0
STIMULATED
CONTROL
80
100
120
140
160
DURATION OF INCUBATION
(min)

Therefore, acute changes in the cytoplasmic level of free Ca^{2+} ions may be detected by following the efflux of radioactive calcium from preloaded tissues. In muscle cells, another indirect way of detecting increases in the cytoplasmic Ca^{2+} ion level is to measure tension.

In the following, I shall describe the results of experiments in which these two simple methods have been used for the screening of a variety of conditions or factors known to stimulate glucose transport. The permeability to glucose was continuously monitored by following the efflux of the nonmetabolized glucose analogue 3-O-methylglucose from preloaded tissues.

C. Relations Between 3-O-Methylglucose Transport, Exchange of Isotopic Calcium, and Tension Development

From Figure 2 it can be seen that when rat soleus muscles are stimulated electrically, the contraction is associated with a rise in the efflux of ^{45}Ca, indicating that the concentration of free Ca^{2+} ions in the cytoplasm is increased. At about the same time, the efflux of 3-O.methylglucose is markedly stimulated.

In the same preparation we have found that both caffeine and K-rich medium induce a rise in tension and the efflux of ^{45}Ca which preceedes a stimulation of 3-O-methylglucose transport (Clausen et al., 1975).

In analogous experiments, we have also tried to induce a release of calcium from the mitochondria. From Figure 3 it can be seen that the addition of 2,4-dinitrophenol (DNP) elicits a rise in the washout of ^{45}Ca which is followed by a contracture. Again, the efflux of labeled 3-O-methylglucose is clearly stimulated.

The other major target for insulin action, adipose tissue shows very much the same response. In whole epididymal fat pads of the rat, DNP induce a release of ^{45}Ca which correlates in time with a stimulation of 3-O-methylglucose efflux (Fig. 4).

Other metabolic poisons (cyanide, salicylate and ethacrynic acid) which have been shown to mimick the effect of insulin on glucose transport (Randle and Smith, 1958), were also found to stimulate the release of isotopic calcium and the efflux of 3-O-methylglucose from rat soleus muscles (Kohn and Clausen, 1971; Clausen et al., 1975) and epididymal fat pads (Clausen, 1969a, 1970).

As shown by Borle (1972) in kidney cells the cytoplasmic level of Ca^{2+} available for transport may be increased by preincubation in the cold. This will inhibit the accumulation of calcium in the mitochondria and in muscle, the sarcoplasmic reticulum. Furthermore, as can be seen from Figure 5, the efflux of calcium across the plasma membrane is diminished, all resulting in a progressive accumulation of calcium in the cytoplasm. This only becomes apparent when the efflux is suddenly allowed to proceed at a higher temperature. When the soleus muscles are transferred from the cold into the warm buffer, there is a rapid increase in the efflux of ^{45}Ca, and it was found that this was larger the longer the exposure to ice-cold buffer lasted.

If Ca^{2+} were of importance for the activation of the glucose transport system, one would now expect that the efflux of 3-O-methyl-

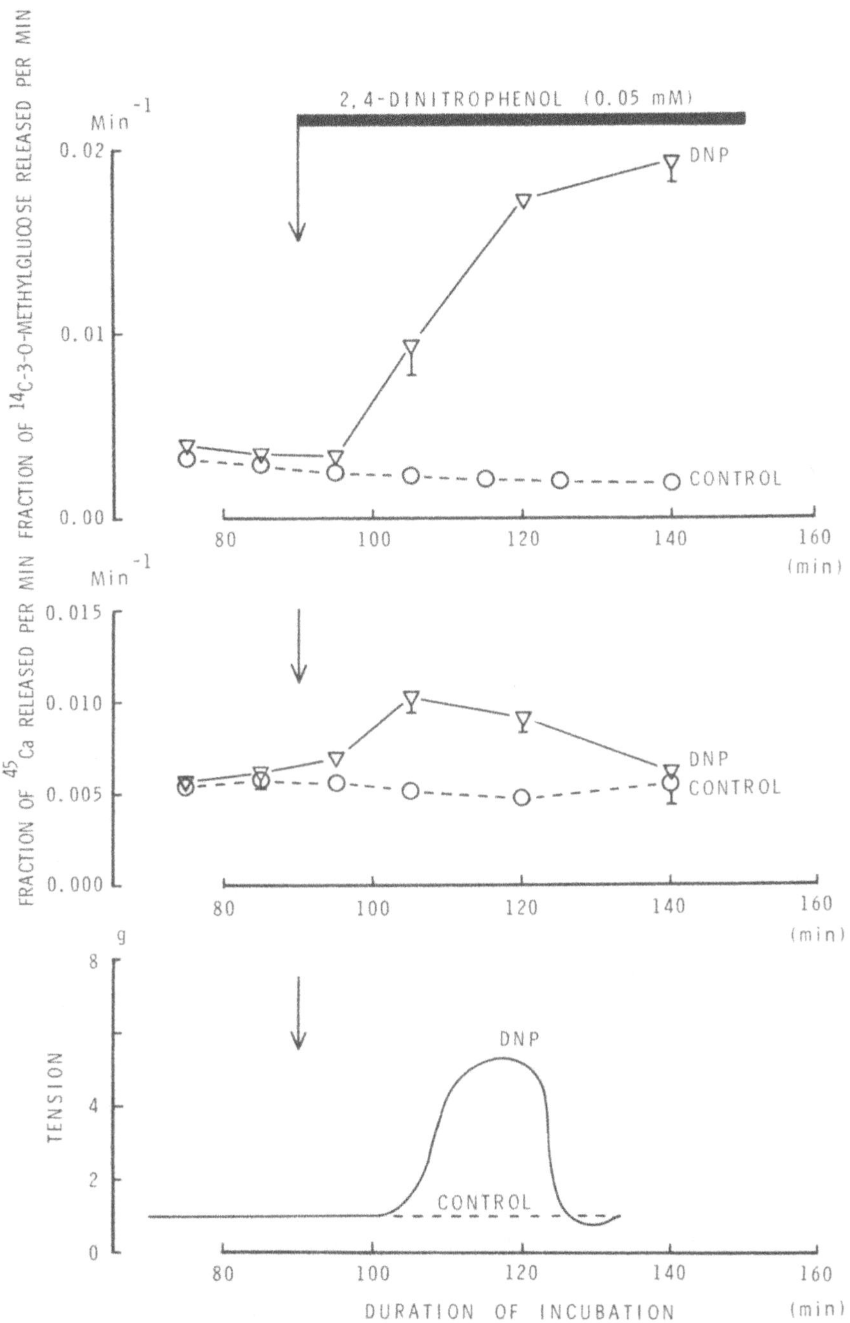

Fig. 3. Effect of 2,4-dinitrophenol on tension and the release of ^{45}Ca and 3-O-^{14}C-methylglucose from rat soleus muscle
Experimental conditions as described in legend to Figure 2. Each curve: mean of 3 observations; bars: S.E.M. (Reprinted, with permission, from Clausen et al., 1975)

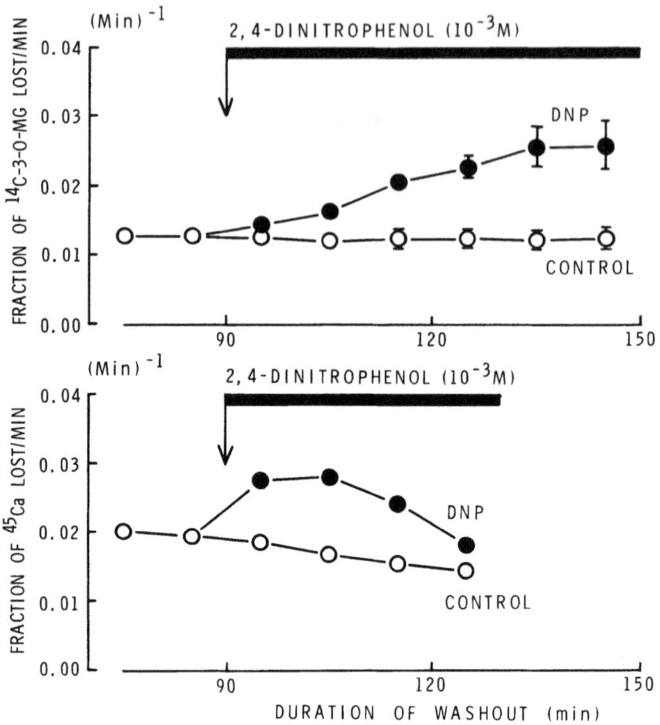

Fig. 4. Effect of DNP on the washout of ^{45}Ca and 3-O-^{14}C-methylglucose from rat epididymal fat pads
Whole epididymal fat pads from fed rats (100-120 g) were loaded for 60 min with ^{45}Ca or 3-O-^{14}C-methylglucose. The radioactivity was then washed out in a series of vials containing unlabeled buffer, and the fraction lost per min calculated for each time interval (For details, see Clausen, 1969a, 1970)

glucose should be stimulated under the same conditions. Indeed, as can be seen from Figure 5, the transfer from 0 to 30°C leads to a large rise in the efflux of 3-O-methylglucose; and again, the increase is related to the duration of cold-exposure. This stimulation of sugar transport (as well as those described before) can be suppressed by phlorizin, indicating that it does not reflect a nonspecific leakage of the plasma membrane.

Finally, the cytoplasmic concentration of Ca^{2+} ions may be increased by exposing the muscles to a hyperosmolar environment (Homsher et al., 1974). This is presumably due to the fact that the intracellular ionic strenght is increased. Thus, the binding of calcium to isolated sarcolemma (Sulakhe et al., 1973), erythrocyte membranes (Gent et al., 1964) or artificial lipid membranes (Dawson and Hauser, 1970) is substantially diminished when the ionic strenght is increased. Furthermore, the accumulation of Ca^{2+} in sarcoplasmic vesicles is impeded by both Na^{+} and K^{+} (de Meis, 1971).

Figure 6 shows that soleus muscles, when exposed to a buffer made hyperosmolar with mannitol (200 or 400 mM) go into a rapid contracture, which coincides with an increase in the efflux of ^{45}Ca. Again, there seems to be some correlation between the increase in the efflux of calcium and the sugar.

Hyperosmolarity also stimulates calcium influx, both in muscles (Table 1) and in fat cells (Martin et al., 1975). Also in fat cells, hyperosmolarity stimulates the transport of glucose and 3-O-methylglucose quite markedly (Kuzuya et al., 1965; Clausen et al., 1970). The kinetics of this effect is very similar to that of insulin.

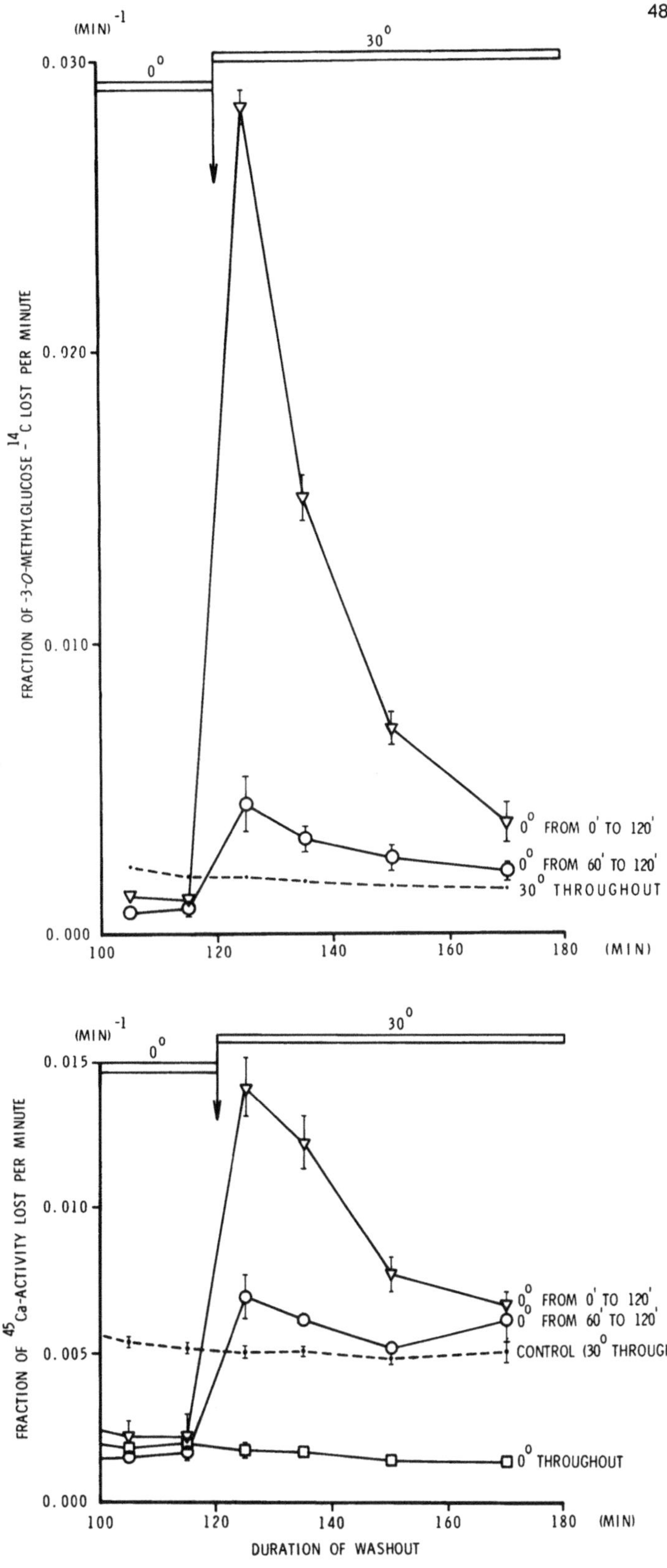

Fig. 5. Effect of cooling on the washout of ^{45}Ca and 3-O-^{14}C-methylglucose from rat soleus muscle Experimental conditions as described in legend to Figure 2. All loading took place at 30°C, but as indicated, the washout of some groups of muscles was performed at 0° for 60, 120 or 180 min. From time indicated by arrow, washout was in some cases allowed to continue at 30°. Each curve: mean of 3-4 observations; bars: S.E.M. Below: washout of ^{45}Ca. Above: washout of 3-O-^{14}C-methylglucose. (Reprinted, with permission, from Clausen et al., 1975)

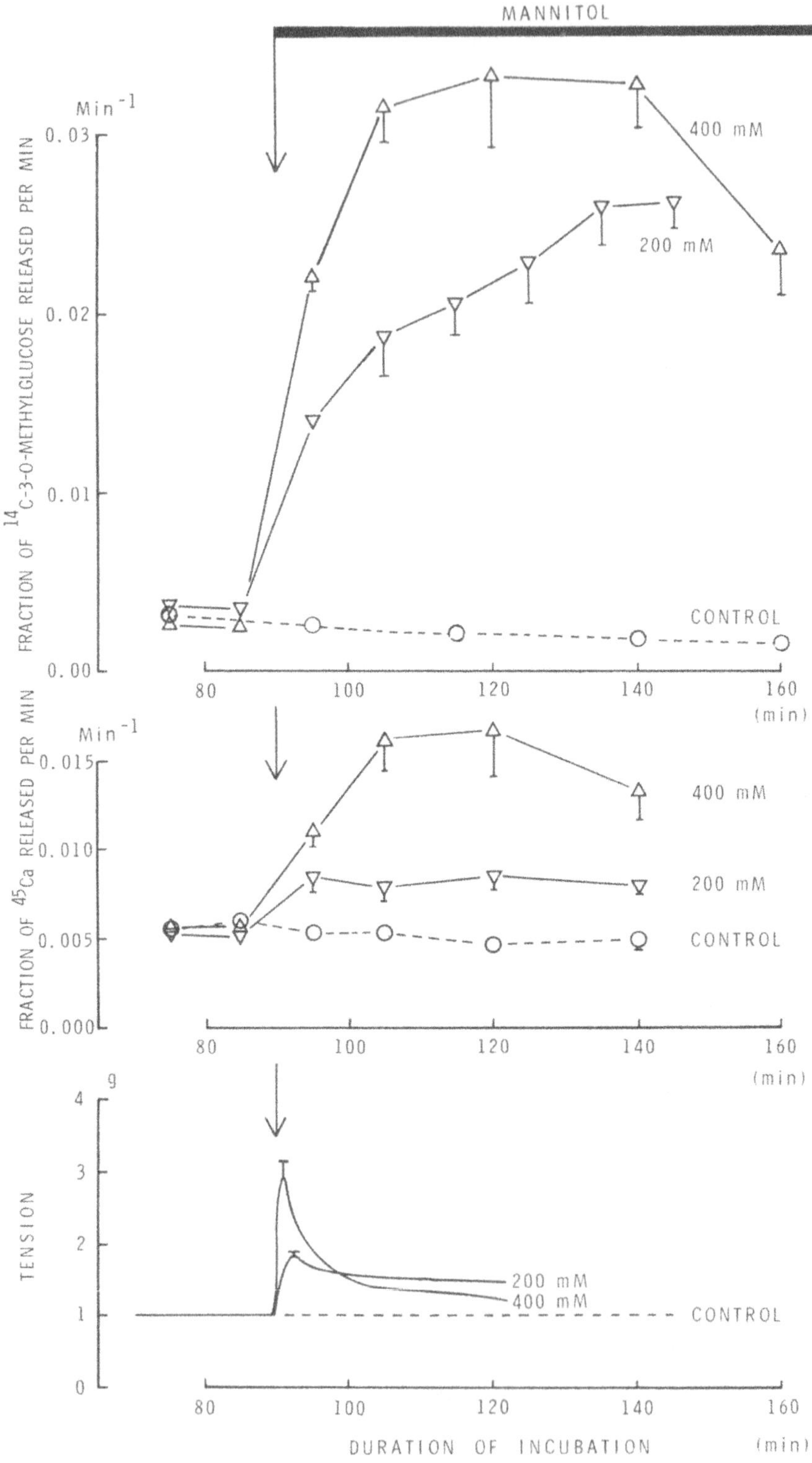
MANNITOL
Min⁻¹
FRACTION OF 14C-3-O-METHYLGLUCOSE RELEASED PER MIN
0.03
0.02
0.01
0.00
400 mM
200 mM
CONTROL
80
100
120
140
160
(min)
Min⁻¹
FRACTION OF 45Ca RELEASED PER MIN
0.015
0.010
0.005
0.000
400 mM
200 mM
CONTROL
80
100
120
140
160
(min)
g
TENSION
4
3
2
1
0
200 mM
400 mM
CONTROL
80
100
120
140
160
DURATION OF INCUBATION
(min)

Table 1. Effect of A 23187 and hyperosmolarity on ^{45}Ca-influx in rat soleus muscle

Additions	^{45}Ca-uptake $\mu mol.g$ wet weight^{-1}x.15 min^{-1}	P
Control	0.114 ± 0.007 (4)	
A 23187 (10 $\mu g.ml^{-1}$)	0.173 ± 0.014 (4)	< 0.01
Control	0.129 ± 0.008 (5)	
Mannitol (200 mM)	0.178 ± 0.012 (5)	< 0.01
Mannitol (400 mM)	0.241 ± 0.012 (5)	<0.001

Intact soleus muscles (for details of preparation, see Kohn and Clausen, 1971) were incubated for 15 min in Krebs-Ringer bicarbonate buffer containing 1.27 mM Ca and 0.4 $\mu Ci.ml^{-1}$ of ^{45}Ca, with or without the additions indicated. Thereafter, the muscles were washed for 3 x 10 min in ice-cold Ca-free buffer containing 0.5 mM EGTA. Both tendons were resected and a trichloroacetic acid homogenate of the muscles counted for ^{45}Ca. On the basis of the specific activity of the incubation medium, the amount of isotopic calcium taken up was calculated and expressed as $\mu mol.g$ tissue wet weight^{-1}.15 min^{-1} ± S.E.M. The number of observations is given in brackets. (Elbrink and Clausen, unpublished observations)

In fat cells, calcium release may be induced by lipolytic hormones (Clausen, 1969b, 197o; Kissebah et al., 1975). This is associated with a stimulation of 3-O-methylglucose efflux (Clausen, 1969a), which, as exemplified in Figure 7, appears with the same time lag.

Apart from inducing release of calcium from cellular pools, inhibition of efflux or stimulation of influx across the plasma membrane may be used as means of increasing the cytoplasmic Ca^{2+} ion level. Thus, when extracellular sodium is replaced by Li or choline, the efflux of ^{45}Ca from epididymal fat pads is decelerated (Clausen, 1970), and in fat cells the influx of isotopic calcium is stimulated (Martin et al., 1972). In fat cells, Na-substitution stimulates glucose metabolism (Letarte and Renold, 1969), and in fat pads, the efflux of 3-O-methylglucose is accelerated in such media (Clausen, 1972).

Ouabain or the omission of K from the incubation medium was found to favor the retention of ^{45}Ca in fat pads (Clausen, 1970). Again, this was associated with a slow rise in the rate coefficient of 3-O-methylglucose efflux (Clausen, 1969a). In rat diaphragm muscles, ouabain was found to stimulate 3-O-methylglucose transport (Bihler, 1968). In rat soleus muscle, a much smaller and delayed effect was observed and this was associated with contractures (Kohn and Clausen, 1971). Cardiac glycosides potentiate the effects of insulin, metabolic inhibitors (Bihler, 1968) hyperosmolarity, cold-exposure and caffeine (Clausen, unpublished observations) on

◁ Fig. 6. Effect of hyperosmolarity on tension and the washout of ^{45}Ca and 3-O-^{14}C-methylglucose from rat soleus muscle
Experimental conditions as described in legend to Figure 2. Osmolarity of washout medium was augmented by the addition of mannitol at indicated concentrations. Each curve: mean of 3-4 observations; bars: S.E.M. (Reprinted, with permission, from Clausen et al., 1975)

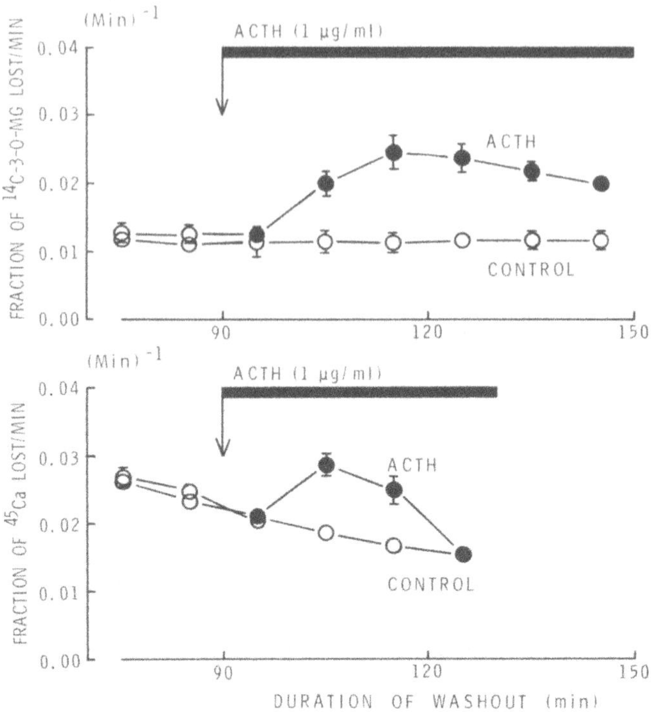

Fig. 7. Effect of ACTH on the washout of ^{45}Ca and 3-O-^{14}C-methylglucose from rat epididymal fat pads Experimental conditions as described in legend to Figure 4. (Reprinted, with permission, from Clausen, 1969a, 1970)

3-O-methylglucose transport in muscle. However, too little is known about the effects of cardiac glycosides on Ca-exchange in skeletal muscle to define their possible relation to sugar transport.

D. Direct Effects of Calcium on the Glucose Transport System?

From the examples given till now, it can only be concluded that in a number of different situations, where there is good reason to believe that the concentration of free Ca^{2+} ions in the cytoplasm is increased, the transport of glucose is accelerated. However, this may be sheer coincidence, and it is desirable to obtain a more direct demonstration of a stimulating effect of calcium on sugar transport.

In principle, this should be possible by assessing the effect of a calcium ionophore on sugar transport. A 23187 is relatively specific for calcium, and in rat soleus muscle, it was found to stimulate ^{45}Ca influx (Table 1). This effect was associated with a rapidly developing contracture which was only seen when the incubation medium contained calcium. However, as can be seen from Figure 8, the efflux of 3-O-methylglucose is not stimulated until much later. Even though experiments performed in Ca-free buffer containing 0.5 mM EGTA showed that the presence of calcium in the extracellular milieu was essential for the stimulation, it seems unlikely that a rise in the cytoplasmic level of Ca^{2+} constitute the sole adequate stimulus for the glucose transport system.

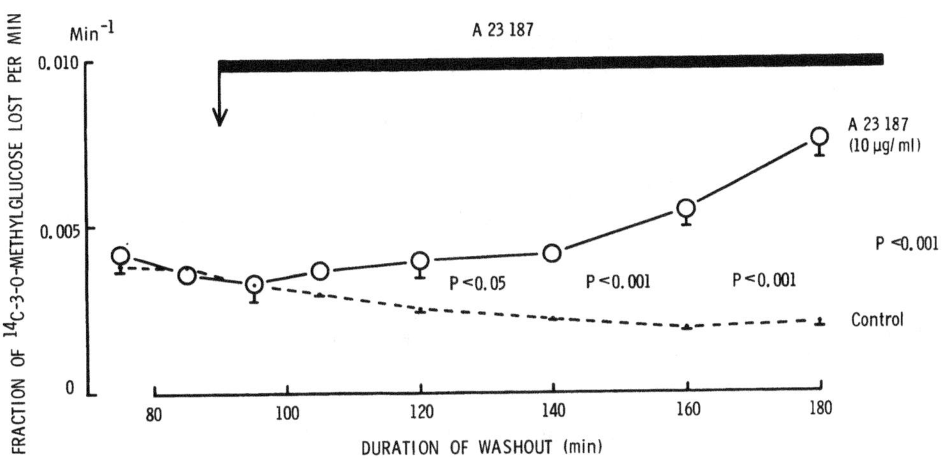

Fig. 8. Effect of A 23187 on the washout of 3-O-^{14}C-methylglucose from rat soleus muscle
Experimental conditions as described in legend to Figure 2. (Clausen, unpublished observations)

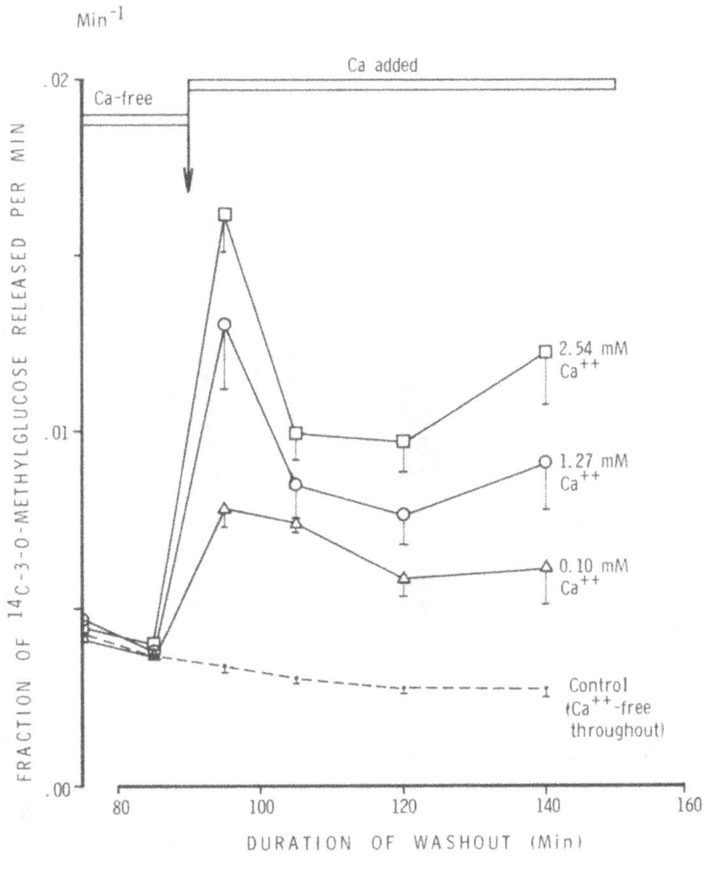

Fig. 9. Effect of A 23187 and calcium on the washout of 3-O-^{14}C-methylglucose from rat soleus muscle. Experimental conditions as described in legend to Figure 2. 90 min of washout took place in Ca-free Krebs-Ringer bicarbonate buffer containing 0.5 mM EGTA and 10 $\mu g.ml^{-1}$ of A 23187. At time indicated by arrow, EGTA was omitted from the washout medium and calcium added at the indicated concentration. A 23187 was present throughout. (Clausen, unpublished observations)

The effect of calcium on 3-O-methylglucose transport may also be demonstrated by preexposing the muscles to A 23187 in a Ca-free medium and then add calcium. As shown in Figure 9, with this procedure the response to calcium appears to be much more rapid and pronounced. In addition, it is clearly dose-dependent. However, these effects can only partially be suppressed by phlorizin or cytochalasin B, and since A 23187 may also affect Mg transport and (presumably as an outcome of its effect on calcium transport) induce K-depletion, nonspecific changes in plasma membrane permeability cannot entirely be ruled out.

E. Effect of Insulin on the Transport and Distribution of Cellular Calcium

Insulin stimulates the release of isotopic calcium from whole epididymal fat pads (Clausen, 1969b) and isolated fat cells (Martin et al., 1973). As can be seen from Figure 10, this effect is not very large. In free fat cells, the response was somewhat more pronounced and statistically significant down to the concentration of insulin found in plasma (10 $\mu U.ml^{-1}$) (Clausen and Martin, in preparation, 1976).

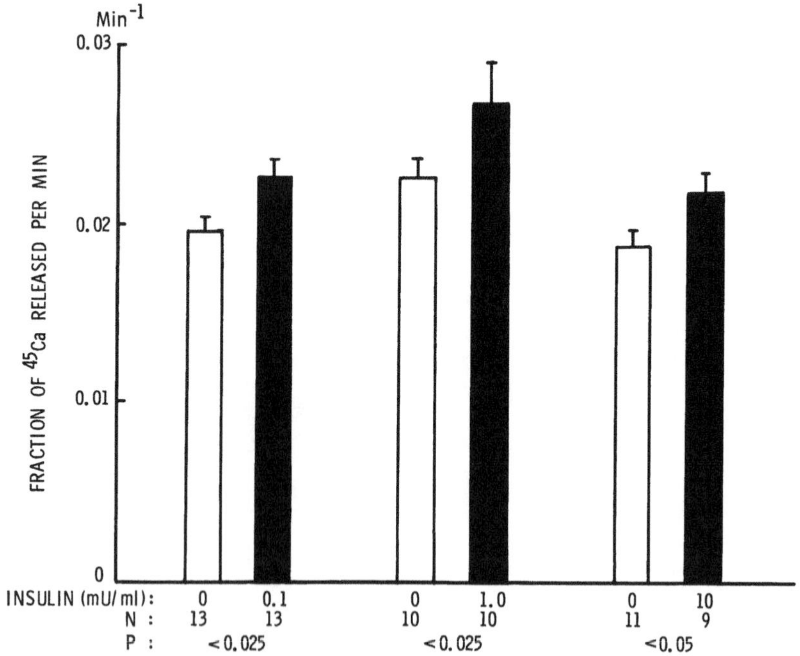

Fig. 10. Effect of insulin on washout of ^{45}Ca from fat pads
Experimental conditions as described in legend to Figure 4. Ca was omitted from efflux medium. 90 min after onset of washing, insulin was added to efflux medium. Each column: mean of rate coefficients determined in the interval from 90 to 100 min after the onset of washout; bars: S.E.M.

Table 2. Effects of insulin, albumin and $ZnCl_2$ on the washout of ^{45}Ca from rat soleus muscles

Experimental conditions	Fraction of ^{45}Ca released per min ± S.E.M.	P
Krebs-Ringer bicarbonate buffer	(180-200 min after onset of washout)	
Control	0.00456 ± 0.00017 (9)	
Insulin (100 $mU.ml^{-1}$)	0.00566 ± 0.00017 (12)	< 0.001
Control	0.00470 ± 0.00022 (6)	
Bovine serum Albumin (0.005%)	0.00471 ± 0.00026 (6)	> 0.95
Krebs-Ringer bicarbonate buffer in which all K was replaced by Na	(110-130 min after onset of washout)	
Control	0.00508 ± 0.00019 (4)	
Insulin (1 $mU.ml^{-1}$)	0.00561 ± 0.00009 (4)	< 0.05
Control	0.00477 ± 0.00023 (7)	
Insulin (5 $mU.ml^{-1}$)	0.00591 ± 0.00019 (7)	< 0.005
Control	0.00597 ± 0.00028 (9)	
Insulin (100 $mU.ml^{-1}$)	0.00797 ± 0.00033 (9)	< 0.001
Control	0.00499 ± 0.00011 (6)	
$ZnCl_2$ (10^{-1}M)	0.00471 ± 0.00006 (3)	> 0.10

Experimental conditions as described in the legend to Figure 2. The fraction of isotopic calcium released per min was determined in the time-interval from 20 to 40 min after the addition of compounds indicated and compared with the control values obtained within the same experiments. The number of observations is given in brackets

In rat soleus muscles, the effect of insulin on the washout of ^{45}Ca is not easy to detect, and previous attempts have failed, possibly because the late phase of the efflux was not sufficiently analyzed (Clausen et al., 1975). When added 160 min after the onset of washout, insulin stimulates ^{45}Ca efflux by around 30% (Table 2). In a K-free medium the effect may be detected earlier in the washout, and as can be seen from Figure 11, it shows some increase with time. Bovine serum albumin at a molar concentration corresponding to 100 mU/ml of insulin did not produce any significant increase. The addition of Zn at a concentration corresponding to that contained in a solution with 100 $mU.ml^{-1}$ of insulin did not produce any effect either (Table 2).

These experiments did not yield any information about where this calcium comes from, but it has been shown that insulin inhibits the binding of calcium to plasma membranes isolated from the liver (Marinetti et al., 1972), and in ghosts of fat cells even very low concentrations of insulin diminish the binding of calcium to a pool with high affinity (Kissebah et al., 1975).

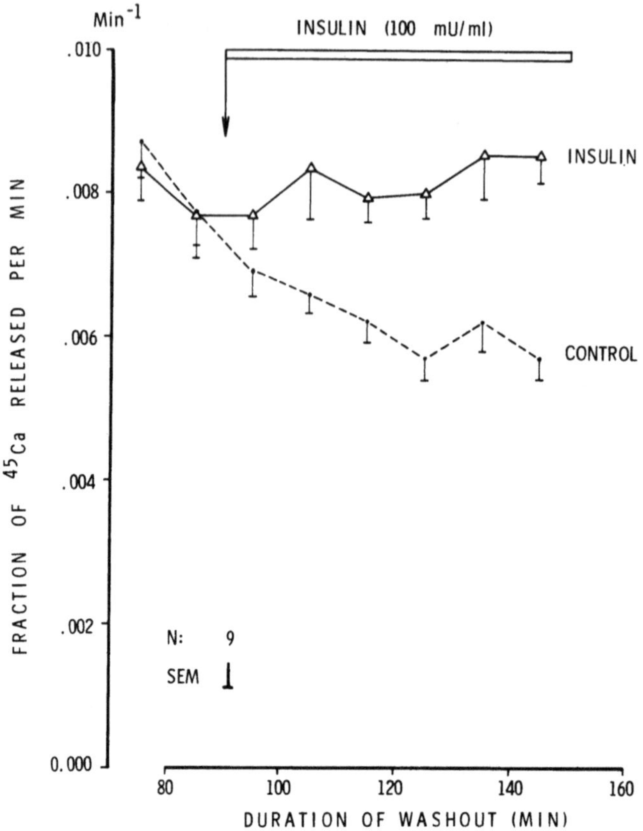

Fig. 11. Effeft of insulin on the washout of ^{45}Ca from rat soleus muscle
Experimental conditions as described in legend to Figure 2. In the washout medium, all K was replaced by an equimolar amount of Na. Each point: mean of 9 observations; bars: S.E.M. (Clausen, unpublished observations)

The next problem is to determine whether the insulin-induced mobilization of calcium can lead to a rise in the cytoplasmic level of free Ca^{2+} ions. In muscle cells, it is obvious to look for a response in the contractile apparatus. From Figure 12 it can be seen that when this has already been partially activated by a hyperosmolar environment, the addition of insulin elicits a pronounced further rise in resting tension. This effect was blocked by insulin antibody, and it could not be produced by $ZnCl_2$ or albumin.

If, as this experiment indicates, insulin really produces a rise in the cytoplasmic Ca^{2+} level, it is reasonable to expect that the mitochondria will take up more calcium. Ca^{2+} ions stimulate mitochondrial respiration and the uptake of inorganic phosphate. Indeed, in skeletal muscle, insulin increases O_2 consumption (Gourley, 1961). Both in fat cells (Hepp et al., 1968) and muscle (Walaas et al., 1969) insulin favors the accumulation of inorganic phosphate. The observation that DNP blocks this effect is compatible with the idea that it is dependent on mitochondrial function (Clausen, 1976). As can be seen from Figure 13, the uptake of labeled inorganic phosphate in rat soleus muscle is clearly suppressed by calcium depletion, both in the absence and in the presence of insulin. This effect is almost completely reversible following reestablishment of the normal extracellular calcium concentration.

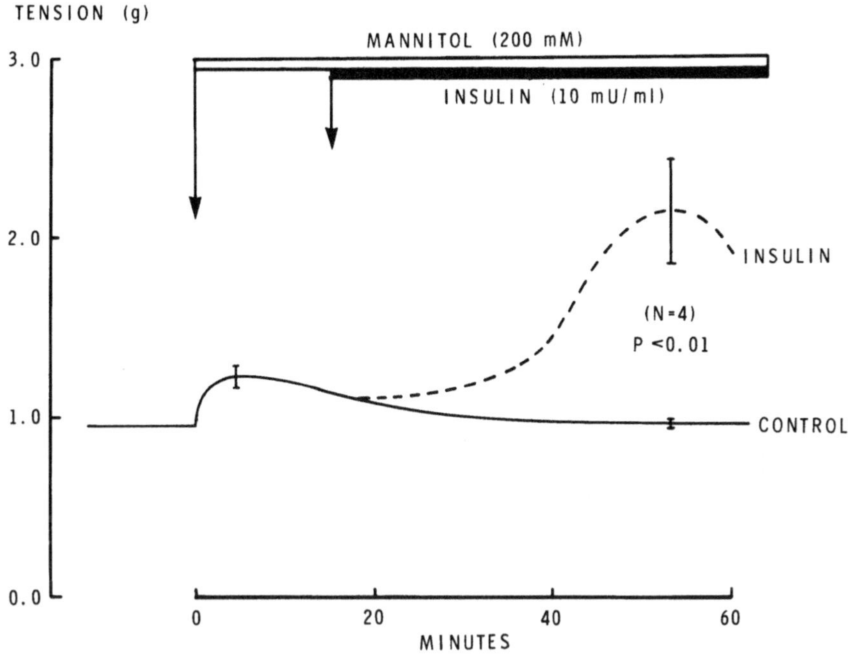

Fig. 12. Effect of hyperosmolarity and insulin on tension in rat soleus muscle Soleus muscles were prepared from 48-h fasted rats and incubated in Krebs-Ringer bicarbonate buffer without metabolizable substrate. Tension was recorded as described elsewhere (Clausen et al., 1975). Two groups of muscles were exposed to hyperosmolarity by addition of mannitol (200 mM); one serves as control, the other was given insulin as indicated. Each curve: mean of 4 observations; bars: S.E.M. (Reprinted, with permission, from Clausen et al., 1975)

But is there any evidence that insulin stimulates the uptake of calcium into the mitochondria? Again, this cannot be detected directly, but as shown in the next figure, there is some indirect support for this idea. In fat cells, the mitochondria contains pyruvate dehydrogenase, which is activated by a Ca^{2+}-sensitive phosphatase (Severson et al., 1974). When fat cells are exposed to hyperosmolarity, Li-substituted buffer or ouabain, the uptake of calcium is increased. This is associated with an increased pyruvate dehydrogenase activity. Thus, the stimulating effect of insulin on pyruvate dehydrogenase may be the result of increased availability of Ca^{2+} ions in the mitochondria (Martin et al., 1972).

F. The Mode of Action of Insulin - A Working Hypothesis

On the basis of these and other observations, I have tried to formulate a working hypothesis for the mode of action of insulin, which is illustrated in Figure 14:

1. Insulin interacts with its specific membrane receptor.

2. Either via the receptor or directly, insulin may induce a con-

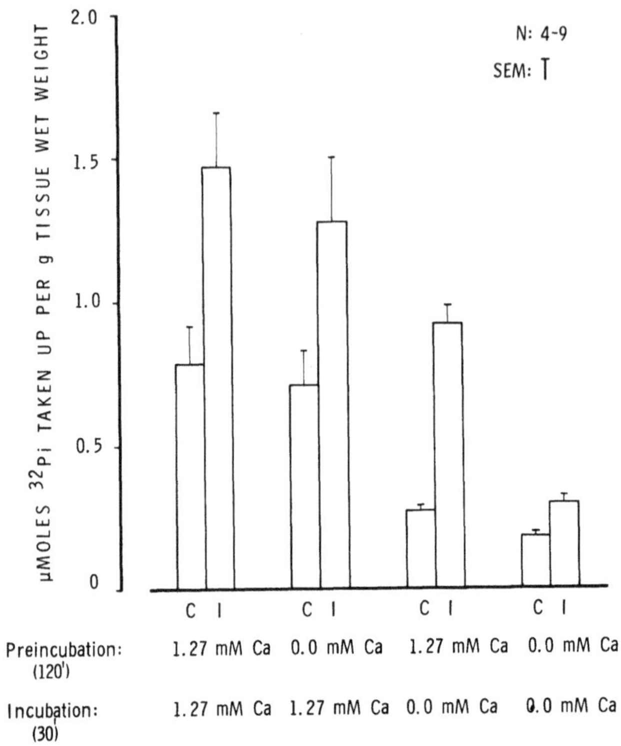

Fig. 13. Effect of Ca-depletion and insulin on the uptake of $^{32}PO_4$ in rat soleus muscle
Soleus muscles were preincubated for 120 min in Krebs-Ringer bicarbonate buffer containing either 1.27 mM Ca or 0 mM Ca and 0.5 mM EGTA. Then the muscles were incubated for 30 min in Ca-free or Ca containing buffer with 0.2 µCi.ml^{-1} of $^{32}PO_4$. Following 60-min wash in ice-cold buffer (so as to remove extracellular radioactivity) the tendons were resected, and after blotting, the muscles were homogenized in 5% trichloroacetic acid. The amount of $^{32}PO_4$ taken up was calculated on the basis of the specific activity of the incubation medium and expressed as µmol.g tissue wet weight^{-1}.30 min^{-1}. Each column represents the mean of 4-9 observations with bars indicating S.E.M. C: Control: I: Insulin (100 mU-ml^{-1})

formational change in the plasma membrane which is decrementally propagated around the receptor area.

3. It is postulated that this change leads to a release of calcium from the polar heads of membrane lipids.

4.
a) The Ca^{2+} ions liberated may be transported out of the cells.

b) Alternatively, the Ca^{2+} may be taken up by the mitochondria, and there induce the activation of pyruvate dehydrogenase described above.

c) The coupling between the uptake of calcium and phosphate in mitochondria (Lehninger, 1974) may account for the stimulating effect of insulin on phosphate uptake.

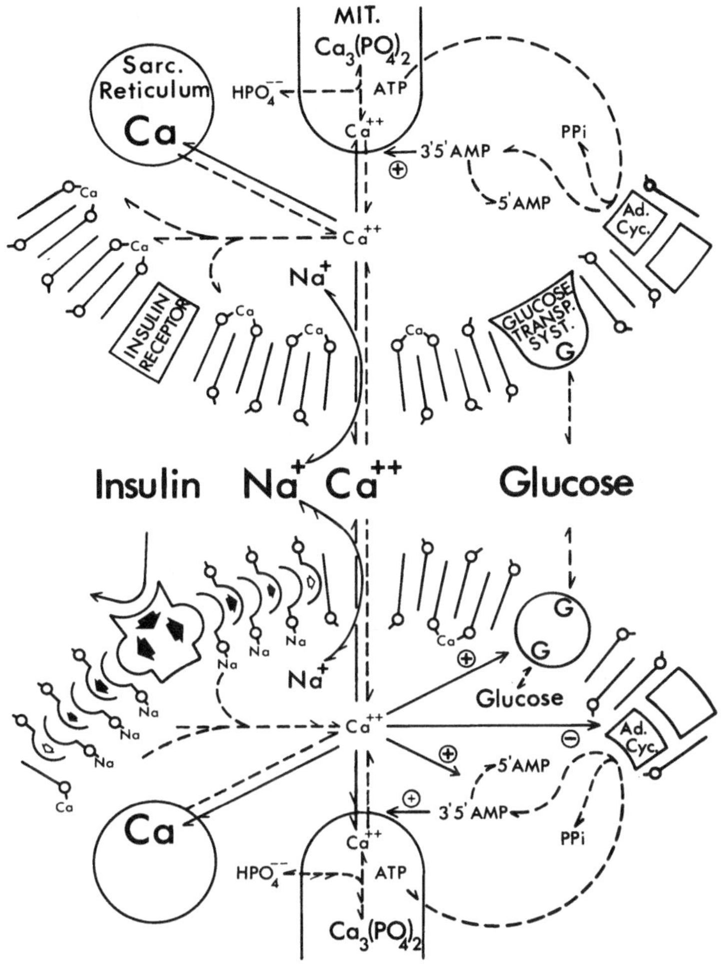

Fig. 14. Hypothetical diagram of cellular elements involved in the action of insulin

Upper part: basal state; lower part: changes supposed to occur during exposure to insulin. Segments of the plasma membrane with insulin receptor, glucose transport system, adenyl cyclase (Ad. Cyc.) and lipids with polar heads binding Ca or Na are shown. Details of the cellular events are described in the text. (Reprinted, with permission, from Clausen, 1975)

d) The observation that Ca^{2+} ions elicit activation of phosphodiesterase (Kakiuchi and Yamazaki, 1970) is consistent with the idea that the stimulating effect of insulin on this enzyme is mediated by calcium (Loten and Sneyd, 1970).

e) Filaments, which resemble contractile proteins in muscle have been detected in plasma membranes isolated from a variety of cells (Jahn and Bovee, 1969), and Ca ions have been shown to trigger their contraction.

These filaments are inactivated by cytochalasin B, a drug which can abolish the effect of insulin on sugar transport in fat cells (Loten and Jeanrenaud, 1974) and muscle (Clausen, 1976). Ca^{2+} ions liberated in the microenvironment of the plasma membrane may trigger a contraction of membrane-associated filaments. As long as Ca^{2+} remains dissociated from the polar heads of the membrane lipids, the mobility of the membrane components may be sufficient to allow a conformational change by which the glucose transport system is activated.

References

Bihler, I.: The action of cardiotonic steroids on sugar transport in muscle, in vitro. Biochim. Biophys. Acta 163, 401-410 (1968).

Borle, A.B.: Kinetic analysis of calcium movements in cell culture. V. Intracellular calcium distribution in kidney cells. J. Membrane Biol. 10, 45-66 (1972)

Clausen, T.: The relationship between the transport of glucose and cations across cell membranes in isolated tissues. V. Stimulating effect of ouabain, K^+-free medium and insulin on efflux of 3-O-methylglucose from epididymal adipose tissue. Biochim. Biophys. Acta 183, 625-634 (1969a)

Clausen, T.: Role of insulin in ion transport in muscle and adipose tissue. Jugoslav. Physiol. Pharmacol. Acta 5, 363-368 (1969b)

Clausen, T.: Electrolytes and the hormonal control of organic metabolism in adipocytes. Horm. Metab. Res. Suppl. 2, 66-70 (1970)

Clausen, T., Gliemann, J., Vinten, J., Kohn, P.G.: Stimulating effect of hyperosmolarity on glucose transport in adipocytes and muscle cells. Biochim. Biophys. Acta 211, 233-243 (1970)

Clausen, T.: Cations, glucose metabolism, and insulin action. Thesis, Århus Univ. (1972)

Clausen, T.: The effect of insulin on glucose transport in muscle cells. In: Current Topics in Membranes and Transport. San Francisco: Academic Press, 1975, Vol. VI, pp. 169-226.

Clausen, T., Elbrink, J., Dahl-Hansen, A.B.: The relationship between the transport of glucose and cations across cell membranes in isolated tissues. IX. The role of cellular calcium in the activation of the glucose transport system in rat soleus muscle. Biochim. Biophys. Acta 375, 292-308 (1975)

Clausen, T.: The role of ions in the control of intermediary metabolism. In: Ion and Enzyme Electrodes in Biology and Medicine. Kessler, M., Clark, L.C., Lübbers, D.W., Silver, I.A., and Simon, W. (eds.). München-Berlin-Wien: Urband and Schwarzenberg, 1976, pp. 237-250

Dawson, R.M.C., Hauser, H.: Binding of calcium to phospholipids. In: Calcium and Cellular Function. Cuthbert, A.W. (ed.). New York: Macmillan, 1970, p. 17

Gent, W.L.C., Trounce, J.R., Walser, M.: The binding of calcium ion by the human erythrocyte membrane. Arch. Biochem. Biophys. 105, 582-589 (1964)

Gourley, D.R.H.: Separation of insulin effects on K content and O_2 consumption of frog muscle with cardiac glycosides. Am. J. Physil. 200, 1320-1326 (1961)

Hales, C.N., Luzio, J.P., Chandler, J.A., Herman, L.: Localization of calcium in the smooth endoplasmic reticulum of rat isolated fat cells. J. Cell Sci. 15, 1-15 (1974)

Hepp, D., Challoner, D.R., Williams, R.H.: Studies on the action of insulin in isolated adipose tissue cells. I. Stimulation of incorporation of ^{32}P-labeled inorganic phosphate into mononucleotides in the absence of glucose. J. Biol. Chem. 15, 4020-4026 (1968)

Holloszy, J.O., Narahara, H.T.: Studies on tissue permeability. X. Changes in permeability to 3-O-methylglucose associated with contraction of isolated frog muscle. J. Biol. Chem. 240, 3493-3500 (1965)

Homsher, E., Briggs, F.N., Wise, R.M.: Effects of hypertonicity on resting and contracting frog skeletal muscles. Am. J. Physiol. 226, 855-863 (1974)

Jahn, T.L., Bovee, E.C.: Protoplasmic movements within cells. Physiol. Rev. 49, 794-862 (1969)

Kakiuchi, S., Yamazaki, R.: Calcium dependent phosphodiesterase activity and its activating factor (PAF) from brain. Studies on cyclic 3', 5'-nucleotide phosphodiesterase (III). Biochem. Biophys. Res. Commun. 41, 1104-1110 (1970)

Kissebah, A.H., Clarke, P., Vydelingum, N., Hope-Gill, H., Tulloch, B., Fraser, T.R.: The role of calcium in insulin action. III. Calcium distribution in fat cells; its kinetics and the effects of adrenaline, insulin and procaine-HCl. Europ. J. Clin. Invest. 5, 339-349 (1975)

Kohn, P.G., Clausen, T.: The relationship between the transport of glucose and cations across cell membranes in isolated tissues. VI. The effect of insulin, ouabain, and metabolic inhibitors on the transport of 3-O-methylglucose and glucose in rat soleus muscles. Biochim. Biophys. Acta 225, 277-290 (1971)

Kuzuya, T., Samols, E., Williams, R.W.: Stimulation by hyperosmolarity of glucose metabolism in rat adipose tissue and diaphragm in vitro. J. Biol. Chem. 240, 2277-2283 (1965)

Lehninger, A.L.: Role of phosphate and other proton-donating anions in respiration-coupled transport of Ca^{2+} by mitochondria. Proc. Natl. Acad. Sci. U.S. 71, 1520-1524 (1974)

Letarte, J., Renold, A.E.: Ionic effects on glucose transport and metabolism by isolated mouse fat cells incubated with or without insulin. III. Effects of replacement of Na^{+}. Biochim. Biophys. Acta 183, 366-374 (1969)

Loten, E.G., Sneyd, J.G.T.: An effect of insulin on adipose-tissue adenosine 3':5'-cyclic monophosphate phosphodiesterase. Biochem. J. 120, 187-193 (1970)

Loten, E.G., Jeanrenaud, B.: Effects of cytochalasin B, colchicine and vincristine on the metabolism of isolated fat-cells. Biochem. J. 140, 185-192 (1974)

Marinetti, G.V., Shlatz, L., Reilly, K.: Hormone-membrane interactions. In: Insulin Action. Fritz, I.B. (ed.). New York: Academic Press, 1972, p. 207

Martin, B.R., Denton, R.M., Pask, H.T., Randle, P.J.: Mechanisms regulating adipose tissue pyruvate dehydrogenase. Biochem. J. 129, 763-773 (1972)

Martin, B.R., Clausen, T., Gliemann, J.: Calcium and the action of insulin on adipose tissue pyruvate dehydrogenase. 9th Intern. Congr. Biochem. Stockholm (1973). Abstr. No. 8 f 9

Martin, B.R., Clausen, T., Gliemann, J.: Relationships between the exchange of calcium and phosphate in isolated fat-cells. Biochem. J. 152, 121-129 (1975)

de Meis, L.: Allosteric inhibition by alkali ions of the Ca^{2+} uptake and adenosine triphosphatase activity of skeletal muscle microsomes. J. Biol. Chem. 246, 4764-4773 (1971)

Randle, P.J., Smith, G.H.: Regulation of glucose uptake by muscle. Biochem. J. 70, 501-509 (1958)

Severson, D.L., Denton, R.M., Pask, H.T., Randle, P.J.: Calcium and magnesium ions as effectors of adipose-tissue pyruvate dehydrogenase phosphate phosphatase. Biochem. J. 140, 225-237 (1974)

Sulakhe, P.V., Drummond, G.I., Ng, D.C.: Calcium binding by skeletal muscle sarcolemma. J. Biol. Chem. 248, 4150-4157 (1973)

Walaas, O., Walaas, E., Wick, A.N.: The stimulatory effect by insulin on the incorporation of ^{32}P radioactive inorganic phosphate into intracellular inorganic phosphate, adenine nucleotides and guanine nucleotides of the intact rat diaphragm. Diabetologia 5, 79-87 (1969)

Ion Transport in Mitochondria

The Proton Pump of the Mitochondrial Respiratory Chain

S. PAPA, F. GUERRIERI, M. LORUSSO, G. IZZO, D. BOFFOLI, and F. CAPUANO

A. Introduction

Fifteen years after Mitchell (1961) first proposed the chemiosmotic hypothesis of oxidative phosphorylation, it has been experimentally established that redox chains and ATPase complexes of energy-transducing membranes directly convert metabolic energy into a transmembrane thermodynamic potential difference of protons (for review, see: Mitchell, 1972; Papa, 1976). The mechanism of the redox and the hydrolytic pumps and their role in ATP formation remain, however, yet to be understood (Papa, 1976).

The chemiosmotic model of Mitchell (1966, 1972) essentially visualizes the redox proton pump as a physical consequence of the asymmetry in the arrangement and mobility in the membrane of electron-transfer and hydrogen-transfer centers of redox catalysts. This model neglects, on principle, the particular role of the apoproteins.

The prosthetic groups of cytochromes and Fe-S centers are, however, covalently bound to apoproteins. It is feasible, and in fact experimentally evident that, at least in the case of cytochromes, the apoproteins feel the transfer of electrons at the metal in the sense that this triggers conformational transitions in the polypeptide chains (Chance, 1972; Slater, 1974; Nicholls and Chance, 1974).

Of specific interest here is the linkage in cytochromes between the redox state of the metal and protolytic equilibria in the apoproteins (for review, see: Dutton and Wilson, 1974).

This linkage would be of the same type as those designated as Bohr effects in soluble hemoproteins (Wyman, 1968; Kilmartin and Rossi-Bernardi, 1973). When the linkage occurs in a membrane-bound electron-carrying protein, it could be amplified in such a way that electron arrival at and departure from the metal respectively would cause proton binding and release from the apoprotein on the opposite sides of the membrane, thus resulting in a vectorial transmembrane flux. This model is referred to as the vectorial Bohr mechanism (Papa et al., 1973; Papa, 1976).

Definition of the stoichiometry of the redox proton pump represents a crucial test for both its molecular mechanism and function.

A basic postulate of Mitchell's mechanism is that, independent of the actual H^+ activity, $2H^+$ have to be translocated across the membrane fro each pair of electrons transferred through a redox loop, or energy conservation site of the chain (Mitchell, 1966, 1972, 1975a). The vectorial Bohr mechanism, on the other hand, is not bound, in principle, to a rigid H^+/e^- stoichiometry. The stoichiometry will in this case be determined by the number of electron carriers exhibiting redox-linked ionization groups, the number of these groups per metal center, their pKs and the actual H^+ activity (Papa et al., 1973; Papa, 1976).

A fixed stoichiometry of $2H^+/2e^-$ per site is apparently supported by a series of experimental determinations by Mitchell and others (for review, see: Mitchell, 1972; Papa, 1976). However, $H^+/2e^-$ ratios higher than 2 have also been obtained (Papa, 1976; Brand et al., 1976).

Clearly the problem requires further experimental exploration.

B. I. Molecular Mechanism of the Redox Proton Pump

A number of characteristics of the protonmotive function of the mitochondrial cytochrome system were recently described (Mitchell, 1972; Wikstrom, 1972; Papa, 1976) which offer a new insight into the molecular mechanism of the process.

The cytochrome system consists of two complexes: cytochrome c oxidoreductase or complex IV and ubiquinol-cytochrome c oxidoreductase or complex III (see Fig. 1).

Complex IV, which exhibits an oblong shape of 70 x 90 Å (Seki and Oda, 1970; Vanderkooi et al., 1972), is plugged across the mitochondrial membrane, protruding in both aqueous phases (Nicholls and Chance, 1974). Electrons are fed to the complex by cytochrome c at the external or C side of the membrane (Nicholls and Chance,

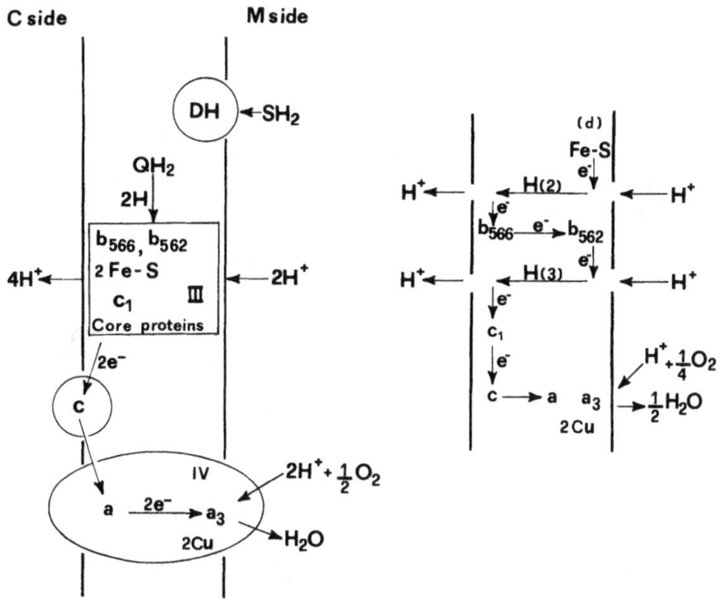

Fig. 1. The cytochrome system of the mitochondrial respiratory chain (see text)

1974). Oxygen protonation on the other hand, takes place at the inner or M side of the membrane (Mitchell and Moyle, 1970; Papa et al., 1974a).

These and other observations indicate that the two hemes and two copper atoms of the oxidase constitute an electron well, which mediate vectorial electron flow from cytochrome c on the C side to oxygen at the M side (Hinkle and Kim, 1973; Papa et al., 1974a; Nicholls and Chance, 1974).

Electrons are fed to cytochromes by dehydrogenases. These oxidize the substrates at the M side of the membrane and donate reducing equivalents to ubiquinone (Klingenberg, 1968).

Kinetic analysis shows that electron flow along complex III, from ubiquinol to cytochrome c, is accompanied by synchronous proton disappearance at the M side and proton appearance at the C side (Papa et al., 1974b, 1975a); this proton translocation is electrogenic.

The right side of Figure 1 illustrates how the redox catalysts of the two complexes would be arranged in the membrane according to the linear redox-loop chemiosmotic model of Mitchell (1972). Inward electron flow at the second redox loop would be mediated by two b cytochromes distributed across the membrane.

The finding that the aerobic oxidation of quinols gives an H^+/e^- ratio of 2 (Lawford and Garland, 1973; Papa et al., 1974b, 1975a) eliminates the proposal that ubiquinone represents the specific hydrogen carrier of the third loop. There is in fact agreement that ubiquinone acts as an hydrogen-transporting pool between dehydrogenases and b cytochromes (Klingenberg, 1968). Thus, there should exist an additional hydrogen carrier on the oxygen side of ubiquinone.

No classical hydrogen carrier is known or suspected to exist on the oxygen side of ubiquinone. Unless a new carrier is discovered having the required kinetics and thermodynamic characteristics to be placed between cytochromes b and c_1, one is faced with two possibilities.

Mitchell (1975a, 1976b) has recently proposed that ubiquinone could act twice as hydrogen carrier on both sides of b cytochromes. He thus changed the two redox loops into a protonmotive ubiquinone cycle. This proposes that semiquinone accepts one hydrogen from the dehydrogenase at the M side. The quinol so formed is oxidized by a b cytochrome to semiquinone and this by cytochrome c_1 to quinone at the C side. The cycle is completed through quinone reduction by a by cytochrome to semiquinone at the M side of the membrane. This model can indeed explain various phenomena exhibited by the b-c_1 region of the chain (Mitchell, 1975a, b). However, it rests upon a number of controversial issues which should be clarified before any conclusion can be drawn on its validity (Papa, 1976).

According to the vectorial Bohr mechanism (Papa et al., 1973; Papa, 1976), an electron carrier of complex III would be responsible for transmembrane proton translocation, because of linkage of oxidoreduction of the metal to protolytic equilibria in the apoprotein. This mechanism requires the midpoint redox potential of the electron carrier(s) responsible for proton pumping to be pH dependent. Furthermore, the protonmotive function of the system should exhibit specific pH dependence.

The minimum functional unit of complex III contains one molecule of cytochrome c_1, two molecules of b cytochromes (Baum et al., 1967; Dutton and Wilson, 1974) and two Fe-S-centers (Albracht and Slater, 1971; Ohnishi et al., 1972). There are also proteins not well characterized. These comprise core proteins (Baum et al., 1967) and, possibly as separate entities, the oxidation factor isolated by Nishibayashi-Yamashita et al. (1972) and the antimycin-binding protein identified by Das Gupta and Rieske (1973). Recently Gellerfors et al. (1975) have fractionated complex III into 8 major polypeptides. Five of these appeared to be associated with the prosthetic groups of the complex.

It has been reported that the midpoint potential of cytochrome c_1 might be pH-independent in the physiological range (Green et al., 1959; Dutton and Wilson, 1974).

The general situation for the pH dependency of the midpoint potential of c cytochromes is however complex (see Fig. 2; (Clark, 1960). Isolated beef heart cytochrome c does not exhibit E_M pH dependence at pHs below 7.8. Above this pH, E_M decreases by 60 mV per pH unit increase (Rodkey and Ball, 1949).

The behavior of c cytochromes of photosynthetic bacteria is quite interesting (Pettigrew et al., 1975).

E_M of Rhodospirillum rubrum cytochrome c_2 shows, for example a slope of 30 mV from pH 5 to 7.5, but no change above this pH (Vernon and Kamen, 1954).

Rhodomicrobium vannielii cyt c_2 exhibits complex pH dependency which can be explained by two ionizations, one of these exhibiting redox linked pK shifts (Pettigrew et al., 1975).

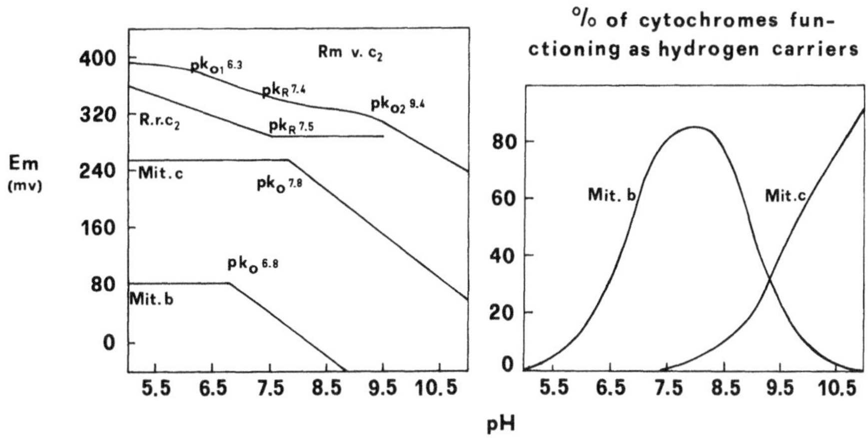

Fig. 2. pH dependency of the midpoint potential of c and b cytochromes (see text)

EM of b cytochromes is pH dependent from pH 6.8 to 9, which indicates that the redox state of the iron is linked to ionization of a group, the pK of which is 6.8 when the iron is oxidized and equal or higher than 9 when the iron is reduced (Straub and Colpa Boonstra, 1962; Urban and Klingenberg, 1969).

Thus coupling of proton transfer to electron transfer is a general property of b and c cytochromes. As a result of this coupling cytochromes switch, at certain pHs, from electron to hydrogen carriers. Thus, in the physiological pH range b cytochromes can act both as electron and hydrogen carriers. In the case of mammalian cytochrome c, on the other hand, hydrogen transfer comes only at alkaline pHs.

Let us now see what the impact is of proton transfer in cytochromes on the redox and protonmotive functions of complex III.

We have studied the pH dependence of the crossover effects exerted on electron flow by the electrical and chemical component of the protonmotive force (Papa et al., 1975b) and the pH dependence of the H^+/e^- stoichiometry of the redox proton pump.

B. II. Crossover Effects Exerted by $\Delta\psi$ on Electron Flow

The rationale of the crossover experiments is that the $\Delta\psi$ component of the protonmotive force should exert a specific control on vectorial electron flow from electron carriers at the C to electron carriers at the M side of the membrane. ΔpH on the other hand should exert selective control on vectorial hydrogen flow from the M to the C side. Thus, collapse of $\Delta\psi$ with valinomycin should specifically promote electron flow from the C to the M side whilst collapse of ΔpH with nigericin should speed up hydrogen flow from the M to the C side. Control of vectorial electron from cytochrome c to oxygen has in fact been demonstrated in this way (Papa et al., 1974a, b).

The crossover experiments led us to favor the possibility (Mitchell, 1972; Wikström, 1972) that two b cytochromes exist, b_{566} located at the C side and b_{562} at the M side of the membrane. b_{566} would be closer to the substrate than b_{562} (see Fig. 3; Papa et al., 1975b; Papa, 1976). With this sequence b cytochromes would mediate vectorial electron flow from the C to the M side. Outward proton pumping would be mediated by an hydrogen carrier located between b and c cytochromes. We found in fact that collapse of $\Delta\psi$,

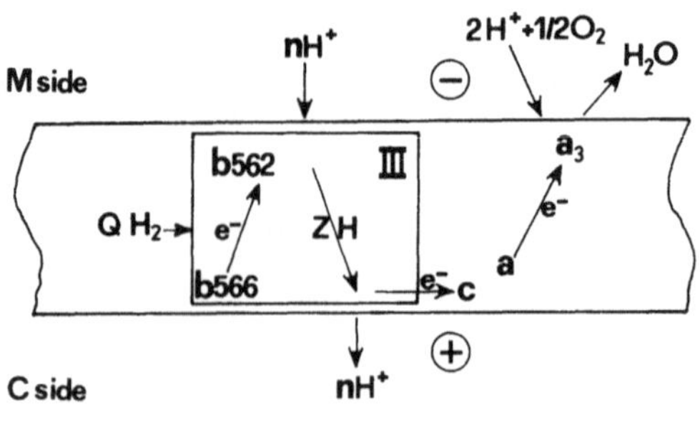

Fig. 3. Model for vectorial electron flow in the cytochrome system of the respiratory chain

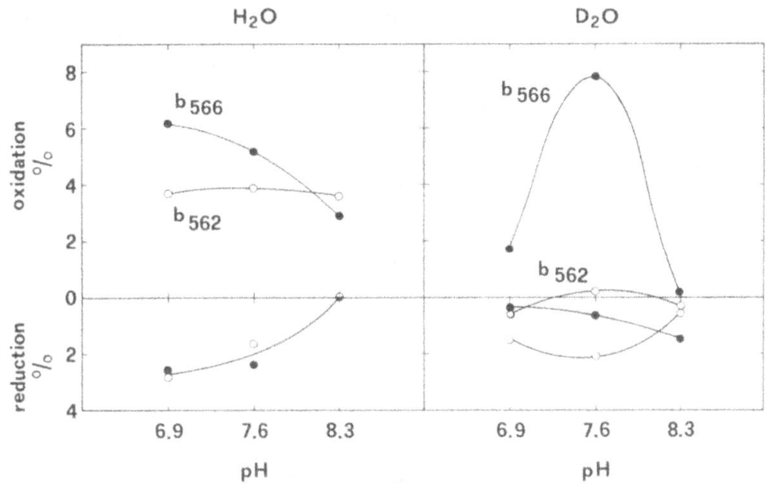

Fig. 4. pH-dependence curves of the effect of valinomycin (plus K^+) on the redox state of b cytochromes of sonic EDTA submitochondrial particles in the respiring steady-state
The redox level of b cytochromes was measured with a double beam dual wavelength spectrophotometer at 562-575 and 566-575 nm. The reaction mixture contained: 200 mM sucrose, 20 mM KCl, 10 mM succinate, 1 µg/mg prot rotenone, 2 µg/mg prot oligomycin, 0.2 µg/ml purified catalase and 2 mg prot/ml EDTA particles. Respiration was activated by adding 15 µl 2% H_2O_2 to anaerobic particles. Valinomycin was added at the concentration of 0.4 µg/mg prot to steady-state respiring particles. Final volume 1.5 ml. Final pH 6.9. Temperature 20°C. Where indicated H_2O was completely replaced by D_2O. The redox transitions of b cytochrome are expressed as % of the total functional amount of b cytochromes

induced by valinomycin addition to steady-state respiring sonic particles, caused rapid oxidation of b cytochromes and simultaneous reduction of c cytochromes. However, as the potential was replaced by extra transmembrane ΔpH, cytochrome b was reduced and cytochrome c reoxidized (Papa et al., 1975b).

As shown on the left of Figure 4, valinomycin-induced oxidation of b cytochromes, was from pH 6.9 to 7.6 larger at 566 than 562 nm. However, as the pH was increased above 6.9 the oxidation of b cytochromes decreased and the difference between b_{566} and b_{562} disappeared. There was, on the other hand, no difference in the extent of re-reduction of b_{566} and b_{562}.

It appears then that $\Delta\psi$exerts a crossover effect between b_{566} and b_{562}. However at pHs around 8, where the redox-linked protonation-deprotonation reactions of b cytochromes become important, electron flow is replaced by hydrogen flow and the electrical-field control disappears.

$\Delta\psi$ also controls electron flow from cytochrome c to oxygen (Papa et al., 1974a, b). Thus, cytochromes c and b_{562} lie between two electrical field crossover sites; this would prevent net reduction of b_{562} when $\Delta\psi$ is collapsed.

Hydrogen flow between b_{562} and cytochrome c should be depressed when water of the medium is replaced by heavy water and the crossover effect exerted by $\Delta\psi$ on the b electron well should be kinetically isolated from that exerted at the oxidase region.

The experiment on the right of Figure 4 shows, in fact, that in the presence of D_2O, collapse of membrane potential caused significant oxidation of b_{566} but not of b_{562}.

It can be noted that the peak for the oxidation curve of b_{566} is shifted in the presence of D_2O to a more alkaline pH. This can be ascribed to the expected decreasing effect of D_2O on the ionization constant of acidic group of the cytochrome (Jencks, 1969).

B. III. pH Dependence of the Redox Proton Pump

The crossover experiments provide evidence that b cytochromes mediate vectorial electron or hydrogen flow from the C to the M side of the membrane. Since hydrogen flow along the b well is opposite in direction to the aerobic proton pump, its switching on at pHs around 8 should result in a lowering of the H^+/e^- ration of the proton pump in the b-c_1 region.

Contrasting results are available as far as the pH dependence of the stoichiometry of the redox proton pump is concerned (Papa, 1976).

According to Mitchell and Moyle (1967) and Brand et al. (1976) the pump is pH-independent in the physiological range.

It has, however, been observed that in sonic particles the H^+/e^- ratio decreases as the pH is increased above 6.5 (Papa, 1976).

On the other hand the H^+/e^- ratio for proton release associated to quinol-cytochrome c electron flow in isolated complex III inlaid in liposomes has been found to increase from 1 to 2 as the pH is raised from 7 to 8 (Guerrieri and Nelson, 1975).

Figure 5 shows experiments on the pH dependence of H^+ release from mitochondria induced by oxidation of duroquinol.

The pH traces at the top illustrate attempts to determine the H^+/e^- ratio with the reductant pulse method (Lawford and Garland, 1973). Valinomycin-supplemented aerobic mitochondria are pulsed with duroquinol and the extent of proton release is calculated from the pH traces by extrapolation from the proton back flow. The extent of proton release decreased as the pH was increased from 6.2 to 9.2. The apparent decrease of the H^+/e^- ratio is, however, mainly due to the fact that as the pH was raised the rate of proton back flow increased so markedly that the required extrapolation became unpracticable.

When the initial rate of proton release was measured with a continuous flow pH-meter, a different pH dependence could be observed (see the curves at the bottom of Fig. 5).

The extent of antimycin or KCN-sensitive active proton extrusion at an interval of 180 msec did not change as the pH was raised from 6.2 to 8.1; however, when further raised to 8.6 and 9.2 it abruptly started to increase.

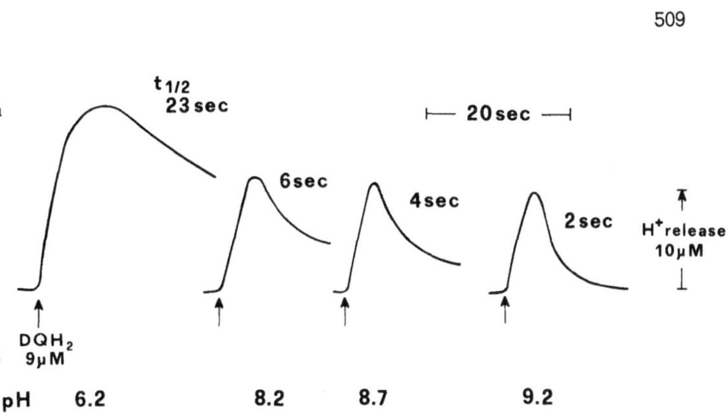

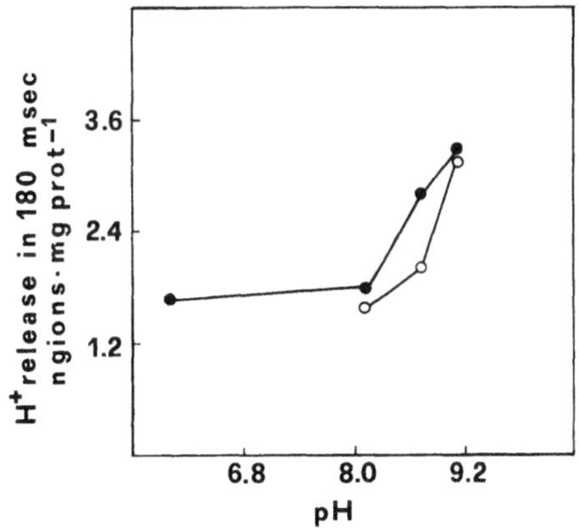

Fig. 5. pH dependence of proton release from aerobic beef-heart mitochondria induced by duroquinol pulses. The top traces indicate static measurements of the proton release obtained by pulsing aerobic beef-heart mitochondria (1 mg prot. ml^{-1}) with 9 μM duroquinol. The curves at the bottom show the pH dependence of the H^+ release measured at 180 msec with a continuous flow pH-meter with a resolution time <2 msec (Papa et al., 1975a). The reaction mixture contained: 150 mM KCL, 5 mM K-malonate, 1 mM glycyl-glycine, 0.2 mM Ethylmaleimide, 1 μg/mg prot oligomycin, 0.5 μg/mg prot vaninomycin, 0.5 μg/mg prot Rotenone. Final volume 1.5 ml. Temperature 25°C. Reaction mixture for the continuous flow pH measurements was same as used for static measurements. Duroquinol was added at a concentration of 100 μM. Experimental values were corrected for antimycin (or KCN) insensitive H^+ release by pulsing the mitochondrial suspension with 100 μM duroquinol in presence of 1 μg/mg prot antimycin (●) or 1 mM KCN (o)

The H^+/e^- ratio can be obtained by dividing the amount of protons translocated at this time by the duroquinol oxidized in the same interval.

The rate of quinol oxidation was determined by measuring polarographically oxygen consumption. Since aerobic oxidation of duroquinol follows in mitochondria Michaelis-Menten kinetics with a K_m of the order of 30-40 μM (Lawford and Garland, 1973), a relatively high concentration (100 μM) of duroquinol was used in these measurements. This maximizes active proton release with respect to passive proton backflow and allows the initial rate of duroquinol oxidation to be measured with good accuracy by the oxygen electrode.

Both H^+ release and duroquinol oxidation were corrected for the antimycin (closed circles) or KCN (open circles) insensitive portion.

Figure 6 shows the pH dependence of the H^+/e^- ratio. The ratio was 2 at pH 6.2, decreased to about 1.6 at pH 8.1 and then sharply increased to 4.5 as the pH was raised to 9.

A possible shortcoming of these experiments might be that when calculating the rate of quinol oxidation from polarographic measurements of oxygen consumption, due to sluggishness of the response of the oxygen electrode, one might fail to detect a transient rapid phase of electron flow immediately following the duroquinol addition.

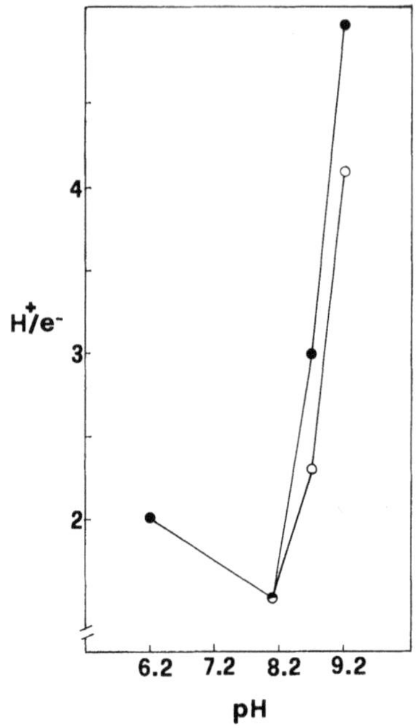

Fig. 6. pH dependance of H^+/e^--ration for duroquinol oxidation in beef-heart mitochondria Ratios obtained by dividing the H^+ released at180 msec by the duroquinol oxidized in same interval. H^+ release was measured with continuous flow pH meter. Amount of duroquinol oxidized was calculated by measuring oxygen consumption with a Clark electrode. Values for curve fitted by closed circles were corrected for antimycin-insensitive H^+ release and duroquinol oxidation. Curve fitted with open circles was corrected for KCN-insensitive H^+ release and duroquinol oxidation. For other experimental conditions and reaction mixtures see legend to Figure 5

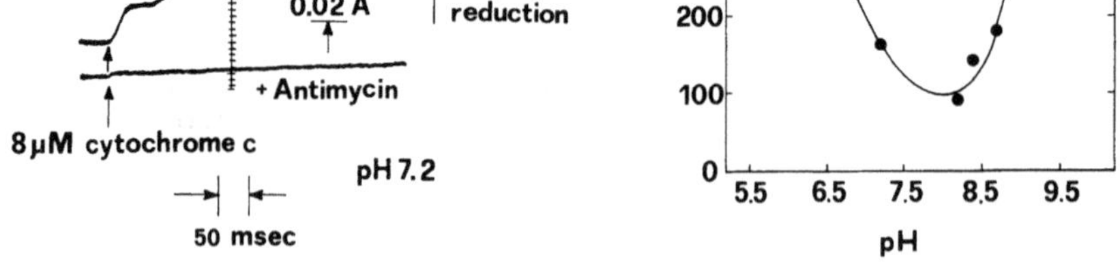

Fig. 7. Kinetics of exogenous cytochrome c reduction by duroquinol in KCN-treated beef-heart mitochondria

Kinetics of cytochrome c reduction was measured at 550-540 nm with stopped flow apparatus (mixing ratio 1:68) applied to a dual-wavelength spectrophotometer (Johnson Foundation Workshop) (Papa et al., 1975a). The main syringe contained 150 mM KCl, 5 mM K-malonate, 1 mM KCN, 2 µg/mg prot oligomycin, 0.5 µg/mg prot valinomycin, 0.5 µg/mg prot rotenone and beef-heart mitochondria (1 mg/ml prot. Duroquinol was added to the mitochondrial suspension at 100 µM final concentration. After 2 min mitochondria were pulsed with cytochrome c, present in side syringe at 0.54 mM solution in 150 mM KCl. Purified beef-heart cytochrome c, mol.wt. 12, 384 (type V Sigma)

We have, therefore, used a second experimental approach by which rapid electron flow from duroquinol to cytochrome c and associated proton release from mitochondria were both directly measured with rapid-flow techniques.

Mitochondria were supplemented with valinomycin plus K^+. Cytochrome oxidase was inhibited with KCN and malonate plus rotenone were also added to inhibit succinate and NADH dehydrogenase. An excess of duroquinol was then added and after a short preincubation to allow redox equilibration mitochondria were rapidly pulsed with 8 μM exogenous cytochrome c.

The stopped-flow double-beam trace presented in Figure 7 shows that added cytochrome c was rapidly and completely reduced with a $t_{1/2}$, at pH 7.2, of 100 msec. This cytochrome c reduction exhibited a small antimycin-insensitive fast phase which amounted to about 1 nano equivalent electron per mg protein. This roughly corresponds to the molar content of the known redox carriers of complex III on the oxygen-side of the antimycin inhibition site. The remaining slower phase of cytochrome c reduction was 95% inhibited by antimycin.

The plot on the right of Figure 7 shows the pH dependence of the rate of antimycin-sensitive cytochrome c reduction by duroquinol. The pH optimum lies around pH 7.5. It can be noted that this bell-shaped pH profile is similar to that exhibited by isolated complex III.

The antimycin-sensitive electron flow from duroquinol to cytochrome was accompanied by synchronous proton release from mitochondria (Fig. 8). To avoid any significant underestimation of active proton ejection an interval of 45 msec was chosen to compute the H^+/e^- ratio proton release associated to the activity of the quinol cyt c oxidoreductase.

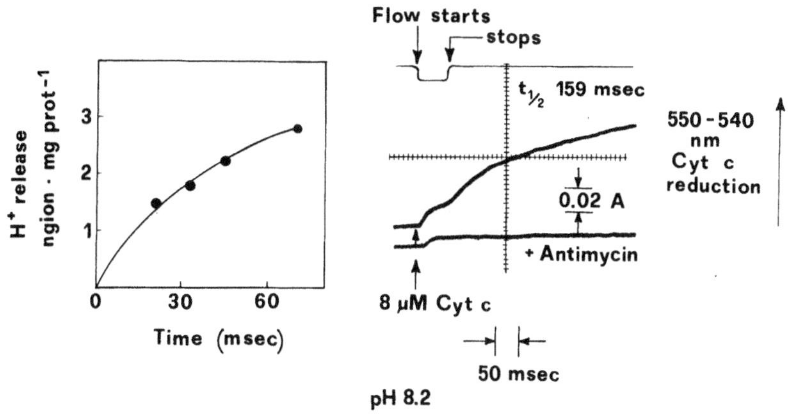

Fig. 8. Kinetics of exogenous cytochrome c reduction and associated proton release in KCN-treated beef-heart mitochondria
Kinetics of proton release obtained with continuous flow apparatus. Kinetics of cytochrome c reduction obtained with stopped flow apparatus. For other experimental details see legend to Figure 7. H^+ release was corrected for antimycin-insensitive reactions

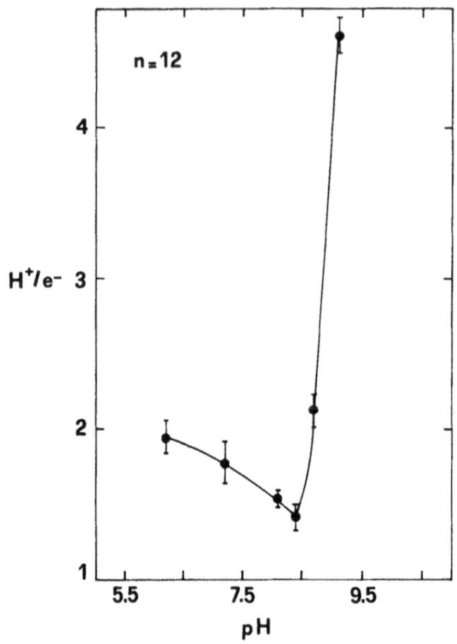

Fig. 9. pH dependance of the H^+/e^- ratio for proton release associated to reduction of exogenous cytochrome c by duroquinol in KCN treated beef-heart mitochondria
H^+/e^- ratios obtained from the measurements indicated in Figure 8 at interval of 45 msec and represent the media of 12 experiments. Vertical bars: S.E.M.
For other details see legend to Figure 8

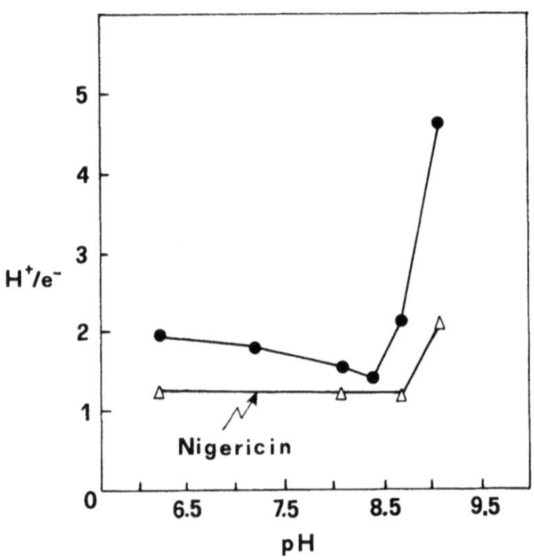

Fig. 10. pH dependence of H^+/e^- ratios for proton release associated to reduction of exogenous cytochrome c by duroquinol in KCN treated mitochondria: Effect of nigericin
For experimental conditions see legend to Figure 8 and 9. ●: Control; Δ: plus 2 μg/mg prot nigericin

Figure 9 shows that the H^+/e^- ratio changed with the pH of the medium with a profile similar to that obtained with the quinol reductant pulse. In fact the ratio was 2 at pH 6.2, decreased to 1.6 as the pH was raised to 8.2, and then sharply increased to 2.5 at pH 8.7 and 4.5 at pH 9.

In the experiment of Figure 10 it was checked with nigericin whether the protons released in excess of net acidification from the overall reaction of duroquinol oxidation by cytochrome c represent vectorial proton translocation rather than scalar deprotonation of acidic groups in the membrane.

It can be seen that in all the pH range tested nigericin lowered the H^+/e^- ratio to around 1. Thus the extra protons appearing in the medium result from vectorial transport across the membrane.

Measurement of the H^+/e^- stoichiometry based upon direct measurement with flow techniques of the rapid initial phase of the elementary steps of the process represents the most direct approach to determine the intrinsic parameters of the proton pump.

This approach should in fact eliminate interferences by secondary proton redistribution across the membrane through leaks and proton-linked solute transporting systems, like the anion/proton symporters (Mitchell, 1961; Papa, 1971) and the alkali metals/proton antiporter (Mitchell and Moyle, 1967; Papa, 1971; Douglas and Cockrell, 1974). The rate of these secondary proton fluxes is in fact negligible with respect to the high rate of electron flow and proton transfer elicited by a pulse of rapid reactants like oxygen, cytochrome c and quinols. On the other hand interference by these reactions is quite important in the experiments as those presented by Brand et al. (1976) where the redox proton pump is activated by a pulse of succinate as reductant. In this case the rate-limiting system for the aerobic proton pump is represented by succinate transport across the membrane, a process whose rate is of the same order as that of proton-anion symports and cation/proton antiport.

Our group has in fact obtained obvious evidence, to be presented elsewhere (Papa et al., 1976) that measurements by Brand et al. (1976) of the H^+/e^- ratios with succinate pulses are affected by secondary solute-linked proton redistribution and cannot be taken as direct measurement of the intrinsic stoichiometry of the redox proton pump.

We checked directly, however, if unrelated secondary proton back flow affected our rapid measurements of the H^+/e^- ratio and its pH dependence.

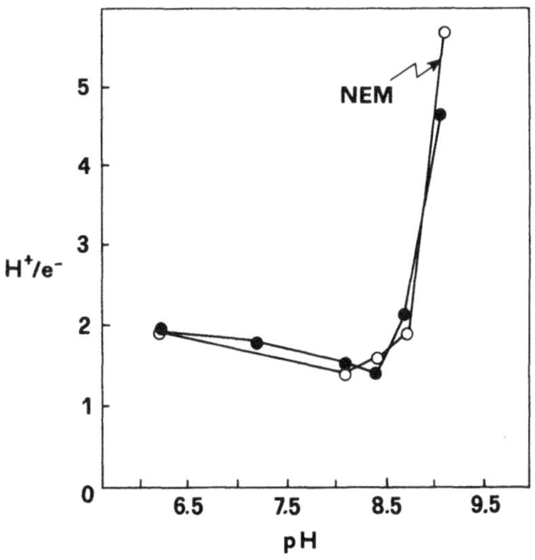

Fig. 11. Effect of N-ethylmaleimide on pH dependence of H^+/e^- ratios for proton release associated to reduction of exogenous cytochrome c by duroquinol in KCN treated beef-heart mitochondria
For experimental conditions see legend to Figure 10. • : Control; o : plus 100 μM N-ethylmaleimide

Table 1. Effect of K^+ Replacement with Li^+ on the H^+/e^- Stoichiometry of the Redox Proton Pump

Expts	Succinate reductant pulse Succinate → O_2 H^+ release	 Electron transfer	 H^+/e^-	Cytochrome c oxidant pulse Duroquinol → cyt. c H^+ release	 Electron transfer	 H^+/e^-
	nEqu.min^{-1}.mg prot^{-1}			nEqu.min^{-1}.mg prot^{-1}		
a	274	194	1.4	3950	2593	1.5
b	373	122	3.1	3224	2205	1.5
c	456	320	1.4			

Succinate reductant pulse:

Expt a) Reaction mixture contained 130 mM KCl, 3 mM Hepes, 0.1 µg.mg prot.$^{-1}$ Valinomycin, 0.5 µg.mg^{-1}prot. rotenone, 30 nmol.mg prot.$^{-1}$N-ethylmaleimide and rat liver mitochondria (2.5 mg prot.ml^{-1}). Final pH 7.2. Final Vol 1.5 ml Temperature 25°C. The reaction was started by addition of 1 mM K-succinate.

Expt b) The same reaction mixture except that 120 mM LiCl plus 10 mM KCl replaced 130 mM KCl. The reaction was started by addition of 1 mM Na succinate.

Expt c) The same reaction mixture of expt (b) except that mitochondria were preincubated 20 min with 1 µg.mg prot^{-1} oligomycin, followed by further 4 min preincubation with 0.5 µg.mg prot^{-1} rotenone. The reaction was started by addition of 1 mM Na-succinate.

Cytochrome reductant pulse:

The experimental conditions are those described in the legend of Figure 8 with the only difference that 130 mM KCl [expt (a)] or 120 mM LiCl plus 10 mM KCl [expt (b)] were used as the main salt component of the reaction medium.

Figure 11 shows that addition of N-ethylmaleimide to block the anion-proton symport systems, had no significant effect on the H^+/e^- ratio. This rules out the possibility that the change of the H^+/e^- ratio is consequent to effect of pH on secondary anion-proton symport.

We have obtained evidence that Li^+ might inhibit the natural cation/proton antiporter (Papa et al., 1973). Thus we tested whether substitution of K^+ with Li in the reaction medium changed the apparent H^+/e^- ratios, measured either with the slow succinate reductant pulse method or the fast cytochrome c pulse method.

The data presented in Table 1 show that suppression of the proton backflow brought about by replacement of K^+ with Li resulted in an increase of the H^+/e^- from 1.4 to 3.1 in the case of the succinate pulse experiment, but had no effect at all on the H^+/e^- ratio measured with the cytochrome c pulse experiment, which in this case amounted with both K^+ or Li, to 1.5.

It is interesting that when mitochondria were preincubated for 20 sec in the presence of oligomycin prior to adding succinate plus rotenone, the H^+/e^- ratio in the presence of LiCl decreased from

3.1 to a value of 1.4 - this ratio being similar to that measured with the cytochrome c pulse. This finding points out that the superstoichiometry of the proton pump obtained with succinate pulses in the Li-medium arises from unrelated proton-release linked to translocation of endogenous solutes.

C. Conclusions

Some interesting indications can be derived from the pH dependence curve of the intrinsic H^+/e^- stoicheiometry of the pump as far its molecular mechanism is concerned.

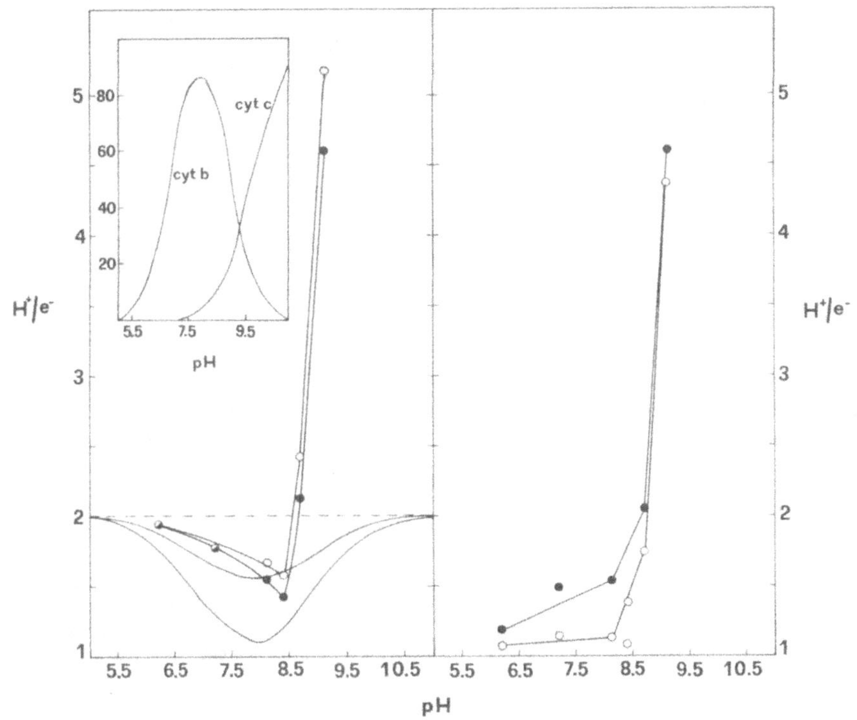

Fig. 12. pH dependence of the H^+/e^- ratios for proton translocation associated to reduction of exogenous cytochrome c by duroquinol in KCN-treated beef-heart mitochondria

Left panel: Closed circles: experimental H^+/e^- ratios; Open circles: H^+/e^- ratios obtained by correcting experimental H^+ release by proton uptake directly associated to reduction of cytochrome c (see insert on upper left). Soft lines at the bottom give the expected pH dependent decrease and subsequent increase of H^+/e^- ratios caused by inward proton pumping by b cytochromes (for other details see text).

Right panel: curves: H^+/e^- ratios calculated on the basis of the extra proton release (in excess with respect to the overall reduction reaction of cytochrome c by duroquinol) corrected for pH dependent H^+ uptake mediated by b cytochromes. Closed circles: 2 hydrogen transported per quinol along the b well; Open circles: transfer of 1 hydrogen

It is very instructive to compare (see Fig. 12) the pH dependence curve of the H^+/e^- stoicheiometry of the proton pump of complex III with the profile, shown in the inset at the upper left, by which b and c cytochromes change from electron to hydrogen carriers as a function of the pH.

First, it has been noted that cytochrome c starts to act as a hydrogen acceptor at pHs above 7. Thus, the measured extent of proton release was corrected for this pH-dependence proton consumption directly associated to cytochrome c reduction.

As shown by the curve fitted by the open circles, this correction did not alter significantly the pH profile of the H^+/e^- ratio.

The soft lines at the bottom show the expected pH dependent decrease and subsequent increase of the H^+/e^- ratio that would be caused by inward proton pumping along the b cytochrome well, in case that one or both of the two hydrogens delivered from the quinol flow along this path from the C to the M side of the membrane. It can be seen that the decrease of the experimental H^+/e^- ratio, going with pH from 6.2 to 8.0, can be explained by inward vectorial hydrogen transfer by b cytochromes. In particular, it corresponds quite well to the curve expected on the basis of only one hydrogen being translocated inward by b cytochromes.

The increase of the experimental H^+/e^- ratios produced by raising the pH above 8.2 cannot, on the other hand, be explained only by the pH dependent decrease of proton linkage to electron transfer on b cytochromes. It is in fact far more steep and leads to H^+/e^- ratios as high as 4.

The H^+/e^- ratios presented on the right of Figure 12 were obtained by correcting the extra H^+ release for pH dependent H^+ uptake mediated by b cytochromes. The curve fitted by closed circles refers to 2 hydrogens transported per quinol along this well, the open circles to transport of 1 hydrogen.

The two curves are not much different. Thus these assumptions are not critical for our main argument. The H^+/e^- ratios so corrected should be intrinsic for proton coupling to electron transfer at the redox carrier of complex III responsible for the proton pump. The pH-dependency curves are indicative of the existence in complex III of multiple ionizable groups whose pKs are linked to the redox state of the electron transfer center of a redox carrier.

Figure 13 shows a diagram which summarizes the observations presented; b cytochromes are apparently arranged so as to form a vectorial redox well across the membrane. Cytochrome b - probably b_{566} - exhibit proton coupling to electron transfer along this vectorial well, and consequent pH dependent inward proton translocation.

It is difficult to explain the specific pH dependence of the H^+/e^- ratio in terms of linear (Mitchell, 1972) or cyclic schemes (Mitchell, 1975a, b) of the chemiosmotic mechanism; nor does this mechanism explain the observation that the H^+/e^- ratio can exceed the value of 2 at alkaline pHs.

These findings are, on the other hand, consistent with the vectorial Bohr mechanism. It might be thought that conformational changes induced in the protein by electron arrival to and removal from the metal of a redox carrier, results in the opening of a channel which

Fig. 13. Scheme describing proton motive function of the mitochondrial cytochrome system

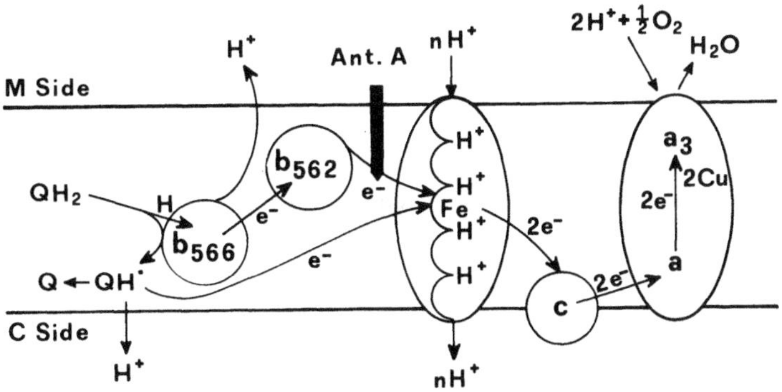

conducts H^+ from the M to the C side of the membrane. It can be conceived that reduction of the electron-accepting center(s) causes the pK values of the aminoacid residues lined along the channel to shift to higher values going from the M to the C side and that oxidation is accompanied by decrease of the pK of acidic group(s) exposed at the C surface. This specific model may be too speculative; however, what is presented in this paper gives an example of the possible functional implications of cooperative linkage between the redox state of the metal, conformational state of apoprotein and protolytic equilibria in membrane bound hemoproteins and metalloproteins.

Aknowledgments. The skillful technical assistance of M. Minuto is gratefully acknowledged. This work has been supported by Grant No. 75.00271.11 of "Consiglio Nazionale delle Ricerche" Rome, Italy.

References

Albracht, S.P.J., Slater, E.C.: EPR studies at 20°K on the mitochondrial respiratory chain. Biochim. Biophys. Acta 245, 503-507 (1971)

Baum, H., Rieske, J.S., Silman, H.I., Lipton, S.H.: On the mechanism of electron transfer in complex III of the electron transfer chain. Proc. Natl. Acad. Sci. U.S. 57, 798-805 (1967)

Brand, M.D., Reynafarje, B., Lehninger, A.L.: The stoichiometry, relationship between energy-dependent proton ejection and electron transport in mitochondria. Proc. Natl. Acad. Sci. U.S. 73, 1-18 (1976)

Chance, B.: The nature of electron transfer and energy coupling reactions. FEBS Lett. 23, 3-20 (1972)

Clark, W.M.: Oxidation-Reduction Potentials of Organic Systems. Baltimore Md.: Waverley Press, 1960

Das Gupta, U., Rieske, J.S.: Identification of a protein component of the antimycin-binding site of the respiratory chain by photoaffinity labeling. Biochem. Biophys. Res. Commun. 54, 1247-1254 (1973)

Douglas, M.G., Cockrell, R.S.: Mitochondrial cation-hydrogen ion exchange - Sodium selective transport by mitochondria and submitochondrial particles. J. Biol. Chem. 249, 5464-5471 (1974)

Dutton, P.L., Wilson, D.F.: Redox potentiometry in mitochondrial and photosynthetic bioenergetics. Biochim. Biophys. Acta 346, 165-212 (1974)

Gellerfors, P., Nelson, B.D.: Analysis of the peptide composition of purified beef-heart complex III by dodecylsulfate electrophoresis. Europ. J. Biochem. 52, 433-443 (1975)

Green, D.E., Järnefelt, J., Tisdale, H.D.: Studies on the electron transport system. XIV. The isolation and properties of soluble cytochrome c. Biochim. Biophys. Acta 31, 34-46 (1959)

Guerrieri, F., Nelson, B.D.: Studies on the characteristics of a proton pump in phospholipid vesicles inlayed with purified complex III from beef-heart mitochondria. FEBS Lett. 54, 339-342 (1975)

Hinkle, P.C., Kim, J.J.: Chemiosmotic coupling in cytochrome oxidase vesicles. In: Mechanisms in Bioenergetics. Azzone, G.F., Ernster, L., Papa, S., Quagliariello, E., Siliprandi, N. (eds.). London: Academic Press, 1973; pp. 53-60

Jencks, W.P.: Catalysts in Chemistry and Enzymology. New York: McGraw Hill, 1969, pp. 243-281

Kilmartin, J.V., Rossi-Bernardi, L.: Interaction of hemoglobin with hydrogen ions, carbon dioxide and organic phosphates. Physiol. Rev. 53, 836-889 (1973)

Klingenberg, M.: The respiratory chain. In: Biological Oxidation. Singer, P.T. (ed.). New York: Interscience (1968), pp. 1-49

Lawford, H.G., Garland, P.B.: Proton translocation coupled to quinol oxidation in ox-heart mitochondria. Biochem J. 136, 711-720 (1973)

Mitchell, P.: Coupling of phosphorylation to electron and hydrogen transfer by a chemiosmotic type of mechanism. Nature (London) 191, 144-148 (1961)

Mitchell, P.: Chemiosmotic coupling in oxidative and photosynthetic phosphorylation. Biol. Rev. 41, 445 (1966)

Mitchell, P.: Structural and functional organisation of energy-transducing membranes and their ion-conducting properties. In: Mitochondria/Biomembranes. In: Proc. 8th FEBS Meeting, Amsterdam. Van Den Bergh, S.G., Borst, P., Van Deenen, L.L.M., Reimersma, J.C., Slater, E.C., Tager, J.M. (eds.). Amsterdam: North Holland-American Elsevier, 1972, Vol. XXVIII, pp. 353-370

Mitchell, P.: Protonmotive redox mechanism of the cytochrome b-c_1 complex in the respiratory chain: Protonmotive ubiquinone cycle. FEBS Lett. 56, 1-6 (1975a)

Mitchell, P.: Protonmotive function of cytochrome systems. In: Electron Transfer Chains and Oxidative Phosphorylation. Quagliariello, E., Papa, S., Palmieri, F., Slater, E.C., Siliprandi, N. (eds.). Amsterdam: North Holland, Publ. Co., 1975b, pp. 305-327

Mitchell, P., Moyle, J.: Respiration-driven proton translocation in rat liver mitochondria. Biochem. J. 105, 1147-1162 (1967)

Mitchell, P., Moyle, J.: The intrinsic anisotropy of the cytochrome oxidase region of the mitochondrial respiratory chain and the consequent vectorial property of respiration. In: Electron Transport and Energy Conservation. Tager, J.M., Papa, S., Quagliariello, E., Slater, E.C. (eds.). Bari: Adriatica Editrice, 1970, pp. 575-587

Nicholls, P., Chance, B.: Cytochrome c oxidase. In: Molecular Mechanisms of Oxygen Activation. Hayaishi, O. (ed.). New York: Academic Press, 1974, pp. 479-534

Nishibayashi-Yamashita, H., Cunningham, C., Racker, E.: Resolution and Reconstitution of the Mitochondrial Electron Transport System. III. Order of reconstitution and requirement for a new factor for respiration. J. Biol. Chem. 247, 698-704 (1972)

Ohnishi, T., Wilson, D.F., Asakura, T., Chance, B.: Studies on iron-sulfur proteins in the site I region of the respiratory chain in pigeon heart mitochondria and submitochondrial particles. Biochem. Biophys. Res. Commun. 46, 1631-1638 (1972)

Papa, S.: Coupling mechanism in ion translocation across the inner mitochondrial membrane. In: Energy Transduction in Respiration and Photosynthesis. Quagliariello, E., Papa, S., Rossi, C.S. (eds.). Bari: Adriatica Editrice, 1971, pp. 173-203

Papa, S., Proton translocation reactions in the respiratory chains. Biochim. Biophys. Acta 456, 39-84 (1976)

Papa, S., Guerrieri, F., Lorusso, M., Simone, S.: Proton translocation and energy transduction in mitochondria. Biochimie 55, 703-716 (1973)

Papa, S., Guerrieri, F., Lorusso, M.: Mechanism of respiration-driven proton translocation in the inner mitochondrial membrane. Analysis of proton translocation associated to oxido-reductions of the oxygen-terminal respiratory carriers. Biochim. Biophys. Acta 357, 181-192 (1974a)

Papa, S., Guerrieri, F., Lorusso, M.: The role of respiratory carriers in the mitochondrial proton pump. In: Membrane Proteins in Transport and Phosphorylation. Azzone, G.F., Klingenberg, M.E., Quagliariello, E., Siliprandi, N. (eds.). Amsterdam: North Holland, Publ. Co., 1974b, pp. 177-186

Papa, S., Guerrieri, F., Lorusso, M.: On the mechanism of the mitochondrial proton pump. In: Dynamics of Energy-Transducing Membranes. Ernster, L., Estabrock, R.W., Slater, E.C. (eds.). Amsterdam: Elsevier, 1974c, pp. 417-432

Papa, S., Guerrieri, F., Capuano, F., Izzo, G., Boffoli, D.: On the stoichiometry of the mitochondrial redox-proton pump. 1976, to be submitted

Papa, S., Lorusso, M., Guerrieri, F.: Mechanism of respiration-driven proton translocation in the inner mitochondrial membrane. Analysis of proton translocation associated with oxidation of endogenous ubiquinol. Biochim. Biophys. Acta 387, 425-440 (1975a)

Papa, S., Lorusso, M., Guerrieri, F. Izzo, G.: On the mechanism of electron flow and proton translocation in the ubiquinone-cytochrome c span of the respiratory chain. In: Electron Transfer Chains and Oxidative Phosphorylation. Quagliariello, E., Papa, S., Palmieri, F., Slater, E.C., Siliprandi, N. (eds.). Amsterdam: Elsevier, 1975b, pp. 317-327

Pettigrew, G.W., Meyer, T.E., Bartsch, R.G., Kamen, M.D.: pH dependence of the oxidation-reduction potential of cytochrome c_2. Biochim. Biophys. Acta 430, 197-208 (1975)

Rodkey, F.L., Ball, E.G.: Oxidation-reduction potentials of the cytochrome c system. Arch. Biochem. 12, 17-28 (1949)

Seki, S., Oda, T.: Studies on cytochrome oxidase. II Ultrastructure of cytochrome oxidase. Arch. Biochem. Biophys. 138, 122-134 (1970)

Slater, E.C.: Electron transfer and energy conservation. In: Dynamics of Energy-Transducing Membranes. Ernster, L., Estabrock, R.W., Slater, E.C. (eds.). Amsterdam: Elsevier 1974, pp. 1-20

Straub, J.P., Colpa-Boonstra, J.P.: The effect of pH on the oxidation-reduction potential of cytochrome b in heart muscle preparations. Biochim. Biophys. Acta 60, 650-652 (1962)

Urban, P.F., Klingenberg, M.: On the redox potentials of ubiquinone and cytochrome b in the respiratory chain. Europ. J. Biochem. 9, 519-525 (1969)

Vanderkooi, G., Senior, A.E., Capaldi, R.A., Hayashi, H.: Biological membrane structure. III. The Lattice structure of membranous cytochrome oxidase. Biochim. Biophys. Acta 274, 38-48 (1972)

Vernon, L.P., Kamen, M.D.: Hematin compounds in photosynthetic bacteria. J. Biol. Chem. 211, 643-675 (1954)

Wikström, M.K.F.: Energy linked change in the redox state and absorption spectrum of cytochrome a in situ. Biochim. Biophys. Acta 283, 385-390 (1972)

Wyman, J.: Regulation in macromolecules as illustrated by haemoglobin. Quart. Rev. Biophys. 1, 35-81 (1968)

The Stoichiometric Coupling of H^+ Efflux During Electron Transport to H^+ Influx During ATP Synthesis by Mitochondria

M. D. BRAND, A. L. LEHNINGER, and B. REYNAFARJE

A. Introduction

It is now widely appreciated that an electrochemical gradient of H^+ generated at the expense of electron transport is the immediate energy source for the transport of various metabolites across the membranes of mitochondria, chloroplasts, and bacteria. Moreover, such proton gradients are also believed to be the immediate driving force for the coupled phosphorylation of ADP. The stoichiometries of H^+ translocation in relation to electron flow and ATP synthesis and hydrolysis are, therefore, fundamental parameters in a great many membrane-associated energy-transformations.

For some years, after the experiments of Mitchell and Moyle (Mitchell and Moyle, 1965, 1967), it has been generally held that the H^+/site ratio (number of H^+ ejected per $2e^-$ per site) of mitochondrial electron transport is 2.0. However, early findings in this laboratory (Rossi and Lehninger, 1963, 1964) which demonstrated that the Ca^{2+} site accumulation ratio of mitochondrial Ca^{2+} transport is 2.0 (i.e. 4 positive charges transported per site), did not appear to be compatible with an H^+/site ratio of 2.0. With the more recent development of knowledge indicating that Ca^{2+} enters respiring mitochondria by an electrophoretic Ca^{2+} uniport (Selwyn et al., 1970; Lehninger, 1974; Rottenberg and Scarpa, 1974), as well as increasing thermodynamic evidence that an H^+/site ratio of 2.0 is inadequate to sustain oxidative phosphorylation against known phosphorylation potentials (Nicholls, 1974; Rottenberg, 1975; Wiechmann et al., 1975; Johnson and Hansford, 1976; Papa, 1976), we have been encouraged to reexamine the H^+/site stoichiometric ratio of mitochondrial electron transport and the H^+/ATP ratio of ATP hydrolysis.

In this paper we describe the results of three different experimental approaches to the question of the stoichiometry of H^+ ejection by the respiratory chain. They show that the H^+/site ratio of 2.0 demonstrated in oxygen pulse measurements is an underestimate, whose origin has been traced to H^+ movements associated with the movements of endogenous phosphate. The true intrinsic H^+/site ratio of mitochondrial electron transport is shown to be at least 3.0 and possibly as high as 4.0.

We also present experiments designed to measure the H^+/ATP ratio in intact mitochondria, and show that this ratio is 2.0, in agreement with other groups (Mitchell and Moyle, 1965, 1968; Moyle and Mitchell, 1973; Thayer and Hinkle, 1973). These values of the H^+/site and H^+/ATP ratios are incorporated in a chemiosmotic scheme in which proton-linked transmembrane movements of phosphate and adenine nucleotides are an integral part of oxidative phosphorylation.

B. Results

I. Measurement of the H^+/site Ratio

1. The H^+/site Ratio Derived from Measurements of Uptake of Ca^{2+} and Weak Acid Anions

A major uncertainty in arriving at the H^+/site ratio of mitochondrial electron transport from measurements of Ca^{2+} uptake by respiring mitochondria is posed by the fact that phosphate has usually been the counteranion in Ca^{2+} transport studies (Rossi and Lehninger, 1963, 1964; Lehninger et al., 1967). Several difficulties attend use of phosphate. Phosphoric acid has multiple ionization stages, the pH of the mitochondrial matrix is not known with accuracy, and H^+ exchanges occur as $Ca_3(PO_4)_2$ is precipitated in the matrix (Lehninger, 1970). We have circumvented these difficulties by replacing phosphate with the anions of various weak aliphatic monocarboxylic acids. It is well known that such anions as acetate, propionate, butyrate, and probably β-hydroxybutyrate pass through biological membranes, including the mitochondrial membrane, in protonated form as the undissociated free acids; no specific membrane transport systems are known for the aliphatic acids. Since these acids also have only a single ionization stage and form relatively soluble salts with Ca^{2+}, there are no complications of the kind described above. Thus each equivalent of weak acid anion entering the mitochondria with Ca^{2+} must take with it a proton. In the alkaline matrix these weak lipophilic acids dissociate to yield their anions; the resulting protons combine with the OH^- generated in the mitochondrial matrix by electron transport.

In our experiments rat liver mitochondria were suspended in an isotonic medium weakly buffered with Tris or Hepes buffer supplemented with $^{45}Ca^{2+}$ and the sodium or potassium salt of the ^{14}C-labeled weak acid. The respiratory substrate was succinate and the system contained rotenone to abolish electron transport from endogenous NAD-linked substrates. Oxygen consumption and proton movements were recorded during the ensuing respiratory stimulation. At various time intervals the mitochondria were recovered by centrifugation and the ^{14}C-anion and $^{45}Ca^{2+}$ in the mitochondrial pellet as well as in the remaining medium were determined. Corrections were applied for the extra-mitochondrial water space of the pellets with ^3H-sucrose (Brand et al., 1976).

The amounts of Ca^{2+} taken up in such experiments were quite large, up to 250 ng-atoms per mg protein, to minimize the contribution of superstoichiometric Ca^{2+} uptake, which is relatively small (Reynafarje and Lehninger, 1974) and which is in any case abolished by oligomycin (Brand and Lehninger, 1975). By this procedure the weak acid anion/Ca^{2+} ratio was found to be 2.0, indicating that for every Ca^{2+} accumulated 2 A^- also entered to yield the soluble CaA_2 salt in the matrix. Table 1 shows the results of such experiments with β-hydroxybutyrate, acetate, propionate, and butyrate as anions. In all cases the anion$^-$/Ca^{2+} ratio was closed to 2.0.

Further experiments have shown that under these conditions 1.7 to 2.0 Ca^{2+} ions are accumulated per site, in agreement with studies carried out in the presence of phosphate (Rossi and Lehninger, 1963, 1964; Lehninger et al., 1967). These data clearly show that 3.5 to 4.0 positive charges in the form of Ca^{2+} ions, as well as an equal number of negative charges in the form of monocarboxylate

Table 1. Stoichiometry of Weak Acid Anion and Ca^{2+} Accumulation by Rat Liver Mitochondria

Mitochondria (2.5 mg protein per ml) were incubated for 10 min in medium containing 120 mM KCl; 3 mM Hepes; 2.5 mM succinate; 2.5 μM rotenone; 150 $\mu g.ml^{-1}$ bovine serum albumin; 400 ng-ion Ca^{2+} mg $protein^{-1}$; 20 mM ^{14}C-labeled acid (10 μCi mM^{-1}); [^{3}H]-sucrose (40 μCi mM^{-1}) at pH 7.2, 25°C. At various times samples were collected by centrifugation, analyzed for radioactivity, and the mitochondrial content of weak acid anion and Ca^{2+} was calculated. The anion/Ca^{2+} ratio was then determined as previously described (Brand et al., 1976)

Weak Acid Anion	Anion/Ca^{2+} Ratio	± S.D.
β-Hydroxybutyrate	2.0	± 0.1
Acetate	1.7	± 0.1
Propionate	2.2	± 0.2
Butyrate	2.0	± 0.2

anions, were transported per $2e^-$ passing through each energy-conserving site. There was no net H^+ movement between the mitochondria and the medium directly associated with Ca^{2+} and BOH^- uptake.

We may now construct vectorial equations to account for the observed stoichiometry, assuming that Ca^{2+} enters by La^{3+}-sensitive eletrophoretic uniport (Selwyn et al., 1970; Lehninger, 1974; Rottenberg and Scarpa, 1974). The observed stoichiometry is most simply accounted for by assuming electron transport causes ejection of 4 H^+ (i.e. 4 positive charges) per site (Fig. 1). Ejection of 4 H^+ can pull in four monocarboxylate anions in the form of the undissociated weak acid (HA), which become 4 A^- on loss of protons to the 4 OH^- generated in the matrix. It will be noted that if 4 H^+ are ejected and used to bring in 4 BOH^- anions there would be no net respiration-dependent H^+ ejection associated with respiration-energized Ca^{2+} and BOH^- uptake, as was observed. The transport of 2 Ca^{2+} per ∿ during mitochondrial electron transport is thus in full agreement with the transport of 2 Ca^{2+} per ∿ for the Ca^{2+}-stimulated ATPase of sarcoplasmic reticulum (MacLennan and Holland, 1975) and for the ATP-dependent Ca^{2+} transport by rat liver mitochondria (Bielawski and Lehninger, 1966). There is, therefore, an essential unity in the stoichiometry of these different Ca^{2+}-transporting systems.

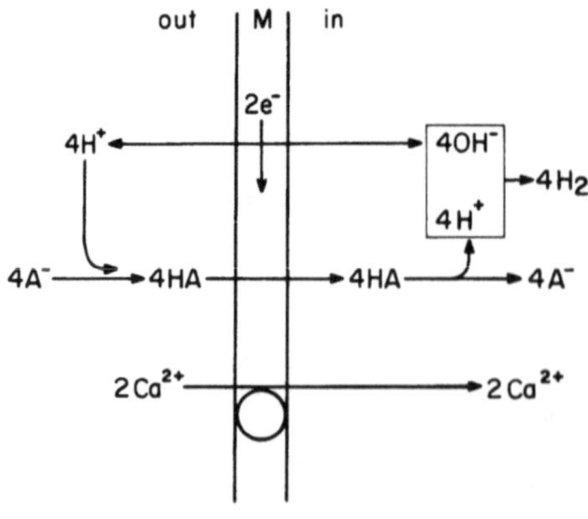

Fig. 1. Scheme to account for the anion/Ca^{2+} ratio of 2.0

Although it has been suggested that Ca^{2+} entry into respiring mitochondria is by means of a Ca^{2+}-proton exchange, such a pathway of entry in aerobic mitochondria has no basis of evidence and is excluded by the osmotic swelling experiments mentioned above (Selwyn et al., 1970; Lehninger, 1974). Moreover, any uncertainty associated with the mechanism of Ca^{2+} transport is in any case removed by the experiments described below, which reveal that the H^+/site ratio exceeds 2.0 under conditions in which Ca^{2+} is not a participant in the transmembrane exchanges.

2. The H^+/site Ratio Derived from Oxygen Pulse Experiments

Recently we decided (Brand et al., 1976a, 1976b) to reevaluate the H^+/site ratio as determined by the oxygen pulse method of Mitchell and Moyle (Mitchell and Moyle, 1965, 1967), in order to determine why the Ca^{2+} method yielded an H^+/site ratio of about 4.0, whereas Mitchell and Moyle reported the value 2.0. Rat liver mitochondria were incubated in a medium of KCl with succinate and rotenone (to inhibit endogenous electron transport) in a glass cell closed from the atmosphere and equipped with a combination glass electrode. The system also contained valinomycin to allow the inward movement of K^+ to compensate electrically for the H^+ ejection. The medium was made anaerobic by flushing with oxygen-free nitrogen. After addition of mitochondria and equilibration of the system for 5-20 min, a small amount of dissolved oxygen was added in the form of a known volume of air-saturated medium. A known amount of electron transport ensues after such an oxygen pulse and is complete within seconds. Simultaneously, the H^+ appearing in the medium is recorded. From the traces the peak H^+ release was extrapolated (Mitchell and Moyle, 1967). Figure 2 shows a typical result of such an experiment, carried out under conditions similar to those described by Mitchell and Moyle (Mitchell and Moyle, 1967). The H^+/site ratio was found to be 2.04, in close agreement with their results. In more than 50 such experiments under these conditions, we have found very close agreement with the results of Mitchell and Moyle.

However, direct measurements of the phosphate concentration of the suspending medium during the anaerobic preincubation revealed that phosphate was lost from the mitochondria during the preincubation period, before the oxygen pulse was added. Inorganic phosphate reached a concentration of 100 μM in the suspension medium in 10 min. It therefore appeared quite probable that reuptake of this phosphate by phosphate-hydroxide transmembrane exchange ensued after addition of oxygen, due to the alkalinization of the matrix, since $H_2PO_4^-$ would enter the mitochondria in exchange for OH^-, a process that is formally equivalent to entry of one H^+ with each $H_2PO_4^-$ ion. The uptake of H^+ coupled to inward movement of phosphate could thus be expected to reduce the observed H^+ ejection.

A simple way of eliminating the phosphate movements was to include in the test system N-ethylmaleimide (NEM), an inhibitor of the phosphate-hydroxide antiporter. Addition of NEM in an amount known to inhibit phosphate transport completely (Johnson and Chappell, 1973; Coty and Pedersen, 1974) caused the observed H^+/site ratio to increase from about 2.0 to about 3.0 (Fig. 2). H^+/site ratios of 3.0 have been consistently observed in a large number of experiments employing NEM with very little deviation. Mersalyl, also an inhibitor of phosphate-hydroxide exchange, gave identical results. Values of 3.0 for the H^+/site ratio have also been observed with the NAD-linked substrates 3-hydroxybutyrate or malate + glutamate.

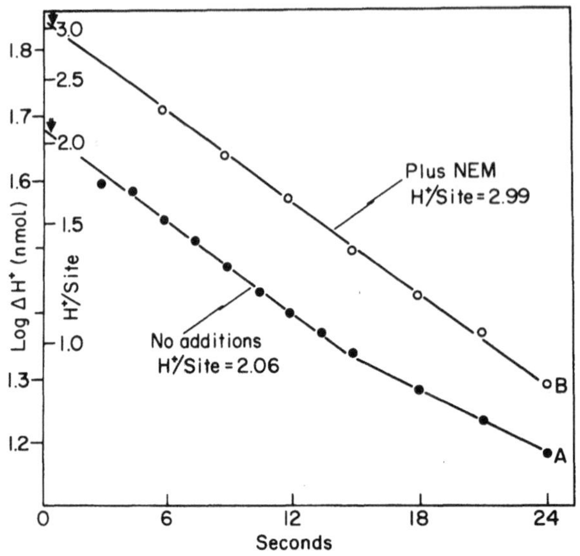

Fig. 2. Determination of H^+/site ratios with the oxygen pulse method. Mitochondria (5 mg protein.mg^{-1}) were incubated anaerobically for 15-20 min in 2 ml of medium containing 120 mM KCl, 3 mM Hepes, 0.5 mM succinate, 5 μM rotenone, and 100 ng valinomycin/mg protein at pH 7.2, 28°C. At zero time H^+ ejection was initiated by injection of 25 μl air-saturated medium (120 mM KCl, 3 mM HEPES). The amount of H^+ ejection (ΔH^+) was calculated from the pH trace obtained. Arrows: time in each experiment at which the added oxygen was calculated to be 50% reduced; the first-order decay of acidity was extrapolated back to this point to calculate the H^+/site ratio. In experiment B N-ethylmaleimide was present at 40 nm.mg protein^{-1}

The ratios obtained either in the presence or absence of NEM were independent of variables such as the amount of oxygen added (below a practical limit), succinate concentration, and pH between 6.5 and 7.5.

That inward movement of phosphate causes spuriously low values of the H^+/site ratio (i.e. about 2.0) has also been demonstrated by depleting mitochondria of endogenous phosphate by incubating them anaerobically, separating and washing them free of released phosphate (and Ca^{2+}) anaerobically, and then adding them to the normal test system for the oxygen pulse measurements. Such phosphate-depleted mitochondria yielded H^+/site ratios approaching 3.0, which were not further increased on addition of NEM. When 100 μM phosphate was added to such phosphate-depleted mitochondria the H^+/site ratios decreased to about 2.0; subsequent addition of NEM returned the H^+/site ratio to near 3.0. We have also found that lowering the temperature to 5°C to slow down the rate of phosphate-hydroxide exchange reaction causes the H^+/site ratio to increase to 3.0 in the absence of NEM. Addition of NEM caused no further increase of the H^+/site ratio at 5°C.

Thus by three different procedures designed to eliminate the movements of endogenous phosphate on the phosphate-hydroxide carrier, the H^+/site ratio is significantly increased from a value of 2.0 to a value of 3.0. Literally hundreds of such experiments have now been carried out and there is little doubt that the intrinsic H^+/

site ratio of mitochondrial electron transport in the absence of phosphate movements must be at least 3.0. We have also excluded the movement of other mitochondrial metabolites by use of appropriate inhibitors. For example, the H^+/site ratio of 3.0 was not influenced by atractyloside, diamox (to block CO_2 accumulation), butylmalonate and other inhibitors to block the movements of dicarboxylate and tricarboxylates, or EGTA to exclude Ca^{2+} uptake.

We are as yet unable to explain why the Ca^{2+} uptake method gives an H^+/site ratio of 3.5 to 4.0 and the oxygen pulse method a value of 3.0. It may be noted, however, that the total proton exchange taking place in the oxygen pulse experiments is of the order of only 5-10 ng-atom H^+ per mg mitochondrial protein, whereas in the Ca^{2+} experiments it is as high as 500 ng-ions H^+ per mg of protein. It is, therefore, conceivable that some minor, but rapid proton-absorbing reaction is causing an additional underestimation of the H^+/site ratio in the oxygen pulse experiments but is not sufficiently large to affect the results of the Ca^{2+} experiments.

3. A Kinetic Method for Measurement of the H^+/site Ratio

In these experiments mitochondria were first preincubated aerobically with rotenone (to block endogenous electron transport), valinomycin, and K^+ (Brand et al., 1976a; Reynafarje et al., 1976). This preincubation results in the oxidation of all electron carriers beyond energy-conserving site 1 and depleted the mitochondria of their energy stores. To initiate electron flow to oxygen succinate is then added and the initial steady-state rates of both oxygen consumption and H^+ ejection are measured from recorder traces made at high chart speeds. Instead of measuring the total amounts of H^+ ejected and oxygen consumed, as in the oxygen pulse experiments, the steady-state rates of H^+ ejection and oxygen consumption were determined and expressed as the H^+/O rate ratio. From the latter the H^+/site ratio was obtained, assuming 2.0 energy-conserving sites for succinate oxidation. Figure 3 shows typical traces of an experiment in a LiCl medium with 10 mM KCl and valinomycin present, in the absence and presence of NEM. The initial velocities of both oxygen consumption and H^+ ejection were linear for at least 8-10 sec in either case. In the absence of NEM the rate of H^+ ejection was close to four times the rate of oxygen consumption, as would be expected for a 2-site substrate and an H^+/site ratio of 2.0. However, in the presence of NEM the rate of H^+ ejection was nearly eight times the rate of oxygen consumption, corresponding to an H^+/site ratio of about 3.8. In such experiments added phosphate depressed the ratio (rate of H^+ ejection)/(rate of oxygen uptake), whereas the addition of NEM increased it. These results are in essential agreement with those obtained by the oxygen pulse method: the H^+/site ratio was about 2.0 in the absence of NEM and increased to 4.0 when phosphate uptake was excluded with NEM.

In these kinetic experiments it is important to emphasize that addition of succinate to rotenone-inhibited, deenergized mitochondria initiated the H^+ ejection and O_2 uptake; there were no preexisting electrochemical gradients that could have compromised the rate ratios observed.

These kinetic experiments therefore yield H^+/site ratio values of almost 4.0, in good agreement with the Ca^{2+} type of measurement described above, but higher than the value of 3.0 observed in the

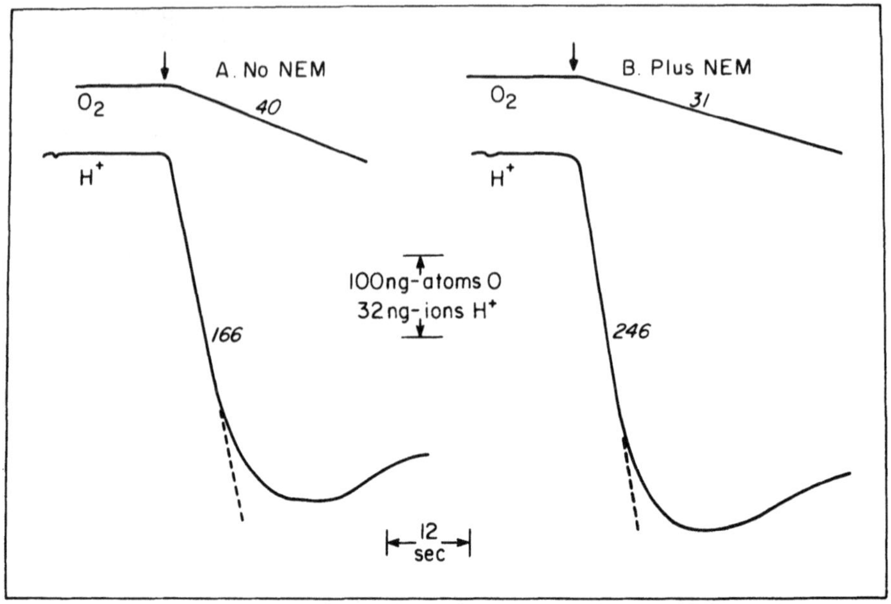

Fig. 3. Determination of H^+/site ratios with the kinetic method. Mitochondria (2.5 mg protein.ml^{-1}) were preincubated aerobically for 2 min at 25°C in 2 ml of medium containing 120 mM LiCl, 10 mM KCl, 3 mM HEPES, pH 7.2, 4 µM rotenone and 1.00 ng valinomycin per mg protein. Where indicated NEM was present at 30 nm.mg protein^{-1}. Respiration was initiated by addition of 1.0 mM sodium succinate, causing oxygen uptake and H^+ ejection. The numbers shown by the traces are rates of O uptake in ng-atoms O min^{-1} and H^+ ejection in ng-ion H^+ min^{-1} mg^{-1}. H^+/site ratios were calculated from the ratio of these rates, as shown

oxygen pulse experiments. Again, no conclusive explanation can be offered for the differences between these methods. However, it may be pointed out that the proton exchanges measured in the kinetic procedure are about twice as great as those taking place in the oxygen pulse method, and also that there is a difference in the redox state of the respiratory chain just before initiation of electron flow.

4. The True Value of the H^+/site Ratio

All three experimental methods described here yield values of the H^+/site ratio of mitochondrial electron transport that are in unmistakable excess over the value of 2.0 originally reported by Mitchell and Moyle (1965, 1967). The Ca^{2+} method and the kinetic method gave values approaching 4.0, the oxygen pulse procedure gave values of 3.0 (Table 2). The common denominator among these three sets of values for the H^+/site ration, which distinguishes them from the earlier measurements, is the exclusion of the movements of endogenous or added phosphate, accomplished by measuring H^+/site ratios under conditions in which phosphate and phosphate movements are avoided.

Our observation that phosphate rapidly passes out of deenergized mitochondria and accumulates in the suspending medium at concentrations of more than 100 µM during the anaerobic preincubation required in oxygen pulse experiments demonstrates the importance

Table 2. Comparison of Observed H^+ and Site Ratios

Method	H^+/site ratio
Ca^{2+}-weak acid	3.5-4.0
Oxygen pulse (NEM)	3.0
Kinetic method (NEM)	4.0
Mitchell and Moyle (23, 24) Oxygen pulse (No NEM) }	2.0

of avoiding or minimizing the participation of phosphate in the ion movements coupled to electron transport. Although Mitchell and Moyle (1967) were not unaware of the potential problems posed by movements of phosphate, their experiments were inadequately controlled for this factor.

II. Measurement of the H^+/ATP Ratio

This ratio has been determined by a variation of the oxygen pulse method described above, in which a small amount of ATP is given in place of the oxygen pulse (Mitchell and Moyle, 1968; Moyle and Mitchell, 1973). Hydrolysis of the ATP is accomplished by ejection of H^+, with an H^+/ATP ratio of 2.0 (Mitchell and Moyle, 1968; Moyle and Mitchell, 1973; Thayer and Hinkle, 1973).

A complication arises, however, owing to the following scalar reaction which causes net H^+ production in addition to the vectorial H^+ movements due to the action of the proton-translocating ATPase:

$$ATP + H_2O \rightarrow ADP + P_i + nH^+$$

The value of n depends on the pH; it is about 0.5 at pH 7.0. It is determined by carrying out an ATP pulse in the presence of an uncoupler, which effectively abolishes vectorial H^+ movements and allows direct measurement of net (scalar) H^+ production.

Mitchell and Moyle (1968, 1973) subtracted the value of n, determined in this way, from the total H^+ ejection observed during an ATP pulse in order to calculate the vectorial component of the observed H^+ ejection. They then used this corrected value to calculate the H^+/ATP ratio.

This correction is invalid because the H^+ produced by the scalar reaction, which occurs in the mitochondrial matrix, will not necessarily appear in the extramitochondrial medium, in which the pH changes are monitored. What will occur, however, is efflux of the $H_2PO_4^-$ produced by ATP hydrolysis, down its concentration gradient in exchange for hydroxide on the phosphate-hydroxide antiporter. This will cause net H^+ ejection which will be uncoupler-sensitive; ionization of the $H_2PO_4^-$ in the extramitochondrial space will cause additional H^+ production which will be uncoupler-insensitive. Such H^+ ejection should be prevented (over short periods of time) by inhibition of the phosphate-hydroxide carrier with NEM or mersalyl; in this case no correction of the observed H^+ ejection should be necessary since the H^+ produced by the scalar reaction will be unable to leave the matrix, and H^+ ejection due to phosphate-hydroxide exchange will be abolished.

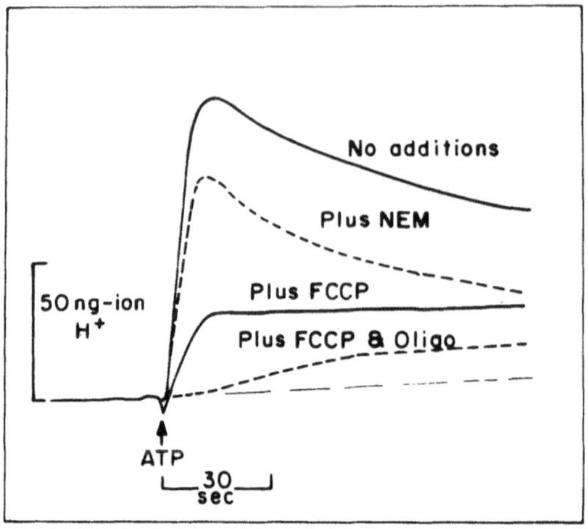

Fig. 4. H^+ ejection following an ATP pulse. Mitochondria (5 mg protein.ml^{-1}) were incubated anaerobically for 15-20 min in 2 ml of medium containing 120 mM KCl, 3 mM HEPES, 1 mM EDTA, 5 μM rotenone, 1 μM antimycin A, 100 ng valinomycin.mg protein^{-1}, and about 2.5 ng aurovertin. mg protein^{-1}, at pH 7.10, 28°C. At zero time H^+ ejection was initiated by injection of 5 μl of 10 mM ATP (50 nm; 5 nm.mg protein^{-1}). The system was calibrated by addition of small amounts of 100 mM HCl. Other additions as shown were 40 nm N-ethylmaleimide/mg protein; 1 μM FCCP (uncoupler); 400 ng oligomycin. mg protein^{-1}

Figure 4 shows experiments designed to measure the H^+/ATP ratio in the presence of NEM. EDTA was present to inhibit adenylate kinase, and aurovertin was added to allow rapid and complete hydrolysis of the ATP (Mitchell and Moyle. 1970; Moyle and Mitchell, 1973). Respiration was prevented by anaerobiosis and respiratory inhibitors, and valinomycin was present to allow K^+ movements to equilibrate charge across the membrane. In the absence of NEM (No additions") injection of a small amount of ATP was followed by rapid appearance of H^+ in the medium, followed by the expected exponential de-

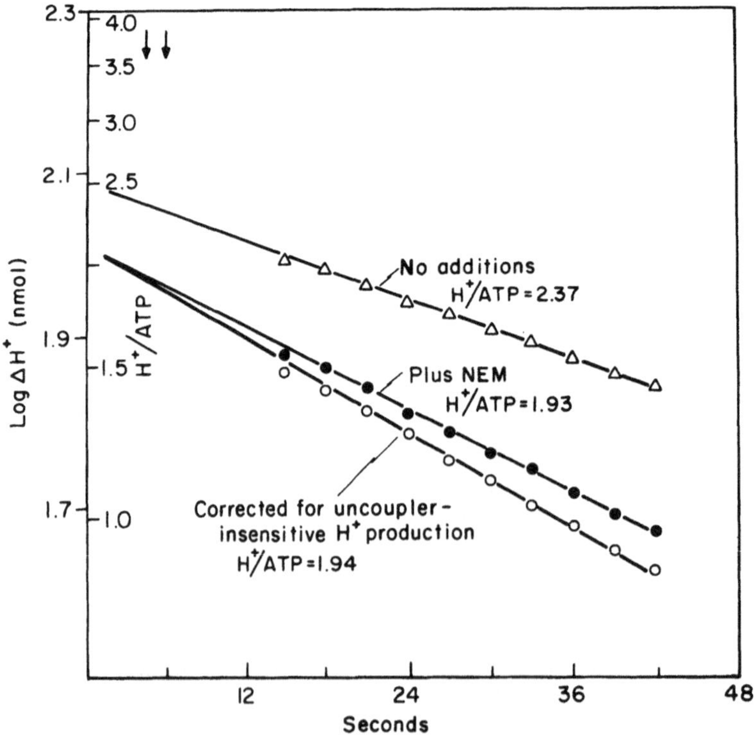

Fig. 5. Determination of H^+/ATP ratio. Data of Figure 4 is plotted to show method of extrapolation. Arrows represent the point at which the ATP hydrolysis was calculated to be half complete in the control (left) and in the presence of NEM (right)

cay of the transmembrane pH gradient. The presence of uncoupler (FCCP) abolished most of the H^+ ejection, and allowed calculation of n. The H^+ ejection was oligomycin-sensitive either in the presence of FCCP or in its absence showing that the H^+ production was due to the action of the mitochondrial ATPase. With NEM present the extent of H^+ ejection was reduced significantly, as predicted above.

The calculation of the H^+/ATP ratios from these curves is shown in Figure 5, in which the exponential decay of acidity was extrapolated back to a time calculated to be halfway through the period of ATP hydrolysis. In the absence of NEM the uncorrected H^+/ATP ratio was 2.4; this was reduced to 1.9 after correction for uncoupler-insensitive H^+ ejection, in confirmation of the results of Mitchell and Moyle (1968, 1973). With NEM present, an experimental design we consider to be more valid, the H^+/ATP ratio was calculated to be 1.9 without any further correction.

Thus the values of about 2.0 we obtain for the H^+/ATP ratio in the presence of NEM are the same as those found using the (invalid) correction of Mitchell and Moyle (1968, 1973), and those obtained for the ATPase in inverted submitochondrial particles (Moyle and Mitchell, 1973; Thayer and Hinkle, 1973) in which the experimental design allows calculation without this correction and in which transmembrane movements of adenine nucleotides and phosphate should not be a factor. The reason that the (invalid) correction applied by Mitchell and Moyle in intact mitochondria gives the same results as addition of NEM, is that, at the pH at which their experiments were carried out, the uncoupler-insensitive scalar H^+ production is of the same magnitude as the H^+ ejection occurring in the absence of uncoupler due to phosphate-hydroxide antiport and subsequent ionizations of extramitochondrial $H_2PO_4^-$. We have found that at lower pH the procedure of Mitchell and Moyle gives H^+/ATP values which rise as high as 3.0 at pH 6 as their correction becomes increasingly inappropriate, whereas in the presence of mersalyl (NEM is less effective at acid pH) to inhibit phosphate-hydroxide exchange, the observed H^+/ATP ratio is constant at 2.0 even at pH 6.

C. Implications

I. Respiratory Chain

The chemiosmotic hypothesis proposes that H^+ ejection by the respiratory chain occurs because of the looped arrangement across the membrane of alternating hydrogen carriers and electron carriers (Mitchell, 1968), although evidence for such an arrangement is available only for the terminal portion of the chain (Hinkle and Mitchell, 1970). In its present formulation such a respiratory chain will eject 2 H^+ at each site. The higher H^+/site ratios of 3.0 or 4.0 that we have demonstrated are not compatible with this arrangement, which must, therefore, be abandoned or modified if our results represent the true stoichiometry of H^+ ejection, and replaced by some as yet unspecified mechanism by which the respiratory carriers may cause H^+ translocation.

II. Coupling Between H^+ Ejection and ATP Synthesis

According to the chemiosmotic hypothesis (Mitchell, 1968) protons ejected from the mitochondria by the respiratory chain subsequently return to the matrix through the reversible proton-translocating ATPase with concomitant synthsis of ATP, which is thus driven by the electro-chemical gradient of H^+. Since (by definition) 1 ATP is synthesized per site it follows that the H^+/site ratio must be equal to the overall H^+/ATP ratio if the hypothesis is correct. If this were not so then net H^+ translocation would occur during oxidative phosphorylation in vivo; this obviously does not occur. Our observation that the H^+/site ratio is 3.0 to 4.0 while the H^+/ATP ratio is 2.0 would thus appear to contradict the chemiosmotic concept of coupling between H^+ ejection by the respiratory chain and H^+ utilization by the ATPase.

The experiments from which we calculate the H^+/ATP ratio are very similar to the oxygen pulse measurements of the H^+/site ratio; any underestimation of the H^+/site ratio due to an undetermined side reaction might also occur in the ATP pulse experiments. Since the experimental approaches are so similar, the H^+/ATP ratio is best compared with the H^+/site ratio derived from oxygen pulse measurements, the problem stated above may then be reformulated "Are H^+/site ratios of 3.0 and H^+/ATP ratios of 2.0 still compatible with a chemiosmotic interpretation of oxidative phosphorylation?" As a note of caution it should be mentioned that in view of the kinetic measurements of the H^+/site ratio we cannot exclude the possibility that both the H^+/ATP (ATP pulse) and H^+/site (oxygen pulse) ratios are underestimated by 1 unit due to some unrecognized side reaction.

A scheme that accommodates this difference between the H^+/site and the H^+/ATP ratios is shown in Figure 6. In essence we propose that 3 H^+ are ejected by the respiratory chain at one site, and 2 H^+ are then used by the ATPase to form 1 ATP. The third H^+ is utilized in transport of the substrates of the reaction (ADP and P_i) into the mitochondria and of intramitochondrial ATP, the product of the reaction, out of the mitochondria. In this way the overall H^+/ATP ratio, including the transport reactions, is 3.0, and equal to the H^+/site ratio.

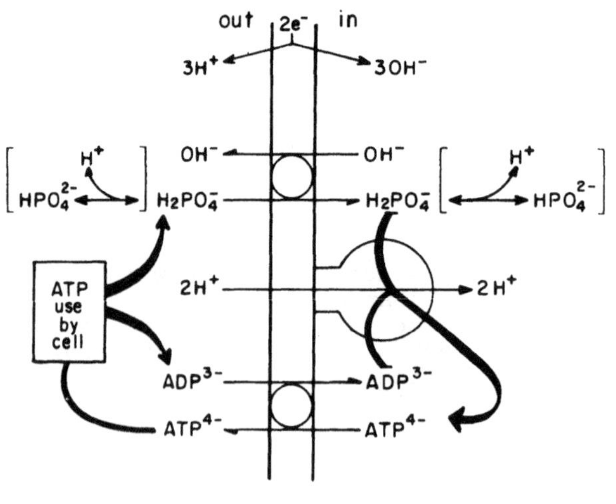

Fig. 6. Functional coupling of the transport of phosphate and adenine nucleotides with ATP synthesis during oxidative phosphorylation in the steady state in vivo

Entry of ADP and P_i in exchange for ATP carries 1 H^+ as follows. $H_2PO_4^-$ exchanges with OH^- on the phosphate-hydroxide antiporter, causing net efflux of 1 OH^-, which is equivalent to net uptake of 1 H^+. This exchange is electroneutral, the charge is carried by the adenine nucleotide carrier allowing entry of ADP^{3-} in exchange for ATP^{4-}. This cycle of transport will occur once for each ATP synthesized in the steady-state. It is important to note that ionization of $H_2PO_4^-$ to give HPO_4^{2-} + H^+ may occur on either side of the membrane, however, the reaction is a "dead end" in the steady state in vivo and cannot contribute to the H^+ fluxes. This is not true in the isolated system where mitochondria are presented with ADP and P_i in this case there will be net loss of HPO_4^{2-} from the medium as ATP is synthesized, with the concomitant alkalinization that is observed under these conditions.

The scheme shown in Figure 6 is based on earlier suggestions of Klingenberg and associates (Klingenberg et al., 1969a, 1969b; Klingenberg, 1975) and Mitchell (Mitchell, 1968). It differs fundamentally in that the H^+/site and H^+/ATP ratios are not identical and thus access of the substrates to the ATPase does not lower the P/O ratio as previously suggested by these authors. The scheme depends on strict $H_2PO_4^-/OH^-$ antiport on the phosphate-hydroxide carrier and strict ADP^{3-}/ATP^{4-} antiport on the adenine nucleotide carrier. These requirements are generally held to be true, although it should be mentioned that both have been questioned (McGivan et al., 1971; Klingenberg, 1975).

The scheme predicts that the overall H^+/ATP ratio will be 3.0 even though the H^+/ATP ratio observed in ATP pulse experiments in the absence of phosphate movements is 2.0. The overall H^+/ATP ratio has been estimated from comparisons of the electrochemical gradient of protons with the phosphate potential with which it should be in equilibrium in state 4. In almost every case a value of close to 3.0 for the H^+/ATP ratio has been observed in this way (Nicholls, 1974; Rottenberg, 1975; Wiechmann et al., 1975; Johnson and Hansford, 1976; Papa, 1976), in good agreement with the predictions of the scheme. In addition the scheme predicts that the H^+/ATP ratio in the mitochondrial matrix should be 1 unit less than that in the medium (i.e. $H^+/ATP_{in} = 2.0$), since 1 of the 3 H^+ and thus 1/3 of the energy for ATP synthesis is invested at the level of translocation of P_i and adenine nucleotides. Several measurements of the difference in phosphorylation potential between the matrix and the extramitochondrial medium are compatible with about this difference between the extramitochondrial and matrix values of the H^+/ATP ratio (Klingenberg, 1970; Heldt et al., 1972; Slater et al., 1973; Baris et al., 1974; Baris and Lumeng, 1975).

D. Conclusions

Our results clearly show that the H^+/site ratio of mitochondrial electron transport is at least 3.0 and possibly as high as 4.0. These experimental findings indicate that the looped arrangement of the respiratory chain proposed by the chemiosmotic hypothesis must be either modified or abandoned in favor of some type of "proton pump."

Our finding that the H^+/ATP ratio is 2.0 even when movements of P_i are prevented leads us to modify the chemiosmotic hypothesis to allow 1/3 of the protonmotive force to be used to translocate phosphate and adenine nucleotides across the mitochondrial membrane during ATP synthesis in the steady state in vivo.

These considerations and modifications strengthen the chemiosmotic hypothesis by removing two of the most pressing criticisms presently leveled at it; (1) the difficulty in identifying a sufficient number of hydrogen carriers in the respiratory chain and in demonstrating a looped arrangement of the respiratory carriers; and (2) the inadequacy of an overall H^+/ATP ratio of 2.0 to maintain or produce the observed phosphorylation potentials for a given protonmotive force. However, they do so at the expense of the loop hypothesis, which must be modified or abandoned in favor of some other type of "proton pump" associated with electron transport.

Acknowledgments. We than Mr Carl Mahle and Mrs Irene Wood for expert technical assistance, and Dr. E.N. Moudrianakis (Johns Hopkins University) for his generous gift of aurovertin. This work was supported by grants from National Institutes of Health (GM 05919) and the National Science Foundation (BMS-75-21923), and a Contract from the National Cancer Institute (NO1-CP-45610).

References

Bielawski, J., Lehninger, A.L.: Stoichiometric relationships in mitochondrial accumulation of calcium and phosphate supported by hydrolysis of adenosine triphosphate. J. Biol. Chem. 241, 4316-4322 (1966)

Brand, M.D., Chen, C-H., Lehninger, A.L.: Stoichiometry of H^+ ejection during respiration-dependent accumulation of Ca^{2+} by rat liver mitochondria. J. Biol. Chem. 251, 968-974 (1976)

Brand, M.D., Lehninger, A.L.: Superstoichiometric Ca^{2+} uptake supported by hydrolysis of endogenous ATP in rat liver mitochondria. J. Biol. Chem. 250, 7958-7960 (1975)

Brand, M.D., Reynafarje, B., Lehninger, A.L.: Stoichiometric relationship between energy-dependent proton ejection and electron transport in mitochondira. Proc. Natl. Acad. Sci. U.S. 73, 437-441 (1976a)

Brand, M.D., Reynafarje, B., Lehninger, A.L.: Re-evaluation of the H^+/site ratio of mitochondrial electron transport with the oxygen pulse technique. J. Biol. Chem. 251, 5670-5679 (1976b)

Coty, W.A., Pedersen, P.L.: Phosphate transport in rat liver mitochondria. J. Biol. Chem. 249, 2593-2598 (1974)

Davis, E. J., Lumeng, L.: Relationships between the phosphorylation potentials generated by liver mitochondria and respiratory state under conditions of adenosine diphosphate control. J. Biol. Chem. 250, 2275-2282 (1975)

Davis, E.J., Lumeng, L., Bottoms, D.: On the relationships between the stoichiometry of oxidative phosphorylation and the phosphorylation potential of rat liver mitochondria as functions of respiratory state. FEBS Lett. 39, 9-12 (1974)

Heldt, H.W., Klingenberg, M., Milovancev, M.: Differences between the ATP/ADP ratios in the mitochondrial matrix and in the extramitochondrial space. Europ. J. Biochem. 30, 434-440 (1972)

Hinkle, P., Mitchell, P.: Effect of membrane potential on equilibrium poise between cytochrome a and cytochrome c in rat liver mitochondria. J. Bioenerg. 1, 45-60 (1970)

Johnson, R.N., Chappell, J.B.: The transport of inorganic phosphate by the mitochondrial dicarboxylate carrier. Biochem. J. 134, 769-774 (1973)

Johnson, R.N., Hansford, R.G.: Protonmotive force (pmf) as an index of energy status in blowfly (Phormia regina) flight muscle mitochondra. Federation Proc. 35, 1556 (1976)

Klingenberg, M.: Metabolite transport in mitochondria: an example for intracellular membrane function. Essays Biochem. 6, 119-159 (1970)

Klingenberg, M.: Energetic aspects of transport of ADP and ATP through the mitochondrial membrane. In: Energy Transformation in Biological Systems. CIBA Found. Symp. Amsterdam: Elsevier, 1975, Vol. XXXI, pp. 105-124

Klingenberg, M., Heldt, H.W., Pfaff, E.: The role of adenine nucleotide translocation in the generation of phosphorylation energy. In: The Energy Level and Metabolic Control in Mitochondria. Papa, S., Tager, J.M., Quagliariello, E., Slater, E.C. (eds.). Bari: Adriatica Editrice, 1969a, pp. 237-253

Klingenberg, M., Wulf, R., Heldt, H.W., Pfaff, E.: Control of adenine nucleotide translocation. In: Mitochondria, Structure and Function. FEBS Symp. Ernster, L., Drahota, Z. (eds.). London-New York: Academic Press, Vol. XVII, pp. 59-77 (1969b)

Lehninger, A.L.: Mitochondria and calcium ion transport. Biochem. J. 119, 129-138 (1970)

Lehninger, A.L.: Role of phosphate and other proton-donating anions in respiration-coupled transport of Ca^{2+} by mitochondria. Proc. Natl. Acad. Sci. U.S. 71, 1520-1524 (1974)

Lehninger, A.L., Carafoli, E., Rossi, C.S.: Energy-linked ion movements in mitochondrial systems. Advan. Enzymol. 29, 259-320 (1967)

MacLennan, D.H., Holland, P.C.: Calcium transport in sarcoplasmic reticulum. Ann. Rev. Biophys. Bioengineering 4, 377-404 (1975)

McGivan, J.D., Grebe, K., Klingenberg, M.: On the coupling between the transport of phosphate and adenine nucleotides in rat liver mitochondria. Biochem. Biophys. Res. Commun. 45, 1533-1541 (1971)

Mitchell, P.: Chemiosmotic Coupling and Energy Transduction. Bodmin: Glynn Research (1968)

Mitchell, P., Moyle, J.: Stoichiometry of proton translocation through the respiratory chain and adenosine triphosphatase systems of rat liver mitochondria. Nature (London) 208, 147-151 (1965)

Mitchell, P., Moyle, J.: Respiration-driven proton translocation in rat liver mitochondria. Biochem. J. 105, 1147-1162 (1967)

Mitchell, P., Moyle, J.: Proton translocation coupled to ATP hydrolysis in rat liver mitochondria. Europ. J. Biochem. 4, 530-539 (1968)

Mitchell, P., Moyle, J.: Influence of aurovertin on affinity of mitochondrial adenosine triphosphatase for ATP and ADP. FEBS Lett. 6, 309-311 (1970)

Moyle, J., Mitchell, P.: Proton translocation quotient for the adenosine triphosphatase of rat liver mitochondria. FEBS Lett. 30, 317-320 (1973)

Nicholls, D.G.: The influence of respiration and ATP hydrolysis on the protonelectrochemical gradient across the inner membrane of rat liver mitochondria as determined by ion distribution. Europ. J. Biochem. 50, 305-315 (1974)

Papa, S.: Proton translocation reactions in the respiratory chains. Biochim. Biophys. Acta 456, 39-84 (1976)

Reynafarje, B., Brand, M.D., Lehninger, A.L.: Evaluation of the H^+/site ratio of mitochondrial electron transport from rate measurements. J. Biol. Chem. (1976, in press)

Reynafarje, B., Lehninger, A.L.: A quantitative analysis of super-stoichiometric H^+ ejection and Ca^{2+} uptake in respiring rat liver mitochondria. J. Biol. Chem. 249, 6067-6073 (1974)

Rossi, C.S., Lehninger, A.L.: Stoichiometric relationships between accumulation of ions by mitochondria and the energy-coupling sites in the respiratory chain. Biochem. Z. 338, 698-713 (1963)

Rossi, C.S., Lehninger, A.L.: Stoichiometry of respiratory stimulation, accumulation of Ca^{2+} and phosphate, and oxidative phosphorylation in rat liver mitochondria. J. Biol. Chem. 239, 3971-3980 (1964)

Rottenberg, H.: The measurement of transmembrane electrochemical proton gradients. J. Bioenerg. 7, 61-74 (1975)

Rottenberg, H., Scarpa, A.: Calcium uptake and membrane potential in mitochondria. Biochemistry 13, 4811-4817 (1974)

Selwyn, M.J., Dawson, A.P., Dunnett, S.J.: Calcium transport in mitochondria. FEBS Lett. 10, 1-5 (1970)

Slater, E.C., Rosing, J., Mol, A.: The phosphorylation potential generated by respiring mitochondria. Biochim. Biophys. Acta 292, 534-553 (1973)

Thayer, W.S., Hinkle, P.C.: Stoichiometry of adenosine triphosphate-driven proton translocation in bovine heart submitochondrial particles. J. Biol. Chem. 248, 5395-5402 (1973)

Wiechmann, A.H.C.A., Beem, E.P., van Dam, K.: The relation between H^+ translocation and ATP synthesis in mitochondria. In: Electron Transfer Chains and Oxidative Phosphorylation. Quagliariello, E., Papa, S., Palmieri, F., Slater, E.C., Siliprandi, E. (eds.). Amsterdam: North-Holland Publ. Co., 1975, pp. 335-342

Mitochondrial Calcium Transport and the Intracellular Calcium Homeostasis

E. Carafoli, M. Crompton, K. Malmström, E. Sigel, M. Salzmann, M. Chiesi, and H. Affolter

Introduction

The concentration of free Ca^{2+} ions in the cytosol is maintained at values which oscillate within narrow limits, and are 3 to 4 orders of magnitude lower than those in the extracellular spaces. The buffering of the Ca^{2+} ions at these low levels is a necessity, dictated by the existence of a large number of reactions which are regulated by Ca^{2+} in the μM concentration range: among them are several enzymatic steps located along key metabolic pathways, various contractile and motile systems (myofibrils, cilia and flagella, microtubules and microfilaments), the release of a large number of hormones, and a long list of membrane-linked functions, ranging from the coupling of excitation to contraction and secretion in muscles and nerve endings, to the exocrine secretion in a variety of cells. The efficacy of the systems responsible for the maintenance of Ca^{2+} ions in the cytosol at these very low levels is remarkable, since Ca^{2+} diffuses continuously from the intracellular spaces into the cell down its concentration gradient. Yet, as first demonstrated by Hodgkin and Keynes (1957), and recently confirmed by Rose and Loewenstein (1975) the penetration of Ca^{2+} into the cell induces only a transient increase of the Ca^{2+}level: the penetrating Ca^{2+} is rapidly sequestered, bound, or extruded. The low level of free Ca^{2+} in the cytosol results from the cooperation of active transport processes across the plasma membrane and across the membrane of intracellular organelles, and/or from the binding of the cation to soluble or insoluble cell components, notably proteins. The extrusion of Ca^{2+} from the cell is carried out by a specific Ca^{2+}-activated ATPase which has been characterized particularly in erythrocytes (Schatzmann, 1966), or by Na^+/Ca^{2+} exchange system which has been studied in detail in heart (Reuter and Seitz, 1968) and axonal (Baker, 1972) membranes. The intracellular sequestration of Ca^{2+} involves the energy-linked transport in the mitochondria of practically all cell types studied (Lehninger et al., 1967; Carafoli and Lehninger, 1971; Carafoli, 1974) and specific ATPases in the sarcoplasmic reticulum of muscle cells (particularly in fast muscles, Weber, 1966; Martonosi, 1972), and in the endoplasmic reticulum of adipose (Hales et al., 1974) and other cells (Moore et al., 1976). Since the membrane area available for transport in the cell organelles greatly exceeds that of the plasma membrane, it is easy to see that cell organelles are the most important components in the rapid regulation of intracellular Ca^{2+}. Their respective quantitative contribution is difficult to evaluate, and is probably variable with the tissues. In general, however, mitochondria deserve the central role, due to their ubiquitous cellular distribution, to their large total mass, and to their well documented ability to store large amounts of Ca^{2+}. In addition, it has been shown that the Ca^{2+} stored in mitochondria can be rapidly released to the extramitochondrial milieu (Drahota et al., 1965; Carafoli, 1967; Carafoli and Lehninger, 1971; Carafoli et al., 1974), presumably in connection with the metabolic Ca^{2+} demands of the cytosol. This observation is of obvious significance from the standpoint of the intracellular Ca^{2+} homeostasis. The maintenance of Ca^{2+} at a constant value (or its oscillations around it) indeed requires that its removal from th- cytosol be a reversible and physiologically modulated process.

The Energy-Linked Uptake of Ca^{2+} by Mitochondria - General Characteristics - The Endogenous Mitochondrial Ca^{2+} Pool

The energy-driven uptake of Ca^{2+} has so far been observed in mitochondria from more than a dozen mammalian tissues, and from some plants (Carafoli and Lehninger, 1971). Although some of the characteristics of the process may vary somewhat with the tissue, the essential properties, which are summarized in Table 1, are constant, and have been described in a number of review articles (Lehninger et al., 1967; Carafoli and Rossi, 1971; Carafoli, 1974). From the standpoint of the intracellular Ca^{2+} homeostasis, some parameters are of particular significance, and will be discussed here in some detail.

The total capacity of the process for Ca^{2+} storage is somewhat variable. As prepared normally in media supplemented with Ca^{2+}-chelators, liver mitochondria contain about 10 nmol of Ca^{2+} per mg of protein, heart mitochondria about twice this value, and smooth muscle mitochondrial up to 150 nmol. It is very unlikely that the endogenous Ca^{2+} pool of isolated mitochondria (or part of it) is due to absorption of Ca^{2+} set free from other organelles and other sources by the homogenization procedure. Indeed, the presence of Ca^{2+} within mitochondria in vivo has been directly demonstrated by microprobe and histochemical studies (Hall, 1975). In addition, studies of the distribution of injected radiostrontium (and presumably radiocalcium) among liver subcellular fractions (Carafoli and Tiozzo, 1967), have shown (Fig. 1) that the percentage of total radioactivity of the homogenate associated with mitochondria decreases from about 60% to about 30% during the first 60 min after the injection. (During the same time, the percentage of radioactivity associated with endoplasmic reticulum increases substantially.) This finding is clearly incompatible with the adventitious association of Ca^{2+} with the mitochondria during the homogenization of the tissue, and indeed suggests that exchanges of Ca^{2+} among subcellular organelles take normally place in vivo.

Table 1. Summary of the Characteristics of Energy-linked Ca^{2+} Transport in Mitochondria

Described in mitochondria from	liver, kidney, heart, skeletal muscle, smooth muscle, chondrocytes, brain, spleen, thyroid, lung tissue, neurospora, yeasts, some plants
Driven by	oxidation of substrates or hydrolysis of ATP
Inhibited by	interruption of energy-flow, low concentrations of lanthanides or ruthenium red. Also by Mg^{2+} (muscle mitochondria)
Stimulated by	simultaneous uptake of permeant anions, particularly phosphate
Stabilized by	adenine nucleotides and Mg^{2+}
Reversed by	interruption of energy-flow, addition of specific Ca^{2+} ionophores

Fig. 1. Association of injected $^{89}Sr^{2+}$ with mitochondria isolated from rat liver
Rats were injected intraperitoneally with a pulse of $^{89}Sr^{2+}$ (Carafoli and Tiozzo, 1967), 5 min before killing. Mitochondrial fraction was separated from the homogenate with a conventional fractionation procedure in 0.25 M sucrose. Data underestimated somewhat the percentage of radioactivity associated with mitochondria, since no correction was applied for the mitochondria associated with the first low-speed sediment, and did not appear in the mitochondrial fraction. Samples were dissolved with aid of a detergent and counted in low-background, gas-flow counter

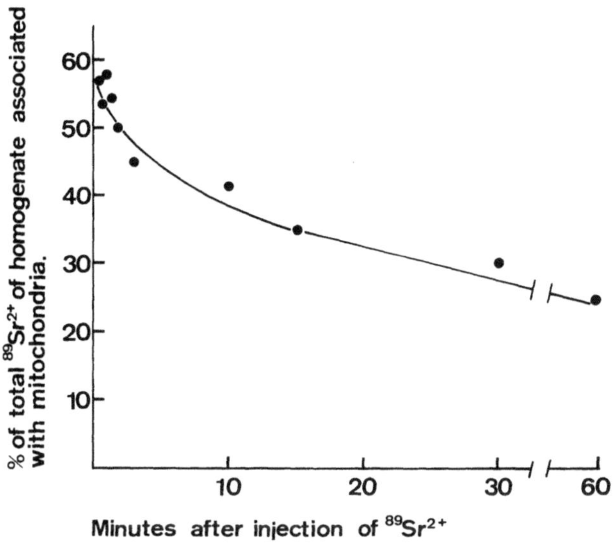

The state of Ca^{2+} in mitochondria in situ is an open matter. The assumption is often found in literature that Ca^{2+} is maintained as a Ca-phosphate precipitate, represented by the dense granules normally observed in the matrix with the electron microscope. However, the presence of Ca^{2+} in the matrix granules has been questioned (Pasquali-Ronchetti et al., 1969), and it has been repeatedly shown (see above) that the endogenous Ca^{2+} can be very rapidly discharged from mitochondria with uncouplers or specific ionophores. It is, therefore, very probable that the endogenous Ca^{2+} pool is not sequestered semi-irreversibly in mitochondria as a salt precipitate, but is maintained in a dynamic state, either in solution, or bound to fixed anionic sites on the membrane. Several questions on the endogenous mitochondrial Ca^{2+} pool, for example the role of adenine nucleotides and Mg^{2+} (Carafoli et al., 1965) still remain unanswered, but are out of scope of this article. Clearly, however, the conclusion that the endogenous mitochondrial Ca^{2+} is rapidly mobile is highly relevant to the problem of the intracellular Ca^{2+} homeostasis.

The total Ca^{2+} storage capacity of mitochondria in some representative tissues is shown in Table 2. It is very probable that the corresponding figures are considerably higher in cells involved in mineralization/demineralization process (osteoblasts, osteoclasts, chondrocytes, egg-shell glands, etc.) but in these tissues at least part of the Ca^{2+} is stored in the form of insoluble deposits, and is thus presumably poorly mobile. On the other hand, the Ca^{2+} pools in the mitochondrial types listed in Table 2, including the very large one present in smooth muscle mitochondria, are promptly and completely mobilized by uncouplers and/or specific ionophores (Fig. 2).

The Ca^{2+} content in mitochondria in vivo rises considerably in nonmineralized tissues exposed to high Ca^{2+} concentrations (Peachey, 1964; Legato et al., 1968), and in pathologically injured

Table 2. Total mitochondrial Ca^{2+} in some representative tissues

Tissue	Mito. content of tissue (mg prot. per gram)	Mito. content of Ca^{2+} (nmol per mg protein	Total mito. Ca^{2+} content in tissue (nmol per g)
Liver	40- 60	10- 12	400 - 720
Heart	80-100	20- 25	1600-2500
Myometrium	14- 20	60-130	840 -2600

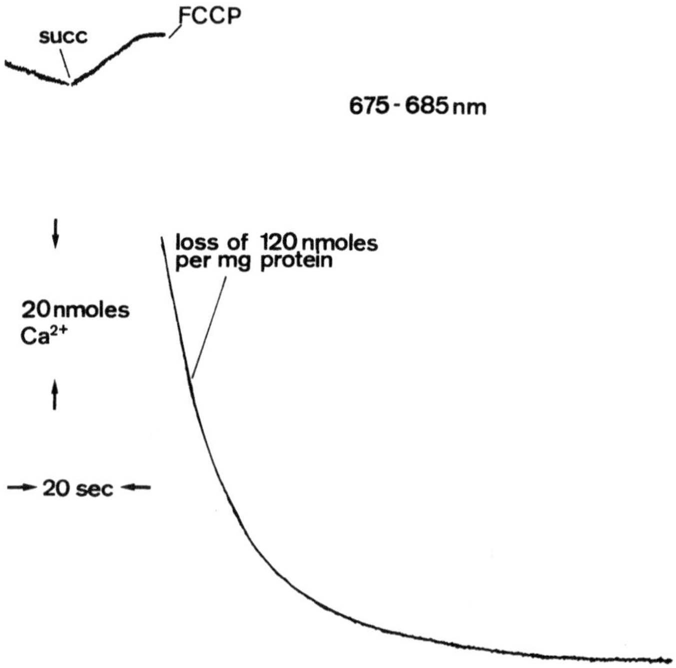

Fig. 2. Rapid release of endogenous Ca^{2+} of human myometrium mitochondria by uncouplers

Mitochondria were prepared from a sample of human myometrium, obtained during a routine hysterectomy operation, in a mannitol (0.21 M), sucrose (0.07 M), Tris-Cl (0.02 M), pH 7.4, Na-EDTA (1.0 mM), Na-EGTA (3.0 mM) medium. The tissue was homogenized in a Polytron PT 20 homogenizer, used for 10 sec with the reostat at position 4, and then with a Polytron PT 10 homogenizer for 5 sec with the reostat at position 5. Mitochondria were collected between 800 g (10 min) and 7000 g (10 min), and washed twice in a MST medium without Ca^{2+}-chelators. They exhibited a respiratory control index by Ca^{2+} of 4 with succinate as substrate, and contained 123 nmoles of endogenous Ca^{2+} per mg of protein. The movements of Ca^{2+} in and out of mitochondria were followed in an Aminco DW-2 Dual Wavelength spectrophotometer, using arsenazo III (Scarpa, 1976) as a Ca^{2+} indicator. The reaction medium contained, in a total volume of 3 ml, 0.12 M KCl, 0.01 M Tris-Cl, pH 7.4, 0.8 ng mitochondrial protein and 80 µM arsenazo. At points indicated 0.005 M Tris-succinate, and 1 µM FCCP were added. The total mitochondrial Ca^{2+} was determined by atomic absorbtion spectrophotometry cells (see Carafoli, 1974, for a review of the pathological conditions leading to mitochondrial calcification), but most of the mitochondrial Ca^{2+} is stored in these cells as an insoluble deposit. In isolated mitochondria, the total amounts of Ca^{2+} can of course increase to impressively high levels after energy-linked uptake, particularly in the presence of phosphate, which penetrates into the matrix to co-precipitate with Ca^{2+}. In the absence of phosphate, the maximal amounts of Ca^{2+} that can be actively accumulated vary considerably with the mitochondrial type. One open problem in the massive loading of mitochondria with Ca^{2+} and phosphate in vitro is the role of adenine nucleotides and Mg^{2+}. Only when they are present in the medium can maximal levels of loading be achieved. In their absence, the accumulated Ca^{2+} and phosphate fail to precipitate optimally, osmotic damage to the mitochondrial structure ensues, and most of the accumulated ions are rapidly lost to the external medium (Carafoli et al., 1965).

The Mechanism of the Energy-Linked Translocation of Ca^{2+}

The mechanism by which Ca^{2+} is actively transported across the inner mitochondrial membrane has been recently the topic of considerable experimentation and discussion. The suggestion by Reed and Bygrave (1975), of a partially charge-compensated Ca^{2+}/H^+ antiport is negated by very recent measurements by Heaton and Nicholls (1976) of the equilibrium distribution of Ca^{2+} between the external medium and the matrix of respiring mitochondria. On the other hand, osmotic swelling studies by Selwyn et al. (1970) suggest penetration as a Ca^{2+} uniport, and recent experiments by Rottenberg and Scarpa (1974) recognize the membrane potential maintained by respiration as the pulling force for the penetration of Ca^{2+}. The conclusion of Rottenberg and Scarpa is based on the comparison of the matrix/medium distributions of $^{45}Ca^{2+}$ and $^{86}Rb^+$ in valinomycin-treated energized mitochondria, and indicate that the electrophoretic uniport system transfers 2 charges per cycle. These conclusions, together with the generally accepted assumption that respiring mitochondria maintain a membrane potential, negative inside, of about 180 mv, lead to some considerations of importance to the problem of the intracellular homeostasis of Ca^{2+}. For a membrane potential of 180 mv and a charge transfer of 2, the Nernst equation predicts a matrix/medium equilibrium distribution ratio of Ca^{2+} of 10^6. A Ca^{2+} content of 10 nmol per mg of protein would on the other hand correspond to a total concentration of Ca^{2+} in the mitochondrial water of about 10^{-2} M, and to a probable concentration of free Ca^{2+} ions not far from 10^{-4} M. Thus, if the electrophoretic uniport of Ca^{2+} were to reach equilibrium uncomplicated by reactions leading to the rapid release of the accumulated cation, the concentration of free Ca^{2+} in the cytosol would decrease to the obviously unacceptably low level of 10^{-10} M. It could of course be proposed that Ca^{2+} penetrates electrophoretically as a monovalent species (for example as Ca^+OH), and in this case the matrix/medium equilibrium distribution ratio would be only 10^3. However, indirect measurements of the levels of free Ca^{2+} in mitochondrial matrix and in the cytosol, based on the activity of Ca^{2+} sensitive enzymes, indicate concentrations of free Ca^{2+} ions between 10^{-7} and 10^{-5} M in the former, and between 10^{-8} and 10^{-5} M in the latter compartment. It is clear, then, that the electrophoretic uniport of Ca^{2+}, particularly if the charge transfer is 2, cannot be permitted to attain equilibrium in vivo. These

Table 3. Gradient of free Ca^{2+} between mitochondria and cytosol in heart

Predicted gradient between mitochondria and sarcoplasm, on the basis of the Nernst equation, for a membrane potential of 180 mV, negative inside, and a charge transfer of 2.

$$0.18 = \frac{RT}{2F} \cdot \ln \frac{[Ca^{2+}]_{in}}{[Ca^{2+}]_{out}}$$

If the species transferred and maintained in mitochondria were monovalent, (e.g., Ca^{+}-OH), the in/out ratio would be predicted to be 10^3

Free Ca^{2+} in mitochondria (from assays of PDHP-phosphatase, PDH-kinase, IDH).	10^{-7}-10^{-5}M
Free Ca^{2+} in sarcoplasm (from assays of α-GP-DH, phosphorylase-kinase, myofibrillar ATPase).	10^{-8}-10^{-5}M

concepts are summarized in Table 3, and emphasize the necessity of mechanisms for the efficient release of the accumulated Ca^{2+} from mitochondria. They will be discussed in detail in the next session.

The Affinity of Mitochondria for Ca^{2+}

The affinity of the energy-linked system for Ca^{2+} has also been the subject of extensive investigations and some conflicting conclusions. As Table 4 shows, most measurements carried out with indirect methods on different mitochondrial types indicate a K_m between 1 and 10 µM. The measurements in which K_ms in excess of 50 µM have been observed are unfortunately invalidated by the presence of inhibitory concentrations of Mg^{2+} in the reaction medium (Sordahl, 1974; Jacobus et al., 1975; Carafoli et al., 1975, 1976). The desirability of a direct measurement of the affinity of mitochondria for Ca^{2+}, and the above-mentioned discovery of the strong inhibitory effect of Mg^{2+}, have lead to the development of an inhibitor-stop method, which is described in Figure 3, and which has permitted the direct measurement of the K_m for Ca^{2+}. Direct measurements of the affinity of the energy-linked system have also been performed with the aid of specific Ca^{2+}-sensitive electrodes. The two methods, which have been applied so far only for heart and myometrium mitochondria, have yielded equivalent results, and have also produced some new interesting information on the kinectic parameters of energy-linked Ca^{2+} uptake (Crompton et al., 1976a).

Curves relating the velocity of uptake by heart mitochondria to the concentration of free Ca^{2+} are presented in Figure 4. Both the inhibitor-stop method (employing buffered Ca^{2+} solutions), and the electrode method, indicate half saturation at about 10 µM external Ca^{2+}, and indicate a maximal velocity of Ca^{2+} removal of about 0.3 nmol per mg of protein per second. It is of interest that phosphate stimulates the rate of uptake at least 2 times, a finding which was rather unexpected, since phosphate (see above) is normally considered to stimulate the extent, not the rate, of Ca^{2+} uptake. Human myometrium mitochondria (Fig. 5), studied with the Mg^{2+} inhibitor-stop method in the presence of phosphate, exhibit a maximal velocity of uptake roughly equivalent to that of heart mitochondria, and an even higher affinity (K_m about 5 µM). As seen

Table 4. A survey of the measurements of the affinity of mitochondria for Ca^{2+} (1971-1975)

Tissue	K_m (μM)	Method	Reference
Liver	50-70	Direct (murexide)	Vinogradov and Scarpa (1973)
"	2	Direct (EGTA buffers)	Bygrave et al. (1971)
"	2-3	Indirect (redox shift of cyt b)	Carafoli and Azzi (1972)
"	4	Direct (EGTA buffers, ruthenium red quenching)	Reed- and Bygrave (1975)
"	8	Indirect (O_2) consumption)	Reynafarje and Lehninger (1973)
Lucilia flight muscle	5	Direct (EGTA buffers, ruthenium red quenching)	Bygrave et al. (1975)
Heart	15	Indirect (redox shift of cyt b)	Chance and Schoerner (1966)
"	55-65	Direct (murexide)	Sordahl (1974)
"	50-100	Direct (murexide)	Scarpa and Graziotti (1973)
"	1	Indirect (redox shift of cyt b)	Jacobus et al. (1975)
Myometrium	25	Direct (murexide)	Wikström et al. (1975)
Vascular smooth muscle	17	Direct (murexide or arsenazo)	Vallières et al. (1975)

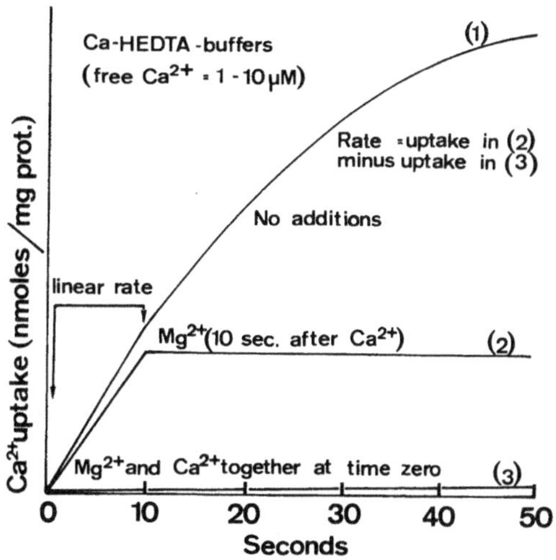

Fig. 3. The Mg^{2+}-inhibitor stop methods for measuring the affinity of heart mitochondria for Ca^{2+}
The reaction medium contained 0.13 M KCl, 0.02 M Tris-Cl, pH 7.0, 0.002 M K-succinate, 1 μg rotenone, 0.5 mg mitochondrial protein, and various amounts of labeled Ca^{2+}, added as a 70 μM total $^{45}Ca^{2+}$-HEDTA complex. Uptake of Ca^{2+} was measured after rapid Millipore filtration (0.45 μ pore size), or after rapid sedimentation of mitochondria in an Eppendorf 3200 centrifuge. The amount of $MgCl_2$ added to stop the uptake of Ca^{2+} was 0.1 M, and the temperature was 25°C

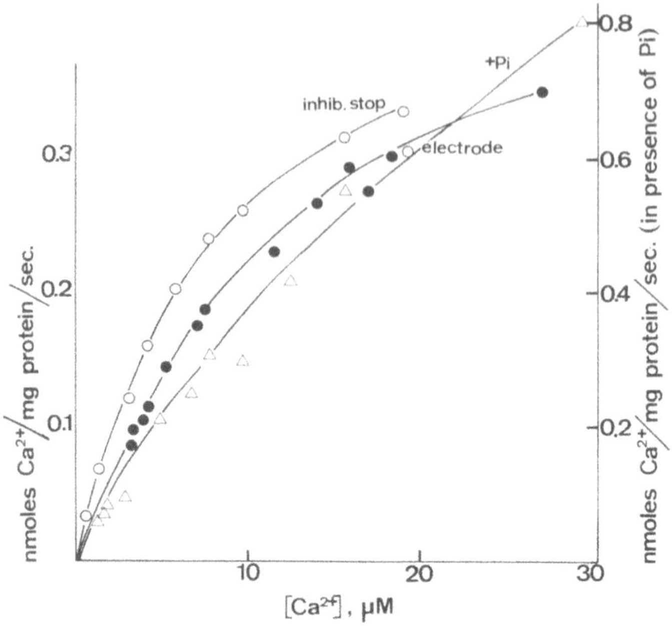

Fig. 4. Kinetics of energy-linked Ca^{2+} uptake by heart mitochondria
Rat heart mitochondria were prepared in a Polytron homogenizer (Sordahl and Schwartz, 1967), essentially according to Scarpa and Graziotti (1973). The conditions for the inhibitor-stop measurements were as described in Figure 3 (the uptake was measured isotopically, following rapid sedimentation of mitochondria in an Eppendorf 3200 centrifuge). The electrode measurements were carried out with the Ca^{2+}-selective electrode developed by Simon and his coworkers (Amman et al., 1975). The reaction medium contained 0.13 M KCl, 0.01 M Tris-Cl, pH 7.0, 1 μg rotenone, 0.5-1.0 mg mitochondrial protein, 0.002 M K-succinate and, when present, 0.001 M inorganic phosphate. $^{45}Ca^{2+}$-HEDTA buffers (total Ca^{2+} concentration 70 μM) were used for the inhibitor-stop method, unbuffered Ca^{2+}-solutions for the electrode experiments. The temperature was 25°C

in Figures 4 and 5, the uptake of Ca^{2+} by both heart and myometrium mitochondria exhibits hyperbolic kinetics. This is at variance with the results obtained by other authors on liver (Vinogradov and Scarpa, 1973; Reed and Bygrave, 1975) and heart mitochondria (Scarpa and Graziotti, 1973), where an evident sigmoidal character for the rate of uptake/versus Ca^{2+} concentration curves was found. The key factor in the discrepancy, at least in the case of heart mitochondria, is probably Mg^{2+}, which was carefully excluded from the media used in the experiments of Figures 4 and 5, and which induces a transition of the curves from hyperbolic to sigmoidal when added to the media in concentrations as low as 1 mM (Fig. 6). The mechanism of the effect induced by Mg^{2+} is currently under investigation, but it seems likely that it may have physiological significance, as a means to control the rate of Ca^{2+} uptake. It must be pointed out, however, that the inhibitory effect of Mg^{2+} against the energy-linked uptake of Ca^{2+} is very evident in heart, but far less pronounced in liver mitochondria. From the data of Figures 4 and 5, one can extrapolate the probable maximal total Ca^{2+} transport ability of mitochondria in situ. The calculations are presented in Table 5, and are based on the observation made in this laboratory that the maximal velocity of uptake is approxi-

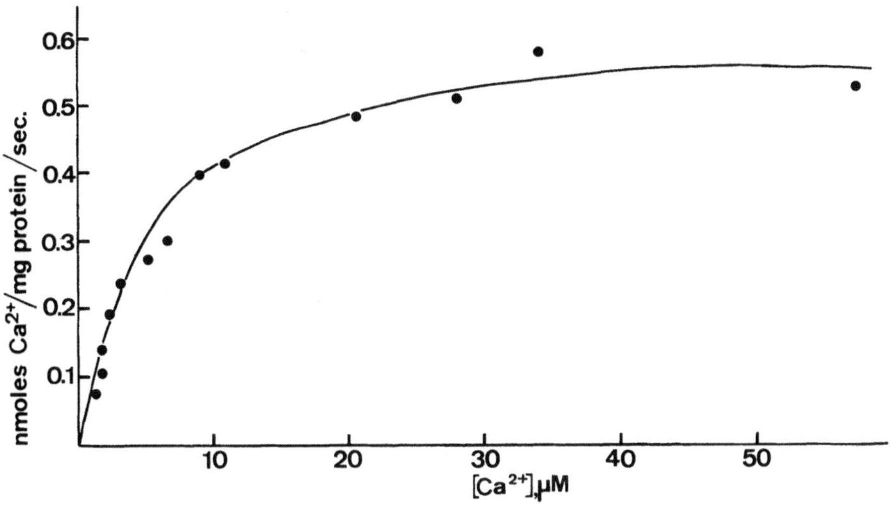

Fig. 5. Kinetics of energy-linked Ca^{2+} uptake by human myometrium mitochondria
Mitochondria were prepared as described in the legend for Figure 2. 0.4-0.6 mg protein was incubated in 1.0 ml medium, containing 0.12 M KCl, 0.01 M Tris-Cl, pH 7.4, 0.002 M inorganic phosphate, 0.005 M Tris-succinate and 2 µg rotenone, at 25°C. Uptake was started by adding the amounts of free Ca^{2+} (added as 70 µM total $^{45}Ca^{2+}$-HEDTA buffers, see legend for Fig. 3), and stopped 10 sec later by adding 0.1 M $MgCl_2$. The uptake was corrected for nonspecific binding by adding the $^{45}Ca^{2+}$ and Mg^{2+} together at time zero. Mitochondria were collected on Millipore filters (0.45 µ pore size) and counted for $^{45}Ca^{2+}$ in a liquid scintillation counter

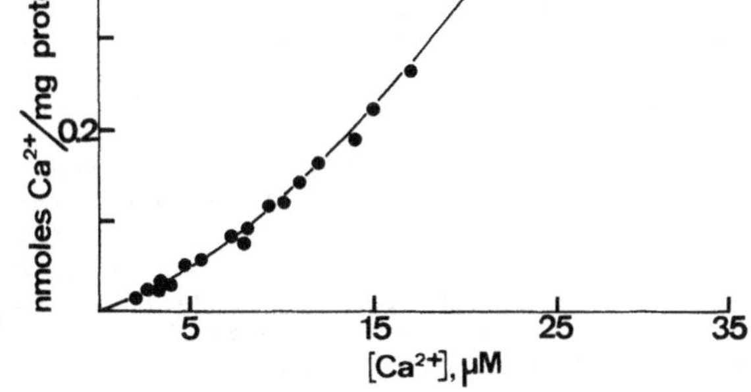

Fig. 6. Effect of Mg^{2+} on energy-linked uptake of Ca^{2+} by heart mitochondria
Conditions as described for Figure 4 (Ca^{2+}-specific electrode). $MgCl_2$, when present, was 0.001 M. The temperature was 25°C

Table 5. Calculated total Ca^{2+} transport capacity of mitochondria in heart and myometrium[1]

Tissue	Mitochondrial content of tissue (mg	Rate of uptake at 1 μM Ca^{2+} (nmol sec^{-1})		Rate of uptake at 10 μM Ca^{2+} (nmol sec^{-1})	
		per mg prot.	per gram tissue.	per mg prot.	per gram tissue.
Heart	80-100	0.12	9.6-12	0.8	64-80
Myometrium	14- 20	0.10	1.4-2.0	0.6	8.4-12

[1] at 38°C in the presence of phosphate, and in the absence of Mg^{2+}.

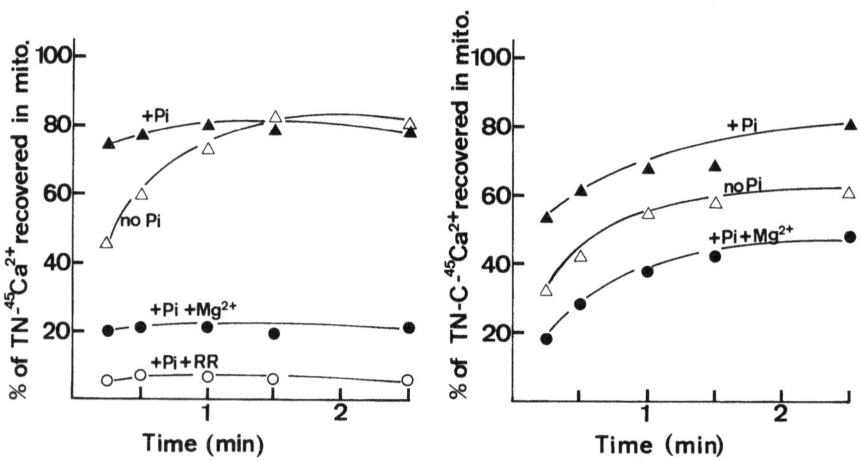

Fig. 7. Interaction of dog heart mitochondria with troponin-bound Ca^{2+} Preparation of mitochondria, of troponin and TN-C, their labeling with $^{45}Ca^{2+}$, and removal of the loosely-bound $^{45}Ca^{2+}$, are described in Affolter et al. (1976). The interaction between mitochondria and Troponin (or TN-C) was measured by the Millipore filtration technique, described by Affolter et al. (1976). The incubations were carried out in small flasks, at 25°C, in a final volume of 0.5 ml. The medium contained 0.13 M KCl, 0.01 M HEPES, pH 7.4, 0.001 M K-succinate, 0.5 μg rotenone and 1 mg mitochondrial protein. The reaction was started by the addition of 4.5 μM troponin (or TN-C). When added, inorganic phosphate was 0.005 M, $MgCl_2$ 0.0025 M, and ruthenium red 1.2 μM

mately doubled when the temperature of the reaction medium is raised from 25°C to 37°C. It may be interesting to relate the rate of uptake at about 1 μM external Ca^{2+} (the presumed average free Ca^{2+} concentration in the sarcoplasm) to the amount of Ca^{2+} mobilized from heart myofibrils during relaxation (about 25 nmol per gram of tissue per 200 msec, Solaro and Briggs, 1974). The calculated rate of uptake is defective with respect to the requirement, at least at 1 μM Ca^{2+}. If the free Ca^{2} concentration were to rise to 5-10 μM (as is probably possible at peak activation) it might however be adequate. It is also likely that isolated mitochondria are not as efficient as mitochondria in situ, but on the other hand in the conditions prevailing in heart sarcoplasm the uptake of Ca^{2+} by mitochondria may be controlled (and possibly limited) by some physiological factor (e.g. Mg^{2+}).

The problem of whether heart mitochondria can extract Ca^{2+} from the myofibrils has been tackled directly by interacting heart mitochondria with $^{45}Ca^{2+}$-labeled troponin, the Ca^{2+} receptor in the myofibril. In the experiment shown in Figure 7, troponin (TN) or its Ca^{2+}-binding subunit (TN-C), were labeled with $^{45}Ca^{2+}$, and then treated with ion-exchange resins to remove the loosely-bound isotope. When interacted with heart mitochondria, both proteins rapidly lost up to 80% of their tightly-bound Ca^{2+} to mitochondria. As expected, the reaction required the supply of energy, was stimulated by phosphate and inhibited by ruthenium red, and was also substantially inhibited by mM concentrations of Mg^{2+}.

Reversibility of Ca^{2+} Transport in Mitochondria

As mentioned above, theoretical considerations, and the evaluation of the free Ca^{2+} concentrations inside and outside the mitochondria (see Table 3) in vivo, strongly suggest that the electrophoretic uniport system responsible for the uptake of Ca^{2+} is not permitted to attain equilibrium in the conditions prevailing in situ. The problem then becomes that of identifying natural compounds (or conditions) capable of releasing the accumulated Ca^{2+} from mitochondria. Recent research in this laboratory (Malmström and Carafoli, 1975) has shown that prostaglandins, added to liver mitochondria after the accumulation of variable Ca^{2+} loads, induce a prompt and extensive release of the accumulated Ca^{2+}. Figure 8 shows an experiment with PGA_1, which induces release when added to mitochondria in concentrations of 10 μM or higher. In the experiment shown in the figure, ruthenium red was added to mitochondria immediately prior to the addition of PGA_1 to stop the reuptake of the lost Ca^{2+}. Indeed, Ca^{2+} would otherwise recycle across the inner membrane, as shown by the polarographic traces also presented in the figure, which demonstrate that PGA_1 acts as an uncoupler, but only when mitochondria contain accumulated Ca^{2+}. As expected, ruthenium red stops the uncoupling by PGA_1. It is interesting that other fatty acids different from PG, for example oleic acid, uncouple respiration irrespective of whether mitochondria contain accumulated Ca^{2+}. Evidently, PG is able to penetrate mitochondria undissociated, as any other uncoupler, but can only leave the organelle as a Ca^{2+} complex. However, even if the demonstration that PG may act as specific Ca^{2+} ionophores is interesting, the high concentrations required for the effect limit the physiological significance of the observation.

Some recent observations in this laboratory (Crompton et al., 1976b) on the release of mitochondrial Ca^{2+} by Na^+ may on the other hand be highly significant physiologically. The experiments have been carried out on heart mitochondria, and have been triggered by previous observations in this laboratory (Carafoli et al., 1973) that Na^+ could under certain conditions release Ca^{2+} from heart mitochondria. Figure 9 shows a Ca^{2+} electrode experiment, in which, after the uptake of Ca^{2+} has been stopped by the addition of ruthenium red, the addition of 5 mM Na^+ induces a rapid release of all the accumulated Ca^{2+}. Also in this case, ruthenium red was added to prevent the re-uptake of the lost Ca^{2+}. The release, however, was still visible (but slower) in the absence of ruthenium red. It is important to stress that the experiment was carried out in an ambient of 120 mM K^+. The effect, therefore, was highly specific for Na^+, and was indeed not duplicated by Rb^+ or Cs^+ (Li^+,

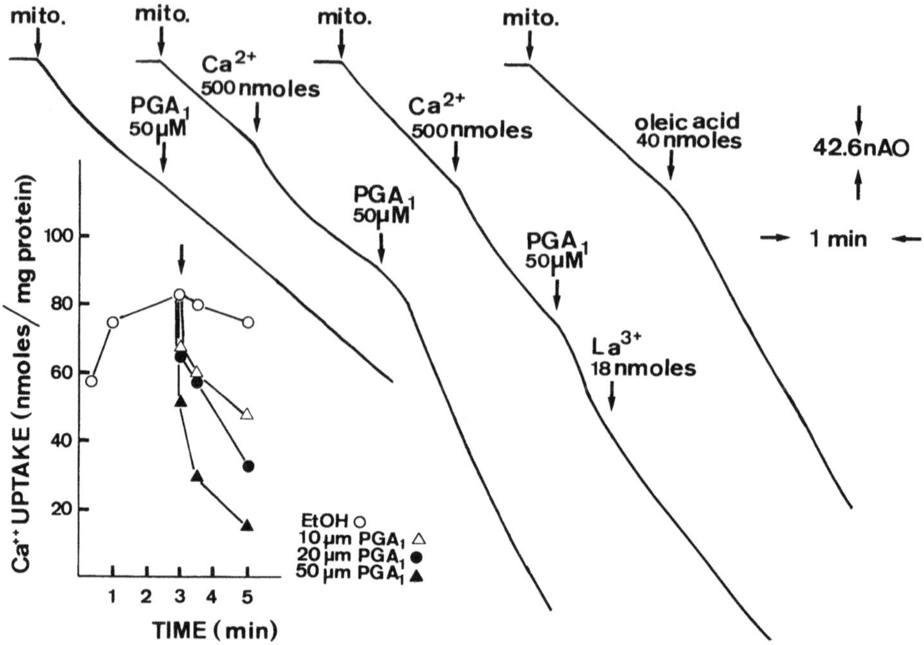

Fig. 8. Release of Ca^{2+} from liver mitochondria by prostaglandins
Liver mitochondria were prepared in a mannitol (0.21 M), sucrose (0.07 M), Tris-Cl (0.005 M) pH 7.4, Na-EDTA (0.5 mM) medium, and washed once in the absence of EDTA. In experiment on left, medium contained 0.15 M NaCl, 0.01 M Na-succinate, 0.001 M imidazole buffer, pH 6.4, 0.5 mM $^{45}Ca^{2+}$, and 12.5 mg mitochondrial protein in a final volume of 5 ml, at 25°C. At arrow, the amounts of PGA indicated were added. Uptake of Ca^{2+} was determined by Millipore filtration, followed by liquid scintillation counting.
Polarographic traces were obtained using a Clark electrode, used in an air-sealed vessel containing a total volume of 1.8 ml, at 25°C. The medium contained 0.15 M sucrose, 0.01 M Na-succinate, 0.001 M imidazole buffer, pH 6.4, and 3.6 mg mitochondrial protein

however, induced some release). Thus, the uptake of Ca^{2+} is inhibited by ruthenium red, but the release induced by Na^+ is not. A logical explanation is that the influx and the (Na^+-induced) efflux of Ca^{2+} proceed by different routes, implying that in the presence of Na^+, Ca^{2+} would tend to cycle across the inner membrane, and in the absence of Na^+ it would not. Figure 10 shows that the rate of induced release is linked to the Na^+ concentration by a sigmoidal relation, with a strong positive cooperativity between 5 and 8 mM Na^+. Calculations of the maximal velocity of induced Ca^{2+} efflux have yielded values of 14.3 nmol of Ca^{2+} per mg of protein per minute.

The obvious explanation for these observations is that Na^+ exchanges for Ca^{2+} in promoting its release. Moreover, the large amounts of Ca^{2+} which can be released in the process (at least 40 nmol per mg of protein), the specificity of the process for Na^+, and its sigmoidal dependence of the Na^+ concentration, suggest the existence of a transmembrane exchange process, rather than the displacement of superficial Ca^{2+} from the external surface of the membrane. It follows, then, that the release of Ca^{2+} should promote penetration of Na^+ into the mitochondrial matrix. The demonstra-

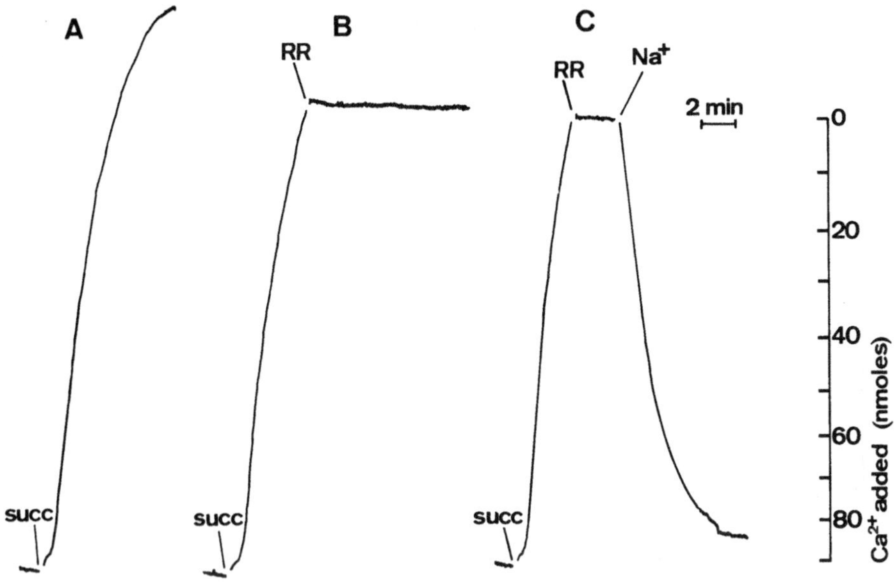

Fig. 9. Release of Ca^{2+} from heart mitochondria by Na^{+}
Rat heart mitochondria were prepared in a Polytron homogenizer, essentially according to Scarpa and Graziotti (1973), in a medium containing 0.21 M mannitol, 0.07 sucrose, 0.01 M Tris-Cl, pH 7.4, 0.1 mM EDTA. They were washed twice, the second time without EDTA, and were incubated 7 min at 25°C in 10 ml of a medium containing 0.12 M KCl, 0.01 M HEPES, pH 7.2, and 30 µg rotenone (to deplete them of endogenous Ca^{2+}). After incubation, suspension was diluted 10 times with cold MST-medium, and washed twice by centrifugation. Reaction medium contained 0.12 M KCl, 0.01 M HEPES, pH 7.2, 2 µg rotenone, and 2 mg mitochondrial protein in 4.0 ml at 25°C. The fluxes of Ca^{2+} were measured with a F 2112 Ca^{2+}-Radiometer specific electrode, and a model 26 Radiometer pH meter, linked to a recorder. Electrode response was calibrated after mitochondria were allowed to accumulate the Ca^{2+} of the medium, and ruthenium red was added to prevent further uptake. Known amounts of Ca^{2+} were added, and the rates of Ca^{2+} fluxes were obtained by plotting calibration curves of the mV against Ca^{2+} added, and replotting the data on a linear scale. Additions: 0.00125 M K-succinate, 0.2 µM ruthenium red, 0.013 M NaCl

tion of influx of Na^{+} into mitochondria in exchange for Ca^{2+} was a difficult task, due to technical problems connected with the use of Na^{+}-specific electrodes to monitor minimal concentration changes against a background of at least 5 mM Na^{+}. Nevertheless, it has been possible to demonstrate in this laboratory that Na^{+} enters the mitochondria in exchange for Ca^{2+}. Moreover, preliminary measurements of the Ca^{2+}/Na^{+} stoichiometry indicate that between 1.8 and 3.3 Na^{+} ions enter for each Ca^{2+} ion that leaves (the sigmoidal nature of the kinetics of the Na^{+}-induced efflux already suggested that more than one Na^{+} was involved per cycle of exchange.) A general scheme that summarizes the observations on the interrelations between Na^{+} and Ca^{2+} is presented in Figure 11.

It is important to stress that the Na^{+}/Ca^{2+} antiporter postulated in this article has so far been found in heart, but not in liver mitochondria. This is interesting, in view of the fact that Na^{+} penetrates into heart cells during the action potential consequent upon the stimulation-dependent depolarization of the sarcoplasm.

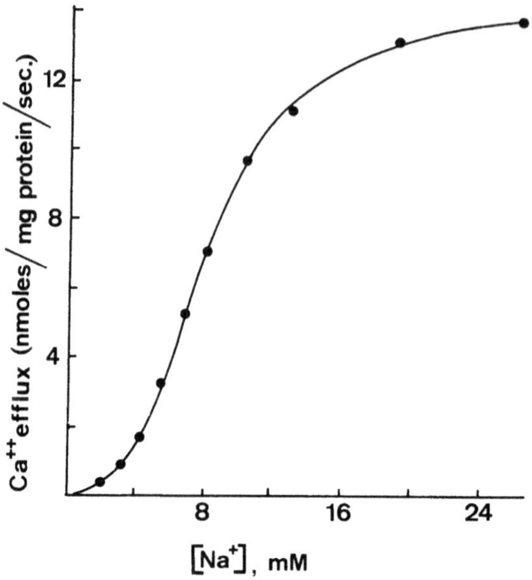

Fig. 10. Dependence of the initial rate of Ca^{2+} efflux on the extra-mitochondrial concentration of Na^+ The preparation and the treatment of mitochondria are described in Figure 9. Incubations (2.9 mg protein) in the medium described in Figure 9, containing 0.001 M K-succinate. After 4 min, 0.12 μM ruthenium red was added, followed by various concentrations of NaCl

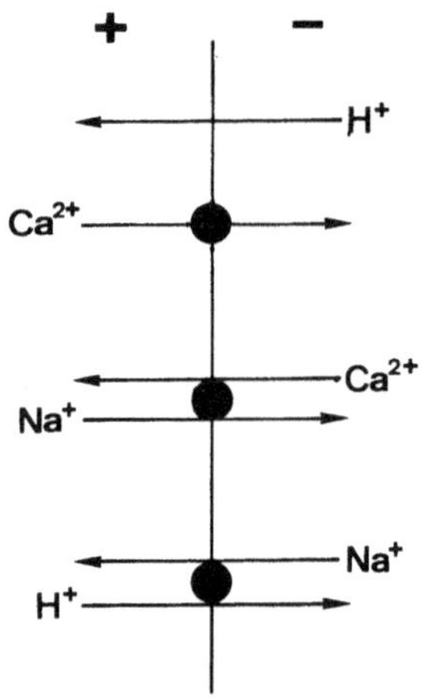

Fig. 11. Fluxes of H^+, Na^+, Ca^{2+}, across the inner membrane of heart mitochondria

It is difficult to decide whether the Na^+ influx during action potential induces an increase of the intracellular Na^+ of sufficient magnitude to affect the distribution of Ca^{2+} between mitochondria and sarcoplasm. Approximate estimates (Niedergerke, personal communication) indicate very small increases, probably smaller than 0.5 mM. However, the most recent measurements of the sarcoplasmic level of free Na^+ show values of about 6 mM (Lee and Fozzard, 1975), which are within the range of optimal sensitivity of Ca^{2+} efflux to Na^+. Even a suboptimal releasing effect could, therefore, have dramatic consequences considering the very large amounts of Ca^{2+} stored endogenously in heart mitochondria. Along slightly different lines, it is perhaps possible to suggest that the positive inotropic effect of cardiac glycosides is related to the Na^+-induced release of Ca^{2+} from mitochondria. Indeed, cardiac glycosides are known to increase the intracellular Na^+ concentration by inhibiting the sarcolemmal Na^+/K^+-ATPase responsible for the extrusion of Na^+, and to increase the exchangeable Ca^{2+} (with the external medium) without increasing the total intracellular Ca^{2+}. This last observation suggests the mobilization of Ca^{2+} normally sequestered within the cell, and could be easily explained with a sustained efflux of Ca^{2+} from mitochondria induced by the increased intracellular level of Na^+.

One last observation made on a completely different cell system, but which could nevertheless be easily rationalized with a Na^+-induced mobilization of mitochondrial Ca^{2+}, is worth mentioning. Pancreatic islets of Langerhans (Lowe et al., 1976) could be induced to release insulin when Na^+ influx into β-cells was permitted by bathing them in a Na^+ and EGTA-containing medium, and by keeping the Na^+ channel in the plasma membrane permanently opened with veratridine. Since the release of insulin is known to require Ca^{2+} (Rasmussen, 1970), and since no Ca^{2+} could penetrate into the cells from the extracellular space due to the presence of EGTA, Na^+ evidently mobilized Ca^{2+} from intracellular stores. The suggestion that the stores could be the mitochondria is very tempting.

Acknowledgments. Parts of the original research described have been supported by the financial contribution of the Swiss National fonds (Grant No. 3.1720.73). Thanks are due to Prof. W. Simon and Dr. D. Amman, Zurich, for valuable discussions on the Ca^{2+}-specific electrodes.

Abbreviations

RR ruthenium red; FCCP carbonyl cyanide p-trifluoromethoxy phenylhydrazone; EDTA ethylene dinitrilo tetraacetic acid; EGTA ethylene glycol-bis-(aminoethylether)-N,N'-tetraacetic acid; HEDTA N-(2 hydroxyethyl)-ethylenediamine tetraacetic acid; HEPES N-2-hydroxyethylpiperazine-N-2-ethanesulphonic acid; PG prostaglandin; TN troponin; TN-C the Ca-receptor polypeptide of troponin; PDHP pyruvate dehydrogenase phosphate; PDH pyruvate dehydrogenase; α-GDPH α-glycerolphosphate dehydrogenase; IDH isocitric dehydrogenase

References

Affolter, H., Chiesi, M., Carafoli, E.: The interaction of mitochondria and sarcoplasmic reticulum with troponin-bound calcium. Europ. J. Biochem. 67, 389-396 (1976)

Amman, D., Gueggi, M., Pretsch, E., Simon, W.: Improved calcium ion-selective electrodes based on a neutral carrier. Anal. Lett. 8, 705-720 (1975)

Baker, P.F.: Transport and metabolism of calcium ions in nerve. Progr. Biophys. Molec. Biol. 24, 177-223 (1972)

Bygrave, F.L., Reed, K.C., Spencer, T.: Sigmoidal kinetics associated with Ca^{2+} uptake and related ATPase in rat liver mitochondria. In: Energy Transduction in Respiration and Photosynthesis. Quagliariello, E., Papa, S., Rossi, C.S. (eds.). Bari: Adriatica Editrice, 1971, pp. 981-988

Bygrave, F.L., Daday, A.A., Doy, F.A.: Evidence for a calcium-ion-transport system in mitochondria isolated from flight muscle of the developing sheep blowfly, Lucilia cuprina. Biochem. J. 146, 601-608 (1975)

Carafoli, E.: In vivo effect of uncoupling agents on the incorporation of calcium and strontium into mitochondria and other subcellular fractions of rat liver. J. Gen. Physiol. 50, 1849-1864 (1967)

Carafoli, E.: Mitochondrial uptake of calcium ions and the regulation of cell function. Biochem. Soc. Symp. 39, 89-109 (1974)

Carafoli, E., Rossi, C.S., Lehninger, A.L.: Uptake of adenine nucleotide by respiring mitochondria during active accumulation of Ca^{2+} and phosphate. J. Biol. Chem. 240, 1154-2261 (1965)

Carafoli, E., Tiozzo, R.: Time-course of the distribution of in vivo administered $^{89}Sr^{2+}$ in rat liver subcellular fractions. Experientia 23, 1017-1020 (1967)

Carafoli, E., Rossi, C.S.: Calcium transport in mitochondria. Advan. Cytopharmacol. 1, 209-227 (1971)

Carafoli, E., Lehninger, A.L.: A survey of the interaction of calcium with mitochondria from different tissues and species. Biochem. J. 122, 681-690 (1971)

Carafoli, E., Azzi, A.: The affinity of mitochondria for calcium. Experientia 27, 906-908 (1972)

Carafoli, E., Tiozzo, R., Crovetti, F., Lugli, G., Kratzing, C.: The release of calcium from heart mitochondria by sodium. J. Molec. Cell. Cardiol. 6, 361-371 (1974)

Carafoli, E., Malmström, K., Capano, M., Sigel, E., Crompton, M.: Mitochondria and the regulation of cell calcium. In: Calcium Transport in Contraction and Secretion. Carafoli, E., Clementi, F., Drabikowski, W., Margreth, E. (eds.). Amsterdam: North Holland Publ. Co. 1975, pp. 53-64

Carafoli, E., Malmström, K., Sigel, E., Crompton, M.: The regulation of intracellular calcium. Clin. Endocrinol. 5, 49S-59S (1975)

Chance, B., Schoerner, J.: High and low energy state of cytochromes. J. Biol. Chem. 241, 4577-4587 (1966)

Crompton, M., Sigel, E., Salzman, M., Carafoli, E.: A kinetic study of the energy-linked influx of Ca^{2+} into heart mitochondria. Eur. J. Biochem. 69, 429-434 (1976a)

Crompton, M., Capano, M., Carafoli, E.: The sodium-induced efflux of Ca^{2+} from mitochondria. A possible mechanism for the regulation of mitochondrial Ca^{2+}. Europ. J. Biochem. 69, 453-469 (1976b)

Drahota, Z., Carafoli, E., Rossi, C.S., Gamble, R.L., Lehninger, A.L.: The steady state maintenance of accumulated Ca^{2+} in rat mitochondria. J. Biol. Chem. 240, 2712-2720 (1965)

Hales, C.N., Luzio, J.P., Chandler, J.A., Herman, L.: Localisation of calcium in the smooth endoplasmic reticulum of rat isolated fat cells. J. Cell. Sci. 15, 1-15 (1974)

Hall, T.A.: The applicability of microprobe analysis to calcium studies. In: Calcium Transport in Contraction and Secretion. Carafoli, E., Clementi, F., Drabikowski, W., Margreth, A. (eds.). Amsterdam: North Holland Publ. Co. (1975), pp. 9-13

Heaton, G.M., Nicholls, D.G.: The calcium conductance of the inner membrane of rat liver mitochondria and the determination of the calcium electrochemical gradient. Biochem. J. 156, 635-646 (1976)

Hodgkin, A.L., Keynes, R.D.: Movements of labelled calcium in squid giant axons. J. Physiol. 138, 253-281 (1957)

Jacobus, W.E., Tiozzo, R., Lugli, G., Lehninger, A.L., Carafoli, E.: Studies of energy-linked Ca^{2+} accumulation by heart mitochondria. J. Biol. Chem. 250, 7863-7870 (1975)

Lee, C.O., Fozzard, H.A.: Activities of potassium and sodium ions in rabbit heart muscle. J. Gen. Physiol. 65, 695-708 (1975)

Legato, M.J., Spiro, D., Langer, G.A.: Ultrastructural alterations produced in mammalian myocardium by variations in perfusate ionic composition. J. Cell. Biol. 37, 1-12 (1968)

Lehninger, A.L., Carafoli, E., Rossi, C.S.: Energy-linked ion movements in mitochondrial systems. Advan. Enzymol. 29, 259-321 (1967)

Lowe, D.A., Richardson, B.P., Taylor, P., Donatsch, P.: Increasing intracellular sodium triggers calcium release from bound pools. Nature (London) 260, 337-338 (1976)

Malmström, K., Carafoli, E.: Effects of prostaglandins on the interaction of Ca^{2+} with mitochondria. Arch. Biochem. Biophys. 171, 418-423 (1975)

Martonosi, A.: Biochemical and clinical aspects of sarcoplasmic reticulum function. Current Topics Membranes Transport. Bronner, F., Kleinzeller, A. (eds.) 3, 83-197 (1972)

Moore, L., Davenport, G.R., Landon, E.J.: Calcium uptake of a rat liver microsomal subcellular fraction in response to in vivo administration of carbon tetra-chloride. J. Biol. Chem. 251, 1197-1201 (1976)

Pasquali-Ronchetti, I., Greenawalt, J.W., Carafoli, E.: On the nature of the dense matrix granules of normal mitochondria. J. Cell. Biol. 140, 565-568 (1969)

Peachey, L.D.: Electronmicroscopic observations on the accumulation of divalent cations in intramitochondrial granules. J. Cell. Biol. 20, 95-111 (1964)

Rasmussen, H.: Cell communication, calcium ion, and cyclic adenosine monophosphate. Science 170, 404-412 (1970)

Reed, K.C., Bygrave, F.L.: A kinetic study of mitochondrial calcium transport. Europ. J. Biochem. 55, 497-504 (1975)

Reuter, H., Seitz, N.: The dependence of calcium efflux from cardiac muscle on temperature and external ion composition. J. Physiol. 195, 451-470 (1968)

Reynafarje, B., Lehninger, A.L.: Ca^{2+} transport by mitochondria from L-1210 mouse ascites tumor cells. Proc. Natl. Acad. Sci. U.S. 70, 1744-1748 (1973)

Rose, B., Loewenstein, W.R.: Permeability of cell junction depends on local cytoplasmic calcium activity. Nature (London) 254, 250-252 (1975)

Rottenberg, H., Scarpa, A.: Calcium uptake and membrane potential in mitochondria. Biochemistry 13, 4811-4817 (1974)

Scarpa, A.: Kinetic and thermodynamic aspects of mitochondrial calcium transport. In: Mitochondria, Bioenergetics, Biogenesis and Membrane Structure. Packer, L., Gómez-Puyou, A. (eds.). New York: Academic Press 1976, pp. 31-45

Scarpa, A., Graziotti, P.: Mechanism for the intracellular regulation of Ca^{2+} in heart. J. Gen. Physiol. 62, 756-772 (1973)

Schatzmann, H.: ATP-dependent Ca^{2+} extrusion from human red cells. Experientia 22, 364-365 (1966)

Selwyn, M.J., Dawson, A.P., Dunnett, S.J.: Calcium transport in mitochondria. FEBS Lett. 10, 1-5 (1970)

Solaro, R.J., Briggs, F.N.: Calcium and the control of enzymatic and mechanical activity in muscle. In: Calcium Binding Proteins. Drabikowski, W., Strzeleska-Golezewska, H., Carafoli, E. (eds.). North Holland Publ. Co., Elsevier 1974, pp. 584-607

Sordahl, L.: Effects of Mg^{2+}, ruthenium red, and the antibiotic ionophore A 23187 on initial rates of Ca^{2+} uptake and release by heart mitchondria. Arch. Biochem. Biophys. 167, 104-115 (1974)

Sordahl, L., Schwartz, A.: Effects of dipyridamole on heart muscle mitochondria. Molec. Pharmacol. 3, 509-515 (1967)

Vallières, J., Scarpa, A., Somlyo, A.P.: Subcellular fractions of smooth muscle. Isolation, substrate utilization, and Ca^{2+} transport by main pulmonary artery and mesentric vein mitochondrial Arch. Biochem. Biophys. 170, 659-669 (1975)

Vasington, F.D., Murphy, J.: Calcium uptake by rat kidney mitochondria and its dependence on respiration and phosphorylation. J. Biol. Chem. 237, 2670-2677 (1962)

Vinogradov, A., Scarpa, A.: The initial velocities of calcium uptake by rat liver mitochondria. J. Biol. Chem. 248, 5527-5531 (1973)

Weber, A.: Energized calcium transport and relaxing factors. Current Topics Bioenerg. 1, 203-254 (1966)

Wikström, M., Ahonen, P., Luukainen, T.: The role of mitochondria in uterine contraction. FEBS Lett. 56, 120-123 (1975)

In vivo Measurements of Ionized Mg^{2+} and Ca^{2+} in Single Cells

A. Scarpa, T. Tiffert, and F. J. Brinley Jr.

The importance of Ca^{2+} and Mg^{2+} in regulation of cellular properties and enzymic function is well established. Metabolism of carbohydrates, proteins and mucleoproteins, photosynthesis and oxidative phosphorylation, cell adhesion, muscle contraction, hormones and neurotransmitters secretion, all exemplify cellular functions where the two divalent cations are regarded as major regulators. During the last decades, the transport of Ca^{2+} and Mg^{2+} by isolated subcellular structures has been studied from various standpoints, often with considerable effort and new experimental design. The detailed knowledge of Ca^{2+} and Mg^{2+} transport and interaction at a subcellular and molecular level is in sharp contrast with the poor understanding of the overall intracellular homeostasis of these two cations. Heretofore, quantitative and sensitive methods for detecting ionized Ca^{2+} and Mg^{2+} in the cytosol of single cells were unavailable, which accounts for the lack of quantitative data on the value of intracellular ionized Ca^{2+} and Mg^{2+} at rest and under various physiological conditions.

In this paper we present a nondestructive technique for measuring quantitatively cytosolic Ca^{2+} and Mg^{2+} and their changes in response to metabolic and electrical events. The technique is based upon the introduction of a metallochromic indicator sensitive to either Ca^{2+} or Mg^{2+} in the cytosol of a single large cell through microinjection and upon the kinetic measurements of the dye absorbance in situ by sensitive pulsed multiwavelength microspectrophotometry.

A. Metallochromic Indicators of Ionized Mg^{2+} and Ca^{2+}

The metallochromic indicators used in the present study were eriochrome blue SE and arsenazo III for the measurements of Mg^{2+} and Ca^{2+}, respectively. The differential spectra of eriochrome blue and eriochrome blue plus various concentrations of either Mg^{2+} or Ca^{2+} are shown in Figure 1. The addition of $MgCl_2$ to eriochrome blue produces an increase of absorbance with a ΔA_{max} at 551 nm and a decrease in absorbance with ΔA_{max} at 580 nm, with an isosbestic point at 563 nm. In contrast, the addition of $CaCl_2$ produces a much broader absorbance decrease from 500 to 600 nm. Other divalent cations, with the exception of Mn^{2+}, produce changes in absorbance similar to that of Ca^{2+}. The different spectral response of eriochrome blue in the presence of Ca^{2+} and Mg^{2+} makes possible the selection of an area of the spectrum where Mg^{2+} can be detected specifically without Ca^{2+} interference. This is evident in Figure 1 (left) which illustrates the time course of the differential absorbance changes of eriochrome blue at 592-554 nm due to Ca^{2+} and Mg^{2+} addition.

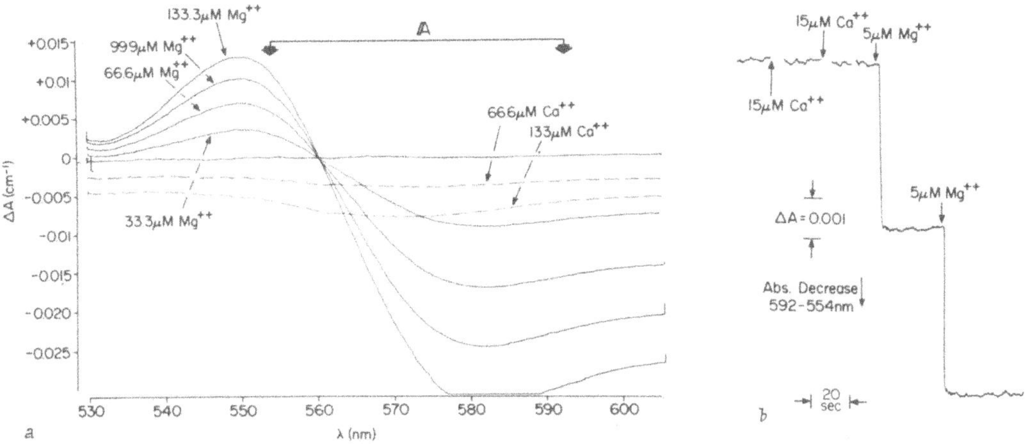

Fig. 1A,B. Difference spectra of eriochrome blue vs. eriochrome blue plus Ca^{2+} and Mg^{2+}. Differential spectrum was obtained by adding to measuring cuvet the concentrations of $CaCl_2$ or $MgCl_2$ reported in figure. Both cuvets contained 100 mM KCl, 30 mM morpholinopropanesulfonate (MOPS) pH 7.1 and 100 μM eriochrome blue. Arrows: regions of the spectra where suitable couples of wavelength pairs may be selected for (Mg^{2+}) measurements without Ca^{2+} interference. (B) illustrates a typical dual-wavelength measurement of eriochrome blue absorbance changes as a function of time due to the addition of $CaCl_2$ and $MgCl_2$

Arsenazo III is an indicator of ionized Ca^{2+} particularly suitable for measurements of ionized Ca^{2+} in nM concentration ranges. The differential spectra of arsenazo III and various amounts of Ca^{2+} versus arsenazo alone are shown in Figure 2. Although several divalent cations produce absorbance changes in arsenazo III in various areas of the spectrum, Ca^{2+} can be measured almost without interference by Mg^{2+} and other divalent cations at 675-685 nm. This is illustrated on the right part of Figure 2 which shows the

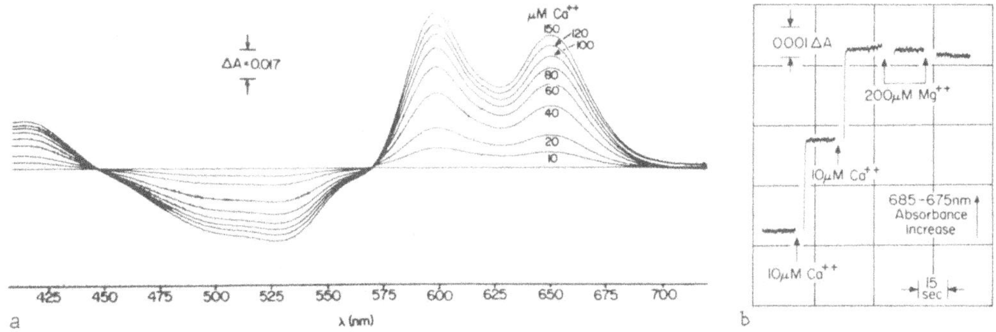

Fig. 2A,B. (A) illustrates differential spectra of arsenazo III vs. arsenazo III plus various Ca^{2+} concentrations. Both cuvets contained 500 mM KCl, 5 mM imidazole buffer, pH 6.8 and 100 μM arsenazo III. Differential spectra were obtained by adding concentrations of Ca^{2+} indicated in figure in measuring cuvet. (B) shows time course of differential absorbance changes due to addition of $MgCl_2$ and $CaCl_2$

TABLE I SOME PROPERTIES OF THE METALLOCHROMIC INDICATORS ERIOCHROME BLUE AND ARSENAZO III.

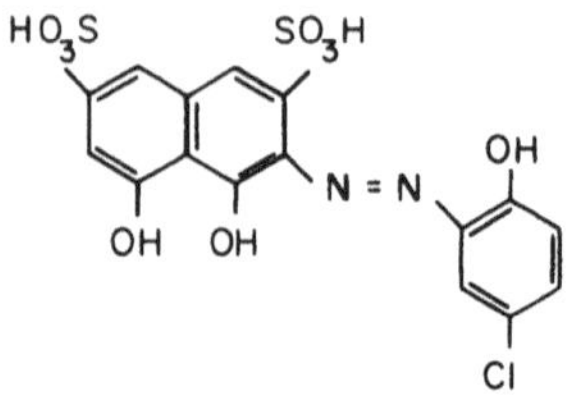

ERIOCHROME BLUE SE

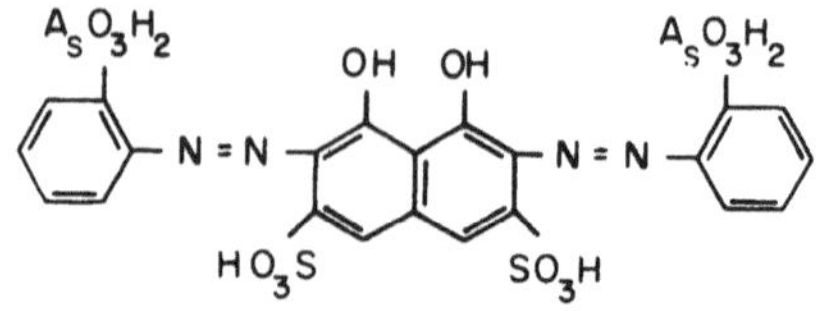

ARSENAZO III

	ERIOCHROME BLUE SE	ARSENAZO III
M.W.	475	776
Water solubility	>20 mM	>20 mM
Binding to cell and cell fractions	- - - +	- - - -
Side effects on cell structure and functions	- - - -	- - - -
Eriochrome blue - Mg^{++} and Arsenazo III - Ca^{++} complexes (pH 7.4, high ionic strength)		
K_D	> 12 mM	~ 60 μM
λ_{max} (nm)	551	695 and 658
suitable differential λ_s (nm)	592-554	585-675
relaxation time, τ (msec)	93	2.4
Δ_ϵ (mM^{-1}. cm^{-1})	~ 5	>25
References	Scarpa, 1974 Scarpa, 1976	Burger, 1973 Di Polo et al, 1976 Scarpa, 1976

absorbance changes of arsenazo III measured as a function of the time by dual wavelength spectroscopy after addition of Ca^{2+} and Mg^{2+}.

Other characteristics of these two indicators are illustrated on Table 1. Of particular importance are the following properties which render these indicators particularly suitable for measuring ionized Ca^{2+} and Mg^{2+} in cytosol in situ: (1) water solubility and lack of binding to or transport across cells and cell fraction; (2) lack of side effects on electrical and metabolic properties of cell and organelles; (3) a relatively low affinity for Mg^{2+} and Ca^{2+}, which results in a minimal disturbance of free Mg^{2+} (in nM range) or free Ca^{2+} (in nM and μM range); (4) a high Δ_ε between the dye and the dye complexed with the divalent cations; (5) a relatively fast complex formation rate between the dye and the cation which permits time resolutions of 1-3 msec for Ca^{2+} transients and 100-200 msec for Mg^{2+} transients (cf. Scarpa, 1974, 1976 for more details).

B. Microspectrophotometric Measurements of the Indicator Absorbance in Situ

Squid giant axons (400-800 μm diameter) were dissected by usual methods from living specimens of Logilo pealei, supplied by the Marine Biological Laboratory, Woods, Hole, Mass., during May 1974, 1975 and 1976, and mounted in a dialysis chamber containing circulating sea water as described by Brinley and Mullins (1967). A similar chamber was used for the experiments using barnacle muscle fibers (800-1500 μm) dissected from specimens of Balanus aquila. Eriochrome blue or arsenazo III were injected by displacement microinjection and were allowed to uniformly distribute within the axoplasm or the sarcoplasm (15-20 min) to reach a concentration ranging from 200 to 1500 μM. For some experiments, a dialysis capillary of 100-150 μm diameter was inserted longitudinally in the fiber: the center of this capillary consisted of a porous area (10-20 Å porous size) through which Ca^{2+} and Mg^{2+} freely diffused, but through which the diffusion of ATP, dyes and larger molecules was limited. The application of the dialysis technique for these measurements has been recently descrobed (Brinley and Scarpa, 1975; Di Polo et al., 1976). The small changes in the absorbance of the indicator, which under certain conditions are a linear function of the ionized Ca^{2+} and Mg^{2+}, were measured kinetically in situ with the apparatus shown in Figure 3, designed and built at the Johnson Research Foundation, University of Penn-

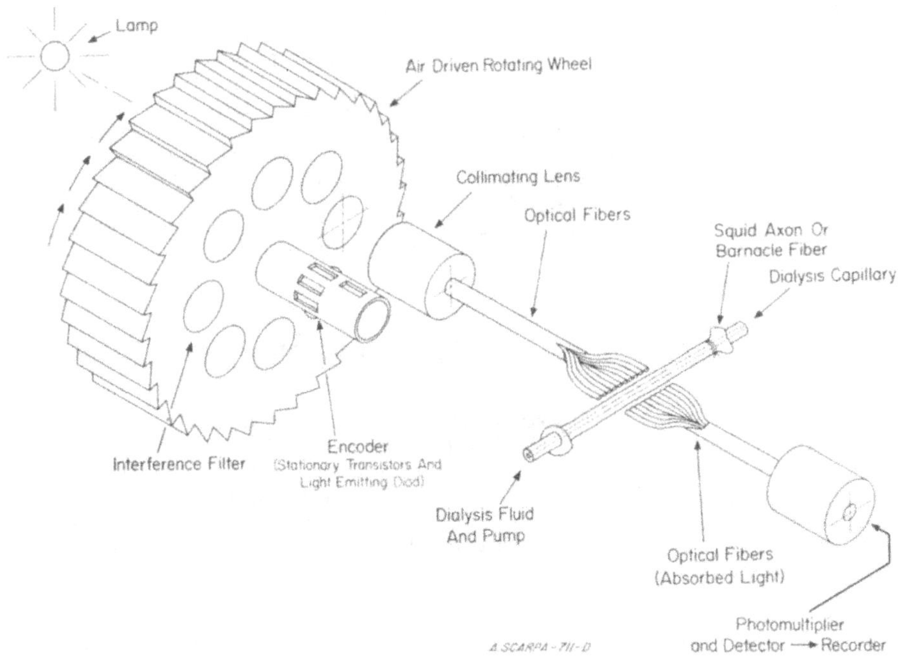

Fig. 3. Diagram of the optical assembly used for microspectrophotometric measurements of eriochrome blue or arsenazo III differential absorbance changes within barnacle sarcoplasms or squid axoplasms. More details in the text

sylvania. The major features of the apparatus consist of: (1) a light source, provided by a stabilized 48 W tungsten iodine lamp (6 amperes); (2) an 11 cm diameter disk containing light 2.2 cm interference filters (40-60% T, 1-2 nm halfband width), driven at variable speeds up to one KHz by compressed air directed on the vanes. The wheel is mounted on a shaft which is machined at one end to form light apertures for a GaAsP light-emitting diode contained inside. Stationary axial rings, each housing phototransistors are mounted over the enclosed shaft which are illuminated during the rotation of the shaft by the light apertures, to produce pulses every 45th rotation of the disk; (3) a collimating lens which focuses the light of the lamp after the passage through each filter on a bundle of minute optical fibers which diverge on a horizontal row at the level of a plexiglas bridge mounted across the barnacle fiber; (4) a photomultiplier tube (EMI 9425 B); (5) electronic circuitry to generate a gate for each filter synchronized to the rotation of the disk shaft position as described above. Other features of the circuit include an 8-channel amplifier, 8 gain equalization gates, 8 detector gates and a dark current clamp gate.

With this instrument, it was possible to maintain continuous readout of the changes in absorbance of the indicator in situ: (1) at the wavelength corresponding to the isosbestic point for the absorbance of the dye vs. dye plus cation; (2) at several wavelength pairs, where the increase in free cation concentration produces either decrease or increase in dye absorbance. Measurements of the change in absorbance at the isosbestic point were a useful measure of changes in total dye concentration and overall absorbance of the fiber (position, light scattering). The differential read out at two closely adjacent wavelengths permits specific measurements of the absorption of the dye with either no or minimal interference by nonspecific absorbance changes such as swelling or changes in refactivity indexes. Furthermore, the simultaneous read out of several sets of wavelength pairs, selected in close areas of the spectrum where the same changes in Ca^{2+} or Mg^{2+} concentration produce differential absorbance changes in opposite direction, reveals the occurrence of possible artifacts and make the interpretation of the results obtained easier.

C. Free Mg^{2+} in Single Isolated Squid Axon and Barnacle Fiber

The magnesium content of axoplasm measured by atomic absorption spectroscopy averages 6.4 nmol.kg $axoplasm^{-1}$ (Baker and Crawford, 1972), but little and indirect information exists in the literature on the state of this intracellular magnesium. Due to the large amount of adenine nucleotides in squid axoplasm it is assumed that a substantial fraction of the total Mg^{2+} exists in a form of complex.

Figures 4 and 5 show the results of an experiment where the ionized Mg^{2+} in the axoplasm were measured by the changes in absorbance of eriochrome blue in the axoplasm by pulsed multiwavelengths microspectrophotometry.

Although in principle it is possible to calculate the ionized magnesium concentration either from the three parameters of K_D, dye concentration and path length or from a calibration curve relat-

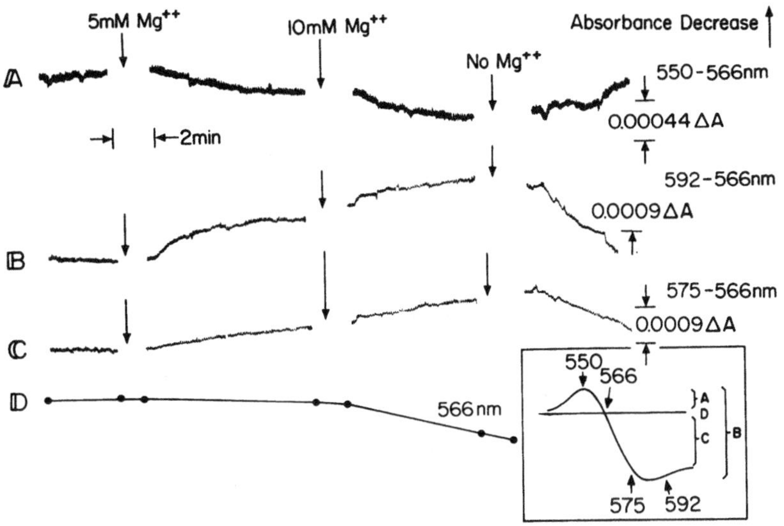

Fig. 4. Simultaneous measurements of dual-wavelength absorbance changes undergone by eriochrome blue at three wavelength pairs and at an isosbestic point during the dialysis of a squid axon. Squid axon was prepared and mounted in the dialysis chamber according to procedure of Brinley and Mullins (1967). Eriochrome blue was introduced into axon by microinjection (Hodgkin and Keynes, 1956) 20 min after dialysis to give an axoplasm concentration of about 4 mM. The axon was then dialyzed with concentration of Mg^{2+} indicated in figure, dissolved in reaction mixture containing final osmolarity of 0.74 made up from the following species (mM):K(350)Na(40) Cl(88), Isothianate(150) D-Aspartate(150), TES(5), Taurine(275). Temperature 21 (Brinley and Scarpa, 1975)

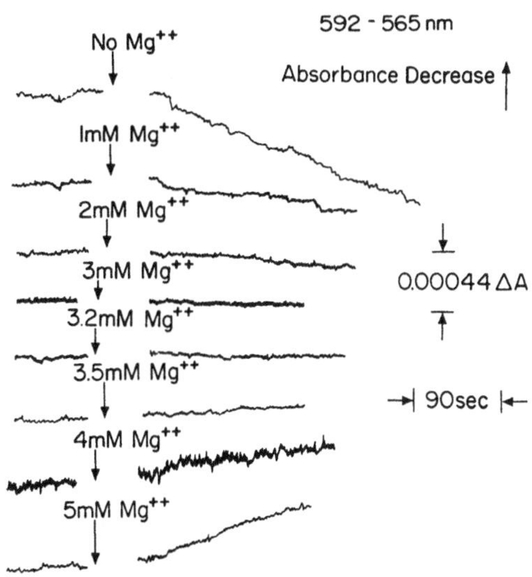

Fig. 5. Recording of changes in absorbance undergone by eriochrome blue with axoplasm of various axons dialyzed with various concentrations of Mg^{2+}. Experimental conditions similar to those in Figure 4 (From Brinley and Scarpa, 1975)

ing absorbance change to ionized magnesium concentration, in these pre-experiments a null point method was used. We determined the internal ionized Mg^{2+} by noting that concentration of Mg^{2+} which, when dialyzed into the axoplasm, produced no changes in the observed differential absorbance. Because the dye binds magnesium very weekly, (K_D = 25 mM in our solutions), the ionized Mg^{2+} was reduced less than 5% by the concentration of eriochrome blue which was used (approx. 500 μM). The response to dialysis of the axon with various concentrations of Mg^{2+} is shown in Figure 4. After microinjecting eriochrome blue into the axoplasm, and waiting for at least 15 to 20 min for the diffusional equilibration of the dye, a base line signal was obtained. At the vertical arrow, dialysis with medium containing 5 mM Mg^{2+} was begun. The dialysis solution evidently contained a higher ionized Mg^{2+} concentration than did the axoplasm because all three differential absorbances deflected in the direction of greater magnesium binding to the dye, reflecting diffusion of Mg^{2+} from the capillary into the axoplasm. After several minutes of dialysis, the concentration was changed to medium containing 10 mM Mg^{2+} and a further change in absorbance was found. The final procedure in the experiment was to replace the dialysis media with one containing no ionized Mg^{2+}. In this case all three differential absorbance changes reversed direction, indicating that ionized Mg^{2+} was being washed out of the axoplasm by the dialysis.

Unfortunately, it was not possible to continue dialysis until Mg^{2+} reached diffusional equilibration, because the dialysis capillary was not entirely impermeable to the dye. As indicated by curve D in Figure 4, which measures the absolute isosbestic absorbance at 566 nm, there was a slow but continuous loss of dye from the axoplasm during dialysis which counted to about 3-4% during the course of the experiment.

In order to avoid this drowback, only the initial deflections were measured for 24 different axons, in which the initial Mg^{2+} concentration in the dialysate ranged from 0 to 10 mM. Eight representative experiments are collected in Figure 5 which shows the initial absorbance changes measured differentially at 592-565 nm. There was essentially no initial deflection when the ionized Mg^{2+} concentration in the dialysate ranged from 3 to 3.5 mM, whereas an initial deflection was obvious when the initial ionized Mg^{2+} was outside this range. Therefore, these results indicate that the level of ionized Mg^{2+} in squid axons, isolated and impaled, but electrically excitable and not metabolically poisoned, is equivalent to that present in our dialysis solution containing between 3.0 and 3.5 mM Mg^{2+}.

A similar technique was used for the in situ measurements of free Mg^{2+} in single muscle fibers isolated from _Balanus aquila_. The null point titration with different concentrations of Mg^{2+} in the dialysate (cf. Fig. 5) is relatively less sensitive in large fibers, such as barnacle muscle fibers. Therefore, in these fibers additional sensitivity was obtained by internally dialyzing the sarcoplasm with EGTA-Mg buffers. The in situ effective dissociation constant for Mg-EGTA is high enough (∿30 mM) to buffer Mg^{2+} concentrations effectively in the mM range with reasonable concentrations of total Mg^{2+} and total EGTA.

When such a mixture is dialyzed through the porous capillary and allowed to diffuse into the sarcoplasm, the relative volume of the sarcoplasm subject to buffer stabilization of ionized Mg^{2+}, is greater than if the dialysate contained an unbuffered magnesium

Table 2. Free and bound Mg^{2+} in the sarcoplasm of Balanus aquila barnacle fiber

	mM	nmol.Kg^{-1} mM.Kg^{-1}	Comments
Mg Total		10.5	atomic absorption measurements
Mg^{2+} Ionized	6	4.2	based upon measurements of internal dialysis using eriochrome blue. The value in nmol.Kg^{-1} were calculated using 20-30% wet weight coupled by solids
Mg bound to ATP		4.2	based upon the measured ATP content (4.9 nmol.Kg^{-1}) and the dissociation constant mM.kg^{-1} for ATP and Mg^{2+} (0.7 mM) obtained under similar conditions (P. De Weer, unpublished observation).
Mg bound to myofibrils		1	(Page et al., 1971)
Mg bound to phosphoarginine and Phosphate		?	

solution. Therefore if the magnesium concentration of the Mg-EGTA in the dialysate does not exactly match the ionized Mg^{2+} in the sarcoplasm, a greater fraction of the dye will show a change in differential absorbance, during the initial minutes of the dialysis. Thus the use of Mg-EGTA in the dialysate increases the amount of absorbance change produced by a given mismatch between ionized Mg^{2+} in the dialysate and in the sarcoplasm.

The results of various measurements of free Mg^{2+} and the calculation of Mg^{2+} bound to intracellular structures in barnacle fibers are summarized in Table 2.

D. Ionized Ca^{2+} Concentration in Squid Axons

It is well established that the intracellular Ca in squid axoplasm is much lower than the Ca in sea water, and that the fraction of the intracellular Ca which exists in ionized form is only a small fraction of the total. The original measurement of ionized Ca (Hodgkin and Keynes, 1956) indicated that it was in the micromolar range and subsequent investigation has indicated that is is somewhat lower, in the range of 350 nM (Baker et al., 1971).

In the present study, the internal concentration of ionized Ca in the resting axoplasm of Logilo pealei giant axons was measured by recording the changes in Δ absorbance of arsenazo III in situ.

The measurement of ionized Ca in the axoplasm of resting axons using arsenazo III depends upon the differential absorbance change produced when the ionized Ca present in axoplasm is chelated by the sudden introduction of an excess of EGTA. In order to validate this technique, it is necessary to demonstrate that (1) the absorbance change is linear with ionized Ca in the range of interest, and (2) chelation of Ca by EGTA is complete under the experimental conditions.

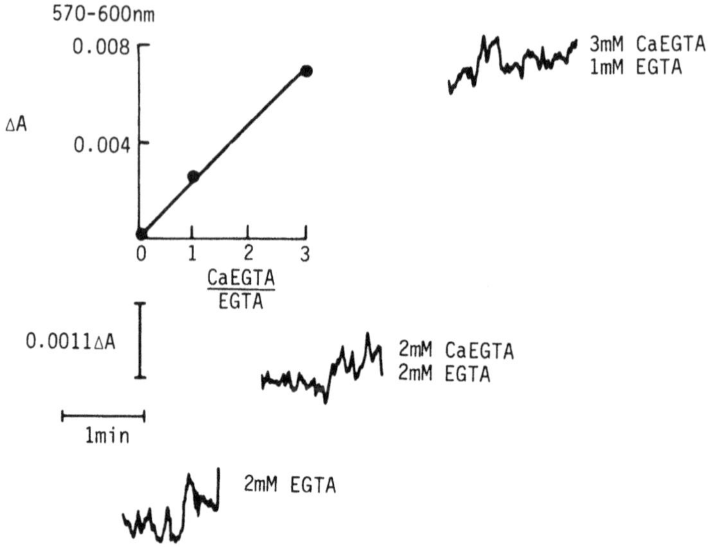

Fig. 6. Dual-wavelength recording of the change in absorbance of arsenazo III in a glass capillary. The calibration curve shows a linearity in Δ absorbance changes (570-600 nm) when dialysis solutions containing 500 μM arsenazo III and increasing amounts of free Ca^{2+} were flushed inside a 500 μm ID glass capillary. Insert: absorbance changes plotted against the ratio of free to bound calcium. The Δ absorbance in 2 mM EGTA was taken as 0 free Ca^{2+}. The calibration curve for differential absorbance was similar at various wavelength pairs (660-685 nm or 675-685 nm) (From Di Polo et al., 1976)

A typical calibration is shown in Figure 6. In this experiment, a glass capillary with a lumen comparable to the usual diameter of a squid axon (500 μm) was positioned in the dialysis apparatus and then flushed with calibrating solutions containing 500 μM arsenazo III and the indicated amounts of EGTA and CaEGTA. The lower part of the figure shows the absorbance changes produced at the wavelength pair 570-600 nm. The concentration of free Ca in the solution containing 2 mM EGTA but no added Ca was taken as zero. The insert shows the data replotted to demonstrate that the absorbance change is linear with ionized Ca. Using an apparent dissociation constant of 0.15 μM for CaEGTA (pH 7.3, ionic strength 0.3) (Di Poloet al., 1976) a calculated ionized Ca of 0.15 μM for the middle point is produced. Calibration for the differential absorbance measured at the wavelength pair 660-685 nm was also linear.

Figure 7 illustrates a combined in vitro calibration and dialysis experiment to demonstrate that the differential absorbance change produced by chelation of Ca with EGTA can be used to measure the amount of ionized Ca present initially in the solution. The principle behind the measurement is diagrammatically illustrated in the top part of the figure. A glass capillary was mounted in the apparatus and flushed with dialysis medium containing 2 mM EGTA. The differential absorbance observed with this solution served as a reference for the 0 Ca base line. The capillary was then flushed with solutions containing the indicated ratios of CaEGTA and EGTA to establish a linear relation between ionized Ca and absorbance. The vertical arrow indicates the onset of dialysis with an isos-

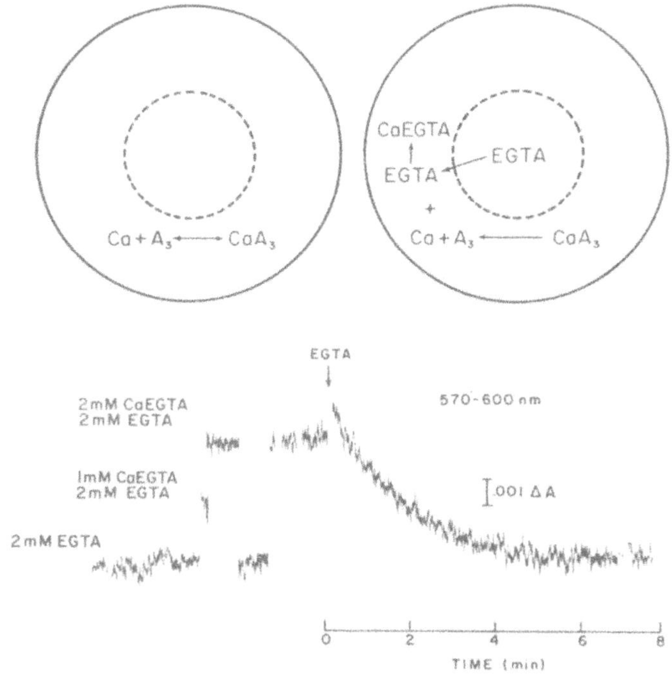

Fig. 7. In vitro calibration of arsenazo III and in vitro dialysis. A dialysis capillary was placed inside a 500 μm ID glass capillary, containing 500 μM arsenazo III dissolved in dialysis medium (for more details, Di Polo et al., 1976). Left hand position shows the absorbance changes produced when dye solutions containing EGTA buffer with various ratios of CaEGTA/EGTA were flushed through the glass capillary. Nominal concentration, based on the determination of effective dissociation constant for Ca-EGTA and Ca-arsenazo III in similar reaction mixtures were 0 (2 mM EGTA) and 150 nm (2 mM EGTA + 2 mM CaEGTA). Right side: absorbance changes when a pulse of EGTA solution (267 mM EGTA, 79 mM KTES, pH 7.3) was pushed through the porous dialysis capillary. The absorbance returns to the level of 0 Ca^{2+} within a time course appropriate for the dimensions of porous plastic capillary and glass capillary

motic solution of EGTA and KTES (267 mM K EGTA and 79 mM KTES). As the figure illustrates, when the EGTA diffuses through the porous capillary into the lumen of the glass tube, the residual ionized Ca is chelated and the differential absorbance returns to the baseline level corresponding to 0 Ca, with a time-course appropriate for the dimensions of the porous plastic and glass capillaries.

Therefore, for calibration purposes, the axon was replaced at the end of an experiment with squid axon by a glass capillary whose lumen was comparable in size to the diameter of the fiber. A sufficient amount of dye previously used for injection into the axon was added to the calibrating solution to give a dye concentration comparable to that of the axoplasm (0.5-1 mM). Aliquots of this dye solution were then treated with appropriate volumes of Ca-EGTA buffers to give solutions with free ionized Ca^{2+} in the range of 0-200 nM. These solution were then flushed through the glass capillary in sequence and the differential absorbance changes noted. A linear dependence of absorbance upon path length and dye con-

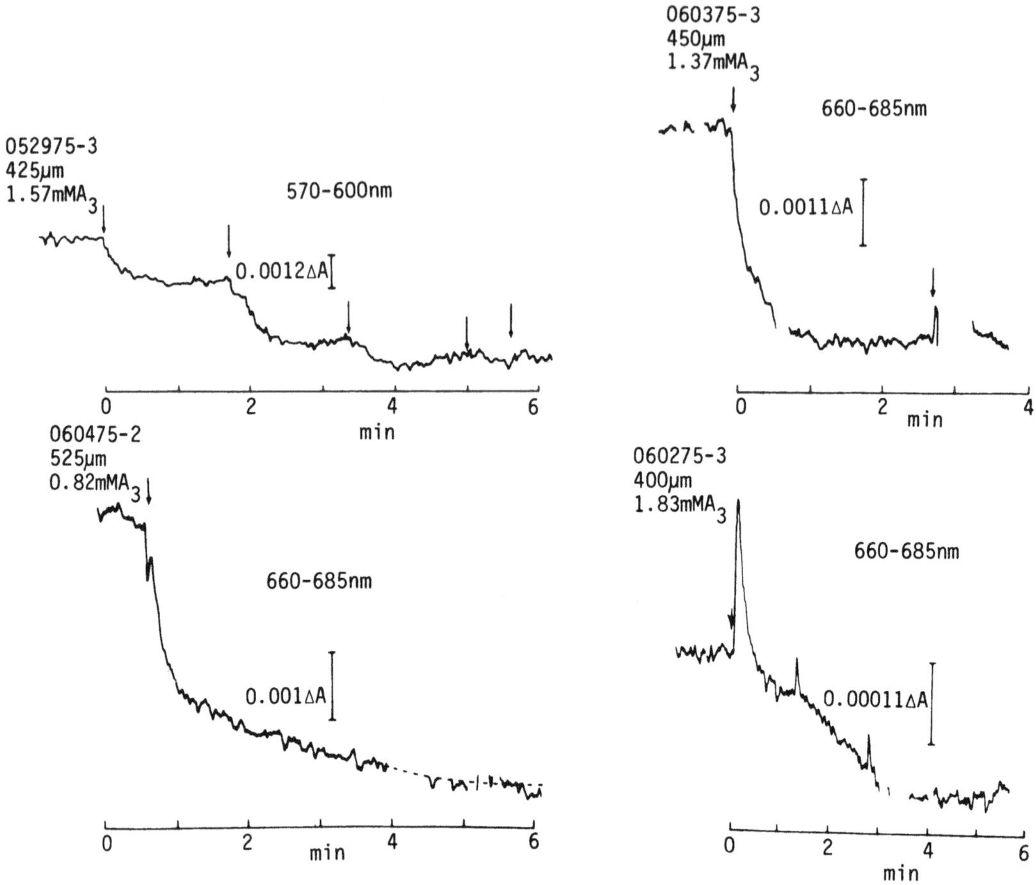

Fig. 8. Determination of internal ionized Ca^{2+} in four axons. Vertical arrows: where the axons containing the concentrations of arsenazo III indicated in the figure were flushed with EGTA through the dialysis capillary. Blank regions of the traces result from deleted artifacts due to external light interference or ink pen failure of the recorder. Vertical bars in each figure: calibration of differential absorbance change

centration was assumed when correcting these in vitro calibration factors for the exact diameter of the axon and in vivo concentration of arsenazo III. The correction factors were about 1.15-1.30.

The result of dialysis in four axons is shown in Figure 8. The vertical arrows indicate the point at which the chelating solution was flushed through the capillary. In the two experiments shown at the top of Figure 8 dialysis was not continuous; instead, a 1-2 sec pulse of chelating solution was pushed through the capillary. In these experiments a subsequent short flush was given, as indicated by the arrows, to confirm that complete chelation of the ionized Ca^{2+} had occurred.

The collected data from nine experiments indicates that the ionized Ca^{2+} in nine squid axons which had been treated with arsenazo III but otherwise untreated, was 40-60 nM.

E. Control of Axoplasmic Ionized Ca^{2+} by Mitochondria in Situ

Microspectrophotometric measurements of arsenazo III absorbance in situ can be successfully used not only to determine the resting concentrations of ionized Ca^{2+} in squid axons, but also the kinetics changes of axoplasmic ionized Ca^{2+} in response to metabolic or electrical events. For this series of experiments, the axon was dissected, positioned in the chamber and microinjected with arsenazo III as described previously. Insertion of the dialysis capillary was omitted.

Electrical stimulation of the isolated axons results in a influx of Ca^{2+} from the bathing fluid to the axoplasm (Hodgkin and Keynes, 1956). Measurements of axoplasmic Ca^{2+} loading under our conditions (500 µm axon, 100 stimulation.sec^{-1}, 112 mM $CaCl_2$ in sea water) showed that Ca $^{+}$ entrance across the axolemma was 15 pmol. cm^{-2}.sec^{-1} and that the uptake of $^{45}Ca^{2+}$ was linear with the time up to 60 min of continuous electrical stimulation (not shown). This Ca^{2+} loading corresponds to an increase in the axoplasmic free Ca^{2+} of about 75 µM.min^{-1} if all the Ca^{2+} entering the axon is assumed to be free in solution.

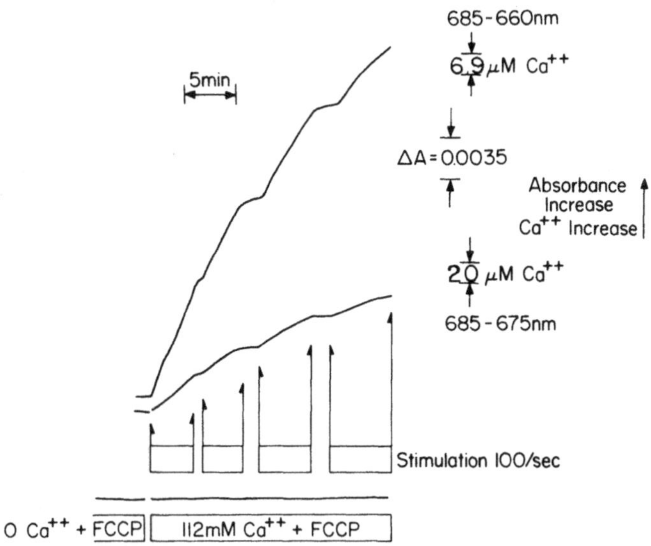

Fig. 9. Increase in ionized Ca^{2+} concentrations in squid axons electrically stimulated in the presence of high concentration of Ca^{2+} and an uncoupler of the oxidative phosphorylation. The axon (480 µm) was microinjected with arsenazo III (240 µM after diffusional equilibration). Where indicated, bathing solution was changed to sea water containing 112 mM Ca^{2+} and the axon was electrically stimulated 100/sec. FCCP (carbonyl cyanide p-trifluoromethyoxyphenylhydrazone)dissolved in dimethyl sulfoxide at a concentration of 2 mg.ml^{-1} was superfused with a motor-driven Hamilton syringe at a rate of 0.2 µl.min^{-1}. Under those conditions the effect of FCCP was maximal since further additions of FCCP to the axons had no additional effect on ionized Ca^{2+} changes

Figure 9 shows the results of an experiment where the increase in axoplasmic ionized Ca^{2+} concentration from electrical stimulation with high levels of extracellular Ca^{2+} was detected as the differential absorbance changes undergone by arsenazo III in situ. The absorbance changes of arsenazo in the axoplasm were measured (1) at the single wavelength 564 nm, (the isosbestic point of the dye and the dye complexed with Ca^{2+}) where no appreciable changes in absorbance were found during the experiment (not shown) and (2) at the pairs of wavelength 685-675 nm and 685-660 nm, where increases in ionized Ca^{2+} produced increases in ΔA to various degrees (cf. Fig. 2). Throughout the experiment the energy-dependent Ca^{2+} uptake by the mitochondria in situ was effectively prevented by the presence of the uncoupler of oxidative phosphorylation FCCP (carbonyl cyanide p-trifluoromethoxyphenylhydrazone). Both Δ absorbance read outs clearly showed an increase in axoplasmic ionized Ca^{2+} concentration immediately after electrical stimulation with a linear increment for several minutes. The figure also shows a clear correlation between the increase in ionized Ca^{2+} in the axoplasm and the electrical stimulation of the axon. The continuous recording of axoplasmic ionized Ca^{2+} increase can be correlated at any time with the values of axoplasmic $^{45}Ca^{2+}$ loading obtained under the identical conditions. Therefore, the buffering capacity of the axoplasm can be calculated as the difference between (1) the value of ionized axoplasmic Ca^{2+} expected from the data of $^{45}Ca^{2+}$ loading under the assumption that all the Ca^{2+} entering the fiber is free in solution and (2) the value of axoplasmic free Ca^{2+} directly measured by arsenazo absorbance. The result indicates that in the absence of energy-dependent Ca^{2+} uptake by mitochondria, only approximately one tenth of all the Ca^{2+} entering the axon remains free in the axoplasm. Figure 10 shows two superimposed traces of the changes in absorbance of arsenazo III in situ at 685-660 nm in two axons of comparable size. In the axon continuously superfused with FCCP, the axoplasmic ionized Ca^{2+} increased almost linearly during the electrical stimulation. On the other hand, in the axon electrically stimulated in the absence of FCCP, the axo-

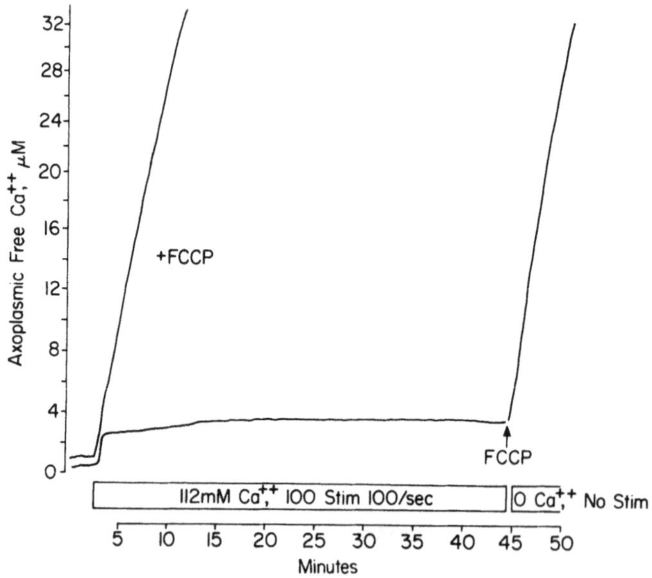

Fig. 10. Energy-dependent Ca^{2+} uptake by mitochondria of squid axons in situ. The experimental conditions were similar to those of Figure 9 except that the axons sizes were 520 and 535 mμ. The arsenazo III concentration was around 320 μM. Only one axon was superfused with FCCP and FCCP was added to the other axon only at the end of the experiments

plasmic concentration of ionized Ca^{2+} rapidly increased from 50-100 nM to about 3 μM, where it remained relatively constant for 30 min in spite of the continuous Ca^{2+} entrance during the electrical stimulation of the axon in high extracellular Ca^{2+}. At the end of the experiment the addition of FCCP produced a dramatic increase in axoplasmic ionized Ca^{2+}. This ionized Ca^{2+} increase was certainly due to Ca^{2+} release from axoplasmic stores, since the extracellular bathing fluid was shifted from one containing 112 mM Ca^{2+} to one containing 0 Ca^{2+} several minutes before FCCP addition.

These results, and several others obtained with various mitochondrial inhibitors, clearly indicate that the mitochondria of squid axons are a powerful Ca^{2+} buffering system which can maintain the axoplasmic Ca^{2+} at relatively low levels (3-5 μM) in spite of the enormous Ca^{2+} loading across the axolemma. On the other hand, the mitochondria seems to be unable to effectively accumulate Ca^{2+} (even when the axoplasmic Ca^{2+} loading was reduced by decreasing extracellular Ca^{2+} or the frequency of stimulation) if the concentration of ionized Ca^{2+} in the axoplasm is below 2-3 μM. These results are in line with the kinetic data available on mitochondria Ca^{2+} uptake in vitro and with the proposed physiological role of mitochondria in controlling intracellular Ca^{2+} homeostasis (Scarpa, 1975).

These data also suggest that two distinct mechanisms are operative in controlling the intracellular Ca^{2+} homeostasis of squid axons: the axolemmal Ca^{2+} pump and the energy-dependent Ca^{2+} transport by mitochondria. The Ca^{2+} pump should effectively control the level of axoplasmic ionized Ca^{2+} during limited Ca^{2+} loading, when the axoplasmic ionized Ca^{2+} concentrations remain confined to the nM levels. On the other hand, the mitochondrial Ca^{2+} uptake seems to be the more powerful Ca^{2+} buffering system in the axoplasm when physiological or pathological events result in a massive Ca^{2+} loading and the concentrations of ionized Ca^{2+} in the axoplasm reach levels much greater than the resting levels.

F. Conclusions

In summary, the absorbance measurements of suitable indicators by pulsed multiwavelength microspectrophotometry can be regarded as a powerful technique for measuring quantitatively ionized Mg^{2+} and Ca^{2+} concentrations in the cytosol of large cells and their kinetic changes under various physiological conditions. The advantage of this technique for measuring cytosolic ionized Mg^{2+} is self-evident, since no other methods have been described in the literature to measure ionized Mg^{2+} with high selectivity and sensitivity. On the other hand ionized Ca^{2+} can be measured in biological systems electrochemically with specific electrodes or by recording the light emission of the photoluminescent Ca-indicator aequorin. To date, in spite of recent development, Ca electrodes lack both sensitivity and rapidity for the measurements of intracellular Ca^{2+}. Aequorin measures Ca^{2+} with high sensitivity and selectivity and has been successfully used to obtain informations on the ionized Ca^{2+} in a variety of cells (Ashley, 1970; Baker et al., 1971; Blinks et al., 1975). However, the interpretation of the data obtained and the future applicability of aequorin to more complete studies of intracellular Ca^{2+} homeostasis are severely limited by the difficulties in calibrating light emission and Ca^{2+} concentration, by the relatively slow response to Ca^{2+} transients and by the practical unfeasibility of measuring ionized Ca^{2+} in the μM range.

Acknowledgments. This work was supported by grants NS 08336, hl 15835 and HL 18708 from N.I.H. T.T. is on leave of absence from the Universidad Peruana Cayetano Heredia and A.S. is an Established Investigator of the American Heart Association.

References

Ashley, C.C.: An estimate of calcium concentration changes during the contraction of single muscle fibers. J. Physiol. 210, 133-134 (1970)

Baker, P.F., Crawford, A.C.: Mobility and transport of magnesium in squid giant axons. J. Physiol. 227, 855-874 (1972)

Baker, P.F., Hodgkin, A.L., Ridgway, E.B.: Depolarisation and calcium entry in squid axons. J. Physiol. 218, 709-755 (1971)

Blinks, J.R., Prendergast, F.G., Allen, D.G.: Photoproteins as biological calcium indicators. Pharmacol. Rev. 27, 438-472 (1975)

Brinley, F.J., Mullins, L.J.: Sodium extrusion by internally dialyzed squid axons. J. Gen. Physiol. 50, 2303-2331 (1967)

Brinley, F.J., Scarpa, A.: Ionized magnesium concentration in axoplasm of dialyzed squid axons. FEBS Lett. 50, 82-85 (1975)

Burger, K.: Organic reagents in metal analysis. In: International Series of Monographs in Analytical Chemistry. Oxford: Pergamon Press, 1973, Vol. LIV

Hodgkin, A.L., Keynes, R.D.: Experiments on the injection of substances into squid axons by means of a microsyringe. J. Physiol. 131, 592-616 (1956)

Hodgkin, A.L., Keynes, R.D.: Movement of labeled calcium in squid axons. J. Physiol. 138, 253-281 (1957)

Di Polo, R., Requena, J., Brinley, F.J., Mullins, L.J., Scarpa, A., Tiffert, T.: Ionized calcium concentrations in squid axons. J. Gen. Physiol. 67, 433-467 (1976)

Scarpa, A.: Indicators of free magnesium in biological systems. Biochemistry 13, 2789-2794 (1974)

Scarpa, A.: Kinetics and energy coupling of Ca^{2+} transport in mitochondria. In: Calcium Transport in Contraction and Secretion. Carafoli, E., Clementi, F., Drabikowski, W., Margreth, A. (eds.). Amsterdam: North-Holland Publ. Co. 1975, pp. 65-76

Scarpa, A.: Metallochromic indicators. Methods Enzym. (1976 in press)

The ADP, ATP Translocation and its Catalyst

M. Klingenberg, H. Aquila, R. Krämer, W. Babel, and J. Feckl

Introduction

The specific transport carriers of biomembranes have become in the last few years one of the most discussed challenges of present-day biochemistry. Studies on the ADP, ATP translocator of the inner mitochondrial membrane have contributed considerably to enlarge our understanding of the translocation mechanism on the molecular basis. Nature has provided unusually favorable conditions for elucidating the mechanism and molecular properties of this system: (1) it is the most active transport system in aerobic eukaryotic cells and, therefore, present in high concentrations in the mitochondria membranes; (2) two groups of highly specific inhibitors have been provided by nature which seem to be engineered exactly for the purpose of tracking a translocation mechanism for the first time on a molecular level. Not enough, these inhibitors then turned out to be instrumental in the first isolation of an undenatured metabolite carrier.

The following mechanistic questions have been discussed with respect to the ADP, ATP translocation. The exchange process with multiple or single carrier sites (Klingenberg, 1967; Klingenberg et al., 1976); the asymmetry of the carrier on both sides of the membrane (Klingenberg et al., 1974); mobile, diffusable, rotatory or fixed gate mechanism; the stages of the translocation process (mobilizing and translocating steps); activation mechanism of the translocator as the result of protein lipid interaction (Klingenberg, 1974, 1975). Some of these questions will be discussed both as the carrier in situ in the membrane, and more conclusively as the isolated carrier protein.

Reorientation Mechanism

Binding studies of the ADP, ATP translocator in the mitochondrial membrane with the substrate and with the inhibitor ligands produced an unexpected insight into the carrier mechanism (Klingenberg et al., 1971, 1973, 1974; Erdelt et al., 1972; Klingenberg, 1974). These binding data were at first puzzling and unexplained, if one resorts to classical interpretations based on an open interaction between ligand and binding site. In particular the introduction of regulatory binding sites in addition to substrate binding sites, according to some sort of fashion in enzymology (allostery), further increased the confusion (Vignais, 1976). Solutions emerged by recognizing that one deals with binding sites of a translocator which according to the mobile carrier model can be distributed both on the inner and outer face of the membrane and can switch back and forth between both sides. On ligand addition the carrier sites might be redistributed depending on the nature and concentration of the ligands. In this manner, the at

REORIENTATION MECHANISM - POSTULATES

- SINGLE BINDING SITE, ORIENTED EITHER IN - OR OUTSIDE
- BINDIND SITE ONLY "MOBILE" WHEN ACTIVATED BY SUBSTRATE (ADP, ATP)
- IDENTICAL BINDING SITES FOR SUBSTRATE AND INHIBITORS
- ATR, CAT AS IMPERMEANT LIGANDS BIN ONLY FROM OUTSIDE, BKA AS PERMEANT LIGAND BINDS FRON INSIDE

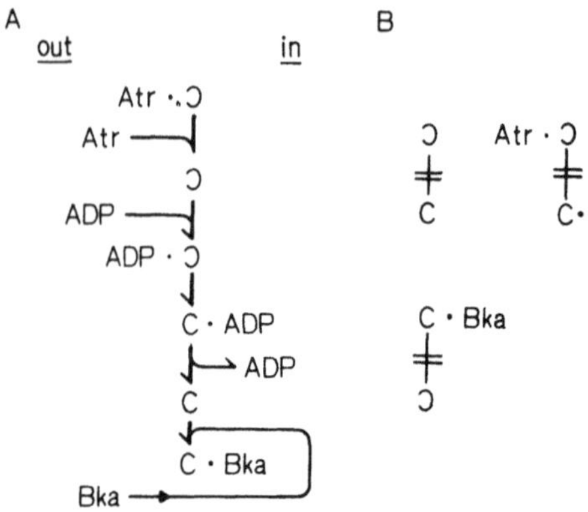

Fig. 1. The reorientation mechanism and its postulates

first apparently contradictory binding data afforded an insight into the mechanism and the dynamics of the translocation process (Klingenberg, 1976).

The most important results came from the interaction between the three types of ligands, the ADP, ATP; atractylate (ATR) and carboxy-atractylate (CAT); and bongkrekate (BKA) and isobongkrekate (IBKA) (Lauquin et al., 1975). Particularly informative were the opposite effects of ATR and BKA on ADP binding which gave the major clue for a unifying explanation of the binding data.

With the reorientation mechanism (Fig. 1) of the ADP, ATP exchange translocator for the first time on a molecular level, a translocation process was demonstrated, Here is not the place for the detailed postulates of the reorientation mechanism. Essentially the binding site of the exchange carrier is assumed to be oriented either to one or the other side of the membrane. The binding site is activated for the translocation only by the substrates ADP and ATP, thus assuming a counterexchange. ATR and CAT as impermeant ligands bind only from the outside of the mitochondrial membrane, BKA as permeant ligand binds also from the inside. Because of the high affinity and specificity with these inhibitor ligands, it is possible in accordance with the reorientation mechanism to fix all carrier sites either on the outside with ATR or CAT or on the inside with BKA.

With sonic particles where the inner m-side of the membrane is turned outside, evidence is provided for an asymmetry of the membrane side with respect to its ligand specificity since there BKA binds directly from the outside as an effective inhibitor whereas CAT and ATR are ineffective. As a result two different states of the carrier were postulated according to whether the binding site is oriented to the c- or m-side of the membrane (Klingenberg et al., 1974):

"c-state:" high affinity for CAT, ATR, low affinity for BKA, medium affinity for ADP, ATP.
"m-state:" high affinity for BKA, low affinity for CAT, ATR, medium affinity for ADP, ATP.

The Asymmetric Gated Pore

On the basis of these data the model of the asymmetric gated pore was derived (Klingenberg et al., 1976) (Fig. 2). This was based on the realization that translocation of the substrate results in an asymmetry of the carrier sites since the substrate on one side binds with another face to the carrier than from the other side. In contrast in a rotatory mechanism, protein and substrate will be orientated equally to each other on both sides of the membrane. The extreme asymmetry of the binding site found in the ADP, ATP carrier such that high affinity ligands lose their affinity almost completely when the carrier site faces the other side of the membrane, has been correlated to the asymmetric substrate translocation.

The process is illustrated in Figure 2. The binding site consists of a central binding area and binding areas which are specifically active in the c- or the m-state. Thus, the single site has a dual specificity by exposing different regions of binding surface with accordingly different specificities. In the c-state the substrate comes in contact only with one specific binding site and in the m-state with the other. The inhibitor ATR binds only with the c- and central region and BKA only with the m- and central region.

The scheme also implies that on binding substrates the protein attains activated conformations (c* or m*) which will be discussed at the end. Furthermore, it indicates that the carrier may be a dimer which forms the central protein. Actual evidence on the isolated protein is in favor of the concept that the ADP, ATP carrier is a dimer consisting of two subunits of each 30,000 mol wt (Riccio et al., 1975; Klingenberg et al., 1976).

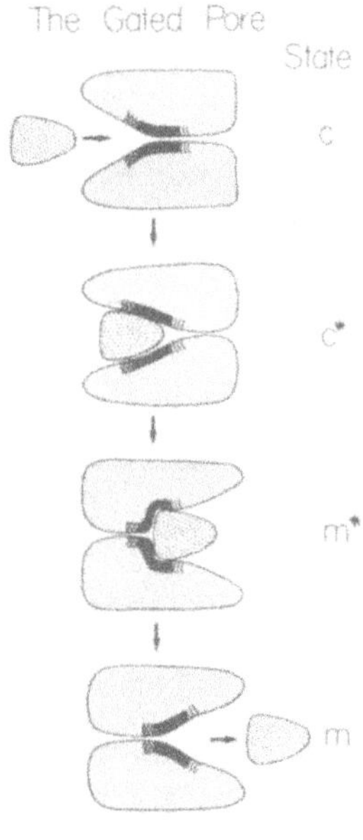

Fig. 2. The asymmetric gated pore. The model stresses that the binding site has a region common to both orientation states (c-,m-region) and that for each state, additional specific regions (c or m-region) are operative in binding the substrate.
c* and m* are the activated states

Reorientation in the Mitochondrial Membrane

Another indication for reorientation of the carrier sites came from studies concerninf the shrinkage, swelling of mitochondria linked specifically to binding of the ligands to the ADP, ATP carrier. In detailed studies we were able to show that on contraction of the mitochondria induced by BKA the carrier is pulled to the m-side and on decontraction by ATR to the c-side (Fig. 3) (Scherer and Klingenberg, 1974). The configurational changes of the mitochondria were interpreted as a spreading of one half larger of the membrane as compared to the other in concurrence with the orientation of the carrier. Since the amount of carrier protein in the membrane is very high (up to 12%) a conformational change and possibly even a shift of this protein from one to the other side of the membrane can be visualized to result in the larger spreading due to the conformation change between the c- and m-state, changing the accessibility of the carrier protein.

For this purpose, the lactoperoxidase catalyzed iodination (Philipps and Morrison, 1970) of the carrier protein in the mitochondrial membrane was examined under the influence of these ligands. The main prerequisite for these studies has been the identification in the mitochondrial membrane of a major protein of mol.wt. of 29,000 as the ADP, ATP carrier (Riccio et al., 1975). This band stands out in SDS-PAGE and has been shown to be the major protein of mitochondria to incorporate iodine (Boxer, 1975), sug-

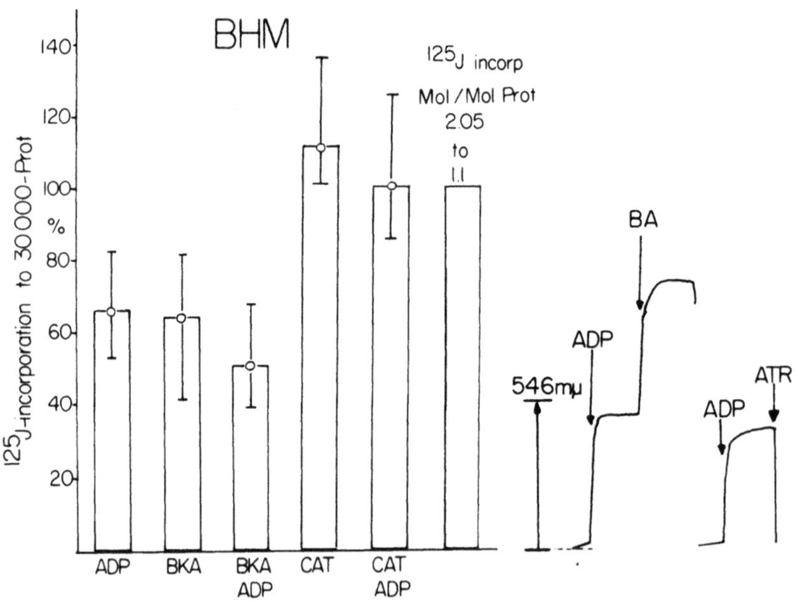

Fig. 3. The influence of ligands and the accessibility of the ADP, ATP carrier to the outer surface of the inner mitochondrial membrane, as assessed by the lactoperoxidase catalyzed ^{125}I-incorporation. Beef heart mitochondria are exposed under the prior addition of ADP, BKA, CAT, etc. to lactoperoxidase and ^{125}I. The mitochondria were dissolved in SDS and the polypeptides separated by PAGE. The coomassie blue stained band at 30,000 mol.wt. was excised from the gels and ^{125}I incorporation measured

gesting that the ADP, ATP carrier is the only major protein of the mitochondrial membrane which surfaces to the outer side. The addition of various ligands to the ADP, ATP carrier changes the incorporation of iodine as shown in Figure 3, such that with CAT there is a slight increase, with ADP and BKA a certain decrease and with BKA plus ADP together the incorporation is inhibited by 50%. This behavior is in good agreement with the prediction of the reorientation mechanism which says that the site in the absence of external ADP is primarily in the c-state with external ADP partially in the m- and c-state and with BKA plus ADP only in the m-state. The difference in the iodination goes parallel to the effect of ligands observed on the absorbance (i.e. contraction) of the mitochondria: no effect with ATR, partial contraction with ADP and full contraction with BKA plus ADP.

The Catalytic Role of ADP on the Transition Between c and m-state in Mitochondria

The most intriguing and revealing information in the studies of ligand binding is the role of ADP and ATP. The replacement of one ligand inhibitor by the other can be directly demonstrated since all these inhibitors are available radioactively labeled. Fortu-

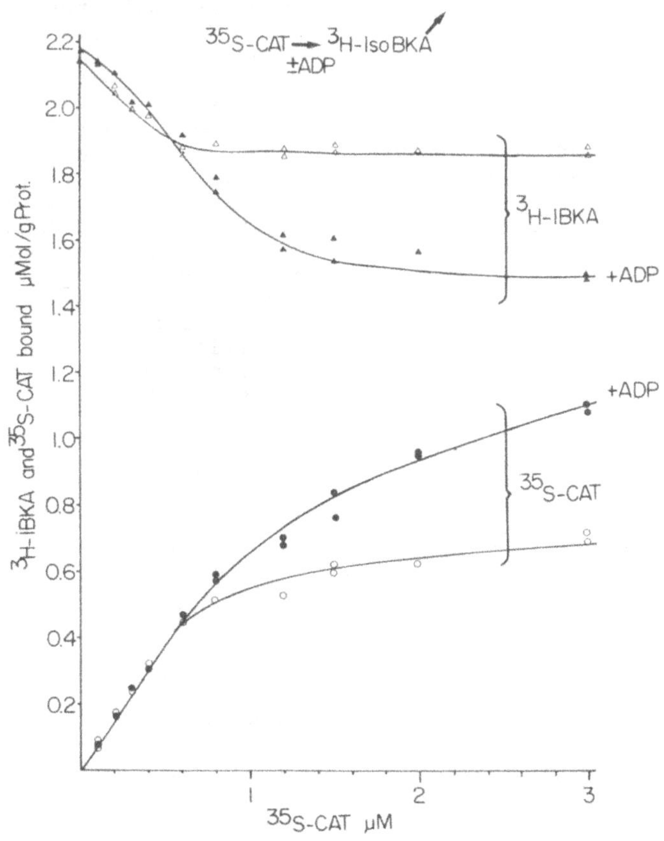

Fig. 4. ADP catalyzed displacement of ^{3}H-IBKA by ^{35}S-CAT (transition m → c state) in mitochondria
Beef heart mitochondria are first loaded with ^{3}H-IBKA, washed and then incubated with increasing concentration of ^{35}S-CAT, in the presence or absence of 10 μM ADP

nately, in both inhibitor groups there is a more tightly and loosely binding form available, CAT and ATR; BKA and iso-BKA. Thus, it is possible to displace the more weakly binding species with the tightly binding compound of the other group and to demonstrate the reversibility of the transition between the two states, in both directions:
Transition c → m state corresponds to displacement of ^{35}S-ATR by ^{3}H-BKA and transition m → c state corresponds to removal of ^{3}H-IBKA by ^{35}S-CAT.

An example for the latter case is shown in Figure 4. It must be noted that ^{3}H-BKA and ^{3}H-IBKA bind also to a large extent nonspecifically to the membrane because of their lipophilic nature whereas ^{35}S-ATR and ^{35}S-CAT bind exclusively to carrier sites. The effect of ADP on promoting the displacement ^{3}H-IBKA is clearly demonstrated. Dependence on the concentration of ^{35}S-CAT reveals that the binding is increased by ADP and this increment corresponds to the displacement of ^{3}H-IBKA. These results demonstrate that the reorientation is catalyzed by the transport active substrates ADP and ATP, and the displacement is part of the translocation process postulated in the scheme of Figure 1.

The Carrier Isolated in the c or m State

The possibility to fix the ADP, ATP translocator in the c-state with CAT and in the m-state with BKA, facilitated the isolation of the translocator in an undenatured, functionally defined form. The high affinity of these ligands prevents denaturation of the carrier by nonionic detergents such as Triton X-100 or alkyl-aminoxide. The CAT-protein complex was isolated two years ago and characterized in the meantime in several respects (Riccio et al., 1975; Klingenberg et al., 1975). It is stable in Triton X-100 for several weeks whereas the BKA protein complex is much less stable and can undergo rapid degradation by proteolysis (Aquila and Klingenberg, unpublished). For our present purpose it is important to note that the content of BKA in the isolated protein is equal to that of CAT. That the isolated CAT protein and BKA protein originate from the same protein is shown by their identical properties and amino-acid composition (Table 1). Furthermore, it may be stated that with the addition of CAT and subsequent loading with ^{3}H-BKA, no ^{3}H-BKA can be isolated.

Table 1. Comparison of purified BKA- vs. CAT-protein

	BKA-Protein	CAT-Protein
Apparent mol.wt. in SPAGE	30 28(5-20%)	30 -
Inhibitor content μM.g prot^{-1}	17.6	18.6
SH- content Mol/30,000 (carboxymethylation)	1.5	1.7
N- Terminal	none detectable	
Amino Acid comp.	identical	
Polarity	39%	

The isolated CAT-protein and BKA-protein can be considered to be a pure undenatured form of the carrier which are completely in a functionally defined state. They can be used for functional studies which relate to the translocator process. A recognition site in the isolated CAT protein for ADP or ATP was shown by the specific displacement of ^{35}S-CAT with ADP and thus indicates that the CAT binding protein is in fact the ADP, ATP carrier, again in agreement with the orientation mechanism. It contrasts to speculations in the literature that CAT binds to separate protein as ADP or ATP (Winkler and Lehninger, 1968).

Of particular interest is the displacement of CAT by BKA. The kinetics of this reaction are shown in Figure 5, in a kinetically resolved microdialysis. Maximum rate of displacement is of course obtained by addition of excess unlabeled CAT. BKA alone is rather inefficient but displace CAT faster in synergism with ADP. This is in most remarkable agreement with effects in the intact membrane, moreover, since the ADP has been postulated in the reorientation mechanism to accomplish the transition from the c- to the m-state by migrating through the carrier. It would appear then that something like a translocation might be achieved in the isolated protein in an apparently slow manner.

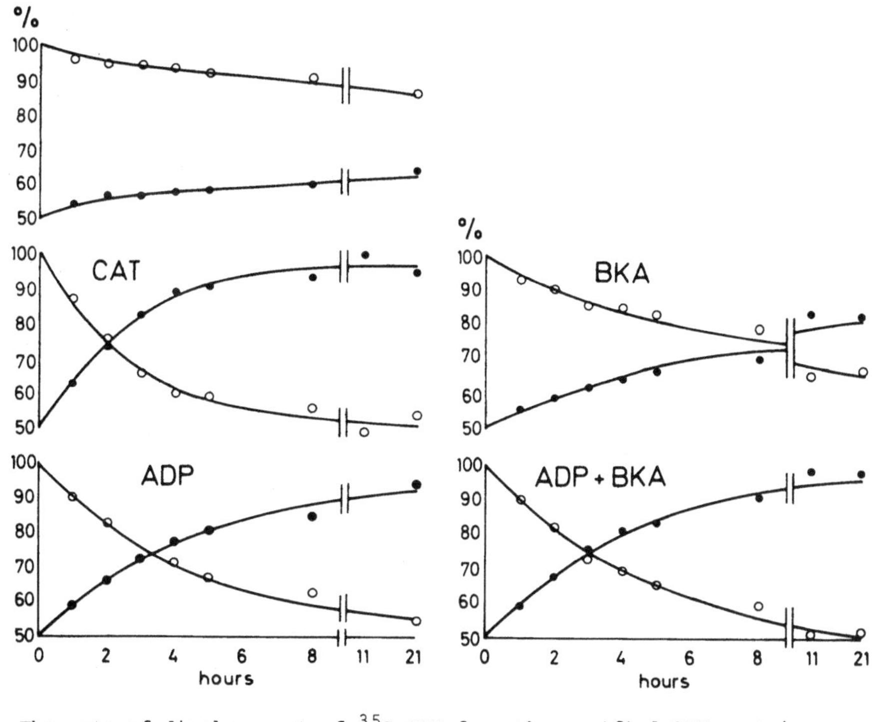

Fig. 5. The rate of displacement of ^{35}S-CAT from the purified CAT-protein complex by various ligands. Demonstration of synergistic effect of ADP and BKA. Measurement by microdialysis, removal of aliquotes from the dialysis chamber at the time indicated. At time = 0 excess of the displacing ligands is added

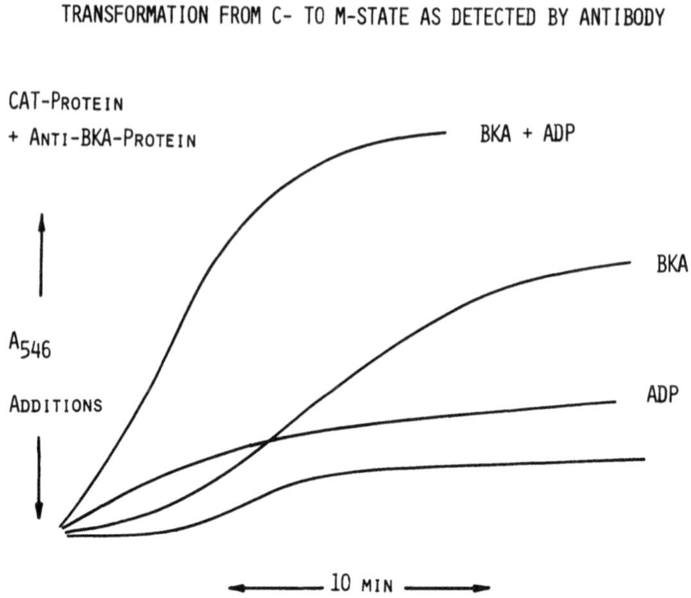

Fig. 6. The interconversion of the CAT protein into the BKA protein as indicated by a BKA specific antibody. Purified CAT protein is incubated with added ligands as indicated, in the presence of anti-BKA-protein. Recording of the absorbance increase due to the immuno-precipitation

Antibodies against the isolated undenatured CAT-protein and also BKA-protein turned out to be a useful tool in studying these conformations. Antibody against the CAT protein and the BKA protein are largely conformation-specific, i.e. they had little cross reaction with the other protein form (Klingenberg et al., 1976; Buchanan et al., 1976). The specificity of the antibodies for the c- and m-conformation are at the present one of the best indicators for the differences in the conformation of the same polypeptide in the two states. In fact, they could be used in trapping the interconversion of the conformation.

The antibody against the BKA protein was used in order to detect the generation of BKA-protein from the CAT-protein on the addition of BKA and ADP (Fig. 6). For this purpose the precipitation due to the antibody complex formation was followed. In fact on addition of BKA and ADP, after a certain lag period, an absorption increase is observed, indicating the formation of the BKA-ADP protein complex with its specific immunogenic properties. Control experiments shown however, that a small portion of the protein has been converted into the m-conformation. The largest part disintegrates to the denatured form after the removal of CAT, due to the lability of the unliganded protein and the BKA-complex.

Because of the conformational specificity of the antibodies it might be expected that in the antigen-antibody complex the transition to the other conformation is inhibited and as a result of the conformational fixation, it should not be possible to remove CAT by BKA + ADP. In fact, it has been shown (Klingenberg et al., 1975; Buchanan et al., 1976) that the removal of ^{35}S-CAT by BKA + ADP is inhibited by antibody. At the same time the displacement of the ^{3}S-CAT is not inhibited by the antibody demonstrating that the antibody binds to the surface different from the CAT binding site. These results may be interpreted to show that the conformations transition from the c- to m-side is so extensive that it is prevented by the protein-protein interaction in the antigen-antibody complex.

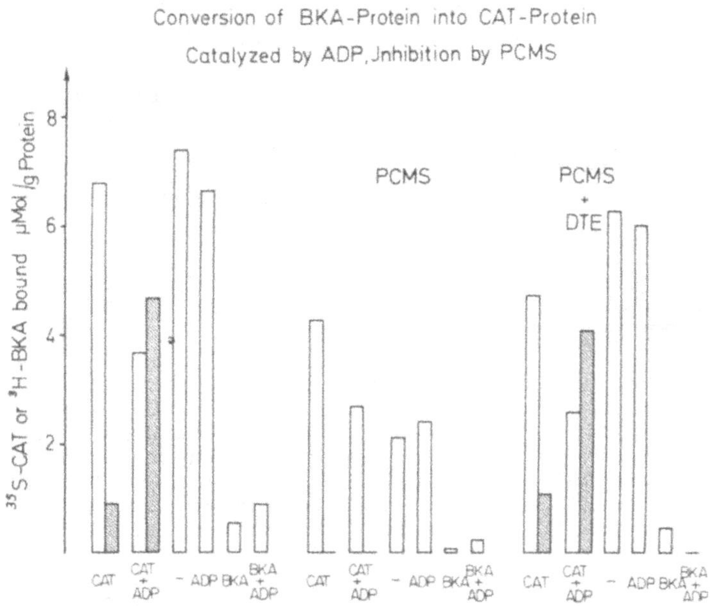

Fig. 7. Direct demonstration of the conversion of the purified BKA protein into the CAT protein, as catalyzed by ADP.
Purified ^{3}H-BKA loaded protein is exposed to ^{35}S-CAT, BKA, ADP as indicated. The binding is measured by equilibrium dialysis. Addition of p-chloro-phenylmercuri-sulfonate (PCMS) and dithioerythrol as indicated (H. Aquila, unpublished)

The opposite conversion BKA protein CAT protein has become possible more recently after the BKA protein became available. This reaction is easier demonstrated because of the greater stability of the produced CAT protein, which, therefore, can be measured in sizeable amounts. The lability of the H-BKA protein requires, however, that these measurements are made within at least 20 h after isolation of the ^{3}H-BKA protein. With double-labeling experiments and the availability of both ^{3}H-BKA and ^{35}S-CAT it has been possible to demonstrate directly the removal and corresponding binding of CAT to the isolated protein. As shown in Figure 7 addition of CAT alone to the protein leads to a relatively small displacement and binding of CAT. Only the addition of CAT + ADP together decreases ^{3}H-BKA by 50% and results in a remarkable ^{35}S-CAT binding which equals the amount of ^{3}H-BKA removed. It also demonstrates that ADP is required as a catalyst for the transition from the m- to the c-state. ADP alone is ineffective in removing BKA whereas BKA alone replaces ^{3}H-BKA completely. Obviously the homo-displacement does not require ADP in contrast to the heterodisplacement (BKA-CAT). This represents the hitherto most convincing experiment for the interconversion between the c- and m-state and its catalysis by ADP, in concurrence with the reorientation mechanism.

In experiments to be reported separately, it could be shown that only in the m-state (with BKA + ADP) a SH-group is available to specific alkylation which then prevents binding of transition to the c-state (Klingenberg et al., 1975). So far, in mitochondria NEM has been applied for this purpose because of its lipophilicity. In the isolated protein also PCMS can be applied. As shown in

Figure 7, PCMS prevents the removal of BKA by CAT and the concomitant binding of CAT. Its level of binding is here somewhat decreased, as seen by the control. At the same time the home-displacement by BKA is not inhibited. After removal of the PCMS by dithioerythrol (DTE), the displacement by BKA by CAT is again possible. These results demonstrate that only in the specific conformation of the m-state is the SH-group exposed, and explain that the inhibition by NEM of the ADP, ATP transport is possible only in the m-state. This SH-group is obviously involved in the conformational transition.

Reconstitution

Over the last years we have also endeavored to "reconstitute" the isolated ADP, ATP carrier protein. The main subject of these studies was at first to see whether the binding of ligands to the protein can be increased by reconstitution of carrier into protein. The studies became more meaningful in particular because it became possible to isolate the undenatured carrier protein with another nonionic detergent alkylaminoxide, even in the uncomplexed state. This form is very labile and a short-cut procedure for a partial purification has been applied. On mixing the protein in the presence of relatively low detergent concentrations with liposomes, the protein becomes completely incorporated. The quantitative uptake of the protein into liposomes reflects the extreme hydrophobic nature of this protein and furnishes a favorable basis for a series of quantitative reconstitution studies.

On liposome incorporation, the binding properties with the various ligands, by which we define at this state the functional state of the carrier, are largely reconstituted. Since at the start the protein is unloaded, binding can be studied by addition of the ligands, by comparing the free and liposome-incorporated protein. The binding of CAT is increased about two-fold (Table 2). More remarkable is the strong increase of the affinity by nearly two orders of magnitude. There is nearly no effect of ADP on the binding of CAT.

Table 2. Reconstitution

Preparation		ADP	Binding affinity high C_o ($nmol.mg^{-1}$)	high K_d (nM)	low C_o ($nmol.mg^{-1}$)	low K_d (nM)
			CAT binding			
BHM	-	-	14-18	12	-	-
solubilized	-	-	1.6	310	-	-
binding	+ liposomes	-	3.3	9	-	-
protein	+ liposomes	+	3.2	10	-	-
			BKA binding			
BHM	-	+	10-18			
solubilized	-	-	0	-	0.2	2000
binding	+ liposomes	-	0.2	100	0.2	1500
protein	-	+	0.7	80	3.75	1500
	+ liposomes	+	6.0	80	0	-

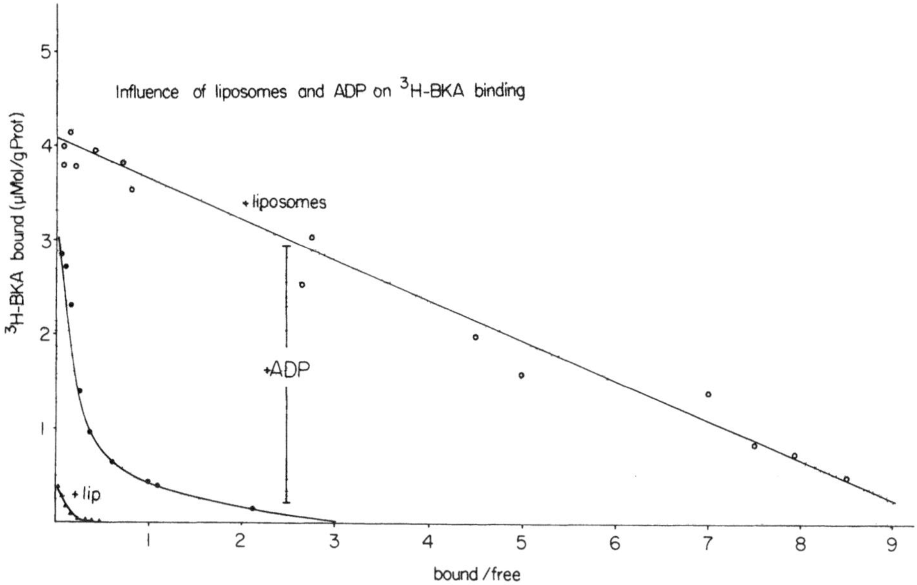

Fig. 8. The reconstitution of BKA binding to isolated, unloaded carrier protein by liposome incorporation.
Mass action plots of ^{3}H-BKA binding demonstrating the number of binding sites and dissociation constants. The unloaded protein is isolated with the detergent alkylaminoxide and mixed with liposomes prepared by sonication from egg lecithine. The binding is measured, after addition of varying amounts of ^{3}H-BKA by Millipore filtration (R. Krämer, unpublished)

The binding of ^{3}H-BKA is much more dependent on the incorporation to the liposomes. Moreover it is nearly fully dependent on the addition of ADP (Fig. 8). The nonincorporated protein has no measurable binding except with ADP. Only on addition of ADP is a small amount of high affinity and a larger amount of low affinity binding $K_d > 10^{7}M$ observed. After the incorporation to liposomes, the binding is only slightly improved but is increased more than ten-fold by ADP reaching nearly 5 μmol.g protein^{-1} ^{3}H-BKA_{bound}. The affinity to ^{3}H-BKA is strongly enhanced, nearly 30-fold, with a uniform class of binding sites. Thus the constitution achieves not only an increase in the number of binding sites but also a strong improvement of the affinity to its ligands reaching nearly that found in the intact membrane. In particular, the requirement for ADP for the transition to the m-state, is most strongly expressed in the reconstituted system.

The fact, that the binding of CAT does not require ADP as compared to binding of BKA, can be rationalized in terms of the reorientation mechanism. In the unloaded state the carrier can be expected to be mainly in the c-state trapped on the outside. Therefore, the isolated protein should be largely in the c-state and would require the ADP catalyzed transition to another state only for the binding of BKA. There is independent evidence based on the immunological properties that the isolated unloaded protein resembles the c-state since it reacts readily with the anti-CAT protein.

Conclusions

In conclusion it can be summarized that the binding studies on the ADP, ATP carrier on the molecular level have given a penetrating insight into the major catalytic effect of biomembranes, the facilitated specific transport. The experimental demonstration on the intact membrane of a reorientation sequence of translocation steps has been extended with a remarkable consistency to the isolated undenatured carrier protein forms. The transition between the c- and m-state was made possible by trapping these states with two different types of inhibitors and is facilitated only by of the transported substrates, ADP and ATP. This shows that the actual transition between the c- and m-state includes the ADP dependent translocation process. The reconstitution of these steps opens further possibilities to forward evidence for the translocation mechanism.

Acknowledgments. This work was supported by the Deutsche Forschungsgemeinschaft, Schwerpunktprogramm: Struktur und Funktion biologischer Membranen.

Abbreviations

ATR, atractylate; BKA, bongkrekate; IBKA, isobongkrekate; NEM, N-ethylmaleimide; PAGE, polyacrylamide electrophoresis; PCMS, p-chloro-phenylmercuri-sulfonate; CAT, carboxy atractylate.

References

Boxer, D.H.: The location of the major polypeptide of the ox heart mitochondrial inner membrane. FEBS Lett. 59, 149-152 (1975)

Buchanan, B.B., Eiermann, W., Riccio, P., Aquila, H., Klingenberg, M.: Antibody evidence for different conformational states of the ADP, ATP translocation protein isolated from mitochondria. Proc. Natl. Acad. Sci. U.S. Vol. 73, 2280-2284 (1976)

Erdelt, H., Weidemann, M.J., Buchholz, M., Klingenberg, M.: Some principle effects of bongkrekic acid on the binding of adenine nucleotides to mitochondrial membrane. Europ. J. Biochem. 30, 107-122 (1972)

Klingenberg, M.: Some speculations on the nature of the adenine nucleotide exchange reaction. In: Mitochondrial Structure and Compartmentation. Quagliariello, E., Papa, S., Slater, E.C., and Tager, J.M. (eds.). Bari: Adriatica Editrice, 1967, pp. 317-319

Klingenberg, M.: The mechanism of the mitochondrial ADP, ATP carrier as studied by the kinetics of ligand binding. In: Dynamics of Energy-Transducing Membranes. Ernster, L., Estabrock, R.W., and Slater, E.C. (eds.). Amsterdam: Elsevier, 1974, pp. 511-528

Klingenberg, M.: Energetic aspects of transport of ADP and ATP through the mitochondrial membrane. In: Energy Transformation in Biological Systems, Ciba Foundation Symposium. Amsterdam: Associated Scientific Publ., 1975, Vol. XXXI, pp. 105-124

Klingenberg, M.: The ADP-ATP carrier in mitochondrial membranes. In: The Enzymes of Biological Membranes: Membrane Transport. Martonosi, A.N. (ed.). New York-London: Plenum Press 1976, Vol. III, pp. 383-438

Klingenberg, M., Aquila, H., Riccio, P., Buchanan, B.B., Eiermann, W., Hackenberg, H.: Characterization of the isolated ADP, ATP carrier. In: Electron Transfer Chains and Oxidative Phosphorylation. Quagliariello, E., Papa, S., Palmieri, F., Slater, E.C., and Siliprandi, N. (eds.). Amsterdam_Oxford: North Holland, 1975, pp. 341-348

Klingenberg, M., Buchholz, M., Erdelt, H., Falkner, G., Grebe, K., Kadner, H., Scherer, B., Stengel-Rutkowski, L., Weidermann, M.J.: The adenine nucleotide carrier: Study of its translocating mechanism by binding with adenosine diphosphate, atractyloside and bongkrekic acid. In: Biochemistry and Biophysics of Mitochondria Membranes. Azzone, G.F., Carafoli, E., Lehninger, A.L., Quagliariello, E., and Siliprandi, N. (eds.). New York-London: Academic Press, 1971, pp. 465-486

Klingenberg, M., Riccio, P., Aquila, H., Buchanan, B.B., Grebe, K.: Mechanism of carrier transport and the ADP, ATP carrier. In: The Structural Basis of Membrane Function. Hatefi, Y., Djavadi-Ohanianc, L. (eds.). New York-San Francisco-London: Academic Press, 1976, pp. 293-311

Klingenberg, M., Riccio, P., Aquila, H., Schmiedt, B., Grebe, K., Topitsch, P.: Characterization of the ADP/ATP carrier in mitochondria. In: Membrane Proteins in Transport and Phosphorylation. Azzone, G.F., Klingenberg, M.E., Quagliariello, E., and Siliprandi, N. (eds.). Amsterdam: North-Holland, Publ. Co., 1974, pp. 229-243

Klingenberg, M., Scherer, B., Stengel-Rutkowski, L., Buchholz, M., Grebe, K.: Experimental demonstration of the reorienting (mobile) carrier mechanism exemplified by the mitochondrial adenine nucleotide translocator. In: Mechanisms in Bioenergetics. Azzone, G.F., Ernster, L., Papa, S., Quagliariello, E., and Siliprandi, N. (eds.). New York-London: Academic Press, 1973, pp. 257-284

Lauquin, G., Duplaa, A.M., Rousseau, A., Vignais, P.V.: Radioactively labeled bongkrekic acid. Isolation and identification of isomers. Studies of their biological properties. Abstr. 1160. 10th FEBS Meet. Paris. 1975

Phillips, D.R., Morrison, M.: The arrangement of proteins in the human erythrocyte membrane. Biochem. Biophys. Res. Commun. 40, 284-289 (1970)

Riccio, P., Aquila, H., Klingenberg, M.: Purification of the carboxy-atractylate binding protein from mitochondria. FEBS Lett. 56, 133-138 (1975)

Scherer, B., Klingenberg, M.: Demonstration of the relationship between the adenine nucleotide carrier and the structural changes of mitochondria as induced by adenosine 5'-diphosphate. Biochemistry 13, 161-170 (1974)

Vignais, P.V.: Molecular and physiological aspects of adenine nucleotide transport in mitochondria. Biochim. Biophys. Acta 456, 1-38 (1976)

Winkler, H.H., Lehninger, A.L.: The atractyloside-sensitive nucleotide binding site in a membrane preparation from rat liver mitochondria. J. Biol. Chem. 243, 3000-3008 (1968)

Bacterial Transport Systems

The Transport of Sugars Across Bacterial Membranes (Aharon-Katzir-Katchalsky Memorial Lecture)*

S. Roseman

In addition to being an especially gifted and imaginative scientist, Aharon Katzir-Katchalsky was a truly unusual human being. It is a great honor to be asked to present this lecture, and I wish to thank the organizers of this meeting for the invitation. I could well spend the remainder of my time trying to discuss Aharon Katzir's many contributions including those to the area of membrane function, but I believe that a real appreciation of his work lies in the future. The complexities of biological membranes and the ways in which they function are only beginning to be realized, and we are only in the early descriptive phase of this work. It is when this phase has been completed, and when these complex organelles have been dissected and reassembled that this field will become a science. Then, we will be able to apply Katzir's ideas and concepts more fully to these systems. Then, we will begin to understand the full significance of his work. He was a man far ahead of his time.

The complexity of membranes and membrane systems is exemplified by my subject, the bacterial phosphotransferase system, or the PTS. As you will see, the PTS itself is not only very complex, but it interacts by unknown mechanisms with a variety of other systems so that it plays a key role in many important physiological processes.

The PTS was accidentally discovered in 1964 (1, 2, 3), and was found to catalyze the phosphorylation of N-acetylmannosamine, yielding the corresponding 6-phosphate ester, a key intermediate in the biosynthesis of the sialic acids (4).

At the same time, the PTS was found to catalyze the phosphorylation of other sugars. Two unique features of this system were immediately obvious. The first was that the phosphoryl donor was not a nucleotide triphosphate, but was phosphoenolpyruvate, or PEP. The second unique feature was the complexity of the system; theoretically, sugar phosphorylation with PEP could be catalyzed by a single protein. Before describing the system and its biochemical complexity, I would like to summarize some of its known physiological functions.

The best defined function is the transport of PTS sugars by the process of group translocation across the plasma membrane. Some of the known regulatory roles of the PTS are as follows: Adler and Epstein (5) and more recently Melton and Hartman (6) have shown that the PTS is involved in chemotaxis of bacterial cells towards sugar substrates of the system. Peterkofsky and his coworkers (7) and others have demonstrated that the PTS regulates adenylate cyclase. Originally, Makman and Sutherland (8) had shown that sugars such as glucose now known to be substrates of the PTS, in some way regulated intracellular levels of cyclic AMP. Pastan and Perlman (9) also suggested that the PTS regulates intracellular levels of cyclic AMP, and thereby the synthesis of certain inducible catabolic enzyme systems. As will be shown, this type of regulation is not entirely dependent on cyclic AMP, but on the fact that the PTS can regulate the transport of at least some solutes

*Contribution 887 from the McCollum-Pratt Institute

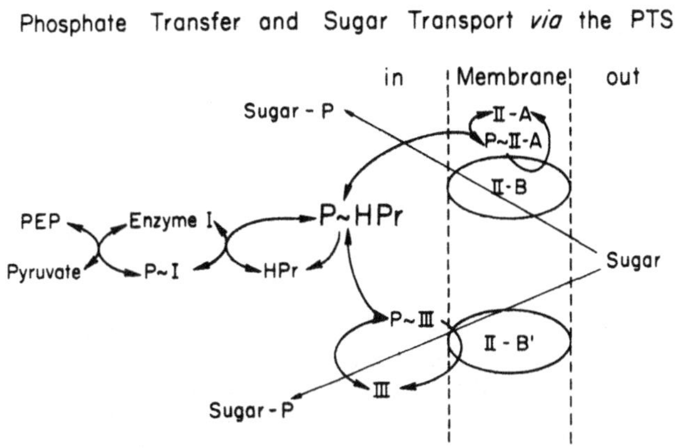

Fig. 1. Sugar translocation and phosphorylation by the phosphoenolpyruvate: sugar phosphotransferase system. Product of vectorial translocation is the sugar phosphate, and the reaction is catalyzed by the sugar binding protein in the membrane (II-B or II-B'), the phosphoryl group being derived from the other sugar-specific component, a soluble protein (phospho-III) or a membrane protein (phospho-II-A). The sugar-specific proteins are phosphorylated by sequential transfer of the phosphoryl group from phosphoenolpyruvate through the general (not sugar-specific) proteins of the system, Enzyme I, and HPr

which are not substrates of the system. In other words, the PTS can regulate at least certain non-PTS transport systems. These regulatory properties of the PTS and perhaps others lead in turn to its involvment as a major component of the complex class of phenomena classed under the name catabolite repression (10, 11).

The fact that the PTS participates in so many cellular functions may be the teleological reason for its unexpected complexity. Various aspects of the PTS have been reviewed (2, 3), and some of these will be briefly discussed here.

The overall reaction, the transfer of a phosphoryl group from PEP to sugar to yield pyruvate and sugar phosphate generally requires four proteins, lipid, and divalent cations. Variants of the basic system exist, but the usual PTS system is that shown in Figure 1. Two of the four proteins (Enzyme I and HPr) are designated as general proteins in that they are not sugar-specific, but are required for the phosphorylation of all sugar substrates of the PTS or, of PTS-sugars. The remaining two proteins are sugar-specific. Thus, a protein duplex is required for a given sugar. Many such sugar-specific protein duplexes can be generated by a single cell, most of them being induced by exposing the cell to the appropriate sugars. As indicated in Figure 1, some PTS sugars (e.g. glucose) are translocated by two sugar-specific duplexes; most sugars appear to be transported by only one system.

With the isolation of purified or homogenous proteins it is possible to show that the overall transfer of the phosphoryl group from PEP to sugar proceeds by a sequential transfer from one protein to another as shown in Figure 1. The phosphoryl group is transferred from PEP to the first general protein of the system, Enzyme I, than to the next general protein called HPr, then to one of the sugar-specific proteins shown as either III (cytoplasmic) or II-A (membrane-bound), and finally to the sugar. The last step is catalyzed by the second sugar-specific protein, called II-B, which is invariably a membrane component. Where a II-B protein has been obtained in purified form, it has been shown to be the sugar-binding protein, or the sugar receptor in the membrane. Again, recent work has shown that variants of this basic system exist. For example, by using appropriate mutants, it has been found that HPr is not always essential but that at least one other protein (perhaps more) can substitute for HPr, and that this variant protein is not normally expressed in the wild type cell. Other work suggests the possibility that PEP, Enzyme I and HPr may not be required but that phosphorylation of some sugar-specific proteins by other high energy phosphoryl compounds may be possible. A full appreciation of these and other potential PTS variants may be realized as we learn more about its regulatory functions, alluded to above. Finally, before leaving Figure 1, it should be emphasized that the PTS comprises two enzymes, Enzyme I and Enzyme II-B. The other two proteins of the system, HPr and Factor III are substrates of these enzymes. The kinetic complexities of such a system are immediately and intuitively obvious; these kinetic complexities have made it very difficult to work with the PTS in crude extracts, and have led to some confusion in the published literature. A brief description of some PTS proteins follows. Enzyme I, isolated from Escherichia coli, Salmonella typhimurium, and Staphylococcus aureus is approximately 70,000 molecular weight. The active enzyme is apparently a dimer (140,000), very sensitive to sulfhydryl inhibitors, labile in the purified state. In all that follows, the E. coli and S. typhimurium proteins appear to be identical and freely substitute for each other in the in vitro assay, whereas the S. aureus proteins are distinctly different and do not crossreact with the comparable proteins from the enteric bacteria (12, 15). HPr is of low molecular weight, 8-10,000 daltons (12, 13, 16). It has been crystallized, but we have not yet succeeded in obtaining crystals large enough for X-ray analysis. On the other hand, the amino acid sequence of S. typhimurium, HPr has been completed (17). In addition, two mutant forms of HPr have been isolated in homogenous form (18) and the location of the amino acid substitutions defined.

In phospho-HPr, the phosphoryl group is linked to the protein at the N-1 position of an imidazole ring of a histidine moiety (16). This is the second phospho-histidine protein isolated, the first being the phosphoryl derivative of succinic thiokinase (19) where the phosphoryl group is linked at the N-3 position. The N-3 linkage has been identified in most of the other more recently isolated proteins of this type. Both of these phosphorimidates are extremely labile to acid hydrolysis, the N-1 phosphoryl derivatives being about ten times more labile than the N-3 derivative. This characteristic lability to hydrolysis is explained by the fact that phospho-HPr is a true high energy phosphoryl derivative as shown in Figure 2 (14). By determining the equilibrium constant of the reaction shown at the bottom of the figure, it is possible to estimate the apparent standard free energy of hydrolysis of phospho-HPr to be -13.4 kilocalories per mole, being slightly below the value for PEP, and almost twice the value for the pyrophosphate bonds in ATP. Thus, phospho-HPr is a true high energy compound, and the appropriate squiggle is used to indicate the linkage.

The sugar-specific proteins transfer the phosphoryl group from phospho-HPr to their sugar substrates. The nomenclature is cumbersome, but was developed purely in an operational manner as differ-

Fig. 2. Apparent standard free energies for the hydrolysis of phosphorylated compounds. The value for the hydrolysis of phospho-HPr was obtained by measuring the equilibrium constant for the enzymatic reaction shown at the bottom of the table

P-HPR: A "HIGH ENERGY PHOSPHATE" COMPOUND

Compound	Standard Free Energy of Hydrolysis $\Delta G^{\circ\prime}$ (kcal)
PEP	−14.8
P~HPR	**−13.4***
1,3-di P-Glycerate	−11.8
P-creatine	−10.3
Acetyl-P	−10.1
ATP	−7.3
Glucose-6-P	−3.3

*Based on Keq(app)=10 (±5) for reaction:
PEP+HPR ⇌ P-HPR + Pyruvate

ent proteins were isolated, before their functions were known. As already indicated, a pair of sugar-specific proteins or a duplex is required for phosphorylating a given sugar. In some cases, both of these proteins are found to be firmly bound to the membrane and cannot be solubilized except by rather drastic techniques (detergents, etc.). In these cases, the sugar-specific protein pair is designated II-A and II-B. In other cases, one protein of the pair is found in the cytoplasmic supernatant of cell-free homogenates, and such proteins are called III. The corresponding membrane component is still designated II-B. Again, it is the II-B proteins which are the sugar receptors or sugar binding proteins in the membrane.

Three types of sugar-specific complexes will be discussed, each to illustrate a different point. Exhaustively washed membranes of E. coli and S. typhimurium obtained from cells grown on any carbon source will phosphorylate three sugars when supplemented with phospho-HPr. The three sugars are glucose, mannose, and fructose and their analogues. These may be regarded as the constitutive II-A/II-B systems of the enteric bacteria. The second system to be disccused is the inducible lactose Enzyme II system from S. aureus. Finally, the third system is again found in enteric bacteria, and is a second glucose specific Enzyme II. It is designated the $III^{Glc}/II\text{-}B^{Glc}$ system of E. coli or S. typhimurium.

The first system was dissociated into its components. When exhaustively washed E. coli membranes were treated with a mixture of urea and butanol (20, 21) giving a soluble fraction and a pellet. Both were required for phosphorylation of the three sugars, fructose, glucose, and mannose, when phospho-HPr was added. Further purification of the soluble fraction led to its separation into three protein components, $II\text{-}A^{Glc}$, $II\text{-}A^{Man}$, and $II\text{-}A^{Fru}$. These appear to be homogenous by conventional disc gel electrophoresis. The pellet fraction was extracted with detergent and purified, ultimately yielding a protein fraction called II-B. The only evidence for homogeneity of this fraction was the fact that it gave a single band on SDS-disc gel electrophoresis. Its tendency to aggregate in the absence of detergent made it difficult to study by conventional physical methods. However, genetic evidence also suggests that II-B is a single protein. In any case, the protein was obtained lipid-free. Reactivation of the II-B protein required both lipid and divalent cation (21) and the most effective lipid by far was a minor E. coli membrane lipid, phosphatidylglycerol. Activation to a lesser extent was also obtained with two other anionic lipids, phosphatidyl-serine and cardiolipin. Inactive lipids included the major E. coli membrane lipid, phosphatidyl-ethanolamine.

One important result of these studies has not yet been fully explored. Reactivation of the II-B protein required both lipid and a divalent cation, but activity was only restored when these were added in a particular sequence. That is, the sequence of mixing of the three components determined whether or not activity was obtained. These results offer an obvious possibility for studying the structure of a membrane lipoprotein, and also of reconstituting the transport system in synthetic vesicles.

The second PTS system to be described is the S. aureus lactose system. In this organism, unlike E. coli, lactose is concomitantly transported and phosphorylated by the PTS. Enzyme I and HPr are constitutive and the HPr protein was purififed to homogeneity (12-15). The sugar-specific duplex, III^{Lac} and $II\text{-}B^{Lac}$ proteins are inducible, and the $II\text{-}B^{Lac}$ protein is a membrane constituent; in fact we did not succeed in solubilizing it despite intensive study. Finally, the III^{Lac} protein has been obtained in homogenous form and shows a molecular weight of 35,000 daltons. It consists of three apparently identical subunits, each of which can accept a phosphoryl group.

The $II\text{-}B^{Lac}$ protein is an integral membrane component and it is this component of the PTS which recognizes and binds the lactose. The important concept is that lactose is simultaneously translocated and phosphorylated by the $II\text{-}B^{Lac}$ protein, the phosphoryl group being derived from phospho-III^{Lac}. This scheme (Fig. 1) represents our thinking on the transport functions of the PTS for its sugar substrates. The general scheme for PTS sugars is shown in Figure 3. Each sugar is translocated and conconmitantly phosphorylated by its Enzyme II complex, the energy for driving this process being derived from phospho-HPr and ultimately from PEP. The model is supported by evidence obtained by different techniques, for example, by the use of membrane vesicles as described

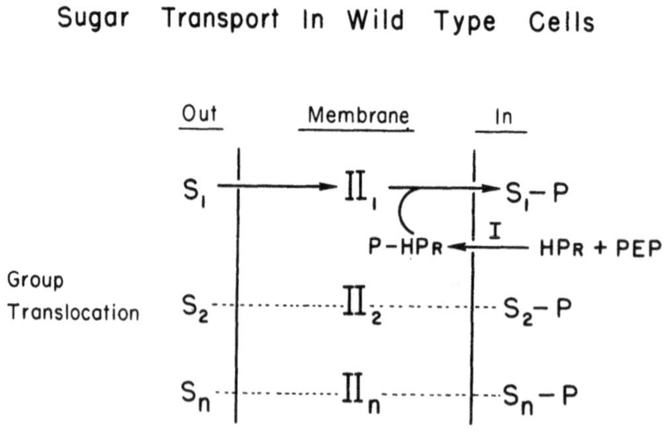

Fig. 3. General scheme for PTS-mediated group translocation. Sugars S_1 through S_n are translocated and concomitantly phosphorylated by their corresponding Enzyme II complexes, the phosphoryl group being derived from phospho-HPr. As indicated in the text, variants of the general system exist

by Kaback (22). There is a great deal of evidence to support this model, and to my knowledge none that argues against it. One experiment illustrates the point. The model shown in Figure 3 makes certain predictions and the predictions are that mutants defective in a sugar-specific protein should be unable to transport that sugar, but should not be affected in the transport of other PTS sugars. Further, mutants defective in a general protein, such as Enzyme I or HPr, should be defective in transporting all PTS sugars, S_1 through S_n. It is conceivable that an Enzyme I or HPr mutant could transport a PTS sugar by facilitated diffusion.

In the case at hand, the lactose system of *S. aureus*, appropriate mutants were, therefore, isolated and tested for their ability to utilize and to transport PTS sugars. The fermentation properties of the mutants agreed with the predictions. Enzyme I mutants were defective in their ability to ferment all sugars, while III^{Lac} and $II\text{-}B^{Lac}$ mutants were unable to ferment either lactose or galactose (galactose is transported via the lactose system in this organism).

The transport properties of the mutants explained their fermentation properties. Methyl α-glucoside was used as a glucose analog to illustrate how a sugar not transported by the lactose system behaves in these mutants. The III^{Lac} and $II\text{-}B^{Lac}$ mutants transported methyl α-glucoside at about the same rate as the parental strain. However, the Enzyme I mutant was unable to take up this sugar at a detectable rate. The same cells were tested with a nonmetabolizable analog of lactose, thiomethyl β-galactoside, and neither the sugar-specific protein mutants, nor the Enzyme I mutant were able to take up the sugar over an extended period of time. Furthermore, the mutants were unable to equilibrate internal with external concentrations of TMG. Thus, studies with these and similar mutants obtained from other organisms meet the predictions made earlier, and fit the concept that the PTS catalyzes a concomitant translocation and phosphorylation of its sugar substrates.

The transport of non-PTS sugars will now be considered. PTS and non-PTS sugars cannot be defined in absolute terms, but instead depend on the organism. This is illustrated in Figure 4. One group

Sugars Transported *via* the **PTS**
(Staphylococcus aureus; Escherichia coli)

S. aureus and *E. coli*
PTS -sugars

Glucose, fructose, mannose (N-Ac-mannosamine), mannitol, sorbitol, β-glucosides, glucosamine, N-acetylglucosamine.

S. aureus PTS-sugars	*E. coli* non-PTS-sugars*
Lactose, sucrose, galactose, trehalose, melibiose, melezitose, pentoses.	**Lactose**, maltose, melibiose, glycerol, pentoses, hexose-P, galactose.

*PTS-sugars are taken up by group translocation, non-PTS-sugars by other processes.

Fig. 4. PTS and non-PTS sugars in *S. aureus* and *E. coli*. *S. typhimurium* displays the same properties as *E. Coli*, except that it does not contain the lactose operon of sugars are PTS sugars in both S. aureus and E. coli. These include glucose, fructose, mannose, etc. In S. aureus, Egan and Morse (23-25) and later Hengstenberg et al., (26), Kennedy and Scarborough (27), and Simoni et al. (12-15) showed that essentially all sugars are PTS sugars except for maltose (28) and perhaps glycerol. In the enteric bacteria, E. coli, and S. typhimurium, the sugars at the lower right of the figure are non-PTS sugars. For example, lactose, which appears as a PTS sugar in S. aureus is a non-PTS sugar in E. coli.

How can a single organism, such as E. coli and S. typhimurium, contain both the PTS system and yet take up sugars such as lactose by mechanisms which do not appear to involve the PTS? There are two obvious possibilities. One, the systems have evolved independently, even in the same organism. Second, that they are related, one having evolved from the other. I will first present evidence that the systems are indeed different, and then, that the PTS does in fact regulate at least some non-PTS sugar uptake systems. For our purpose, we chose melibiose (29), rather than lactose because we wished to use S. typhimurium which does not contain the lactose operon, but where we could apply the genetic expertise of our colleague, Prof. Philip Hartman. Melibiose is a dissaccharide, D-galactopyranosyl-(α,1→6)-D-glucose, and in a sense is an analogue of lactose (β,1→4). In E. coli, two separate transport systems have been demonstrated for these dissaccharides. They enter the cell as free sugars and are split by the corresponding α- and β-galactosidases to galactose and glucose. As already indicated, S. typhimurium does not contain the lactose operon and therefore, only generates the melibiose permease system. One further point of confusion relates to the nonmetabolizable substrate used in these studies, TMG (thiomethyl β-galactoside). It is taken up by both transport systems.

The primary finding obtained in our laboratory was that the uptake of TMG via the melibiose permease system was greatly stimulated by sodium chloride. At the time these studies were performed, bacterial transport systems were generally thought to be quite different from animal transport systems, and thus, this effect of sodium was quite unexpected. About half maximum stimulation of TMG uptake was obtained at about 1 mM sodium ion concentration. The effect was quite specific, the only cation that served in place of sodium ion was lithium, and the kinetics of the transport system showed a remarkable similarity to the results obtained in the transport of dissaccharides and other sugars by animal intestinal cells. It was further shown, as indicated in Figure 5, that the uptake of the melibiose analog TMG was sodium dependent, but, likewise, that the sodium uptake was dependent on TMG. Uninduced cells did not take up either solute to a significant extent, and sodium was not taken up in the absence, but only in the presence of TMG. Furthermore, there was an over-shoot in sodium uptake, where excess sodium was excreted from the cell shortly after uptake.

To this point, we had considered bacterial cells to be primary transport systems. That is, a solute was taken up independently of any other solute. The translocation process was extremely rapid, much more rapid than passive diffusion, stereochemically specific, and catalyzed by specific permease type proteins in the membrane. Facilitated diffusion was defined as movement down an electrochemical gradient, active transport against the gradient, requiring the expenditure of metabolic energy, and group translocation being derivatization of the solute as it crossed the membrane as in the case of the PTS. The results on the melibiose

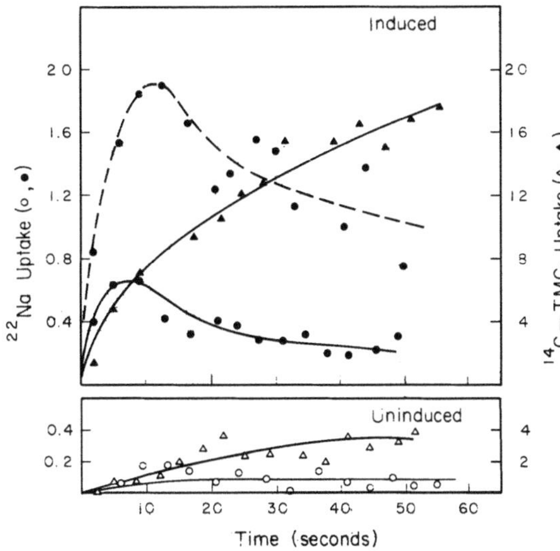

Fig. 5. Uptake of labeled sodium and methyl β-thiogalactoside. *S. typhimurium* (wild type, LT-2) was grown in salts medium containing either 0.5% melibiose (induced) or 0.2% glycerol (uninduced) to midexponential phase, harvested, washed, and transport experiments conducted. Units on the ordinates represent nmoles of solute per mg, dry weight, of cells. Upper curve for sodium uptake in the presence of TMG was performed by dilution into 1 mM NaCl just prior to filtration, while the lower curve (solid line, filled circles) was obtained by dilution into 10 mM NaCl. If the labeled solutes had equilibrated between the external medium and cell water, the value for Na^+ would be 2, and the value for TMG would be 20

Fig. 6. Melibiose, Na^+co-transport

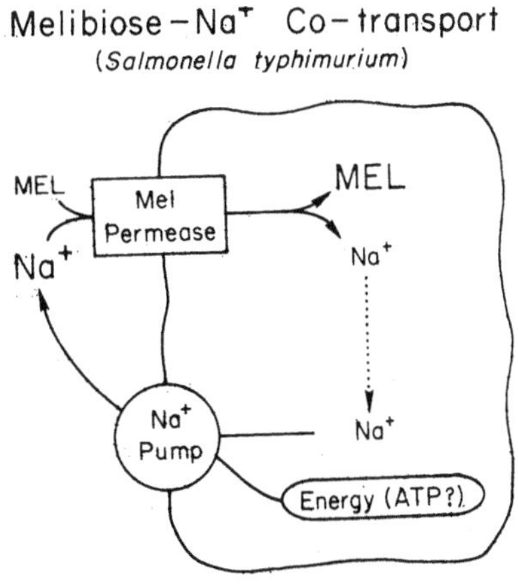

transport system, however, lead to quite a different conclusion, and the conclusion is shown in Figure 6. Melibiose and sodium are co-transported by the appropriate permease, and the sodium ion is then pumped out of the cell, this against an electrochemical gradient. The pump required metabolic energy. This type of co-transport (symport) system had long been recognized in animal cells

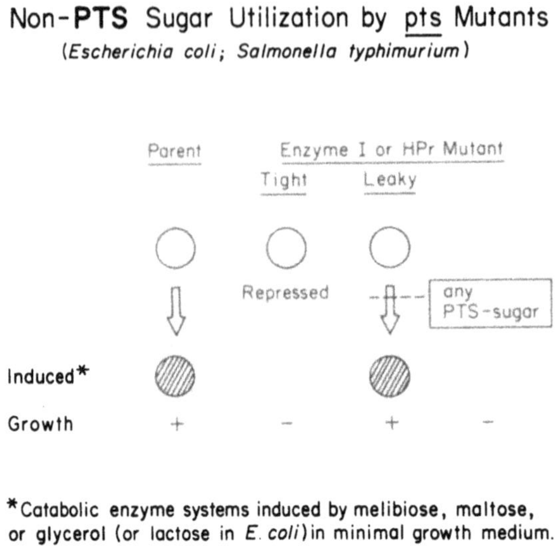

Fig. 7. PTS-mediated repression of inducible enzyme synthesis. Wild type cells are induced to synthesize the catabolic systems required for utilization of the non-PTS sugars melibiose, maltose, glycerol. and lactose, and therefore grow on these sugars. Tight pts mutants cannot be induced. Leaky mutants are normally induced, but are repressed by the presence of low concentrations of any PTS sugar in the inducing medium

(30, 31). Countertransport or antiport is a similar event, except that here the movement of the cation is in the opposite direction from the movement of the solute.

Mitchell (32) has made enormous contributions to the thinking in this field, and in the case of coupled transport systems he suggests that protons are the cations which are either co- or countertransported with the solute. Proton motive force is thus used in a translocation of protons coupled to a solute. There is considerable evidence in support of Mitchell's model, some of it presented at this meeting. Thus, at this point, we would conclude that non-PTS solutes such as melibiose are transported by coupling to the transport of cation, either sodium ions or protons, and such transport systems are different and distinct from PTS-driven systems. I would now like to examine this conclusion.

We have tested this conclusion by studying the transport of the following non-PTS sugars in E. coli and S. typhimurium: melibiose, lactose, maltose and glycerol (33-35). The basic finding is that shown in Figure 7. If parental cells are placed in a mineral salts medium, containing any of the four non-PTS sugars, melibiose, lactose, maltose or glycerol, the cells are induced to synthesize the corresponding catabolic enzyme systems and therefore grow on these sugars. On the other hand, if a tight Enzyme I or HPr mutant is placed in the same medium, no induction and no growth occurs. If a leaky Enzyme I or HPr mutant is placed in the inducing medium, normal induction and growth occur. However, the addition of any PTS sugar at very low concentrations inhibits and prevents induction and growth. The PTS, therefore, represses induction of synthesis of these catabolic systems. How?

This form of repression is related to a series of complex phenomena that have been studied over many, many years, variously called the glucose effect, catabolite repression, diauxie and so on. The repression illustrated in Figure 7 is called PTS-mediated repression. It is clearly one form of so-called catabolite repression.

Fig. 8. Two explanations for PTS-mediated repression as described in Figure 7

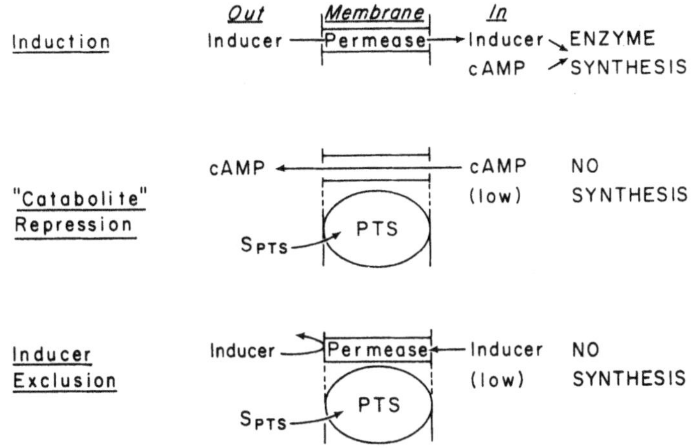

Fig. 9. Inducer exclusion via the PTS. For details, see text

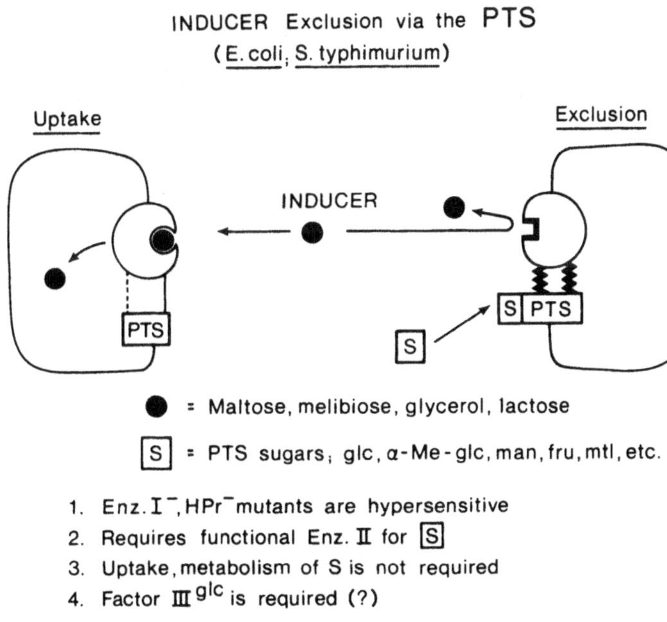

The explanations for PTS-mediated repression are shown in Figure 8. Induced enzyme synthesis is an extremely complex process, but for our purposes can be simplified; normal induction requires minimal concentrations of inducer and cyclic AMP in the cell. If either of these fall below the minimal level, enzyme synthesis is not induced. Pastan and Perlman (9) suggested that PTS-mediated repression occurs by loss of cyclic AMP from the cell when a PTS sugar is added to the inducing medium. We believe that this in fact does occur in some strains of bacteria, and that in these strains it is a component of PTS-mediated repression. We do not believe that it is the entire explanation. In our opinion (33-35) a major component of PTS-mediated repression is inducer exclusion, shown at the bottom of figure 8. The PTS in some way regulates non-PTS per-

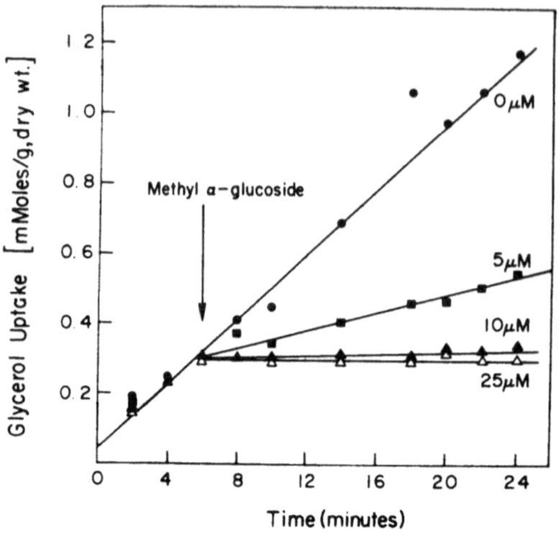

<u>Fig. 10.</u> PTS-mediated repression of uptake and growth on glycerol. Methyl α-glucoside was added at <u>arrow</u> at the indicated concentrations to a leaky <u>S. typhimurium</u> Enzyme I mutant (fully induced cells)

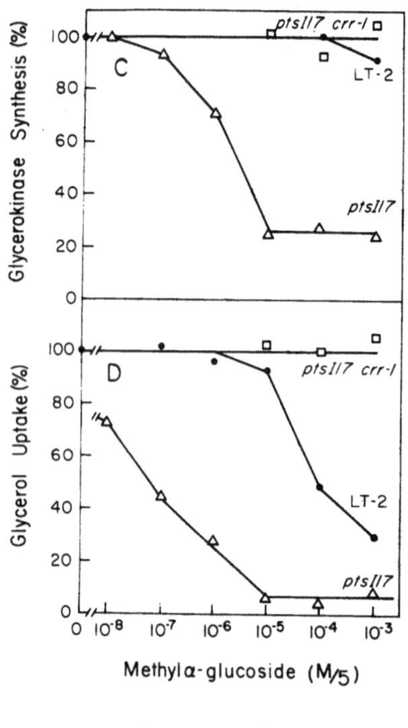

<u>Fig. 11.</u> Effect of methyl α-glucoside on glycerokinase induction and on glycerol uptake in <u>S. typhimurium</u> strains: LT-2, wild type cells; <u>ptsI17</u>, leaky Enzyme I mutant; <u>ptsI17 crr-1</u>, double mutant

mease systems, and prevents uptake of inducer. The inducer concentration in the cell falls below the minimal required levels and no enzyme synthesis occurs.

Inducer exclusion is schematically illustrated in Figure 9 in somewhat greater detail. Normal induction of the maltose, melibiose, glycerol and lactose catabolic systems occurs when the corresponding sugars are taken up through their permeases (present at low levels in the cells prior to induction). PTS-mediated repression is effected when any PTS sugar interacts with the PTS. This in turn regulates the non-PTS permease so that it cannot take up the inducer.

An example of PTS-mediated repression is shown in Figure 10. Cells (a leaky Enzyme I mutant) growing on glycerol so that they are fully induced for this catabolic system are exposed to extremely low concentrations of methyl α-glucoside. As can be seen, there is immediate growth stasis and an immediate cessation in the uptake of glycerol When this phenomenon is examined in more detail one obtains the results shown in Figure 11. The upper portion of the figure presents the induction of glycerokinase as a function of methyl α-glucoside concentration in the inducing medium, while the lower portion of the slide shows glycerol uptake under the same conditions. As can be seen, the parental strain is reasonably resistant to methyl α-glucoside, but the Enzyme I leaky mutant is extremely sensitive. More important, uptake of the inducer, glycerol, is equally sensitive to the glucoside. Exactly the same correlations are found with the melibiose, maltose and lactose systems.

A summary of these kinds of studies is given in Figure 9. The non-PTS sugars maltose, melibiose, glycerol and lactose were studied in E. coli and S. typhimurium, along with a variety of PTS sugars. Enzyme I and HPr mutants are hypersensitive to the phenomenon of inducer exclusion. In order to observe inducer exclusion with a PTS sugar, a functional Enzyme II for this sugar must be present; inducer exclusion by other PTS sugars is not affected in such a mutant. If uptake of the PTS sugar is required, it is at such low levels that it is questionable. The last point in Figure 9 is discussed below.

These results lead to the conclusion that the PTS regulates at least these non-PTS systems. What is the mechanism underlying this kind of regulation? We do not know, but some clues have come from the following series of experiments.

It is possible to overcome PTS-mediated repression by introducing a second site mutation into Enzyme I and HPr mutants (33-35) either leaky or tight or into deletions of both (36, 37).

The mutation which reverses the phenotypic behavior of Enzyme I and HPr mutants towards non-PTS (but not towards PTS) sugars is called crr (carbohydrate repression resistant). Such mutants were selected for by their ability to grow on the non-PTS sugars in the presence of high levels of methyl α-glucoside. The crr gene maps adjacent to (36, 37) but not as part of the pts operon, which contains the structural genes for Enzyme I and HPr.

Typical results with an Enzyme I, crr double mutant are shown in Figure 11. The sensitivity of the Enzyme I mutant to methyl α-glucoside, with respect to both glycerokinase induction and to glycerol uptake, is reversed by the crr mutation. Identical results were obtained with the maltose, melibiose (and lactose in E. coli) systems. This single mutation, therefore, affects PTS-mediated repression by all PTS sugars, and it affects all of the non-PTS sugars tested. What is the biochemical nature of the crr mutation?

In order to consider this question, a third Enzyme II system must first be described, the $III^{Glc}/II\text{-}B^{GLC}$ system. Several PTS sugars such as glucose are taken up by more than one PTS system, as indicated in Figure 1. Glucose can be transported by the II-A/II-B system that I have already discussed, but also by a separate $III^{Glc}/II\text{-}B^{Glc}$ system.

We have recently shown that methyl α-glucoside is transported only by the $III^{Glc}/II\text{-}B^{Glc}$ system, and when used at less than 1 mM concentration, it is phosphorylated only by this system.

After extensive effort, $II\text{-}B^{Glc}$, a membrane protein, has recently been solubilized and purified close to homogeneity (38). As in the case of the purified II-B protein, $II\text{-}B^{Glc}$ requires lipid and divalent cation for reactivation. It represents less than 1% of the protein in exhaustively washed membranes, but must have a very high turnover number in view of the rapidity of phosphorylation of methyl α-glucoside by crude membrane preparations. It is $II\text{-}B^{Glc}$ which is defective in ptsG mutants (38, 39), and thus this is a key protein in PTS-mediated repression by methyl α-glucoside and by glucose.

The III^{Glc} protein has also been extensively purified to homogeneity. The available data indicate that it is a protein of about 20,000 daltons (gel filtration), containing 3 to 4 subunits; the subunits appear to be active in the in vitro phosphorylation assay. Phospho-III^{Glc} appears to be an acyl-phosphate, similar to the Na^+, K^+ and Ca^{2+} ATPases of eukaryotic cells, but unlike the other characterized phosphoproteins of the PTS, all of which are phosphohistidinyl proteins.

Based on studies with methyl α-glucoside (7), and on the work described above, it is evident that the $III^{Glc}/II\text{-}B^{Glc}$ system regulates adenylate cyclase.

III^{Glc} may also be of critical importance in regulating the glycerol, melibiose, maltose and lactose transport systems. This conclusion is based on the fact that the only biochemical defect thus far detected in crr mutants is the III^{Glc} protein; the mutants contain 0-30% of wild type levels. The critical question yet to be resolved is whether the crr gene is the structural or a regulatory gene for III^{Glc}. If it is the structural gene, then III^{Glc} must be a key protein in regulating at least the four non-PTS transport systems listed above. Assuming that crr is the structural gene for III^{Glc}, how can the results described above be explained?

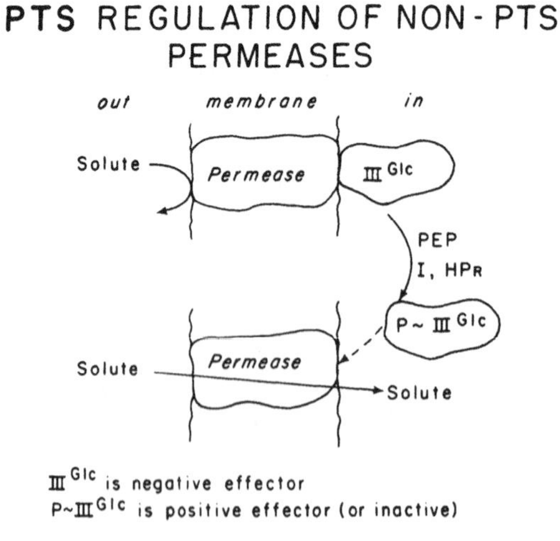

Fig. 12. Scheme for mechanism for inducer exclusion. The solute is a non-PTS compound

At the moment, we can only speculate. A model consistent with all of the data, only some of which I have presented here, is shown in Figure 12. According to this scheme, it is the ratio of phospho-III^{Glc} to free III^{Glc} which may be the determining factor. III^{Glc} is proposed to be a negative effector of non-PTS permease systems. Phospho-III^{Glc} is either inactive or is a positive effector. Mutants defective in Enzyme I or HPr would contain high levels of the inhibitor, III^{Glc}. This type of inhibition would be relieved by a crr mutation, which would lead to low levels of III^{Glc}. Exactly the same kind of thinking can be applied to the published results on the regulation of adenylate cyclase.

The last question to be considered is whether non-PTS transport systems have evolved directly from, or independently of the PTS. It is for this reason that we have been searching for variants of the PTS. Some have already been found, and it is likely that more exist. Furthermore, the distinction between PTS and non-PTS sugars is not always maintained, in spite of the rather strict specificity one observes in bacterial transport systems. For example, galactose is not a PTS sugar in enteric bacteria, while glucose and its analogue methyl α-glucoside are. However, under certain conditions there is good evidence (3, 40, 41) for overlapping specificities between the systems. The galactose system takes up glucose and methyl α-glucoside, while the glucose-mannose system (II-A/II-B) takes up galactose.

It therefore seems possible from these and other data that non-PTS systems have evolved from the PTS, as suggested in Figure 13. The PTS in its most easily recognizable form catalyzes group translocation. The Enzyme II portion of the system could catalyze facilitated diffusion, as shown at the top of the slide, and there are indeed reports that Enzyme II complexes can do this (3). Finally, in active transport, the driving force would be cleavage of the phospho-Enzyme II complex which would be accompanied by translocation of the solute against its electrochemical gradient. Good examples of this type are the sodium, potassium and the calcium stimulated ATPases of higher organisms.

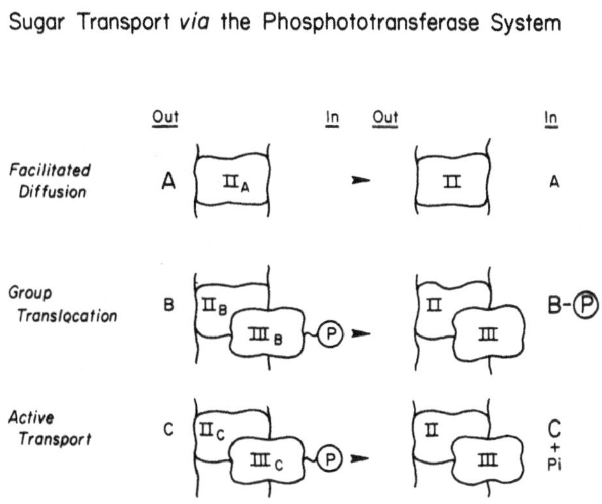

Fig. 13. Possible functions of Enzyme II complexes. Solute A is transported down its electrochemical gradient, solute B by group translocation (as in the case of PTS sugars), solute C against its electrochemical gradient. The energy in the last case is derived by hydrolysis of the phosphoprotein, as in the case of the Na^+, K^+ pump of eukaryotic cells

In conclusion, the PTS has been directly implicated in diverse and important physiological phenomena. Its inherent complexity may be required to permit the flexibility required for these different functions. While it has been shown to regulate at least some other transport systems, the molecular mechanisms underlying this type of regulation remain to be established. Such studies should ultimately show whether or not other solute transport proteins have been derived from the PTS.

References

1. Kundig, W., Ghosh, S., Roseman, S.: Phosphate bound to histidine in a protein as an intermediate in a novel phosphotransferase system. Proc. Natl. Acad. Sci. U.S. 52, 1067-1074 (1964)
2. Roseman, S.: Carbohydrate transport in bacterial cells. In: Metabolic Pathways. Hokin, L.E. (ed.). New York: Academic Press, 1972, Vol. VI, pp. 41-89
3. Postma, P.W., Roseman, S.: The bacterial phosphoenolpyruvate: sugar phosphotransferase system. Biochim. Biophys. Acta 457, 213-257 (1976)
4. Roseman, S.: Metabolism of sialic acids and D-mannosamine. Federation Proc. 21, 1075-1083 (1962)
5. Adler, J., Epstein, W.: Phosphotransferase system enzymes as chemoreceptors for certain sugars in Escherichia coli chemotaxis. Proc. Natl. Acad-Sci. U.S. 71, 2895-2899 (1974)
6. Melton, T.: Ph.D. Thesis, Johns Hopkins University, 1976
7. Harwood, J.P., Gazdar, C., Prasad, C., Peterkofsky, A., Curtis, S.J., Epstein, W.: Involvement of the glucose enzymes II of the sugar phosphotransferase system in the regulation of adenylate cyclase by glucose in Escherichia coli. J. Biol. Chem. 251, 2462-2468 (1976)
8. Makman, R.S., Sutherland, E.W.: Adenosine 3',5'-phosphate in Escherichia coli. J. Biol. Chem. 240, 1309-1314 (1965)
9. Pastan, I., and Perlman, R.L.: Repression of β-galactosidase synthesis by glucose in phosphotransferase mutants of Escherichia coli. J. Biol. Chem. 244, 5836-5842 (1969)
10. Paigen, K., Williams, B.: Catabolite repression and other control mechanisms in carbohydrate utilization. Advan. Microb. Physiol. 4, 251-324 (1970)
11. Magasanik, B. In: The Lactose Operon. Beckwith, J.R., Zipser, D. (eds.). Cold Spring Lab., 1970, pp. 189-219
12. Simoni, R.D., Nakazawa, T., Hays, J.B., Roseman, S.: Isolation and characterization of the lactose phosphotransferase system in Staphylococcus aureus. J. Biol. Chem. 248, 932-940 (1973)
13. Hays, J.B., Simoni, R.D., Roseman, S.: A trimeric lactose specific phosphocarrier protein of the Staphylococcus aureus phosphotransferase system. J. Biol. Chem. 248, 941-956 (1973)
14. Simoni, R.D., Hays, J.B., Nakazawa, T., Roseman, S.: Phosphoryl transfer in the lactose phosphotransferase system of Staphylococcus aureus. J. Biol. Chem. 248, 957-965 (1973)
15. Simoni, R.D., Roseman, S.: Lactose transport in Staphylococcus aureus. J. Biol. Chem. 248, 966-976 (1973)
16. Anderson, B., Weigel, N., Kunding, W., Roseman, S.: Purification and properties of a phosphocarrier protein (HPr) of the phosphoenolpyruvate-dependent phosphotransferase system of Escherichia coli. J. Biol. Chem. 246, 7023-7033 (1971)
17. Weigel, N., and Powers, D.: Studies on the primary structure of a phosphocarrier protein of the bacterial phosphotransferase system. Federation Proc. 34, 491 (1975)
18. Beneski, D.: Ph.D. Thesis, Johns Hopkins University, in preparation

19. DeLuca, M., Ebner, K.E., Hultquist, D.E., Kreil, G., Peter, J.B., Moyer, R.W., Boyer, P.D.: The isolation and identification of phosphohistidine from mitochondrial protein. Biochem. Z. 338, 512-525 (1963)
20. Kundig, W., Roseman, S.: Isolation of a phosphotransferase system from Escherichia coli. J. Biol. Chem. 246, 1393-1406 (1971)
21. Kundig, W., Roseman, S.: Characterization of constitutive membrane-bound Enzymes II of the Escherichia coli phosphotransferase system. J. Biol. Chem. 246, 1407-1418 (1971)
22. Kaback, H.R.: The role of the phosphoenolpyruvate-phosphotransferase system in the transport of sugars by isolated membrane preparations of Escherichia coli. J. Biol. Chem. 243, 3711-3724 (1968)
23. Egan, J.B., Morse, M.L.: Carbohydrate transport in Staphylococcus aureus I. Biochim. Biophys. Acta 97, 310-319 (1965)
24. Egan, J.B., Morse, M.L.: Carbohydrate transport in Staphylococcus aureus II. Biochim. Biophys. Acta 109, 172-183 (1965)
25. Egan, J.B., Morse, M.L.: Carbohydrate transport in Staphylococcus aureus III. Biochim. Biophys. Acta 112, 63-73 (1965)
26. Hengstenberg, W., Egan, J.B., Morse, M.L.: The accumulation of phosphorylated carbohydrate derivatives and evidence for a new enzyme splitting lactose phosphate. Proc. Natl. Acad. Sci. U.S. 58, 274-279 (1967)
27. Kennedy, E.P., Scarborough, G.A.: Mechanism of hydrolysis of O-nitrophenyl-β-galactoside in Staphylococcus aureus and its significance for theories of sugar transport. Proc. Natl. Acad. Sci. U.S. 58, 225-228 (1967)
28. Button, D.K., Egan, J.B., Hengstenberg, W., Morse, M.L.: Carbohydrate transport in Staphylococcus aureus. Maltose accumulation and metabolism. Biochem. Biophys. Res. Commun. 52, 850-855 (1973)
29. Stock, J., Roseman, S.: A sodium-dependent sugar co-transport system in bacteria. Biochem. Biophys. Res. Commun. 44, 132-138 (1971)
30. Stein, W.D.: The Movement of Molecules Across Cell Membranes. New York: Academic Press, 1967
31. Schultz, S.G., Curran, P.F.: Coupled transport of sodium and organic salts. Physiol. Rev. 50, 637-718 (1970)
32. Harold, F.M.: Conservation and transformation of energy by bacterial membranes. Bacteriol. Rev. 36, 172-230 (1972)
33. Saier, M.H., Simoni, R.D., Roseman, S.: Properties of mutant bacteria defective in proteins of the phosphoenolpyruvate: sugar phosphotransferase system. J. Biol. Chem. 251, 6584-6597 (1976)
34. Saier, M.H., Roseman, S.: The crr mutation: its effect on repression of enzyme synthesis. J. Biol. Chem. 251, 6598-6605 (1976)
35. Saier, M.H., Roseman, S.: Inducer exclusion and regulation of the melibiose, maltose, glycerol, and lactose transport system by the phosphoenolpyruvate: sugar phosphotransferase system. J. Biol. Chem. 251, 6606-6615 (1976)
36. Cordaro, J.C., Roseman, S.: Deletion mapping of the genes coding for HPr and Enzyme I of the phosphoenolpyruvate: sugar phosphotransferase system in Samonella typhimurium. J. Bacteriol. 112, 17-29 (1972)
37. Cordaro, J.C., Anderson, R.P., Grogan, E.W., Wenzel, D.J., Engler, M., Roseman, S.: Promoter-like mutation affecting HPr and Enzyme I of the phosphoenolpyruvate: sugar phosphotransferase system in Salmonella typhimurium. J. Bacteriol. 12C, 245-252 (1974)
38. Waygood, B., Stock, J., Meadow, N., Roseman, S.: Kinetic characterization of bacterial phosphotransferase system (PTS) mutants. Federation Proc. 35, 1357 (1976)
39. Curtis, S.J., Epstein, W.: Phosphorylation of D-glucose in Escherichia coli mutants defective in glucose phosphotransferase, mannosephosphotransferase, and glucokinase. J. Bacteriol. 122, 1189-1199 (1975)
40. Simoni, R.D., Levinthal, M., Kundig, F.D., Kundig, W., Anderson, B., Hartman, P.E., Roseman, S.: Genetic evidence for the role of a bacterial phosphotransferase system in sugar transport. Proc. Natl. Acad. Sci. U.S. 58, 1963-1970 (1967)
41. Postma, P.W.: Involvement of the phosphotransferase system in galactose transport in Salmonella typhimurium. FEBS Lett. 61, 49-53 (1976)

Molecular Biology and Energetics of Membrane Transport *

H. R. KABACK

Bacterial cytoplasmic membrane vesicles have provided a unique model system for the study of active transport (Kaback, 1970, 1972, 1974b), and have led to the development of similar experimental systems from the cells of higher organisms (Murer and Hopfer, 1974; Colombini and Johnstone, 1974; Sigrist-Nelson et al., 1975; Quinlan et al., 1976; Lever, 1976a, 1976b). Vesicles are prepared by lysis of osmotically-sensitized cells (i.e., protoplasts or spheroplasts), and consist of osmotically-intact, unit-membrane bound sacs which are approximately 0.5 to 1.0 μ in diameter (Kaback, 1971; Short et al., 1975). The sacs are devoid of internal structure, their metabolic activities are restricted to those provided by the enzymes of the membrane itself, and a number of observations demonstrate that the vesicle membrane has the same orientation as the membrane in the intact cell (see Stroobant and Kaback, 1975, for a summary of these observations). Transport by membrane vesicles per se is practically nil, and the energy source for the transport of a particular substrate can be determined by studying which compounds or experimental manipulations drive solute accumulation. Moreover, metabolic conversion of the transport substrate and the energy source is minimal. These properties of the vesicle system constitute a considerable advantage over intact cells with respect to the study of active transport, as the system provides a clear definition of the reactions involved in the transport process.

This discussion is not intended as a general review, but is concerned primarily with respiration-dependent active transport in membrane vesicles isolated from *Escherichia coli*, and emphasis is placed upon very recent observations to be presented at this meeting. For a more general discussion the reader is referred to numerous reviews of this area, two of the most recent of which are those by Konings (1976) and by Harold (1976). Studies with this bacterial system are relevant not only to active transport and other membrane-related phenomena, but to the general problem of energy transduction. As opposed to mitochondria and chloroplasts where respiratory energy and light are converted to another form of chemical enery (i.e., ATP), respiratory energy in isolated bacterial membrane vesicles is converted primarily to work in the form of solute concentration against an electrochemical or osmotic gradient.

* This manuscript has also been submitted to *Harvard Books in Biophysics* and to the *Journal of Cellular Physiology*.

Transport Properties of Bacterial Membrane Vesicles

Transport assays with membrane vesicles may be performed in various ways (Kaback, 1971, 1974a). The most widely used method is a filtration assay in which vesicles are incubated with a radioactive transport substrate in the presence of an appropriate energy source, and at a given time, the reaction mixtures are diluted to terminate the uptake reaction. The vesicles are then immediately separated from the medium by means of rapid membrane filtration. Although this method is rapid and convenient, it suffers from the disadvantages that the reaction mixtures must be diluted prior to filtration, and in many instances, this operation results in significant losses of accumulated solutes (Ramos et al., 1976). The vesicles can also be separated from the reaction mixtures by centrifugation, but in this case, samples cannot be assayed rapidly, corrections must be made for radioactive solute trapped in the pellet, and the reaction mixtures become anaerobic no matter how rapidly centrifugation is carried out (Ramos et al., 1976). Very recently, the flow dialysis technique devised by Colowick and Womack (1969) has been adapted to measure transport (Ramos et al., 1976, Ramos and Kaback, 1977a,b), and it is presently clear that this is the method of choice. By this means, changes in the external concentration of solute are determined continuously, accurately, and highly reproducibly under conditions which require no manipulation of the experimental system.

Evidence was presented as early as 1960 (Kaback, 1960) which indicated that bacterial membrane vesicles would provide a useful model system for the study of active transport. Subsequent studies have demonstrated that the vesicles catalyze the transport of a plethora of metabolites in the presence of appropriate energy sources (Kaback, 1974b), and that initial rates of transport of many of these metabolites are comparable to those of the intact cell (Lombardi and Kaback, 1972; Short et al., 1972). Moreover, the vesicles accumulate many solutes to concentrations markedly in excess of those in the external medium. Initial progress with the system was slow primarily because of preconceived ideas regarding the physical nature of the vesicles and the energetics of active transport. It was generally thought that transport would be driven by ATP or another nucleoside triphosphates, and considerable effort was expended in an effort to provide evidence for this notion. However, as it turns out, the energy source for transport in isolated membrane vesicles varies with the organism, with the substance transported, and ATP or other nucleoside triphosphates do not drive transport in the vesicle system (see below).

In general, the transport systems elucidated in the vesicle system fall into three main categories: (1) group translocation mechanisms in which a covalent change is exerted upon the transported molecule so that the reaction itself results in the passage of the molecule through the diffusion barrier; (2) active transport in which solute is accumulated against an electrochemical or osmotic gradient; and (3) passive diffusion of certain weak acids and lipophilic ions, followed by equilibration with the pH gradient and the electrical potential, respectively, across the membrane.

In E. coli membrane vesicles, the uptake of glucose, fructose, mannose, mannitol, sorbitol, and a few additional sugars occurs by vectorial phosphorylation via the phosphoenolpyruvate-phosphotransferase system (PTS) [a group translocation mechanism

(Kaback, 1970)]. The PTS, described initially in 1964 (Kundig et al., 1964), catalyzes the transfer of phosphate from phosphoenolpyruvate through certain proteins to the appropriate carbohydrates in such a manner that the carbohydrate is translocated across the membrane and accumulated as a result of phosphorylation. Although the system is crucial to the bacteria that employ it, the PTS is not ubiquitous among bacterial species, and has not been found thus far in organisms phylogenetically higher than bacteria. It is also noteworthy that evidence for other group translocation reactions has been presented in vesicles from E. coli and other bacteria (Kaback, 1974b).

As opposed to group translocation mechanisms, transport of a whole variety of other solutes (i.e., sugars, amino acids, organic acids, and ions) by E. coli membrane vesicles occurs by active transport (Kaback, 1974b). These transport systems are coupled primarily to the oxidation of D-lactate to pyruvate, catalyzed by a flavin adenine dinucleotide-linked, membrane-bound D-lactate dehydrogenase (D-LDH) which has been purified to homogeneity (Kohn and Kaback, 1973; Futai, 1973). Electrons derived from D-lactate are passed to oxygen via a membrane-bound respiratory chain, and in the sequence of reactions between the primary dehydrogenase and cytochrome b_1, the first cytochrome in the E. coli respiratory chain, respiratory energy is converted into work in the form of active transport (Barnes and Kaback, 1971; Kaback and Barnes, 1971; Lombardi and Kaback, 1972; Kaback, 1972; Stroobant and Kaback, 1975). Although other oxidizable substrates such as L-lactate, succinate, α-glycerol-P, and NADH also stimulate transport to some extent, they are not nearly as effective as D-lactate unless ubiquinone is added to the vesicles (Stroobant and Kaback, 1975). It should be emphasized, however, that D-lactate is not the electron donor for active transport in all bacterial membrane vesicles (Kaback, 1974b). Active transport in the vesicle system can also be driven by non-physiological electron donors such as reduced phenazine methosulfate (PMS) (Konings et al., 1971) or pyocyanine (Stroobant and Kaback, 1975), both of which donate electrons to the respiratory chain at a site prior to the cytochromes. The use of these nonphysiological electron donors has allowed the generalization of the vesicle system to many other bacteria in addition to E. coli (Konings et al., 1971; Kaback, 1974b).

Active transport in E. coli membrane vesicles also functions in the absence of oxygen when the appropriate anaerobic electron transfer systems are present (Konings and Kaback, 1973; Boonstra et al., 1975; Konings and Boonstra, 1976). Lactose and amino acid transport under anaerobic conditions can be coupled to the oxidation of α-glycerol-P with fumarate as an acceptor or to the oxidation of formate utilizing nitrate as electron acceptor. Both of these anaerobic electron transfer systems are induced by growth of the organism under appropriate conditions, and components of both systems are loosely bound to the membrane.

Finally, active transport can be driven by artificially-induced potassium gradients of appropriate polarity (Hirata et al., 1973; 1974). When membrane vesicles prepared in potassium-containing buffers are diluted into solutions lacking this cation, and the ionophore valinomycin is added, the diffusion of potassium out of the vesicles creates an electrical potential across the vesicles membrane (interior negative), and uptake of certain substrates is observed. Although the extent of transport observed under these conditions is considerably less than that observed with D-lactate and especially reduced PMS, the finding that solute accumulation occurs under these conditions has great implications with respect to the energetics of respiration-linked active transport (see below).

Obligate and facultative anaerobes growing in the absence of anaerobic electron acceptors transport nutrients. Moreover, mutants of E. coli which require δ-aminolevulinic acid or heme do not manifest transport defects despite an absence of cytochromes (Devor et al., 1974). These observations and others (Abrams and Smith, 1971; Klein and Boyer, 1972; Schairer and Haddock, 1972; Berger, 1973; Butlin, 1973; Parnes and Boos, 1973; Or et al., 1973) suggest that ATP or other high-energy phosphate intermediates may drive transport in intact cells under certain conditions. However, it is extremely important to recognize that respiration-linked active transport in isolated membrane vesicles does not involve either the generation or hydrolysis of ATP or other high-energy phosphate compounds. This contention is supported by several lines of evidence (see Kaback, 1974b): (1) Neither ATP nor ATP generating systems support transport of any substrate, even when they are included during lysis and the subsequent steps in the preparation of vesicles. Under the same conditions, when colloidal gold is added during lysis, it becomes trapped within the intravesicular space. (2) Vesicles do not catalyze oxidative phosphorylation to a significant extent, and transport occurs in the presence of high concentrations of arsenate. (3) Very little ATP can be detected in the vesicle preparations whether or not oxidizable substrates are present. (4) Conditions that stimulate or inhibit ATPase activity have no effect on transport in the vesicles. (5) Transport occurs in a completely normal fashion in membrane vesicles prepared from uncA mutants of E. coli. These mutants are deficient in calcium, magnesium-stimulated ATPase activity and uncoupled for oxidative phosphorylation. Thus, although it appears likely that ATP or some other high-energy phosphate intermediate may be able to drive transport under some conditions, it is likely that most of the energy for active transport in E. coli under aerobic conditions is derived from respiration without the mediation of high-energy phosphate intermediates.

Energetics of Active Transport

Although the precise means by which energy released from the oxidation of D-lactate or reduced phenazine methosulfate is coupled to transport are unknown, an increasing accumulation of experimental evidence indicates that chemiosmotic phenomena (Mitchell, 1961, 1966, 1967, 1968, 1970a, b, 1973; Greville, 1969; Harold, 1972) play a central role in the process. Quite recently, as a result of the "discovery" that flow dialysis can be applied to the study of active transport (Ramos et al., 1976; Ramos and Kaback, 1977a,b), a number of important observations have been made which provide convincing experimental support for this theory. In addition, the new findings extend certain important aspects of the chemiosmotic concept, and finally, they provide explanations for earlier observations which seemed to exclude the importance of the concept in the mechanism of active transport in this experimental system.

As visualized by Mitchell, oxidation of electron donors via a membrane-bound respiratory chain or ATP hydrolysis catalyzed by the membraneous calcium, magnesium-stimulated ATPase is accompanied by the expulsion of protons into the external medium, leading to the generation of an electrochemical gradient of protons ($\Delta\bar{\mu}_{H^+}$) which is composed of electrical and chemical parameters according to the following relationship:

$$\Delta\bar{\mu}_{H^+} = \Delta\Psi - \frac{2.3RT}{F}\Delta pH \qquad (1)$$

Where $\Delta\Psi$ represents the electrical potential across the membrane, and pH is the chemical difference in proton concentrations across the membrane (2.3RT/F is equal to 58.8 mV at room temperature). Moreover, according to this hypothesis, it is the electrochemical gradient of protons or one of its components which is postulated to be the immediate driving force for the accumulation of transport substrates. Transport of organic acids is postulated to be dependent upon the chemical gradient of protons (ΔpH) (i.e., the undissociated acid is transported through the membrane and is presumed to accumulate in the ionized form due to the relative alkalinity of the internal milieu), while the transport of positively charged compounds such as lysine or potassium is purportedly coupled to the electrical component ($\Delta\Psi$), and the uptake of neutral substrates such as lactose or proline is thought to be coupled to $\Delta\bar{\mu}_{H^+}$ and to occur via co-transport with protons (i.e., symport in Mitchell's terminology). In addition, the inhibitory effects of uncoupling agents on transport are attributed to the ability of these compounds to conduct protons across the membrane, thus short-circuiting the so-called protonmotive force ($\Delta\bar{\mu}_{H^+}$).

Clearly, as opposed to an earlier model which attempted to explain the coupling between respiration and active transport in a direct manner (Kaback and Barnes, 1971), the chemiosmotic formulation stipulates that transport is the result of an indirect process whereby respiration is coupled to active transport through the mediation of the electrochemical proton gradient. As such, the experimental approach to the problem at this level of resolution is dependent upon the development of techniques which allow the measurement of $\Delta\bar{\mu}_{H^+}$ in a system which is not amenable to a direct electrophysiological approach and upon the demonstration that changes in $\Delta\bar{\mu}_{H^+}$ and its components correlate with changes in the transport of appropriate solutes.

Determination of $\Delta\Psi$.

In 1971, Grinius et al. initiated the use of lipophilic ions to measure electrical potentials in biological systems which are not amenable to the introduction of microelectrodes. These ions are constructed in such a fashion that they are sufficiently lipophilic to enter the hydrophobic core of the membrane, and apparently they must also be able to delocalize their charge in order to allow them to passively equilibrate with the electrical potential across the membrane (Haydon and Hladky, 1972). Using isotopically labeled cations of this type [i.e., dimethyldibenzylammonium (DDA^+) in the presence of tetraphenylborate, or triphenylmethylphosphonium ($TPMP^+$)], or a fluorescent lipophilic cation (safranine-O) or rubidium in the presence of valinomycin, it has been demonstrated that E. coli membrane vesicles generate a $\Delta\Psi$ (interior negative) of approximately -75 mV in the presence of reduced PMS or D-lactate (Hirata et al., 1973; Altendorf et al., 1975; Schuldiner and

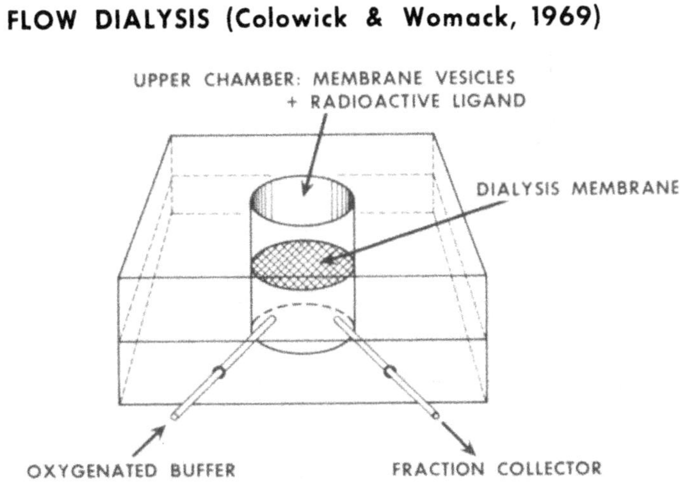

Fig. 1. Schematic representation of flow dialysis apparatus (Colowick and Womack, 1969) as modified by Ramos et al. (1976). Although not shown, upper and lower chambers are stirred by means of small magnetic stirring bars and upper chamber is gassed with a stream of oxygen

Kaback, 1975). Furthermore, it was shown that there is a direct relationship between the magnitude of $\Delta\Psi$ and the accumulation of lactose and that lactose accumulation partially dissipates $\Delta\Psi$ (Schuldiner and Kaback, 1975). Although these findings lend strong support to the chemiosmotic hypothesis, it was apparent that $\Delta\Psi$ in itself was insufficient to account for the magnitude of solute accumulation by the vesicles (Schuldiner and Kaback, 1975). This deficiency, in addition to the apparent absence of a transmembrane pH gradient, left reasonable doubt as to the quantitative relationship between the electrochemical proton gradient and solute accumulation. Recent experiments utilizing flow dialysis to measure ΔpH (Ramos et al., 1976; Ramos and Kaback, 1977a,b) have resolved this problem.

Determination of ΔpH.

Recently, Padan et al. (1976) have shown that intact E. coli generate a ΔpH (interior alkaline), and that the magnitude of the ΔpH is very dependent upon external pH, exhibiting a maximal value of about 2 pH units at pH 6.0 or below. In addition, Rottenbarg (1975) has demonstrated that acetate may be utilized to determine ΔpH in mitochondria, and suggested (personal communication) that this weak acid might be more useful than 5,5-dimethyloxazolidine-2,4-dione (DMO) (Waddel and Butler, 1959) in certain systems because it might be less permeant. Initial experiments with E. coli membrane vesicles utilizing the usual assay techniques revealed a small amount of acetate uptake which was sensitive to external pH, but the quantity of acetate taken up was so small that the putative ΔpH appeared to be thermodynamically insignificant. It was not until Sofia Ramos, Shimon Schuldiner and I utilized flow dialysis, a technique that Schuldiner and I had been using to measure binding of dansylgalactosides to vesicles (Schuldiner et al., 1976a), that the puzzle was resolved (Ramos et al., 1976).

Figure 1 is a schematic representation of a flow dialysis cell as modified by Ramos et al. (1976). The apparatus consists of a lucite block containing two chambers separated by a dialysis membrane. A suspension of membrane vesicles in an appropriate buffer is placed in the upper chamber and oxygenated buffer at the same pH as the

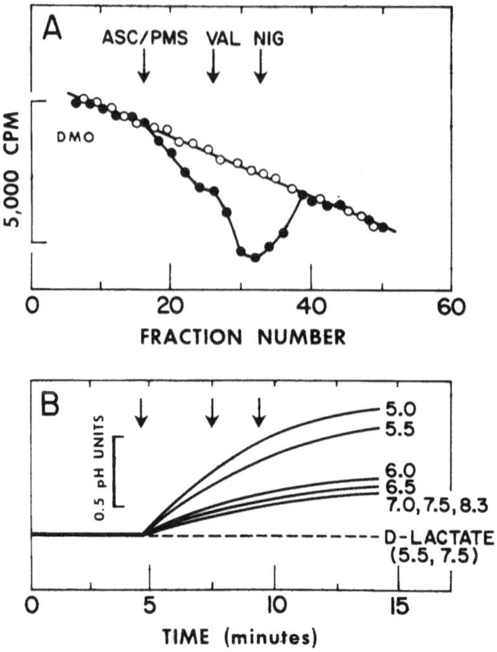

Fig. 2A,B. (A) Flow dialysis determinations of DMO uptake by membrane vesicles prepared from *E. coli* ML 308-225. Assays shown were carried out with membrane vesicles (about 2.5 mg of protein per ml) suspended in 0.05 M potassium phosphate at pH 5.5 as described by Ramos et al. (1976). [2-^{14}C]DMO (11 mCi mmol^{-1}) was used as indicated at a final concentration of 200 μM. As shown by arrows, sodium ascorbate and phenazine methosulfate (ACS/PMS), valinomycin (VAL), and nigericin (NIG) were added to the upper chamber at final concentrations of 20 mM, 0.1 μM, and 0.1 μM, respectively (closed symbols). Open symbols: obtained from an identical experiment carried out in the absence of ascorbate and PMS. (B) Change in external pH during oxidation of ascorbate-PMS. The tracings shown are recordings of the pH of the upper chamber of the flow dialysis apparatus containing ML 308-225 membrane vesicles (about 2.5 mg of protein per ml) suspended in 0.05 M potassium phosphate. Potassium phosphate (0.05 M) at the same pH as the initial pH of the upper chamber was pumped through the lower chamber. At time indicated by arrow, sodium ascorbate (20 mM) and PMS (0.1 mM) or lithium D-lactate (20 mM) were added to the upper chamber, and the pH was recorded continuously with a radiometer 2323C pH electrode connected to a radiometer pH meter (Model PHM64). The signal was recorded with a double channel Corning recorder (Model 845). The pH values given represent the initial pH of the suspension prior to addition of ascorbate-PMS and D-lactate

suspension in the upper chamber is pumped from the lower chamber at a constant rate to a fraction collector. Both chambers are stirred by means of small magnetic stirring bars, and the reaction mixture in the upper chamber is gassed with a stream of oxygen. Solute added to the upper chamber diffuses into the lower chamber through the dialysis membrane, and by this means, the concentration of solute in the medium surrounding the vesicles can be monitored easily and continuously. If the vesicles accumulate or bind solute, its diffusible concentration in the upper chamber will diminish, and the change in concentration in the upper chamber will be reflected by a decrease in the concentration of the solute in the lower chamber.

The data presented in Figure 2A illustrate a typical flow dialysis experiment carried out in 0.05 M potassium phosphate buffer at pH 5.5 with the weak acid DMO (Ramos et al., 1976). Shortly after addition of isotopically labeled DMO to the upper chamber containing membrane vesicles, radioactivity appears in the dialysis, increases linearly for about 2-3 min (not shown), and reaches a maximum which then decreases at a slow rate (open symbols). When ascorbate and PMS (closed symbols) or D-lactate (not shown) are added to the upper chamber, the vesicles accumulate DMO, and the concentration in the dialysate decreases relative to the level observed in the absence of electron donor. Addition of valinomycin, an ionophore which specifically increases the potassium permeability of the membrane, causes the vesicles to accumulate DMO more, and the concentration in the dialysate decreases still further. This effect of valinomycin is dependent upon the presence of potassium. Finally, when nigericin is added (Fig. 2A), DMO is released from the vesicles, and the external concentration returns to the control levels, an observation which is consistent with the notion that nigericin catalyzes an exchange of potassium for protons, and thus collapses ΔpH.

Using the equilibrium concentration of DMO in the dialysate subsequent to the addition of ascorbate-PMS (i.e., fraction 25 or 7 min), the amount of DMO taken up by the vesicles can be calculated (6.9 nmoles per mg of membrane protein). With these values, together with the external concentration of DMO and its pK value (6.30) and the intravesicular volume of 2.2 μl per mg of membrane protein, it can be calculated (Waddel and Butler, 1959; Rottenberg, 1975) that the intravesicular pH is 7.8. Similarly, using the equilibrium concentration recovered in the dialysate after addition of valinomycin (i.e., fraction 25 or 8.5 min), it can be calculated that the intravesicular pH is 8.3. Given the external pH values at the analogous times (Fig. 2B), the ΔpH obtained after addition of ascorbate-PMS is about 2.0 pH units (7.8 inside minus 5.8 outside), corresponding to -118 mV, and the ΔpH obtained after addition of ascorbate-PMS and valinomycin is about 2.2 pH units (8.3 inside minus 6.1 outside), corresponding to -130 mV. It is also noteworthy that similar results are obtained with D-lactate as electron donor and with acetate, propionate, and butyrate as permeant acids, that ΔpH values obtained by flow dialysis are constant over at least a 100-fold range of weak acid concentrations and over a range of membrane protein concentrations from 1 to 3 mg per ml (Ramos et al., 1976). It is apparent from these studies that the previous failures to detect ΔpH in the vesicles using DMO and standard assay techniques are due to artifacts inherent in the experimental procedures. DMO is not accumulated to any extent when uptake is assayed by filtration or rapid centrifugation, only a small amount of acetate uptake is observed, and filtration assays yield increasingly lower values for propionate and butyrate relative to acetate (Ramos et al., 1976). Presumably, the low levels of uptake observed with filtration and centrifugation are due to the high passive permeability of the vesicles to weak acids (filtration) and to the sensitivity of ΔpH to oxygen tension (centrifugation).

Although each weak acid utilized yields the same value for intravesicular pH, the relative decrease in the external concentration of each acid after addition of ascorbate-PMS (or D-lactate) varies with its pK value. Acetate which has a pK of 4.75 exhibits the most marked change on addition of valinomycin (i.e., 145% of the value obtained after addition of ascorbate-PMS), while propionate (pK 4.87) exhibits intermediate properties, and DMO (pK 6.3) exhibits the least marked change on addition of ascorbate-PMS and the most dramatic effect on addition of valinomycin (i.e., 230% of the value obtained after addition of ascorbate-PMS). These differences can be explained by theoretical considerations (Waddel and Butler, 1959; Rottenberg, 1975). Weak acids equilibrate with a transmembrane pH gradient (interior alkaline) in the following manner: the undissociated species of the acid is passively permeant and enters the intramembranal pool in this form. If the intramembranal compartment is sufficiently alkaline relative to the pK of the permeant weak acid, the acid dissociates, and the anionic species accumulates in the intramembranal pool. It follows, therefore, that the greater the differential between the internal pH and the pK of the acid, the higher the percentage of the acid in the ionized form in the intramembranal pool, and the higher the level of accumulation. Thus, at an intramembranal pH of 7.8 (the intramembranal pH after the addition of ascorbate-PMS), essentially all of the acetate and propionate and significantly less DMO will be ionized in the intravesicular pool. Therefore, DMO will accumulate to a relatively lesser extent than acetate or propionate. It is also clear that an increase in the intravesicular pH (such as occurs on addition of valinomycin) will have a greater relative effect on DMO accumulation because it will increase the concentration of the anionic form of DMO inside of the vesicles to a significant extent relative to acetate and propionate which are already highly dissociated.

Effect of External pH on ΔpH, $\Delta\Psi$, Internal pH, $\Delta\bar{\mu}_{H^+}$, and Respiration

The data presented in Figure 3 demonstrate that ΔpH in ML 308-225 membrane vesicles varies markedly with external pH, as reported by Padan et al. (1976) for intact cells. From pH 5.0 to 5.5, ΔpH remains constant at -115 to -120 mV, decreases drastically above pH 5.5, and is negligible at pH 7.5 and above. Significantly, despite marked variation in ΔpH as a function of external pH, internal pH and $\Delta\Psi$ remain essentially constant at pH 7.8 and -75 mV, respectively. As a result of the variation in ΔpH, $\Delta\bar{\mu}_{H^+}$ exhibits a maximal value of about -195 mV at pH 5.5 (-120 mV ΔpH + -75 mV $\Delta\Psi$) and a minimal value of about -75 mV at pH 7.5 and above (0 ΔpH + -75 mV $\Delta\Psi$). Strikingly, when the rate of oxidation of reduced PMS is studied as a function of external pH, a relationship similar to that observed for ΔpH and $\Delta\bar{\mu}_{H^+}$ is obtained. That is, the rate of reduced PMS oxidation is relatively high at pH 5.0, increases slightly from pH 5.0 to 5.5, decreases sharply above pH 5.5, and remains constant at about 35% of maximal activity from pH 7.5 to pH 8.3.

The variation in ΔpH (and as a result, $\Delta\bar{\mu}_{H^+}$) with external pH is clearly due in part to variations in the rates of oxidation of reduced PMS. Since the relatively low rates of reduced PMS oxidation observed above pH 5.5 produce increasingly lower ΔpH values, but constant $\Delta\Psi$ values, it seems reasonable to suggest that relatively low rates of electron flow through the energy-coupling site are sufficient to generate $\Delta\Psi$, while relatively high rates of electron flow are necessary to support a significant ΔpH. It is also important to realize however that in order to generate a significant ΔpH across the vesicle membrane without producing an increase in $\Delta\Psi$, an anion(s) must move with the protons that are ejected from the vesicles or alternatively, a cation(s) must move in the oppo-

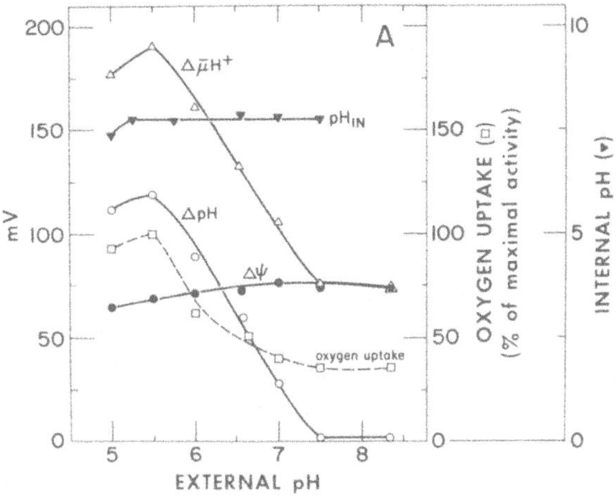

Fig. 3. Effect of external pH on internal pH, ΔpH, ΔΨ, $\Delta\bar{\mu}_{H^+}$, and oxidation of reduced PMS. Experiments shown were carried out as described in Fig. 2 using *E. coli* ML 308-225 membrane vesicles. Internal pH (▽——▽) was calculated from flow dialysis experiments with [1-^{14}C]acetate performed at each pH given using sodium ascorbate and PMS at final concentrations of 20 mM and 0.1 mM, respectively. ΔpH values (o——o) were calculated as described (Ramos et al., 1976) after correcting the external pH for the change induced by ascorbate-PMS oxidation (see Fig. 2B). The external pH prior to addition of ascorbate-PMS. ΔΨ values (●——●) were calculated from filtration assays carried out with [^{3}H]TPMP$^+$ (1.33 Ci mmol^{-1}) at a final concentration of 24 μM (Schuldiner and Kaback, 1975; Ramos et al., 1976). Assays were carried out after 10 min incubation using ascorbate and PMS. $\Delta\bar{\mu}_{H^+}$ values (Δ——Δ) were calculated from ΔpH and ΔΨ using Equation (1). Rates of oxidation of reduced PMS (□——□) were determined as described previously (Barnes and Kaback, 1971; Konings and Kaback, 1971) using sodium ascorbate and PMS at final concentrations of 20 mM and 0.1 mM, respectively. Values were corrected for nonenzymatic oxidation of reduced PMS by subtracting control values obtained at each pH in the absence of membrane vesicles. The maximum rate (100% observed at pH 5.5) was about 480 ng atoms of oxygen taken up per min per mg of membrane protein

site direction (i.e., into the vesicles). Although these possibilities have not been investigated exhaustively, a number of observations (Ramos and Kaback, unpubl.) suggest that the compensatory ion movement(s) that must take place at pH 5.5 are probably not related to the movement of a particular ionic species. Thus, the relatively high rates of oxidation and the high rate of proton extrusion observed at relatively acid pH (see Fig. 12 in Lombardi et al., 1973) may cause an increase in ΔpH with no change in ΔΨ because of compensatory movements of ions in both directions across the membrane. It also seems likely that ion movements may play an important role in the diminution and ultimate loss of ΔpH at progressively alkaline pH, since similar effects of external pH on ΔpH at progressively alkaline pH, since similar effects of external pH on ΔpH have been observed in intact *E. coli* (Padan et al., 1976), *Halobacterium halobium* (Bakker et al., 1976), and in vesicles from *Pseudomonas capsulata* (Rottenberg, pers. comm.) without corresponding changes in respiration. In this regard, recent experiments (Eisenbach et al., 1977; Ramos, Rottenberg, and Kaback, unpubl. information) suggest the operation of a mechanism which catalyzes the exchange of external protons for intravesicular sodium or potassium at relatively alkaline pH. It seems clear, in any event, that the effect of external pH on ΔpH may be a complicated phenomenon and will require further study.

Effect of Various Electron Donors on ΔpH, ΔΨ, and $\Delta\bar{\mu}_{H^+}$

Although there is little relationship between the ability of the vesicles to oxidize an electron donor and the ability of that electron donor to drive active transport, a qualitative correlation

Table 1. Effect of various electron donors on ΔpH, ΔΨ, and $\Delta\bar{\mu}_{H^+}$

Electron donor	ΔpH[a]	ΔΨ[b] (mV)	$\Delta\bar{\mu}_{H^+}$[c]
Ascorbate-PMS	-115	-74	-189
D-lactate	-102	-70	-172
Succinate	0	-64	-64
NADH	0	0	0
NADH + Q1	-59	-62	-131

[a] ΔpH was calculated from flow dialysis experiments carried out with sodium [^{3}H]acetate (685 mCi.mmol^{-1}) at a final concentration of 18 μM as described in Figure 2. Ascorbate and PMS, lithium D-lactate, sodium succinate, NADH (sodium salt), and ubiquinone-1 (Q1) were used at final concentrations of 20 mM and 0.1 mM, 20 mM, 20 mM, 5 mM, and 0.08 mM, respectively.

[b] ΔΨ was determined from filtration assays performed after 5- and 10-min incubations (Schuldiner and Kaback, 1975) except that [^{3}H]TPMP$^+$ (1.33 Ci.mmole^{-1} was used at a final concentration of 24 μM. Electron donors were used at the concentrations given above; and the data were corrected for the amount of TPMP$^+$ taken up in the absence of exogenous electron donors.

[c] $\Delta\bar{\mu}_{H^+}$ was calculated from Equation (1)

exists between the ability of various electron donors to drive transport and their ability to generate a ΔΨ (interior negative) (Schuldiner and Kaback, 1975). The data presented in Table I corroborate the latter observation and demonstrate further that a similar relationship exists with respect to ΔpH and $\Delta\bar{\mu}_{H^+}$. Clearly, ascorbate-PMS and D-lactate produce maximal relative values for each parameter, while succinate and especially NADH produce much weaker effects. Moreover, when ubiquinone-1 is added to the vesicles, NADH generates much higher values. This observation is highly significant in that NADH oxidation under these conditions drives transport in the presence of ubiquinone-1, but not in its absence (Stroobant and Kaback, 1975).

Effect of Valinomycin, Nigericin, and CCCP on ΔpH, ΔΨ, $\Delta\bar{\mu}_{H^+}$.

As increasing concentrations of valinomycin are added to vesicles at pH 5.5 or 7.5, ascorbate-PMS-dependent TPMP$^+$ uptake decreases to approximately 10% of the control value at 2.5 to 5 μM valinomycin (Fig. 4), corresponding to a decrease in ΔΨ from about -70 mV to -20 mV (insert). Strikingly, there is an increase in acetate uptake to about 130% of the control value at 1 μM valinomycin [an increase in ΔpH of about 10 mV (insert)] which remains constant at higher valinomycin concentrations. The effect of valinomycin on the total driving force at pH 5.5 is relatively small, producing only about a 20% loss at 5 μM valinomycin [from a $\Delta\bar{\mu}_{H^+}$ of about -190 mV to -150 mV (insert)].

As shown in Figure 4B, nigericin induces effects which are opposite to those of valinomycin. Acetate uptake decreases to 0 as the nigericin concentration is increased from 0 to 0.1 μM [from a ΔpH of -110 mV to 0 (insert)], while TPMP$^+$ uptake increases almost 2.5-fold over the same concentration range [from a ΔΨ of about -60 mV to -90 mV (insert)]. Despite marked effects on ΔpH and ΔΨ, 0.1 μM nigericin produces only a 45% decrease in the total driving force at pH 5.5 [from a $\Delta\bar{\mu}_{H^+}$ of -170 mV to -90 mV (insert)]. It is also

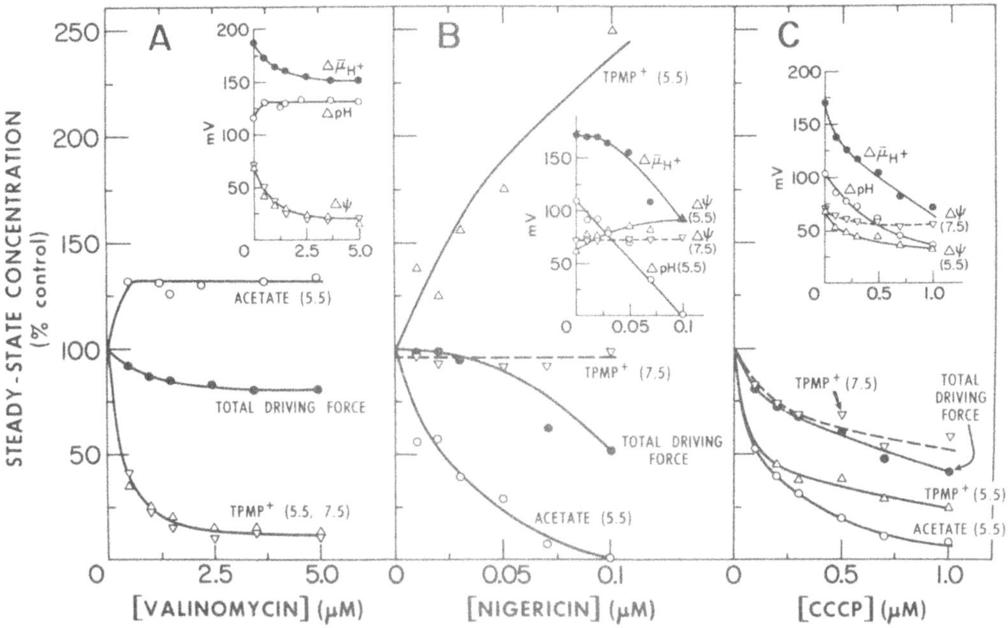

Fig. 4A-C. Effect of valinomycin (A), nigericin (B), and CCCP (C) on ΔpH, $\Delta\Psi$, and $\Delta\bar{\mu}_{H^+}$. ΔpH was determined by flow dialysis in the presence of acetate and ascorbate-PMS at pH 5.5, and the indicated concentrations of valinomycin (panel A), nigericin (B), or CCCP (C) as described in Figure 2. Steady-state levels of $TPMP^+$ accumulation ($\Delta\Psi$) were determined at pH 5.5 (Δ——Δ) and pH 7.5 (∇——∇) in the presence of given concentrations of valinomycin, nigericin, or CCCP by filtration assays as described in Figure 3. The effect of the inhibitors on the total driving force at pH 5.5 ($\Delta\mu_{H^+}$, ●——●) was calculated from ΔpH and $\Delta\Psi$ at each inhibito concentration. Percentage of the total driving force remaining at each inhibitor concentration is percentage of $\Delta\bar{\mu}_{H^+}$ measured in the absence of inhibitors. Following control values (100%) were obtained in the absence of inhibitors (nmoles per mg membrane protein): acetate uptake: 5.4; $TPMP^+$ uptake: 0.80 (at pH 5.5); 0.91 (at pH 7.5). The mV values plotted in the inserts were calculated from the experimental values presented. (From Ramos et al., 1976)

apparent that this ionophore has no effect on $TPMP^+$ uptake at pH 7.5, a finding which is consistent with the absence of a ΔpH at this external pH (cf. Fig. 3).

The proton conductor carbonylcyanide-m-chlorophenylhydrazone (CCCP) inhibits both $TPMP^+$ and acetate uptake at pH 5.5 over a concentration range from 0 to 1 μM CCCP, and diminishes the total driving force at pH 5.5 by approximately 60% (a decrease in $\Delta\bar{\mu}_{H^+}$ from -170 mV to about -65 mV at 1 μM CCCP) (Fig. 4C). It is also noteworthy that CCCP inhibits more markedly at pH 5.5 than at pH 7.5, an effect that is presumably due to the pK of the proton conductor itself. The observation that CCCP dissipates both ΔpH and $\Delta\Psi$, provides strong evidence that protons are the electrogenic species.

Finally, it should be emphasized that valinomycin, nigericin, and CCCP do not elicit significant changes in the ability of the vesicles to oxidize either reduced PMS or D-lactate at pH 5.5 or pH 7.5 (Ramos and Kaback, 1977a). Thus, as opposed to mitodondria and chloroplasts where these compounds generally cause an increase in respiration (i.e., proton pumping), this is not the case with bacterial membrane vesicles.

The explanation for the valinomycin effects may be conceptualized by considering a theoretical membrane which is completely impermeable to ions. In such a membrane, a small amount of proton extrusion will result in a large membrane potential with an immeasurably small proton gradient. If such a membrane is made permeable to ions (e.g., potassium by the addition of valinomycin), the result will be a diminution in the electrical potential coincident with the appearance of a measurable pH gradient. With bacterial membrane vesicles at pH 5.5, as discussed above, there must be a finite permeability to ions in order to account for the ΔpH that is present at this external pH. Moreover, if this suggestion is correct, the ion permeability of the membrane is still limiting for ΔpH, as evidenced by the observation that valinomycin increases ΔpH in the presence of potassium. Thus, valinomycin, by allowing potassium to move as a counterion to protons could cause an increase in ΔpH at the expense of $\Delta\Psi$ without an increase in respiratory activity (i.e., proton extrusion). The finding that the ionophore does not cause the appearance of ΔpH at pH 7.5 may result from relatively low rates of oxidation at this pH which are not increased when the vesicles are made permeable to potassium or other ions (i.e., the vesicles do not exhibit respiratory control).

Nigericin, on the other hand, presumably catalyzes an electrically silent exchange of protons for potassium across the membrane, and it is not immediately obvious why the decrease in ΔpH caused by this ionophore should result in an increase in $\Delta\Psi$ without a commensurate increase in respiration (i.e., proton extrusion). However, returning to the analogy of the membrane which is completely impermeable to ions, it is apparent that very small increases in proton extrusion in such a theoretical membrane would result in large increases in $\Delta\Psi$. Therefore, it seems possible that a small increase in reduced PMS oxidation which is within the limits of detection of the oxygen electrode utilized could account for the increase in $\Delta\Psi$ observed with nigericin at pH 5.5.

Relationship Between $\Delta\bar{\mu}_{H^+}$ and Active Transport

In addition to providing direct support for the chemiosmotic theory, the work discussed above establishes a powerful experimental framework within which to test other more specific aspects of the concept.

Effect of Solute Accumulation on ΔpH and $\Delta\Psi$

If accumulation of solute is coupled to the electrochemical proton gradient across the membrane ($\Delta\bar{\mu}_{H^+}$), and if the passive accumulation of weak acids and lipophilic cations reflects the individual components of $\Delta\bar{\mu}_{H^+}$ (ΔpH and $\Delta\Psi$, respectively), it follows that acetate and/or $TPMP^+$ accumulation might be diminished in the presence of transport substrates such as lactose which are accumulated in relatively large amounts by the vesicles. Both predictions are borne out by the data presented in Figure 5. As shown in the upper panel, when ML 308-225 membrane vesicles are allowed to accu-

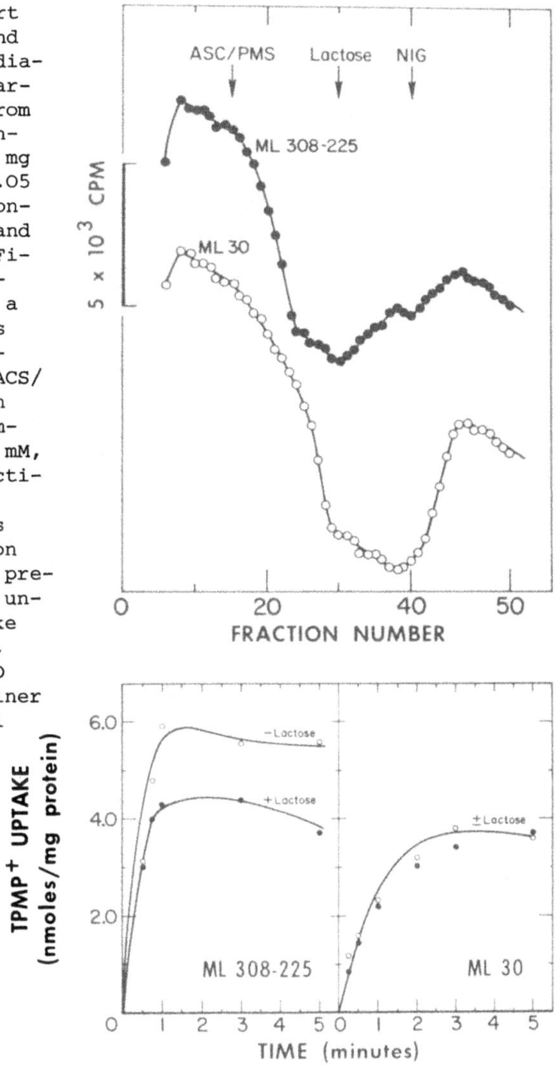

Fig. 5. Effect of lactose transport on acetate uptake (upper panel) and TPMP$^+$ uptake (lower panel). Flow dialysis assays (upper panel) were carried out with membrane vesicles from *E. coli* ML 308-225 (●——●) and uninduced ML 30 (o——o) (about 2.5 mg of protein per ml) suspended in 0.05 M potassium phosphate at pH 5.5 containing 0.01 M magnesium sulfate and 1 μM valinomycin as described in Figure 2 except that sodium [^{3}H]acetate (685 mCi mmol^{-1}) was used at a final concentration of 18.5 μM. As indicated by arrows, sodium ascorbate and phenazine methosulfate (ACS/PMS), lactose (LAC), and nigericin (NIG) were added to the upper chamber at final concentrations of 20 mM, 0.1 mM, 2.5 mM, and 0.1 μM, respectively.
The data shown in the lower panels illustrate the effect of lactose on TPMP$^+$ uptake in membrane vesicles prepared from *E. coli* ML 308-225 and uninduced ML 30. o——o, TPMP$^+$ uptake in the absence of lactose; ●——●, TPMP$^+$ uptake in the presence of 10 mM lactose. (See Fig. 9 in Schuldiner and Kaback, 1975, for experimental details)

mulate acetate in the presence of ascorbate-PMS and 1 μM valinomycin at pH 5.5 [at this valinomycin concentration, there is only about 60% inhibition of lactose transport (Ramos and Kaback, 1977b)], addition of lactose causes the release of about 50% of the accumulated acetate, and the concentration of the weak acid in the dialysate increases appropriately. When the same experiment is carried out with uninduced ML 30 vesicles which do not transport lactose, acetate accumulation is not affected by addition of lactose. Similarly, as shown in the bottom panels of Figure 5, when lactose is added to vesicles which transport this dissaccharide, the uptake of TPMP$^+$ is diminished (lower left), and no effect is observed on addition of lactose to vesicles which do not contain this transport system (lower right). Clearly, a good explanation for these effects is that lactose is taken up by co-transport with protons, as suggested by Mitchell (1973). It should also be emphasized that earlier experiments carried out with intact cells are also consistent with

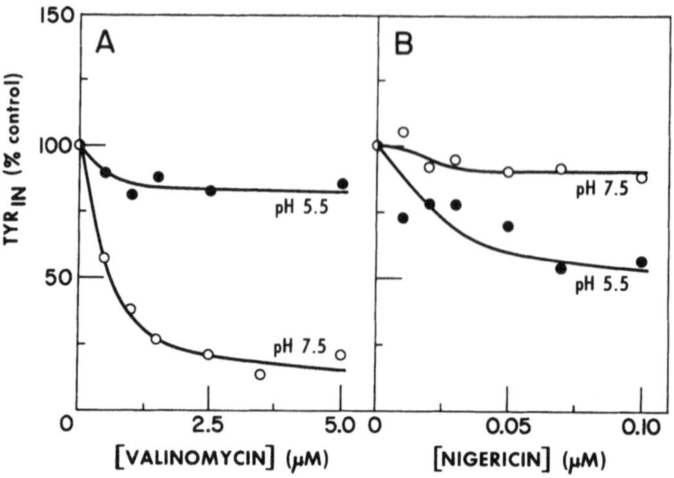

Fig. 6. Effect of valinomycin (panel A) and nigericin (panel B) on the steady-state level of tyrosine accumulation by ML 308-225 membrane vesicles. Aliquots of membrane vesicles prepared from E. coli ML 308-225 (25 μl containing about 0.1 mg of protein) were diluted to a final volume of 50 μl, containing, in final concentrations, 0.05 M potassium phosphate [pH 5.5 (Δ——Δ) or pH 7.5 (V——V)] and 0.01 M magnesium sulfate, and valinomycin (A) or nigericin (B) were added to given final concentrations. The reaction vessels were fitted with perforated rubber stoppers so that each sample could be gassed with oxygen by means of a hypodermic needle inserted through the stopper. Additions to the reaction mixtures were made with a Hamilton microsyringe by insertion through the perforation in the stopper (Kaback, 1974a). The samples were incubated at 25°C, and gassed with pure oxygen for about 1 min prior to addition of ascorbate-PMS and a given transport substrate and throughout the incubation. Sodium ascorbate (20 mM) and PMS (0.1 mM) were added, and immediately thereafter, U-[^{14}C]-tyrosine (404 mCi mmol^{-1}) at a final concentration of 4 μM. The incubations were continued for 5 min, terminated, and assayed as described previously (Kaback, 1974a)

this explanation (West, 1970; West and Mitchell, 1973; West and Wilson, 1973; Kashket and Wilson, 1973).

Effect of Valinomycin and Nigericin on the Steady-State Level of Accumulation of Transport Substrates

Before proceeding to the experiments to be presented in this section a few important points derived from the data presented in Figs. 3 and 4 will be reemphasized: (1) $\Delta\bar{\mu}_{H^+}$ is maximal at pH 5.5 where approximately two-thirds of the total driving force is ΔpH and one-third is ΔΨ; (2) At pH 7.5, $\Delta\bar{\mu}_{H^+}$ is reduced by approximately two-thirds, and consists solely of a ΔΨ component; (3) At pH 5.5, ΔpH and ΔΨ can be manipulated reciprocally by valinomycin and nigericin with little or no effect on $\Delta\bar{\mu}_{H^+}$; (4) Nigericin has no effect on ΔΨ (and therefore no effect on $\Delta\bar{\mu}_{H^+}$) at pH 7.5.

The data presented in Figures 6 and 7 are representative data derived from experiments (Ramos and Kaback, 1977b) in which the steady-state level of accumulation of 14 individuals solutes in the presence of ascorbate-PMS at pH 5.5 (●——●) and pH 7.5 (o——o) was studied as a function of increasing concentrations of valinomycin (panels A) and nigericin (panels B). Results for tyrosine (Fig. 6) and gluconate (Fig. 7) only are presented, and the data are plotted as a percentage of control values obtained in the absence of valinomycin and nigericin.

Direct quantitative correlations between variations in the accumulation of a particular substrate and variation in $\Delta\bar{\mu}_{H^+}$, $\Delta\Psi$, or ΔpH are observed in only a few instances at pH 5.5. Nevertheless, certain qualitative statements are clearly justified when the experiments are considered as a whole:

1. Accumulation of tyrosine (Fig. 6), lactose, proline, serine, glycine, leucine, and surprisingly, lysine, glutamate, and succinate at pH 5.5 responds to increasing concentrations of valinomycin and nigericin in a manner that correlates reasonably well with the effect of these ionophores on $\Delta\bar{\mu}_{H^+}$. Accumulation of each of these solutes is progressively and mildly inhibited or relatively unaffected by increasing concentrations of both ionophores.

2. Accumulation of gluconate (Fig. 7), glucose-6-P, D-lactate, and glucoronate at pH 5.5 is stimulated by valinomycin and inhibited by nigericin in a manner clearly reminiscent of the effects of the ionophores on ΔpH.

3. Regardless of whether the accumulation of a particular solute is stimulated, inhibited, or unaffected by valinomycin or nigericin at pH 5.5, in each and every case, valinomycin causes marked inhibition of accumulation at pH 7.5 and nigericin has no effect whatsoever at this external pH (Figs. 6 and 7). It is generally clear therefore that at pH 5.5, the transport systems fall into two categories: Those that are driven preferentially by $\Delta\bar{\mu}_{H^+}$ and those that are driven preferentially by ΔpH. Moreover, it is patently obvious that all of the systems are driven by $\Delta\Psi$ at pH 7.5 where $\Delta\Psi$ comprises the totality of $\Delta\bar{\mu}_{H^+}$.

Since the coupling between the accumulation of a particular substrate and $\Delta\bar{\mu}_{H^+}$, ΔpH, and $\Delta\Psi$ apparently varies with external pH, it is not surprising that strict correlations between $\Delta\bar{\mu}_{H^+}$ or ΔpH are observed in only a few instances at pH 5.5. With tyrosine (Fig. 6), leucine, lysine and succinate, for example, there is reasonably good quantitative correlation between the effects of valinomycin and nigericin on $\Delta\bar{\mu}_{H^+}$ and the effects of the ionophores on the accumulation of these solutes (Ramos and Kaback, 1977b). However, lactose and glycine accumulation are also coupled preferentially to $\Delta\bar{\mu}_{H^+}$ at pH 5.5, but there is an apparent bias towards $\Delta\Psi$ in both instances, since valinomycin inhibits lactose and glycine accumulation more effectively than it inhibits $\Delta\bar{\mu}_{H^+}$ at pH 5.5. Similarly, although the accumulation of gluconate (Fig. 7), glucose-6-P, D-lactate, and glucuronate is coupled to ΔpH at pH 5.5, in only one case (i.e., D-lactate) is the accumulation of these substrates in complete equilibrium with ΔpH (Ramos and Kaback, 1977b). It is also clear that nigericin does not inhibit the accumulation of these solutes as effectively as it dissipates ΔpH at pH 5.5 (Fig. 7 with Fig. 4B).

Although these results support the conclusion that chemiosmotic phenomena play an important role in active transport, many of the observations are not readily explained by the chemiosmotic hypothesis as it is presently formulated. For instance, according to current notion, accumulation of organic acids is obligatorily dependent on the relative alkalinity of the internal pH. Organic acids are presumably transported in their undissociated form and

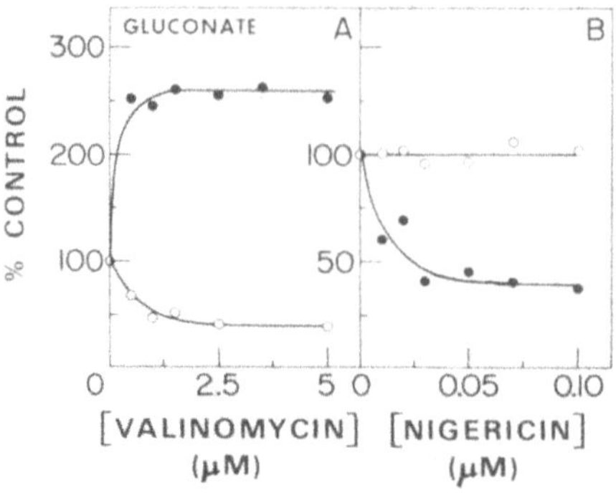

Fig. 7A,B. Effect of valinomycin (A) and nigericin (B) on the steady-state level of gluconate accumulation by membrane vesicles prepared from E. coli ML 30 grown on gluconic acid. Experiments were performed as described in Figure 6 using U-[^{14}C]gluconate (4.1 mCi mmol^{-1}) at a final concentration of 0.4 mM

accumulate in the internal pool in the ionized form. Since there is no ΔpH across the membrane at pH 7.5 in either whole cells (Padan et al., 1976) or isolated membrane vesicles (cf. Fig. 3), the putative mechanism cannot account for acid accumulation at high external ΔpH. A possible explanation for these and other unexpected findings will be discussed subsequently.

Effect of External pH on Accumulation of Specific Transport Substrates

As shown in Fig. 3, $\Delta\bar{\mu}_{H^+}$ is extremely sensitive to external pH, exhibiting a maximum at pH 5.5, decreasing dramatically above pH 5.5, and remaining constant at pH 7.5 and above. These variations in $\Delta\bar{\mu}_{H^+}$ are due primarily to changes in ΔpH which is also maximal at pH 5.5 but negligible at pH 7.5 and above. Given these observations, pH titrations of various transport systems take on new and important significance. With a stoichiometry of 1:1 between protons and most transport substrates, as suggested by West and Mitchell

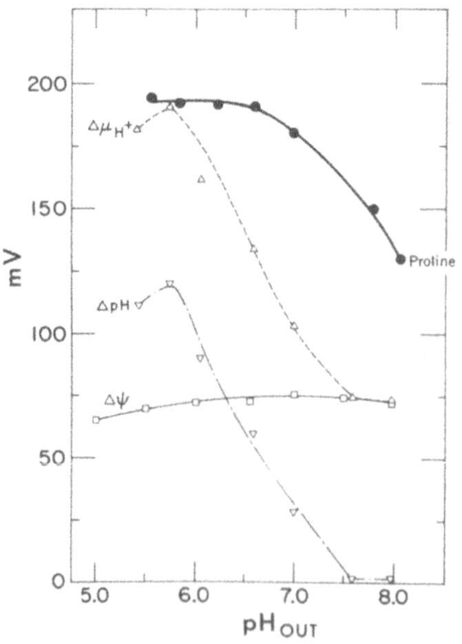

Fig. 8. Effect of external pH on ΔΨ, ΔpH and $\Delta\bar{\mu}_{H^+}$ and steady-state level of proline accumulation in E. coli ML 308-225 membrane vesicles. Data for ΔΨ, ΔpH, and $\Delta\bar{\mu}_{H^+}$ are taken from Figure 3, except that the external pH values given are the actual values for the pH of the upper chamber at the time the measurements were made. Concentration gradients for proline were determined by flow dialysis as described in Figure 2 using U-[^{14}C]proline (236 mCi mmol^{-1}) at a final concentration of 8 μM

(1972, 1973) and Mitchell (1973) those transport systems which are driven by $\Delta\bar{\mu}_{H^+}$ or ΔpH should exhibit maxima for accumulation at approximately pH 5.5, and those systems which are driven by ΔpH should exhibit little or no activity at pH 7.5 and above. The data presented in Figure 8 illustrate the effect of external pH on the accumulation of proline and on $\Delta\bar{\mu}_{H^+}$, ΔpH, and ΔΨ. Each measurement was made under the identical conditions using the flow dialysis technique. Clearly, the predictions are not fulfilled. Although the shape of the proline titration curve resembles that of $\Delta\bar{\mu}_{H^+}$, a quantitative correlation between the accumulation of proline and $\Delta\bar{\mu}_{H^+}$, ΔpH, or ΔΨ at all values of external pH is not evident. Strikingly, moreover, although the magnitude of $\Delta\bar{\mu}_{H^+}$ is sufficient to account thermodynamically for proline accumulation at pH 5.0 and pH 5.5, it is clear that above pH 5.5, proline is accumulated to an extent which is in excess of the value determined for $\Delta\bar{\mu}_{H^+}$. Thus, proline is accumulated to an intravesicular concentration which is approximately 2000-fold higher than that of the external medium at pH 5.5 (i.e., equivalent to about -195 mV), and is therefore apparently in complete equilibration with $\Delta\bar{\mu}_{H^+}$. However, at pH 7.5, where $\Delta\bar{\mu}_{H^+}$ (i.e., ΔΨ) is only about -80 mV, proline is concentrated to the extent of about -160 mV. Similar observations have been made with lactose, lysine, and succinate (Ramos and Kaback, 1977b).

It is apparent from these data that if the stoichiometry between transported solutes and protons remains constant at 1:1 at all values of external pH, the concentration gradients observed at relatively high external pH are impossible to accommodate thermodynamically with the present chemiosmotic model. However, the model is inadequate to explain the accumulation of organic acids at pH 7.5 where there is no discernible ΔpH or the inhibition of acid accumulation by valinomycin observed at this external pH (Fig. 7). However, there is at least one reasonably simple explanation for these observations within the bounds of the chemiosmotic framework. Possibly, the stoichiometry between protons and accumulated solute varies as a function of external pH in such a manner that it is 1:1 at pH 5.5, but increases to higher values as external pH is increased. If, for instance, the stoichiometry between protons and proline were 2:1 at pH 7.5 rather than 1:1, the concentration gradient of proline would be thermodynamically compatible with $\Delta\bar{\mu}_{H^+}$ at pH 7.5 (i.e., if the stoichiometry is 2:1, the concentration gradient varies as the square of the charge gradient). In a similar vein, it is conceivable that at pH 5.5, transport of organic acids occurs in a manner consistent with the standard chemiosmotic mechanism (i.e., one proton is taken up per mole of undissociated acid), while at pH 7.5, two protons are taken up per mole of acid, one in association with the substrate itself and one in association with the carrier molecule. By this means, anion uptake at pH 7.5 would become electrogenic. Such changes in stoichiometry might result from the dissociation of important functional groups in the carriers at relatively high pH values.

These suggestions are not only necessary on theoretical grounds, but they are testable experimentally using the techniques that have been developed in this laboratory (Ramos and Kaback, in preparation). At pH 5.5, imposed variations in ΔpH (using a fixed concentration of valinomycin and increasing concentrations of nigericin) should be reflected by a linear change in the concentrations gradient established with a given transport substrate, and the function should have a slope of one. Alternatively, at pH 7.5, imposed variations in ΔΨ (using a fixed concentration of nigericin and increasing concentrations of valinomycin should be reflected by a linear change in the concentration gradient, and the function should have a slope of two ore more (see Fig. 8 in Schuldiner and Kaback, 1975). In any case, it is important to point out that in order to account for these observations in any manner, it must be conceded that the carriers mediate the response of transport substrates to $\Delta\bar{\mu}_{H^+}$ and its components in an unexpected manner.

Resolution of Inconsistencies

A number of initial observations from this laboratory (see Lombardi et al., 1974 for a review) appear to be in direct conflict with the proposal that chemiosmotic phenomena play a central role in active transport in isolated bacterial membrane vesicles. Although some of these inconsistencies were resolved in subsequent publications (Kaback, 1974b; Schuldiner and Kaback, 1975; Stroobant and Kaback, 1975), some of the more inexplicable observations can be resolved only in light of these recent findings. One of the most difficult of the observations to accommodate is the variation in the ability of different electron donors to drive different transport systems relative to D-lactate or reduced PMS. In chemiosmotic theory, oxidation of any particular electron donor (i.e, reduced PMS, D-lactate, succinate, or NADH) would presumably be associated with a unique value of $\Delta\bar{\mu}_{H^+}$. Accordingly, the order of effectiveness of the various electron donors in stimulating steady-state uptake would be expected to be the same for all respiration-linked transport systems, this order reflecting their relative effectiveness in generating $\Delta\bar{\mu}_{H^+}$. As shown previously (Lombardi et al., 1974), the observations do not support the theory. However, the inconsistency can now be resolved. Since some transport systems are driven primarily by $\Delta\bar{\mu}_{H^+}$ with and without biases towards $\Delta\Psi$ and others are driven primarily by ΔpH, and since some electron donors generate a significant $\Delta\Psi$ but no ΔpH (Table 1), it is clear that those transport systems with a bias towards $\Delta\Psi$ will be stimulated more effectively by these electron donors that other transport systems which are coupled to $\Delta\bar{\mu}_{H^+}$ or ΔpH.

Another apparent inconsistency that is resolved by the present finding is the variation in the inhibitory potency of valinomycin with different transport systems. As shown by Lombardi et al. (1973), the half-maximal inhibitory concentrations of this ionophore vary over three orders of magnitude when its effect on various transport systems is examined. As demonstrated here, depending on the particular transport system and the external pH, valinomycin may stimulate, inhibit, or have essentially no effect. It is apparent, therefore, that in light of the current observations, most of the earlier contradictory findings, if not all, can be accommodated within the confines of the chemiosmotic formulation.

Molecular Aspects of Carrier Function

It seems apparent from the foregoing discussion that although many of the details are not yet completely clear, a unifying concept which explains the energetics of respiration-dependent active transport in bacterial systems is emerging. However, it should be emphasized that this is only the beginning, that ultimately, carrier molecules must be solubilized, purified, and reconstituted so that this concept can be studied on a molecular level. Initial

FLUORESCENT β-GALACTOSIDES

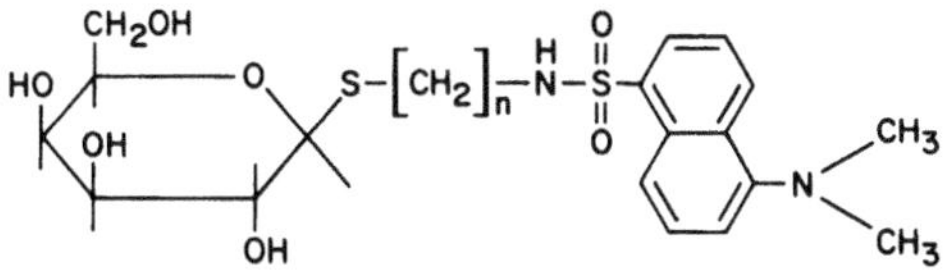

R=	
	DG_0
$S(CH_2)_2$	DG_2
$S(CH_2)_3$	DG_3
$S(CH_2)_4$	DG_4
$S(CH_2)_5$	DG_5
$S(CH_2)_6$	DG_6
$O-CH_2CH_2$	OXY DG_2

Fig. 9. Structural formulae of various dansylgalactosides

PHOTOREACTIVE β-GALACTOSIDES

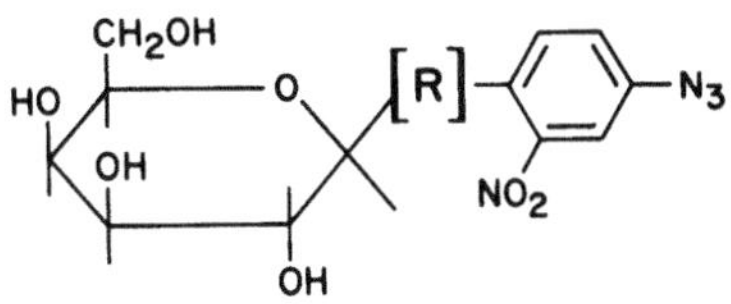

R=	
S	APG_0
SCH_2CH_2-N(H)-	APG_2

Fig. 10. Structural formulae of azidophenyl-galactosides

promising developments in this respect have already been reported (Hirata et al., 1976). In any case, within the past few years, certain important insights into carrier functions at a more refined level of resolution have been achieved. These studies (see Schuldiner et al., 1976b for a review) have involved a collaborative effort between the author's laboratory and that of Dr. Rudolf Weil of Sandoz Forschungsinstitut in Vienna, Austria.

In addition to the wealth of information available with regard to the β-galactoside transport system, and the ability to manipulate it genetically, this transport system has another notable property which makes it particularly advantageous for study. As opposed to most other bacterial transport systems, the β-galactoside system is relatively nonspecific with respect to the non-galactosyl moiety of the dissaccharide. Thus, reporter groups or chemically reactive species can be incorporated into this class of sugars without abolishing their affinity for the lac carrier protein.

Substituted Galactopyranosides

The structures of the compounds to be discussed are shown in Figs. 9 and 10. As shown, they fall into two general categories (N-dansyl)aminoalkyl-β-D-galactopyranosides [dansylgalactosides] (Reeves et al., 1973; Schuldiner et al., 1975a,b,c, 1976a) and (2-nitro-4-azidophenyl)-β-D-galactopyranosides [azidophenylgalactosides] (Rudnick et al., 1975a,b). In both cases, a β-D-galactopyranoside (usually 1-thio-β-D-galactopyranoside) is linked directly or through an alkyl chain of varying length to the appropriate probe.

In one instance, a fluorescent moiety commonly referred to as a dansyl group, is attached to the galactopyranoside (Fig. 9). The dansyl fluorophore is sensitive to solvent polarity, exhibiting an increase in quantum yield and a blue shift in the emission spectrum as the solvent becomes less polar. Thus, the dansyl moiety is useful as a reporter group, yielding information about the polarity of its environment.

The second group of compounds contains an arylazide moiety linked to the galactosyl portion of the molecule (Fig. 10). The rationale behind the use of this class of compounds is that irradiation with visible light causes photolysis of the azido group to form molecular nitrogen and a highly reactive nitrene which then reacts covalently with the macromolecule to which the azido-containing ligand is bound, resulting in irreversible inactivation.

Dansylgalactosides are not Transported

Each of the dansylgalactosides shown in Figure 9 is a competitive inhibitor of lactose uptake in membrane vesicles prepared from strains of *E. coli* which contain a functional lac carrier protein (K_i values for $Dns^{0,2,3,4,5,}$ and ^{6}-Gal are approximately 550, 30, 12, 6, 3, and 5 μM, respectively). Strikingly however, an exhaustive number of experiments indicates that the dansylgalactosides are not transported to a significant extent (Reeves et al., 1973; Schuldiner et al., 1975a,b).

Binding is Energy Dependent

The fluorescent properties of the dansylgalactosides in aqueous solution are similar to those of other dansyl derivatives, exhibiting emission maxima at approximately 540-550 nm and excitation maxima at approximately 340 nm. Significantly, no change in these parameters is observed in the presence of membrane vesicles. On addition of D-lactate, however, there is a marked increase in dansylgalactoside fluorescence which is absolutely dependent upon the presence of functional *lac* carrier protein in the vesicles. Moreover, no change in the fluorescence of 2'-(N-dansyl)aminoethyl 1-thio-β-D-glucopyranoside (dansyl*gluco*side) is observed, indicating that the effects observed are specific for the galactosyl configuration of the ligand. It is also highly significant that the effect of D-lactate can be mimicked by the imposition of an ionic diffusion potential (interior negative) or by a lactose diffusion gradient (inside→outside).

The fluorescence increase observed with each of the dansylgalactosides under the conditions described exhibits an emission maxima at 500 nm and excitation maxima at 345 and 292 nm. The blue shift in the emission maximum is equivalent to that observed when the dansylgalactosides are dissolved in 85% dioxane, indicating that the affected molecules find themselves in a relatively hydrophobic environment. The appearance of the new peak in the excitation spectrum at 292 nm indicates moreover that the bound dansylgalactoside molecules are excited by energy tranfer from tryptophanyl residues in the membrane proteins.

The increase in dansylgalactoside fluorescence induced by D-lactate is blocked or rapidly reversed by addition of β-galactosides, sulfhydryl reagents, inhibitors of D-lactate oxidation, and proton conductors or other reagents which collapse the membrane potential. On the other hand, the fluorescence increase induced by imposition of ionic diffusion potentials is blocked by β-galactosides, sulfhydryl reagents, and proton conductors, but not by respiratory poisons. Moreover, the degree to which dansylgalactoside fluorescence is increased under these conditions is dependent upon the magnitude of the applied ionic diffusion potential. Since D-lactate oxidation leads to the generation of a membrane potential (interior negative), it seems clear that the increase in dansylgalactoside fluorescence observed with D-lactate or artificially imposed ion gradients occurs by a similar mechanism. Finally, the increase in dansylgalactoside fluorescence observed with a lactose diffusion gradient is inhibited by β-galactosides and sulfhydryl reagents, both of which block the binding site of the carrier, but not by respiratory poisons or reagents which collapse the membrane potential. Thus, it seems clear that the fluorescence changes induced under these conditions occur by a mechanism which is independent of the membrane potential.

Titration of vesicles containing the lac carrier protein with each dansylgalactoside demonstrates that the vesicles bind 1-2 nm of each dansylgalactoside per mg of membrane protein, a value almost identical to that obtained from direct binding measurements with $[^3H]DG_6$ (Schuldiner et al., 1976a). Assuming that 1 dansylgalactoside molecule is bound per molecule of lac carrier protein and that the molecular weight of the lac carrier protein is approximately 30,000, 1-2 nm/mg membrane protein is equivalent to 3-6% of the membrane protein. This value is very similar to that reported by Jones and Kennedy (1969) who used a completely independent method. In addition, these titration studies demonstrate that the affinity of the lac carrier for ligand is directly related to the length of the alkyl chain between the galactosyl and dansyl moieties of the dansylgalactosides. There is excellent agreement, moreover, between the affinity constants of the various dansylgalactosides as determined by fluorimetric titration or flow dialysis (Reeves et al., 1973; Schuldiner et al., 1975a, 1976a,b) and their apparent K_i values for lactose transport as given above.

There are at least three possible mechanisms by which energy might lead to dansylgalactoside binding to the lac carrier protein. These mechanisms are shown schematically in Figure 11. In the first scheme (I), the carrier is accessible to the external medium and binding occurs spontaneously. In this case, energy coupling results in partial translocation of the bound ligand, resulting in its exposure to the hydrophobic interior of the membrane, and thus to the fluorescence changes observed. This scheme appears to be

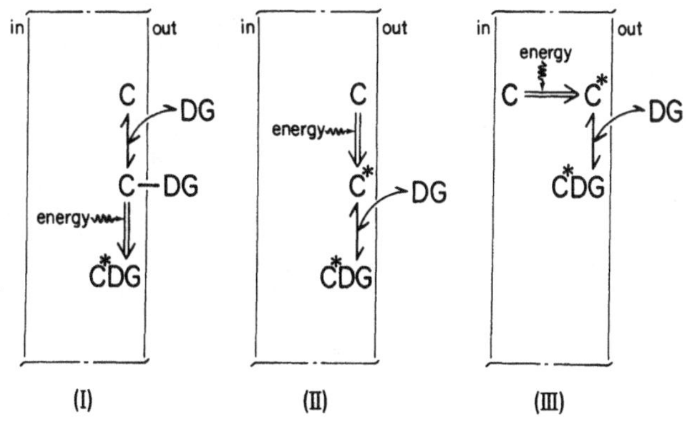

Fig. 11. Possible mechanisms for the increase in dansylgalactoside fluorescence observed on energization of membrane vesicles. C, β-galactoside carrier protein; DG, dansylgalactoside; *, activated state

unlikely because no changes in the emission or excitation spectra of the dansylgalactosides are observed in the absence of D-lactate and for reasons to be discussed below. The second scheme (II) suggests that the carrier is accessible to the external medium in the absence of energy coupling, but that its affinity is increased when energy is supplied. In the third scheme (III), the carrier is inaccessible to the external medium, and energy coupling causes a conformational change in the carrier such that high affinity binding sites appear on the external surface of the membrane. If it is postulated, moreover, that the *lac* carrier protein (or part of it) has a negative charge, the appearance of binding sites on the exterior surface of the membrane can be more easily conceptualized. Imposition of a membrane potential (interior negative) would cause "movement" of the negatively charged carrier to the external surface of the membrane and binding of dansylgalactosides. It should be emphasized that schemes II and III are not mutually exclusive and that energy-coupling may well increase the accessibility of the carrier and its affinity for ligand simultaneously. In any case, the data suggest that energy is coupled to one of the initial steps in transport, and that facilitated diffusion therefore cannot represent the first step in the active transport of β-galactosides.

Further evidence for the proposition that the fluorescence changes observed upon "energization" of the membrane are due to binding of the dansylgalactosides per se rather than binding followed by translocation into the hydrophobic interior of the vesicle membrane has been provided by the use of two other independent techniques. Anisotropy of fluorescence can be used to assess binding specifically since changes in this parameter reflect alterations in the rotation of molecules in solution. Studies with Dns^2-Gal and Dns^6-Gal (Schuldiner et al., 1975c) demonstrate that there is a marked increase in fluorescence anisotropy in vesicles containing the lac carrier protein on addition of D-lactate. In the absence of this electron donor, anisotropy values are minimal and identical in vesicles with or without the *lac* carrier protein, and no increase in anisotropy is observed in vesicles devoid of the *lac* carrier protein when D-lactate is added. In addition to these changes in anisotropy, fluorescence lifetime studies with Dns^2-Gal and Dns^6-Gal are also consistent with the proposition that the changes in dansylgalactoside fluorescence observed on "energization" of the appropriate vesicles reflect binding of the fluorescent probes to the *lac* carrier protein (Schuldiner et al., 1975c). Furthermore, it can be calculated from the anisotropy and lifetime values that the rotational relaxation time of Dns^2-Gal increases dramatically when the probe is bound to the *lac* carrier protein.

Finally, high specific activity $[^3H]Dns^6$-Gal has been synthesized, and its binding to membrane vesicles has been studied directly by flow dialysis (Schuldiner et al., 1976a). With vesicles containing the *lac* carrier protein, little, if any, binding is detected in the absence of D-lactate or reduced PMS. In the presence of these electron donors, binding is observed and the binding constant and number of binding sites are approximately 5 μM and 1.5 nm/mg membrane protein, respectively. Both values are in excellent agreement with those obtained by fluorescence titration. These results provide virtually unequivocal evidence in support of the hypothesis that changes in dansylgalactoside fluorescence observed on energization of membrane vesicles containing the β-galactoside transport system reflect binding of the probe to the *lac* carrier protein.

Energy Dependent Photoinactivation

Studies with the photoreactive azidophenylgalactosides (fig. 10) have provided completely independent support for the conclusions derived from the dansylgalactoside experiments. 2-Nitro-4-azidophenyl-1-thio-β-D-galactopyranoside (APG_0) is a competitive inhibitor of lactose transport in ML 308-225 membrane vesicles, exhibiting an apparent K_i of 75 μM (Rudnick et al., 1975a). The initial rate and steady-state level of $[^3H]APG_0$ accumulation are markedly stimulated by the addition of D-lactate to vesicles containing the β-galactoside transport system, and kinetic studies reveal an apparent K_m of 75 μM. Vesicles devoid of the β-galactoside transport system do not take up significant amounts of APG_0 in the presence or absence of D-lactate. When exposed to visible light in the presence of D-lactate, APG_0 irreversibly inactivates the β-galactoside transport system. Strikingly, APG_0-dependent photoinactivation is not observed in the absence of D-lactate. Kinetic studies of the inactivation process yield a K_D of 77 μM, a value which is almost identical to the K_m and K_i values obtained with this compound. Moreover, lactose protects against APG_0 photoinactivation and significant inactivation of amino acid transport is not observed with APG_0. Thus, it is clear that these effects are specific for the lac carrier protein.

Analogous studies carried out with 2'-N(2-nitro-4-azidophenyl) aminoethyl-1-thio-β-D-galactopyranoside demonstrate that this compound behaves similarly with respect to photoinactivation of the β-galactoside transport system with two important exceptions (Rudnick et al., 1975b). Like its analogue DG_2, APG_2 is not accumulated by the vesicles in the presence of D-lactate or ascorbate-PMS and it exhibits a higher affinity for the lac carrier protein than APG (i.e., the K_i for competitive inhibition of lactose transport and the K_D for photoinactivation in the presence of D-lactate are 35 μM). In addition, it has been demonstrated that an artificially imposed membrane potential (interior negative) also leads to APG_2-dependent photoinactivation of the lac carrier protein.

Apparent Discrepancy

As discussed above, it seems clear from studies with dansyl- and azidophenylgalactosides that binding of these ligands by the lac carrier protein is dependent upon energy-coupling. However, Kennedy et al. (1974) have reported that binding of β-D-galactopyranosyl-1-thio-β-D-galactopyranoside (TDG) and p-nitrophenyl α-D-galactopyranoside (NPG) to membrane particles prepared by ultrasonic disruption is independent of the presence of an energy source and is not inhibited by sodium azide. It should be emphasized though that the amount of binding observed under these conditions is approximately 10-20 times less than the total amount of lac carrier protein (i.e., M protein) present in the membrane as determined by Jones and Kennedy (1969) and as determined by titration studies with the dansylgalactosides. In any case, a small but significant amount of nonspecific binding of dansylgalactosides is detected by fluorescence anisotropy and lifetime studies, allowing the possibility that the techniques described thus far are not sufficiently sensitive to detect less than 10% of the binding observed under energized conditions. For this reason, high specific activity $[^3H]NPG$ was synthesized and its binding to membrane vesicles investigated by means of flow dialysis (Rudnick et al.,

1976). These studies corroborate the findings of Kennedy et al. (1974), but they also confirm the observations discussed above. There is a small amount of NPG binding by vesicles containing the lac carrier protein (about 0.2 nmoles per mg of membrane protein at saturation) which is abolished by p-chloromercuribenzenesulfonate (p-CMBS) but not by proton conductors, and this binding is not dependent upon the presence of D-lactate. On addition of D-lactate, however, approximately 2.3 nmol of NPG are bound per mg of membrane protein (i.e., an amount virtually identical to that observed with the dansylgalactosides), and all of the bound ligand is displaced by p-CMBS. Moreover, the K_D for binding observed under both conditions is very similar (i.e., 6-9 μM). Thus, although there is a small amount of binding in the non-energized state, it seems quite clear that the great bulk of the lac carrier protein is unable to bind ligand unless the membrane is energized.

Translocation and Accumulation of β-Galactosides

The observation that a small but significant amount of NPG binding to membrane vesicles containing the lac carrier protein occurs in the absence of an electrochemical gradient of protons or artificially applied ion diffusion potentials suggests that the lac carrier protein my exist in two forms which are in a state of dynamic equilibrium (Fig. 12): (1) a high affinity form which is accessible on the external surface of the membrane; and (2) a low affinity, cryptic form. In the absence of D-lactate or reduced phenazine methosulfate, 90% or more of the carrier is in the low affinity, cryptic form and only 10% or less is in the high affinity, accessible form. Upon generation of an electrochemical proton gradient across the vesicle membrane (interior negative and alkaline), one or more negatively charged groups in the low affinity, cryptic form of the protein might be influenced, resulting in a conformational change and a shift in the equilibrium. According to such a model, active transport would occur by binding of ligand to the high affinity form of the carrier on the external surface of the membrane, followed by conversion of the carrier to the low affinity, cryptic form and release of ligand from the inner surface of the membrane. However, since the carrier is presumed to be negatively charged, translocation of ligand would require neutralization of the high affinity form of the carrier on the external surface of the membrane. This might be accomplished

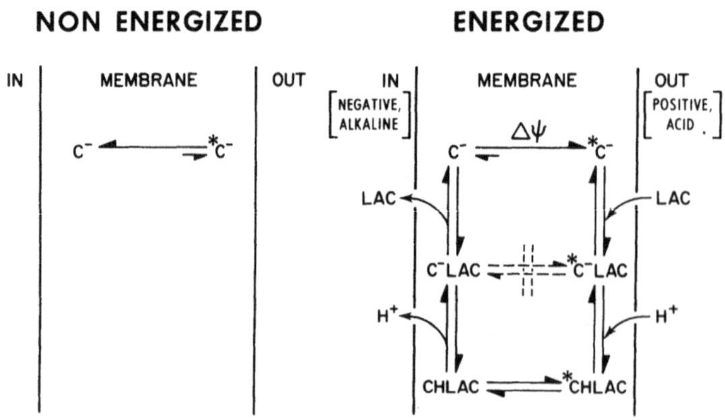

Fig. 12. Kinetic model for lactose accumulation. C: low affinity, cryptic form of the lac carrier protein; *C: high affinity, accessible form of the lac carrier protein; C⁻: unprotonated lac carrier protein; CH: protonated form of the lac carrier protein; LAC: lactose

if ligand binding increased the pK_a of a negatively charge functional group(s) in the carrier. The protein would then be uncharged and no longer under the influence of the electrochemical gradient. The protonated carrier-ligand complex would "relax" to the cryptic form and release proton(s) and ligand on the inner surface of the membrane, regenerating the charged form of the carrier, and the cycle could then be repeated. Clearly, this reaction sequence is consistent with many of the observations discussed above. It is also noteworthy that according to this model, the carrier would not translocate protons across the membrane in the absence of ligand, a stipulation that is necessary lest the carriers themselves dissipate the electrochemical gradient without performing work. It should also be emphasized that this formulation might account for low rates of facilitated diffusion without necessitating that the process represent the initial step in the active transport of β-galactosides.

Abbreviations

The abbreviations used in this paper are: PTS: phosphoenolpyruvate phosphotransferase system; D-LDH: D-lactate dehydrogenase; PMS: phenazine methosulfate; DDA^+: dimethyldibenzylammonium; $TPMP^+$: triphenylmethylphosphonium; Dns^0-Gal: 1-(n-dansyl)amino-1-thio-β-D-galactopyranoside; Dns^2-Gal: 2'-(N-dansyl)aminoethyl-1-thio-β-D-galactopyranoside; Dns^3-Gal: 3'-(N-dansyl)aminopropyl-1-thio-β-D-galactopyranoside; Dns^4-Gal: 4'-(N-dansyl)aminonutyl-1-thio-β-D-galactopyranoside; Dns^5-Gal: 5'-(N-dansyl)aminopentyl-1-thio-β-D-galactopyranoside; Dns^6-Gal: 6'-(N-dansyl)aminohexyl-1-thio-β-D-galactopyranoside; TDG: β-D-galactopyranosyl-1-thio-β-D-galactopyranoside; NPG: p-nitrophenyl-α-D-galactopyranoside; APG_0: 2-nitro-4-azidophenyl-1-thio-β-D-galactopyranoside; APG_2: 2'-N-(2-nitro-4-azidophenyl)aminoethyl-1-thio-β-D-galactopyranoside; CCCP: carbonyl-cyanide-m-chlorophenylhydrazone; p-CMBS: p-chloromercuribenzenesulfonate.

References

Abrams, A., Smith, J.B.: Biochem. Biophys. Res. Commun. 44, 1418 (1971)
Altendorf, K., Hirata, H., Harold, F.M.: J. Biol. Chem. 249, 4587 (1975)
Bakker, E.P., Rottenberg, H., Caplan, S.R.: Biochim. Biophys. Acta 440, 557 (1976)
Barnes, E.M., Jr., Kaback, H.R.: J. Biol. Chem. 246, 5518 (1971)
Berger, E.A.: Proc. Natl. Acad. Sci. U.S. 70, 1514 (1973)
Boonstra, J., Huttunen, M.T., Konings, W.N., Kaback, H.R.: J. Biol. Chem. 250, 6792 (1975)
Butlin, J.D.: Thesis, Australian Nat. Univ., Canberra, Australia, 1973
Colombini, M., Johnstone, R.M.: J. Membrane Biol. 18, 315 (1974)
Colowick, S.P., Womack, F.C.: J. Biol. Chem. 244, 774 (1969)
Devor, K.A., Schairer, H.U., Renz, D., Overath, P.: Europ. J. Biochem. 45 451 (1974)
Eisenbach, H., Cooper, S., Garty, H., Johnstone, R.M., Rottenberg, H., Caplan S.R.: Biochim. Biophys. Acta, in press (1977)
Futai, M.: Biochemistry 12, 2468 (1973)
Greville, G.D.: Current Topics Bioenerg. 3, 1 (1969)
Grinius, L.L., Jasaitis, A.A., Kadziauskas, Yu P., Liberman, E.A., Skulachev, V.P., Topali, V.P., Tsofina, L.M., Vladimirova, M.A.: Biochim. Biophys. Acta 216, 1 (1971)
Harold, F.M.: Bacteriol. Rev. 36, 172 (1972)
Harold, F.M.: Current Topics Bioenerg. (in press, 1976)
Haydon, D.A., Hladky, S.B.: Quart. Rev. Biophys. 5, 187 (1972)

Hirata, H., Altendorf, K.H., Harold, F.M.: Proc. Natl. Acad. Sci. U.S. 70, 1804 (1973)
Hirata, H., Altendorf, K., Harold, F.M.: J. Biol. Chem. 249, 2939 (1974)
Hirata, H., Sone, N., Yoshida, M., Kagawa, Y.: Biochem. Biophys. Res. Commun. 69, 665 (1976)
Jones, T.H.D., Kennedy, E.P.: J. Biol. Chem. 244, 5981 (1969)
Kaback, H.R.: Federation Proc. 19, 130 (1960)
Kaback, H.R.: Ann. Rev. Biochem. 39, 561 (1970)
Kaback, H.R.: In: Jakoby, W.B. (ed.). Methods in Enzymol. New York: Academic Press, 1971, Vol. XXII, p. 99
Kaback, H.R., Barnes, E.M., Jr.: J. Biol. Chem. 246, 5523 (1971)
Kaback, H.R.: Biochim. Biophys. Acta 265, 367 (1972)
Kaback, H.R.: 1974a. In: Fleischer, S., Packer, L. (eds.). Methods in Enzymol. New York: Academic Press, Vol. XXXI, p. 698
Kaback, H.R.: Science 186, 882 (1974b)
Kashket, E.R., Wilson, T.H.: Proc. Natl. Acad. Sci. U.S. 70, 2866 (1973)
Kennedy, E.P., Rumley, M.K., Armstrong, J.B.: J. Biol. Chem. 249, 33 (1974)
Klein, W.L., Boyer, P.D.: J. Biol. Chem. 247, 7257 (1972)
Kohn, L.D., Kaback, H.R.: J. Biol. Chem. 248, 7012 (1973)
Konings, W.N., Barnes, E.M., Jr., Kaback, H.R.: J. Biol. Chem. 246, 5857 (1971)
Konings, W.N., Kaback, H.R.: Proc. Natl. Acad. Sci. U.S. 70, 3376 (1973)
Konings, W.N., Boonstra, J.: In: Kleinzeller, A., Bronner, F. (eds.). Current Topics in Membranes and Transport (in press, 1976)
Konings, W.N.: 1976. Advan. Microbiol. Physiol (in press, 1976)
Kundig, W., Ghosh, S., Roseman, S.: Proc. Natl. Acad. Sci. U.S. 52, 1607 (1964)
Lever, J.: Proc. Natl. Acad. Sci. U.S. (in press, 1976)
Lever, J.: J. Cell. Physiol. (in press, 1976)
Lombardi, F.J., Kaback, H.R.: J. Biol. Chem. 247, 7844 (1972)
Lombardi, F.J., Reeves, J.P., Kaback, H.R.: J. Biol. Chem. 248, 3551 (1973)
Lombardi, F.J., Reeves, J.P., Short, S.A., Kaback, H.R.: Ann. N.Y. Acad. Sci. 227, 312 (1974)
Mitchell, P.: Nature (London) 191, 144 (1961)
Mitchell, P.: Biol. Rev. Cambridge Phil. Soc. 41, 445 (1966)
Mitchell, P.: Advan. Enzymol. 29, 33 (1967)
Mitchell, P.: Chemiosmotic Coupling in Oxidative and Photosynthetic Phosphorylation. Bodmin, England: Glynn Research, 1968
Mitchell, P.: In: Membranes and Ion Transport. Bittar, E.E. (ed.). New York: Wiley, 1970a, Vol. I, p. 192
Mitchell, P.: In: Organization and Control in Prokaryotic and Eukaryotic Cells. Charles, H.P., Knight, B.C.J.G. (eds.). London-New York: Cambridge Univ., 1970b, Vol. XX, p. 121
Mitchell, P.: J. Bioenerg. 4, 63 (1973)
Murer, H., Hopfer, H.: Proc. Natl. Acad. Sci. U.S. 71, 484 (1974)
Or, A., Kanner, B.I., Gutnick, D.L.: FEBS Lett. 35, 217 (1973)
Padan, E., Zilberstein, D., Rottenberg, H.: Europ. J. Biochem. 63, 533 (1976)
Parnes, J.R., Boos, W.: J. Biol. Chem. 248, 4429 (1973)
Quinlan, D.C., Parnes, J.R., Shalom, R., Garvey, T.Q., Isselbacher, K.J., Hochstadt, J.: Proc. Natl. Acad. Sci. U.S. 73, 1631 (1976)
Ramos, S., Schuldiner, S., Kaback, H.R.: Proc. Natl. Acad. Sci. U.S. 73, 1892 (1972)
Ramos, S., Kaback, H.R.: Biochemistry, in press (1977a)
Ramos, S., Kaback, H.R.: Biochemistry, in press (1977b)
Reeves, J.P., Shechter, E., Weil, R., Kaback, H.R.: Proc. Natl. Acad. Sci. U.S. 70, 2722 (1973)
Rottenberg, H.: J. Bioenerg. 7, 61 (1975)
Rudnick, G., Schuldiner, S., Kaback, H.R.: Biochemistry, in press (1976)
Rudnick, G., Weil, R., Kaback, H.R.: J. Biol. Chem. 250, 1371 (1975a)
Rudnick, G., Weil, R., Kaback, H.R.: J. Biol. Chem. 250, 6847 (1975b)
Schairer, H.U., Haddock, B.A.: Biochem. Biophys. Res. Commun. 48, 54 (1972)
Schuldiner, S., Kaback, H.R.: Biochemistry 14, 5451 (1975)
Schuldiner, S., Kerwar, G.K., Weil, R., Kaback, H.R.: J. Biol. Chem. 250, 1361 (1975a)
Schuldiner, S., Kung, H., Weil, R., Kaback, H.R.: J. Biol. Chem. 250, 3679 (1975b)

Schuldiner, S., Spencer, R.D., Weber, G., Weil, R., Kaback, H.R.: J. Biol. Chem. 250, 8893 (1975c)
Schuldiner, S., Weil, R., Kaback, H.R.: Proc. Natl. Acad. Sci. U.S. 73, 109 (1976a)
Schuldiner, S., Rudnick, G., Weil, R., Kaback, H.R.: Trends Biochem. Sci. 1, 41 (1976b)
Short, S.A., Kaback, H.R., Kohn, L.D.: J. Biol. Chem. 250, 4291 (1975)
Short, S.A., White, D.C., Kaback, H.R.: J. Biol. Chem. 247, 7452 (1972)
Sigrist-Nelson, K., Murer, H., Hopfer, H.: J. Biol. Chem. 250, 5674 (1975)
Stroobant, P., Kabackm H.R.: Proc. Natl. Acad. Sci. U.S. 72, 3970 (1975)
Waddel, W.J., Butler, T.C.: J. Clin. Invest. 38, 720 (1959)
West, I.C.: Biochem. Biophys. Res. Commun. 41, 665 (1970)
West, I.C., Mitchell, P.: Bioenergetics 3, 445 (1972)
West, I.C., Mitchell, P.: Biochem. J. 132, 587 (1973)
West, I.C., Wilson, T.H.: Biochem. Biophys. Res. Commun. 50, 551 (1973)

Solubilisation and Partial Purification of Alanine Carrier from Membranes of a Thermophilic Bacterium and its Reconstitution into Functional Vesicles

H. HIRATA, N. SONE, M. YOSHIDA, and Y. KAGAWA

Summary

An alanine transport carrier was solubilized from membranes of thermophilic bacterium PS3 with cholate-deoxycholate mixture and then partially purified by DEAE cellulose and gel filtration in the presence of Triton X-100. For assay of alanine carrier activity it was reconstituted into vesicles with phospholipids and the transport energy was supplied as a membrane potential introduced by K^+-diffusion via valinomycin. The partially purified carrier had no ATPase or NADH dehydrogenase activity and the apparent molecular weight was between 150,000 and 200,000. The Km of the carrier in reconstituted vesicles for alanine transport was 10 μM which coincided with that in original membranes. PS3 P-lipids were strictly required for functional reconstitution and particularly phosphatidyl-ethanolamine or phosphatidylglycerol were found to be essential under the condition studied.

A. Introduction

Energy coupling to active transport of various nutrients across bacterial membrane is still one of the unsolved problems in the transport studies. Recent results showed, however, that coupling of metabolic energy to transport can best be interpreted by Mitchell's chemiosmotic theory (Mitchell, 1966; Hirata et al., 1974; Altendorf et al., 1975; Harold, 1976). But on the other hand, of course, there remain some conflicting evidences and in some cases quite different transport mechanisms have been suggested other than chemiosmotic (Berger, 1973; Berger and Heppell, 1974). In any event, more precise information about the molecular mechanism of transport is eagerly required. For this purpose, one of the general ideas would be to resolve the transport system in the membranes and reconstitute them from each component. Unfortunately, no transport carrier has as yet been isolated from bacterial membranes. This may be because all known transport carriers are too unstable to be solubilized from membranes. Thus for further studies a more stable material is required.

It seems possible that this could be obtained from the membrane of the thermophilic bacterium PS3, which are resistant to various drugs, organic solvents and detergents (Yoshida et al., 1975). Indeed, recently a dicyclohexyl carbodiimide sensitive ATPase complex was purified from this bacterium using ionic and nonionic detergents (Sone et al., 1975) and further characterization of the ATPase complex was carried out under a variety of conditions (Yoshida et al., 1976).

This communication describes the solubilization and partial purification of an alanine carrier from this bacterium and its reconstitution with P-lipids into vesicles capable of active transport of alanine. Part of the work has already been published (Hirata et al., 1976a).

B. Solubilization and Patial Purification of Alanine Carrier

There are naturally some difficulties in this kind of research, namely, how to solubilize the carriers from membrane, how to reconstitute them into functional vesicles, the assay method of the carrier activity and finally purification procedures. Among these difficulties the assay method of carrier activity might be a crucial point, since we do not know whether the carrier has any other catalytic activity such as sarcoplasmic Ca^{2+}-ATPase or Na^{+}-K^{+}-ATPase, or whether the carrier binds substrate without input of energy. Therefore, we had to use an appropriate method as following: the aimed protein(s) was reconstituted into vesicles with phospholipid and then the alanine carrier was defined as protein(s) catalyzing active transport of radioactive alanine into the reconstituted vesicles depending on an artificial membrane potential imposed by K^{+}-diffusion via valinomycin. The method for reconstitution and K^{+}-loading of proteoliposomes were carried out as reported by Hirata et al. (1976).

Membranes of PS3 were solubilized by Na-cholate-deoxycholate mixture, centrifuged at 140,000 g for 1 h, and proteoliposomes were reconstituted from both supernatant and precipitate protein. On ad-

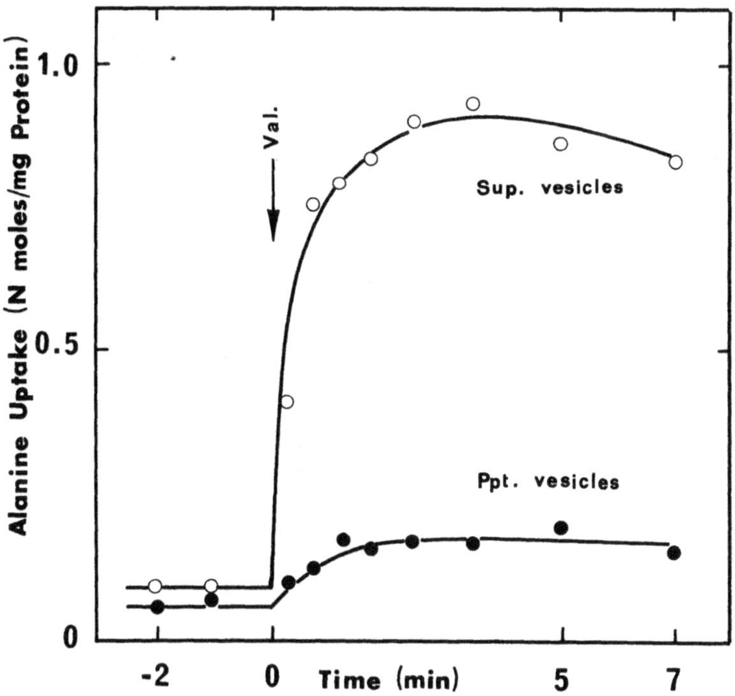

Fig. 1. Alanine uptake by reconstituted, K^{+}-loaded vesicles. Vesicles containing the detergent-soluble (O) or -insoluble (●) proteins were reconstituted and loaded with K-phosphate (pH8). For details, see Hirata et al. 1976

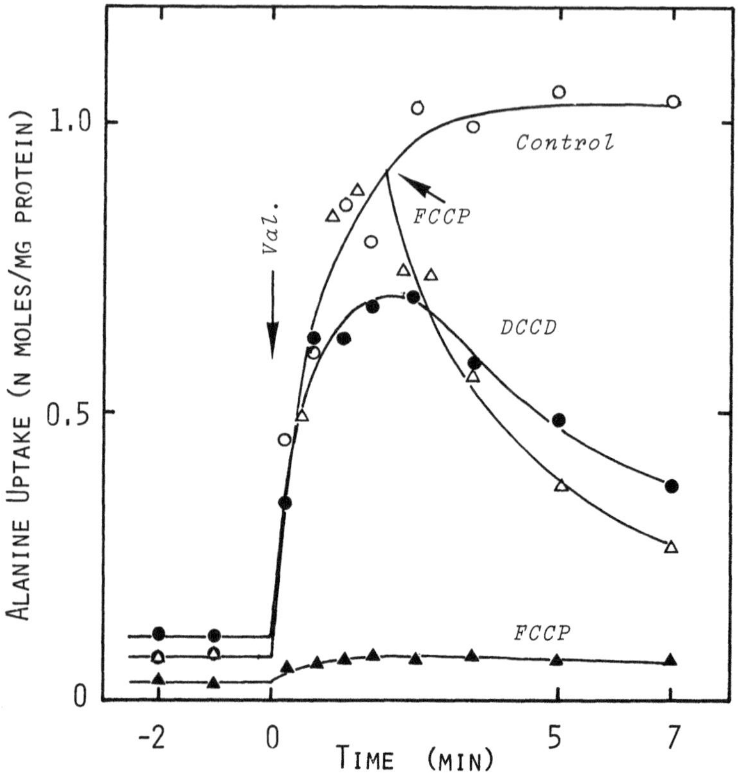

Fig. 2. Effect of inhibitors on alanine uptake by reconstituted-K^+-loaded vesicles. Symbols were: control (o, Δ before addition of FCCP); FCCP (▲, Δ after addition of FCCP as indicated by arrow); DCCD (●)

dition of valinomycin, vesicles reconstituted from the detergent-soluble proteins showed rapid uptake of alanine which reached a maximum in 2 min. Vesicles reconstituted from the insoluble proteins showed less active uptake (Fig. 1).

The validity of this assay method was confirmed since the initial rate of alanine uptake was proportional to the amount of protein (up to 600 μg) added to a fixed amount of phospholipids (10 mg) (Hirata et al., 1976a). Figure 2 shows that the uptake of alanine by these reconstituted vesicles was completely blocked by the uncoupler, carbonylcyanide p-trifluoromethoxyphenyl-hydrazone, while DCCD gave a slight effect.

The alanine carrier was purified using DEAE-cellulose and Sepharose 6B column chromatography in the presence of 0.1% Triton X-100 as described by Hirata et al. (1976a).

The purification of alanine carrier is summarized in Table 1. Fibal preparation (Sepharose 6B Fr. 3) was approximately 13-fold purification in specific activity; however, its yield is too small to proceed to further purification.

Table 1. Purification of alanine carrier

	Protein (mg)	Sp. Act. (nmol.min^{-1}.mg^{-1})
Membrane vesicles	1,096	0.76
Cholate-DOC ext.	96.3	1.85
Triton ext. sup	49.7	1.89
ppt	37.8	0.50
DEAE 50 mM Na_2SO_4	2.5	5.12
200 mM Na_2SO_4	16.0	0.77
Sepharose Fr. 1	0.65	9.93
Fr. 2	0.64	5.88
Fr. 3	0.18	1.84

Characterization of Alanine Carrier

The final preparation obtained as above contained neither ATPase activity nor NADH dehydrogenase activity, but a minute amount of cytochromes was still present.

Figure 3 shows an elution pattern of gel filtration of DEAE eluate of another batch in the presence of 0.5% Triton X-100. Although the presence of higher concentration of the detergents was rather destructive, the activity was found in the second peak which was eluted between urease and yeast alcohol dehydrogenase. Thus, an apparent molecular weight of the alanine carrier would lie between 150,000 and 200,000. However, this does not necessarily mean a true value, since the possibility of aggregation or micelle formation was not excluded.

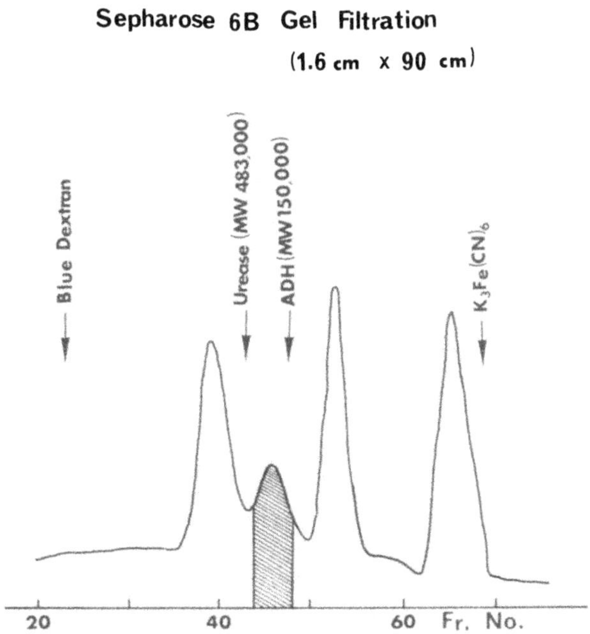

Fig. 3. Gel filtration pattern. Eluates from DEAE cellulose were applied on to a Sepharose 6B column (1,6 cm x 90 cm) equilibrated with 25 mM Tris SO_4 (pH8) containing 0.5% Triton X100, 0.2 M Na_2SO_2 and 0.25 mM dithiothreitol. UV absorbing fractions (4 peaks) were collected and combined. After removal of Triton X100 by ammonium sulfate fractionation, proteins were reconstituted into vesicles with P-lipids and alanine transport activity was assayed

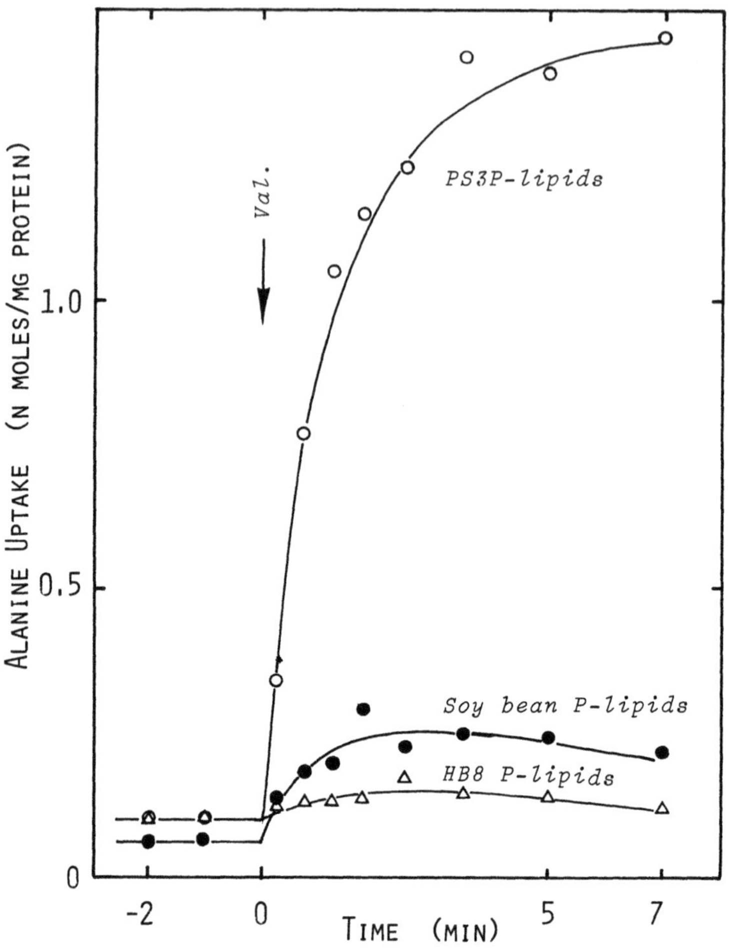

Fig. 4. Phospholipid specificity for reconstitution. Proteins of cholate-DOC extract were reconstituted into vesicles with various P-lipids and alanine transport activity was assayed. PS3 P-lipids are strictly required for functional reconstitution

The apparent Km of the carrier in reconstituted vesicles for alanine transport was 10 μM, which coincided very well with that in original membranes (Hirata et al., 1976b). Among various amino acids tested, the reconstituted vesicles transported glycine as well as alanine. Furthermore, glycine transport seemed to be competitive with alanine transport (not shown). The apparent Ki of glycine for alanine transport was 5.5 μM.

Phospholipid Requirement for Functional Reconstitution

As shown in Figure 4, only PS3 P-lipids were able to reconstitute besides with alanine carrier. Other P-lipids including soy bean P-lipids and those from Thermus thermophilus HB8, ATCC27634 were slightly able to form functional vesicles. As described previously, PS3 P-lipids contained virtually no unsaturation in their fatty acyl groups (Kagawa, 1976).

Table 2 shows that more than 80% of total lipid fraction labeled with ^{14}C-glycerol were phospholipids. From ^{32}P distribution in the

Table 2. Distribution of ^{14}C or ^{32}P in lipid fraction (% of total cpm)

Nuclide	^{14}C	^{32}P	^{32}P
Method	TLC[a]	TLC[a]	PC[b]
Neutral lipids	14.1	-	-
Cardiolipin	23.0	19.2	18.0
Phosphatidyl-ethanolamine	35.7	57.6	58.4
Phosphatidyl-glycerol	27.2	12.9	23.6
Unknown		10.3	
Total	100	100	100

[a] Thin layer chromatography of total lipids
[b] Paper chromatography of Dawson's hydrolysate

phospholipid fraction, more than 50% of phospholipids was phosphatidylethanolamine and the rest was cardiolipin and phosphatidylglycerol.

In Table 3 P-lipid specificity for functional reconstitution was summarized. Among components of PS3 P-lipids, phosphatidylethanolamine and phosphatidylglycerol were found to be the most effective P-lipids, while cardiolipin was unable to form functional vesicles or rather inhibitory when mixed with PE and PG. Table 3 also indicates that any P-lipids from other sources including a synthetic P-lipid could hardly reconstitute functional vesicles.

However, we do not know at the moment whether the requirement of specific P-lipids is due to the intrinsic necessity of the carrier or the artifact in the assay procedures - for example, reconstitution, K^+-loading, membrane filters, etc...

Table 3. Alanine uptake

	Initial rate ($nmol.min^{-1}.mg^{-1}$)	Maximum ($nmol.mg^{-1}$)
PS3-PL	1.25	1.45
PS3-PE	4.97	3.73
PS3-PG	1.95	1.50
PS3-CL	0.07	0.26
PS3-PE+PG+CL	0.41	0.27
Soy Bean PL	0.33	0.20
Egg York PL	0.37	0.45
HB8 PL	0.00	0.00
DPPC[a]	0.14	0.16

[a] dipalmitoyl phosphatidyl choline

Acknowledgments. The authors are grateful to Dr. Tairo Oshima of Mitsubishi, Life Science Institute, Machida City, Tokyo, for kindly providing the bacterial strain used. They also thank members of Kyowa Hakko Co., Ltd for large-scale culture of the bacterium. The technical assistance of Miss Toshiko Kanbe and Miss Keiko Ikeba is acknowledged.

References

Altendorf, K., Hirata, H., Harold, F.M.: Accumulation of lipid-soluble ions and of rubidium as indicators of the electrical potential in membrane vesicles of Escherichia coli. J. Biol. Chem. 250, 1405-1502 (1975)

Berger, E.A.: Different mechanisms of energy coupling for the active transport of proline and glutamine in Escherichia coli. Proc. Natl. Acad. Sci. U.S. 70, 1514-1518 (1973)

Berger, E.A., Heppel, L.A.: Different mechanisms of energy coupling for the shock-sensitive and shock-resistant amino acid permeases of Escherichia coli. J. Biol. Chem. 249, 7747-7755 (1974)

Harold, F.M.: Membranes and energy transduction in bacteria. In: Current Topics in Bioenergetics. New York: Academic Press, 1976

Hirata, H., Altendorf, K., Harold, F.M.: Energy coupling in membrane vesicles of Escherichia coli. J. Biol. Chem. 249, 2939-2945 (1974)

Hirata, H., Sone, N., Yoshida, M., Kagawa, Y.: Solubilization and partial purification of alanine carrier from membranes of a thermophilic bacterium and its reconstitution into functional vesicles. Biochem. Biophys. Res. Commun. 69, 665-671 (1976a)

Hirata, H., Sone, N., Yoshida, M., Kagawa, Y.: Active transport of alanine by thermostable membrane vesicles isolated from a thermophilic bacterium. J. Biochem. (Tokyo) 79, 1157-1166 (1976b)

Kagawa, Y., Sone, N., Yoshida, M., Hirata, H., Okamoto, H.: Proton translocating ATPase of a thermophilic bacterium. J. Biochem. (Tokyo) 80, 141-151 (1976)

Mitchell, P.: Chemiosmotic coupling in oxidative and photosynthetic phosphorylation. Biol. Rev. 41, 445-502 (1966)

Sone, N., Yoshida, M., Hirata, H., Kagawa, Y.: Purification and properties of a dicydohexylcarbodiimide-sensitive adenosine triphosphatase from thermophilic bacterium. J. Biol. Chem. 250, 7917-7923 (1975)

Yoshida, M., Sone, N., Hirata, H., Kagawa, Y.: A highly stable adenosine triphosphatase from a thermophilic bacterium. J. Biol. Chem. 250, 7910-7916 (1975)

Yoshida, M., Sone, N., Hirata, H., Okamoto, H., Kagawa, Y.: Proton translocating ATPase from a thermophile: Components, characterization and reconstitution. 10th Intern. Congr. Biochem. Abstr. (in press, 1976)

Some Ion Transport Systems in *E. coli.* (Transport of Potassium and of Anionic Sugars)

A. KEPES, J. MEURY, A. ROBIN, and J. JIMENO

A. Introduction

I. Physiology of K^+ Transport

K^+ ion is essential for the growth of E. coli and is the major intracellular cation. The bacteria can attain optimal growth rates in media containing a few mM to a few hundred mM K^+ salts (1, 2), and in spite of a 10^5-fold range of K^+ concentrations in the medium, intracellular K^+ is maintained in a narrow range of 100-300

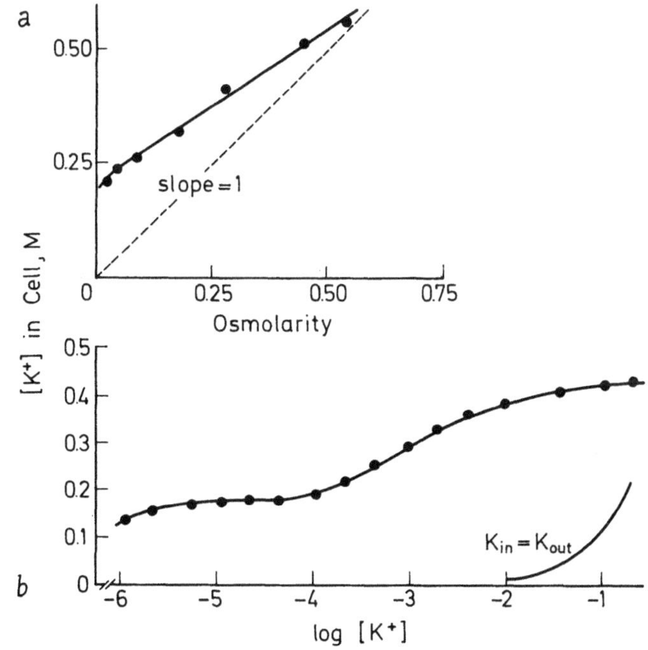

Fig. 1a,b. Variation of the potassium content of *E. coli* cytoplasm as a function of the osmolarity (a) and the K^+ concentration of the incubation medium. (a) The osmolarity of the culture medium was 0.3 osmol, essentially Na phosphate pH 7 and the K concentration was 1 mM. Higher osmolarities were obtained by addition of NaCl, lower osmolarities by dilution of the growth medium. K^+ was kept at 1 mM. The increase of intracellular K^+ is approximately 0.5 mM per increase of 1 milliosmol in the medium. (b) K^+ in the sodium based medium was varied from 10^{-6} M to 0.2 M. When K^+ concentration was in excess of 1 mM, Na^+ concentration was decreased accordingly to keep osmolarity constant. The smooth line indicates theoretical equilibrium in the absence of transmembrane potential difference

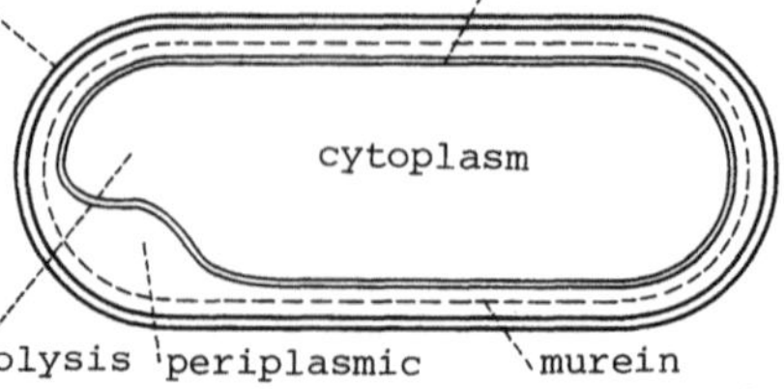

Fig. 2. Schematic representation of the envelopes of *E. coli*. The murein layer (dotted line) is a rigid sacculus made by a single cross-linked macromolecule. It limits the swelling of the cell but allows its shrinkage

mM (3). Even these minor variations of intracellular K^+ are primarily determined by the osmotic pressure of the medium rather than by the K^+ concentration in the medium. Figure 1 shows the variation of intracellular K^+ as a function of these two parameters.

The increase of intracellular K^+ is between 0.5 and 1.0 mM for an increase of 1.0 milliosmol in the medium.

From the diagrammatic representation of the envelopes of E. coli on Figure 2 it appears that the inner membrane of E. coli, the osmotic barrier of the cell, is prevented from osmotic stretching by the rigid murein layer in which it is enclosed and against which it is pressed by the internal hydrostatic pressure under physiological conditions. The inner membrane is capable nevertheless of a fast response to an increase of the external osmotic pressure by an osmotic efflux of water which is completed in about 0.1 sec (4). This amounts to a limited shrinkage which can be monitored turbidimetrically. This shrinkage is corrected by a slower process extending to several seconds or a few minutes.

This strongly suggests that the essential role of cellular K^+ is to maintain the osmolarity of the cytoplasm above the osmolarity of the medium so that the plasma membrane is constantly pressed against the rigid murein envelope and the adjacent outer membrane by the turgor pressure. It is noteworthy that laboratory techniques to obtain lasting plasmolysis, such as practiced for the shock of Neu and Heppel (5) or for the preparation of lysozyme spheroplasts (6) selected empirically K^+ free media to prevent the correction of plasmolysis.

II. Genetics of K^+ Transport

This important homeostatic mechanism is subject to double security and to high-precision control. Genetic data give a glimpse into the biological importance of these mechanism(s). Four independent K^+ uptake systems are coded by at least 9 genes (7, 8). These are represented on the genetic map of Figure 3. Four Kdp genes code for a high affinity transport with a K_m around 1 μM, which is normally repressed when bacteria are growing on high K^+. When K^+ of the medium is nearly exhausted, the Kdp transport system is derepressed and the remaining traces of K^+ are used up, while intracellular K^+ concentration is held constant. When no more K^+ is available, the cells are capable of one doubling, whereby the intracellular K^+ concentration drops to half of its optimal value, and growth ceases.

Fig. 3. Potassium transport loci on the genetic map of *E. coli*. (From Epstein and Davies, 1970)

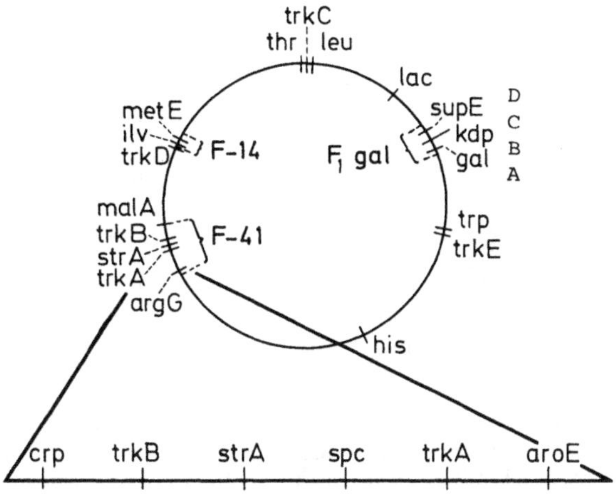

Mutations in the Kdp genes render E. coli unable to grow on micromolar concentrations of K^+, instead millimolar concentrations are required. In these conditions, mutations in the genes Ktr result in even higher K^+ requirements, while such mutations have no major phenotypic effects in Kdp^+ strains. TrK A codes for a major potassium transport system synthesized constitutively and having a K_m around 1 mM, TrK D codes for another system of smaller capacity and TrK B, C and E mutants are leaky for K^+ retention. This picture emerged from the work of Epstein and his associates (1970, 1971). The kinetic description which follows is based on experiments with wild type cells grown on high K^+ and therefore are believed to refer predominantly to the transport system coded by TrK A gene.

B. Experimental

I. Potassium Transport System(s)

1. Potassium Depletion of the Cell

A complete exploration of net fluxes and of unidirectional fluxes of K^+ became possible when a K^+-free condition of E. coli became available. This was most conveniently obtained by a sudden osmotic down shock, which leads to an instantaneous, but transient breakdown of the permeability barrier and a leak of intracellular pools of small molecules, either neutral or electrically charged (9, 10). One simple way is to filter a bacterial suspension on millipore filter and when drained, to rinse with a solution hypoosmotic in comparison with the original medium (Fig. 4). It was shown that a difference of 200 to 400 milliosmol between the two solutions was necessary to release most or all of a pool, depending on the bacterial strain, growth conditions and the size and electrical charge of the solute. The actual osmotic pressure of the suspending and washing solutions have no major importance, nor is the nature of the osmoactive solute critical for the final effect, only the difference of the osmotic pressures and the suddenness of the change from the first to the second are essential.

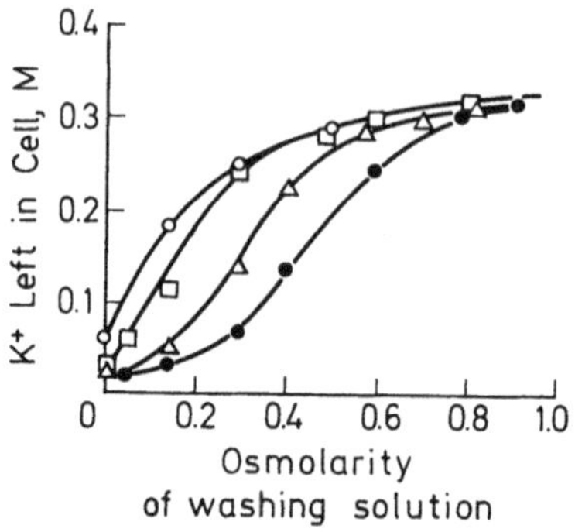

Fig. 4. The effect of hypoosmotic wash on the potassium content of *E. coli*. Aliquots of a culture were preincubated in media of the following osmolarities: 0.9 osM (hexagones); 0.7 osM (triangles); 0.5 osM (squares) and 0.3 osM (circles). Samples were filtered on millipore filters and rinsed with buffers of the osmolarity indicated in abscissa. The potassium retained on the millipore was measured by flame photometry

Very often K^+ depletion was obtained by a distilled water wash 90-95% of the total cell potassium is washed out and the eluate is at neutral pH. This means that intracellular K^+ is nearly completely balanced by small molecular anions, since macromolecules are not released by the distilled water shock. This observation rules out all theories where the bacterial cytoplasm plays the role of an ion exchange resin to a significant extent with respect to intracellular K^+.

2. Uptake of Potassium

In most routine experiments the K^+-depleted cells were suspended in a Na^+-based medium and K^+ was added to observe net uptake. But net uptake can also be obtained with cells already "equilibrated" with external K^+ by increasing the osmotic pressure of the medium. The main cation mouvements in these routine conditions are schematically described in Table 1.

K^+ uptake is upon addition of KCl (Line 2) electrically compensated to the extent of 70-80% by Na^+ efflux, for 5-10% by H^+ efflux and for the remainder presumably by influx of anions.

On the other hand, when the extra K^+ uptake was obtained as a response to the increase of external osmolarity by an electrically neutral molecule like sucrose (Line 3), it could be safely assumed that the inward flux of K^+ was accompanied by an equivalent inward flux of anions, usually Cl^-.

Table 1. Summary of potassium uptake in two steps

Step	Medium	Cytoplasm
0	100 mM NaCl	200 mM Na^+
1	100 mM NaCl 1 mM KCl	50 mM Na^+ 180 mM K^+
2	100 mM NaCl 1 mM KCl 200 mM sucrose	300 mM K^+

(O). Starting with cells emptied by distilled water shock resuspended in a sodium-based electrolyte
(1). the addition of KCl causes a first step of potassium uptake and
(2) the addition of sucrose by elevating the osmotic pressure of the medium causes a further potassium uptake.

3. Requirements for Electric Neutrality

The question could be raised whether any of the electrically compensating ion movements played a specific and essential role. To test this, a cell suspension depleted from intracellular K^+ and also presumably of nearly all small molecular ionic solutes was incubated in a medium buffered with ampholine at pH about 7.0. Ampholines are a mixture of carbon chains of mol.wt. 1000 and above, with a variable proportion of COO^- and NH_3^+ side chains. Appropriate mixtures of an initial pH near 7 or near 4 or near 9 are commercially available. Ampholines were assumed to be nonpenetrating solutes since the outer membrane excludes hydrophilic solutes of mol.wt. > 10^3.

When an acidic mixture of ampholines is titrated with KOH to neutrality, it contains only K^+ as a penetrating species besides the ions of water H^+ and OH^-. Conversely, when an alkaline mixture of ampholines is titrated to neutrality with HCl, it contains only a penetrating anion: chloride.

When depleted E. coli suspended in neutral ampholine was presented with K^+ as the sole ionic species in the form of potassium ampholine very little K^+ uptake occurred. When ampholine chloride was added to reach (Cl^-) concentrations of 7, 50 and 100 mM, the K^+ uptake curves of Figure 5 were obtained.

Similar results were obtained by adding after K^+ alone, $SO_4^=$, $PO_4^=$, NO_3^- ions singly. Any one of these ions was able to support K^+ uptake with small differences in the respective half optimal concentrations. From this, we can rather safely conclude, that K^+ transport is electrogenic and must be compensated by ion movements, such as efflux of cations or influx of anions, but no single ion movement is essential or specific to the K^+ transport mechanism.

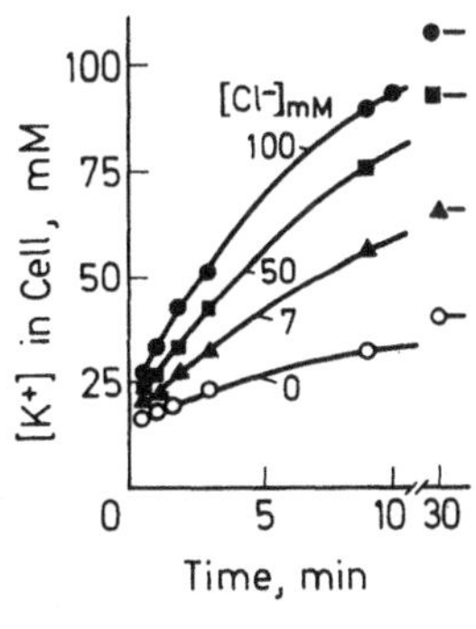

Fig. 5. Dependence of potassium uptake from the presence of a penetrating anion in the medium. Cells were depleted by a distilled water wash and resuspended in a neutral ampholine buffer. K^+ ampholine was added at zero time. Aliquots were supplemented with variable amounts of ampholine chloride

4. Steady-State Fluxes Compared to Uptake

Most active transport systems in bacteria concentrate nonmetabolizable transport substrates optimally 10^2-10^3-fold, some are able to concentrate 10^5-fold (11, 12). All those so far explored reach a steady-state concentration during which influx and efflux compensate each other and influx is of the same magnitude as during the initial accumulation period (13). This situation is often described as a pump and leak (Fig. 6a). Sometimes the unidirectional

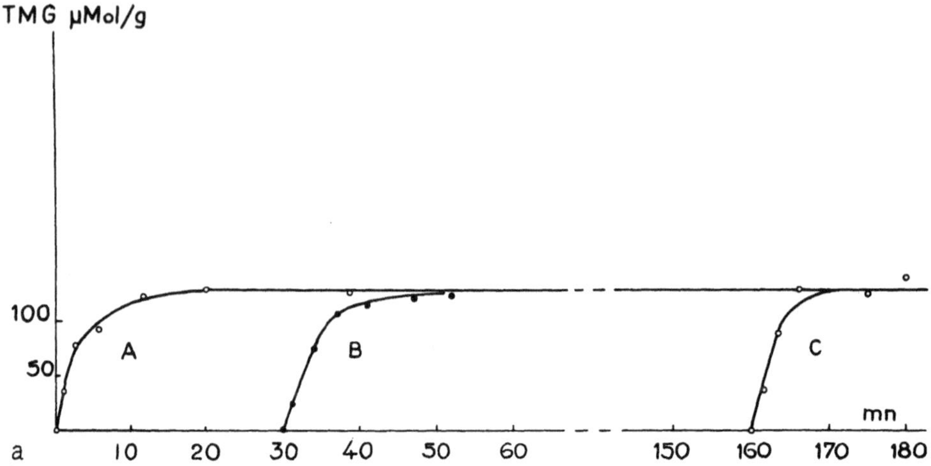

Fig. 6a,b. Uptake and exchange of non metabolized intracellular solutes. (a) Uptake and turnover of methyl thiogalactoside TMG in lactose permease positive bacteria. Uptake was started by 1 mM TMG at zero time. In A, ^{14}C TMG was added, in B and C nonradioactive TMG. These received tracer ^{14}C-TMG at 30 and 160 min respectively. (From Kepes and Monod, 1956)

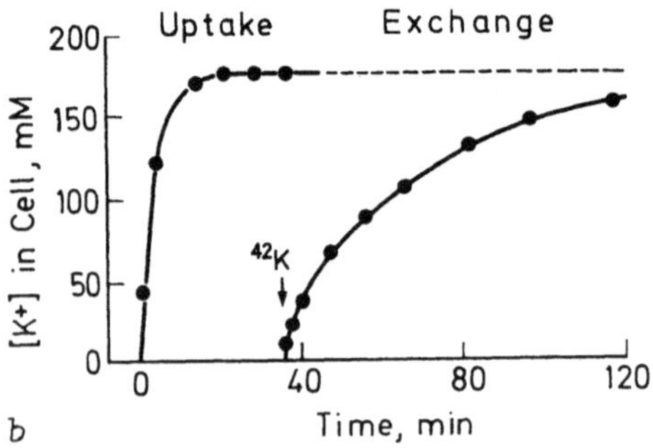

(b) Potassium uptake in cells depleted by a distilled water wash and suspended in a Na based medium was started at zero time by the addition of 1 mM KCl. One suspension had ^{42}K immediately, the other 35 min later

influx during steady state is somewhat higher than during uptake and this is described as trans-stimulation and often supposed to reflect a mobile carrier mechanism.

In all these cases, the steady-state level is itself a saturable function of extracellular transport substrate concentration and proportional to the rate of uptake which means that efflux is first-order with respect to intracellular concentration. Potassium uptake differs in this respect from all other transport systems as shown by Figure 6b.

Influx was found to be a saturable function of (K^+) in the medium. Since as shown above (Fig. 1), steady-state level is virtually independent of the extracellular concentration, a first sharp difference with all other uptake systems appears.

It was found moreover that the maximal rate of exchange was five times lower than the maximal rate of uptake. This is a second difference from all other uptake systems hitherto described. Thus, the uptake system is inhibited at least 80% from carrying maximal fluxes, by the accumulated level of K^+. This can be qualified a feedback inhibition.

5. Regulation of the Fluxes

During an uptake experiment, when internal K^+ concentration is increasing, the relationship between this and the feedback inhibition can be explored. To do this, K^+ uptake was monitored by ^{42}K added at zero time, while unidirectional K^+ influx was monitored by adding trace amounts of ^{42}K at close intervals to aliquots of a suspension during uptake of nonradioactive potassium (Fig. 7a). Rate of uptake remained constant over several minutes while K^+ concentration reached about half the final level. Unidirectional flux was equal to net flux during this period, i.e., efflux was 0. At later times uptake slowed down and unidirectional influx became larger than net flux.Efflux reached its steady-state value around 2/3 completion of the final K^+ level and remained constant while influx slowly reached the same values as efflux, its steady-state value (Fig. 7b).

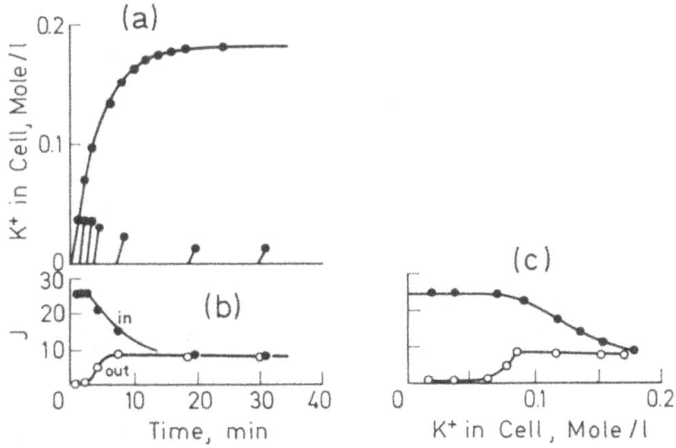

Fig. 7a-c. Variation of unidirectional potassium fluxes during potassium uptake. (a) Uptake as in Figure 6. Small aliquots received tracer ^{42}K at close intervals during uptake of non radioactive K^+ to measure unidirectional influx for periods of 45 s. (b) Unidirectional influx J_{in} expressed as $mM.min^{-1}$ was plotted vs. the midpoint of the period of measurement. Unidirectional efflux J_{out} was calculated as the difference between net flux (a) and unidirectional influx. (c) Unidirectional fluxes of Figure 7b were plotted vs. the intracellular K^+ content of the cells at midinterval

The variation of fluxes is a function of the intracellular K^+ level. Figure 7c shows its high-precision rigid control. This experiment shows that influx and efflux are regulated separately by some feedback mechanism. The regulation of efflux is sharply cooperative. The signal probably comes from a differential osmometer or manometer sensing the difference in hydrostatic pressure inside and outside the cell. Its mechanism is completely unknown.

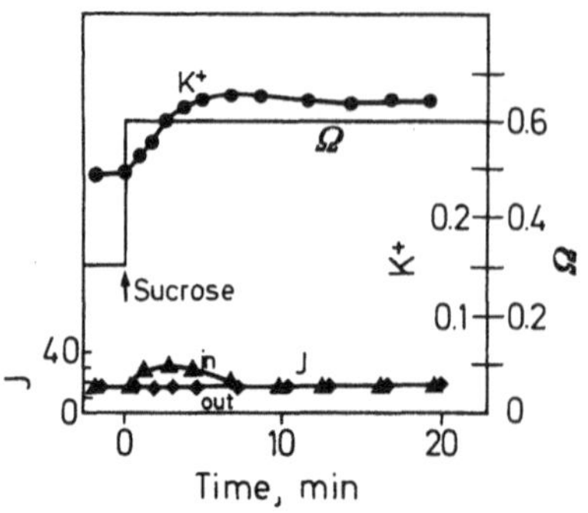

Fig. 8. Increase of intracellular K^+ in response to an osmotic upshock with sucrose. Net flux and unidirectional fluxes J_{in} and J_{out} were measured as in Figure 7a and b

Even more striking is the separate regulation of the two fluxes, when, starting from a steady state, an extra uptake is provoked by a sudden increase of the external osmotic pressure. Influx and efflux were equal during the previous steady state, and upon the osmotic signal, influx suddenly increased, while efflux remained constant. When the new steady-state level was closely approached, influx decreased to its steady-state value, efflux still remained constant (Fig. 8).

6. Steady State is not an Equilibrium State

One important conclusion which stems from this observation and is relevant to the behavior of the cell membrane, is that the final ratio of K_{in}/K_{out} does not reflect a Gibbs-Donnan equilibrium of any kind, since fluxes are sharply stopped by signals where ions may play no role.

Figure 9 illustrates special circumstances where K^+ level clearly does not reflect an equilibrium but rather a metastable state. When energy inhibitors were added to cells depleted of intracellular K^+ before uptake or to cells at the steady state after uptake, the same final conditions led to different final results, i. e., final K^+ level is different depending on the order of addition of the inhibitor and K^+. In particular, CCCP completely inhibits K^+ uptake whereas it only moderately lowers the steady-state K^+ level. At the slightly lower steady state, exchange goes on. In summary, exchange fluxes are not shut off in loaded cells while net uptake flux by empty cells is completely abolished. In contrast, cyanide only moderately inhibits net K^+ uptake, moderately lowers the steady-state K^+ level so that the K_{in}/K_{out} ratio is still about 100/1, but shuts off K^+ exchange sharply at the new

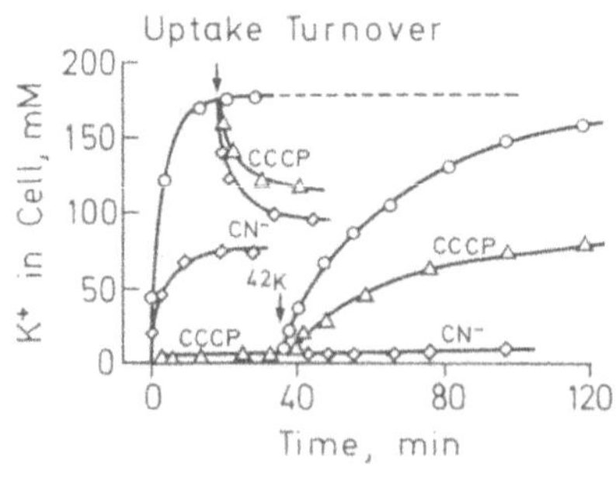

Fig. 9. Uptake and exchange of potassium in the presence of a respiratory inhibitor and an uncoupling agent. Both cyanide and CCCP moderately lower the steady-state level of potassium but CCCP completely inhibits uptake and cyanide completely abolishes exchange

lowered steady state. Here the final level of K^+ does not depend on the order of addition of K^+ and CN^- as if a new equilibrium value would result from the respiratory inhibition, but this seeming equilibrium reflects a complete stillstand of transmembrane K^+ fluxes and therefore this equilibrium is not maintained by the mass action relationship.

7. Metabolic Conditions of the Potassium-Potassium Exchange

Figure 10a and b further illustrate the metabolic conditions which abolish potassium-potassium exchange. In Figure 10a, aerobically grown cells with glycerol as the carbon source with a high steady-state K^+ content were submitted to anoxia by bubbling N_2. This stopped K^+-K^+ exchange as shown by the nonpenetration of added ^{42}K. Bubbling air immediately reversed this inhibition, which was instantly reestablished by adding cyanide or phenazine methosulfate.

In Figure 10, bacteria with no carbon source but at high steady-state K^+ content fail to exchange internal K^+ against external ^{42}K. Addition of glycerol, the carbon source used for growth or of succinate or fructose, which can support a high respiratory rate, immediately restored K^+ exchange. Anoxia, cyanide or phenazine methosulfate immediately stopped this exchange.

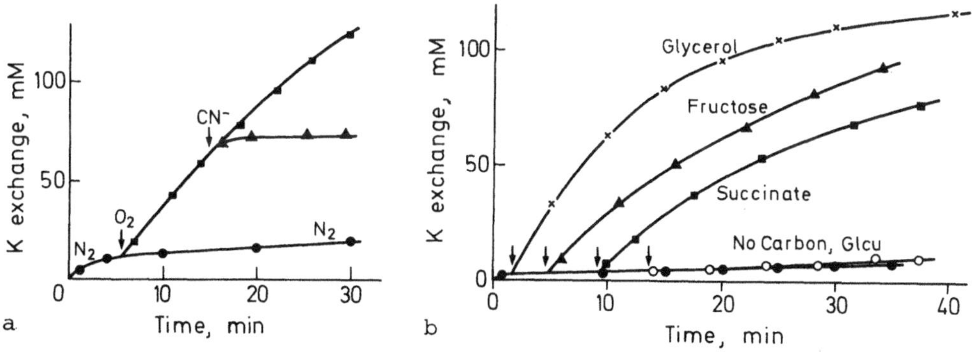

Fig. 10a,b. Inhibition of K^+ - K^+ exchange. (a) Inhibition by anoxia is immediately releived by bubbling O_2 and inhibition is reestablished by adding cyanide. (b) K^+ - K^+ exchange is negligible in the absence of a carbon source. Addition of any carbon source capable to support respiration reestablished normal exchange, while a nonmetabolizable sugar, here glucuronate in mutant AJ_{19} was without effect

8. Membrane Permeability

The virtual abolition of potassium-potassium exchange is noteworthy. This sets a highest limit to the potassium permeability of the cell membrane, which is astonishingly low. Half-times of turnover of the intracellular potassium as long as 100 min or more were repeatedly observed.

This would represent a negligible leak and could not explain turnover times of 20-40 min in physiological conditions when oxidizable substrate, oxygen, and an intact respiratory chain are present.

Therefore, this steady-state exchange depends upon a special transport system apparently dependent on oxidative metabolism, although it does not carry out thermodynamic work.

The impermeability is intrinsic to the bulk of the membrane and is comparable to the often quoted impermeability of the halophilic bacteria (14), which live in 4 M sodium salt solutions and contain up to 5 M K salts in the cytoplasm. The maintenance of such a gradient requires a very efficient barrier to reduce losses.

9. The Effect of Thiol Reagents

Only one class of inhibitors was found which is able to abolish the potassium gradient between the cell and its environment: thiol reagents. N ethyl maleimide is the best example when added at steady state, intracellular K^+ leaks out completely within a few minutes and K^+ uptake is completely abolished. The residual intracellular K^+ corresponds to a gradient of 2-3-fold and it is freely exchangeable with potassium in the medium. The time constant of exchange is an order of magnitude smaller than the time constant of exchange at the steady state of unpoisoned cells, and two orders of magnitude smaller than the exchange during respiratory inhibition.

This fast K^+ exchange strongly suggests that a Gibbs-Donnan equilibrium is approached and that it corresponds to a transmembrane potential difference of 20-30 mV negative inside.

A class of TrK mutations render the cells leaky to K^+ (15). The mutants have a normal intracellular potassium level but its maintenance required constant intensive work of the potassium pump. A decrease of the external K^+ concentration, a limited supply of O_2 to the respiratory chain cause a rapid decrease of intracellular K^+ and even under the optimal conditions of K^+ transport, the turnover of intracellular K^+ is fast. The system is transformed by the mutation into a pump and leak type. The fast leak of K^+ from these cells seems to indicate that the K^+ gradient is not maintained by a high transmembrane electric field, or else, that the primary pump which creates the transmembrane potential difference is damaged.

The evidence that K^+ transport is electrogenic and therefore presumably a primary transport system, the fact that the steady-state K^+ gradient does not normally correspond to an equilibrium state and particularly not to a Gibbs-Donnan equilibrium and the phenomenal impermeability of the cell membrane to K^+ under physiological or nearly physiological conditions raises the question of the role of the K^+ pump and/or the K^+ gradient in the energy conversions of the cell.

II. The role of Potassium in the Energization of Other Transport Systems

A number of reports describe the energization of several transport systems in *E. coli* membrane vesicles by a K^+ diffusion potential generated by the electrogenic efflux of intravesicular potassium upon addition of valinomycin (16). The direction of the potassium gradient in vivo could suggest that a similar mechanism might be operative. In this respect, it was checked that a number of active transport systems for neutral or anionic molecules able

Fig. 11. Uptake of methyl thiogalactoside TMG by cells depleted from intracellular potassium. Depletion by a distilled water shock was followed by resuspension in sodium phosphate buffer and 10 min preincubation with or without 1 mM KCl

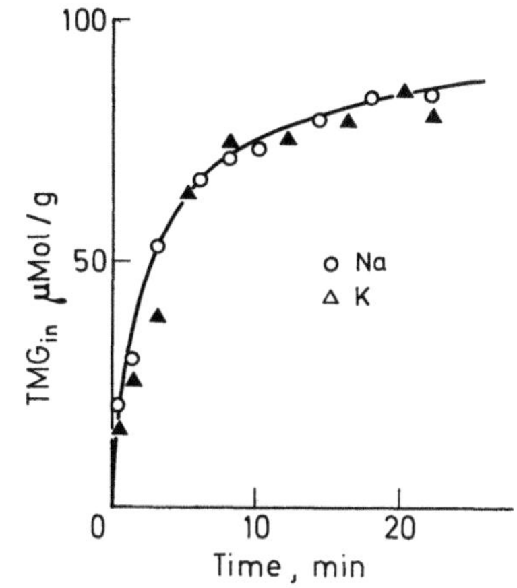

to establish 10^2-10^3-fold concentration gradient

1. did not require the presence of K^+ in the medium (replaced by Na^+)
2. did not require a K^+ gradient or any intracellular pool of free potassium, i.e. did function normally in K^+ depleted cells (Fig. 11)
3. did not require the K^+ pump and the K^+ permeability barrier, i. e., functioned normally in the presence of N ethyl maleimide (Fig. 12)
4. did not require a normal K^+ pump, i.e. continued to perform active transport during the early period of recovery of the K^+ pump after inactivation by NEM
5. were not correlated with any increase of K^+ flux either a net flux or an exchange flux even if the latter was close to zero (see Fig. 10b).

These results are briefly summarized in Table 2.

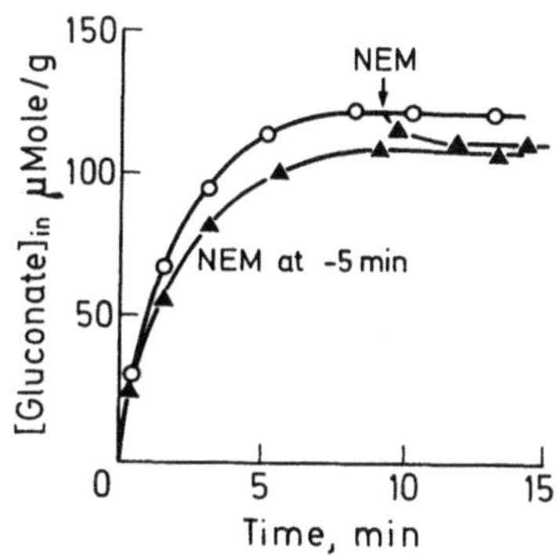

Fig. 12. Gluconate uptake by *E. coli* mutant AR_{13} with or without N-ethyl maleimide added 5 min before or after accumulation of gluconate

III. Relationship Between the Transport of Anionic Sugars and Potassium Transport

Scrutinizing the transport systems of two anionic sugars, gluconate and glucuronate is particularly instructive in connection with the K^+ transport system.

Table 2. Potassium dependence of a variety of active transport systems in E. coli

	Required K^+ in medium	Required K^+ in cell	Active transport in the presence of NEM	Active transport after treatment with NEM	K^+ flux increase[d]
Lactose	no	no		+[a]	
Melibiose	no	no			
Galactose (Pmgl)	no	no	+		
Glucuronic acid	no	no		+[b]	no
Gluconic acid[c]	no	no	+		

a NEM treatment inactivated the K^+ pump while the lac permease was protected by thiodigalactoside (TDG). Transport assay was performed after washing out NEM and TDG.

b NEM treatment in strain AJ_{19} did not inactivate glucuronic acid transport unless glucuronic acid was present. Transport assay was performed after washing out NEM.

c This transport is accompanied by an H^+ symport as shown by alcalinization of the medium and acidification of the cytoplasm.

d ^{42}K was added to a cell suspension with no carbon source. Addition of glucuronate did not increase K^+ exchange whereas addition of glycerol reestablished optimal exchange flux.

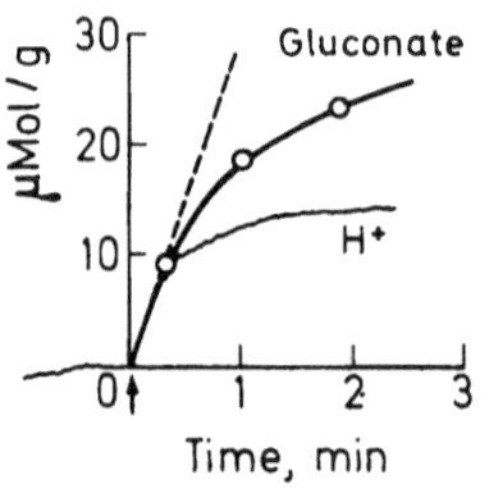

Fig. 13. Hydrogen ion uptake during the initial transient period of gluconate uptake. H^+ uptake was monitored by a pH recorder in an unbuffered suspension of *E. coli* AR_{13} spontaneously stabilized around pH 5.6. After an initial period of 1 to 1 uptake during about 20 sec, the uptake of H^+ ceased while the uptake of gluconate continued

1. Gluconate Transport

Gluconate uptake is accompanied during its first 20 sec by a one-to-one influx of H^+ ions which can be observed to disappear from the medium and also to cause acidification of the cytoplasm (17, 18). This substantiates a proton symport during active transport while the often quoted observation of a flux of H^+ ion driven by a sugar gradient when all presumed primary proton pumps have been inhibited supports the existence of a symporter (19). It also shows that during transient states rapid fluxes can be observed even if their continuation is made impossible by a build-up of an electric field or excessive acidity (Fig. 13). These rapid and limited fluxes indicate direct coupling as opposed to indirect, secondary couplings. Gluconate transport occurs unchanged if potassium is omitted from the medium, with cells emptied of cytoplasmic potassium or in the presence of thiol reagents such as N ethyl maleimide, which collapse the potassium gradient, but do not inhibit gluconate permease (see Fig. 12).

2. Glucuronate Transport

Glucuronate permease can also perform accumulative transport in the absence of extra or intracellular potassium, but it is inactivated by -SH reagents. Its functional integrity in uphill transport can be observed in the presence of a nearly inactive K^+ pump, when the latter was inactivated with NEM while glucuronate permease was protected by the absence of its transport substrate (20). Removing unreacted NEM by washing or by addition of excess 2 mercaptoethanol makes possible the transport assay for glucuronate while the potassium pump is in its early stage of recovery. The observation of glucuronate transport provides two other pieces of information. One concerns the permeability of the bacterial membrane to anions. When glucuronate is accumulated in the cells to about 10^{-1} M and these are resuspended in a glucuronate-free medium, exit is very slow, half times ranging between 30-60 min. This could be due to a very efficient recapture of the anion which leaks out since excess glucuronate in the medium, as well as an energy poison such as sodium azide, accelerate the efflux of intracellular glucuronate by a factor of 10 to 50. That the retention of glucuronate actually reflects the impermeability of the cell membrane can be shown by addition of a thiol reagent such as N ethyl maleimide which slows down glucuronate efflux even more and at the same time prevents the possibility of a chase by external glucuronate (Fig. 14). Half-times of leak now range beyond 60-100 min. This contrasts with the rapid leak of intracellular potassium and the collapse of the K^+ gradient upon addition of the inactivating agent. Thus, potassium exit under its own gradient cannot drive glucuronate anion out of the cell when glucuronate permease is inactivated.

Glucuronate uptake with a maximal velocity of approximately 100 μ mol.g dry weight^{-1}.min^{-1} is independent of the presence of a carbon source in the medium. When the intracellular K^+ is at the steady state e.g., 200 mM, in the absence of a carbon source, its exchange flux is very slow, about 5 μmol.g^{-1}.min^{-1}, and of course, its net flux is zero.

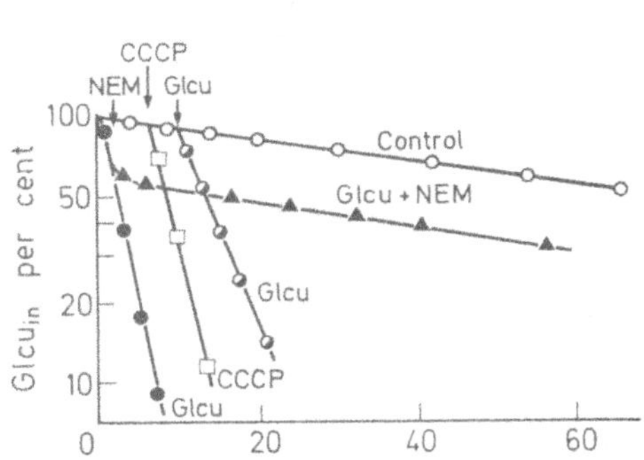

Fig. 14. Efflux of glucuronate from *E. coli*. E. coli strain AJ_{19} was loaded with ^{14}C glucuronate, centrifuged and resuspended in a large volume of glucuronate free medium. The efflux in this control suspension (open circles) was modified by addition of saturating non radioactive glucuronate at zero time (full circles) or half saturating glucuronate at 8 min (half-filled circles), 5 μM carbonyl cyanide m chlorophenyl hydrazone (CCCP, open squares) and the glucuronate containing sample was supplemented at 1 min with 0.5 mM N-ethyl maleimide (NEM, closed triangles)

These fluxes are not significantly increased upon the addition of glucuronate (in the non metabolizing mutant). When the inward flux is monitored with ^{42}K, the assay is very sensitive, so that an increase of K^+ influx matching 10% of the active glucuronate uptake could have been easily detected. This was not observed even for a short transient period (see Fig. 10b).

C. Conclusions

K^+ transport is due to a primary pump, it is electrogenic with a number of optional pathways of electrical compensation. K^+ gradient does not reflect an equilibrium state but obeys an at least two-fold feedback regulation in a rather rigid way, the signal of which is the turgor pressure and the output an inhibition of the inward flux on one hand and of the exchange flux on the other.

The permeability of the cell membrane to K^+ is about 10^5 times lower than its permeability to water, passive "equilibration" of K^+ has time constants of 100 min or more i.e., more than a bacterial generation time. K^+ conductance is sharply increased by thiol reagents and as a result of certain mutations.

If transport systems for neutral or anionic solutes are driven by cotransport with a cation e.g., H^+, K^+ cannot serve as an intermediate or as a secondary electrical compensation to an electrogenic transport system. No K^+-H^+ or K^+-Na^+ exchange systems could be detected.

The actual transmembrane potential difference was not measured in the present study. It cannot be ruled out that the distribution of Na^+ in K^+-depleted cells or the distribution of K^+ in NEM-treated cells, which are both compatible with a transmembrane potential difference of 15-30 mV inside negative, represent a near equilibrium. It must be stressed that in both of these conditions the "energized state" of the membrane remains of normal magnitude as judged by the concentration gradients which a variety of active transport systems are able to establish. This low value of the potential difference (at pH near 7.0) combined with the known ΔpH of 0.5 to 1.0 pH unit, inside alkaline (21, 22), would result in a total "proton motive force" insufficient to drive most active transport systems with a one-to-one stoichiometry.

Special mention has to be made of the high resistance of the cell membrane to the passage of K^+ ions. The highest impermeability is characteristic of the bulk membrane structure and any increased permeability is due either to a perturbation of this basic structure or to a specialized ion transport channel or carrier. Since increased permeability can be due to at least three genetically distinct mutations and to a variety of -SH reagents, the concept of perturbations leading to a leak is suggested. The specificity of the leak channels thus created remains to be explored, we could only show in the present work that NEM creates channels permeable to K^+, but not to glucuronate.

References

1. Damadian, R.: Ion metabolism in a potassium accumulation mutant of E. coli. B. J. Bacteriol. 95, 113 (1968)
2. Lubin, M., Kessel, D.: Preliminary mapping of the genetic locus for potassium transport in E. coli. Biochem. Biophys. Res. Commun. 2, 249 (1960)
3. Schultz, S.G., Solomon, A.K.: Cation transport in E. coli. I. Intracellular Na and K concentrations and net cation movement. J. Gen. Physiol. 45, 355 (1961)
4. Matts, T.C., Knowles, C.J.: Stopped-flow studies of salt induced turbidity changes of Escherichia Coli. Biochem. Biophys. Acta 249, 583-587 (1971)
5. Neu, H.C., Heppel, L.A.: The release of enzymes from Escherichia coli by osmotic shock and during the formation of spheroplasts. J. Biol. Chem. 240, 3685-3692 (1965)
6. Repaske, R.: Lysis of Gram-negative organisms and the role of versene. Biochem. Biophys. Acta 30, 225-229 (1958)
7. Epstein, W., Kim, B.S.: Potassium transport loci in E. coli K12. J. Bacteriol. 108, 639 (1971)
8. Epstein, W., Davies, M.: Potassium-dependent mutants of E. coli K12. J. Bacteriol. 101, 836 (1970)
9. Leder, I.: Interrelated effects of cold shock and osmotic pressure on the permeability of the E. coli membrane to permease accumulated substrates. J. Bacteriol. 111, 211 (1972)
10. Tsapis, A., Kepes, A.: Transient breakdown of the permeability barrier of the membrane of Escherichia coli upon hypoosmotic shock, submitted to Biochem. Biophys. Acta (1976)
11. Kepes, A.: The place of permeases in cellular organization. In: The Cellular Functions of Membrane Transport. Hoffman, J.F. (ed.). Englewood Cliffs, N.J.: Prentice Hall, 1964, p. 155
12. Vorisek, J., Kepes, A.: Galactose transport in E. coli and the galactose binding protein. Europ. J. Biochem. 28, 364 (1972)
13. Kepes, A., Monod, J.: Etude du fonctionnement de la galactoside perméase d'Escherischia coli. Compt. Rend. Acad. Sci. (Paris) 244, 809 (1957)
14. Ginzburg, M., Sachs, L., Ginzburg, B.Z.: Ion metabolism in a Halobacterium. J. Membrane Biol. 5, 78-101 (1971)
15. Lubochinsky, B., Meury, J., Stolkowski, J.: Cinêtique des èchanges de potassium chez E. coli, souche B 207, qui ne peut croître normalement qu'en prêsence de concentrations élevêes en potassium. Compt. Rend. Acad. Sci. (Paris) 258, 5106 (1964)
16. Hirata, H., Altendorf, K., Harold, F.M.: Energy coupling in membrane vesicles of E. coli. Accumulation of metabolites in response of an electrical potential. J. Biol. Chem. 249, 2939 (1974)
17. Robin, A., Kepes, A.: The mechanism of maintenance of electroneutrality during the transport of gluconate by E. coli. FEBS Lett. 36, 133 (1973)
18. Robin, A., Kepes, A.: Inducible gluconate permease in gluconate kinase deficient mutants of Escherichia coli. Biochem. Biophys. Acta 406, 50-59 (1975)
19. West, J.C., Mitchell, P.: Proton-coupled β-galactoside translocation in nonmetabolizing E. coli. J. Bioenerget. 3, 445 (1972)
20. Jimeno-Abendano, J., Kepes, A.: Sensitization of D-glucuronic acid transport system of E. coli to protein group reagents in presence of substrate or absence of energy source. Biochem. Biophys. Res. Commun. 54, 1342 (1973)
21. Kahsket, E.R., Wong, P.T.S.: The intracellular pH of E. coli. Biochim. Biophys. Acta 193, 212 (1969)
22. Haguenauer, R., Kepes, A.: Na F inhibition of phosphorylation and dephosphorylation involved in α-methyl glucoside transport in E. coli K12. A pH-dependent phenomenon sensitive to uncoupling agents. Biochimie 34, 505-512 (1972)

The Role of Periplasmic Proteins and Outer Membrane Receptors in the Process of Active Transport in *E. coli.*

W. Boos

Summary

Two types of proteins are discussed in their role of facilitating the transport of maltose and sn-glycerol-3-phosphate in Escherichia coli. The first protein is the receptor for phage λ, known to be an outer membrane protein. By facilitating the diffusion of maltose and the higher maltodextrins through the outer membrane the effect of the λ receptor is to decrease the K_m of the transport system without influencing the V_{max} of substrate flux.

The second protein is a periplasmic protein that is induced by growth on glycerol and is essential for transport of sn-glycerol-3-phosphate in whole cells but not in membrane vesicles. This protein has solely been identified by the use of a two-dimensional polyacrylamide gel electrophoresis of periplasmic proteins in wild-type and mutants defective in L-α-glycerolphosphate transport.

A. Introduction

It is convenient to classify the active transport systems in E. coli in three groups according to their sensitivity against osmotic shock, to the type of proteins involved, and to the mode of energy coupling for the accumulation of their substrate. Figure 1 ia a schematic representation of these systems, exemplified by the lactose, the β-methylgalactoside, and the maltose transport system.

1. The lactose type systems are generally insensitive against the classical cold osmotic shock procedure of Neu and Heppel (1965) and are still operating in isolated membrane vesicles. These vesicles are freed of periplasmic binding proteins, outer membrane components and cytoplasmic constituents. Genetically the best understood system, lactose (Kennedy, 1970), consists of only one component. This component is tightly membrane-bound and the product of the lacY gene of the lactose operon. General agreement exists about the nature of the energy coupling of this system. As a proton symporter the lactose carrier makes use of the membrane potential (inside negative) and the pH gradient (inside alkaline) across energized membranes according to the proposal of Mitchell (Kaback, this vol.).

2. The β-methylgalactoside type systems (Boos, 1974a) are highly sensitive to the treatment with the cold osmotic shock procedure, since they are dependent on periplasmic substrate binding proteins which are removed from the cell envelope during this process. Where genetic data are available it has been found that binding proteins are essential but not sufficient for a

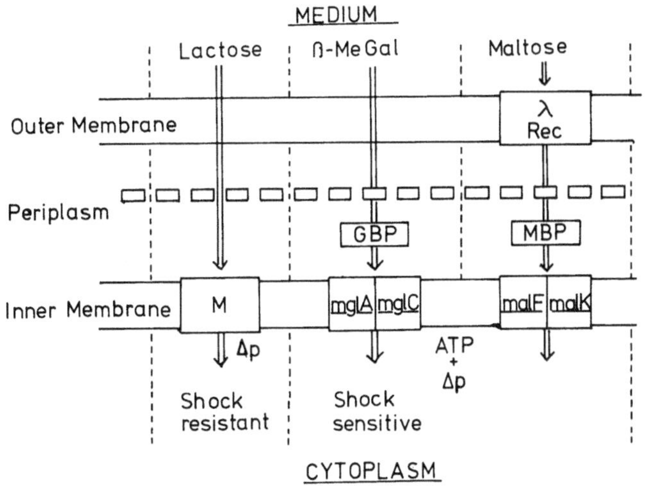

Fig. 1. Schematic representation of three different types of active sugar transport systems in Escherichia coli. β-MeGal: 1-O-β-methyl-galactopyranoside; λRec: receptor for phage λ; GBP: galactose-binding protein; MBP: maltose-binding protein; M: M-protein, gene product of the lacY gene; *mglA*, *mglC*, *malF*, and *malK*: symbolize the as yet unknown proteins of the corresponding genes; Δp: proton-motive force equal to sum of the membrane potential $\Delta\Psi$ and the chemical potential of the pH gradient expressed in electrical dimensions, $-2.3\frac{RT}{F} \cdot \Delta pH$

functioning transport system. In the case of the β-methylgalactoside transport systems two additional genes besides that for the galactose-binding protein are necessary for the complete transport system (Ordal and Adler, 1874). We must postulate a membrane location for these as yet unknown gene products (Johnson et al., 1975). It is not clear how energy coupling occurs in binding protein dependent systems. Several observations indicate that Mitchell's protonmotive force is not the main source of energy, while phosphate-bound energy, possibly ATP itself, might play a key role (Berger and Heppel, 1974). Difficulties arise from the observation that entry and exit might occur via different routes (Parnes and Boos, 1973) and exit appears to be energy dependent (Wilson, 1976). Similarly, exit of maltose from maltose loaded cells is dependent on metabolic energy and prevented by energy uncouplers of the proton conducting type (Ferenci, 1976). Moreover, other lines of evidence indicate at least a participation of the proton-motive force, Δp, on the energetics of binding protein dependent transport systems. Colicin K affects the "energized state of the membrane" but does not decrease ATP levels provided the membrane bound, Ca^{2+}, Mg^{2+} activated ATPase is defective. Under these conditions both types of transport system (shock-sensitive and resistant) are equally reduced in their energy coupling (Plate et al., 1974). Also, Lieberman and Hong (1976) reported the isolation of a temperature sensitive energy uncoupled mutant that behaved at the non-permissive temperature similarly to an ATPase negative mutant treated with colicin K. Both, binding protein dependent and independent transport systems were equally strong reduced even though ATP levels were high. Therefore, both the proton motive force as well as ATP appear to be required for the energy coupling of binding protein dependent systems.

It is not clear how binding proteins participate in the transport mechanism. One favored idea is to endow them with a K_m-factor like function for a tightly membrane-bound system of low affinity (Robbins and Rotman, 1975). Other possibilities have been discussed elsewhere (Silhavy et al., 1974).

3. The maltose-type systems contain in addition to their components in the cytoplasmic membrane and possibly periplasmic proteins also proteins in the outer membrane. In the case of the maltose transport system it is the λ receptor (Szmelcman and Hofnung, 1975). Other transport systems make use of colicin and other phage receptors (Wang and Newton, 1971; DiMasi et al., 1973; Hantke and Braun, 1975).

B. The Maltose Transport System and the Phage λ Receptor (Szmelcman et al., 1976)

Wiesmeyer and Cohn first demonstrated a transport system which was energy-dependent and inducible by maltose (Wiesmeyer and Cohn, 1960). Genetic studies performed later on the maltose system revealed that three genes malE, malF, malK located in the malB region, are essential in the maltose transport capacity (Hofnung, 1974) and are under the positive control of a malT gene. It is clear that this transport system is the only in E. coli taking up maltose, since malE, malF, or malK mutants do not grow on maltose. A maltose-binding protein was extracted by osmotic shock, purified to homogeneity, and shown to be an essential element in maltose transport (Kellerman and Szmelcman, 1974), as well as in maltose chemotaxis (Hazelbauer, 1975). The synthesis of this protein is induced by maltose and is under the same positive control as the other proteins of the maltose system. It is coded for by the malE gene. The maltose-binding protein is a monomer of 40,000 daltons and it recognizes maltose and the α, 1→4 linked higher maltodextrins.

Gene lamB, located in the same operon as malK, is the structural gene for an outer membrane protein which acts as a receptor for bacteriophage λ (Randall-Hazelbauer and Schwartz, 1973). lamB mutants have recently been shown to be impaired in maltose transport when the concentration of this sugar in the medium is less than 0.1 mM. Furthermore, treatment of wild-type bacteria ($lamB^+$) with antibodies directed against purified λ receptor strongly reduced the initial rates of maltose transport, thus demonstrating that the phage receptor is an element of the maltose uptake machinery (Szmelcman and Hofnung, 1975).

In the following the kinetic parameters of maltose and maltotriose transport in wild-type and λ resistant mutants are compared with the binding constants of the maltose-binding protein for these substances.

I. Active Transport of Maltose

Since a nonmetabolizable analogue of maltose is not yet available all transport experiments were performed with pop. 1000, an amylomaltase negative strain which is unable to utilize maltose or maltotriose as carbon source. Figure 2 shows a typical uptake curve

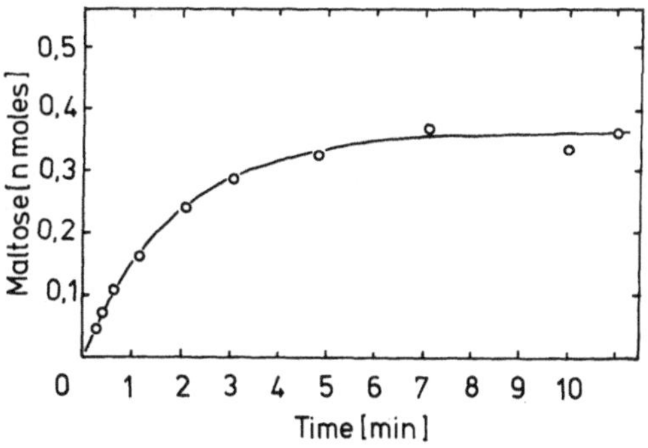

Fig. 2. Uptake of maltose by pop 1000 (malQ7, $lamB^+$). Cells were grown in glycerol containing 50 mM maltose as inducer. Washed cells ($5x10^7$ cells. ml^{-1}) and an initial [^{3}H]maltose concentration of 1 μM were used. Plotted are the amounts of maltose accumulated per 2.5 . 10^7 cells. (Adapted from Szemelcman et al., 1976 with permission of the author and the publisher)

of maltose at 1 μM initial external concentration using 5 . 10^7 cells.ml^{-1}. Under these conditions accumulation reaches saturation after 6 min. Assuming a cellular volume of 10^{-12} ml (Winkler and Wilson, 1966) one can estimate the ratio of intra- to extracellular radioactivity at the equilibrium state of accumulation is in the order of 8 . 10^4. However, chromatographic analysis of the remaining external radioactivity revealed that it is no longer only maltose but contains in addition a compound that runs faster than maltose on paper chromatography. Moreover, the accumulated radioactivity consists of about 70% maltose and 30% of the same unknown compound which is found in the external medium. Despite these unusual features the initial uptake of maltose still gives the kinetic parameters of the transport system. Time-dependent chromatography of the accummulated compounds reveals that this unknown compound is formed only internally, i.e. not during the transport step.

II. Kinetic Constants for maltose and Maltotriose Transport via the Complete Transport System

From the double reciprocal plots of the initial rate of maltose uptake versus different maltose concentrations (Fig. 3) one obtains a straight line with an apparent K_m of 0.9 μM and a V_{max} of 2.0 nmol per min per 10^8 cells. The same plot at higher as well as lower substrate concentrations remains linear. This indicates that maltose uptake is due to a homogeneous system. Moreover, since the K_D of the maltose-binding protein for maltose shows the same K_D as the K_m for maltose transport one might conclude that the maltose-binding protein is the only recognition site for mal-

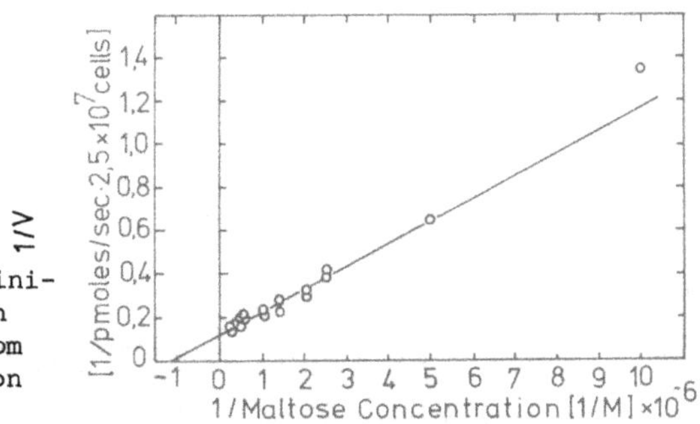

Fig. 3. Lineweaver-Burk plot of the initial rate of maltose uptake in strain pop 1000 (malQ7, $lamB^+$). (Adapted from Szmelcman et al., 1976 with permission of the author and the publisher)

tose during the transport step. However, this is not true when the next higher maltodextrin is used as substrate. Double reciprocal plots for the initial rate of maltotriose uptake versus maltotriose concentration (not shown) results again in a straight line with an apparent K_m of 2 μM and a V_{max} of 1.1 nmol per min per 10^8 cells. However, the K_D of the maltose-binding protein for maltotriose as measured by fluorescence quenching is 0.15 μM (Fig. 6). Yet, despite the discrepancy of K_m of transport and K_D of binding to the maltose-binding protein the uptake system appears kinetically homogeneous and not mediated via two different recognition sites.

III. Kinetic Constants of Maltose Transport in Two Mutants Defective in the Receptor of Phage λ

From strain pop 1000 (malQ7) a mutant was isolated which showed resistance against phage λ as well as against host range mutants of λ. The mutation was shown to be in lamB, the structural gene of the receptor protein of phage λ. Double reciprocal plots of the initial rate of maltose transport in this mutant against maltose concentrations gave an apparent K_m of 0.5 mM and a V_{max} of 0.9 nmol per min per 10^8 cells. Furthermore, a previously characterized missence lamB mutation was introduced in a strain containing the malQ7 (amylomaltase defective) mutation. The determination of K_m and V_{max} of maltose transport in wild-type and the isogenic lamB mutant is shown in Figure 4. The mutant exhibited a K_m of 0.1 mM in contrast to the wild-type where the K_m was found to be 1 μM. Despite the differences in K_m the V_{max} for both strains was identical. Thus, the absence of an intact λ receptor results in an apparent increase of the K_m for maltose uptake above the K_D of the maltose-binding protein. In the lamB mutants no transport of maltotriose could be measured even at substrate concentrations of 1 mM. This finding is corroborated by the observation that nonsense lamB mutants grow very poorly on maltodextrins (unpublished results).

IV. Substrate Induced Fluorescence Quenching of the Maltose-Binding Protein

The addition of substrate to the maltose-binding protein drastically alters the intrinsic fluorescence properties of this protein. Figure 5B shows the emission spectrum (excitation at 280 nm) and

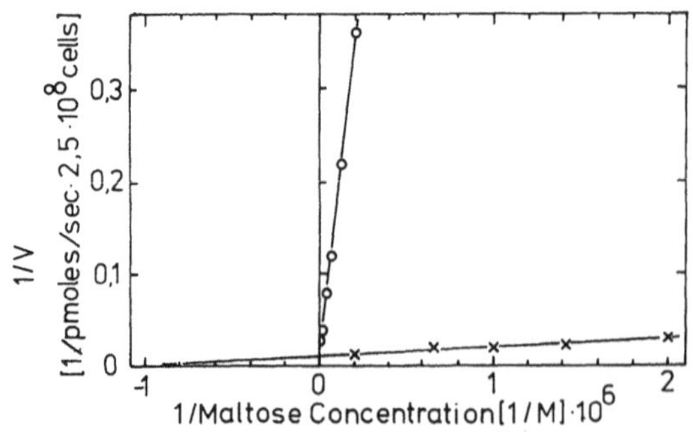

Fig. 4. Lineweaver-Burk plot of the initial rate of maltose uptake in strains x, pop 1010 (malQ7, lamB+); o pop 1101 (malQ7, lamB 103). (Adapred from Szmelcman et al., 1976 with permission of the author and the publisher)

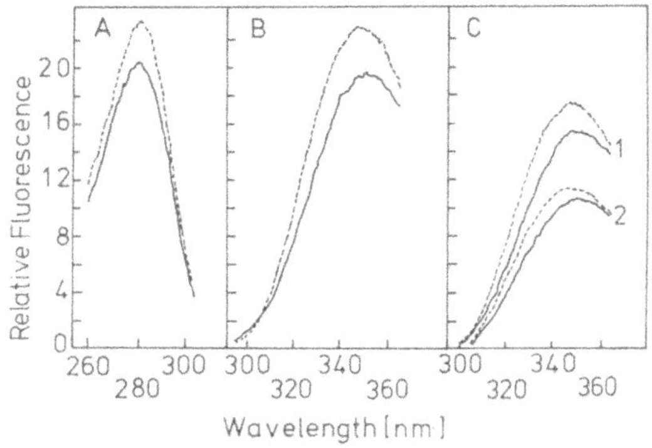

Fig. 5A-C. Fluorescence excitation and emission spectra of the maltose-binding protein. In this and all subsequent figures the excitation and emission slits were 7.5 nm, the temperature was 22°C, and the buffer was buffer A, pH 7.1. Dotted lines: relative fluorescence of the protein (3.5 µg/ml) alone; Solid lines: fluorescence in the presence of 0.1 mM maltose.
(A) excitation spectra with an emission wavelength of 350 nm; (B) emission spectra with an excitation wavelength of 280 nm; (C) emission spectra with excitation wavelength of 290 nm (upper curves), (1) and 295 nm (lower curves) (2). (Adapted from Szmelcman et al., 1976 with permission of the author and the publisher)

Figure 5A the excitation spectrum (emission at 350 nm) of the maltose-binding protein in the presence and absence of saturating concentrations of maltose (0.1 mM). Excitation maximum occurs at 281 nm and emission maximum at 348 nm. The addition of maltose results in quenching of the fluorescence emission with a maximal effect at an emission wavelength of 333 nm. The emission spectra remain unchanged in relative shape when the protein is excited at 280, 285, or 290 nm (Fig. 5C); indicating that emission is due to mainly one type of chromophore, probably tryptophan (Udenfriend, 1962). Addition of maltose not only results in an emission quenching but also in a slight but reproducible shift of 2 nm to longer wavelength. The fluorescence quenching of the maltose-binding protein is specific: The sugars glucose, α-methylglucoside, lactose as well as isomaltose in 1 mM concentrations have no effect and do not inhibit the fluorescence change caused by maltose. On the other hand higher maltodextrins (maltotriose, maltotetraose, maltopentaose, maltoheptaose, and cyclic heptaose) which have previously been shown to interfere with binding of maltose to the maltose-binding protein (Kellerman and Szmelcman, 1974) exhibit fluorescence quenching of the maltose-binding protein that is in the same order of magnitude as that produced by maltose.

V. Titration of the Fluorescence Quenching Effect of the Maltose-Binding Protein by Maltose and Maltotriose

Figure 6 shows the dependence of fluorescence quenching of the maltose-binding protein on the total concentration of maltose. Maximum quenching occurs at concentration above 10 µM with a half maximal response at 1 µM. Similar experiments performed with maltotriose gave a maximal response above concentration of 2 µM with a half maximal response at 0.15 µM (data no shown). Therefore, maltose-binding protein has a higher affinity for maltotriose than

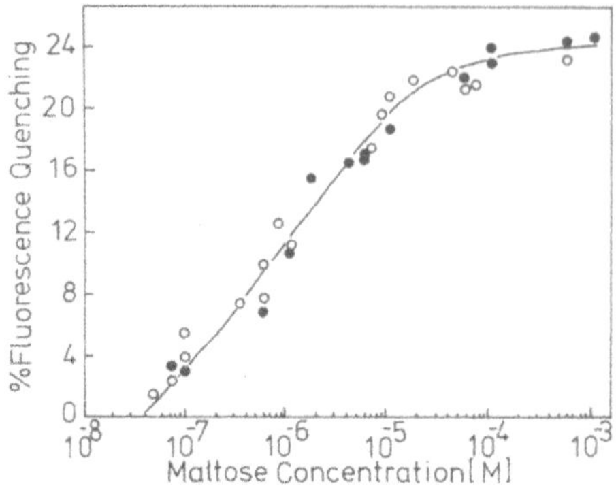

Fig. 6. Concentration dependence of maltose induced fluorescence quenching. Maltose was added to a protein containing cuvette in 5 µl aliquots and the fluorescence emission was measured at 350 nm, excitation was at 280 nm. The same volume of buffer was added to an identical control cuvette and the % quenching was corrected for the observed dilution. Maltose-binding protein concentration was -o-, 7 µg/ml; -•-, 14 µg/ml. (Adapted from Szmelcman et al., 1976 with permission of the author and the publisher)

for maltose. However, the affinity for longer maltodextrins such as maltotetraose, pentaose, or heptaose is the same as for maltotriose.

By comparing the kinetic parameters of transport in vivo with the binding parameters of the maltose-binding protein in vitro the following points can be made:

1. Reciprocal plots of the initial rate of uptake versus substrate concentration with all substrates and mutants were linear over large concentration ranges. Thus, there is no evidence for the independent functioning of two recognition sites; λ receptor must therefore function in sequence with the maltose-binding protein rather than in parallel with it.

2. With maltose as substrate the apparent K_m of transport into the wild-type and K_D of binding to the binding protein was identical, 1 µM. Thus, in the presence of λ receptor the access of maltose, to the maltose-binding protein is not limited.

3. Mutations in the receptor for phage λ increased the K_m for maltose transport by a factor of 100-500 without altering the maximal rate of transport at saturating substrate concentrations. Since λ receptor is known to be located in the outer membrane Randall-Hazelbauer and Schwartz, 1973) this result suggests that in absence of this protein diffusion of maltose through the outer membrane is the rate limiting step in transport at low concentrations of this substrate.

4. With maltotriose as a substrate the K_m of transport (2 µM) in the wild-type is thirteen fold higher than the K_D of binding to the binding protein. This suggests that even in the presence of λ receptor diffusion of maltotriose through the outer membrane is the limiting step at low concentrations of substrate.

5. The two lamB strains tested do not transport maltotriose. Therefore, in the absence of λ receptor the outer membrane is essentially impermeable to trisaccharides, as suggested by others (Nakae, 1975; Nakae and Nikaido, 1975).

6. The very bulky substrate cyclic heptaose is bound to the binding protein with a K_D that is in the same order of magnitude as maltotriose but it cannot be transported via the maltose transport system, not even in the wild-type, nor does it inhi-

bit maltose transport, presumably because it cannot pass the outer membrane barrier.

How does the λ receptor facilitate the permeability through the outer membrane for maltose and higher maltodextrins? Purified λ receptor did not bind maltose after isolation with cholate (unpublished results). Until now attempts to detect binding activity of the λ receptor either in vitro or in vivo using malE mutants defective in maltose-binding protein were equally negative. Under the experimental conditions a binding affinity with a K_D of 5 μM and lower would have been detected.

In analogy to Inouye's model regarding the formation of pores by lipoprotein in the outer membrane (Inouye, 1974) it is tempting to speculate that the λ receptor protein also plays the role of a "pore" for maltose and higher maltodextrins. According to such a model the λ receptor may have no binding affinity for its substrates. One may speculate that the passage of other molecules, particularly of carbohydrates of similar size as maltose or smaller, would equally well be facilitated by the λ receptor. Evidence that this so would require the availability of a system in which the rate of transport is limited by diffusion across the outer membrane. Other transport systems have recently been reported to require, in addition to their intrinsic transport proteins, outer membrane components otherwise recognized to be phage or colicin receptors (Wang and Newton, 1971; DiMasi et al., 1973; Hantke and Braun, 1975). Whether the role of these proteins is similar to that proposed here for the λ receptor is not clear at present.

Quite interestingly the maltose-binding protein exhibits a higher affinity for maltotriose and other maltodextrins than for maltose. This suggests that the binding site was selected to recognize maltotriose rather than maltose. Taken together with the existence of the lamB protein, which is more stringently needed for growth on maltodextrins than on maltose, this observation suggests that the maltose system in fact evolved to utilize the mixture of linear oligosaccharides originating form the hydrolysis of starch and glycogen.

C. A Periplasmic Protein Related to the sn-Glycerol-3-Phosphate Transport System of Escherichia coli (Silhavy et al., 1976)

The sn-glycerol-3-phosphate (GLP) transport system in *E. coli* appears to be ideally suited for a combined genetic-biochemical analysis. There exists an easy selection procedure for transport-negative (glpT) strains by selecting for resistance against phosphonomycin (Hendlin et al., 1969; Lin, 1970; Venkateswasan and Wu, 1972). Transport positive revertants can be obtained by growth on GLP. Also, constitutive (glpR) strains can be selected (Lin et al., 1962). glpT strains have been found 50% cotransducible by phage P1 to the nalA marker and the map position of 43 min on the *E. coli* chromosome has been proposed for the glpT mutation (Cozzarelli et al., 1968; Kistler and Lin, 1971). From these studies it cannot be determined whether the glpT gene consists of one or more cistrons. Unlike most transport systems in *E. coli* the GLP transport system is homogeneous, i.e. the substrate is transported by only one system. This system has been characterized as an energy dependent active transport system (Km 12 μM) translocating the substrate without chemical alteration (Hayashi et al., 1964).

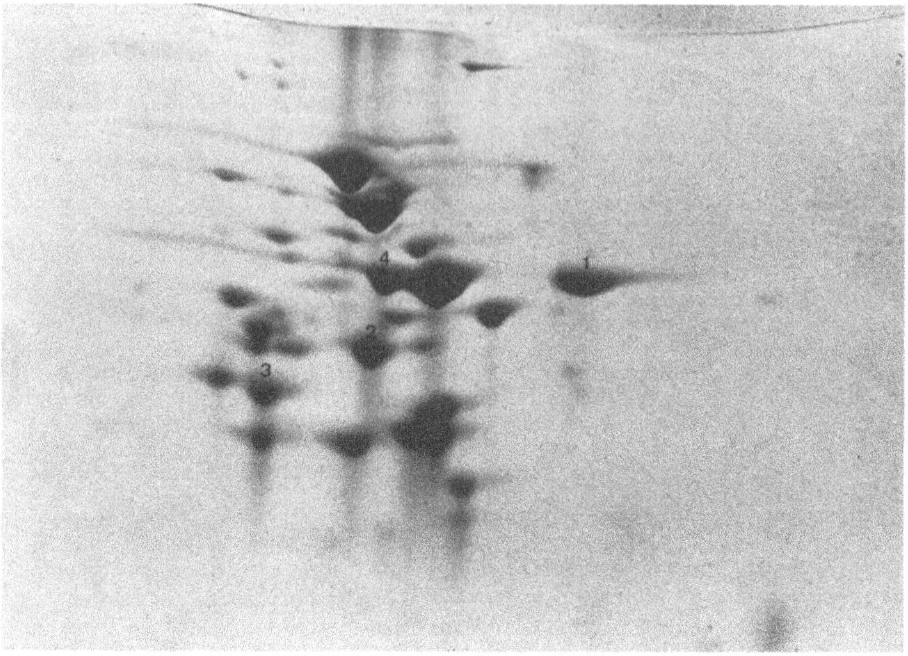

Fig. 7. Two dimensional polyacrylamide gel electrophoresis of shock fluid of strain LA 3400 (wild-type) grown in the presence of glycerol.
First dimension is electrophoresis in 8 M urea, pH 8.4 followed by electrophoresis in 0.2% SDS, pH 6.48 as described (Johnson et al., 1975). About 300 μg protein were applied. The numbers and molecular weights correspond to the following proteins: 1: GLPT; 2: galactose-binding protein (Boos and Gordon, 1971); 3: ribose-binding protein (Willis and Furlong, 1974); 4: maltose-binding protein (Kellerman and Szmelcman, 1974). (Adapted from Silhavy et al., 1976 with permission of the author and the publisher)

Despite these advantages no attempts have been made to study the system with biochemical methods. In particular, it is not clear whether the system is mediated via a periplasmic binding protein (Boos, 1974b) or a typical membrane bound system active in membrane vesicles of the Kabackosome type (Kaback, 1972).

In the following we describe the identification of a periplasmic protein that is related to the sn-glycerol-3-phosphate transport system.

I. Two-Dimensional Gel Electrophoresis of Shock Fluid from *E. coli*

Figure 7 shows the pattern of a Comassie Blue stained two dimensional polyacrylamide gel slab of the periplasmic proteins obtained by osmotic shock from strain LA-3400 (wild-type) after growth with 0.4% glycerol as carbon source. Separation of the proteins occurs in the first dimension (left to right) in 8 M urea predominantly according to their electrical charge. The second dimension (top to bottom) occurs in SDS emphazising differences in molecular weight. With this technique nearly all proteins in the osmotic shock fluid of *E. coli* can be separated. Therefore, if the synthesis of a yet

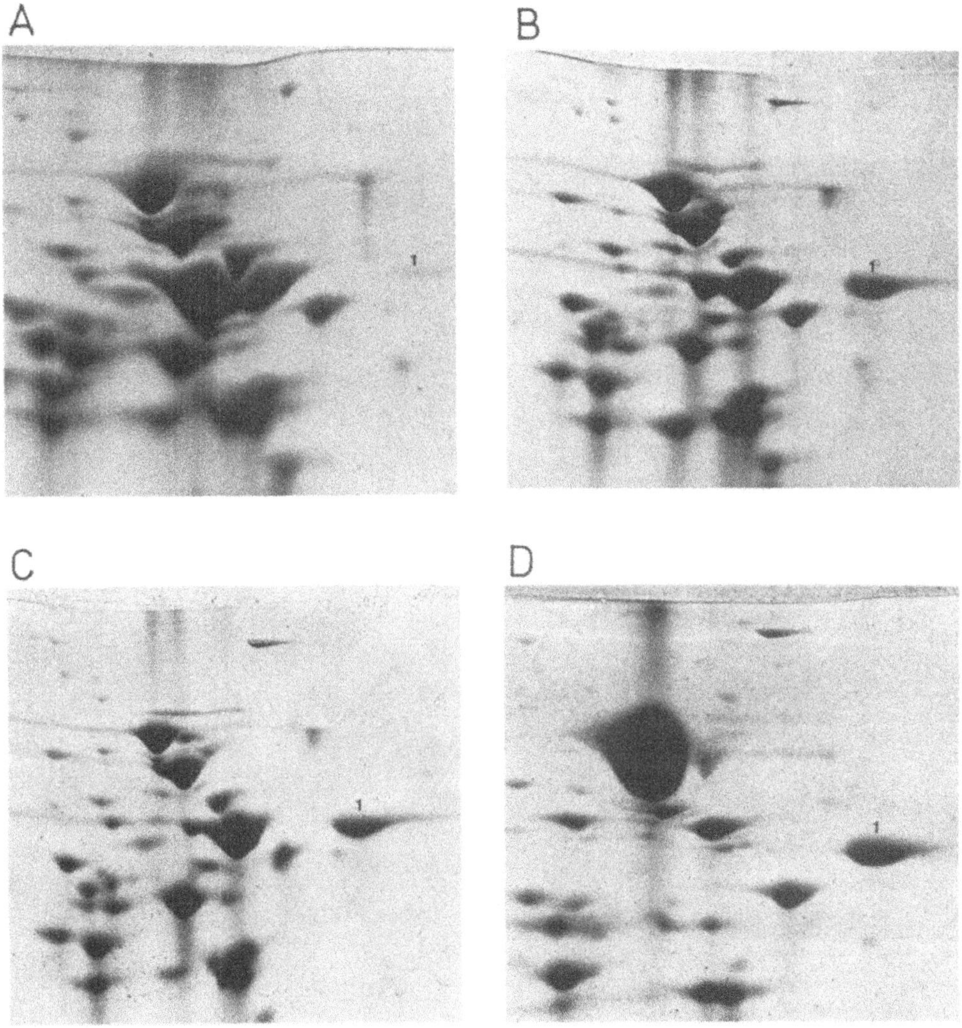

Fig. 8A-D. Detailed pictures of two dimensional polyacrylamide gel slabs of shock fluids from several strains: (A) LA 3400 (wild-type) grown on trehalose; (B) LA 3400 grown on glycerol; (C) strain 72 (glpR) grown on succinate; (D) strain 72 grown on glycerol in the presence of fumarate under anaerobic conditions. Experimental conditions as in Figure 7. (Adapted from Silhavy et al., 1976 with permission of the author and the publisher)

unknown protein belonging to this compartment can be influenced by induction, repression, or mutation it should easily be recognized by this analytical technique.

Based on this rationale we observed a protein among the periplasmic proteins that appeared to be inducible during growth in the presence of glycerol.

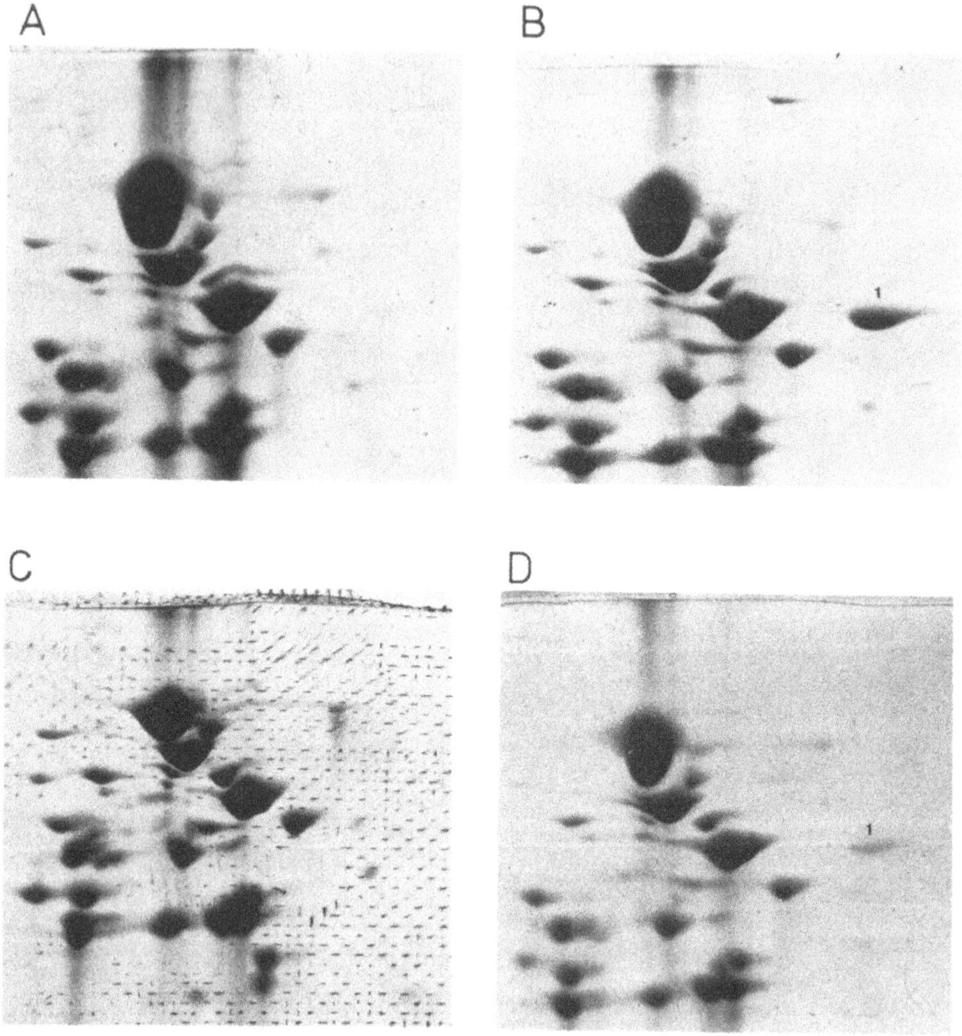

Fig. 9A-D. Detailed pictures of two dimensional polyacrylamide gel slabs of shock fluids from the following glpT mutants and their revertants. (A) LA 3401 (glpT) grown on glycerol; (B) LA 3406 ($glpT^+$ revertant of LA 3401) grown on glycerol; (C) LA 3402 (glpT) grown on glycerol; (D) LA 3407 ($glpT^+$ revertant of LA 3402) grown on glycerol. The number 1 designates GLPT. Experimental conditions as in Figure 7. (Adapted from Silhavy et al., 1976 with permission of the author and the publisher)

II. The sn-Glycerol-3-Phosphate Transport Protein (GLPT)

Figure 8 A and B show the detailed area of the two dimensional gels from shock fluids of strain LA-3400 grown with trehalose and glycerol as carbon source. One protein spot (No. 1) appears inducible by glycerol. The protein exhibits under the denaturing conditions of the first dimension an unusual high negative charge. In the second dimension (SDS) an approximate molecular weight of 40,000 daltons can be estimated. The same analysis of the shock protein of a glycerol constitutive glpR strain (Fig. 8C) grown in the presence of succinate demonstrates that the protein in question is under the control of the glpR regulatory gene. In addition, as seen in Figure 8D growth of the glpR strain under anaerobic conditions in the presence of glycerol and fumarate (Freedberg and Lin, 1973) GLPT appears fully induced. From the availability of this protein in the periplasmic shock fluid it seemed likely that it might be either the gene product of the glpF (glycerol facilitator) (Richey and Lin, 1972) or the glpT gene, known to be involved in GLP transport (Cozzarelli et al., 1968).

III. glpT Mutant Analysis

By growth in the presence of phosphonomycin an inhibitor of cell wall synthesis and substrate of the GLP transport system we isolated spontaneous glpT mutants (Venkateswasan and Wu, 1972). The two-dimensional polyacrylamide gel electrophoresis of the shock proteins or two independently isolated glpT strains, derivatives of the constitutive glpR strain 72 are shown in Figure 9A and C. Both mutant preparations miss the protein entirely. Simultaneously both strains fail to grow on GLP but grow normally on glycerol. Their defect must therefore be confined in the transport of GLP. Measurements of GLP uptake in these mutants in comparison to its wild-type parent is shown in Figure 10. Using a substrate concentration of 0.3 μM, far below 12 μM, the Km of the system (Hayashi et al., 1964), the mutants exhibit less than 5% transport activity of their parent. Transport of GLP in the wild-type is competitively

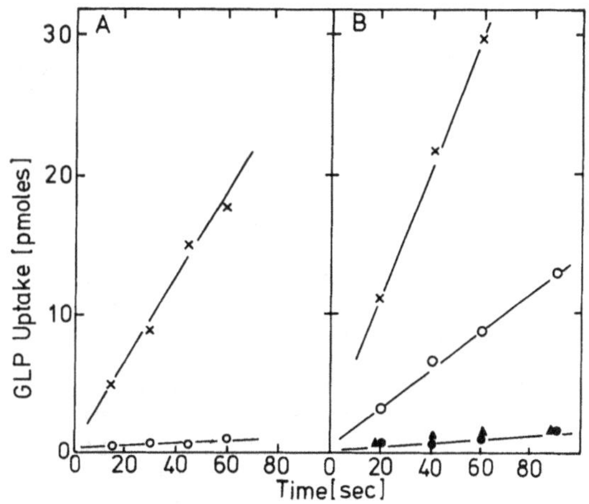

Fig. 10A,B. GLP transport activities of wild-type mutant and revertant strains. The bacterial cultures were resuspended in growth medium minus carbon source at an optical density of 0.5 (576 nm). [^{14}C]GLP was added at an initial concentration of 0.3 μM. The data are expressed in amount GLP taken up per 0.5 ml samples. All operations were done at room temperature.
(A) strain LA 3400 before (x-x-x) and after (o-o-o) osmotic shock (1); (B) mutant strain LA 3401 (●-●-●), revertant LA 3406 (o-o-o); mutant strain LA 3402 (▲-▲-▲); revertant strain LA 3407 (x-x-x). (Adapted from Silhavy et al., 1976 with permission of the author and the publisher)

inhibited by orthophosphate with a Ki of 20 mM (not shown). This value agrees well with the Ki determined previously (Hayashi et al., 1964) and indeed characterizes the measured GLP uptake as being mediated by the GLP transport system. Application of the cold osmotic shock procedure of Neu and Heppel (1965) to the wild-type strains clearly shows that the GLP transport system is dependent on a periplasmic component (Fig. 10A). Spontaneous revertants of the two mutants described above to grow again on GLP were isolated and measured for the transport activity as well as their capability to produce GLPT. Figure 9 B and D show that one strain (LA 3406) has fully regained the capability to produce GLPT while strain LA 3407 shows only a partial reversion. The extent by which GLPT reappears in these revertants is reflected in their transport activity of GLP (Fig. 10B).

Shock fluids of several other mutants that fail to grow on GLP after isolation for resistance against phosphonomycin were analyzed by the two dimensional gel electrophoresis (not shown). Some of these mutants (LA 3405) exhibit a reduced amount of GLPT, others (LA 3404) show no apparent alteration neither in the amount of GLPT nor in its position on the gel slab. This is consistent with the observation in several binding protein dependent transport systems: The periplasmic component is essential but not sufficient for transport activity (Boos, 1974b).

IV. Genetic Analysis of the Isolated glpT Mutants

Since phosphonomycin resistant mutants have previously been reported to arise also from mutation of the hexosephosphate transport system as well as in the phosphoenolpyruvate uridine diphospho-N-acetylglucosamine enolpyruvyl transferase (Venkateswasan and Wu, 1972) it was necessary to genetically characterize our GLP transport mutants.

A lysate of phage P1 grown on a nalA strain was introduced into two GLP transport negative strains selecting for growth on GLP. The transductants were scored for nalA. In the first transduction with strain LA 3403 of 42 transductants 18 or 43% were nalA. In the second transduction with strain LA 3401 of 79 transductants 49 or 62% were nalA. These results are similar to the published cotransduction frequency of 50% between glpT and nalA (Cozzarelli et al., 1968; Kistler and Lin, 1971) and establishes the mutation of these strains as glpT. The shock fluids of four $glpT^+$ transductions were checked by two dimensional electrophoresis and found to contain GLPT (not shown).

From the analogy of a series of other transport systems that are mediated via periplasmic components (Boos, 1974b) it seems likely that GLPT is a binding protein specific for GLP. Yet, all attempts to measure in crude shock fluids binding activity for GLP even with the most sensitive method of ultrafiltration (Paulus, 1970) have failed so far to give positive results. However, with an expected binding affinity of 12 μM (the Km of GLP transport in whole cells (Hayashi et al., 1964) and the relative small amount of GLPT in crude shock fluids, binding activity might be difficult to detect. Another complication arises from the observation that all shock fluids independent of their content of GLPT contain an enzymatic activity that splits GLP in glycerol and Pi. This is surprising in view of the fact that all strains used in this study are phoA and therefore do not produce the periplasmic alkaline phosphatase. By paper chromatographic analysis it can be seen that under conditions similar to those used for the binding assay by ultrafiltration 30-60% of the GLP is hydrolyzed in one hour. Therefore, a meaningful test for binding activity or any enzymatic activity of GLPT has to await the purification of GLPT.

Since transport of GLP appears to depend on the presence of the periplasmic GLPT one would not expect that membrane vesicles of the Kabackosome-type are able to transport GLP. However, stimulation of membrane-bound transport systems by GLP as well as its own transport has been reported (Barnes and Kaback, 1971; Heppel et al., 1972). Measurement of possible residual amounts of GLPT in membrane vesicles might explain this discrepancy.

Alternatively periplasmic transport components might be essential for transport in whole cells but not in membrane vesicles, a phenomenon reported for the E. coli dicarboxylic transport system (Lo and Sanwal, 1975).

References

Barnes, E.M., Kaback, H.R.: Mechanism of active transport in isolated membrane vesicles. I. The site of energy coupling between D-lactic dehydrogenase and β-galactoside transport in Escherichia coli membrane vesicles. J. Biol. Chem. 246, 5518-5522 (1971)

Berger, E.A., Heppel, L. A.: Different mechanisms of energy coupling for the shock-sensitive and shock-resistant amino acids permease of Escherichia coli. J. Biol. Chem. 249, 7747-7755 (1974)

Boos, W.: Pro and contra transport carriers; the role of the galactose-binding protein in the β-methylgalactoside transport system of Escherichia coli. Current Topics Membranes Transport 5, 51-136 (1974a)

Boos, W.: Bacterial transport. Ann. Rev. Biochem. 43, 123-146 (1974b)

Boos, W., Gordon, A.S.: Transport properties of the galactose-binding protein of Escherichia coli. Occurrence of two conformational states. J. Biol. Chem. 246, 621-628 (1971)

Cozzarelli, N.R., Freedberg, W.B., Lin, E.C.C.: Genetic control of the L-α-glycerolphosphate system in E. coli. J. Molec. Biol. 31, 371-387 (1968)

DiMasi, D.R., White, J.C., Schnaitman, C.A., Bradbeer, C.: Transport of vitamin B_{12} in Escherichia coli: Common receptor sites for vitamin B_{12} and the E colicins on the outer membrane of the cell envelope. J. Bacteriol. 115, 506-513 (1973)

Ferenci, T.: this volume

Freedberg, W.B., Lin, E.C.C.: Three kinds of controls affecting the expression of the glp regulon in Escherichia coli. J. Bacteriol. 115, 816-823 (1973)

Hantke, K., Braun, V.: Membrane receptor dependent iron transport in Escherichia coli. FEBS Lett. 49, 301-305 (1975)

Hayashi, S., Koch, J.P., Lin, E.C.C.: Active transport of L-α-glycerophosphate in Escherichia coli. J. Biol. Chem. 239, 3098-3105 (1964)

Hazelbauer, G.L.: Maltose chemoreceptor of Escherichia coli. J. Bacterio. 122, 206-214 (1975)

Hendlin, D., Stapley, E.O., Jackson, M., Wallick, H., Miller, A.K., Wolf, F. J., Miller, T.W., Chaiet, L., Kahan, F.M., Foltz, E.L., Woodruff, H.B.: Phosphonomycin, a new antibiotic produced by strains of streptomyces. Science 166, 122-123 (1969)

Heppel, L.A., Rosen, B.P., Friedberg, J., Berger, E.A., Weiner, J.H.: Studies on binding proteins, periplasmic enzymes and active transport in Escherichia coli. In: The Molecular Basis of Biological Transport. Woessner, J.F., Jr., Huijing, F. (eds.). New York-London: Academic Press, 1972, pp. 133-150

Hofnung, M.: Divergent operons and the genetic structure of the maltose B region in Escherichia coli K-12. Genetics 76, 169-184 (1974)

Inouye, M.: A three-dimensional molecular assembly model of a lipoprotein from the Escherichia coli outer membrane. Proc. Natl. Acad. Sci. U.S. 71, 2396-2400 (1974)

Johnson, W.C., Silhavy, T.J., Boos, W.: Two-dimensional polyacrylamide gel electrophoresis of envelope proteins of Escherichia coli. Appl. Microbiol. 29, 405-413 (1975)

Kaback, H.R.: Transport across isolated bacterial cytoplasmic membrane. Biochim. Biophys. Acta 265, 367-416 (1972)

Kellerman, O., Szmelcman, S.: Active transport of maltose in Escherichia coli K12. Involvement of a "periplasmic" maltose-binding protein. Europ. J. Biochem. 47, 139-149 (1974)

Kennedy, E.P.: The lactose permease system of Escherichia coli. In: The Lactose Operon. Beckwith, J.R., Zipser, D. (eds.). New York: Cold Spring Harbor Laboratory, 1970, pp. 49-92

Kistler, W.S., Lin, E.C.C.: Anaerobic L-α-glycerophosphate dehydrogenase of Escherichia coli: Its genetic locus and its physiological role. J. Bacteriol. 108, 1224-1234 (1971)

Lieberman, M.A., Hong, J.-S.: Energization of osmotic shock-sensitive transport system in Escherichia coli requires more than ATP. Arch. Biochem. Biophys. 172, 312-315 (1976)

Lin, E.C.C.: The genetics of bacterial transport systems. Ann. Rev. Genetics 4, 225-262 (1970)

Lin, E.C.C., Koch, J.P., Chused, T.M., Jorgensen, S.E.: Utilization of L-α-glycerophosphate by Escherichia coli without hydrolysis. Proc. Natl. Acad. Sci. U.S. 48, 2145-2150 (1962)

Lo, T.C.Y., Sanwal, B.D.: Isolation of the soluble substrate recognition component of the dicarboxylate transport system of Escherichia coli. J. Biol. Chem. 250, 1600-1602 (1975)

Nakae, T.: Outer membrane of Salmonella typhimurium: Reconstitution of sucrose-permeable membrane vesicles. Biochem. Biophys. Res. Commun. 64, 1224-1230 (1975)

Nakae, T., Nikaido, H.: Outer membrane as a diffusion barrier in Salmonella typhimurium. J. Biol. Chem. 250, 7359-7365 (1975)

Neu, H.C., Heppel, L.A.: The release of enzymes from Escherichia coli by osmotic shock and during the formation of spheroplasts. J. Biol. Chem. 240, 3685-3692 (1965)

Ordal, G.W., Adler, J.: Isolation and complementation of mutants in galactose taxis and transport. J. Bacteriol. 117, 509-516 (1974)

Parnes, J.R., Boos, W.: Energy coupling of the β-methylgalactoside transport system of Escherichia coli. J. Biol. Chem. 248, 4436-4445 (1975)

Paulus, H.: A rapid and sensitive method for measuring the binding of radioactive ligands to protein. Anal. Biochem. 32, 91-100 (1970)

Plate, C.A., Suit, J.L., Jetten, A.M., Luria, S.E.: Effects of Colicin K on a mutant of Escherichia coli deficient in Ca^{2+}, Mg^{2+}-activated adenosine triphosphatase. J. Biol. Chem. 249, 6138-6143 (1974)

Randall-Hazelbauer, L., Schwartz, M.: Isolation of the bacteriophage lambda receptor from Escherichia coli. J. Bacteriol. 116, 1436-1446 (1973)

Richey, D.P., Lin, E.C.C.: Importance of facilitated diffusion for effective utilization of glycerol by Escherichia coli. J. Bacteriol. 112, 784-790 (1972)

Robbins, A.R., Rotman, B.: Evidence for binding protein-independent substrate translocation by the methylgalactoside transport system of Escherichia coli K12. Proc. Natl. Acad. Sci. U.S. 72, 423-427 (1975)

Silhavy, T.J., Boos, W., Kalckar, H.M.: The role of the Escherichia coli galactose-binding protein in galactose transport and chemotaxis. In: 25th Mosbacher Coll. Biochem. Sensory Functions. L. Jaenicke (ed.). Berlin-Heidelberg-New York: Springer Verlag, 1974

Silhavy, T.J., Hartig-Beecken, I., Boos, W.: A periplasmic protein related to the sn-glycerol-3-phosphate transport system of Escherichia coli. J. Bacterio. 126, 951-958 (1976)

Szmelcman, S., Hofnung, M.: Maltose transport in Escherichia coli K12. Involvement of the bacteriophage lambda receptor. J. Bacteriol. 124, 112-118 (1975)

Szmelcman, S., Schwartz, M., Silhavy, T.J., Boos, W.: Maltose transport in Escherichia coli K12. Europ. J. Biochem. 65, 13-19 (1976)

Udenfriend, S.: Fluorescence Assay in biology and medicine. New York: Academic Press, 1962

Venkateswasan, P.S., Wu, H.C.P.: Isolation and characterization of a phosphonomycin resistant mutant of Escherichia coli K12. J. Bacteriol. 110, 935-944 (1972)

Wang, C.C., Newton, A.: An additional step in the transport of iron defined by the trnB of Escherichia coli. J. Biol. Chem. 246, 2147-2151 (1971)

Wiesmeyer, H., Cohn, M.: The characterization of the pathway of maltose utilization by Escherichia coli. III. A description of the concentrating mechanism. Biochim. Biophys. Acta 39, 440-447 (1960)

Willis, R.C., Furlong, C.E.: Purification and properties of a ribose-binding protein. J. Biol. Chem. 249, 6926-6929 (1974)

Wilson, D.B.: Properties of the entry and exit reactions of the β-methylgalactoside transport system of Escherichia coli. J. Bacteriol. 126, 1156-1165 (1976)

Winkler, H.H., Wilson, T.H.: The role of energy coupling in the transport of β-galactosides by Escherichia coli. J. Biol. Chem. 241, 2200-2211 (1966)

Subject Index

Advances in Biochemical Engineering

Volume 1

70 figures. VII, 194 pages. 1971
ISBN 3-540-05400-6

Contents:

Volume 2

70 figures. V, 215 pages. 1972
ISBN 3-540-06017-0

Contents:

Volume 3

119 figures.VI, 290 pages. 1974
ISBN 3-540-06546-6

Contents:

Volume 4 Engineering

87 figures. V, 172 pages. 1976
ISBN 3-540-07747-2

Contents:

Springer-Verlag
Berlin Heidelberg New York

www.ingramcontent.com/pod-product-compliance
Lightning Source LLC
Chambersburg PA
CBHW061149240926
51846CB00006B/963
* 9 7 8 3 6 4 2 6 6 5 6 6 0 *